T0392381

Phytopathogenic Bacteria and Plant Diseases

Phytopathogenic Bacteria and Plant Diseases

B.S. Thind

CRC Press is an imprint of the
Taylor & Francis Group, an **informa** business

CRC Press
Taylor & Francis Group
6000 Broken Sound Parkway NW, Suite 300
Boca Raton, FL 33487-2742

Printed on acid-free paper

International Standard Book Number-13: 978-0-367-19715-5 (Hardback)

Library of Congress Cataloging-in-Publication Data

Names: Thind, B. S., author.
Title: Phytopathogenic bacteria and plant diseases / B.S. Thind.
Description: Boca Raton : Taylor & Francis, 2020. | Includes bibliographical references and index.
Identifiers: LCCN 2019012214| ISBN 9780367197155 (hardback : alk. paper) | ISBN 9780429242786 (ebook)
Subjects: LCSH: Bacterial diseases of plants. | Phytopathogenic bacteria.
Classification: LCC SB734 .T445 2020 | DDC 632/.32--dc23
LC record available at https://lccn.loc.gov/2019012214

Visit the Taylor & Francis Web site at
http://www.taylorandfrancis.com

and the CRC Press Web site at
http://www.crcpress.com

Printed and bound in Great Britain by
TJ International Ltd, Padstow, Cornwall

Dedication

Dedicated to the memory of my mentor and teacher late Dr. M. M. Payak, Ex-Head, Division of Plant Pathology, Indian Agricultural Research Institute, New Delhi, who guided and trained me in the discipline of Plant Pathology and Plant Bacteriology.

Contents

Abbreviations of Bacterial Genera Used in This Book

Genus	Abbreviation	Genus	Abbreviation
Acetobacter	*Ace.*	*Listeria*	*Li.*
Acidovorax	*Aci.*	*Lonsdalea*	*Lo.*
Acinetobacter	*Acin.*	*Neisseria*	*Ne.*
Agrobacterium	*Ag.*	*Nocardia*	*No.*
Arthrobacter	*Ar.*	*Paenibacillus*	*Pae.*
Bacillus	*Ba.*	*Pantoea*	*Pa.*
Bacterium	*Bac.*	*Paraburkholderia*	*Par.*
Bradyrhizobium	*Bra.*	*Pectobacterium*	*Pe.*
Brenneria	*Br.*	*Phytomonas*	*Ph.*
Burkholderia	*Bu.*	*Pseudoalteromonas*	*Psa.*
Campylobacter	*Cam.*	*Pseudomonas*	*Psm.*
'*Candidatus* Liberibacter'	'*Ca.* Li.'	*Ralstonia*	*Ral.*
'*Candidatus* Phlomobacter'	'*Ca.* Phl.'	*Rathayibacter*	*Rat.*
'*Candidatus* Phytoplasma'	'*Ca.* Phy.'	*Rhizobacter*	*Rhba.*
Citrobacter	*Cit.*	*Rhizobium*	*Rhbi.*
Clavibacter	*Cla.*	*Rhizorhapis*	*Rhr.*
Clostridium	*Clo.*	*Rhodococcus*	*Rho.*
Corynebacterium	*Cor.*	*Salmonella*	*Sal.*
Curtobacterium	*Cur.*	*Samsonia*	*Sam.*
Dickeya	*D.*	*Serratia*	*Se.*
Enterobacter	*En.*	*Shigella*	*Sh.*
Enterococcus	*Enc.*	*Sphingomonas*	*Sph.*
Erwinia	*Er.*	*Spiroplasma*	*Sp.*
Escherichia	*Es.*	*Staphylococcus*	*Sta.*
Gibbsiella	*Gi.*	*Streptococcus*	*Stc.*
Gluconobacter	*Gl.*	*Streptomyces*	*St.*
Herbaspirillum	*He.*	*Tatumella*	*Ta.*
Janibacter	*Jani.*	*Thermus*	*Th.*
Janthinobacterium	*Jant.*	*Vibrio*	*V.*
Lactobacillus	*Lactob.*	*Xanthomonas*	*Xa.*
Lactococcus	*Lactoc.*	*Xylella*	*Xyl.*
Leifsonia	*L.*	*Xylophilus*	*Xyp.*
Leucobacter	*Leb.*	*Yersinia*	*Y.*
Leuconostoc	*Len.*		

Preface

The study of phytopathogenic bacteria and the plant diseases caused by them is of utmost importance to mankind as these diseases cause huge economic losses. Some of these diseases, such as bacterial wilt of solanaceous plants, soft rot of fleshy vegetables and fruits, crown gall of plants, citrus canker, citrus huanglongbing, fire blight of pome fruit trees, and bacterial blight of rice, are of international importance. Plant diseases caused by phytopathogenic bacteria are also disastrous because many of these diseases cannot be effectively controlled due to the lack of effective chemicals. Moreover, the secondary spread of these diseases is very fast compared to fungal diseases.

The classification and nomenclature of bacteria has undergone a sea change during the last three decades. The sequence analysis of highly conserved regions of the bacterial genome such as small sub unit rRNA gene has provided very valuable information concerning numerous taxonomic changes and realignment of different taxonomic groups. The sequencing of genomes of many phytopathogenic bacteria has certainly stabilized their ever fluctuating nomenclature and taxonomy. Since the 1st edition of *Bergey's Manual of Systematic Bacteriology* was published in 1984, the number of published bacterial species has more than tripled, i.e., 390 new genera and 2,200 new species have been described. Many new genera containing plant pathogenic bacteria have been recognized, and the rearrangement of the species among the genera has also occurred. During 1970s, there were only 6 genera of bacteria that contained plant pathogenic bacteria, but now this number is 53, including 3 candidatus categories.

In contrast to direct monetary losses and other harmful effects of phytopathogenic bacteria, they have a great deal of commercial value. They are used for the commercial production of many industrial products as well as tools for various scientific studies. Their use in industrial fermentation, production of antibiotics, xanthan gums, and polysaccharides such as curdlans are the important applications. Production of pectin-degrading enzymes produced by soft-rotting erwinias and restriction nucleases produced by *Agrobacterium* and *Xanthomonas* species are other notable examples of their commercial use. *Gluconobacter* species have various biotechnological applications. Their biotransformed strains are used for the production of rare and special sugars like L-ribose and miglitol, both very promising pharmaceutical lead molecules. *Ag. tumefaciens* (the causal agent of crown gall of many plants), apart being a plant pathogen, is one of the best known vehicles for gene transfer and most useful genetic engineering tool for developing transgenics.

Our knowledge of phytopathogenic bacteria, plant diseases caused by them, host-pathogen interactions, molecular mechanisms involved in virulence and pathogenesis, and disease management strategies developed has now reached a stage where it would be useful to integrate the concepts and ideas into a book. Therefore, the book entitled, *Phytopathogenic Bacteria and Plant Diseases* is proposed.

There are 7 chapters included in the book. Chapter 1 includes economic importance of phytopathogenic bacteria and historical review of phytobacteriology. Chapter 2 contains basic information about the bacterial cell. Chapter 3 deals with diagnosis of bacterial diseases of plants. The up-to-date classification of plant pathogenic bacteria is given in Chapter 4. The bacterial names included are based on the information published in *International Journal of Systematic Bacteriology* (*IJSB*), *International Journal of Systematic and Evolutionary Microbiology* (*IJSEM*), and related journals up to December issues of 2018, and supplemented with the information given in *Names of Plant Pathogenic Bacteria, 1864–2004* prepared by Young et al. (2004), *Comprehensive List of Names of Plant Pathogenic Bacteria, 1980–2007* prepared by Bull et al. (2010), *List of New Names of Plant Pathogenic Bacteria (2008–2010)* prepared by Bull et al. (2012), and *List of New Names of Plant Pathogenic Bacteria (2011–2012)* prepared by Bull et al. (2014). Chapter 5 deals with information on molecular mechanisms of phytopathogenic bacteria involved in virulence and pathogenesis. In Chapter 6, an altogether a new topic, i.e., plants acting as carriers of human enteric bacterial pathogens has been included. It needs an immediate attention of the scientists to decontaminate plant and plant products used for human consumption.

Chapter 7 contains descriptions of specific bacterial plant diseases. Although it was not possible to include all the bacterial plant diseases, 70 specific diseases, including economically important diseases of international importance and representative diseases of different crops and regions, have been described. In order to make discussion of phytobacteriology even more effective, full-color photographs of symptoms of most of these diseases are given. Information published in plant pathological and related biological journals until December issues of 2018 have been included in the book.

The abbreviations of names of bacterial genera, included in the book, are given on page xix. Because a majority of the genera included in this book start with the same letter, a deviation has been made from the tradition of abbreviating the genus name by using only the first letter of the genus. In Chapter 7, the names of causal agents of different diseases are given along with author citations. In author citations, when there are more than two authors, the name of the first author is given followed by "et al." One or two and, in a few cases, up to three recent synonyms of each causal organism are also given. In the corresponding references, the names of up to five authors are given, followed by "et al." when there are additional authors. The names of all the authors/editors are given for book references.

At the undergraduate level, students have limited exposure to phytobacteriology. However, at the post-graduate level, students are taught phytobacteriology in detail. State agricultural universities and the universities teaching biological sciences have courses dealing exclusively with phytobacteriology. PhD students specializing in phytobacteriology are taught advanced courses. This book will serve as a textbook for advanced undergraduate and post-graduate students and also as a reference book for research workers engaged in the discipline of plant pathology and phytobacteriology.

I am indebted to numerous colleagues and friends for suggestions made during the preparation of the manuscript. Also, in the preparation of this manuscript, I have referred to a number of publications, and I would like to acknowledge all the authors of these excellent and valuable publications/articles.

There might be some errors, mistakes, and shortcomings in this publication. I will greatly appreciate the healthy criticism and suggestions for the improvement of this publication in future.

B.S. Thind

Acknowledgments

It is my pleasure to acknowledge the help of all those who contributed in the preparation of this publication. At the outset, I am highly thankful to my friend and ex-colleague, Dr. G.S. Miglani, formerly Professor of Genetics, Punjab Agricultural University (PAU), Ludhiana, for his frequent counseling and help. I am also grateful to my friend and ex-colleague, Dr. R.S. Kahlon, Ex-Prof.-cum-Head, Department of Microbiology, PAU, Ludhiana, for reviewing Chapter 2, "Bacterial cell," and making critical suggestions.

I owe my special thanks to Dr. Parminder Singh, Assistant Plant Pathologist, PAU, Ludhiana, who helped me immensely in downloading a large number of pictures and photographs from the Internet and also for editing and arranging these photographs. His help in typing and setting the text material is also acknowledged. I am also indebted to Dr. M.S. Hujan, Assistant Professor of Plant Pathology, PAU, Ludhiana, for going through Chapter 6, "Plants as carriers of human enteric bacterial pathogens," and making useful suggestions.

The generous help rendered by my ex-colleagues and friends, namely, Dr. Bipin Kumar Sharma, Ex-Director, PAU Regional Research Station, Kapurthalla; Dr. Surinder Kumar Thind, Prof. of Plant Protection, PAU, Ludhiana; Dr. Kuldip Singh, Ex-Senior Molecular Geneticist, PAU, Ludhiana; Dr. R.P. Maharshi, Ex-Head, Department of Plant Pathology, SKN College of Agriculture, Jobner, Rajasthan; and the late Dr. Ram Kishun, ex-Principal Scientist and Head, Crop Protection Division, Central Institute for Subtropical Horticulture, Lucknow, for lending me the photographs of the symptoms of different plant diseases.

I am also indebted to Dr. N.W. Schaad, Agricultural Research Service-United States Department of Agriculture, Foreign Disease-Weed Science Research Unit, Ft. Detrick, MD, USA, for his expert advice and frequent counseling. My sincere thanks are also due to Professor Ian K. Toth, The James Hutton Institute Invergowrie, Dundee, DD2 5DA, UK; Dr. N.W. Schaad; Dr. M.K. Mondal, Sr. Scientist, Division of Plant Pathology, I. A. R. I., New Delhi; Dr. R. Sridhar, ex-Head, Department of Plant Pathology, Central Rice Research Institute, Cuttack; and Dr. R.P. Maharshi for agreeing to act as referees/reviewers for this book. The frequent counseling and help rendered by Dr. Paramjit Singh Randhawa, President, California Plant and Seed Labs, USA, is thankfully acknowledged.

I owe a special gratitude to my wife, Mrs. Ravinder Kaur Thind for her continuous and untiring encouragement during this entire period and also for her help in the preparation of subject index. My sincere thanks also go to my ex-colleagues and friends who helped me directly or indirectly in the preparation of this manuscript, but whose names have not been mentioned here.

B.S. Thind

Author

Dr. B.S. Thind retired as Professor-cum-Head, Department of Plant Pathology, Punjab Agricultural University (PAU), Ludhiana, India after serving the university for 32 years. After retirement, he became an Emeritus Scientist (Indian Council of Agricultural Research [ICAR], New Delhi) in the Department of Plant Pathology, PAU, Ludhiana from July 2003 to July 2005 on the scheme entitled, "Development, mass culturing, and formulation of effective and economical biocontrol agents for the control of *Xanthomonas oryzae* pv. *oryzae*, the causal agent of bacterial blight of rice."

Dr. Thind completed his PhD in Plant Pathology from Indian Agricultural Research Institute, New Delhi in 1970. His first encounter with plant bacteriology or phytobacteriology sub discipline occurred during his PhD program when he worked on bacterial stalk rot of maize caused by *Dickeya zeae*. To this day, he continues to work on bacterial diseases of plants and has made significant contributions in this field.

He has been a principal investigator of an Indian Council of Agricultural Research (ICAR)-funded research scheme entitled, "Biological control of bacterial blight, sheath blight, sheath rot, and brown leaf spot of rice," two ICAR-funded research schemes entitled, "Perpetuation, variability, and control of *X. oryzae* pv. *oryzae*, the causal agent of bacterial blight of rice," and "Detection and control of phytopathogenic bacteria from cowpea and mung bean seeds."

He has also taught many courses of plant pathology to undergraduate and post-graduate students, including 600 series advanced courses of plant bacteriology to PhD students and has guided 10 PhD and 13 MSc students. All these students conducted the research work on bacterial diseases of different crops.

Dr. Thind attended the 7th, 8th, and 9th International Conferences on Plant Pathogenic Bacteria held in Budapest in 1989, in Versailles in 1992, and in Chennai in 1996, respectively. As member of the National Organizing Committee, he organized the 9th International Conference on Plant Pathogenic Bacteria held in Chennai, India during August 26–29, 1996 and also chaired a session. He also attended I, II, and III International Symposia of Plant Pathology and International Conference in New Delhi in 1967, 1971, 1981, and 1997, respectively; the Global Conference in Plant Pathology in Udaipur in 1995 and Asian Biotechnology Rice Network Workshop at Ludhiana in 1998; and 25 national and eight zonal meetings of different professional societies.

Dr. Thind has been a member of Indian Phytopathological Society, New Delhi, since 1966 and is now designated as a life member. He has been a member of the Indian Society of Mycology and Plant Pathology, Udaipur, since 1976, of the Indian Society of Plant Pathologists, Ludhiana, since 1984, and of the Indian Science Congress Association, Kolkata, since 1992.

Dr. Thind has served as paper setter and external examiner of four universities and Agricultural Scientists' Recruitment Board, New Delhi; as an expert of selection committees for the post of Associate Professors/Professors and as an external examiner of PhD and MSc students of different universities; and as reviewer/referee of various national and international research journals.

He has also published 127 research papers in scientific journals of international and national repute, has authored one manual entitled, *Plant Bacteriology* and one textbook entitled, *Phytopathogenic Procaryotes and Plant Diseases* published by Scientific Publishers (India).

Section I

General Aspects

1 Introduction

1.1 PHYTOBACTERIOLOGY

Phytobacteriology or plant bacteriology is a subdiscipline of plant pathology, which deals with plant-associated bacteria and their interactions with each other and with their hosts (Dumenyo et al. 2001). The subdiscipline covers all aspects of plant-associated bacteria including basic as well as of applied nature. The morphology, physiology, taxonomy, genetics, and serology of the bacteria comprise the basic aspects. Host-pathogen interactions, including pathogenesis and infection process, epidemiology, effect of environment on disease development, and disease management are the applied aspects dealt with in this subdiscipline. In the past, the emphasis has been mainly on the investigation of plant diseases caused by bacteria due to heavy losses caused by these diseases. However, lately, the field of phytobacteriology has expanded considerably to include beneficial bacterial-plant interactions such as nodulation and nitrogen fixation; promotion of plant growth by phyllosphere, rhizosphere, and soil-inhabiting bacteria, and control of ice-nucleation active bacteria by phyllosphere bacteria. Phytobacteriology also includes the management and control of bacterial plant diseases using various methods such as cultural practices, resistant cultivars, chemical and biological control, and molecular genetics. A basic knowledge of various branches of science like microbiology, genetics, chemistry, biochemistry, plant physiology, agronomy, soil science, and molecular biology is essential for better understanding of plant diseases caused by microorganisms. The molecular biology is now becoming increasingly important in every biological science including phytobacteriology.

1.2 PROCARYOTES

On the basis of cellular organization, the living organisms can be divided into two groups, namely, procaryotes and eucaryotes. The procaryotes have a set of characteristics, which differentiate them from eucaryotes. The main characteristic of the procaryotes is that with a few exceptions, nucleoplasm or genophore (earlier called nucleoid) is not separated from the cytoplasm by a unit-membrane system, i.e., the nuclear membrane is absent. The other characteristic features are the presence of 70S ribosomes (with the exception of archaea, whose ribosomes have a slightly higher S value) dispersed in the cytoplasm, no cytoplasmic streaming, and absence of endoplasmic reticulum with attached ribosomes. They take the nutrients in liquid form. Most of them possess a rigid cell wall except one group, i.e., **mollicutes**. They occur as single cells or simple association of similar cells. The cells may be motile or non-motile; in case of motile, the motility is mediated by bacterial flagella or gliding motility on solid surface.

The procaryotes are inhabitants of moist environments and predominantly occur as unicellular microorganisms. However, they also occur in other forms such as, filaments, mycelia, or colonies. In some cases, differentiating structures like holdfasts and resting cells are formed. The gene transfer and recombinations occur, but in these processes gametogenesis and zygote formation does not ever occur.

The above given characteristics of procaryotes are sufficient to distinguish them from eucaryotic microorganisms, but some common features are found in some procaryotes and eucaryotes. For example, the hyphae formed in actinomycetes look similar to those of molds, and ability of spirochetes to twist and contort their shape matches the flexibility shown by certain protozoa. Some large bacteria may approach the size of eucaryotic cells, while some eucaryotic cells are as small as bacteria. However, procaryotic and eucaryotic cells can be easily differentiated on the basis of fluorescent-labeled genes.

The procaryotes comprise two different groups, namely, the ***Bacteria*** and the ***Archaea***, which have been recognized as different domains for the purpose of classification (Krieg 2001).

1.2.1 Bacteria

On the basis of Gram stain reaction and presence or absence of cell wall, the bacteria have been divided into the following three phenotypic subgroups:

1. Gram-stain-negative bacteria having cell walls
2. Gram-stain-positive bacteria having cell walls
3. Bacteria lacing cell walls.

1.2.1.1 Gram-Stain-Negative Bacteria Having Cell Walls

These bacteria possess a Gram-stain-negative type of cell wall, which is composed of an outer membrane, and an inner, relatively thin peptidoglycan layer containing muramic acid; the latter may be absent in a few organisms. Reproduction is usually by binary fission, some groups show budding and a very few divide by multiple fission. The cells may occur in the form of spheres, ovals, straight or curved rods, helicals, or filaments. The cells may be non-motile or motile showing swimming or gliding motility. In relation to oxygen requirement, the members may be aerobic, anaerobic, facultatively anaerobic, or microaerophilic.

1.2.1.2 Gram-Stain-Positive Bacteria Having Cell Walls

These bacteria have a Gram-stain-positive type of cell wall, in which there is no outer membrane and the peptidoglycan layer is relatively thicker than that of Gram-stain-negative type of cell wall. The cell shapes include rods, spheres, or filaments; the rods and filaments may show true branching.

The reproduction is mostly by binary fission, but actinomycetes and their relatives produce sporogenous hyphae. Some members produce spores, also called **endospores**, which help to tide over the adverse conditions, but are not a mean of multiplication. The oxygen requirement of the members is the same as that of Gram-stain-negative bacteria.

1.2.1.3 Bacteria Lacking Cell Walls

They are commonly called the "**mycoplasmas**" (also called mollicutes or soft-skin microorganisms). They lack the cell wall, but are enclosed by a unit membrane, plasma membrane. As they lack the cell wall (peptidoglycan layer), they are not sensitive to β-lactam antibiotics or other antibiotics, which inhibit cell wall synthesis. They are highly pleomorphic, have different cell forms varying in size and some of these forms are filterable. Cells stain Gram-stain-negative, no resting forms are produced, and usually, they are non-motile, but some members show gliding motility. Different modes of reproduction such as, budding, fragmentation, and/or binary fission occur. Most members require complex media and all require cholesterol or related sterols for growth. The typical colony under adequate growth conditions is biphasic having **fried-egg appearance**, i.e., consisting of an opaque, granular central area which grows down into the medium and a flat, translucent peripheral zone. The mol% G + C contents of rRNA (43–48) and of DNA (23–46) are lower than that of walled bacteria.

1.2.2 Archaea

Archaea, predominantly inhabitants of terrestrial and aquatic habitats, are found in anaerobic, hypersaline, or hydrothermally and geothermally heated environments. Some members also live symbiotically in the digestive tracts of animals. Archaea grow as mesophiles and thermophiles, but some species grow at very high temperatures, i.e., up to 110°C. They grow as aerobes, anaerobes, and facultative anaerobes. Archaea are not sensitive to β-lactam antibiotics as their cell walls do not contain murein, peptidoglycan containing muramic acid. All archaea contain glycerol isopranyl ether lipids, which is their unique biochemical property.

Gram-stain-reaction varies depending on the composition of cell envelopes. It may be Gram-stain-positive or -negative within the same group. The cells are of different shapes, which include spherical, spiral, plate, or rod. Unicellular or multicellular forms also occur in filaments or aggregates. The modes of reproduction include binary fission, budding, fragmentation, constriction, or mechanisms not yet known. The cell masses exhibit a variety of colors including red, pink, purple, green, yellow, grey, white, etc. Five major groups recognized in archaea are given below.

1. Methanogenic archaea
2. Sulphate-reducing archaea
3. Extremely halophilic archaea
4. Cell wall-lacking archaea
5. Extremely thermophilic S°-metabolizing archaea.

1.3 PHYTOPATHOGENIC BACTERIA

Phytopathogenic bacteria include cell-walled bacteria, including both Gram-stain-negative and Gram-stain-positive, and cell wall-lacking bacteria (mollicutes). The general characteristics of these phytopathogens are that majority of them are non-spore-forming, rod-shaped organisms. Only *Streptomyces* species produce spores and are filamentous. Most plant pathogenic bacteria are nutritionally non-fastidious. Their optimal growth temperature is around 23°C–28°C. However, a few of them, such as *Ralstonia solanacearum* and *Burkholderia cepacia*, can grow at 42°C. Plant pathogenic bacteria can be either strictly aerobic or facultatively anaerobic.

Phytopathogenic and photosymbiotic procaryotes have one unique feature, i.e., their ability to multiply in the plants that separates them from all other procaryotes. Pathogens can increase from an initial low number of cells to a higher number in a short period of time within a host plant. No other feature clearly distinguishes them from the myriad of saprophytes that occupy every other plant-associated habitat (Kennedy and Lacy 1982). They have evolved to grow well in their plant hosts, often in specific hosts as well as in a specific plant part, such as blossoms, leaves, roots, tubers, or corms. Inside the plant hosts, phytopathogenic procaryotes multiply in the intercellular spaces and penetration of the cells generally occurs after the disruption and disorganization of the cell walls. Genetic studies have shown the involvement of numerous genes in virulence and pathogenesis. A number of plant pathogenic bacteria harbor virulence and pathogenicity genes on plasmid (an extrachromosomal element), whereas other pathogens bear virulence and pathogenicity genes in their chromosomes. The conservation of virulence and pathogenicity genes in pathogenicity islands strongly suggests interchangeable movement of these genes across species boundaries via lateral gene transfer flow (Hacker and Kaper 2000; Syvanen and Kado 2002). Many of these genes provide an inherent ability to the pathogen to grow in its host and hence, are essential for successful colonization of the host.

Phytopathogenic procaryotes cause plant diseases on a wide range of host plants. Some of these diseases are highly devastating and cause serious economic losses. Bacterial plant pathogens develop new survival strategies to thwart efforts to control and eliminate plant diseases caused by them. Kado (2009) reported that acquisition of novel genes through lateral gene transfer is a powerful means for pathogen development. The generation, through selective mutations, of novel and functional genes whose products enhance successful colonization can also facilitate pathogen fitness. Irrespective of the mechanism, genetic changes that are beneficial to pathogen survival occur due to selection pressure put on the pathogen by various control measures adopted in farming practices.

As the procaryotes include both bacteria and archaea, and among them only bacteria are pathogenic to plants, the terms "**phytopathogenic prokaryotes**" and "**phytopathogenic bacteria**" have been used interchangeably in this book.

1.3.1 Economic Importance of Phytopathogenic Bacteria

There is not enough data available on losses due to plant diseases caused by phytopathogenic bacteria in spite of the fact that some of these diseases are of international importance. The loss caused by a plant disease varies from region to region due to variation in environmental factors, which influence the disease development and also variation in susceptibility of the varieties under cultivation. For example, bacterial blight of rice is severe mainly in Southeast Asia. Moreover, the importance of a crop also varies in a particular region or country. Chemical control of bacterial plant diseases is less effective than those of caused by fungi due to the availability of less number of effective chemicals and low efficacy of the available chemicals. Moreover, many antibiotics, which are effective against many plant pathogenic bacteria, are also used for the control of human and animal bacterial pathogens. Hence, their use in plant disease control in many countries is not permitted due to the risk of transfer of resistance to human and animal pathogens. The crop losses due to plant pathogenic procaryotes in the United States in 1976 are given in Table 1.1 (Kennedy and Alcorn 1980).

Bacterial wilt of solanaceous plants is considered the world's single most destructive bacterial plant disease. According to Champoiseau and Allen (2009), brown rot bacterium is responsible for an estimated loss of one billion US dollars worldwide every year, mostly to poor small-scale farmers in tropical highlands. Bacterial wilt of potato is also considered the second most important disease of potato worldwide after late blight.

Bove (2006) has rightly stated that citrus Huanglongbing (greening) is the most important, serious, severe, destructive, and devastating disease of citrus in the world. The reduction in yield can range from 30% to 100% depending on proportion of affected canopy. The affected orchards become economically unviable in 7–10 years after planting. About 100 million trees have been destroyed in many countries of South and Southeast Asia, Indonesia, the Philippines, India, Arabian Peninsula, and South Africa. Since 2004, more than 500 thousand trees have been officially destroyed in Brazil due to this disease and around 300–400 thousand trees unofficially destroyed by commercial citrus growers (Gottwald et al. 2007). Catastrophic losses due to the disease from India during 1960s were reported by Fraser et al. (1966).

Incomplete eradication of citrus canker from the United States, particularly from the state of Florida, has taken a very heavy toll of citrus trees. After its first introduction in Florida in 1912, it took about 20 years to eradicate the disease from Florida by destroying and burning more than three million nursery and a quarter million fruit bearing trees. The second introduction of disease in Florida in 1986 resulted in the destruction of more than 20 million nursery and orchard trees, and the eradication program continued until 1992. The third introduction of the disease in Florida in 1995 and subsequent years led to the destruction of 1.56 million commercial trees and 6,000,000 dooryard trees leaving the eradication target incomplete. The successful eradication of the disease from Australia, South Africa, and New Zealand and partially successful eradication (incidence below 0.22% in São Paulo) from Brazil was achieved after destroying citrus trees on a large scale.

Pierce's disease of grapevine caused by *Xyl. fastidiosa* subsp. *fastidiosa* was noticed in California State of the United States in the 1880s, but the lack of appropriate vector of casual agent and other non-favoring factors kept the disease

TABLE 1.1
Estimates of US Crop Losses Caused by Phytopathogenic Procaryotes in 1976[a]

Pathogen[b]	Disease	Loss (millions of US dollars)
Ralstonia solanacearum	Bacterial wilt of tomato and tobacco	9.4
Agrobacterium tumefaciens	Crown gall of fruits and nuts	23.0
Erwinia amylovora	Fire blight of pear	4.7
Er. tracheiphila	Cucumber wilt	0.9
Pantoea stewartii subsp. *stewartii*	Stewart's wilt of corn	0.2
Pectobacterium atrosepticum and/or *Pe. carotovorum* subsp. *carotovorum*	Blackleg and/or soft rot of potato	14.0
Clavibacter michiganensis	Bacterial wilt of alfalfa	17.0
Pseudomonas savastanoi pv. *glycinea*	Bacterial blight of soybean	65.0
Psm. syringae pv. *syringae*	Bacterial leaf blight of wheat	18.0
Xanthomonas citri subsp. *malvacearum*	Angular leaf spot of cotton	15.0
Xa. axonopodis pv. *phaseoli*	Common bacterial blight of bean	5.0
Leifsonia xyli subsp. *xyli*	Ratoon stunting disease of sugarcane	10.0
Xylella fastidiosa subsp. *fastidiosa*	Pierce's disease of grapevine	3.0
Spiroplasma citri	Citrus stubborn disease	1.0
'*Candidatus* Phytoplasma spp.'	Pear decline	1.6
	Lethal yellowing of coconut	3.0

[a] Based on data summarized from Kennedy and Alcorn (1980).
[b] Recent names of the pathogens are given.

under check. The introduction of glassy-winged sharpshooter (*Homalodisca vitripennis*), a highly efficient vector of the bacterium, in Southern California has changed the status of the disease from minor to severe with enormous increase in losses. It has threatened not only the cultivation of the European grape (*Vitis vinifera*), but also of the many ornamental crops in the southeastern United States.

In 1918, 75% of the bean fields in New York State of the United States were affected with common blight (*Xa. axonopodis* pv. *phaseoli*), and the disease caused heavy losses. In the subsequent years, losses ranged from 20% to 50%. In 1953, the disease was widespread in Nebraska (USA), and it caused an estimated loss of over one million US dollars. In 1976, in the United States, common blight was economically the most important disease among the bacterial diseases of beans, and it caused an estimated loss of four million US dollars. According to Gilbertson and Maxwell (1992), in 1967, the disease caused yield losses up to 38% in dry bean in Ontario (Canada) and 10%–20% in navy bean in Michigan (United States). Yield losses ranging from 17% to 22% were also reported from Colombia. Besides loss in yield, the disease causes reduction in quality of seeds through discoloration of infected seeds.

Although the bacterial blight of rice, caused by *Xa. oryzae* pv. *oryzae,* occurs in many countries, it is most destructive in Southeast Asia. The losses are generally high in an early-infected crop and occurrence of kresek phase may lead to the total failure of the crop under certain situations. In Japan, prior to introduction of resistant varieties, losses generally ranged from 20% to 30%, but in some cases went as high as 50% (Ou 1972). Reports from Philippines, India, and Indonesia estimate that losses due to kresek phase have reached 60%–75% depending on weather conditions, location, and rice variety (Reddy et al. 1979; Ou 1985). In addition to reduction in yield, the disease may also affect grain quality by interfering with maturation.

Bacterial canker (black spot) of mango causes heavy losses because most of the commercial varieties of mango are highly susceptible to the pathogen, and the infection results in drastic yield losses. The fruit infection ranging from 50% to 80% is common on highly susceptible varieties. In 1966 and 1967, severe epiphytotics of the disease caused almost 100% fruit loss on most susceptible cultivars in most of the mango growing areas of South Africa. The fruit loss was mainly due to the fruit drop and non-marketability of the harvested fruit. The estimated loss alone in 1966 was approximately one million US dollar (Gagnevin and Pruvost 2001). Kishun (1981) observed 20%–80%, 10%–70%, 10%–40%, and 10%–55% fruit drop in Alphonso, Pairi, Totapuri, and local cultivars grown in India, respectively, and the maximum fruit drop in all the cultivars occurred 30–60 days after fruit set. He further observed that the rot in storage in these varieties was 5%–80%, 11%–67%, 10%–100%, and 5%–80%, respectively.

Lethal yellowing of coconut palm trees has killed several palm trees in Florida State of the United States and some other countries. First observed in Florida in 1971, it killed 40,000 trees by 1974, and by 1975, 75% of palm trees in Dade County of the state were either dead or dying due to this disease. In Jamaica, 5.4 million trees were killed during a span of 20 years starting from 1961. In Mexico and Tanzania, palm trees in thousands of hectares and in Ghana over a million coconut palm trees were killed within 30 years. The disease also killed more than 60 thousand palm trees in Togo by 1964 (Agrios 2004).

Common scab of potato is one of the most economically important diseases of the crop. It costs the Tasmanian potato industry in excess of 3.6 million Australian dollars per annum (Wilson 2004).

An estimated loss of $4.7 million due to fire blight occurred in California in 1976, while in Southwest Michigan alone, the loss in 1991 was $3.8 million. Again in 2000, a severe outbreak of the disease occurred on apples in Michigan, resulting into removal and destruction of 35,000–45,000 apple trees from approximately 1550–2300 acres and amounting to a loss of $42 million. In 1998, apple and pear growers in Washington and northern Oregon suffered a loss of $68 million due to fire blight damage.

Stewart's wilt of corn, caused by *Pa. stewartii* subsp. *stewartii*, is a good example of a disease that continues to have enormous implications on trade and commerce. The disease has been a major threat to corn producers in the eastern corn belt of the United States for over 100 years. Severe outbreaks of the disease occurred in Nebraska in 1999 and 2000. In 1999, it was found to occur in 762 out of 1317 fields in Iowa (58% prevalence). The impact of the disease was that the seed produced from 58% of the seed corn fields in Iowa in 1999 could not be exported to the countries having phytosanitary (quarantine) restrictions for this pathogen.

In addition to causing direct loss in crop yield and quality, phytopathogenic bacteria cause other harmful effects to mankind. *Rathayibacter toxicus* causes gumming disease in ryegrass and other fodder grasses, and produces toxins, which when consumed often result in fatal poisoning of livestock. Various species of *Anguina* (nematode) act as vector of this bacterium. Glycolipid toxins, known as corynetoxins, are produced by the bacterium as the infected grass matures and becomes senescent. In Australia, *Rat. toxicus* is found most commonly in *Lolium rigidum* (annual ryegrass) with *Anguina funesta* as a vector and in *Polypogon monspeliensis* (rabbit-foot grass) and *Lachnagrostis filiformis* (syn. *Agrotis avenacea*, annual blown grass) with an undescribed *Anguina* vector. In Australia, more than 100,000 sheep and thousands of cattle have died from this disease in certain years (Edgar 2004). Besides affecting farmers or growers, bacterial diseases also affect transport and other industries dependent on the affected crops.

In contrast to direct monetary losses and other harmful effects caused by phytopathogenic bacteria, they have a great deal of commercial value (Starr 1984). The studies of these bacteria have yielded very useful products and tools. Several phytopathogenic bacteria are used in industrial fermentation. The most prominent case is the large-scale use of *Xa. campestris* to make xanthan gums, extracellular polysaccharides of considerable economic importance. Xanthan gums are used for various purposes including oil-well drilling muds, tertiary recovery of petroleum from spent wells, ceramic glazes, thickening agents in foods, and antidrip agents in paints. Another class of economically important polysaccharides produced by phytopathogenic bacteria includes curdlans, which are produced by various species of *Agrobacterium* and other bacteria. On heating, the curdlans form very elastic and resilient gums. The curdlans have a potential use in various food products as non-caloric gelling, thickening, binding, and stabilizing additives. A pseudomonad isolated from an aquatic plant produces an agar-like polysaccharide that has the potential to be used as a substitute for agar in bacteriological media.

Besides vinegar production, *Gluconobacter* species have many biotechnological applications. In the last decades, many bioconversion routes for rare and special sugars involving *Gluconobacter* spp. have been developed. Among the recent ones, are the biotransformations involved in the production of L-ribose and miglitol, both very promising pharmaceutical lead molecules. Most of these processes make use of *Gluconobacter*'s membrane-bound polyol dehydrogenases.

Various enzymes produced by phytopathogenic bacteria have various applications and are produced commercially. Pectin-degrading enzymes have found an important application in the food industry, where they are used in the clarification of fruit juices. An enzyme produced by an *Agrobacterium* strain is used for converting super coiled, covalently closed, circular DNA into a simple, covalently closed, circular form. Various *Xanthomonas* species and other phytopathogenic bacteria produce restriction endonucleases having much utility in genetic engineering and basic research. An L-asparaginase (protein synthesis inhibitor drug) produced by soft rot bacterium, *Pe. chrysanthemi* has a great potential for use as an antileukemic agent. The potential of Ti-plasmid of *Ag. tumefaciens* to move genetic material from bacteria to plants has made it an important tool in plant genetic engineering and it is widely used for this purpose. This technology is largely responsible for the development of plant biotechnology industry and the latter is revolutionizing agriculture and health care. The production of agrocin by an avirulent bacterium, *Ag. radiobacter* (strains K84 and K1026) and its role in controlling crown gall disease on commercial scale are well documented. At present, there are more than 16 commercial biocontrol products based on different bacteria including *Ag. radiobacter*, which are used to control plant diseases caused by different pathogens. The detailed studies of the nodulation process by *Rhizobium and Bradyrhizobium* species have led to the preparations of bacterial inoculum for application to legume crops (Dumenyo et al. 2001).

1.4 HISTORICAL REVIEW OF PHYTOBACTERIOLOGY

Antony van Leeuwenhoek discovered the bacteria for the first time in 1676 and called them animalcules, i.e., small animals. In 1876, Robert Koch showed that anthrax of animals is caused by a bacterium, *Bacillus anthracis*. Robert Koch also devised pour plate method for the purification of bacteria and gave Koch's postulates to prove the pathogenicity of causal agents of pathogenic diseases.

Woronin, a Russian worker, was the first person to show association and multiplication of bacteria in root nodules of leguminous plants in 1866. However, the multiplication of bacteria in root nodules is considered a symbiotic relationship and not a strictly pathogenic action. Professor T.J. Burrill of the University of Illinois was the first to show a bacterium as a causal agent of a plant disease. In a series of experiments starting from 1878 till 1884, he showed that the fire blight of pear was caused by a bacterium, which he identified as *Micrococcus amylovorus*. As the pure culture technique was in early stage of development in the laboratory of Robert Koch in Germany, Burrill was unaware of this modern microbiological method. Therefore, he did not use the pure culture of the bacterium in his work. J.C. Arthur working at New York Experiment Station confirmed Burrill's work during 1885–1887 using pure culture of the bacterium. In 1883, J.H. Wakker from Holland showed that yellow-slime disease of hyacinth was caused by a bacterium. Luigi Savastano, an Italian, demonstrated in 1887 that a bacterium incited the knot disease of olive. In 1893, L.H. Pammel found that bacteriosis of cabbage is caused by a yellow bacterium, *Bacillus campestris*, a name that was later changed to *Xa. campestris* pv. *campestris*.

Frediano Cavara (1857–1929), an Italian botanist who was the director of the Royal Botanical Gardens at the University of Naples, had for some time keenly observed galls on grapevines. He made the successful isolations of a bacterium from the galls and demonstrated that it caused an identical gall disease on healthy grapevines (Cavara 1897). Later on, Erwin F. Smith identified this bacterium and named it as *Bacterium tumefaciens* (Smith and Townsend 1907), which was changed to *Phytomonas tumefaciens* and then again changed to *Agrobacterium tumefaciens*. In the reports by Cavara (1897), it is clear that Smith, who visited Cavara in 1904, was shown the bacterial cause of the crown gall disease. Although Erwin F. Smith and C.O. Townsend (Smith and Townsend 1907) are cited as the discoverers of the bacterial nature of crown gall, credit should also be given to Cavara for the initial discovery and demonstration of the bacterial etiology of crown gall. In fact, Smith and Townsend confirmed the earlier findings of Cavara. Smith became intrigued with this disease and used it as a paradigm of cancer in animals and humans (Smith 1911). In 1901, L.R. Jones showed that the soft rot of vegetables is caused by *Bacillus carotovorus*.

Soon after 1890, Erwin F. Smith started work on bacterial diseases and the first disease investigated by him was bacterial wilt of cucurbits. Later on, he worked on many bacterial diseases of plants including bacterial wilt of solanaceous plants and black rot of crucifers. Erwin F. Smith's monumental contributions are included in his monograph entitled, *Bacteria in Relation to Plant Diseases* published in three volumes from 1905 to 1914. In 1896, Smith postulated that there are in all probability as many bacterial diseases of plants as there are of animals (Smith 1896). As late as 1920, Smith predicted that there will be bacterial diseases found on plants of every plant family.

A number of plant pathologists and authors of many books have given credit to T.J. Burrill for providing first credible proof that bacteria could cause plant diseases. However, Kennedy et al. (1979), in their critical appraisal of research work on bacterial diseases of plants carried out by T.J. Burrill and others, have opined that the credit for this discovery should be shared by E. Mitscherlich, a German chemist, T.J. Burrill, and J.C. Arthur and not given to Burrill alone. They have further stated that Burrill's place in history with respect to these studies is perhaps based on the fact that he was proved by others to have been right, and that he had a strong advocate in Erwin F. Smith.

Most of the work on bacterial diseases of plants from 1878 to 1900 and even later was carried out in the United States because there was a distinct air of disbelief on the part of European botanists regarding the role of bacteria in causing plant diseases. This was mainly due to the influence of De Bary, who considered bacteria as of minor importance as pathogens of plants. De Bary was aware of Wakker's work on the hyacinth disease, but not of Burrill's work on fire blight when he published his *Comparative Morphology and Biology of Fungi, Mycetozoa and Bacteria* in 1884. In this publication in the section of bacteria, he devoted just one short paragraph to those parasitic on plants and also pointed out that they had scarcely ever been observed. De Bary in his *Lectures on Bacteria* published in 1887 devoted only 2 out of 146 pages to bacteria parasitic to plants and stated that saprophytic bacteria may also under special conditions, attack the tissues of living plants as facultative parasites, produce disease in them, and destroy them.

Alfred Fischer, a professor of botany at Leipzig, Germany, who was a student of De Bary in 1881, challenged the work of Erwin F. Smith and others concerning the role of bacteria in causing plant diseases in 1897. A. Fischer in his book entitled, *Lectures on Bacteria* refused to accept the findings of Smith and others terming them unreliable. Smith replied to Fischer in 1899 pointing out his complete ignorance of the subject. Fischer wrote back the same year, and Smith replied in two rebuttal papers, which seemed to end the controversy. Smith was the clear winner of the debate. This debate has been duly recorded in one of the Phytopathological Classics published by APS Press (Campbell 1981). Walker (1969) has rightly commented on the controversy created by Alfred Fischer, stating that while this debate is now chiefly of historical importance, it is significant that more than 20 years after the work of Robert Koch and of T.J. Burrill; a rather prominent botanist still challenged the existence of bacterial plant pathogens. By 1900, the concept that bacteria could cause plant diseases was solidly established (Starr 1984), thereby creating **phytobacteriology** as a new subdiscipline of plant pathology. Erwin F. Smith's contributions to the field of plant bacteriology and his efforts to put it on sound footing earned him the title of "**father of plant bacteriology**." Goto (1992) has given a brief account of research work done on bacterial diseases of plants in Japan from the late 1880s to 1970s.

Although Erwin F. Smith made highly significant contributions to the field of plant bacteriology, he did not establish a dynasty or a school where young scientists could nurture their enthusiasm and interest in the newly discovered science of phytobacteriology as remarked by Sequeira (2000) in his review article. Therefore, after his death in 1927, phytobacteriology began to wane. Around the middle of twentieth century, plant pathology was dominated by mycology and virology.

In the initial stage of plant bacteriology, the emphasis was on reporting new bacterial diseases of plants and description and identification of their causal bacteria. Then the emphasis shifted to the investigations on the physiology of diseased plants, and during 1940s to 1960s, the significant contributions were made in this field along with ecology of plant pathogenic bacteria. The work of R.M. Klein (1954) and of A.C. Braun and Ross Pringle (1959) provided the evidence that the crown gall bacterium, *Ag. tumefaciens* caused permanent transformation of host cells leading to their autonomous and rapid growth in culture. This work along with pioneer work of A.C. Braun and his colleagues on the structure and mode of action of wildfire toxin produced by *Psm. syringae* pv. *tabaci* greatly impacted many plant bacteriologists to determine the pathogenicity determinants in other diseases (Sequeira 2000).

The discovery of mycoplasma-like organisms by Doi et al. (1967) gave the correct etiology of yellow-type diseases, which were earlier thought to be caused by viruses. The discovery of mycoplasma-like organism also resulted in the subsequent discovery of spiroplasmas and fastidious procaryotes. The mycoplasma-like organisms are now called phytoplasmas. The demonstration of conjugative transfer of bacterial genes in *Erwinia* by Chatterjee and Starr (1972) shifted the emphasis of research to molecular genetics of bacterial plant pathogens. Another notable contribution during 1970s was the success achieved in the biological control of crown gall with *Ag. radiobacter* strain K84 (New and Kerr 1972). Subsequently, an effective and viable biological control of fire blight has also been achieved. The introduction of pathovar system by Young et al. (1978) has greatly streamlined the taxonomy of plant pathogenic bacteria. Rapid advances in DNA sequencing technology and sequence analysis of highly conserved regions of the bacterial genome, such as the small subunit rRNA gene have led to the natural classification of procaryotes that reflects the evolutionary history of bacteria and archaea. The widespread application of new methods of classifying procaryotes has led to an explosive growth in the number of validly published species and higher taxa. Since the completion of first edition of *Bergey's Manual of Systematic Bacteriology* in 1989, the number of published species has more than tripled and accompanied with numerous taxonomic rearrangements and changes in nomenclature. The number of bacterial genera containing phytopathogenic bacteria, which was only six during 1970s, now stands at 53 including three '*Candidatus* genera'.

The first avirulence gene (*avrA*) characterized was cloned from a race 6 strain of the soybean pathogen, *Psm. syringae* pv. *glycinea* (recent name, *Psm. savastanoi* pv. *glycinea*) by Staskawicz et al. (1984). It elicits a resistant reaction on soybean cultivars having resistant gene *Rpg2*. Avirulence genes determine the race specificity of pathogen in a compatible host pathogen interaction. The products of *avr* genes apparently function as recognition factors in resistant hosts (i.e., carrying the *R* gene) and virulence determinants in susceptible hosts (i.e., carrying the *r* allele). In most cases, bacterial avirulence gene function is dependent on interactions with hypersensitive reaction and pathogenicity (*hrp*) genes. The *hrp* genes which control the ability of phytopathogenic bacteria to cause disease on susceptible plants and elicit hypersensitive reaction on resistant or non-host plants were first discovered for *Psm. syringae* pv. *phaseolicola* by Lindgren and associates in 1986. The role of avirulence genes and of *hrp* genes in host parasite interaction have been reviewed by Leach and White (1996) and Lindgren (1997), respectively.

Investigations carried on *Ag. tumefaciens*, the causal agent of crown gall, during the last 100 years (since its report by Smith and Townsend in 1907) have made this bacterium from merely a plant pathogen to a house-hold name in plant genetic engineering and plant molecular genetics (Nester 2008). The basic research carried out with this bacterium has provided very useful insights into host-parasite interactions and the mechanism of gene transfer from a procaryote to a eucaryote. The main problem of using Ti-plasmid as a gene vector, due to its larger size, has been solved by separating it into two parts; the 25 bp borders flanking the T-DNA between which the genes of interest are inserted and the vir region that is necessary for the processing and transfer of the T-DNA. This binary vector system is used in most laboratories. *Agrobacterium* is truly a natural genetic engineer.

Lately, atmospheric pressure plasmas have received much attention as a promising physical tool for biological decontamination and sterilization. Several types of plasma devices have been described for sterilization of animal and human tissues, and the system can also be used for reducing the population of various plant microbes that cause infectious diseases. Mráz et al. (2014) used *Cla. michiganensis* subsp. *michiganensis*, *Er. amylovora*, and *Escherichia coli* (as a control) to study the bacterial growth mechanism after exposure to the atmospheric pressure GlidArc plasma. It was found that low-temperature plasma treatment affected all the three bacterial species by slowing down growth and reproduction rate. After plasma treatment, both *Es. coli* and *Er. amylovora* reached their maximum growth sooner than the *Cla. michiganensis* subsp. *michiganensis*. This is, however, caused by a longer cultivation period of *Clavibacter* genus. The application of GlidArc plasma is a suitable treatment that can be used in agriculture for plant protection. Further studies of low-temperature atmospheric pressure plasmas' effect on bacteria could be focused on its practical use in seed treatment of various crops.

A brief account of development of phytobacteriology in India is given below. In India, Cappel (1892) reported bangle blight or bungdi disease of potato from Pune, Maharashtra. Butler in 1903 also reported this disease and stated that it was similar to brown rot of potato caused by *Psm. solanacearum*. Coleman in 1909 proved the bacterial nature of the disease. Later on, Dr. M.K. Patel started work on bacterial diseases in 1948. He along with his associates worked for nearly 15 years and reported nearly 40 bacterial diseases, most of which were new records. Based on his contribution to the field of plant bacteriology in India, Dr. M.K. Patel is regarded as the "**father of plant bacteriology**" in India. A review of bacterial plant disease investigation published by Patel and Kulkarni (1953) included 27 species of bacterial pathogens reported from India. The first book on bacterial plant diseases in India was written by G. Rangaswami in 1962, in which he listed 73 bacterial plant pathogens. In 1965, Mathur and associates published a list of bacterial plant pathogens found in India. Phytopathogenic bacteria of India and bibliography published by Chakravarti et al. (1973) contains a list of 483 references on plant pathogenic bacteria and 148 bacterial plant pathogens belonging to six genera, namely, *Agrobacterium*, *Bacillus*, *Corynebacterium*, *Erwinia*, *Pseudomonas*, and *Xanthomonas*. A summer institute on Plant Bacteriology under the directorship of Dr. P.N. Patel was held in the Division of Mycology and Plant Pathology, Indian Agricultural Research Institute, New Delhi in 1972, and its proceedings were published in three volumes edited by Dr. P.N. Patel. First National Symposium on Plant Bacterial Diseases was held at Sri Venkateswara University, Tirupati in 1980 under the directorship of Dr. M.V. Nayudu and several important recommendations were made. Ninth International Conference on Plant Pathogenic Bacteria was held in 1996 at University of Madras, Chennai under the chairmanship of Dr. A. Mahadevan.

At present, phytobacteriology is one of the leading areas in plant pathology and has attracted a large number of scientists from areas other than plant pathology. It is evident from the large number of papers on phytobacteriology published in *Molecular Plant-Microbe Interactions*, *The Plant Cell*, and *Applied and Environmental Microbiology*.

REFERENCES

Agrios, G. N. 2004. *Plant Pathology*. 5th ed. Elsevier Academic Press, San Diego, CA, 922 pp.

Bove, J. M. 2006. Huanglongbing: A destructive, newly emerging century-old disease of citrus. *J. Plant Pathol.* 88: 7–37.

Braun, A. C., and Pringle, R. B. 1959. Pathogen factors in the physiology of disease-toxins and other metabolites. pp. 88–99 in: *Plant Pathology: Problems and Progress*. S. Holton, G. W. Fischer, R. W. Fulton, H. Hart, and S. E. A. McCallan eds. University of Wisconsin, Madison, WI.

Campbell, C. L. ed. 1981. *The Fischer-Smith Controversy: Are There Bacterial Diseases of Plants*? Phytopathological Classics No. 13, APS Press, St. Paul, MN, XVIII + 65 pp.

Cappel, E. L. 1892. A note on a potato disease prevalent in Poona district and elsewhere. Bombay Dir. Department of Land Records and Agriculture (Abstr. Butler, 1903).

Cavara, F. 1897. Eziologia di alcune malattie de piante coltivate. *Staz. Sper. Agr. Ital.* 30: 482–509.

Chakravarti, B. P., Hegde, S. V., Ahmed, S. R., Gupta, D. K., and Rangarajan, M. 1973. Phytopathogenic bacteria of India and bibliography (1892–1972). Bulletin, Department of Plant Pathology, University of Udaipur, Udaipur, India.

Champoiseau, P. G., and Allen, C. 2009. *Ralstonia solanacearum* race 3 biovar 2: Tropical losses, temperate anxieties. APSnet Feature Story, http://www.apsnet.org/online/ feature/ralstonia/
Chatterjee, A. K., and Starr, M. P. 1972. Genetic transfer of episomic elements among *Erwinia* species and other Enterobacteria: F'lac+. *J. Bacteriol.* 111: 169–176.
Doi, Y. M., Teranaka, M., Yora, K., and Asuyama, H. 1967. Mycoplasma- or PLT group-like microorganisms found in the phloem elements of plants infected with mulberry dwarf, potato witches' broom, aster yellows, or paulownia witches' broom. *Ann. Phytopathol. Soc. Jpn.* 33: 259–266.
Dumenyo, C. K., Chatterjee, A., and Chatterjee, A. K. 2001. Phytobacteriology. pp. 762–769 in: *Encylopedia of Plant Pathology.* Vol. II, O. C. Maloy, and T. D. Murray eds. John Wiley & Sons, New York.
Edgar, J. 2004. Future impact of food safety issues on animal production and trade: Implication for research. *Aust. J. Exp. Agri.* 44: 1073–1078.
Fraser, L. R., Singh, D., Capoor, S. P., and Nariani, T. K. 1966. Greening virus, the likely cause of citrus dieback in India. *FAO Plant Prot. Bull.* 14: 127–130.
Gagnevin, L., and Pruvost, O. 2001. Epidemiology and control of mango bacterial black spot. *Plant Dis.* 85: 928–935.
Gilbertson, R. L., and Maxwell, D. P. 1992. Common blight of beans. pp. 18–39 in: *Plant Diseases of International Importance. Vol. II, Diseases of Vegetables and Oilseed Crops.* H. S. Chaube, U. S. Singh, A. N. Mukhopadhyay, and J. Kumar eds. Prentice Hall, Englewood Cliffs, NJ.
Goto, M. 1992. *Fundamentals of Bacterial Plant Pathology.* Academic Press, San Diego, CA, 342 pp.
Gottwald, T. R., da Graca, J. V., and Bassanezi, R. B. 2007. Citrus huanglongbing: The pathogen and its impact. APSnet Feature Story, http://www.apsent.org/online/feature/huanglongbing/
Hacker, J., and Kaper, J. B. 2000. Pathogenicity islands and evolution of microbes. *Annu. Rev. Microbiol.* 54: 641–679.
Kado, C. I. 2009. Horizontal gene transfer: Sustaining pathogenicity and optimizing host-pathogen interactions. *Mol. Plant Pathol.* 10: 143–150.
Kennedy, B. W., and Alcorn, S. M. 1980. Estimates of U.S. crop losses to procaryote plant pathogens. *Plant Dis.* 64: 674–676.
Kennedy, B. W., Widin, K. D., and Baker, I. S. F. 1979. Bacteria as the cause of disease in plants: A historical perspective. *ASM News* 45: 1–5.
Kennedy, R. W., and Lacy, G. H. 1982. Phytopathogenic procaryotes: An overview. pp. 3–17 in: *Phytopathogenic Procaryotes.* Vol. 1, M. S. Mount, and G. H. Lacy eds. Academic Press, New York.
Kishun, R. 1981. Loss in mango fruit due to bacterial canker *Xanthomonas mangiferaeindicae. Proceedings of the 5th International Conference on Plant Pathogenic Bacterica,* Cali, Colombia, pp. 181–184.
Klein, R. M. 1954. Mechanisms of crown-gall induction. pp. 97–114 in: *Abnormal and Pathological Plant Growth. Brookhaven Symposia in Biology.* No. 6, Brookhaven National Laboratory.
Krieg, N. R. 2001. Procaryotic domains. pp. 21–25 in: *Bergey's Manual of Systematic Bacteriology.* 2nd ed. Vol. I, *The Archaea and the Deeply Branching and Phototrophic Bacteria.* D. R. Boone, and R. W. Castenholz eds. Springer Verlag, New York.
Leach, J. E., and White, F. F. 1996. Bacterial avirulence genes. *Annu. Rev. Phytopathol.* 34: 153–179.
Lindgren, P. B. 1997. The role of *hrp* genes during plant bacterial interactions. *Annu. Rev. Phytopathol.* 35: 129–152.
Mráz, I., Beran, P., Šerá, B., Gavril, B., and Hnatiuc, E. 2014. Effect of low-temperature plasma treatment on the growth and reproduction rate of some plant pathogenic bacteria. *J. Plant Pathol.* 96: 63–67.
Nester, E. 2008. *Agrobacterium:* The natural genetic engineer: 100 years later. APSnet Feature Story, http://www.apsnet.org/online/feature/Agrobacterium/
New, P. B., and Kerr, A. 1972. Biological control of crown gall: Field observations and glasshouse experiments. *J. Appl. Bacteriol.* 35: 279–287.
Ou, S. H. 1972. *Rice Diseases.* 1st ed. Commonwealth Mycological Institute, Kew, UK, 368 pp.
Ou, S. H. 1985. *Rice Diseases.* 2nd ed. C. A. B. International, Slough, UK, 380 pp.
Patel, M. K., and Kulkarni, Y. S. 1953. A review of bacterial plant disease investigation in India. *Indian Phytopathol.* 5: 131–140.
Reddy, A. P. K., MacKenzie, D. R., Rouse, D. I., and Rao, A. V. 1979. Relationship of bacterial leaf blight severity to grain yield of rice. *Phytopathology* 69: 967–969.
Sequeira, L. 2000. Legacy for the millennium: A century of progress in plant pathology. *Annu. Rev. Phytopathol.* 38: 1–17.
Smith, E. F. 1896. The bacterial diseases of plants: A critical review of the present state of our knowledge. *Am. Nat.* 30: 797–804.
Smith, E. F. 1911. Crown-gall and sarcoma. *U.S. Dep. Agric. Bur. Plant Ind. Circ.* 85: 1–4.
Smith, E. F., and Townsend, C. O. 1907. A plant-tumor of bacterial origin. *Science* 25: 671–673.
Starr, M. P. 1984. Landmarks in the development of phytobacteriology. *Annu. Rev. Phytopathol.* 22: 169–188.
Staskawicz, B. J., Dahlbeck, D., and Keen, N. T. 1984. Cloned avirulence gene of *Pseudomonas syringae* pv. *glycinea* determines race-specific incompatibility on *Glycine max* (L.) Merr. *Proc. Natl. Acad. Sci. USA* 81: 6024–6028.
Syvanen, M., and Kado, C. I. eds. 2002. *Horizontal Gene Transfer.* 2nd ed. Academic Press, San Diego, CA.
Walker, J. C. 1969. *Plant Pathology.* 3rd ed. McGraw-Hill Book Co., New York, 819 pp.
Wilson, C. R. 2004. A summary of common scab disease of potato research from Australia. *Proceedings of the International Potato Scab Symposium 2004.* S. Naito, N. Kondo, S. Akino, et al. eds. Hokkaido University, Sapporo, Japan, pp. 198–214.
Young, J. M., Dye, D. W., Bradbury, J. F., Panagopoulos, C. G., and Robbs, C. F. 1978. A proposed nomenclature and classification for plant pathogenic bacteria. *N. Z. J. Agric. Res.* 21: 153–177.

2 Bacterial Cell

On the basis of cellular organization, the living organisms are divided into two groups, namely, procaryotes and eucaryotes. The members of the procaryotic world comprise a vast heterogeneous group of very small, mostly unicellular organisms. Stanier and van Niel (1962) proposed that bacteria are distinguished from other forms of life, including viruses, protists, fungi, algae, plants, and animals, by their procaryotic cell structure. They defined the procaryotic cell by three major criteria: the absence of internal membranes that compartmentalize the nuclear material and the enzymatic machinery for respiration and photosynthesis, nuclear division that occurs by fission and not mitosis, and the presence of peptidoglycan in the cell wall.

During the last 50 years, a wealth of new information has been produced that greatly enriches our understanding of procaryotes. These organisms have proven to be of enormous abundance and diversity, the product of complex evolutionary processes over billions of years. They dominated life on the earth prior to the appearance of eucaryotes. Their distant past implies that many of the most salient features of modern life evolved in an entirely procaryotic world, including most of the organizing principles of the cell; the basic mechanisms of replication, transcription, and translation; the major catabolic and anabolic pathways; and the biogeochemical cycles that maintain the biosphere (Whitman 2009). Although Stanier and van Niel (1962) precisely defined the term procaryote, it has occasionally been misused, usually as synonymous with bacteria.

Procaryotes include two types of organisms, namely, bacteria and archaea. The majority of procaryotes, including the photosynthetic cyanobacteria are included in the bacteria. The bacteria look quite simple when we look at size, shape, and arrangement of their cells with an ordinary light microscope. However, the modern electron microscopy has revealed an amazing complexity and details of external and internal cell structures, which were not possible with an ordinary light microscope.

2.1 SIZE

Being microscopic, the bacteria are measured in micrometers (μm), which are equivalent to 1/1000 mm (10^{-3} mm). The cell size of bacteria varies with the species, but most cells are approximately 0.5–1.0 μm in diameter or width and 2.0–5.0 μm in length. The cells of *Staphylococcus* and *Streptococcus* species show a slightly larger diameter measuring 0.75–1.25 μm and some filamentous forms may be as long as 100 μm.

A few cases of bacteria having extremely large size have come to the knowledge of bacteriologists. *Epulopiscium fishelsoni* found in the gut of surgeon-fish measures 50–100 μm in width and 0.5 mm in length. Another bacterium, *Thiomargarita nomibiensis* measures 100–300 μm in diameter. This appears to be the largest bacteria discovered so far. The eucaryotic cells on an average measure from 2 to 200 μm in diameters.

The size of the bacterial cells is influenced by age and cultural conditions. Actively growing cells are slightly larger than the cells in their stationary phase. In stained preparations, the cell size is also reduced due to shrinkage of cells.

The extent of small size of a bacterium is evident from the fact that approximately 1 trillion (1,000,000,000,000 or 10^{12}) bacterial cells weigh a mere 1 g.

The importance of extremely minute size of a bacterial cell is reflected due to the very high ratio of its surface area to the volume. This highlights the fact that there is a large surface through which nutrients can enter for a relatively small volume of cell substance to be nourished. This is partly responsible for the high rate of metabolism and growth of bacteria. Due to their rapid growth and multiplication, the bacteria are used more frequently in biological research.

2.2 SHAPE

Every bacterial species is associated with a constant shape. There are three basic shapes of bacteria, namely, spherical, cylindrical, or spiral. Most commonly found bacteria are either cocci or bacilli.

2.2.1 Spherical

Spherical cells are called **cocci** (singular, **coccus**). They are usually round, but they can be ovoid or ellipsoidal (Figure 2.1a). Their size is measured in diameter.

2.2.2 Cylindrical

They are rod-shaped and have length and breadth. They are called **bacilli** (singular, **bacillus**). The ends of the cells may be flat, rounded, pointed, or tapered (Figure 2.1b). Their size is measured in length and breadth. There is not much difference in the width of different species, but differences in length are considerable.

2.2.3 Spiral

The cells are helical in shape and are called **spirilla** (singular, **spirillum**). Spiral bacteria have one or more twists, and they are never straight (Figure 2.1c). Certain spiral bacteria are curved short rods and resemble distinctive commas. These are called **vibrios**. *Vibrio cholerae*, the causal agent of cholera, is a typical example of vibrios. The other spiral bacteria are long rods

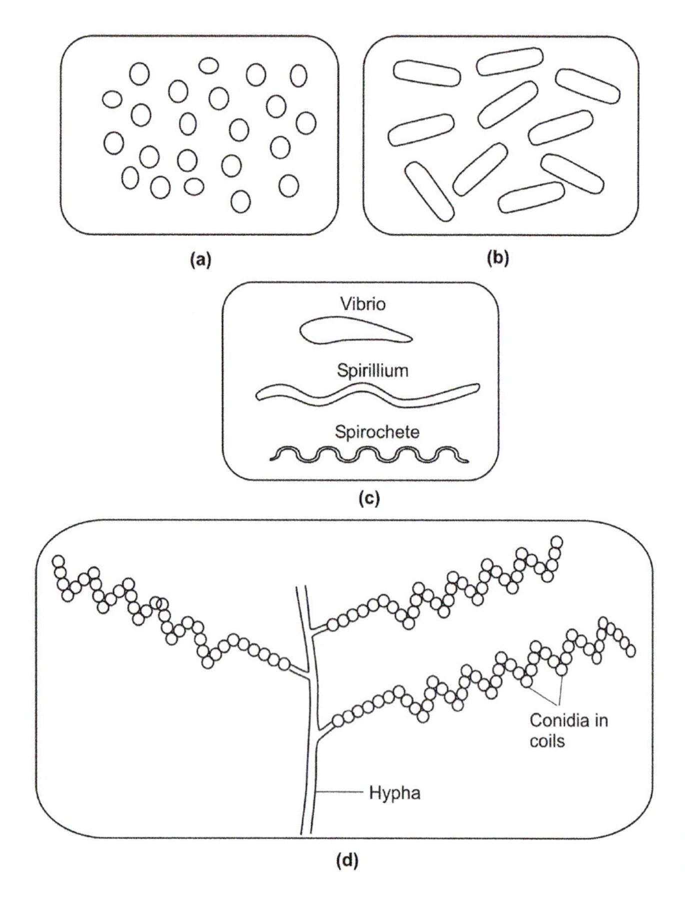

FIGURE 2.1 Shapes of bacterial cells: (a) cocci, (b) bacilli, (c) spirals, and (d) filamentous hyphae.

twisted into spirals; if the rods are rigid having thick cell walls, they are called **spirilla**, and **spirochetes** when cells have thin flexible cell walls (Tortora et al. 2007). *Treponema pallidum*, the causal agent of syphilis, is a typical example of a spirochete.

There are many modifications of these three basic shapes. The cells of *Pasteuria* are pear-shaped, while those of *Caryophanon* are disc-shaped. Square and star-shaped forms also occur. The genus *Caulobacter* contains appendaged bacteria.

The members of Actinomycetes produce long **filamentous hyphae** that may branch to produce a mass of hyphae called **mycelium** (Figure 2.1d). Although they appear like fungi, but the cells are typically procaryotic, hence classed with eubacteria.

Although the shape of cells of most bacterial species is constant under a definite set of environmental conditions, a few species have a variety of cell shapes and are called **pleomorphic**. Mycoplasmas are pleomorphic.

2.3 ARRANGEMENT

The majority of bacteria exist in unicellular forms; however, some species show arrangements or attachment of cells. This arrangement is usually found in spherical and cylindrical forms. In spherical cells, the following five types of arrangement of cells are found.

2.3.1 Diplococci

A coccus cell divides in one plane forming a pair of two cells, e.g., some spp. of *Neisseria* (Figure 2.2a).

2.3.2 Streptococci

The cells divide in one plane forming a chain of cells, e.g., *Streptococcus* and *Lactococcus* spp. (Figure 2.2b).

2.3.3 Tetrads

The cocci divide at a right angle to first plane of division forming a group of four cells in the shape of a square, e.g., *Pediococcus* and *Micrococcus* spp. (Figure 2.2c).

2.3.4 Sarcinae

In this case, the third division occurs in a plane different to the first two divisions resulting in cuboidal or packet arrangement of eight cells, e.g., *Sarcina* spp. (Figure 2.2d).

2.3.5 Staphylococci

When the divisions in the three planes occur in an irregular pattern, the arrangement of cells look like a bunch of grapes, e.g., *Staphylococcus* spp. (Figure 2.2e).

In cylindrical bacteria, only the following two types of cell arrangement are found.

2.3.6 Diplobacilli

Two cells are arranged from end to end (Figure 2.2f).

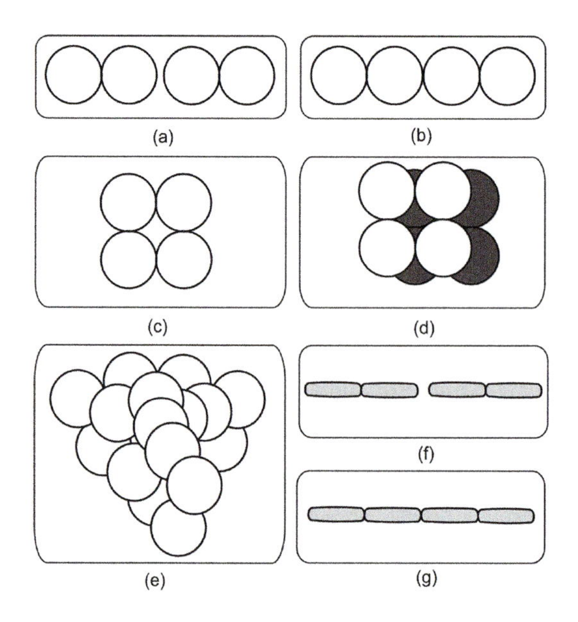

FIGURE 2.2 Arrangement of cells: (a) diplococci, (b) streptococci, (c) tetrads, (d) sarcina, (e) staphylococci, (f) diplobacilli, and (g) streptobacilli.

2.3.7 Streptobacilli

Three or more cells are arranged from end to end forming a chain, e.g., *Bacillus* spp. (Figure 2.2g).

Each of these arrangements is typical of a particular species and is helpful in its identification. However, rarely all the cells of a given species are arranged exactly in the same pattern. It is the predominant arrangement that is taken into account while studying bacteria. The size, shape, and arrangement of bacteria are important parameters of their gross morphology and are used in their identification.

2.4 CELL STRUCTURE

The cell structures described below are not common to all the bacteria. Some of these structures such as cytoplasmic membrane, cytoplasm, ribosomes, and nucleoid, region containing all or most of genophore, are found in all the bacteria, while the others are associated with certain particular species. Some of these structures are present inside the cell, while others are present outside the cell wall (Figure 2.3).

2.4.1 Flagella and Pili

Many bacteria are motile, and this motility is due to the presence of flagella (singular, flagellum). Many species of bacilli and spirilla possess flagella, while cocci rarely have these organelles. Flagella are thin hair-like structures with a helical shape. These originate from the cytoplasmic membrane and extend to the surface after penetrating the cell wall. Flagella are many times longer than the cells, measuring 15–20 μm in length. However, the diameter of a flagellum is only a fraction of width of the cells, i.e., 12–20 nanometers (nm). Being too thin in width, unstained flagella cannot be seen with an ordinary light microscope. The staining procedures using a mordant make flagella thicker due to the deposition of dye and thus making them visible by light microscopy.

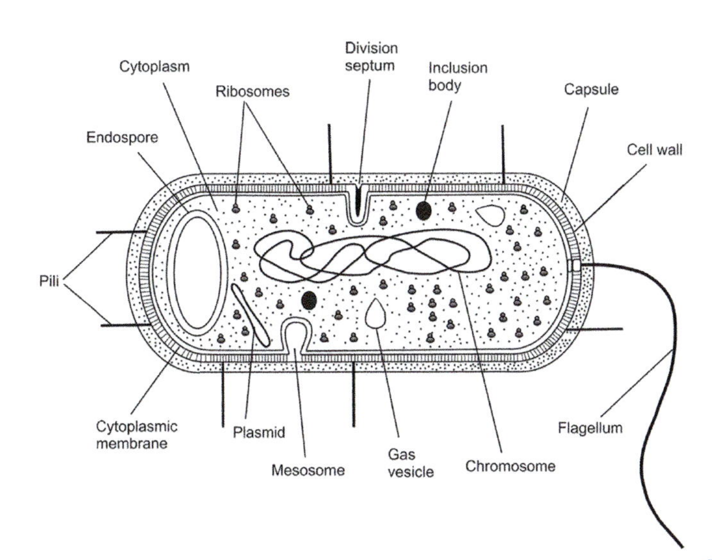

FIGURE 2.3 Bacterial cell.

The number of flagella and the pattern of flagellar attachment on a bacterial cell are used to classify bacteria. Four types of flagellar attachment found in bacteria are given below:

Monotrichous: One flagellum at one pole of the cell, e.g., *Xanthomonas* spp. (Figure 2.4a).

Lophotrichous: A cluster of flagella at one pole of the cell, e.g., *Pseudomonas* spp. (Figure 2.4b).

Amphitrichous: One flagellum or cluster of flagella at both poles of the cell, e.g., *Spirillum* spp. (Figure 2.4c).

Peritrichous: The flagella scattered over the entire surface of the cell, e.g., *Erwinia* and *Escherichia* spp. (Figure 2.4d).

Some bacteria, e.g., spirochetes have specialized flagella called **periplasmic flagella** (also called **axial filaments**). The helical filaments of these flagella are not free and loose like that of normal flagella. These arise at the poles and twine around the protoplasmic cylinder beneath the outer membrane of the cell wall. The spirochetes move in corkscrew-like manner with the help of these flagella.

A flagellum consists of three parts, the basal body, hook, and a long filament. The basal body is embedded in the cell and consists of a central rod surrounded by rings. Gram-stain-positive bacteria have only one pair of rings surrounding the central rod, one ring present in each of cytoplasmic membrane and cell wall. In Gram-stain-negative bacteria, two pairs of rings are present,

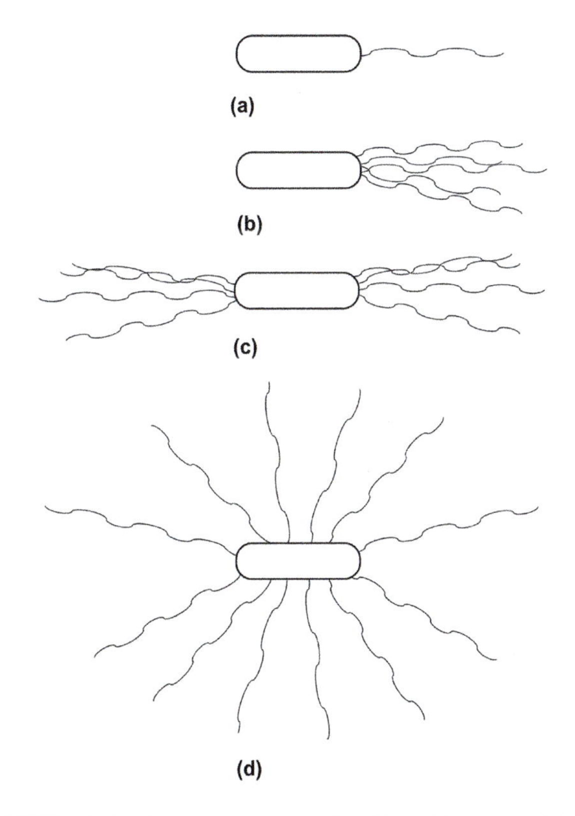

FIGURE 2.4 Arrangement of flagella: (a) monotrichous, (b) lophotrichous, (c) amphitrichous, and (d) peritrichous.

outer rings attached to the cell wall and inner rings lying in the cytoplasmic membrane. The helical filament is made of protein molecules called flagellin. These molecules are made within the cell and then transported along the hollow core of the flagellum for adding to the distal end, thereby confining the growth to the distal end of the flagellum (Pelczar et al. 2004).

Flagella function by rotating in a corkscrew-like fashion and propel the bacteria through liquid. The rings of the basal body rotate the flagella through chemical reaction. The bacteria having polar flagella show more rapid movement than those having peritrichous flagella. The environment affects the motility of the bacteria, i.e., they move towards a favorable environment or away from a harmful environment. Swimming bacteria also exhibit chemotaxis, which is movement in response to chemicals in the environment. Bacteria usually swim towards increasing levels of nutrients and away from increasing levels of harmful substances.

Much is known about the flagellum, i.e., its basic structure, the function of its individual motor components, and the regulation of its synthesis. However, there is only a beginning to identify the dynamics of flagellar proteins and to understand how the motor structurally adapts to environmental stimuli. In their review, Baker and O'Toole (2017) discuss the external and cellular factors that influence the dynamics of stator complexes (the ion-conducting channels of the flagellar motor). There is also focus on recent discoveries suggesting that stator dynamics are a means for controlling flagellar function in response to different environments.

The flagella, being the organs of locomotion, play a role in the virulence of some plant pathogenic bacteria. The bacteria swim towards their host plants often after sensing a chemical signal from the host. In *Agrobacterium tumefaciens,* aflagellated mutants show about 38% reduction in virulence and produce small and less number of tumors on the host in comparison to their wild-type parent. Also, Rvi$^-$ (reduced in virulence) mutants of *Pectobacterium atrosepticum* that are deficient in flagella biosynthesis and motility, but produce extracellular enzymes levels similar to the wild-type, are less virulent than the parents.

2.4.1.1 Pili

The pili (singular, pilus), also called **fimbriae**, are short, straight, non-helical structures present on the surface of bacterial cells. These are thinner (3–10 nm in diameter) than flagella and can be seen only with electron microscope. The pili are more common in Gram-stain-negative bacteria and are no way associated with the motility of the cells. These are hollow like flagella and are made of protein subunits called pilin, which are arranged in a spiral fashion around the central core. These originate beneath the cell wall, but no anchoring structures analogous to the basal bodies of flagella are found. They are present all along the cell, and their number can range from one to several hundred per cell. In *Es. coli*, there are 100–300 pili arranged peritrichously around a cell, each pilus measuring 2–20 nm in length and 7 nm in diameter. Different morphological types of pili occur, each associated with a different function, but attachment to the substrate or host cell is the primary function. A special type of pilus called **F** or **sex pilus** is associated with the bacterial conjugation. Bacteria that have this pilus are called F$^+$ (donor cells), and those without it are called F$^-$ (recipient cells). In animal pathogenic bacteria, pili help bacteria to attach to host cells lining the respiratory and intestinal tracts. *Ne. gonorrhoeae*, the causal agent of gonorrhea, bears pili that recognize and adhere to receptor on human cells. Pili also provide the site of attachment to some bacteriophages.

As adhesion fimbriae are a major virulence factor for many pathogenic Gram-stain-negative bacteria, they are also potential targets for antibodies. Fimbriae are commonly required for initiating the colonization, which leads to disease, and their success as adhesion organelles lies in their ability to both initiate and sustain bacterial attachment to epithelial cells. The ability of fimbriae to unwind and rewind their helical filaments presumably reduces their detachment from tissue surfaces with the shear forces that accompany significant fluid flow. Hence, the disruption of functional fimbriae by inhibiting this resilience should have high potential for use as a vaccine to prevent disease. Singh et al. (2017) showed that two characteristic biomechanical features of fimbrial resilience, namely, the extension force and the extension length, were significantly altered by the binding of antibodies to fimbriae. The fimbriae that were studied are normally expressed on enterotoxigenic *Es. coli*, which are a major cause of diarrheal disease. This alteration in biomechanical properties was observed with bivalent polyclonal antifimbrial antibodies that recognize major pilin subunits, but not with the antigen-binding fragments (Fab) of these antibodies. The authors proposed that the mechanism by which bound antibodies disrupt the uncoiling of natural fimbria under force is by clamping together layers of the helical filament, thereby increasing their stiffness and reducing their resilience during fluid flow. Another proposal was that antibodies tangle fimbriae via bivalent binding, i.e., by binding to two individual fimbriae and linking them together. Use of antibodies to disrupt physical properties of fimbriae may be generally applicable to the large number of Gram-stain-negative bacteria that rely on these surface-adhesion molecules as an essential virulence factor.

2.4.2 Surface Layers (Capsule and Slime Layer)

Mainly two types of surface layers, namely, capsule and slime layer are found in bacteria. If the layer is definite in outline and shape and is attached firmly to the cell wall, it is called a **capsule**. Capsule may be found in many bacteria, but is not universal in occurrence. Capsule is generally made of polysaccharides, but in some cases it may consist of polypeptides. It can be best seen by negative staining or through electron microscopy. The capsule is a protective layer and often plays a major role in pathogenicity of pathogenic bacteria. The capsules of *Stc. pneumoniae* protect the bacterial cells from being engulfed by the white blood cells (phagocytosis) of eucaryotic cells, thereby increasing the chances of infection. In many pathogenic species, the loss of capsule may result in the loss of pathogenicity, but not the viability.

When the surface of external layer is not of definite shape, disorganized, and is attached loosely to the cell wall, it is called a **slime layer** (e.g., *Rhizobium trifolii*). The slime layer being soluble in water creates many problems in the industry. The slime producing bacteria clog filters in sugar industry by increasing the viscosity of the fluids and coat pipes and other equipment.

The slime layer and capsule (also termed as **glycocalyx**) perform other functions also depending on the bacterial species. Among these, adherence of cells to rocks in fast-moving water, plant roots, teeth, etc. is a major function. Capsule may prevent attachment and lysis of cells by bacteriophages. They may also act as a reservoir of stored food and protect against temporary drying of cells by binding water molecules.

2.4.3 Cell Wall

Barring a few groups of bacteria, e.g., mycoplasmas, cell wall is the universal outer most layer of bacterial cells. It is a rigid structure responsible for the shape of the cell. It accounts for 20%–30% dry weight of the cell. Besides giving a characteristic shape to the cell, the cell wall serves as a selective barrier, allowing the influx of desired nutrients and liquids and preventing the escape of certain enzymes. It also prevents the movement of dyes, bile salts, antibiotics, heavy metals, and degradative enzymes into the cell.

In bacteria, peptidoglycan (also called murein or mucopeptide) is mainly responsible for determining the shape of cell wall. The cell wall of bacteria is also unique in the sense that peptidoglycan is neither found in archaea nor in eucaryotes. The basic structure of peptidoglycan in all the bacterial species consists of the following three building blocks: (1) *N*-acetyl glucosamine (NAG), (2) *N*-acetylmuramic acid (NAM), and (3) a peptide made of four amino acids or tetrapeptide.

The alternating units of NAG and NAM linked together by glycosidic bonds form the backbone and give strength to the cell wall. The small peptides attached to NAM give an additional rigidity to the cell wall. In Gram-stain-positive bacteria and actinomycetes, these peptides consist of four amino acids: L-alanine, D-glutamic acid, L-lysine, and D-alanine. Diaminopimelic acid (DAP) replaces L-lysine in Gram-stain-negative bacteria, myxobacteria, cyanobacteria, and some Gram-stain-positive bacteria. The D-configuration of glutamic acid and alanine makes the peptide bond resistant to pepsin and trypsin enzymes. The peptide chains are cross-linked with each other either through a direct bond (Gram-stain-negative bacteria) or through a pentaglycine bridge that is formed between terminal D-alanine of one peptide chain and the third amino acid (lysine or DAP) of the other.

The work of more than five decades supports the idea that cell envelope synthesis, including the inward growth of cell division, is tightly coordinated with DNA replication and protein synthesis through central metabolism. Remarkably, no unifying model exists to account for how these fundamentally disparate processes are functionally coupled. Recent studies show that proteins involved in carbohydrate and nitrogen metabolism can moonlight as direct regulators of cell division, coordinate cell division and DNA replication, and even suppress defects in DNA replication. In their minireview, Sperber and Herman (2017) focus on studies illustrating the intimate link between metabolism and regulation of peptidoglycan synthesis during growth and cell division, and identify the following three recurring themes. (i) Nutrient availability, not growth rate, is the primary determinant of cell size. (ii) As the degree of gluconeogenic flux is likely to have a profound impact on the metabolites available for cell envelope synthesis, the selection of a growth medium is a critical consideration when designing and interpreting experiments related to morphogenesis. (iii) Perturbations in pathways relying on commonly shared and limiting metabolites, like undecaprenyl phosphate, can lead to pleotropic phenotypes in unrelated pathways.

The enzyme lysozyme can completely destroy the cell wall of Gram-stain-positive bacteria leaving a spherical cell called protoplast. However, the cell walls of Gram-stain-negative bacteria are relatively resistant to this enzyme and are not completely destroyed; the spheroplast produced in this manner is equally susceptible to osmotic lysis as the protoplast.

Penicillin is a potent inhibitor of cell wall synthesis of Gram-stain-positive bacteria. It acts by preventing the formation of cross-linkages. Penicillin does not affect the preexisting cross-linkages, but does not allow the formation of fresh cross-linkages. Therefore, the non-growing population of bacteria is not inhibited. Mycoplasmas are also not affected by penicillin because they lack the cell wall.

The cell walls of bacteria and archaea differ in both chemical function and structure. The cell walls of archaea as already stated, lack peptidoglycan, but contain proteins, glycoproteins, or complex polysaccharides. Some species contain pseudomurein, which consists of *N*-acetyltalosaminuronic acid in place of *N*-acetylmuramic acid and lacks D-amino acid.

In spite of the fact that peptidoglycan is common to the cell walls of Gram-stain-positive and Gram-stain-negative bacteria, the cell walls of both the groups differ in thickness and chemical composition. These differences help in the classification of some bacteria and also explain the basis of Gram's staining reaction. The cell walls of Gram-stain-positive species are generally thicker (20–25 nm), consisting of several layers of peptidoglycan and showing high degree of cross linking up to 90% than those of Gram-stain-negative species (10–15 nm), showing low degree of cross linking, up to only about 10%. The similar situation occurs with respect to Gram-stain-positive and Gram-stain-negative archaea.

2.4.3.1 Cell Wall of Gram-Stain-Positive Bacteria

Their cell walls have much greater amount of peptidoglycan and hence much thicker than those of Gram-stain-negative bacteria. In some species, the weight of this polymer may go up to 50% or more of the dry weight of the cell wall compared to 10% in Gram-stain-negative species. Teichoic acids (acidic polysaccharides) are also found in walls of many species. Teichoic acids, which are polymers of a carbohydrate like glucose, an alcohol (glycerol or ribitol), and phosphate, are attached to the peptidoglycan and to the cytoplasmic membrane. These may help in the storage of phosphorus and being negatively charged, also help in the transport of positive ions into and out of the cell. Gram-stain-positive bacteria have neither outer membrane nor periplasmic space as found in Gram-stain-negative species.

2.4.3.2 Cell Wall of Gram-Stain-Negative Bacteria

These are more complex, but thinner than the cell walls of Gram-stain-positive bacteria. The cell walls of these bacteria are bilayered structures consisting of an outer membrane and a peptidoglycan layer; the latter is present in the periplasmic space (space between outer membrane and cytoplasmic membrane). Peptidoglycan layer of Gram-stain-negative bacteria constitutes only 5%–10% of the dry weight of the cell wall. Outer membrane is responsible for the distinguishing character of these bacteria.

The outer membrane also acts as a selective barrier for the movement of some substances into and out of the cell. This selective permeability is attributed to the presence of proteins called porins that form diffusion channels for the passage of molecules. The permeability is also related to the electric charge and size of molecules. Both specific and non-specific porins are present. Outer membrane is attached to the underlying peptidoglycan layer by a lipoprotein.

Outer membrane, a bilayered membrane contains phospholipids, lipopolysaccharides (LPSs), integral membrane proteins, and lipoproteins. These components are synthesized in the cytoplasm or at the inner leaflet of the inner membrane and have to be transported across the inner membrane and through the periplasm to assemble eventually in the correct membrane (Bos et al. 2007). In the outer membrane, lipopolysaccharides are present exclusively in the upper layer and phospholipids almost entirely in the lower layer.

Lipopolysaccharides found only in Gram-stain-negative bacteria consist of lipid A, core polysaccharide, and O antigen. The lipid part of a lipopolysaccharide called an endotoxin is poisonous in nature. It can cause diarrhea, fever, and destruction of red blood cells in the host.

The O antigens present in the outer membrane are responsible for many serological reactions of these bacteria. They also act as sites for the attachment of bacteriophages. Some outer member proteins also act as receptor sites for the attachment of bacteriocins and bacteriophages.

2.4.3.3 Gram-Staining/Reaction

The differences in chemical composition and structures of cell walls of Gram-stain-positive and Gram-stain-negative bacteria are the bases for differential Gram reaction. The treatment of bacterial cells with crystal violet followed by Lugol's iodine solution results in the formation of crystal violet-iodine (CVI) complex within the cells. The subsequent washing of Gram-stain-negative bacterial cells with ethanol dissolves the lipids and removes them. This disrupts the outer membrane, increases its permeability, and results in washing away of the dye complex. Hence, decolorized cells are stained pink with the counter stain, safranin. On the other hand, in the Gram-stain-positive bacteria, ethanol makes the pores in peptidoglycan layer to shrink, thus trapping the CVI dye complex inside the cells.

2.4.4 Cytoplasmic Membrane

Cytoplasmic membrane present below the cell wall is the seat of many enzymatic activities. It also acts as a selective barrier for the transport of water soluble molecules into and out of the cell and is much more selective than the cell wall. It contains various enzymes, some of which are involved in energy production and cell wall synthesis. Permeases proteins present in the membrane transport small molecules into the cell.

Cytoplasmic membrane is a bilayered 7.5 nm thick structure as seen under electron microscope. It is mainly composed of phospholipids and proteins. The phospholipids form a bilayer in which most of the proteins are partially or totally buried (intrinsic proteins). Some of the proteins called extrinsic proteins are also present on the surface of the phospholipid bilayer. The fluidity nature of the membrane allowing both proteins and lipids to move within phospholipids matrix is called the **fluid mosaic model**.

Cytoplasmic membranes of most procaryotes do not contain sterols such as cholesterol, and hence they are less rigid than those of eucaryotes. However, among the procaryotes, the cytoplasmic membranes of the mycoplasmas contain the sterols, which impart rigidity to these membranes. In the mycoplasmas, cytoplasmic membrane is the outer most membrane of the cells.

The electron micrographs of thin sections of bacterial cells reveal extensive invaginations of the cytoplasmic membranes, called **mesosomes**. They are more prominent in Gram-stain-positive bacteria. Mesosomes may be present near the cytoplasmic membrane or deep inside the cytoplasm. Earlier, these were thought to be associated with various functions like cell division, metabolic process, enzyme secretion, etc. But now, these are considered to be more of an artifact than the real structure.

In the absence of membrane-bounded organelles in bacterial cells, the cytoplasmic membranes act as multiple structures. The bacteria, in which the metabolism is based on the use of light energy or change of gases, their cytoplasmic membranes show extensive and elaborate intracellular extensions. Such type of extensions increases the surface area available for these activities. In phototrophic bacteria, the photosynthetic pigments are attached to these membrane systems. The type of these membrane systems varies with different groups of phototrophic bacteria. In purple bacteria, these modifications are quite elaborate and often occupy a sizable internal cell area. In *Ectothiorhodospira*, the systems consist of flat sheets called **lamellae**, while in *Chromatium,* these occur as round tubes or vesicles. In *Pelodictyon*, a green bacterium, chlorosomes constitute the photosynthetic apparatus which are cylindrical structures attached to the lower side of the cytoplasmic membrane. The lamellar membrane systems of oxygenic cyanobacteria are often complex and consist of multi-layered structures called **thylakoids**.

Cytoplasmic membrane plays an important role in regulating the transport/movement of water soluble molecules. This involves simple diffusion or facilitated diffusion down a concentration gradient or energy-based active transport.

2.4.5 Cytoplasm

Below the cytoplasmic membrane is the cytoplasmic area. The cytoplasmic area contains the cytoplasm, ribosomes, enzymes, genetic material (chromosomes and plasmids), granules, vacuoles, etc. The cytoplasm is a matrix of gel-like consistency and consists of water (about 80%), nucleic acids, carbohydrates, proteins, lipids, inorganic ions, etc. Many chemical reactions, involving the synthesis of cell components from nutrients, take place in cytoplasm. Contrary to eucaryotic cells, there is no cytoplasmic streaming in procaryotic cells, i.e., cytoplasm does not flow around within the cell.

The **cytoskeleton** is a structure that helps cells maintain their shape and internal organization, and it also provides mechanical support that enables cells to carry out essential functions like division and movement. It is a network of protein fibers supporting cell shape and anchoring organelles within the cell. The three main structural components of the cytoskeleton are microtubules (formed by tubulins), microfilaments (formed by actins), and intermediate filaments.

Only in the past two and a half decades, it has been widely accepted that bacteria are internally organized and possess a complex multi-faceted cytoskeleton similar to eucaryotic cells. The proteins that make up the bacterial cytoskeleton can largely be grouped into families based on homology to the major types of eucaryotic cytoskeletal filaments, i.e., tubulin-like, actin-like, and intermediate filament-like. Additionally, bacteria are host to some filament systems that lack recognizable homology to eucaryotic cytoskeletons (Barry and Gitai 2011).

Tubulin-related bacterial protein FtsZ and actin-related bacterial proteins MreB/Mbl have recently been described as constituents of bacterial cytoskeletons. Genes coding for MreB/Mbl could only be found in elongated bacteria, but not in coccoid forms. It was speculated that constituents of today's eucaryotic cytoskeleton (tubulin, actin) may have evolved from procaryotic precursor proteins closely related to today's bacterial proteins FtsZ and MreB/Mbl. Prior to the description of proteins MreB/Mbl, evidence had been obtained for the existence of a shape-preserving cytoskeleton ubiquitously present in all bacteria (Mayer 2003).

The presence of actin, tubulin, and intermediate filament homologs in bacteria suggests that primitive cytoskeletons first arose in bacteria. Busiek and Margolin (2015) have reviewed the function and direct interactions between these actin and tubulin homologs. Actin and tubulin homologs are prominent in the bacterial cytoskeleton, but other proteins such as RodZ, DivIVA, and intermediate filament homologs are emerging as important supporting players. Though we know many of these players and how some of them interact, the mechanisms by which cytoskeletal components dictate cell shape still need to be worked out. As it is generally true that cell architecture is inherited, it is also true that rod-shaped cells can emerge from round cells or from round dormant spores, which must depend on *de novo* assembly of cytoskeletal structures (Margolin 2009).

2.4.6 Ribosomes

The ribosomes, the site of protein synthesis, are densely packed throughout the cytoplasm. During protein synthesis, many ribosomes may be held together to form polyribosomes or polymers. In procaryotic cells, the ribosomes are 70S (Svedberg unit) type (compared to 80S found in eucaryotic cells) each having two subunits; the smaller, squash-shaped 30S, and the larger, somewhat spherical 50S. The Svedberg unit is a coefficient of sedimentation in the ultra centrifugation. High levels of Mg^{2+} ions are required to keep both the subunits of ribosomes associated. The antibiotics like streptomycin, neomycin, tetracycline, etc. that inhibit protein synthesis cause malfunctioning of the ribosomes. In *Es. coli*, the ribosomes are made up of 66% RNA, known as ribosomal RNA (r RNA) and 34% proteins.

2.4.7 Inclusions and Storage Products

Different kinds of chemical substances often accumulate forming insoluble deposits in the cytoplasm. Although they differ in their chemical composition and other aspects, they function as the reservoir of energy and other structural building blocks. Most of them are bounded by a non-unit, thin, lipid-based membrane and are called inclusion bodies. They can be easily seen with light microscope in suitably stained preparations.

Some phototrophic (purple and green) bacteria store large amounts of elemental sulphur in the form of granules by oxidizing sulphur compounds such as hydrogen sulphide, thiosulphate, etc. These granules may be present inside the cells (e.g., *Chromatium* spp.) or on the surface of cells (e.g., *Chlorobium* spp.) These granules may serve as stored energy for the bacteria. Another commonly stored substance in bacteria is a lipid material called poly-β-hydroxybutyrate (PHB). Poly-β-hydroxyalkanoate (PHA), a derivative of PHB, also accumulates as granules in bacteria. The plastic-like consistencies of many of these PHAs and biodegradable nature of their plastics have generated a lot of interest in the recent years. These granules can be stained with fat soluble dyes like Sudan black and Nile blue. Glycogen, a polymer of glucose subunits, is another stored product found in bacterial cells. Being polysaccharide, glycogen granules stain brown with iodine solution. Both PHB and glycogen granules serve as reservoir for carbon and energy.

Volutin granules or metachromatic granules composed of polyphosphates are also found in some bacteria. In electron microscopy, they appear as round, dark areas and stain intense reddish purple with dilute methylene blue dye.

Intracellular crystals of iron oxide, magnetite, and Fe_3O_4 are found in some aquatic bacteria. These crystals are bounded by a membrane and are called **magnetosomes**. They vary in shape from square to rectangular or look like a spike. As such bacteria possess ability of orienting and migrating along geomagnetic or exhibit magnetotaxis, they are called magnetotactic bacteria.

2.4.8 Gas Vesicles

The bacteria living in water bodies possess gas vesicles, with the help of which they can move up and down in response to different environmental conditions. The large blooms produced by some cyanobacteria are endowed with these gas vesicles. They are hollow, but rigid, spindle-shaped structures and proteinaceous in nature. They vary in length and width and also in number from a few to several hundred per cell. The vesicle membrane is permeable to gasses, but does not allow the water and solutes to enter the vesicle.

2.4.9 Nucleoid (Genophore)

The genetic material in bacteria, comprising of single haploid chromosome and extrachromosomal genetic elements, is called nucleoid or genophore. The earlier belief that the nuclear material of bacteria is dispersed throughout the cell was due to the inadequacy of the staining techniques. With advancement of science, the exact nature of bacterial hereditary material can be easily demonstrated. It is a fairly compact clump, organized into a definite body, and occupies roughly one-third volume of the cell. Unlike nucleoplasm of eukaryote, nucleoid is **not surrounded by a membrane**. In contrast, the nuclear material in eucaryotes or higher organisms is surrounded by a well-defined nuclear membrane. Fluorescent microscopy has been found very helpful for delineating the gross boundaries of the nuclear region. The fluorescent probes like 4,6-diamidino-2-phenylindole dihydrochloride (DAPI) capable of entering the cell and binding to DNA, at a non-toxic concentration, greatly help in visualization of nucleoid boundaries. The electron micrograph of *Es. coli* revealed that its nucleoid has a highly clefted appearance and is confined to a ribosomes-free zone in the cell. Nucleoid is largely composed of DNA (60%) with small amounts of RNA and proteins. The nucleoid proteins or nucleoid-associated proteins are distinct from histones of eucaryotic nuclei and do not form nucleosomes. Common nucleoid associated proteins are HU, histone-like nucleoid-structuring protein (H-NS), Fis, curved DNA-binding protein (cbpa), and DNA-binding proteins from starved cells (DPS) that help in compaction of DNA by bending, bridging, and aggregation. Proteins like HU and Fis probably play role in transcription.

2.4.9.1 Chromosome

Bacterial genome primarily comprises of single, haploid, circular chromosome, and in addition there may be present some extrachromosomal genetic elements, referred as plasmids, which may account for about 1% of the total genome. However, as recently reported, some bacteria such as *V. cholerae* and *Rohodobacter sphaeroides* have two chromosomes. The chromosome of *Es. coli* is a closed circular DNA molecule and is approximately 1 mm long. Taking into account 1–2 μm length of this bacterial cell, the chromosome must have condensed about 500–1000 times to fit into the cell. The condensation of the chromosome occurs by organizing the DNA into a series of loops and super coiling of these loops in an underwound fashion. The transcription, recombination, and other functions are not hindered by the folding pattern (Srivastava and Srivastava 2003). In most bacteria, the chromosome is circularly closed, but in a few cases linear chromosomes have also been reported, e.g., *Streptomyces lividans* and *Rhodococcus fascians*. Among the procaryotes, *Mycoplasma genitalium* possesses the smallest genome containing only 470 genes (~10%–15% of *Es. coli*).

2.4.10 Plasmids

Plasmids are extra chromosomal genetic material found in most bacteria, and they play a significant role in their adaptation and evolution. Most plasmids are covalently closed circular structures consisting of double stranded DNA, and the DNA is negatively super coiled. However, some bacteria such as *Borrelia burgderferi* and some species of genus *Streptomyces* possess linear plasmids. The plasmids have the ability to replicate independently. The number of copies of a plasmid per cell varies with different plasmids, F-plasmid or P1 prophage-plasmid exists as a single copy, while 40–300 copies of plasmid PIJ101 are found in *St. coelicolor*. Many bacteria may contain more than one type of plasmids. Initially, the plasmids were named on the basis of functions which they performed, e.g., F-plasmid as it mediates fertility and Ti-plasmid of *Ag. tumefaciens* induces tumor formation in some vascular plants. Now a more standardized method to name plasmids has been worked out. In this method, "p" letter is placed before the capital letters that describe either the plasmid or denote the initials of person(s) who isolated or constructed the plasmid, e.g., $pPATH_{Pag}$, pathogenicity plasmid of *Pantoea agglomerans* pv. *gypsophilae*. The methods for the detection and isolation of plasmids have also been devised.

Plasmids are of common occurrence in phytopathogenic bacteria and contribute significantly to host evolution in a multifaceted manner. Plasmids tend to encode determinants of virulence and ecological fitness that can enhance adaptation to a specific niche or can influence niche expansion. Many of these determinants appear to have been acquired from other bacteria via horizontal transfer, showing an important function of plasmids in the acquisition of sequences that enable rapid evolution. Ultimately, these genes are transferred to the host chromosomes through plasmid integration events; thereby resulting in stabilization of important acquired determinants within the genome. Most plasmids characterized in phytopathogenic bacteria are self-transmissible and possess suites of genes encoding type IV secretion systems. Sundin (2007) has given an excellent account of genomic insights into the contribution of phytopathogenic bacterial plasmids to the evolution of their hosts. The role of plasmids in the evolution of phytopathogenic bacteria was also reviewed earlier by Coplin (1989).

Most genera of plant pathogenic bacteria harbor one or more plasmids containing genes that encode traits of importance to ecological fitness and/or virulence. Till 2007, complete genome sequence information for 22 medium- to large-sized plasmids belonging to six plant pathogenic species was obtained (Sundin 2007).

Bacterial plasmids are well known carriers of antibiotic and heavy metal resistance genes in all ecosystems. This fact is also true for plasmids of plant pathogenic bacteria. Plasmids-encoded resistance determinants to copper and streptomycin in *Psm. syringae* pv. *tomato* and *Xa. vesicatoria* were some of the first plasmid-encoded determinants to be cloned and characterized in these bacteria. With few exceptions, all strains of *Er. amylovora* isolated from diseased tissues contain a non-conjugative plasmid called the ubiquitous plasmid pEA29 (plasmid of *Erwinia amylovora*). This plasmid can be cured in the laboratory, and the cured strains are reduced in virulence and show thiamine auxotrophy. The presence of pEA29 influences the expression of the chromosomal amylovoran biosynthetic operon during infection. The exopolysaccharide amylovoran is a pathogenicity determinant of the fire blight pathogen, and thiamine biosynthesis is required for the high level expression of the amylovoran biosynthetic operon during infection. Of the other plasmids found in *Er. amylovora*, the plasmid pEL60 (Plasmid of *Erwinia amylovora*) is of interest because it is an IncL/M plasmid with a high degree of similarity to the backbone of Plasmid of *Citrobacter freundii* (pCTX-M3), a multiple antibiotic resistance plasmid identified in the opportunistic human pathogen, *Citrobactor freundii*. pEL60 is a cryptic plasmid, encoding only genes involved in replication, maintenance, and conjugative transfer, while pCTX-M3 contains two additional insertions totaling 29.6 kb that encode a number of antibiotic resistance genes.

The acquisition and expression of novel sequences may confer pathogenicity on a saprophytic strain. A classical example is the transformation of *Pa. agglomerans*, a widespread epiphytic and commensal bacterium into a host specific tumourigenic pathogen by acquiring a plasmid called pPATH (plasmid governing pathogenicity of *Pa. agglomerans*) containing pathogenicity island (PAI). The PAI was evolved on an interon plasmid of IncN family, which is distributed among genetically diverse populations of *Pa. agglomerans*. The horizontal gene transfer was the major contributing force in the creation of PAI, although a pathoadaptive mechanism might also be involved.

Vivian et al. (2001) found that most strains of *Psm. syringae* analyzed, regardless of pathovar, contained at least one indigenous plasmid. Genetic studies of plasmids from different pathovars reveal that the large majority of plasmids in *Psm. syringae* share a major replication gene (*repA*). Sequencing and functional characterization of *repA* from 100 kb plasmid pPT23A of *Psm. syringae* pv. *tomato* PT 23 and analysis of the distribution of *repA* in diverse pathovars of *Psm. syringae* led to the recognition of the pPT23A plasmid family as a group, called pPT23A family plasmids (PFPs). Most plasmids of this family range from approximately 35 to 100 kb and are believed to have originated from a common ancestor because they share *repA* gene that encodes an essential replication protein. The pPT23A plasmid family is considered to be universally distributed among *Psm. syringae* pathovars. The *repA* replication gene is the only gene conserved in all PFPs studied by Zhao et al. (2005). Genes encoding virulence factors such as type III effectors, coronatine toxin biosynthesis, plant hormone production, and determinants important for epiphytic fitness including UV radiation tolerance are commonly found on PFPs (Vivian et al. 2001). Genes of potential ecological importance or virulence genes of general importance to *Psm. syringae* are more widely distributed among pathovars. The loss of virulence in *Psm. savastanoi* pv. *phaseolicola* to bean is associated with the loss of plasmid pAV511.

The plasmids act as carriers of virulence/avirulence genes in *Xanthomonas* species. Additional avirulence/effector genes such as those of *avrBs3/pthA* family and *avrBs4* are plasmid encoded in *Xanthomonas* spp., and in some cases these genes are located on conjugative plasmids. The *avrBs3/pthA* factors play a role in pathogen aggressiveness and virulence. Fujikawa et al. (2006) have shown that *avrBs3/pthA* effectors function as suppressors of host defense responses.

The integration of plasmids into the host chromosomes occurs, and it represents one delivery method for horizontally acquired sequences. Integration of plasmid pMC7105 (plasmid of *Pseudomonas syringae*) from *Psm. savastanoi* (=*syringae*) pv. *phaseolicola* was first demonstrated in mid-1980s. It was shown to replicate autonomously or integrate into the host chromosome. Integration was achieved via recombination between copies of the insertion sequence IS*801* on the plasmid and chromosome. An extensive sequence homology found between pMC7105 and plasmid DNA from *Psm. savastanoi* (=*syringae*) pv. *glycinea* indicates that integration of plasmids may occur in other *Psm. syringae* pathovars. Plasmid pFKN, a PFP from *Psm. syringae* pv. *maculicola*, has also been shown to integrate into the chromosome through recombination between alleles of the effector gene *avrPphE* (avirulence gene). This integration was considered to comprise an evolutionary event resulting in the formation of a new pathogenicity island in the *Psm. syringae* pv. *maculicola* chromosome. The integration of plasmids into chromosomes can be viewed as the first step in the generation of new island sequence.

Insertion sequence (IS) elements play a role in the evolution of bacteria in various ways. These sequences move within plasmids, between plasmids, and between plasmid and chromosome. IS elements can facilitate recombination between plasmids and chromosomes. Plasmid-encoded IS elements effect pathogen virulence and evolution. The element IS*801* appears prominently in the evolution of certain *Psm. syringae* pathovars while absent in other pathovars. IS*801*-like sequences occur close to effector genes in *Psm. savastanoi* pvs. *glycinea* and *phaseolicola* and downstream of plant hormone biosynthetic genes in *Psm. savastanoi* pv. *savastanoi*. On account of this close proximity, IS*801* has been considered to be involved in the mobility of virulence genes. IS*3* family is one of the largest families of IS element. IS*476* encoded on plasmids of *Xa. campestris* pv. *campestris* and *Xa. vesicatoria*, IS*1389* of *Xa. campestris* pv. *amaranthicola*, IS *Xac3* of *Xa. citri* subsp. *citri*, IS *Psy26* of *Psm. syringae* pv. *tomato,* and an unnamed element of pEA29 of *Er. amylovora* are members of the IS*3* family. IS elements like IS*801* serve as sites for recombination facilitating integration of plasmids into chromosomes, and these elements can also mobilize adjacent sequences during the transposition process.

The plasmids also play a role in other plant pathogenic bacteria. *Psm. syringae* pv. *syringae*, incitant of bacterial canker of stone fruits, contains plasmid-mediated genes for copper resistance. Patro and Jindal (2000) have shown the existence of plasmids in *Xa. axonopodis* pv. *vignaeradiatae*, incitant of bacterial leaf spot of green gram, and their correlation with virulence. The highly virulent isolates contained the maximum number of plasmids. The plasmid-borne genes were shown to govern virulence, exopolysaccharide production, antibiotic resistance, and pigmentation.

The plasmids have played a very important role as vectors in genetic engineering. They have been used to transfer required genes into the strains of the same species, into different species, or even into organisms of different kingdoms. Their small size, easy isolation, and introduction, more number of copies, possession of an easily selectable marker, presence of sites for restriction endonucleases, and amenability to desired modifications have greatly enhanced their utility in genetic engineering. Tumor inducing (Ti) plasmid of *Ag. tumefaciens* is commonly used for transferring genetic material into different bacterial species or even into plants. *Ag. tumefaciens* is an excellent genetic engineering tool. The DNA transformation capabilities of the bacterium have been extensively exploited in biotechnology as means of inserting foreign genes into plants. Therefore, Nester (2008) has appropriately called it a natural genetic engineer and also a household name in plant genetic engineering and plant molecular genetics.

2.4.11 Dormant Forms

Some bacteria produce dormant forms or resting stages called spores and cysts. These dormant forms help the bacteria to tide over or withstand the adverse conditions like heat or drying. They are metabolically inactive, i.e., not growing but under favorable conditions, they can germinate and grow into metabolically active vegetative cells.

2.4.11.1 Spores

Spores are thick-walled, dormant, and highly refractile structures. As they are formed inside the cells, they are also called **endospores**. The spores are highly resistant to heat, drying, and chemicals toxic to the vegetative cells. The endospores of *Clostridium botulinum*, the cause of botulism (food poisoning), can resist boiling for several hours, while vegetative cells of most bacteria are killed above 70°C. As the spores are resistant to adverse environmental conditions, they help the bacteria to tide over the adverse environmental conditions.

Among the rod-shaped bacteria, the genera *Bacillus* and *Clostridium* are exclusively spore formers. Among the spherical bacteria, the members of genus *Sporosarcina* form the spores. Normally, one spore is produced per cell, and they are usually produced in old cultures. The spores vary in shape (spherical or ovoid) and location within the cell. The location in the cell can be central, subterminal, or terminal (Figure 2.5a–c, respectively). In some cases, terminal spores

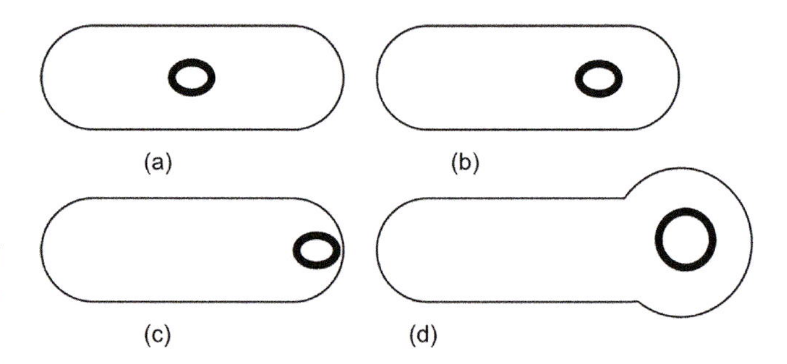

FIGURE 2.5 Location of spores in the cell: (a) central, (b) subterminal, (c) terminal, and (d) terminal spore with swollen sporangium.

are formed in swollen sporangia (Figure 2.5d). Spore position in the mother cell or sporangium frequently differs among species and is helpful in the identification of bacteria. It is essential to apply heat to the endospores when staining for light microscopy. In bacteria, endospores are not the means of multiplication, but means to tide over the adverse conditions because one vegetative cell produces one spore and on germination one spore produces one cell.

During sporulation, dehydration process occurs that expels most of the water from the spore. All endospores contain large amounts of dipicolinic acid (DPA), a compound not found in vegetative cells. DPA accounts for 5%–10% of the endospore's dry weight, and it occurs in combination with large amounts of calcium, both accounting for 15% of the spore's dry weight. Earlier, it was thought that DPA was directly responsible for resistance of spores to heat. However, the occurrence of heat-resistant mutants lacking DPA has seriously weakened this view-point. Calcium does aid in resistance to wet heat, oxidizing agents, and sometimes dry heat. It may be that calcium-dipicolinate complex often stabilizes spore nucleic acids. The calcium-dipicolinate complex in combination with acid-soluble protein stabilization of DNA, protoplast dehydration, DNA repair, the spore coat, and the greater stability of cell proteins in bacteria adapted to growth at high temperatures are considered responsible for imparting resistance to spores against heat and other lethal agents (Prescott et al. 2005).

The outermost exosporium layer of *Clo. difficile* spores is likely to play a major role in spore-host interactions during recurrent infections, contributing to the persistence of the spores in the host. Pizarro-Guajardo et al. (2016a) demonstrated by transmission electron microscopy the presence of two ultra structural morphotypes of the exosporium layer in spores formed from the same sporulating culture. However, it was not clear whether these distinct morphotypes appeared due to purification techniques or they appeared during biofilm development. Pizarro-Guajardo et al. (2016b) demonstrated through transmission electron microscopy that these two exosporium morphotypes are formed under sporulation conditions and are also present in spores formed during biofilm development. This work provides definite evidence that in a population of sporulating cells, spores with a thick outermost exosporium layer and spores with a thin outermost exosporium layer are formed.

Another type of spores produced by bacteria is **conidia** (singular, conidium), produced by actinomycetes. These spores are not much resistant to heat, but are resistant to drying. As one thallus of actinomycetes produces large number of conidia, these are the means of reproduction, but not the means to tide over the adverse conditions.

2.4.11.2 Cysts

The cysts are also dormant, thick-walled structures that resist drying. However, they are not highly resistant to heat like endospores. The vegetative cells give rise to cysts, and under favorable conditions cysts germinate to give rise to vegetative cells. The vegetative cells of several bacterial genera have been reported to produce cysts, but those produced by the members of *Azotobacter* are considered the typical bacterial cysts.

REFERENCES

Baker, A. E., and O'Toole, G. A. 2017. Bacteria, rev your engines: Stator dynamics regulate flagellar motility. *J. Bacteriol.* 199: e00088–e00117.

Barry, R., and Gitai, Z. 2011. Self-assembling enzymes and the origins of the cytoskeleton. *Curr. Opin. Mirobiol.* 14: 704–711.

Bos, M. P., Robert, V., and Tommassen, J. 2007. Biogenesis of the Gram-negative bacterial outer membrane. *Ann. Rev. Phytopathol.* 61: 191–214.

Busiek, K. K., and Margolin, W. 2015. Bacterial actin and tubulin homologs in cell growth and division. *Curr. Biol.* 25: R243–R254.

Coplin, D. L. 1989. Plasmids and their role in the evolution of phytopathogenic bacteria. *Ann. Rev. Phytopathol.* 27: 187–212.

Fujikawa, T., Ishihara, H., Leach, J. E., and Tsuyumu, S. 2006. Suppression of defence response in plants by the *avrBs3/pthA* gene family of *Xanthomonas* spp. *Mol. Plant-Microbe Interact.* 19: 342–349.

Margolin, W. 2009. Sculpting the bacterial cell. *Curr. Biol.* 19: R812–R822.

Mayer, F. 2003. Cytoskeletons in prokaryotes. *Cell Biol. Int.* 27: 429–438.

Nester, E. 2008. *Agrobacterium*: The natural genetic engineer: 100 years later. APSnet Feature Story, http://www.apsnet.org/online/feature/Agrobacterium/

Patro, T. S. S. K., and Jindal, J. K. 2000. Expression of plasmid-borne characters in *Xanthomonas axonopodis* pv. *vignaeradiatae*. *Trop. Agri. Res.* 12: 86–94.

Pelczar, M. J. Jr., Chan, E. C. S., and Krieg, N. R. 2004. *Microbiology: Concepts and Applications*. 5th ed. Tata McGraw-Hill, New Delhi, India, 918 pp.

Pizarro-Guajardo, M., Calderón-Romero, P., Castro-Córdova, P., Mora-Uribe, P., and Paredes-Sabja, D. 2016a. Ultrastructural variability of the exosporium layer of *Clostridium difficile* spores. *Appl. Environ. Microbiol.* 82: 2202–2209.

Pizarro-Guajardo, M., Calderón-Romero, P., and Paredes-Sabja, D. 2016b. Ultrastructure variability of the exosporium layer of *Clostridium difficile* spores from sporulating cultures and biofilms. *Appl. Environ. Microbiol.* 82: 5892–5898.

Prescott, L. M., Harley, J. P., and Klein, D. A. 2005. *Microbiology*. 6th ed. McGraw-Hill Companies, New York, 1130 pp.

Singh, B., Mortezaei, N., Savarino, S. J., Uhlin, B. E., Bullitt, E., et al. 2017. Antibodies damage the resilience of fimbriae, causing them to be stiff and tangled. *J. Bacteriol.* 199: e00665–16.

Sperber, A. M., and Herman, J. K. 2017. Metabolism shapes the cell. *J. Bacteriol.* 199: e00039–e00117.

Srivastava, S., and Srivastava, P. S. 2003. *Understanding Bacteria*. Kluwer Academic Publishers, Dordrecht, the Netherlands, 469 pp.

Stanier, R. Y., and van Niel, C. B. 1962. The concept of a bacterium. *Arch. Microbiol.* 42: 17–35.

Sundin, G. W. 2007. Genomic insights into the contribution of phytopathogenic bacterial plasmids to the evolutionary history of their hosts. *Ann. Rev. Phytopathol.* 45: 129–151.

Tortora, G. J., Funke, B. R., and Case, C. L. 2007. *Microbiology: An Introduction*. 9th ed. Benjamin Cummings, San Francisco, CA, 958 pp.

Vivian, A., Murillo, J., and Jackson, R. W. 2001. The roles of plasmids in phytopathogenic bacteria: Mobile arsenals? *Microbiology* 147: 763–780.

Whitman, W. B. 2009. The modern concept of the procaryote. *J. Bacteriol.* 191: 2000–2005.

Zhao, Y., Ma, Z., and Sundin, G. W. 2005. Comparative genomic analysis of the pPT23A plasmid family of *Pseudomonas syringae*. *J. Bacteriol.* 187: 2113–2126.

3 Diagnosis of Bacterial Diseases of Plants

For an effective disease management, the correct diagnosis of the disease based on the identification of its causal agent is very essential. The diagnosis of some diseases on the basis of symptoms can be made to a satisfactory level, but the identification of the causal agent is necessary to confirm the cause of the disease. For the diagnosis of a plant disease, a comprehensive host list that covers a known disease, its typical symptoms, and known potential pathogens for a specific host are required. The compendia of different crop diseases published by American Phytopathological Society are very valuable publications for this purpose. A list of plant diseases found on the website of American Phytopathological Society (http://www.apsnet.org/online/common/top.asp) is also a source of valuable information. *Westcott's Plant Disease Handbook* is also useful because specific symptoms associated with each disease are given (Horst 2001). The photographs of disease symptoms, especially the colored ones, also aid in the identification of plant diseases. A list of known pathogens for a given crop greatly reduces the choices to one or two suspected genera for the given symptoms.

3.1 ASSESSMENT OF SYMPTOMS

The assessment of symptoms and the diagnosis of the disease based on the symptoms in a laboratory depend on the sample sent to the laboratory. It must consist of plant material showing different stages/phases of the disease, i.e., young and old infections. If the disease affects different plant parts, i.e., leaves, stems, bulbs, roots, etc., the sample should contain all the plant parts showing symptoms. In completely diseased tissue, it is difficult to find the primary cause of the disease as in many cases it is generally replaced by secondary pathogens or saprophytes. Therefore, the symptoms in the sample must contain borderline between healthy and diseased tissue. The samples should be wrapped in paper and put in a plastic bag, leaving some space for air, to prevent rotting and decay by secondary organisms during transit. The information including name of cultivar, date of sowing, date of sampling, place, weather conditions, soil type, etc. should accompany the sample. Finally the sample should be statistically representative.

The symptoms of plant diseases caused by different groups of pathogens, namely, fungi, bacteria, viruses, phytoplasmas, and nematodes have many similarities and overlapping of the symptoms usually occurs. However, two signs, i.e., **water-soaking** and **bacterial exudation**, are associated only with most, but not all the bacterial plant diseases. The amount of bacterial ooze varies with environmental conditions and the nature of disease. The ooze is generally more forceful in vascular diseases and in freshly infected lesions. The intensity of ooze is also more under humid conditions. In a dry atmosphere, the bacterial ooze is converted into scales/flakes, beads, often becoming detectable only with the help of a magnifying glass, especially when it is sparse. Riley et al. (2006) have also emphasized the importance of signs by stating that signs are much more specific to disease-causing agents than the symptoms and are extremely useful in the diagnosis of a disease and identification of the agent causing the disease.

In bacterial wilt of solanaceous plants, the surface of the transverse section is generally moister in appearance in comparison to a fungal wilt. White droplets of bacterial ooze generally appear on the cut surface, and the ooze intensifies on squeezing the infected plant tissue. A diagnostic feature of bacterial wilt of cucurbits is appearance of bacterial ooze from the cut ends of the infected stems. This sticky bacterial mass can be drawn into fine threads by touching the cut end of the stem with a finger or a knife blade and gently pulling them away from the cut end. The fine threads of bacterial ooze can also be seen by pulling away the cut ends of the infected stem (Figure 3.1).

Microscopic examination of diseased plants is extremely helpful for establishing bacterial nature of the diseases. The bacterial streaming is found in almost all the bacterial diseases except the crown gall. A portion of the lesion along with an adjoining healthy tissue is cut, mounted in a drop of water, and placed under the microscope, without covering it with a cover slip. Clouds of bacteria streaming out of the cut end of the mounted tissue can be seen. The intensity of ooze is more in vascular diseases like bacterial blight of rice and bacterial wilt of cucurbits. Phase contrast microscopy is more suitable for this purpose because starch or latex particles can be distinguished from the bacterial cells. Bacterial ooze test can also be performed in the field by cutting an infected stem (bacterial wilt of cucurbits) or leaf (bacterial blight of rice) and placing it in a glass/beaker of clean water. After a few minutes, fine threads of bacterial ooze coming out of the cut end can be seen and the water turns milky.

3.1.1 Koch's Postulates

To prove that a particular organism/pathogen is a cause of the disease, it is essential to prove the Koch's postulates given below. These postulates were given by Robert Koch, a renowned German bacteriologist.

1. An organism/pathogen is always associated with a disease
2. Isolation of the pathogen or causal agent and its identification
3. Inoculation of the pathogen on the host and production of symptoms identical to the symptoms observed in nature
4. Reisolation of the pathogen from inoculated plants, and it should be identical in all respects to the inoculated pathogen.

FIGURE 3.1 Bacterial wilt of cucurbits: Strands of viscous bacterial ooze drawn from cut stem ends. (Courtesy of M.P. Hoffman.)

3.2 ISOLATION OF BACTERIA

The method of isolation described below is applicable only to the culturable plant pathogenic bacteria. It is not applicable to the spiroplasmas, phytoplasmas, and fastidious xylem- or phloem-limited bacteria as they are non-culturable and the methods to study them are different. In diseases caused by fastidious bacteria such as spiroplasmas, phytoplasmas, and *Xylella fastidiosa*, the vascular bundles of the sectioned tissues are examined with an electron microscope to detect the presence of these pathogens.

Take a fresh disease sample and wash it thoroughly with tap water. Cut small pieces/bits of diseased tissue bordering the healthy tissue and place them in sodium hypochlorite solution for 1–2 min for surface sterilization. These bits can also be disinfected with 70% alcohol. After surface sterilization, the bits are washed four times with sterile water to remove the residue of chlorine/alcohol. Then, these bits are placed in a tube containing sterile water or phosphate-buffered saline (PBS, pH 7.2) and left for 30 min to allow the bacteria to ooze out of the tissue. Alternatively, a few bits of diseased tissue can be crushed in a sterile small dish containing phosphate-buffered saline or sterile water to make the cell suspension. For the isolation of *Xanthomonas citri* subsp. *malvacearum*, leaf tissue showing vein infection should be preferred over the other symptomatic tissue.

The isolation of bacteria is more difficult from seed samples, host tissue having very low number of bacteria or plant samples having latent infection or epiphytic populations. In traditional isolation, plant tissue is washed and macerated, often filtrated and centrifuged; and then a small aliquot of the suspension is streaked onto a suitable non-selective or semi-selective medium. It is also advisable to inoculate a part of this suspension into a test plant, which may act as a selective medium. However, these methods are labor-intensive and require a lot of time. Selective media have a fairly high detection level ($10^{3\text{-}5}$ cells mL^{-1}). The following two methods are used for plating the cell suspension on the solid medium.

1. Streak method
2. Pour plate method.

3.2.1 Streak Method

In streak method, the dilute cell suspension is streaked on the solidified culture medium poured in the plates. An aliquot of 20–100 μL of the cell suspension is placed on the solid media in a plate and spread with a sterile spreader. The same spreader is used to further spread bacteria on two more plates. In this way, a dilution of approximately 1:10 on second plate and 1:100 on third plate is obtained, yielding separate colonies that can be examined for the desired colony type. The cell suspension can also be streaked on the solid media using an inoculating needle. The loop of an inoculating needle, charged with cell suspension, is dragged from one end to the other end of the plate; brought back to the first end and again moved to the other end. This way the whole plate is streaked in a zigzag fashion. The same loop, without charging with cell suspension, is streaked on two to three more plates. After streaking, the plates (inverted position) are incubated at the required temperature. An advantage of streaking with inoculating needle is that colonies growing on the streaks are supposed to originate from the streaked cell suspension and hence, are picked up.

For most plant pathogenic bacteria, a general, non-selective medium is generally used. These media contain usually water and non-defined nutrients (e.g., peptones, beef extract, yeast extract) or defined inorganic salts and organic nutrients, making them suitable for growth of most bacterial species. A general purpose, non-selective, but differential medium containing tetrazolium chloride is useful for distinguishing a potential pathogen from saprophytes during the initial isolation and for this purpose the concentration of the tetrazolium chloride should be reduced to 0.001% to avoid inhibitory effects on members of *Xanthomonas* genus. Semi-selective or selective media are used for specific bacterial pathogens. Semi-selective media contain substances to enhance production of pigments and/or substances which can only be used by certain bacteria and/or inhibitors like antibiotics for non-desired bacteria. Semi-selective media are more helpful for isolations from soil, but generally not from plant tissues because the latter are generally surface sterilized before isolation. Selective media contain nutrients and inhibitors, which by their quantity and quality allow only a specific bacterium to form colonies in/on the medium in an ideal situation. To enhance the isolation of target bacteria and to exclude non-pathogenic bacteria, semi-selective or selective isolation media are often used. However, many of the non-desirable bacteria can also form colonies that may be very similar to those of the target bacterium. Therefore, the isolated bacterium must be compared with the bacterium known to be pathogenic. Culture media are solidified by adding agar-agar, a polysaccharide, i.e., sulphuric acid ester of linear galactan, obtained from red algae of the genera *Gracilaria*, *Gelidium*, *Pterocladium*, and *Acanthopeltis*. Media are sterilized in the autoclave at 115°C or 121°C for 20 min.

3.2.2 Pour Plate Method

In this method, 20–40 µL of cell suspension is poured in each plate and then a molten, but cool solid medium is poured in these plates. The plates are rotated clock and anti-clock wise to mix the cell suspension with the medium. After solidification of the medium, the plates (inverted position) are incubated at the required temperature.

Most plant pathogenic bacteria grow well in an oxygen-rich atmosphere at 20°C–28°C. In most cases, the colonies appear on the isolation plates after 2–5 days of incubation. The colony morphology is very important for the identification of a bacterium. Based on colony characteristics of the bacterium under investigation, isolated single colonies of the desired bacterium are transferred on medium slants in the tubes and incubated for growth. The characteristics of the colonies and of mass growth on the slants are recorded. Many plant pathogenic bacteria produce pigmentation in the culture media that helps greatly in their identification. *Xanthomonas* species produce yellow, water insoluble color (xanthomonadins). Some bacteria produce water-soluble pigments, and these diffuse into the medium, e.g., the brown melanoid pigment of *Streptomyces scabiei* and *Ralstonia solanacearum* and the fluorescent pigments (pyoverdins) of many *Pseudomonas* species.

The pure culture of the bacterium is obtained using single colony culture technique. It can be obtained by both streak and pour plate methods. In streak method, a loop-full of a very dilute cell suspension of the isolated bacterium is streaked continuously (without recharging) in 5–6 plates. If the isolated colonies appearing in the last streaked plates appear similar in morphology including shape, size, color, etc., it means that the colonies formed are descendants of a single colony and the culture is pure. To make it sure, this process is repeated 3–4 times, and the single colony culture obtained from the last repeat is labeled as pure culture and used for further studies. A pure culture forms the basis for all further studies. In pour plate method, the serial dilutions of bacterial cell suspension are mixed (1 mL per plate) with the molten, but cool solid medium and plates are incubated. Isolated colonies are supposed to appear in the highest dilutions. The remaining procedure is same as in streak method.

3.3 IDENTIFICATION OF BACTERIA

The identification of the bacteria is very time consuming as it involves a large number of tests to be conducted, and some of the tests take long time. The following morphological, physiological, and biochemical characteristics can be determined using conventional methods of identification.

3.3.1 Morphological Characteristics

These include Gram-stain reaction, cell shape and size, spore formation, and presence or absence of flagella and capsule. Most plant pathogenic bacteria are Gram-stain-positive or -negative rods except *Streptomyces* spp., which are filamentous. The colony morphology is also very important as form, color, and smell of a colony may be an indication of a bacterial pathogen, but many other non-pathogenic bacteria may also form similar type of colonies. Diffusible pigment production also plays an important role in the identification of a bacterium.

3.3.2 Physiological Characteristics

These include the temperature range for growth, thermal death point (temperature at which the bacterial cells in dilute cell suspension die, when kept for 10 min at that temperature), and tolerance to NaCl concentration. The temperature range for growth of plant pathogenic bacteria is approximately from 4°C to 37°C; thermal death point usually ranges from 50°C to 55°C. The determination of antibiotic resistance may be useful because the antibiotics are generally used to control bacterial pathogens in the field (e.g., streptomycin against *Erwinia amylovora*). Furthermore, screening for toxin production and ice-nucleation activity will yield additional characteristics useful in the identification of plant pathogenic bacteria, especially those belonging to the *Psm. syringae* group and *Clavibacter* spp.

3.3.3 Biochemical Characteristics

The determination of biochemical tests of plant pathogenic bacteria is important as the reactions of these tests are based on the diverse enzyme systems of the bacteria, and many of these enzymes are involved in plant pathogenic action. These tests include the ability of the bacteria to utilize certain carbon sources like sugars, alcohols, organic acids, glycosides, and/or nitrogen sources, e.g., amino acids. The determination of oxidative and fermentative metabolism of glucose is important for differentiation of *Pseudomonas* and *Xanthomonas* from *Erwinia*. The former two genera are oxidative positive and fermentative negative, while the latter is positive for both the tests. Some biochemical tests give the information regarding the formation of certain end products by the bacteria, e.g., formation of H_2S from cysteine, indole from tryptophan, and NO_2 or N_2 from nitrate. Action of enzymes, namely, pectinases, gelatinase, amylases, and casease is judged by the hydrolysis of pectin, gelatin, starch, and casein, respectively. Cellulases and lipases (esterases) cause hydrolysis of cellulose and fats/fatty acids, respectively.

The diagnosis of an undescribed or a new disease is certainly more difficult and requires more skill and expertise than confirming the presence of an already known disease. Therefore, the entire set of diagnostic principles should be used to narrow down the choices, as no simple tests are suitable. More emphasis should be put on microscopic examination and the initial association of the pathogen with the disease syndrome. Some basic information about the pathogen should be obtained to decide which tests are to be used for further identification before spending lot of time on numerous phenotypic and genotypic tests for pathogen identification. The tests for basic information include Gram-staining reaction, oxidative versus fermentative metabolism, and presence or absence of spores and flagella. To further characterize the bacterium, the relatively inexpensive tests include metabolic tests (analytical profile index, API, strip tests, bioMerieux, Inc.), analysis of fatty methyl esters (MIDI, Newark, DE), metabolic substrate analysis (MicroLog TM, Biolog, Inc., Hayward, CA), or 16S ribosomal (r)DNA sequence analysis. An extreme caution must be exercised in interpreting results of any one test, especially for identification of an unknown disease caused by an unidentified bacterial pathogen. The past experience emphasizes the importance of performing basic bacteriological tests before accepting a name, which comes from a database, regardless of the refined analytical methods used to generate a similarity index.

Besides, routine morphological, physiological, and biochemical tests, serological and molecular techniques have been developed, which are more specific and sensitive. These techniques are routinely used in various laboratories throughout the world for the identification of bacteria and diagnosis of bacterial plant diseases.

A polyphasic analysis still appears to be the most reliable approach for identification of new pathogens. After the pathogenicity has been proved, and the genus determined with relatively simple basic bacteriological tests, the final identification of the bacterium is interpreted from the results of various genotypic methods. Hu et al. (2001) found that comparison of 16S rDNA sequence data with phenotypic data for type strains assisted in selection of determinative tests that may discriminate distinct taxa for simplified laboratory analyses.

3.4 SEROLOGICAL TECHNIQUES

Serology is an important tool for providing additional information for the identification of bacteria, but a bacterium cannot be identified on the basis of its serological reactions alone. Many times antisera are not specific enough, and they may cross-react with non-pathogenic bacteria possessing some of the antigenic determinants of the target bacterium in common, especially at lower dilutions of the serum. In serology, usually specific antisera are preferred to react with the target bacterium or its products only. However, sometimes antisera are too specific and may miss certain strains of the pathogen.

Agglutination test or latex agglutination test (antibodies are coated to sensitized latex beads, as a result of which the reaction is more readily visible and more sensitive) and precipitation test are not very sensitive and are also liable to disturbing cross reactions. However, immunofluorescence (IF) test is a very robust and sensitive serological test (can detect 10^{3-4} cells mL^{-1} in plant extract), and the primary reaction of antigen and antibody is clearly visible. Moreover, binding reactions can be observed at very high titers. In IF test, the antibodies are marked with a chemical dye, usually fluorescein isothiocyanate (FITC) that fluoresces in blue light. A light microscope fitted with epi-fluorescent light having suitable excitation and barrier filters for FITC is required. Indirect IF is slightly more sensitive and less specific than direct IF (Anonymous 1998). Immunofluorescence is widely used in Europe to detect bacterial pathogens in seed and propagating materials. In the Netherlands, Van der Wolf and Schoen (2004) reported its use to screen 60,000 potato seed pieces annually to determine the presence of *Ral. solanacearum*. It is also used to detect *Cla. sepedonicus* in potato seed pieces, and in France, it is used to screen tomato seed lots for the presence of *Cla. michiganensis*.

3.4.1 Monoclonal Antibodies

With the development of hybridoma technology, considerable improvement has been made in immunodiagnostic techniques. Monoclonal antibodies (MAbs) produced with this technology have a defined specificity to a single epitope. Reactivity of monoclonal antisera is usually lower than that of polyclonal antisera, because only one epitope of the bacterium will show the binding reaction. To enhance reactivity of monoclonal antisera, mixtures of monoclonals can be made, if available (De Boer 1987).

Hybridomas produce consistent antisera that give consistent result in different laboratories. Several MAbs produced by plant pathogenic bacteria and used in their epidemiological studies are listed in Table 3.1. Many of these antibodies have been characterized by testing their specificities with large number of strains of the target pathogens. Some bacterial taxa are relatively homozygous in the sense that they possess common antigenic determinants and one antibody generally reacts with all or nearly all strains of the taxon. *Cla. michiganensis* and *Xa. axonopodis* pv. *pelargonii* fall in this category. However, many plant pathogenic bacteria like *Pectobacterium atrosepticum* and *Pe. carotovorum* subsp. *carotovorum* are serologically heterogeneous, therefore, not all members of the population react with a polyclonal antiserum or a taxon-specific MAb. Pathogen specific MAbs that react with heat stable lipopolysaccharides enable development of robust detection kits for rapid use in laboratory and field diagnosis. MAbs directed towards capsule and/or extracellular polysaccharides in pathogens, such as *Cla. michiganensis* and *Ral. solanacearum*, are useful in a number of immunodiagnostic formats.

Most of the MAbs produced for different bacterial plant pathogens have been tested using enzyme-linked immunosorbent assay (ELISA). The ability of ELISA assays to detect 10^{5-6} cfu mL^{-1} is adequate for the identification of bacterial pathogens from symptomatic plants and colonies on the selective media. Moreover, the sensitivity of ELISA can be enhanced 10-fold by using an extraction buffer containing ethylenediaminetetra acetic acid and lysozyme, which releases lipopolysaccharides into solution, thereby enhancing the antibody-antigen reaction without increasing background readings. ELISA techniques using both polyclonal and monoclonal antibodies are available for many taxa of phytopathogenic bacteria, and rapid detection kits are also commercially available. Multitarget ELISA, which can detect more than one species in the same ELISA plate wells using different enzyme labels,

TABLE 3.1
Taxon-Specific Monoclonal Antibodies Produced for Plant Pathogenic Bacteria[a]

Genus, Species/Sub species, or Pathovar	MAb Designation	No. of Target Strains Tested	Reference
Acidovorax avenae subsp. *citrulli*	3D1F3	26	M. Bandla (personal communication)
Agrobacterium tumefaciens biovar 3	A6F21-1D3G7C8		Bishop et al. (1989)
Clavibacter michiganensis subsp. *michiganensis*	Cmm1	88	Alvarez et al. (1993b)
Cla. michiganensis subsp. *sepedonicus*	McAb1 to McAb-5	19	De Boer and Wieczorek (1984)
Erwinia carotovora subsp. *atroseptica*	14/18.6, 14/2, 14/8.6	3	Vernon-Shirley and Burns (1992)
Er. chrysanthemi	6A6	36	Singh et al. (1999)
Er. stewartii	C/G7/B2	43	Lamka et al. (1991)
Pseudomonas avenae	Pa1 to Pa-5	20	Alvarez et al. (1993a)
Psm. syringae pv. *phaseolicola*	Six MAbs[b]	223	Ovod et al. (1995)
	AG-1, AG-2	9	Wong (1990)
Ralstonia solanacearum species-specific strain-specific	Rs1, Rs1a	75	Alvarez et al. (1992)
	MAb		Griep et al. (1998)
Xanthomonas albilineans	12 MAbs	1	Tsai et al. (1990)
	7 MAbs	38	Alvarez et al. (1996)
Xa. axonopodis pv. *begoniae*	Xcb-1	26	Benedict et al. (1990)
Xa. axonopodis pv. *citri*	X-4600	30	Permar and Gottwald (1989)
Xa. axonopodis pv. *dieffenbachiae*	Xcd1, Xcd3, Xcd7,	329	Lipp et al. (1992)
	Xcd108	302	Norman and Alvarez (1994)
Xa. campestris pv. *campestris*	X9, X13, X17	200	Alvarez et al. (1985)
	10C5, 20H6, 16B5, 17C12, 10H12, 11B6	37	Franken (1990)
Xa. hortorum pv. *pelargonii*	Xcp-1	76	Benedict et al. (1990)
Xa. hortorum pv. *phaseoli*			
	XP2	18	Wong (1990)
Xa. oryzae pv. *oryzae*	Xco1, Xco2, Xco5	178	Benedict et al. (1989)
Xa. oryzae pv. *oryzicola*	Xco1a	8	Benedict et al. (1989)
Xa. translucens pv. *undulosa*	AB3-B6	44	Bragard et al. (1993)
Xa. campestris pv. *mangiferaeindicae*	XCM-1-XCM-6	4	Sanders et al. (1994)
Xylophilus ampelinus		63	Gorris et al. (1989)

[a] Data summarized from Alvarez (2004).
[b] Two MAbs (Ps core-1 and Ps core-2) specific to core lipopolysaccharide and four O-chain-specific MAbs (Ps-O:2-1, Ps-O:2-2, Ps-O:2-3, and Ps-O:3-1) were used to classify 223 strains belonging to 19 pathovars of *Psm. syringae*.

have been developed by Agdia, Inc. A multiple ELISA for detection of *Cla. michiganensis* subsp. *michiganensis* and *Xa. axonopodis* pv. *vesicatoria* is currently available (Alvarez 2004). A brief description of ELISA technique is given below.

3.4.2 Enzyme-Linked Immunosorbent Assay

In this test, the marker is an enzyme responsible for the reaction. The antibodies are first adsorbed (coated to) in the wells of a plastic ELISA plate. Subsequently, bacterial antigens in buffered sample solution are pipetted in the wells, which are trapped by the coated antibodies. After incubation and washing, an enzyme-labeled antiserum (conjugate) is added. Only wells where antigens react with the enzyme change in color after incubation with a suitable enzyme substrate. The enzyme used is generally alkaline phosphatase and substrate *p*-nitrophenylphosphate. A major disadvantage of ELISA is that in many cases bacterial products and not the whole cells are detected as the cells are washed out due to insufficient binding capacity of coating antibodies. Moreover, there is no information about the viability of cells and the dead cells may give false reading.

3.4.3 Flow Cytometry

The development of flow cytometry has enhanced the immunodiagnostic detection of bacteria. In this technique, the bacterial cells or other particles that pass individually through a sensor in a liquid stream, are rapidly identified and quantified. Cells are identified by conjugation of fluorescent dyes to specific antibodies and multiple cellular parameters are determined simultaneously based on cell's fluorescence and its ability to scatter light. The cells may be sorted electronically, permitting purification and/or culturing of subpopulations of selected cells for further confirmatory tests (Alvarez and Adams 1999; Alvarez 2001). Flow cytometry has been used for the detection of *Cla. michiganensis* subsp. *michiganensis* in tomato seed extracts, of *Xa. axonopodis* pv. *dieffenbachiae* in anthurium (Alvarez and Adams 1999), of *Xa. campestris* pv. *campestris* in seed extracts of *Brassica* sp. (Chitarra et al. 2002), and to know the viability of *Ral. solanacearum* in seed potatoes (Van der Wolf et al. 2004).

Other methods such as immune-magnetic separation and lateral flow devices are also used for the detection of plant pathogenic bacteria. Lateral flow devices use the principles, primarily those of ELISA, but different types of filters used act as solid support for the initial binding reaction (Danks and Barker 2000). A lateral flow device test kit, developed by Central Science Laboratory, UK, detects *Ral. solanacearum* in a 3-minute single step process. Rapid ImmunoStrip® kits for detection of *Ral. solanacearum*, *Cla. michiganensis*, and *Xa. hortorum* pv. *pelargonii* are sold by Agdia, Inc. In immune-magnetic separations, target cells are isolated from a mixed solution using paramagnetic polystyrene beads coated with specific antibodies. After washing, bound cells can be used for polymerase chain reaction or as they are viable, they can also be grown on semi-selective media.

It is emphasized here that commercially available modern serological kits can only be used for presumptive identification in a presumptive diagnosis. As these systems use only one characteristic or one set of similar characteristics, no authentic identification can be made. There are too many risks of errors resulting from cross-reactions, background errors, etc. Till now, only the fulfilling of **Koch's postulates** has led to satisfactory identifications and a correct diagnosis.

3.5 GENOMIC TECHNIQUES

During the last two decades, many genomic techniques have been developed for the detection of known pathogens in plant samples. The brief description of four techniques is given below.

3.5.1 Polymerase Chain Reaction

At the start of polymerase chain reaction (PCR), the DNA preparation from which the desired segment is to be amplified, an excess of two primer molecules, the four deoxyriboside triphosphates, and the DNA polymerase are mixed together in a reaction mixture. The multiplication of nucleic acid is achieved by repeated cycles of the following steps performed sequentially.

1. *Denaturation* (*melting*): The reaction mixture is heated usually to 95°C that assures denaturation
2. *Annealing*: The mixture is now cooled to 68°C (depending on the primers used) that permits annealing of the primer to the complementary sequences in the DNA; these sequences are located at the 3′ ends of the two strands of the desired segment
3. *Extension of nucleic acid strands*: The temperature is now so adjusted (72°C, if Taq polymerase is used) that the DNA polymerase synthesizes the complementary strands by utilizing 3′ OH of the primers. The primers are extended towards each other so that the DNA segment lying between two primers is copied; this is ensured by employing primers complementary to the 3′ ends of the segment to be amplified. After multiplication, the PCR products can be visualized on an ethidium bromide-stained agarose gel via electrophoresis.

Taq polymerase (originally isolated from the hot-spring thermophilic bacterium *Thermus aquaticus*, but now also available as an artificially synthesized product) is generally used and the temperature is adjusted to 72°C. At this temperature, the base pairing between about 20 bases long primers and the DNA is much more specific than at 37°C, the optimal temperature for *Escherichia coli* DNA polymerase. This minimizes the chances of annealing of primers to imperfectly matched sequences and, thereby, amplification of unwanted DNA.

The sensitivity of PCR is very high, and, theoretically, it can detect one copy of the target DNA in a sample. However, in practice, usually a lower sensitivity is reached due to the presence of inhibitory substances in the plant extract. Hybridization with probes via blotting is often used to verify identity of PCR products. If specific restriction sites are present in the product, restriction enzyme analysis can be performed as well. If there is enough information about the specificity of the primers used, the method can be used simultaneously for detection and identification. The negative PCR results can be the fallout of the following factors:

1. Absence of the target or target sequence in the sample; Rico et al. (2003) reported that when non-toxigenic strains are important under field infections and the PCR uses a sequence of DNA coding for the toxin, the pathogen will not be detected
2. Degradation of the target DNA in the sample
3. Failure of reaction due to wrong experimental conditions or presence of inhibitory compounds.

For successful completion of PCR, an internal control such as primers against plant DNA (which is always present in the plant extract) that will multiply this DNA should also be included in the same PCR experiment. Unfortunately false-positive reactions in a bacteriological context are still possible due to homology of non-target organism DNA with the primers thought to be specific for the target. In order to increase sensitivity and/or specificity of PCR, it can also be performed in combination with pre-enrichment in selective liquid medium, immuno-magnetic capture, and fluorescent labels. Lopez et al. (2003) reported an achievement of an ELISA plate set up in some tests.

3.5.2 Real-time (TaqMan®) PCR

The development of detectors (sequence detection systems) that can measure fluorescence that is emitted during the PCR cycle has led to the invention of a quantitative PCR called real-time (TaqMan®) PCR. The method is based on the 5′→3′ exonuclease activity of the *Taq* DNA polymerase, which results in cleavage of fluorescent dye-labeled TaqMan® probes during PCR. In this technique, as the multiplication of DNA during early stages can be made visible; the positive samples can be detected before the inhibitory compounds in the extract block the reaction. Moreover, the detection is rapid compared to conventional PCR; it is more accurate and has several other advantages.

3.5.3 DNA/RNA (Dot/Slot-Blot) Hybridization

In dot-blot hybridization, genetic material (DNA, RNA) of the bacterium (under investigation) is used to bind with the target bacterium. It is a very specific reaction, and due to its high specificity, it can be used simultaneously for detection and identification. The diseased plant sample is treated to remove all possible non-DNA material. Thereafter, DNA cleavage is allowed to take place under low pH conditions. After fixing a single-stranded DNA of target bacterium on a nitrocellulose filter, a DNA probe is added. This probe contains a large number of specific nucleic acid molecules of the target bacterium, which have been multiplied before in another bacterium, mostly using a virus as a vector. The nucleic acid probe binds to the complementary parts, if present, on the single strand fixed on the filter, under a process called hybridization. After washing, the hybridized DNA or RNA or RNA:DNA complexes can be visualized, using either radioactive or non-radioactive-labeled nucleic acid probe. The detection level of this technique is high (ca. 10^5 cells mL^{-1}) and can be further enhanced by application of enrichment culture.

Alvarez et al. (2015) devised a portable field pathogen-detection system by developing primers for the loop-mediated isothermal amplification (LAMP) kit from the 16S rDNA gene because it is a tandem region and is polymorphic for different phytoplasma groups. For LAMP reactions, fluorochrome hydroxynaphthol blue that does not need observation under UV light was used. For reactions in the field, a prototype incubator was developed, involving a 5-W solar panel, 12 V 12 Ah battery, voltameter, and mini dry bath with a 35-sample capacity. The kit's sensitivity is equal to that of the quantitative PCR technique and 1000 times that of the nested PCR technique. This LAMP kit can process 35 samples in 90 min without requiring electrical connections or any need to leave the field.

Gétaz et al. (2017) designed a reliable and sensitive LAMP assay using a unique marker, providing a highly specific and rapid detection technique, convenient for on-site detection. The specificity of the designed assay was tested on 37 strains from a culture collection of *Xa. fragariae*, 82 strains of other *Xanthomonas* species and pathovars, and 11 strains of other bacterial genera isolated from strawberry leaves. A detection limit of 10^2 fg was achieved, approximating to 20 genome copies per reaction. A consistent lower detection efficiency of 10^2 cfu mL^{-1} was achieved, when the analysis was performed with crude plant material. The LAMP assay designed in this study was adapted to work on crude plant material without any prior extensive extraction steps or incubation period. Moreover, it does not require advanced analytical knowledge or a fully equipped laboratory. Results were achieved within 7–20 min, depending on the pathogen concentration, thus providing a high-throughput and user-friendly method for detection and screening of plant material for quarantine regulations.

Fuller et al. (2017) designed oligonucleotide primers and probes to target *virD2* (virulence gene D2 of *Agrobacterium tumefaciens*) for use in a molecular diagnostic tool, which relies on isothermal amplification and lateral-flow-based detection. The oligonucleotide tools were tested in the assay and evaluated for detection limit and specificity in detecting alleles of *virD2*. One set of primers that successfully amplified *virD2* when used with an isothermal recombinase, was selected. Both the tested probes had detection limits in picogram amounts of DNA. Probe 1 could detect all tested pathogenic isolates that represented the maximum diversity of *virD2*. Finally, the coupling of lateral-flow detection to the use of these oligonucleotide primers in isothermal amplification made the process simple and easy and alleviated reliance on specialized tools necessary for molecular diagnostics. The assay is an advancement for the rapid molecular detection of pathogenic *Agrobacterium* spp.

3.5.4 Fluorescent *In Situ* Hybridization (FISH) Using rRNA (rDNA) Oligonucleotide Probes

In this method, short (20–30 mers) oligonucleotide probes are used against 16S or 23S (ribosomal) RNA/DNA. These oligonucleotide probes are used for *in situ* hybridization as they are able to diffuse through the cell walls of microorganisms present in thin tissue sections or in plant or soil extracts fixed on a microscopic slide. For Gram-positive bacteria, sometimes degradation of peptidoglycan layer of the cell wall with lysozyme is necessary to enhance penetration of the probe into the cells. Sensitivity depends also on the metabolic activity of the cells, as the dead cells show a positive reaction for a considerable period of time. After labeling the probes with a fluorescent dye or a gold label, the microorganisms can be visualized by incident light (fluorescence) microscopy. The specificity of the method is quite high, but false-positive reactions occur due to homology of RNA of non-target organisms with the specific probe sequence as reported by Moter and Göbel (2000) and Wullings et al. (1998).

The following techniques involving separation of bacterial components, namely, proteins and nucleic acids in an electric field, are also used for identification of plant pathogenic bacteria.

1. Polyacrylamide gel electrophoresis (PAGE), one- or two-dimensional plasmid electrophoresis
2. Immuno-electrophoresis
3. Native protein electrophoresis (enzyme detection)
4. Restriction fragment length polymorphism (RFLP) analysis
5. Random amplified polymorphism DNA (RAPD) analysis
6. Repetitive sequence-based (REP) fingerprinting PCR (REP, ERIC or BOX)
7. Amplified fragment length polymorphisms (AFLP) analysis.

Many workers have reported the development and/or application of pathogen-specific primers for detection of bacterial plant pathogens in heterogeneous mixtures. Most of these involve amplification by PCR coupled with one or more other techniques. The main emphasis has been on the development of specific probes for the detection of target pathogens. Manulis et al. (1994) developed pathovar-specific probes and PCR tests for *Xa. campestris* pv. *pelargonii*, which could detect 10–50 cfu of the bacterium per sample. The purpose for developing such probes was to detect the pathogens in asymptomatic plant tissue also. The primers and a probe were also developed for the detection of *Psm. syringae* pv. *actinidiae* in asymptomatic kiwi fruits (Koh and Nou 2002). Cubero and Graham (2002) developed primer sets for *Xa. axonopodis* pv. *citri* that could distinguish pathotype A from *Xa. aurantifolii* pathotypes B and C. Primers were based on sequence differences in the internal transcribed spacer (ITS) region and *pthA* gene. Primer sets based on ribosomal sequences had high level of specificity for *Xa. axonopodis* pv. *citri* A (of citrus canker type A), whereas those based on *pthA* were universal for pathogens of all types of citrus bacterial cankers. Multiplex PCR is used to identify several pathogens simultaneously (Bertolini et al. 2003). Caruso et al. (2003) developed cooperational PCR (Co-PCR) to detect *Ral. solanacearum* in water. Other genomic methods include DNA-hybridization, dot-blots, and nucleic acid sequence-based amplifications.

3.6 ENRICHMENT TECHNIQUES

The enrichment techniques have frequently been used to increase the sensitivity of immunofluorescence (IF), ELISA, and/or PCR. An enrichment-ELISA assay using pathogen-specific MAbs has been used to detect *Xa. axonopodis* pv. *dieffenbachiae* in leaf samples or *Pe. atrosepticum* or *Ral. solanacearum* in potato seed pieces. Alvarez et al. (1997) reported that even a limited culturing step, not sufficient for colony identification, increases the sensitivity of ELISA by about 10,000-fold. Enrichment assays have the added advantage that only the viable cells multiply to give a positive signal. Immunofluorescence colony-staining (IFC) technique combines an enrichment step in pour plates for colony detection with fluorochrome-labeled antibodies and also increases the sensitivity of IF approximately 1,00,000 times (Veena and Van Vuurde, 2002). The microcolonies formed due to the multiplication of viable cells can be seen with an epifluorescence stereomicroscope at 40–60X, whereas individual dead cells are not visible at this low magnification. IFC is used mostly for epidemiological studies and for confirmation of results achieved with other detection methods, which do not distinguish between living and dead cells. Optimization of buffers and conjugates also improves test results. Antibodies of sub class IgG are highly suitable for IFC formats because their molecules are smaller than those of IgAs (dimers) and IgMs (pentamers), and they diffuse easily through agarose gel.

The enrichment of target bacteria on semi-selective media also improves the sensitivity of PCR reaction. Such tests are often termed “BIO-PCR” because living cells harvested from colonies growing on culture plates are used in the PCR reaction. The major bands result from amplifications of DNA from living cells, although small amplicons may also be formed from dead bacteria. Sakthivel et al. (2001) were able to detect 55 fg of DNA per reaction tube which was equivalent to 70 cfu of *Xa. oryzae* pv. *oryzae* mL^{-1} of original sample and about seven cells per reaction tube. For detection of *Burkholderia cepacia* by a “BIO-PCR” assay, contaminated samples were incubated only for 24 h in broth; DNA was extracted and added to PCR beads with specific primers. The entire process was completed in 27 h in comparison to 5–6 days required for isolation and identification of the target bacterium.

Certain compounds that interfere with DNA amplification have restricted the use of direct PCR for the detection of target bacteria from natural samples. In many such cases, immunocapture or immune-magnetic separation has been employed to solve the problem. The optimum time for immunocapture of *Psm. syringae* pv. *syringae* with Advanced Magnetics™ beads coated with a polyclonal antiserum was just 1 h incubation. The detection limits were significantly improved when immunocapture was followed by a specific PCR test. Walcott and Gitaitis (2000) reported that immune-magnetic separation plus PCR resulted in a 100-fold increase in detection of *Aci. avenae* subsp. *citrulli* compared to direct PCR. Van Overbeek et al. (2002) also supported the polyphasic approach for studying the interaction between *Ral. solanacearum* and its biocontrol agent, *Psm. corrugata*.

Schaad et al. (2003) have reviewed the advances made in molecular-based diagnostic techniques used for the detection of plant pathogens including bacteria and compared the efficiency and sensitivity of conventional PCR, real-time PCR, BIO-PCR, and immunocapture PCR. Real-time BIO-PCR is much more sensitive than conventional real-time PCR, and it can detect 2 cfu mL^{-1} of *Psm. savastanoi* pv. *phaseolicola* in seed extracts and of *Cla. sepedonicus* in potato tuber extracts. Multiplex PCR can be used to detect more than one species of bacteria in the same reaction tube using probes labeled with different fluorescent dyes. To avoid cross absorption, the wavelength of each dye must be well separated.

Newly developed serological and molecular methods have played an important role in diagnosis of bacterial plant diseases and identification of plant pathogenic bacteria. However, after going through the literature carefully, it becomes certain that conventional methods are equally valid, important, and still used by workers. Many merits and plus points of molecular methods have been advocated and those that have proven to be often true are given below:

1. Rapid, sensitive, and cost effective
2. Have been integrated into certification/inspection schemes
3. Standardized test kits are commercially available for use
4. Can be applied to non-culturable organisms such as phytoplasmas/spiroplasmas
5. Genetically modified organisms can be traced in the environment more easily
6. Less sensitive to mutation or variation
7. Differentiation at low taxonomic level, even at strain level is possible.

It is generally argued that molecular methods, being genetic, are of a higher and better level than conventional methods, which are phenotypic. Isolated nucleic acids no longer function in the experiments as genetic material because they are immobilized in gels, cut into pieces, etc., and their analysis is a purely phenotypic one, too. Some possible pitfalls of molecular biological methods are given below:

1. Specificity, sensitivity, and reproducibility have been tested only to a limited extent
2. Negative influence of conditions and biochemicals may lead to experimental errors
3. Inability to discriminate between viable and non-viable cells and free nucleic acid in a sample
4. Difficult to verify false negatives and false positives
5. Change in probes/primers/enzymes/methods/chemicals may yield different or conflicting patterns or no patterns at all
6. As only a small part of structural elements of an organism is used, it may lead to a sampling error
7. Koch's postulates cannot be fulfilled.

3.7 BACTERIOPHAGES

The host range of the bacteriophages (phages) varies greatly. The host range may be narrow, confined to a particular species or a strain or a group of strains within a species. The non-specific or polyvirulent phages have a wide host range, which can infect bacteria belonging to different species, genera, or even families.

The phages have been used to detect phytopathogenic bacteria from diseased plant material. The increase in population of phage added to the substrate indicates the presence of the host bacterium in the substrate. The propagated phage is separated from the microbial complex either by centrifugation or by chloroform treatment. In case of low population of the target bacterium in the substrate, the required enrichment of the bacterium is achieved by growing it in a suitable medium.

Phages have been successfully used to detect phytopathogenic bacteria from seeds of cowpea and green gram (Soni and Thind 1991), rice, soybean, pea, and kidney bean. Two phages, ϕ RSM1 and ϕ RSS1, of *Ral. solanacearum* are highly strain specific and can be effectively used for the detection of specific strains of this bacterium (Yamada et al. 2007). Kawasaki et al. (2007) reported that bacterial cells of *Ral. solanacearum*, transformed with a green fluorescent protein (GFP)-expressing plasmid pRSS12, constructed from a filamentous phage ϕ RSS1, emitted a strong green fluorescence, and the presence of these cells was easily monitored in infected potato tissues and soil samples. These results suggest that this plasmid can serve as an easy-to-use GFP-tagging tool for any given strain of *Ral. solanacearum* in cytological and field studies. A novel technique called, "sensing of phage-triggered ion cascades" has been developed by Dobozi-King et al. (2005) for the rapid detection and identification of live pathogenic bacteria on the scale of minutes, with an unparalleled specificity. The method has a potential ultimate sensitivity of 1 bacterium/microliter (1 bacterium/mm^3) and is applicable to only physiological viable cells infected by mycophages and siphophages. The method combines the specificity and avidity of bacteriophages with fluctuation analysis of electrical noise. The method is based on the massive transitory ion leakage that occurs at the moment of phage DNA injection into the host cell. The ion fluxes require only that the cells should be physiologically viable (i.e., having energized membranes) and can occur within seconds after mixing the cells with sufficient concentration of phage particles.

Phages can also be used for the identification of bacteria including phytopathogenic ones. The phages used for this purpose should have a high specificity for the species or the strains to be identified. The phages attacking different groups of bacteria differ in their host range. Phages infecting *Pe. carotovorum* subsp. *carotovorum* generally have a narrow host range. Similarly, ϕ RSM1 and ϕ RSS1 of *Ral. solanacearum* have a narrow host range, while two other phages, namely, ϕ RSA1 and ϕ RSL1 of the same bacterium have a relatively wide host range, and the phage ϕ RSA1 was able to infect all the 15 strains of the bacterium belonging to different races and biovars tested in the study (Yamada et al. 2007). The ability of two phages, namely, Cp1 and Cp2 to infect roughly 98% strains of *Xa. citri* subsp. *citri* (syn. *Xa. axonopodis* pv. *citri*) occurring in Japan indicates the wide host range of the phages of this pathogen (Wakimoto 1967). Similar results were also obtained in Taiwan (Wu et al. 1993). In some phages, phage virulence is correlated to pathogenic variability or certain biochemical properties of the host bacterium. These phages can be used for quick identification of pathogenic races. Certain phages of *Ral. solanacearum* attack strains of the bacterium characterized by lactose oxidation and pathogenicity to tobacco. Phages specific to a lactose-utilizing strain of *Xa. citri* subsp. *malvacearum* (syn. *Xa. axonopodis* pv. *malvacearum*) and a mannitol-utilizing strain of *Xa. citri* subsp. *citri* occur in nature and are used for quick identification of these bacteria. Another bacteriophage successfully used for strain identification has been reported by Du Plessis et al. (1981).

Three bacteriophages of *Xa. arboricola* pv. *pruni* isolated from soil beneath diseased plum trees were used to type six South African and four other *Xa. arboricola* pv. *pruni* isolates which were previously shown to be serologically closely related. On the basis of patterns of lysis and efficiency of plaque formation by the phages, the bacterial isolates formed three distinct groups. Isolates from South Africa, the United States, and Argentina were lysed by the three phages with a high degree of efficiency, while one isolate from New Zealand showed no lysis, and the other isolate showed lysis with a low degree of efficiency. Both the New Zealand isolates were lysogenic, liberating phages spontaneously or following UV irradiation. The phage typing also provided evidence that peach, apricot, and plum trees in South Africa were infected by the same strain of *Xa. arboricola.* pv. *pruni.*

3.8 PATHOGENICITY TEST

Pathogenicity test is the most important step in the fulfillment of Koch's postulates because there are saprophytic bacteria that resemble pathogens in phenotypic or genetic characteristics (almost) completely. In spite of many molecular and serological tests available to detect the presence of a pathogen in host tissue, it is still a **most reliable test** to prove that a particular organism is the cause of the disease. It is the final proof and confirmation that a particular bacterium isolated from a plant with or without symptoms is really the cause of the disease and can be proved by a pathogenicity test using a susceptible host of the suspected pathogen. Janse (2005) has rightly concluded, "**The confirmatory pathogenicity test is still valid, still going strong, and important to fulfill the Koch's postulates**." In legal cases between government and growers or importers/exporters and between countries, when a pathogen is detected or described for the first time, this test is still indispensable and obligatory in many official testing and diagnostic schemes. An example is the diagnostic scheme for detection of latent infections, identification pathways, and the confirmatory pathogenicity test for *Ral. solanacearum* specified in an official European Union (EU) and European Plant Protection Organization (EPPO) testing scheme. The pathogenicity tests for quarantine organisms should be performed in a special quarantine greenhouse, where insects are excluded and a special quarantine protocol and regime for workers is in place.

However, the pathogenicity test requires time, availability of host plants, and optimum environmental conditions for disease development. Moreover, in some bacterial pathogens, the time required to produce the disease symptoms is too long. Some strains of *Cla. michiganensis* require 21–28 days to produce the symptoms. On the other hand, the confirmation by pathogenicity tests is often quick for leaf spotting pathogens as they take only 3–4 days to produce the symptoms under optimum conditions.

3.8.1 Inoculation of Host

The isolated bacterium is inoculated on the host to prove its pathogenicity. The bacterial cell suspension (made in buffer) used for inoculation should contain 10^{8-9} cells mL^{-1}. There are different methods of inoculation depending on the pathogen and host involved. For leaf spot diseases, leaves are first rubbed with carborundum powder to make small wounds, and then the bacterial suspension is smeared or sprayed onto the leaf surface. For xylem-inhabiting bacteria, like *Xyl. fastidiosa*, cell suspension is infiltrated into the xylem under vacuum through the injured root, stem, or leaf. For bacterial blight of rice, another vascular disease, the inoculation by cutting of leaf tips with scissors dipped in cell suspension works very well. To avoid pitfalls, it is a must to include a negative control (plants inoculated with sterile buffer solution only) in a pathogenicity test. If necessary, also a positive control, i.e., using a known pathogenic strain of the pathogen should be included. To avoid any contamination among controls and test plants in the greenhouse, all the three sets of plants should be well separated from each other.

For some bacterial pathogens, hypersensitivity test on tobacco, as an alternative to pathogenicity test, is quite useful, while it does not work with others. Hypersensitivity test on tobacco is a specific plant test, which differentiates most (fluorescent) plant pathogenic *Pseudomonas* spp., and some *Xanthomonas* and *Erwinia* spp. from non-plant pathogenic bacteria. A dense bacterial suspension containing 10^8 cells mL^{-1} is infiltrated in the mesophyll of a tobacco leaf. A pathogenic bacterium will cause a hypersensitive reaction, in which the cells leak and collapse within 24 h after infiltration, rendering the tissue glassy, and later turning necrotic. Infiltration with non-pathogens only causes some yellowing after several days or no visible change at all.

3.8.2 Reisolation of Pathogen

To be certain that the inoculated bacterium really caused the symptoms observed in the test plants, reisolation from inoculated plants at a distance from the point of original inoculation is necessary. Reisolated bacterium should be purified and pure culture obtained should be identical in all respects to the original inoculated culture. Some rapid confirmatory positive tests based on serology and PCR will make it a conclusive diagnosis. When a bacterium is found for the first time in a country or imported material, the tests for its identification should be repeated to confirm its identity and declare it a cause of the disease. In such cases, the original sample, sample extract, pure culture, and other test material obtained should be kept for reference purposes. Lelliot and Stead (1987) have given a detailed account for the diagnosis of bacterial diseases of plants by giving their detailed symptoms, diagnostic procedures, host inoculation tests, composition of culture media, and procedures for staining techniques and biochemical tests.

3.9 DIAGNOSIS REPORT

At the end, a diagnosis report is made containing the following information and is sent to the original correspondent.

1. Origin and description of the sample
2. Nature of the damage
3. Name of the bacterium isolated

4. Name of disease caused by the bacterium and description of disease symptoms
5. Nature (biology, epidemiology) of the pathogen
6. Preventive and management measures to be adopted.

The merits and demerits of modern serological and molecular methods have been given above. A positive view of these methods is to acknowledge them as a welcome addition to the already existing ones. They should be carefully checked for their specificity, reproducibility, and limitations because false positive reactions with unknown bacteria are easily possible. Besides their demerits, the molecular methods have not been adopted in all the laboratories due to certain constraints. For some laboratories, the cost of chemicals and equipment and the requirement of trained personnel for DNA extraction are the limitations for using PCR in the routine. Therefore, keeping in view the merits and demerits of all the available methods, Alvarez (2004) has rightly advocated the adoption of integrated approaches for the detection of plant pathogenic bacteria and diagnosis of bacterial plant diseases.

Mahlein (2016) has highlighted the potential of optical techniques, such as RGB imaging, multi- and hyperspectral sensors, thermography, or chlorophyll fluorescence in the detection, identification, and quantification of plant diseases at an early stage of epidemics. Recently, 3D scanning has also been added as an optical analysis that supplies additional information on vitality of plants. Different platforms from proximal to remote sensing are also available for multiscale monitoring of single crop organs or entire fields. Accurate and reliable detection of diseases is facilitated by highly sophisticated and innovative methods of data analysis, which lead to new insights derived from sensor data for complex plant pathogen systems. Non-destructive, sensor-based methods support and expand upon visual and/or molecular approaches to plant disease assessment. The most relevant areas of application of sensor-based analyses are precision agriculture and plant phenotyping.

Miller et al. (2009) have stressed the importance of early and accurate diagnoses and pathogen surveillance on local, regional, and global scales to predict outbreaks and allow time for development and application of mitigation strategies. Plant disease diagnostic networks have been developed worldwide to address the problems of efficient and effective disease diagnosis and pathogen detection, engendering cooperation of institutes and experts within countries and across national borders. Internet has opened global avenues for access to databases, communication, and cooperation, which was not possible decades ago. Without this innovation, the organization of effective networks to meet the challenge of invasive pathogens was unlikely. Case studies of regional, national, and international diagnostic networks are presented. The authors have also stressed the importance of morphological identification and systematics for the detection and identification of the pathogens in spite of frequent use of molecular methods in developed countries.

The National Plant Diagnostic Network (NPDN) has become a critical component of the plant biosecurity infrastructure of the United States. The target set forth in 2002 for a distributed, but coordinated system of plant diagnostic laboratories at land grant universities and state departments of agriculture has been realized. NPDN has become a model for cooperation among the public and private entities necessary to protect our natural and agricultural plant resources. Established at five regional networks, NPDN laboratories upload diagnostic data records into a National Data Repository at Purdue University. By facilitating early detection and providing triage and surge support during plant disease outbreaks, NPDN has become an important partner among federal, state, and local plant protection agencies and with the industries that support plant protection (Stack et al. 2014).

REFERENCES

Alvarez, A. M. 2001. Differentiation of bacterial populations in seed extracts by flow cytometry. pp. 393–396 in: *Plant Pathogenic Bacteria*. S. H. De Boer ed. Kluwer, Dordrecht, the Netherlands.

Alvarez, A. M. 2004. Integrated approaches for detection of plant pathogenic bacteria and diagnosis of bacterial diseases. *Annu. Rev. Phytopathol.* 42: 339–366.

Alvarez, A. M., and Adams, P. D. 1999. Flow cytometry: A promising tool for seed health testing. *Proceedings of the Third International Seed Health Symposium*. International Seed Test Association, Ames, Iowa, pp. 110–114.

Alvarez, A. M., Benedict, A. A., and Gananamanickam, S. S. 1993a. Identification of seed-borne bacterial pathogens of rice with monoclonal antibodies. *Proceedings of the Symposium on Seed Health Testing 1st*, Ottawa, Canada.

Alvarez, A. M., Benedict, A. A., and Mizumoto, C. Y. 1985. Identification of xanthomonads and groupings of *Xanthomonas campestris* pv. *campestris* strains with monoclonal antibodies. *Phytopathology* 75: 722–728.

Alvarez, A. M., Berestecky, J., Stiles, J. I., Ferreira, S. J., and Benedict, A. A. 1992. Serological and molecular approaches to identification of *Pseudomnas solanacearum* strains from *Heliconia*. *Presented at Bacterial Wilt International Conference*, Kaohsiung, Taiwan.

Alvarez, A. M., Derie, M., Benedict, A. A., and Gabrielson, R. 1993b. Characteristics of a monoclonal antibody to *Clavibacter michiganensis* subsp. *michiganensis*. *Phytopathology* 83: 1405.

Alvarez, A. M., Rehman, F. U., and Leach, J. E. 1997. Comparison of serological and molecular methods for detection of *Xanthomonas oryzae* pv. *oryzae* in rice seed. *Presented at Seed Health Test: Progress Towards the 21st Century*, Cambridge, UK.

Alvarez, A. M., Schenck, S., and Benedict, A. A. 1996. Differentiation of *Xanthomonas albilineans* strains with monoclonal antibody reaction patterns and DNA fingerprints. *Plant Pathol.* 45: 358–366.

Alvarez, E., Pardo, J. M., and Truke, M. J. 2015. A portable field pathogen-detection system for disease-free planting materials. *Phytopathology* 105: S1.5–S1.6.

Anonymous. 1998. Interim testing scheme for the diagnosis, detection and identification of *Ralstonia solanacearum* (Smith) Yabuuchi et al. in potatoes. Annex. II to the Council Directive 98/57/EC of 20 July 1998 on the control of *Ralstonia solanacearum* (Smith) Yabuuchi et al. Publication 97/647/EC, Official Journal European Communities No. L 235, 8–39.

Benedict, A. A., Alvarez, A. M., Berestecky, J., Imanaka, J., Mizumoto, C. Y., et al. 1989. Monoclonal antibodies specific to *Xanthomonas campestris* pv. *oryzae* and pv. *oryzicola*. *Phytopathology* 79: 322–328.

Benedict, A. A., Alvarez, A. M., and Pollard, L. W. 1990. Pathovar-specific antigens of *Xanthomonas campestris* pv. *begoniae* and *X. campestris* pv. *pelargonii* detected with monoclonal antibodies. *Appl. Environ. Microbiol.* 56: 572–574.

Bertolini, E., Olmas, A., Lopez, M. M., and Cambra, M. 2003. Multiplex nested RT-PCR in a single closed tube for sensitive and simultaneous detection of four RNA viruses and *Pseudomonas savastanoi* pv. *savastanoi* in olive trees. *Phytopathology* 93: 286–292.

Bishop, A. L., Burr, T. J., Millak, V. L., and Katz, B. H. 1989. A monoclonal antibody specific to *Agrobacterium tumefaciens* biovar 3 and its utilization for indexing grapevine indexing material. *Phytopathology* 79: 995–998.

Bragard, C., Mehta, Y. R., and Maraile, H. 1993. Serodiagnostic assays versus the routine techniques to detect *Xanthomonas campestris* pv. *undulosa* in wheat seeds. *Fitopathol. Bras.* 18: 42–50.

Caruso, P., Bertolini, E., Cambra, M., and Lopez, M. M. 2003. A new and sensitive co-operational polymerase chain reaction (Co-PCR) for rapid detection of *Ralstonia solanacearum* in water. *J. Microbiol. Methods* 55: 257–272.

Chitarra, L. G., Langerak, C. J., Bergervoet, J. H., and van den Bulk, R. W. 2002. Detection of the plant pathogenic bacterium, *Xanthomonas campestris* pv. *campestris* in seed extracts of *Brassica* sp. applying fluorescent antibodies and flow cytometry. *Cytometry* 47: 118–126.

Cubero, J., and Graham, J. H. 2002. Genetic relationship among worldwide strains of *Xanthomonas* causing canker in citrus species and design of new primers for their identification by PCR. *Appl. Environ. Microbiol.* 68: 157–164.

Danks, C., and Barker, I. 2000. On-site detection of plant pathogens using lateral flow devices. *Bull. OEPP/EPPO* 30: 421–426.

De Boer, S. H. 1987. Use of monoclonal antibodies to identify and detect plant pathogenic bacteria. *Can. J. Plant Pathol.* 9: 182–187.

De Boer, S. H., and Wieczorek, A. 1984. Production of monoclonal antibodies to *Corynebacterium sepedonicum*. *Phytopathology* 74: 1431–1434.

Dobozi-King, M., Kim, J. U., Young, R., Cheng, M., and Kish, L. B. 2005. Rapid detection and identification of bacteria: Sensing of Phage-Triggered Ion Cascade (SEPTIC). *J. Biol. Phys. Chem.* 5: 3–7.

Du Plessis, H. J., Loos, M. A., and Matthee, F. N. 1981. Bacteriophage typing of *Xanthomonas campestris* var. *pruni* isolates from different stone fruit species. *Phytophylactica* 13: 57–62.

Franken, A. A. J. M. 1990. Production, characterization and application of monoclonal antibodies for detecting *Xanthomonas campestris* pv. *campestris* in cabbage seeds. *Proceedings of the Symposium "Perspectives for Monoclonal Antibodies in Agriculture."* PUDOC, Wageningen, the Netherlands.

Fuller, S. L., Savory, E. A., Weisberg, A. J., Buser, J. Z., Gordon, M. I., et al. 2017. Isothermal amplification and lateral-flow assay for detecting crown-gall-causing *Agrobacterium* spp. *Phytopathology* 107: 1062–1068.

Gétaz, M., Bühlmann, A., Schneeberger, P. H. H., Van Malderghem, C., Duffy, B., et al. 2017. A diagnostic tool for improved detection of *Xanthomonas fragariae* using a rapid and highly specific LAMP assay designed with comparative genomics. *Plant Pathol.* 66: 1094–1102.

Gorris, M. T., Cambra, M., and Lopez, M. M. 1989. Production of monoclonal antibodies specific to *Xylophilus ampelinus*. *Proceedings of the 7th International Conference on Plant Pathogenic Bacteria*, Budapest, Hungary, pp. 913–921.

Griep, R. A., Van Twisk, C., Van Beckhoven, J. R. C. M., Van der Wolf, J. M., and Schots, A. 1998. Development of specific recombinant monoclonal antibodies against the lipopolysaccharides of *Ralstonia solanacearum* race 3. *Phytopathology* 88: 795–803.

Horst, R. K. 2001. *Westcott's Plant Disease Handbook*. 6th ed. Kluwer Academic Publishers, Boston, MA.

Hu, F.-P., Young, J. M., Triggs, C. M., Park, D.-C., and Saul, D. J. 2001. Relationships within the Proteobacteria of plant pathogenic *Acidovorax* species and subspecies, *Burkholderia* species and *Herbaspirillum rubrisubalbicans* by sequence analysis of 16S rDNA, numerical analysis and determinative tests. *Antonie Van Leeuwenhoek J. Microbiol.* 80: 201–214.

Janse, J. D. 2005. *Phytobacteriology: Principles and Practice*. CABI Publishing, Wallingford, UK, 360 pp.

Kawasaki, T., Satsuma, H., Fujie, M., Usami, S., and Yamada, T. 2007. Monitoring of phytopathogenic *Ralstonia solanacearum* cells using green fluorescent protein-expressing plasmid derived from bacteriophage ϕRSS1. *J. Biosci. Bioeng.* 104: 451–456.

Koh, Y. J., and Nou, I. S. 2002. DNA markers for identification of *Pseudomonas syringae* pv. *actinidiae*. *Mol. Cells* 13: 309–314.

Lamka, G. L., Hill, J. H., McGee, D. C., and Braun, E. J. 1991. Development of an immunosorbent assay for seedborne *Erwinia stewartii* in corn seeds. *Phytopathology* 81: 839–846.

Lelliot, R. A., and Stead, D. E. 1987. *Methods for the Diagnosis of Bacterial Diseases of Plants*. Blackwell Scientific Publications, Oxford, UK, 215 pp.

Lipp, R. L., Alvarez, A. M., Benedict, A. A., and Berestecky, J. 1992. Use of monoclonal antibodies and pathogenicity tests to characterize strains of *Xanthomonas campestris* pv. *dieffenbachiae* from aroids. *Phytopathology* 82: 677–682.

Lopez, M. M., Bertolini, E., Olmos, A., Caruso, P., Gorris, M. T., et al. 2003. Innovative tools for detection of plant pathogenic viruses and bacteria. *Int. Microbiol.* 6: 233–243.

Mahlein, A.-K. 2016. Plant disease detection by imaging sensors – Parallels and specific demands for precision agriculture and plant phenotyping. *Plant Dis.* 100: 241–251.

Manulis, S., Valinsky, L., Lichter, A., and Gabriel, D. W. 1994. Sensitive and specific detection of *Xanthomonas campestris* pv. *pelargonii* with DNA primers and probes identified by random amplified polymorphic DNA analysis. *Appl. Environ. Microbiol.* 60: 4094–4099.

Miller, S. A., Beed, F. D., and Harmon, C. L. 2009. Plant disease diagnostic capabilities and networks. *Annu. Rev. Phytopathol.* 47: 15–38.

Moter, A., and Göbel, U. B. 2000. Fluorescence in situ hybridization (FISH) for direct visualization of microorganisms. *J. Microbiol. Methods* 41: 85–112.

Norman, D. J., and Alvarez, A. M. 1994. Rapid detection of *Xanthomonas campestris* pv. *dieffenbachiae* in anthurium plants with a miniplate enrichment/ELISA system. *Plant Dis.* 78: 954–958.

Ovod, V., Ashorn, P. Yakovleva, L., and Krohn, K. 1995. Classification of *Pseudomonas syringae* with monoclonal antibodies against the core and O-side chains of the lipopolysaccharide. *Phytopathology* 85: 226–232.

Permar, T. A., and Gottwald, T. R. 1989. Specific recognition of *Xanthomonas campestris* Florida nursery strain by a monoclonal antibody probe in a microfiltration enzyme immunoassay. *Phytopathology* 79: 780–783.

Rico, A., López, R., Asensio, C., Aizpún, M. T., Asensio-S-Manzanera, M. C., et al. 2003. Nontoxigenic strains of *Pseudomonas syringae* pv. *phaseolicola* are a main cause of halo blight of beans in Spain and escape current detection methods. *Phytopathology* 93: 1553–1559.

Riley, M. B., Williamson, M. R., and Maloy, O. 2006. Plant disease diagnosis. APSnet Feature Story, http://www.apsnet.org/online/feature/plant disease/

Sakthivel, N., Mortensen, C. N., and Mathur, S. B. 2001. Detection of *Xanthomonas oryzae* pv. *oryzae* in artificially inoculated and naturally infected rice seeds and plants by molecular techniques. *Appl. Microbiol. Biotechnol.* 56: 435–441.

Sanders, G. M., Verschoor, J. A., van Wyngaard, S., Korsten, L., and Kotze, J. M. 1994. Production of monoclonal antibodies against *Xanthomonas campestris* pv. *mangiferaeindicae* and their use to investigate differences in virulence. *J. Appl. Bacteriol.* 77: 509–518.

Schaad, N. W., Frederick, R. D., Shaw, J., Schneider, W. L., Hickson, R., et al. 2003. Advances in molecular-based diagnostics in meeting crop biosecurity and phytosanitary issues. *Annu. Rev. Phytopathol.* 41: 305–324.

Singh, U., Trevors, C. M., De Boer, S. H., and Janse, J. D. 1999. Fimbrial-specific monoclonal antibody-based ELISA for European potato strains of *Erwinia chrysanthemi* and comparison to PCR. *Plant Dis.* 84: 443–448.

Soni, P. S., and Thind, B. S. 1991. Detection of *Xanthomonas campestris* pv. *vignaeradiatae* (Sabet et al.) Dye from green gram seeds and *X. campestris* pv. *vignicola* (Burkh.) Dye from cowpea seeds with the help of bacteriophages. *Pl. Dis. Res.* 6: 6–11.

Stack, J. P., Bostock, R. M., Hammerschmidt, R., Jones, J. B., and Luke, E. 2014. The National Plant Diagnostic Network: Partnering to protect plant systems. *Plant Dis.* 98: 708–715.

Tsai, C. C., Lin, C. P., and Chen, C. T. 1990. Characterization and monoclonal antibody production of *Xanthomonas albilineans* (Ashby) Dowson, the causal agent of sugarcane leaf scald disease. *Plant Prot. Bull.* 32: 125–135.

Van der Wolf, J. M., and Schoen, C. D. 2004. *Bacterial Pathogens: Detection and Identification Methods.* Marcel Dekker, New York, pp. 1–5.

Van der Wolf, J. M., Sledz, V., Van Elsas, J. D., Van Overbeek, L., and Van Bergervoet, J. H. W. 2004. Flow cytometry to detect Ralstonia solanacearum and to assess viability. p. 6 in: *Bacterial Wilt: The Disease and the* Ralstonia solanacearum *Species Complex.* C. Allen, P. Prior, and A. C. Hayward eds. APS Press, St. Paul, MN.

Van Overbeek, L. S., Cassidy, M., Kozdroz, J., Trevors, J. T., and Van Elsas, J. D. 2002. A polyphasic approach for studying the interaction between *Ralstonia solanacearum* and potential control agents in the tomato phytosphere. *J. Microbiol. Methods* 48: 69–86.

Veena, M. S., and van Vuurde, J. W. 2002. Indirect immunofluorescence colony staining method for detecting bacterial pathogens of tomato. *J. Microbiol. Methods* 49: 11–17.

Vernon-Shirley, M., and Burns, R. 1992. The development and use of monoclonal antibodies for detection of *Erwinia. J. Appl. Bacteriol.* 72: 97–102.

Wakimoto, S. 1967. Some characteristics of citrus canker bacteria, *Xanthomonas citri* (Hasse) Dowson, and the related phages isolated from Japan. *Ann. Phytopathol. Soc. Japan* 33: 301–310.

Walcott, R. R., and Gitaitis, R. D. 2000. Detection of *Acidovorax avenae* subsp. *citrulli* in watermelon seed using immunomagnetic separation and the polymerase chain reaction. *Plant Dis.* 84: 470–474.

Wong, W. C. 1990. Production of monoclonal antibodies to *Pseudomonas syringae* pv. *phaseolicola* and *Xanthomonas campestris* pv. *phaseoli. Lett. Appl. Microbiol.* 10: 241–244.

Wu, W. C., Lee, S. T., Kuo, H. F., and Wang, L. Y. 1993. Use of phages for identifying the citrus canker bacterium *Xanthomonas campestris* pv. *citri* in Taiwan. *Plant Pathol.* 42: 389–395.

Wullings, B. A., Van Beuningen, A. R., Janse, J. D., and Akkermans, A. D. L. 1998. Detection of *Ralstonia solanacearum*, which causes brown rot of potato, by fluorescent in situ hybridization with 23S rRNA targeted probes. *Appl. Environ. Microbiol.* 64: 4546–4554.

Yamada, T., Kawasaki, T., Nagata, S., Fujiwara A., Usami S., et al. 2007. New bacteriophages that infect the phytopathogen *Ralstonia solanacearum. Microbiology* 153: 2630–2639.

4 Classification of Bacteria

Classification is an arrangement of organisms into groups, called taxonomic ranks, on the basis of similarities or relationships. Two other areas closely related with the classification are nomenclature and identification. Nomenclature is the assignment of names to the taxonomic groups according to the nomenclature code. Identification is the practical use of classification scheme to determine the identity of an organism as a member of an established taxon or as member of a previously unidentified species.

4.1 NOMENCLATURE CODE

For many years, the Botanical Code of Nomenclature was the basis for bacterial nomenclature. In the International Congress of Microbiology held in London in 1936, it was decided to have a separate code for bacterial nomenclature. Therefore, the first separate code for bacteria, i.e., International Bacteriological Code of Nomenclature was published in 1948. The code was emended many times, but a major revision occurred in 1976 (Lapage et al. 1975). The name of the code has also been changed several times, but since 2000, it is known as the **International Code of Nomenclature of Procaryotes** and is governed by the **International Committee on Systematics of Procaryotes**. Before the adoption of abovementioned last name of the code, it was called **International Code of Nomenclature of Bacteria** and was governed by the **International Committee on Systematic Bacteriology**.

The 1976 revision of the International Code of Nomenclature of Bacteria (Lapage et al. 1975) provides that after January 1, 1980, priority of publication shall date from January 1, 1980, and on that date all names published prior to January 1, 1980 and included in the **Approved Lists of Bacterial Names** of ICSB (International Committee on Systematic Bacteriology) shall be validly published for the first time on that date. Those names validly published prior to January 1, 1980, but not included in the Approved Lists will have no further standing in nomenclature. After January 1, 1980, the priority of publication shall date from the date of their publication in the *International Journal of Systematic Bacteriology* (from 2000 in *International Journal of Systematic and Evolutionary Microbiology*). Since 1980, the names effectively published outside the abovementioned two journals can only make a claim to priority, if they are subsequently published in validation lists published in these two journals. The priority shall date from the date of their publication in the validation lists and not from the date of their original publication.

The Bacteriological Code (1990 revision) does not cover taxa above the rank of a class. Therefore, in the second edition of *Bergey's Manual of Systematic Bacteriology*, other codes of nomenclature have been used for naming the higher ranks. As there is no recognition of a rank above kingdom, Bergey's Trust, after considerable deliberations, concluded to incorporate the rank of **domain** into the hierarchy. Further, the rank of kingdom has not been used in this edition to avoid possible conflicts with other codes of nomenclature.

The International Code of Nomenclature of Procaryotes (ICNP; Parker et al. 2015) covers the nomenclature of procaryotes up to the rank of class only. In comparison, the International Code of Nomenclature for algae, fungi, and plants (McNeill et al. 2012) also covers the ranks of kingdom, division, or phylum. The rank of phylum is extensively used for groups of procaryotes, and names of more than 30 phyla are known in the literature for which there are cultured representatives in the literature (http: //www.bacterio.net/-classifphyla.html). In addition, there are many deep lineages equivalent to phyla that are not yet represented by organisms available in pure culture. In spite of the wide use of the term, the rank of phylum does not have standing in the nomenclature as regulated by the ICNP. Oren et al. (2015) have proposed to include the rank of phylum in the International Code of Nomenclature of Procaryotes. Whitman et al. (2018) proposed the suffix *-ota* to denote phyla, replacing the somewhat awkward *-aeota* as an addendum to the earlier proposal of Oren et al. (2015) to include the rank of phylum in the International Code of Nomenclature of Prokaryotes.

4.2 SPECIES

The bacterial species is the basic and most important taxonomic rank in the bacterial systematics. The concept of a bacterial species is poorly understood in comparison to higher organisms. This is mainly due to the reason that bacteria being procaryotic organisms differ markedly from higher organisms. Unlike higher organisms, sexuality is not used in bacterial species classification because relatively few bacteria undergo conjugation. Moreover, most of the procaryotic microorganisms are relatively simple and similar in morphological characters. Therefore, the morphological features alone are usually of little classificatory significance in comparison to higher organisms.

Brenner et al. (2001) defined the bacterial species as "a distinct group of strains that have certain distinguishing features and that generally bear a close resemblance to one another in the more essential features of organization." Each species differs considerably and can be distinguished from all other species. The above given definition of the species was generally loosely followed. It was mainly due to the reasons that one cannot accurately determine "a close resemblance," "essential features," or how many "distinguishing features" are sufficient to create a species.

4.3 CLASSIFICATION

Von Naegeli (1857) placed the bacteria in class **Schizomycetes**. In the 7th edition of *Bergey's Manual of Determinative Bacteriology*, class Schizomycetes was placed in one the five divisions, namely, Protophyta, recognized under the kingdom Plantae (Breed et al. 1957). Kingdom Plantae was proposed by Carolus Linnaeus in 1753. In 1937, E. Chatton, on the basis of cellular organization, divided the living organisms into two groups, namely, **procaryotes** and **eucaryotes** (Chatton 1937). R.G.E. Murray (1968) proposed two kingdoms, namely, Procaryotae and Eucaryotae to include these two groups. During 1970s, the comparison of oligonucleotide catalogs of the 16S ribosomal (r)RNA of procaryotes and of 18S rRNA of eucaryotes of a broad range of living organisms revealed that there are two different kinds of procaryotes: the ***Archaea*** (also called archaebacteria or archaeobacteria) and the ***Bacteria***. On the basis of these findings, Woese et al. (1990) opined that there are three fundamentally different kinds of living organisms: the *Archaea*, the *Bacteria,* and the eucaryotes (now called the ***Eucarya***).

Different criteria have been used to classify bacteria. First of all, it was the morphological approach in which the major emphasis was placed on morphological characters. Later on, the emphasis was also placed on physiological and biochemical characters along with the morphological characters. As a result of this, the classification of bacteria became tedious and time consuming because a large number of tests were to be performed and some of these tests required long time for completion.

With the rapid advances in DNA sequencing technology, sequence analysis of highly conserved regions of the bacterial genome, most notably genes coding for the RNA of the small ribosomal sub unit has now led to a natural classification that reflects the evolutionary history of bacteria. An important development was the discovery by Doi and Igarashi (1965) and Dubnau et al. (1965) that the rRNA citrons in bacterial species were conserved (slower to change) to a greater extent than the bulk of the genome, probably because of their critical function for the life of a cell. This function would allow any slow changes in nucleotide sequence to occur over a long period of time, relative to other genes that were not so critical to cell functions. This in turn led to the idea that rRNA-DNA hybridization might be useful for deducing the broader relationships that DNA-DNA hybridization could not reveal.

Genotyping (whole-genome sequence comparisons) is the most powerful procedure that will likely stabilize the taxonomy of bacteria. Current taxonomic classifications utilize DNA-DNA hybridization values, guanine-plus-cytosine content in mole percent, 16S rDNA sequences, and to a lesser extent, intergenic spacer sequences (16S–23S) and sequence comparisons of a few characterized genes, such as ATPase, the β sub unit of F_1F_0 elongation factor Tu, *recA*, and *gyrA*, to generate new taxonomic proposals. The sequence analyses of genomes and plasmids have revealed complete profiles of gene families and identified pathogenicity islands and specific genes or gene sets that proved more useful and reliable than physiological and biochemical tests. The growing number of whole-genome sequences available provides the needed data to analyze evolutionary trends and define the age of each genus or bacterial family by using molecular clocks.

Nowadays, the polyphasic approach used in the taxonomy and systematics of the bacteria and archaea includes the use of phenotypic, chemotaxonomic, and genotypic data. The use of 16S rRNA gene sequence data has revolutionized our understanding of the microbial world and led to a rapid increase in the number of descriptions of novel taxa, especially at the species level. In many cases, it has allowed for the demarcation of taxa into distinct species, but its limitations in a number of groups have resulted in the continued use of DNA-DNA hybridization. As technology has improved, next-generation sequencing has provided a rapid and cost-effective approach to obtaining whole-genome sequences of microbial strains. The incorporation of genomics into the taxonomy and systematics of the bacteria and archaea coupled with computational advances will boost the credibility of taxonomy in the genomic era (Chun and Rainey 2014).

Bergey's Manuals of Determinative Bacteriology and *Bergey's Manuals of Systematic Bacteriology* are the most comprehensive and authoritative publications in the field of bacterial taxonomy. The first edition of *Bergey's Manual of Determinative Bacteriology* was published in 1923, and the subsequent editions up to 7th were published in 1925, 1930, 1934, 1939, 1948, and 1957, respectively. As the new information became available, it was added in the subsequent editions. The 8th edition of *Bergey's Manual of Determinative Bacteriology* (Buchanan and Gibbons 1974) is different in format, presentation of information, and approach to classification from the 7th edition. This manual is meant mainly for the identification of bacteria, and these are divided into 19 groups. The 9th edition of *Bergey's Manual of Determinative Bacteriology* (Holt et al. 1994) published in 1994 is a departure from past editions that it is intended solely for the identification of those bacteria that have been described and cultured. In this edition, the bacteria have been divided into 35 groups including five groups of Archaeobacteria (archaea). These groups are not formal taxonomic ranks, but easily recognized phenotypic groups useful for identification purposes.

Second edition of *Bergey's Manual of Systematic Bacteriology* is the latest edition of the manual. It represents a major departure from the 1st edition, as well as from the 8th and 9th editions of *Bergey's Manual of Determinative Bacteriology*, in that the organization of the content follows a phylogenetic framework based on analyses of the nucleotide sequence of the ribosomal small sub unit RNA, rather than one based on phenotypic characters. The classification in 8th and 9th editions of *Bergey's Manual of Determinative Bacteriology* and in the 1st edition of *Bergey's Manual of Systematic Bacteriology* was organized in a non-hierarchical scheme because information about higher taxa was insufficient for construction of a formal hierarchical classification such as those used in all the previous editions. Instead, the genera were organized into phenotypic groupings, and these groupings were called parts, sections, or groups.

It is generally agreed that procaryotes (also called ***Procarya***) fall into two major lines of evolutionary descents, the *Archaea* and the *Bacteria*. These have been recognized as two different domains in the 2nd edition of *Bergey's Manual of Systematic Bacteriology*. The domains are sub divided into phyla that represent the main organizational units in this edition. With the exception of two phyla, i.e., Cyanobacteria and Actinobacteria, phyla are further sub divided into classes, classes into orders, orders into families, families into genera, and genera into species. In the case of Cyanobacteria, sub sections replace the families. In the case of Actinobacteria, sub classes and sub orders are included. Different taxonomic ranks and their numbers recognized until 2003 are given in Table 4.1 (Garrity et al. 2005).

The domain *Archaea* is divided into two phyla and nine classes. The domain *Bacteria* is divided into 24 phyla and 32 classes. The domain *Bacteria* is much larger and more complex than the other procaryotic domain, *Archaea*. The size of various phyla of domain *Bacteria* differs considerably. The phylum *Proteobacteria* is the largest and at the end of October 2003, it contained 2279 validly named species belonging to 521 genera. The phylum *Dictyoglomi* is the smallest and contains only a single recognized species. The phyla of the domain *Bacteria* along with their phyla numbers are listed in Table 4.2 (Garrity et al. 2005).

Since the publication of Garrity et al. (2005), Oren et al. (2015) have listed seven more phylla names currently in use, i.e., *Armatimonadetes*, *Elusimicrobia*, *Ignavibacteriae*, *Lentisphaerae*, *Euryarchaeota*, *Thaumarchaeota*, and *Crenarchaeota*, but omitted the phylum name *Cyanobacteria* already recognized.

TABLE 4.2
Phyla of the Domain *Bacteria*

Phylum Number	Phylum Name	Phylum Number	Phylum Name
B1	*Aquificae*	B13	*"Firmicutes"*
B2	*Thermotogae*	B14	*"Actinobacteria"*
B3	*Thermodesulfobacteria*	B15	*"Planctomycetes"*
B4	*"Deinococcus-Thermus"*	B16	*"Chlamydiae"*
B5	*Chrysiogenetes*	B17	*"Spirochaetes"*
B6	*Chloroflexi*	B18	*"Fibrobacteres"*
B7	*Thermomicrobia*	B19	*"Acidobacteria"*
B8	*"Nitrospirae"*	B20	*"Bacteroidetes"*
B9	*Deferribacteres*	B21	*"Fusobacteria"*
B10	*Cyanobacteria*	B22	*"Verrucomicrobia"*
B11	*"Chlorobi"*	B23	*"Dictyoglomi"*
B12	*Proteobacteria*	B24	*Gemmatimonadetes*

" " Used for provisional arrangement, the rest are established.

4.3.1 Classification of Plant Pathogenic Bacteria

The classification of plant pathogenic bacteria has also seen many revisions and rearrangements. Before 1967, i.e., prior to the discovery of role of mycoplasma-like organisms in causing plant diseases, there were only six genera of bacteria, which included plant pathogens. These genera were *Agrobacterium*, *Corynebacterium*, *Erwinia*, *Pseudomonas*, *Streptomyces*, and *Xanthomonas*. The 1976 revision of International Code of Nomenclature of Bacteria (Lapage et al. 1975) had a major effect on the taxonomy of bacterial plant pathogens. As a result of revision of the code, the names of many plant pathogenic bacteria were not accepted and hence, not included in the approved lists. Some of these were not accepted because of their poor descriptions or absence of suitable type cultures. The others were not accepted because some of the plant pathogenic species reported were similar in morphological, cultural, and biochemical characters and could not be separated from each other on the basis of these characters. However, they could be separated from each other on the basis of their pathogenicity on different host plants. Most of the names in the latter category were nomenspecies of *Xa. campestris* and *Psm. syringae* groups. More than 40 species included in the genus *Pseudomonas* were identical to *Psm. syringae* on the basis of cultural, morphological, and biochemical characters. According to the revised code, the species epithets of these species (not included in the approved lists) would thereafter be available for reuse to name new taxa. The loss of species designations for recognized plant pathogens and subsequent potential reuse of these names for new taxa was bound to result in unnecessary confusion and loss of the knowledge previously published about these pathogens.

TABLE 4.1
Different Taxonomic Ranks Used in Classification of Bacteria

Taxonomic Rank	Archaea	Bacteria	Total
Domain	1	1	2
Phylum	2	24	26
Class	9	32	41
Sub class	0	5	5
Order/sub section	13	75	88
Sub order	0	17	17
Family	23	217	240
Genus	79	1,115	1,194
Species	281	6,185	6,466

4.3.1.1 Pathovar Concept

The executive committee of International Society for Plant Pathology (ISPP) set up a Committee on Taxonomy of Plant Pathogenic Bacteria (CTPPB) to see the consequences of 1976 revision of code for plant pathologists and to make necessary recommendations for retaining the names of plant pathogenic bacteria not included in the approved lists. The infrasubspecific taxon "pathovar" records most accurately the differences exhibited by plant pathogenic bacteria in terms of host range or in producing different symptoms (diseases) on the same host. The International Standards define a pathovar as "a strain or a group of strains with the same or similar characteristics, differentiated at the infrasubspecific level from other strains of the same species or sub species on the basis of distinctive pathogenicity to one or more plant hosts." These differences in pathogenicity can be in symptoms or host range. Pathovars may also differ in other characteristics as well as pathogenicity, but such differences must be of minor importance, otherwise there would be grounds for separation at sub species or even species level (Bradbury 1982). CTPPB of ISPP requested the Judicial Commission of International Committee on Systematic Bacteriology to alter the rules of the code to include the taxon pathovar. However, the Judicial Commission denied the request because it could set a precedent for the inclusion of other infrasubspecific sub divisions.

The consensus of opinion of ISPP members and its CTPPB was that the pathovar designation most accurately records the known status of nomenspecies of plant pathogenic bacteria not included in the approved lists and that formal recognition is essential to avoid confusion in the nomenclature. Young et al. (1978) proposed that bacterial plant pathogens which do not have a claim to species rank be classified as pathovars and also listed all validly published names of plant pathogens having authentic cultures as species or pathovars. In order to formalize the use of pathovar names, the ISPP through its executive committee and its CTPPB, decided to publish a list of pathovar names and standards, to be known as the International Standards for Naming Pathovars of Phytopathogenic Bacteria.

4.3.1.2 International Standards for Naming Pathovars of Phytopathogenic Bacteria

The rules for the standards were adopted from the code and were so written that if the code ever does expand to assume authority over the infrasubspecific sub divisions, the pathovar standards would be transferred easily to the code. The standards were adapted to the code in a manner that allows both the standards and the code to be applied without conflict. The principal requirements of the code for legitimate names are essentially the same for pathovars.

Unlike the valid publication of species names, no journal is specified for the valid publication of pathovar names. Although there is requirement of submission of pathotype culture to one international culture collection, the ISPP-CTPPB suggests that researchers should follow the code and submit the new pathotype strains at least to two international culture collections located in two different countries to ensure stability and access. Priority for pathovars included in the list of pathovars names published with the standards (Dye et al. 1980) and also those published before the publication of this list is assigned in a manner similar to those of species. These are treated as though they had been validly published for the first time on January 1, 1980. Priority for pathovar names published after January 1, 1980 is determined according to the date of effective publication of legitimate pathovar names. For more details on nomenclature of bacterial plant pathogens, the readers may consult the review articles by Young (2008) and Bull et al. (2008).

Plant pathologists or plant bacteriologists have to follow two separate, but complementary set of rules: the rules set forth in International Code of Nomenclature of Procaryotes for naming down to sub species level, and the standards in the International Standards for Naming Pathovars of Plant Pathogenic Bacteria to name pathovars. A list of pathovar names and pathotype strains is maintained by the CTPPB of ISPP. The first such list was an appendix to the standards and only contained the names of organisms for which pathovar names and pathotype or neopathotype strains needed to be assigned (Dye et al. 1980). Names of plant pathogenic species in the approved lists were not included in the ISPP-CTPPB list. These two documents together contained the names of all bacterial plant pathogens validly published as of 1980. Since then, the ISPP-CTPPB periodically publishes lists of recently published legitimate names and updates an annotated comprehensive list of names of plant pathogenic bacteria. The subsequent lists published by ISPP-CTPPB are given below.

1. "Names of plant pathogenic bacteria 1864–1995" (Young et al. 1996) is the first comprehensive list of names published by the ISPP-CTPPB in which all the names proposed for plant pathogenic bacteria were included in a single list. This was originally published in the *Review of Plant Pathology* (Vol. 75, pp. 721–863) and in 2000 was placed on the ISPP website, http://www.isppweb.org/names_bacterial_revised.asp.
2. "Names of plant pathogenic bacteria published since 1995" (Young et al. 2004; http://www.isppweb.org/names_bacterial.asp) is a list of names published from 1995 to March 2004. It is published only online and includes all names that were published during this period.
3. "Names of plant pathogenic bacteria, 1864–2004" (Young et al. 2004; http://www. isppweb.org).
4. "Names of plant pathogenic bacteria, 1864–2006" (Bull et al. 2008; http://www. isppweb.org/about_tppb.asp).
5. "Comprehensive list of names of plant pathogenic bacteria, 1980–2007" (Bull et al. 2010).
6. "List of new names of plant pathogenic bacteria, 2008–2010" (Bull et al. 2012).
7. "List of new names of plant pathogenic bacteria, 2011–2012" (Bull et al. 2014).

In the 2nd edition of *Bergey's Manual of Systemic Bacteriology*, phytopathogenic bacteria have been placed in three phyla, namely, *Proteobacteria*, "*Firmicutes*," and "*Actinobacteria*." Among these phyla, *Proteobacteria* is the largest and is divided into five classes, namely, "*Alphaproteobacteria*," "*Betaproteobacteria*," "*Gammaproteobacteria*," "*Deltaproteobacteria*," and "*Epsilonproteobacteria*." Out of these five classes, first three classes contain the plant pathogenic bacteria. The phyla, "*Firmicutes*" and "*Actinobacteria*" contain three and one class, respectively, and all these classes contain plant pathogenic bacteria. The important characteristics of the phyla containing plant pathogenic bacteria are given in Table 4.3.

TABLE 4.3
Phyla of the Domain Bacteria Encompassing Phytopathogenic Bacteria

Phylum Number	Phylum Name/Properties	Class	Genera Containing Phytopathogenic Bacteria
B12	*Proteobacteria* Largest phylum of the domain *Bacteria* and 34 genera of this phylum contain plant pathogenic bacteria. It is divided into five classes.	"*Alphaproteobacteria*"	*Acetobacter*, *Agrobacterium*, '*Candidatus* Liberibacter', *Gluconobacter*, *Rhizobium*, *Rhizorhapis*, *Sphingomonas*
		"*Betaproteobacteria*"	*Acidovorax*, *Burkholderia*, *Paraburkholderia*, *Herbaspirillum*, *Janthinobacterium*, *Ralstonia*, *Xylophilus*
		"*Gammaproteobacteria*"	*Acinetobacter*, *Brenneria*, '*Candidatus* Phlomobacter', *Citrobacter*, *Dickeya*, *Enterobacter*, *Erwinia*, *Gibbsiella*, *Lonsdalea*, *Pantoea*, *Pectobacterium*, *Pseudoalteromonas*, *Pseudomonas*, *Rhizobacter*, *Salmonella*, *Samsonia*, *Serratia*, *Tatumella*, *Xanthomonas*, *Xylella*
		"*Deltaproteobacteria*"	None
		"*Epsilonproteobacteria*"	None
B13	"*Firmicutes*" Contains phenotypically diverse groups of bacteria such as thermophilic and hyperthermophilic bacteria; anaerobic, straight, curved and helical Gram-stain-negative rods; phototrophic and non-phototrophic bacteria; Gram-stain-negative and Gram-stain-positive cocci; Gram-stain-positive and Gram-stain-negative rods; mycoplasmas and thermoactinomycetes.	"*Bacilli*"	*Bacillus*, *Paenibacillus*, *Lactobacillus*, *Lactococcus*, *Leuconostoc*
		"*Clostridia*"	*Clostridium*
		Mollicutes	'*Candidatus* Phytoplasma', *Spiroplasma*
B14	"*Actinobacteria*" A very high level of morphological, physiological, and genomic diversity is found among its members. The members are divided into two major phenotypic groups: unicellular non-sporulating actinobacteria and filamentous sporulating sporoactinomycetes. The unicellular non-sporulating group contains Gram-stain-negative aerobic rods and cocci; budding and/or appendaged bacteria; Gram-stain-positive cocci; regular and irregular non-sporulating Gram-stain-positive rods and mycobacteria. The sporoactinomycetes include nocardioform actinomycetes, actinomycetes having multilocular sporangia, *Streptomyces* and other related genera.	*Actinobacteria*	*Arthrobacter*, *Clavibacter*, *Corynebacterium*, *Curtobacterium*, *Janibacter*, *Leifsonia*, *Leucobacter*, *Nocardia*, *Rathayibacter*, *Rhodococcus*, *Streptomyces*

4.3.1.3 Genera of Phytopathogenic Bacteria

Until now 53 genera of phytopathogenic bacteria have been reported in the literature. These include 50 established genera and three candidatus categories, whose relatedness has been determined and verified by *in situ* probing. A brief description of these genera is given below. The authenticated species, sub species, and pathovars belonging to these genera are also listed alphabetically.

Domain: *Bacteria*
I. Phylum: *Proteobacteria*
1. Class: "*Alphaproteobacteria*"
   i. Order: *Rhodospirillales*
   Family: *Acetobacteraceae*
   Genera: *Acetobacter* and *Gluconobacter*

4.3.1.3.1 Genus Acetobacter *Beijerinck 1898*

The genus was established by Beijerinck (1898). The important characteristics of the genus are given below.

Cells are Gram-stain-negative, non-sporing, straight or slightly curved rods measuring 0.6–0.8 × 1.0–4.0 µm. Cells may occur singly, in pairs, or in chains and may be motile or non-motile; if motile, the motility is by means of peritrichous flagella. Colonies of most strains are non-pigmented, while some produce pale colonies, and a small percentage of strains produce brown water soluble pigments. Obligately aerobic; metabolism is strictly respiratory, never fermentative, with oxygen as the terminal electron acceptor. Tests for oxidase, gelatin liquefaction, indole, and H_2S production are negative. Catalase test is usually positive and some strains reduce nitrates to nitrites. Chemoorganotrophs and some strains need growth factors such as thiamine, niacin, pantothenic acid, or *p*-aminobenzoic acid. Optimal temperature for growth is 30°C and optimal pH is in the acidic range, i.e., 4.0–6.0. Acetate and lactate are oxidized to CO_2 and H_2O, and ethanol to acetic acid. Major quinone is Q-9. *Acetobacter* strains are found in fermented foods, flowers, fruits, vinegar, and palm wine, and cause spoilage of cocoa and grape wine, beer, cider, and fermented foods. The genus does not contain any human and animal pathogen. The G + C content of the DNA varies from 50.5 to 60.3 mol%.

Type species: *Acetobacter aceti*

Beijerinck and Folpmers (1916) described two species, namely, *Ace. aceti* and *Ace. pasteurianus* isolated mainly from flowers, fruits, palm wine, vinegar, kefir, and fermented food. Later on, these species were also found to cause brown rot of apple and pear fruits and pink disease of pineapple (Kontaxis and Hayward 1978; Dhanvantari et al. 1978).

Plant pathogenic species: *Ace. aceti*, *Ace. pasteurianus.*

4.3.1.3.2 Genus Gluconobacter Asai 1935

The genus was established by Asai (1935). At present it contains only one plant pathogenic species, the causal agent of bacterial rot of mature fruits (pear, apple, and grape) and pink disease of pineapple.

The genus has the following characteristic features. Cells are Gram-stain-negative, ellipsoidal to rod-shaped, and measure 0.5–1.0 × 2.6–4.2 μm. Cells may be motile or non-motile and motile cells have 3–8 polar flagella. The cells may occur singly and/or in pairs, but rarely in chains. Non-sporing. Obligately aerobic. Metabolism is strictly respiratory type in which the oxygen acts as a terminal acceptor. The best growth occurs at 25°C–30°C. Optimum pH for growth is 5.0–6.0, though most strains can even grow at pH 3.5. Oxidase test is negative, but catalase test is positive. Indole and hydrogen sulphide are not produced, gelatin is not liquefied, and nitrates are not reduced. Ethanol is oxidized to acetic acid, but acetic acid is not further oxidized to carbon dioxide and water, as the bacterium does not have a full set of Krebs cycle enzymes. Acid is generally produced from maltose, glucose, fructose, and xylose, and growth can occur even up to concentration of 10.0% glucose. Major quinone possessed is ubiquinone of Q-10 type. *Gluconobacter* strains occur in sugar rich environments like fruits, flowers, soft drinks, grapes, wine, palm sap, coca wine, cider, beer, etc. The G + C content of the DNA varies from 52.0 to 64.0 mol%.

In addition to vinegar production, *Gluconobacter* species have many biotechnological applications. In the last decades, many bioconversion routes for rare and special sugars involving *Gluconobacter* spp. have been developed. Among the recent ones, are the biotransformations involved in the production of L-ribose and miglitol, both very promising pharmaceutical lead molecules. Most of these processes make use of *Gluconobacter*'s membrane-bound polyol dehydrogenases. The other enzymes of recent interest in industrial biotechnology include dextran-dextrinase, capable of transglucosylating substrate molecules and intracellular NAD-dependent polyol dehydrogenases involved in coenzyme regeneration.

Type species: *Gluconobacter oxydans*

Plant pathogenic species: *Gl. oxydans.*

ii. Order: "*Rhizobiales*"
   a. Family: *Rhizobiaceae*
      Genera: *Agrobacterium* and *Rhizobium*

4.3.1.3.3 Genus Agrobacterium Conn 1942

The genus was established by Conn (1942) to include *Ag. tumefaciens* and *Ag. rhizogenes* which cause hypertrophies of the affected plant parts. The pathogenicity character is encoded by genes present on the transmissible plasmids. The ease with which the capacity to induce hypertrophies is transferred between species of *Agrobacterium* suggests that a stable taxonomy based on pathogenicity may not sound well. The use of names of *Ag. tumefaciens* and *Ag. rhizogenes* with natural classification recognizing pathogenicity of strains according to their "tumorigenic" or "rhizogenic" and "non-pathogenic state" has been supported earlier. According to Young et al. (2001), this also posed a problem as the epithet "*tumefaciens*" was too closely associated with tumorigenicity to stand independently as a species name. In 1993, it was proposed to replace the name of *Ag. tumefaciens* with *Ag. radiobacter* to resolve this problem, but it could not be adopted because *Ag. tumefaciens* was conserved as the type species of this proposed combination (Sawada et al. 1993). Young et al. (2001) found that three genera of family *Rhizobiaceae*, namely, *Rhizobium*, *Agrobacterium*, and *Allorhizobium* do not differ phenotypically and also do not have unique phenotypic and genetic circumscription. Moreover, their phylogenetic differentiation on the basis of algorithm and sequences of 16S rRNA is also not strong. Therefore, they amalgamated these three genera in a single genus *Rhizobium* as the senior subjective synonym. However, Garrity et al. (2005) have recognized both the genera, *Rhizobium* and *Agrobacterium* in the family *Rhizobiaceae* for the time being. *Agrobacterium* is treated as a separate genus here also. Bouzar and Jones (2001) have included the causal agent of aerial tumors of *Ficus benjamina* in this genus.

The genus has the following important characteristics. Cells are Gram-stain-negative, non-sporing rods, measuring 0.6–1.0 × 1.5–3.0 μm. Cells are found singly or in pairs and are motile by means of 1–6 peritrichous flagella. Aerobic and chemoorganotrophs. Metabolism is strictly respiratory type in which oxygen is used as terminal electron acceptor. Colonies are generally circular, smooth, convex, and colorless to light beige. Growth on media containing carbohydrates is usually copious and slimy. Optimal temperature for growth is 25°C–28°C. Catalase and urease are always positive, while oxidase is generally positive. Growth in mineral salts media containing mannitol and other carbohydrates is accompanied by an acidic reaction. Common inhabitants of soil. Tumor forming strains occur in soils where diseased plants grew earlier or, are growing. The G + C content of the DNA ranges from 57.0 to 63.0 mol%.

Type species: *Agrobacterium tumefaciens*

Ag. tumefaciens causes crown gall of plants. Virulent strains of the bacterium contain a tumor-inducing plasmid, called Ti-plasmid. For successful infection of the host, T-DNA, called transfer DNA is cut from the Ti-plasmid and is transferred to the host cell genome. The penetration of *Ag. tumefaciens* into the host is through wounds. Although *Agrobacterium* species generally infect plants, they have been shown to cause opportunistic infections in humans with weakened immune systems (Hulse et al. 1993), but have not been shown to be a primary pathogen in otherwise healthy persons. The DNA transmission capabilities of *Agrobacterium* have been extensively exploited in biotechnology, as a means of inserting foreign genes into plants. The plasmid transfer DNA that is transferred to the plants is an ideal vehicle for genetic engineering. Tumor inducing genes are deleted from the plasmid and the desired gene sequence is cloned into the transfer DNA.

Plant pathogenic species: *Ag. larrymoorei, Ag. rhizogenes, Ag. rubi, Ag. tumefaciens, Ag. vitis.*

4.3.1.3.4 Genus Rhizobium *Frank 1889*

The salient characteristics of the genus are given below. Cells are Gram-stain-negative, non-sporing, small- to medium-sized rods measuring 0.5–1.0 × 1.2–3.0 µm. Motile by 1–6 flagella, which may be sub polar, bipolar, or peritrichous. Cells contain characteristic granules of polymerized β-hydroxybutyrate, which stain with Sudan Black and appear as highly refractive bodies under phase contrast illumination. Optimal temperature for growth lies between 25°C and 30°C. Optimum pH lies between 6.0 and 7.0. Aerobic, having a respiratory type of metabolism with oxygen as the terminal electron accepter. Colonies are circular, convex, opaque or semi-translucent, raised, and mucilaginous on yeast mannitol/mineral salt agar. Colony color is generally white or beige. Growth on carbohydrate media is usually accompanied by abundant amounts of extracellular polysaccharide slime. *Rhizobium* species can grow on relatively simple synthetic media. Chemoorganotrophs and utilize a wide range of carbohydrates and salts of organic acids as sole carbon sources, without gas formation. Produce an acidic reaction in mineral salts medium containing mannitol or other carbohydrates. Glutamate is much superior to nitrate or the ammonium ions as nitrogen source. Vitamin requirements differ for different species and strains. Thiamine, biotin, and calcium pantothenate are required for growth while nicotinic acid, folic acid, *p*-aminobenzoic acid, inositol, vitamin B12, and riboflavin do not promote growth.

Casein, starch, chitin, and agar are not hydrolyzed. Starch and cellulose are not utilized. An acid reaction is produced in mineral-salts medium having mannitol or other carbohydrates. Strains of all species cause hypertrophies in plants resulting in formation of root nodules with or without symbiotic nitrogen fixation. In root nodules, bacteria live endophytically and reduce or fix atmospheric nitrogen into a form utilizable by the host plant. Plant-host specificity is usually for a few legume genera, but in some strains, may extend to many legume genera and is largely determined by the chemical structure of lipochito-oligosaccharide Nod factors produced. The genes, which code for nodulation (*nod*) and nitrogen fixation (*nif*), are clustered on large plasmids or megaplasmids (pSyms) present in the bacteria. Plasmid transfer occurs between species and results in expression and stable inheritance of characters possessed by the plasmid-donor species. The G + C content of the DNA varies from 57.0 to 66.0 mol%. Members are common inhabitants of arable land.

Three isolates of Gram-stain-negative, rod-shaped, non-sporing bacteria isolated from galls on chrysanthemum and cherry plum were able to cause crown galls on various plant species. On the basis of phylogenetic and phenotypic evidence, Pulawska et al. (2012) considered that the isolates represent a single novel species of the genus *Rhizobium*, for which they proposed the name *Rhizobium skierniewicense* sp. nov.

Type species: *Rhizobium leguminosarum*

Plant pathogenic species: *Rhbi. nepotum, Rhbi. skierniewicense.*

Both *Agrobacterium* and *Rhizobium* have been recognized as separate genera in this book, as done by Garrity et al. (2005). The plant pathogenic species included under *Rhizobium* by Young et al. (2001) are included in the genus *Agrobacterium* in this book.

***Candidatus* category**: There are many bacteria that cannot be cultured by currently available methods. Therefore, these cannot be investigated in ways that provide the description for proposal as legitimate taxa. Keeping this in view, Murray and Schleifer (1994) proposed the rank or category of *Candidatus* for classification of such procaryotes. The International Committee on Systematic Bacteriology accepted the proposal of Murray and Schleifer (1994) and recommended the category '*Candidatus*' as a possible taxonomic rank for the unculturable procaryotic organisms (Murray and Stackebrandt 1995). In addition to genomic information, such as sequences apt to determine the phylogenetic position of the organism, all information, including structural, metabolic, and reproductive features, should be included in the description of a provisional taxon, together with the natural environment in which the organism can be identified by *in situ*-hybridization or other similar techniques for cell identification. *Candidatus* species are differentiated primarily on the basis of the comparative analyses of 16S rRNA sequences, other sequence data, ecological information, and metabolic data where available. To make clear the identification of organisms as *Candidatus*, names are preceded by the word *Candidatus* (italicized), but the name itself is not italicized and the entire designation is written between single quotation marks, e.g., '*Candidatus* Liberibacter asiaticus.' *Candidatus* nomenclature is neither governed by the code nor by the Standards for

Naming Pathovars. However, to prevent any nomenclature confusion, the Phytoplasma/Spiroplasma Working Team of the International Research Project for Comparative Mycoplasmology (IRPCM) has suggested rules for the description of organisms as novel taxa within '*Candidatus* Phytoplasma' (IRPCM 2004) and by extension to, other plant pathogenic '*Candidatus* genera' (Young 2008). As the *Candidatus* microorganisms cannot be cultured, their type strains cannot be isolated or maintained for reference. The reference material for *Candidatus* is either a DNA sample or a specific rDNA sequence. Until now, three *Candidatus* genera have been recognized which are described below.

b. Family: "*Phyllobacteriaceae*"
   Genus: '*Candidatus* Liberibacter'

4.3.1.3.5 'Candidatus *Liberibacter' Jagoueix-Eveillard et al. 1994*

Citrus huanglongbing (earlier called citrus greening) disease was initially considered to be caused by **mycoplasma-like organisms** (MLOs) and **rickettsia-like organisms** (RLOs). Later studies revealed that its causal agent is enclosed by 25 nm thick cell envelop and was called bacterium-like organism (BLO). Jagoueix-Eveillard et al. (1994) characterized two uncultured phloem-restricted bacteria causing citrus huanglongbing in Asia and Africa and named them as '*Candidatus* Liberobacter asiaticum' and '*Candidatus* Liberobacter africanum,' respectively. Following the rules of the International Code of Nomenclature of Bacteria, these two pathogens have been renamed as '*Candidatus* Liberibacter asiaticus' and '*Candidatus* Liberibacter africanus.' Teixeira et al. (2005) have reported '*Candidatus* Liberibacter americanus' associated with citrus huanglongbing in São Paulo state of Brazil.

The below given description of the '*Candidatus* Liberibacter' is based mainly on Jagoueix-Eveillard et al. (1994). Bacteria are Gram-stain-negative, filamentous in shape, and occur in phloem sieve tubes of plants. Round forms are also observed, but they are considered to be degenerating bacteria. They are transmitted by two insect vectors (Psyllidae), namely, *Trioza erytriae* and *Diaphornia citri*. They also multiply in the hemolymph and cells of salivary glands of both the vectors. The above given description of the '*Candidatus* genus' is based on the bacteria present in the phloem sieve tubes of sweet orange trees affected by huanglongbing (ex. greening disease) in Pune, India and in Nelspruit, South Africa. '*Ca.* Li. asiaticus' requires higher temperature than '*Ca.* Li. africanus' for optimum expression of the disease symptoms.

Sechler et al. (2009) designed a medium called Liber A to culture all the three '*Candidatus* Liberibacter spp.' which cause citrus huanglongbing. The medium containing citrus vein extract and a growth factor supported the growth of '*Ca.* Liberibacter spp.' for four or five single colony transfers before the decline of viability. The colonies of '*Ca.* Li. asiaticus' were irregularly shaped, convex, and 0.1–0.3 mm after 3–4 days of incubation. Two strains of '*Ca.* Li. asiaticus' and one of '*Ca.* Li. americanus' grown on this medium proved pathogenic to citrus and were isolated from uninoculated tissues of inoculated trees and seedlings 9 and 2 months later, respectively. The identity of the bacteria was confirmed by RT-PCR and 16S rDNA sequencing. This is the first report of the culturing and pathogenicity of '*Ca.* Li. asiaticus' and '*Ca.* Li. americanus' associated with citrus huanglongbing.

Roberts et al. (2015) reported three novel lineages of '*Ca.* Liberibacter africanus' associated with native rutaceous hosts of *Trioza erytreae* in South Africa.

Type species: '*Candidatus* Liberibacter asiaticus'

Plant pathogenic species: '*Ca.* Li. africanus', **'*Ca.* Li. africanus':** subsp. capensis', subsp. clausenae', subsp. vepridis', subsp. zanthoxyli'; '*Ca.* Li. americanus', '*Ca.* Li. asiaticus', '*Ca.* Li. psyllaurous', '*Ca.* Li. solanacearum'.

iii. Order: "*Sphingomonadales*"
   Family: *Sphingomonadaceae*
   Genera: *Sphingomonas* and *Rhizorhapis*

4.3.1.3.6 *Genus* Sphingomonas *Yabuuchi et al. 1990 emend. Yabuuchi et al. 1999 emend. Takeuchi et al. 2001 emend. Yabuuchi et al. 2002 emend. Feng et al. 2017*

Yabuuchi et al. (1990) created the genus *Sphingomonas* to include strictly aerobic, chemoheterotrophic, yellow pigmented, and Gram-stain-negative rod-shaped bacteria that contain glycosphingolipids as cell envelop components. Later on, Yabuuchi et al. (1999) emended the genus and transferred *Rhizomonas suberifaciens*, the causal agent of corky root disease of lettuce, along with two other species to this genus. On the basis of morphological, physiological, and chemotaxonomic characteristics along with DNA-DNA hybridization and 16S rDNA sequences comparison data, Takeuchi et al. (1995) included four new species in this genus. These species were *Sph. rosa* (isolated from rose plants and formerly named as *Ag. rhizogenes*), *Sph. pruni* (isolated from *Prunus persica*), *Sph. asaccharolytica*, and *Sph. mali* (isolated from apple trees). They also identified two strains, isolated from *Psychotria nairobiensis* and formerly named as *Chromobacterium lividum*, as *Sph. yanoikuyae*. Buonaurio et al. (2002) described a novel pathogen causing brown spot on yellow Spanish melon (*Cucumis melo* var. *indodorus*) fruits and identified it as *Sph. melonis*.

The below given description of the genus is taken mainly from the emended description of the genus proposed by Yabuuchi et al. (2002). Cells are Gram-stain-negative, nonsporing, straight or slightly curved rods or ovoid-shaped cells, which measure 0.2–1.4 × 0.5–4.0 μm. Reproduction is generally by binary fission, but budding or asymmetric division as visualized by electron microscopy has been found in two species, namely, *Sph. natatoria* and *Sph. ursincola*. Motility is usually by a single polar flagellum, but *Sph. trueperi* is peritrichously flagellate. Many species are non-motile, since they do not show spreading growth on 0.3% semi-solid agar plates. Bipolar staining in Gram stain (*Sph. ursincola*) and a rosette-like aggregation caused by polar-fimbriae occur in some species. Aerobic, having a strictly respiratory type of metabolism with oxygen as the terminal electron accepter. Colony color is variable, for instance, deep-yellow, orange, lemon-yellow, or non-pigmented. In certain cases, the pigmentation varies with the medium used. The colonies of some non-pigmented strains such as *Sph. yanoikuyae* become lemon-yellow after incubation of more than 3 days at room temperature. Esculin is hydrolyzed. Two species, namely, *Sph. ursincola* and *Sph. natatoria* containing bacteriochlorophyl A are facultative photoorganotrophs. Catalase positive. Oxidase positive or negative. Cell walls contain sphingolipids instead of lipopolysaccharides. Respiratory quinone is ubiquinone 10. Glucuronosyl-(1→1)-ceramide (SGL-1), galacturonosyl-β(1→1)-ceramide (in several species), and 2-hydroxymyristic acid, but not 3-hydroxy fatty acids are found. Major long-chain bases of SGL-1 are C-18:0, C-20:1, and C-21 cyclopropane. The G + C content of the DNA ranges from 59.0 to 68.0 mol%. Inhabitants of natural and man-made environments. Some species are opportunistic pathogens of humans and a few are plant pathogens.

Type species: *Sphingomonas paucimobilis*

Plant pathogenic species: *Sph. asaccharolytica*, *Sph. mali*, *Sph. melonis*, *Sph. pruni*, *Sph. rosa*, *Sph. yanoikuyae*.

4.3.1.3.7 Genus Rhizorhapis *Francis et al. 2014*

The previous illegitimate name of the genus was *Rhizomonas* (van Bruggen et al. 1990).

The below given description of the genus is based mainly on the publication of Francis et al. (2014). Cells are Gram-stain-negative, straight or slightly curved rods, and motile by one lateral, sub polar or polar flagellum, or are non-motile (van Bruggen et al. 1990). Resting stages are unknown. Cells are obligate aerobes having an oxidative metabolism and multiplication occurs by binary fission. Cells accumulate polyhydroxybutyrate (PHB) granules and are arginine dihydrolase-negative. Colonies are non-fluorescent, white or creamy white, and are smooth or wrinkled. The optimum growth temperature is 28°C–32°C and maximum is 37°C. Ethanol is not converted to acetic acid. Oxidase and catalase are produced. Denitrification to N_2 gas does not occur. Ubiquinone Q10 is present. Whole-cell fatty acids consist mainly of even-numbered unsaturated ($C_{18:1}$ and $C_{16:1}$) and saturated ($C_{16:0}$) straight-chain fatty acids, as well as $C_{14:0}$ 2-OH. The $C_{18:1}$ fatty acids represent at least 50% of the total fatty acids. The G + C content of the DNA ranges from 58 to 60 mol%. The type species is *Rhizorhapis suberifaciens* formerly known as *Rhizomonas suberifaciens* (van Bruggen et al. 1990).

Type species: *Rhizorhapis suberifaciens*

The bacterium causing corky root of lettuce was first classified as *Rhizomonas suberifaciens*. As the genus name *Rhizomonas* was found illegitimate, the bacterium was reclassified as *Sphingomonas suberifaciens* based on 16S rRNA gene sequences and the presence of sphingoglycolipid in the cell envelope. As the genus *Sphingomonas* is so diverse, the further reclassification of corky root of lettuce pathogen was deemed necessary. Twenty new isolates obtained from lettuce or sow thistle roots, or from soil using lettuce seedlings as bait, and previously reported isolates were characterized in a polyphasic study including 16S rRNA gene sequencing, DNA-DNA hybridization, DNA G + C content, whole-cell fatty acid composition, morphology, substrate oxidation, temperature, and pH sensitivity and pathogenicity to lettuce. The isolates causing lettuce corky root belonged to the genera *Rhizorhapis* gen. nov., *Sphingobium*, *Sphingopyxis*, and *Rhizorhabdus* gen. nov. Francis et al. (2014) specifically proposed to reclassify *Rhizomonas suberifaciens* as *Rhizorhapis suberifaciens* gen. nov., comb. nov., and also proposed the novel species *Rhizorhabdus argentea* gen. nov., sp. nov., *Sphingobium xanthum* sp. nov., and *Sphingobium mellinum* sp. nov. Several strains isolated from lettuce roots belonged to the genus *Sphingomonas*, but none of them was pathogenic.

Plant pathogenic species: *Rhiz. suberifaciens*.

2. Class: "*Betaproteobacteria*"
   Order: "*Burkholderiales*"
   a. Family: *Comamonadaceae*
      Genera: *Acidovorax* and *Xylophilus*

4.3.1.3.8 Genus Acidovorax *Willems et al. 1990 emend. Willems et al. 1992*

The genus was created by Willems et al. (1990) to include bacteria isolated from soil, water, activated sludge, and clinical samples. Two years later, Willems et al. (1992) transferred four plant pathogenic species and sub species of the genus *Pseudomonas* to this genus. Two more pathogens, i.e., causal agents of bacterial leaf spot of anthurium (*Anthurium andraeanum*) and bacterial leaf spot of lamb's lettuce (*Valerianella locusta*) have been included in this genus (Gardan et al. 2000, 2003b). On the basis of genetic and phenotypic data, Schaad et al. (2008) proposed an emendation of the species *Aci. avenae* and elevated *Aci. avenae* subsp. *citrulli* and *Aci. avenae* subsp. *cattleyae* to species rank as *Aci. citrulli* and *Aci. cattleyae*, respectively. They also proposed a new taxon, *Aci. oryzae* sp. nov. with FC-143^T = International Collection of Phytopathogenic Bacteria (ICPB) 30003^T = International Collection of Microorganisms from Plants (ICMP) 3960^T = American Type Culture Collection (ATCC) 19882^T as the type strain.

The below given description of the genus is as proposed by Willems et al. (1990) and emended by Willems et al. (1992). Cells are Gram-stain-negative, straight to slightly curved rods, and measure 0.2–1.2 × 0.8–5.0 μm. Cells occur singly, in pairs, or in short chains, and are motile by means of one or rarely two or three polar flagella. Most strains produce non-pigmented growth on nutrient agar, but some phytopathogenic strains produce a yellow to slightly brown diffusible pigment. Optimal temperature for growth is 30°C–35°C. Oxidase test is positive, while urease activity is variable among strains. Aerobic, having a strictly oxidative type of metabolism with oxygen as the terminal electron acceptor. However, some strains of *Aci. temperans* and *Aci. delafieldii* are capable of heterotrophic denitrification of nitrate. Most strains are chemoorganotrophs, although strains of *Aci. facilis* and *Aci. delafieldii* can grow lithoautotrophically, using the oxidation of H_2 as an energy source. Organic acids, amino acids, and peptone support good growth of the bacterium, but only a limited number of sugars support the growth. Two fatty acids, namely, 3-hydroxyoctanoic acid and 3-hydroxydecanoic acid, are always present, while 2-hydroxy-substituted fatty acids are absent. The G + C content of the DNA varies from 62.0 to 70.0 mol% (T_m). *Acidovorax* strains can be obtained from infected plants, soil, water, and clinical samples and activated sludge.

Type species: *Acidovorax facilis*

Plant pathogenic species: *Aci. anthurii*, *Aci. avenae*, ***Aci. avenae***- subsp. *avenae*; *Aci. cattleyae*, *Aci. citrulli*, *Aci. konjaci*, *Aci. oryzae*, *Aci. valerianellae*.

4.3.1.3.9 Genus Xylophilus *Willems et al. 1987*

Willems et al. (1987) found that the causal agent of bacterial necrosis and canker of grapevines, earlier classified as *Xanthomonas ampelina*, was not related to the genus *Xanthomonas*. They created a new genus *Xylophilus* and transferred *Xa. ampelina* to the new genus as *Xylophilus ampelinus*. This being the only species in the genus was designed as type species.

The below given description of the genus is taken from Willems et al. (1987). Cells are Gram-stain-negative, straight to slightly curved rods, measure 0.4–0.8 × 0.6–3.3 μm and are motile by a single polar flagellum. Long filamentous cells (length ≥ 30 μm) may appear in older cultures. Cells occur singly, in pairs, or in short chains. Strictly aerobic, having a respiratory type of metabolism using oxygen as the terminal electron acceptor. The bacteria are chemoorganotrophs with oxidative carbohydrate metabolism; oxidase test negative and catalase test positive. The growth is generally very slow and poor even at the optimal temperature of 24°C. The G + C content of the DNA ranges from 68.0 to 69.0 mol%. The genus belongs to the acidovorans rRNA complex in rRNA superfamily III and is equidistantly related to the other taxa in this complex. *Xylophilus* strains can grow on L-glutamine, but not on calcium lactate, while the reverse is true for *Xanthomonas* strains.

Type species: *Xylophilus ampelinus*

Plant pathogenic species: *Xyp. ampelinus*.

b. Family: "*Burkholderiaceae*"
   Genera: *Burkholderia*, *Paraburkholderia*, and *Ralstonia*

4.3.1.3.10 Genus Burkholderia *Yabuuchi et al. 1993 emend. Gillis et al. 1995*

The genus was created by Yabuuchi et al. (1993) to include some species of the then genus *Pseudomonas*. Gillis et al. (1995) further emended this genus on the basis of polyphasic taxonomy and also proposed a new species for this genus. All the 19 species of this genus belong to the ribosomal RNA similarity group II, which can be differentiated from the other groups of aerobic pseudomonads by rRNA/DNA hybridization or by rDNA sequencing.

Oliveira et al. (2017) reported *Bu. cenocepacia* causing sour skin of onion (*Allium cepa*) in Brazil. As one plant pathogenic species of this genus has been transferred to a newly created genus *Ralstonia*, it contains only seven plant pathogenic species.

The genus was named after W.H. Burkholder, an American bacteriologist, who first discovered the etiological agent of sour skin of onions. The following description of the genus is based on its proposal by Yabuuchi et al. (1992) and emendment by Gillis et al. (1995). Cells are Gram-stain-negative, non-sporing, straight or curved rods, but not helical and measure generally 0.5–1.0 × 1.5–4.0 μm. Cells may occur singly or in pairs and are motile by one or generally several polar flagella. One species, namely, *Bu. mallei* lacks flagella and is non-motile. Most strains accumulate poly-β-hydroxybutyrate as carbon reserve material and are capable of *ortho* cleavage of protocatechuate. Catalase is produced and oxidase activity varies among species. Chemoorganotrophs. Metabolism is strictly respiratory type with oxygen as the terminal electron acceptor. Some species are capable of anaerobic respiration using nitrate. Strains of some species such as *Bu. vietnamiensis* and *Bu. cepacia* are capable of fixing N_2. Most species grow at 40°C. The species are characterized by the presence of hydroxy fatty acids having 14, 16, and 18 carbon atoms and most characteristic of these acids is the $C_{16:0\ 3OH}$. The type strains of several species are characterized by the presence of two types of ornithine lipids. All species can utilize glucose, galactose, glycerol, inositol, sorbitol, and mannitol as sole sources of carbon. More than half of the species are pathogenic on plants, animals, and humans. The species are isolated from plant materials, soil, and clinical samples. The G + C content of the DNA ranges from 59.0 to 69.6 mol%.

The emended description of the genus by Sawana et al. (2014) is as below. The members of the genus are Gram-stain-negative, straight or slightly curved rods, which exhibit motility mediated by one or more polar flagella. Only, *Bu. mallei* lacks flagella and is non-motile. The members do not produce sheaths or prosthecae and do not pass through any resting stages. Most species are able to accumulate and utilize poly-β-hydroxybutyrate for growth. The species are mostly aerobic chemoorganotrophs, but some species are capable of anaerobic respiration using nitrate as the terminal electron acceptor. The members of the genus form a distinct monophyletic clade in phylogenetic trees, and they are distinguished from all other bacteria by the conserved sequence indels reported in this work in the following proteins: periplasmic amino acid-binding protein, 4-hydroxybenzoate 3-monooxygenase, 6-phosphogluconate dehydrogenase, sarcosine oxidase sub unit alpha, a putative lipoprotein, and a putative lyase. The G + C content for the members ranges from 65.7% to 68.5%.

Type species: *Burkholderia cepacia*

Plant pathogenic species: *Bu. andropogonis*, *Bu. caryophylli*, *Bu. cenocepacia*, *Bu. cepacia*, *Bu. gladioli*, ***Bu. gladioli***- pv. *agaricicola*, pv. *alliicola*, pv. *gladioli*; *Bu. glumae*, *Bu. plantarii.*

4.3.1.3.11 Genus Paraburkholderia *Sawana et al. 2015, gen. nov. emend. Dobritsa and Samadpour 2016*

Sawana et al. (2014), on the basis of molecular signatures and phylogenomic analysis of the genus *Burkholderia*, proposed the amendment of genus *Burkholderia* into the emended genus *Burkholderia* containing pathogenic organisms and a new genus *Paraburkholderia* gen. nov. harboring environmental species.

Cells are Gram-stain-negative, straight or slightly curved rods that are motile with one or more polar flagella. Other morphological and metabolic characteristics are similar to those of genus *Burkholderia*. The G + C content for the members of the genus ranges from 61.4% to 65.0%. Unlike *Burkholderia* species, *Paraburkholderia* members are not commonly associated with human infections. *Paraburkholderia* members form a monophyletic clade within the family *Burkholderiaceae*, and this prompted their distinction as a genus independent from *Burkholderia* species, in combination with the finding of robust conserved signature indels (CSIs), which are unique to *Paraburkholderia* species, and are lacking in members of the *Burkholderia* genus. These CSIs distinguish the genus from all other bacteria. Moreover, the CSIs that were found to be shared by *Burkholderia* species are absent in *Paraburkholderia*, providing an evidence of separate lineages.

The CSIs found within the *Paraburkholderia* genus are in parallel with phylogenomic analyses that indicate two monophyletic clades within the genus; one clade harbors unnamed and *Candidatus* Paraburkholderia, while the other clade is inclusive of environmental *Paraburkholderia*, commonly used for agricultural purposes. CSIs have been found exclusive to each of these clades, and have not been found specific for any other combination of *Paraburkholderia* species, providing an additional level of phylogenetic resolution within the genus level.

At present, only one species have been found to be plant pathogenic. Sugiyama et al. (2017) reported *Par. andropogonis* causing bacterial leaf streak of *Strelitzia reginae* in Hawaii detected from imported plants.

Type species: *Paraburkholderia graminis*

Plant pathogenic species: *Par. andropogonis*

4.3.1.3.12 Genus Ralstonia *Yabuuchi et al. 1996*

The genus was created in 1996 by Yabuuchi et al. (1995) to accommodate bacteria from ecological diverse niches that were previously classified in *Burkholderia* and *Alcaligenes.* The genus was established on the basis of phenotypic characterization, cellular lipid, and fatty acids analysis, phylogenetic analysis of 16S rDNA sequences, and rRNA-DNA hybridization within the homology group II of pseudomonads. The genus contains plant pathogens, human pathogens, and metal resistant bacteria. Yabuuchi et al. (1995) transferred *Bu. solanacearum* (syn. *Psm. solanacearum*) at the time of creation of the genus, but *Psm. syzygii*, causal agent of Sumatra disease of clove, was transferred later on by Vaneechoutte et al. (2004).

Ral. solanacearum species complex has long been recognized as a group of phenotypically diverse strains that can be sub divided into four phylotypes. Using a polyphasic taxonomic approach on an extensive set of strains, Safni et al. (2014) have provided enough evidence for a taxonomic and nomenclatural revision of members of this complex. Data obtained from phylogenetic analysis of 16S–23S rRNA ITS gene sequences, 16S–23S rRNA intergenic spacer (ITS) region sequences, and partial endoglucanase (egl) gene sequences and DNA-DNA hybridizations demonstrate that the *Ral. solanacearum* species complex comprises three genospecies. One of these genospecies includes the type strain of *Ral. solanacearum* and consists of strains of *Ral. solanacearum* phylotype II only. The second genospecies includes the type strain of *Ral. syzygii* and contains only phylotype IV strains. This genospecies is further divided into three distinct groups, namely, *Ral. syzygii*, the causal agent of Sumatra disease on clove trees in Indonesia, *Ral. solanacearum* phylotype IV strains isolated from different host plants mostly from Indonesia, and strains of the blood disease bacterium (BDB), the causal agent of blood disease of banana, a bacterial wilt in Indonesia affecting bananas and plantains. The last genospecies consists of *Ral. solanacearum* strains that belong to phylotypes I and III. As these genospecies are also supported by phenotypic data, which allow the differentiation of the three genospecies, the following taxonomic proposals are made.

1. Emendation of the descriptions of *Ral. solanacearum* and *Ral. syzygii*
2. Descriptions of *Ral. syzygii* subsp. *syzygii* subsp. nov., *Ral. syzygii* subsp. *indonesiensis* subsp. nov., *Ral. syzygii* subsp. *celebesensis* subsp. nov., and *Ral. pseudosolanacearum* sp. nov.

The below given description of the genus is based on Yabuuchi et al. (1995). The genus was named after E. Ralston, an American bacteriologist who first described *Psm. pickettii* and suggested a taxonomic relationship of this species to *Psm. solanacearum* based on DNA homology between these species. Cells are Gram-stain-negative, motile or non-motile rods. When motile, the motility is by means of a single polar flagellum or peritrichous flagella. Oxidase activity varies with species. Lysine and ornithine decarboxylases are not produced. Major quinone is ubiquinone Q-8. Cellular lipids of the members contain two kinds of phosphatidylethanolamines, PE-1 and PE-2. PE-2 contains 2-hydroxy fatty acid at *sn*-2 position of the glycerol moiety. Monosaccharides, disaccharides, and polyalcohols are oxidized and assimilated as a sole source of carbon and energy. The inability of *Ral. pickettii*, *Ral. eutropha*, and *Ral. solanacearum* to assimilate galactose, mannose, sorbitol, and mannitol distinguishes these species from *Burkholderia* species. Two-dimensional thin layer chromatography failed to detect two kinds of ornithine-lipids and C19 cyclopropanoic acid in cellular fatty acids in the type strains of the abovementioned three *Ralstonia* species, while the presence of these compounds was clearly demonstrated in six species of *Burkholderia* including *Bu. cepacia*. Species have different flagellar morphology. The members of the genus are plant and human pathogens and knallgas bacteria and metal-resistant bacteria. Strains of *Ral. pickettii* are found in various clinical specimens and hospital environments. *Ral. solanacearum* is taxonomically heterogeneous and infects a large number of host plants. The G + C content of DNA ranges from 64.0 to 68.0 mol%.

Type species: *Ralstonia pickettii*

Plant pathogenic species: *Ral. pseudosolanacearum*, *Ral. solanacearum*, *Ral. solanacearum* complex: banana (*Musca* sp.), ***Ral. syzygii***: subsp. *celebesensis*, subsp. *indonesiensis*, subsp. *syzygii*.

c. Family: "*Oxalobacteraceae*"
   Genera: *Herbaspirillum* and *Janthinobacterium*

4.3.1.3.13 Genus Herbaspirillum *Baldani et al. 1986 emend. Baldani et al. 1996*

The genus was created in 1986 to include a root-associated nitrogen fixing bacterium. Later, a plant pathogen, earlier described as *Psm. rubrisubalbicans* (causal agent of mottled stripe disease of sugarcane) was transferred to this genus.

The below given description of the genus is based on the proposal of Baldani et al. (1986) and emended by Baldani et al. (1996). Cells are Gram-stain-negative, generally vibroid and sometimes helical, measuring 0.6–0.7 µm in width, while the cell length varies from 1.5 to 5.0 µm depending on the culture media. Cells are motile, having one to three or more flagella on one or both the poles. The bacterium has a typical respiratory metabolism, and sugars may be oxidized, but are not fermented. Among the species included in the genus, three species are capable of fixing atmospheric N_2 under microaerobic conditions and grow well with N_2 as the sole nitrogen source, even in the presence of 10% sucrose. Oxidase and urease tests are positive, while catalase test is weak or variable. The salts of certain organic acids such as malate, fumarate, succinate, pyruvate, citrate, and trans-aconitate are favored carbon substrates for both NH_4^+- and N_2-dependent growth. Other carbon sources, such as glycerol, mannitol, sorbitol, D-glucose, galactose, L-arabinose, ribose, D-xylose, D-fructose, D-mannose, but not sucrose, are also utilized. Starch and gelatin are not hydrolyzed. Optimal temperature for growth is between 30°C and 34°C and optimal pH between 5.3 and 8.0. The genus constitutes a separate rRNA cluster within the β-subclass of the *Proteobacteria*. The G + C content of the DNA ranges from 60.0 to 65.0 mol% (T_m). The bacterium occurs mainly in association with graminaceous plants, endophytically colonizing roots, stems, and leaves, and also in soil.

Type species: *Herbaspirillum seropadicae*

Plant pathogenic species: *He. rubrisubalbicans*

4.3.1.3.14 Genus Janthinobacterium *De Ley et al. 1978 emend. Lincoln et al. 1999*

The genus was created in 1978 to include some Gram-stain-negative psychrotrophic and violacein producing bacterial species associated with food spoilage. In 1999, Lincoln and associates described a new species causing soft rot of mushrooms (*Agaricus bisporus*) and assigned it to this genus.

The description of the genus given below is as proposed by De Ley et al. (1978) and emended by Lincoln et al. (1999). Cells are Gram-stain-negative, straight and sometimes slightly curved rods having rounded ends, and measure 0.8–1.5 × 1.8–6.0 µm. Cells occur mostly singly, but sometimes in pairs or short chains. Non-sporing and motile. Strictly aerobic, having a respiratory type of metabolism with oxygen as the terminal electron acceptor. Temperature range for growth is 4°C–30°C, but best growth occurs at 25°C. Optimal pH lies between 7.0 and 8.0. Many strains produce the violet pigment, violacein, hence the genus name *Janthinobacterium*, which is derived from janthinus meaning violet-colored. Chemoorganotrophs. Citrate is used as a sole source of carbon and ammonium as a sole source of nitrogen. The major phospholipids are phosphatidyl ethanolamine, phosphatidylglycerol, and diphosphatidylglycerol as polar lipids. The major respiratory lipoquinone is Q-8. In Hugh and Leifson medium, acid is always produced from galactose, xylose, L-arabinose, and *m*-inositol, and often from lactose, mannitol, cellobiose, sucrose, inulin, and glycerol. Growth occurs on L-leucine, but not on ornithine and phenylalanine. The G + C content of the DNA varies from 61.0 to 67.0 mol%. The bacterium is commonly found in cold and temperate climates in soil and water, but can also be isolated from diseased mushrooms.

Type species: *Janthinobacterium lividum*

Plant pathogenic species: *Jant. agaricidamnosum.*

3. Class: "*Gammaproteobacteria*"
   i. Order: "*Enterobacteriales*"
      Genome-based phylogeny and taxonomy of the Enterobacteriales have been revised by Adeolu et al. (2016). The authors have proposed a new order "*Enterobacteriales*" and divided it into seven families. Out of these seven families, four families, namely, *Enterobacteriaceae*, *Erwiniaceae* fam. nov., *Pectobacteriaceae* fam. nov., and *Yersiniaceae* fam. nov. contain plant pathogenic bacteria.
   a. Family: *Enterobacteriaceae*
      Genera: *Citrobacter*, *Enterobacter*, *Gibbsiella*, and *Salmonella*

4.3.1.3.15 Genus Citrobacter Werkman and Gillen 1932

The below given description of the genus is as given by Sakazaki (1984). Cells are straight rods, measure 1.0 × 2.0–6.0 µm, and occur singly and in pairs. Gram-stain-negative, usually motile by peritrichous flagella, but usually not encapsulated. Facultatively anaerobic, having both a respiratory and a fermentative type of metabolism. Grow readily on ordinary media. Colonies on nutrient agar are generally 2–4 mm in diameter, smooth, low convex, moist, translucent or opaque, and grey with shiny surface and entire edges. Mucoid or rough forms may occur occasionally. Catalase positive, but oxidase negative. Chemoorganotrophs. Nitrate is reduced to nitrite. Citrate can be used as a sole source of carbon. Lysine is not decarboxylated. Methyl red test is positive, but Voges–Proskauer test is negative. Acid and gas is produced from glucose. Phenylalanine deaminase, gelatinase, lipase, and deoxyribonuclease are not produced. Alginate and pectate are not decomposed. The bacterium is found in the feces of man and other animals; probably a normal intestinal inhabitant. Often isolated from clinical specimens as opportunistic pathogens. Also found in food, water, soil, and sewage. The mol% G + C of the DNA is 50–52 (T_m).

Type species: *Citrobacter freundii*

Allahverdi et al. (2016) reported *Citrobacter freundii* as the cause of bacterial canker of mulberry from Iran.

Plant pathogenic species: *Cit. freundii.*

4.3.1.3.16 Genus Enterobacter Hormaeche and Edwards 1960

The genus was established in 1960 by Hormaeche and Edwards. Earlier, the genus did not include any plant pathogen. Now it contains four plant pathogenic species transferred from *Erwinia*. In order to infer an encompassing phylogeny of this genus, 16S rDNA sequences of additional members of *Enterobacter* are required.

The genus has the following important characteristics. Cells are Gram-stain-negative, non-sporing, straight rods, which measure 0.6–1.0 × 1.2–3.0 µm. Some strains are encapsulated. Motile by peritrichous flagella. Facultative anaerobes, and some enterobacter bacteria as facultative anaerobes ferment both glucose and lactose as a carbon source. Both acid and gas are produced from fermentation of glucose. *Enterobacter* species can be distinguished from other Gram-stain-negative rods (*Escherichia coli* and *Klebsiella* spp.) by virtue of being fast fermenter of lactose. Gas is produced from metabolic processes. Nitrate is reduced to nitrite. H_2S is not produced from thiosulphate. Methyl red and Voges–Proskauer tests are positive for many strains. Tween 80 and gelatin are either very slowly hydrolyzed or not hydrolyzed. Optimal temperature for growth is 30°C.

Enterobacter are found on plants and human skin, in soil, water, sewage, and intestinal tracts of humans and animals. Some *Enterobacter* species such as *En. sakajakii* are opportunistic human pathogens. They cause infections in immunocompromised (usually hospitalized) hosts and rarely in otherwise healthy individuals. *En. sakajakii* is well known for infections in infants who were fed milk-based powdered infant formulae. The bacterium survives when the contaminated powdered formula is heated and prepared. Since it has caused several outbreaks of infection in the past, baby formulae are effectively screened for *En. sakajakii* before they are sold. *En. cloacae* A-11 and similar bacteria are found on cucumber, radish, peas, soybean, sunflower, and sweet corn seeds. *En. cloacae* is also used for biological control of plant diseases. The G + C content of the DNA is 52.0–60.0 mol%.

Two isolates of Gram-stain-negative, facultatively anaerobic rods, 0.3–1.0 µm wide and 0.8–2.0 µm long, with peritrichous flagella, showing a DNA G + C content of 55.1 ± 0.5 mol%, isolated from diseased mulberry roots were investigated. Biochemical data revealed that the isolates could be differentiated from their nearest neighbors by the presence of lysine decarboxylase activity and their ability to utilize D-arabitol. DNA-DNA relatedness also distinguished the two isolates from phylogenetically closely related *Enterobacter* strains. Hence, these isolates were named *Enterobacter mori* sp. nov. (Zhu et al. 2011).

Furtado et al. (2012) isolated a facultative anaerobic, Gram-stain-negative bacterium causing brownish, necrotic, irregular spots on leaves of *Mabea fistulifera* Mart. in Brazil and on the basis of phenotypic, biochemical, and molecular tests identified it as *Enterobacter cowanii*. This was the first report of a member of the *Enterobacteriaceae* family causing disease in *M. fistulifera*. Retana et al. (2017) reported *En. hormaechei* causing stem rot on *Hylocereus* spp. in Costa Rica.

Type species: *Enterobacter cloacae*

Plant pathogenic species: *En. cancerogenus*, *En. cloacae*, ***En. cloacae***- subsp. *dissolvens*; *En. cowanii*, *En. hormaechei*, *En. mori*, *En. nimipressuralis*, *En. pyrinus*.

4.3.1.3.17 Genus Gibbsiella Brady et al. 2011 emend. Kim et al. 2013

The genus was named in honor of British forest pathologist John N. Gibbs for his contributions to forest pathology. The below given description of the genus is based on the publication of Brady et al. (2010a).

Cells are Gram-stain-negative, short rods (0.9 × 1–1.5 µm), facultatively anaerobic, oxidase negative, and catalase positive. Cells occur singly, in pairs, or in groups of four, and possess very fine fimbriae, but no flagella. Colonies on nutrient agar are white to creamy, round, convex, and smooth having entire margins. Strains can grow at temperatures between 10°C and 40°C. β-galactosidase activity is positive, but H_2S, urease, indole, acetoin, and gelatinase production are negative. Citrate is utilized and nitrate is reduced to nitrite. Arginine dihydrolase, lysine decarboxylase, ornithine decarboxylase, and tryptophan deaminase activity are negative. The G + C content ranges from 56.0 to 56.4 mol%.

The emended description of the genus is the same as given by Brady et al. (2010a), except that the DNA G + C content ranges from 56.0 to 58.7 mol% (Kim et al. 2013).

Brady et al. (2010a) reported *Gi. quercinecans* gen. nov., sp. nov., the cause of acute oak decline from Britain and Spain.

Type species: *Gibbsiella quercinecans*

Plant pathogenic species: *Gi. quercinecans.*

4.3.1.3.18 *Genus* Salmonella *Lignières 1900*

Joseph Leon Lignières named genus *Salmonella* in honor of Daniel Salmon, a veterinary pathologist of the US Department of Agriculture. *Salmonella* was first visualized in 1880 by Karl Eberth in the Peyer's patches and spleens of typhoid patients. In 1884, Georg Theodor Gaffky successfully grew the pathogen in pure culture. In 1985, Theobald Smith discovered what would be later known as *Salmonella enterica* (var. Choleraesuis). Initially, *Salmonella* Choleraesuis was thought to be the causal agent of hog cholera, so Salmon and Smith named it "Hog-cholerabacillus." The name *Salmonella* was not used until 1900, when Joseph Leon Lignières proposed that the pathogen discovered by Daniel Salmon's group be named *Salmonella* in his honor (Lignieres 1900).

The cells are rod-shaped, measure 0.7–1.5 × 2.0–5.0 μm, and are Gram-stain-negative. Cells are motile having peritrichous flagella and do not form endospores. They are chemotrophs and obtain their energy from oxidation and reduction reactions using organic sources. They are also facultative anaerobes, capable of surviving with or without oxygen. Most sub species of *Salmonella* produce H_2S, which can readily be detected by growing them on culture media containing ferrous sulphate. Most isolates exist in two phases, i.e., a motile phase I and a non-motile phase II. Cultures that are non-motile in primary culture may be switched to the motile phase using a Craigie tube or ditch plate. The bacteria are not killed by freezing, but UV light and heat accelerate their destruction. The cells are killed after being heated to 55°C (131°F) for 90 min or to 60°C (140°F) for 12 min. The heating of food for at least 10 minutes to an internal temperature of 75°C (167°F) is recommended to protect against *Salmonella* infection.

Genus *Salmonella* consists of only two species, *Sal. bongori* and *Sal. enterica*, and the latter is divided into six sub species, namely, *enterica*, *salamae*, *arizonae*, *diarizonae*, *houtenae*, and *indica*. *Sal. enterica* subsp. *enterica* includes more than 1,500 serotypes (while all the six sub species include over 2,500 serotypes), which despite their high genetic similarity vary greatly in their host range and disease outcome ranging from enteritis to typhoid fever (Ohl and Miller 2001). *Sal. enterica* subsp. *enterica* is an important economic and public health problem throughout the world.

The members lead predominantly host-associated lifestyles, but can be found to persist in a bathroom setting for weeks following contamination, and are frequently isolated from water sources, which act as bacterial reservoirs and may help to facilitate transmission between hosts. *Salmonella* species can be found in the digestive tracts of humans and animals, especially reptiles. Food and water can also be contaminated with the bacteria if they come in contact with the feces of infected people or animals. *Sal. enterica* sub species are found worldwide in all warm blooded animals, and in the environment. *Sal. bongori* is restricted to cold blooded animals, particularly reptiles. Strains of *Salmonella* cause illnesses such as typhoid fever, paratyphoid fever, and food poisoning (salmonellosis).

Type species: *Salmonella enterica*

Strictly speaking, *Salmonella* species are not plant pathogens. Two reports by Klerks et al. (2007) and Schikora et al. (2008) have suggested that plants show chlorosis or reduced vigor following *Sal. enterica* inoculation, but Koch's postulates were not proved in either case. Therefore, *Sal. enterica* remains characterized as plant-associated bacteria. However, Kado (2010) has listed *Sal. enterica* as a plant pathogen, responsible for causing systemic colonization of thale cress [*Arabidopsis thaliana* (L.) Heynh.]. There are numerous produce-implicated outbreaks, which are testament to *Sal. enterica* and pathogenic *Escherichia coli*'s ability to persist and use plants as vectors to humans (Taormina et al. 1999).

Plant pathogenic species: *Sal. enterica.*

4.3.1.3.19 'Candidatus *Phlomobacter' Zreik et al. 1998*

Garnier (2015) included this genus in the family *Enterobacteriaceae,* but Adeolu et al. (2016) excluded it from this family. Zreik et al. (1998) created 'Candidatus Phlomobacter' to include bacterium-like organisms responsible for causing marginal chlorosis of strawberry. At present, it contains only one plant pathogenic bacterium.

The below given description of the '*Candidatus* Phlomobacter' is taken from Zreik et al. (1998). Gram-stain-negative, walled bacterium, restricted to the sieve tubes of the phloem tissue of strawberry plants. The bacterium is 0.20–0.27 μm in diameter and 4.0 μm long. The bacterium is unable to grow in culture media, could not be transmitted to periwinkle plants through dodder, and insect vectors are not known. Sequence analysis of the cloned rDNA revealed that it is a new bacterium within group 3 of class 'Gammaproteobacteria,' while citrus huanglongbing agents belong to class "Alphaproteobacteria." '*Candidatus* Phlomobacter fragariae' name was proposed for strawberry marginal chlorosis agent.

Type species: '*Candidatus* Phlomobacter fragariae'

Plant pathogenic species: '*Ca.* Phl. fragariae'.

b. Family: *Erwiniaceae*
   Genera: *Erwinia*, *Pantoea*, and *Tatumella*

4.3.1.3.20 *Genus* Erwinia *Winslow et al. 1920 emend. Hauben et al. 1999*

The genus was coined to contain all plant pathogenic bacteria that were Gram-stain-negative, peritrichously flagellate, and facultatively anaerobic rods. The genus was named in honor of Erwin F. Smith, a pioneer plant bacteriologist.

The heterogenicity of this genus was generally acknowledged right from its inception. Several proposals were put forth to emend or revise this genus. Waldee (1945) suggested that *Erwinia* should be emended to contain pathogens that cause only necrosis or wilt diseases (*Er. amylovora*, *Er. salicis*, and *Er. tracheiphila*), utilize a restricted range of carbon compounds for growth, and the biochemically more active soft rotting pathogens (*Er. carotovora* and *Er. chrysanthemi*) should be placed in a new genus, *Pectobacterium*. Several workers supported Waldee's suggestion. However, in 8th edition of *Bergey's Manual of Determinative Bacteriology*, three clusters of organisms, namely, *amylovora*, *herbicola*, and *carotovora* groups were recognized under *Erwinia*. According to Kim et al. (1999), the phylogenetic group of seven species of the then genus *Erwinia* cluster *amylovora* (cluster-I), based on the complete 16S rDNA sequences, represents the true authentic erwinias. Some of the species and pathovars of *Erwinia* have been transferred to other genera, namely, *Brenneria*, *Dickeya*, *Enterobacter*, *Pantoea*, and *Pectobacterium*. However, Gardan et al. (2004) added one species, the causal agent of pink canker of papaya, to this genus.

The below given description of the genus is based on the emended description of the genus published by Hauben et al. (1999). Cells are Gram-stain-negative, non-sporing, straight rods which measure 0.5–1.0 × 1.0–3.0 µm. Cells occur mostly singly or in pairs and are motile by peritrichous flagella. They are facultative anaerobes, but anaerobic growth of some species is poor. Optimal temperature for growth is between 27°C and 30°C. Strains do not produce indole and do not oxidize gluconate. Catalase is positive, but oxidase and pectinase are negative. Strains utilize L-alanine, L-serine, glycylglycine, and L-glutaminic acid, but not kynureninic acid and trigonelline as nitrogen sources. The members are sensitive to chloramphenicol, furazolidone, oxytetracycline, tetracycline, and nalidixinic acid. Strains do not possess arginine dihydrolase, caseinase, urease, and phenylalanine deaminase.

The species of the genus comprise a distinct phylogenetic group, as determined by 16S rRNA gene sequence comparisons and are characterized by 14 generic signature nucleotides. The major fatty acids include 12:0, 14:0, and 16:0. The G + C content of the DNA ranges from 49.8 to 54.1 mol%. Members of the genus occur as saprophytes, epiphytes, and plant pathogens. Plant pathogens cause plant diseases that include mainly wilts and blights. The entry of the pathogens into host is mainly through wounds and natural openings.

The phylogenetic analyses of the 16S rRNA gene, *gyrB*, and *rpoD* gene sequences of a bacterium, isolated from black lesions on shoots of European pear trees (*Pyrus communis* L.) in an orchard in Japan, indicated that it could not be assigned to any recognized species of the genus *Erwinia*. DNA-DNA hybridization confirmed that the bacterial strains represented a novel species. Hence, the name *Erwinia uzenensis* sp. nov. was proposed (Matsuura et al. 2012).

Phenotypic analyses of eight *Erwinia* strains, isolated from necrotic pear blossoms in València, Spain, clustered them into one phenon, distinct from other species of the genus *Erwinia*. Phylogenetic analysis of 16S rRNA, *gpd*, and *recA* gene sequences showed that the eight novel strains could not be assigned to any recognized species. Based on these results, the novel isolates represent a novel species of the genus *Erwinia*, for which the name *Erwinia piriflorinigrans* sp. nov. is proposed (Lopez et al. 2011). Uribe-Lorío and Wang (2017) identified *Er. billingiae* as a causal agent of bacterial canker on mango (*Mangifera indica*) trees in Costa Rica. Luo et al. (2018) reported for the first time *Er. aphidicola* causing pepper fruit brown spot from China.

Type species: *Erwinia amylovora*

Plant pathogenic species: *Er. amylovora*, *Er. aphidicola*, *Er. billingiae*, *Er. cacticida*, *Er. mallotivora*, *Er. papayae*, *Er. persicina*, *Er. piriflorinigrans*, *Er. psidii*, *Er. pyrifoliae*, *Er. rhapontici*, *Er. tracheiphila*, *Er. uzenensis*.

4.3.1.3.21 Genus Pantoea *Gavini et al. 1989 emend. Mergaert et al. 1993*

Gavini and associates established this genus in 1989 to include some species isolated from diverse geographical and ecological sources. Kageyama et al. (1992) described a new species from citrus fruits and transferred it to this genus. Later, some species of the then genus *Erwinia*, based mainly on the differences in DNA-DNA hybridization were also transferred to this genus. On the basis of multi-locus sequence analyses and amplified fragment length polymorphism analysis, Brady et al. (2009) identified bacteria isolated from eucalyptus leaves and shoots, showing symptoms of blight and dieback, collected in Uganda, Uruguay, and Argentina; and from maize displaying brown stalk rot symptoms in South Africa as *Pantoea vagans* sp. nov., *Pantoea eucalypti* sp. nov., and *Pantoea deleyi* sp. nov. The below given description of the genus is as given by Gavini et al. (1989) and emended by Mergaert et al. (1993).

The cells are Gram-stain-negative, non-encapsulated, non-sporing, straight rods, and measure 0.5–1.3 × 1.0–3.0 µm. Cells are peritrichously motile or non-motile. Cultures grow well on nutrient agar at 30°C. Colonies on nutrient agar are smooth, translucent, and more or less convex having entire margins. The colonies are yellow, pale beige to pale reddish-yellow, or non-pigmented. Facultatively anaerobic and oxidase test is negative. No lysine decarboxylase, tryptophan deaminase, or urease activity. Pectate is not degraded, and hydrogen sulphide is not produced from thiosulphate. Glucose dehydrogenase and gluconate dehydrogenase are produced. The bacterium has been isolated from plant surfaces, seeds, fruits, soil, and water as well as from humans (wounds, blood, urine, and internal organs) and animals in several parts of the world. The G + C content of the DNA ranges from 49.7 to 60.6 mol% (T_m).

Ma et al. (2016) has reported *Pa. beijingensis* as an emerging pathogen of *Pleurotus eryngii*, causing soft rot disease in China and Korea.

Type species: *Pantoea agglomerans*

Plant pathogenic species: *Pa. agglomerans*, ***Pa. agglomerans***: pv. *betae*, pv. *gypsophilae*, pv. *millettiae*; *Pa. allii*, *Pa. ananas*, *Pa. ananatis*, ***Pa. ananatis***: pv. *ananatis*, pv. *uredovora*; *Pa. beijingensis*, *Pa. cypripedii*, *Pa. deleyi*, *Pa. dispersa*, *Pa. eucalypti*, *Pa. stewartii*, ***Pa. stewartii***: subsp. *indologenes*, subsp. *stewartii*; *Pa. vagans*.

4.3.1.3.22 Genus Tatumella *Hollis et al. 1982 emend. Brady et al. 2010b*

Hollis et al. (1981) named a new genus *Tatumella* in honor of Harvey Tatum, an American bacteriologist who made many contributions to our understanding of the classification and identification of fermentative and non-fermentative bacteria of medical importance. Hollis et al. (1981) proposed the name *Tatumella ptyseos* gen. nov., sp. nov., for a group of organisms (previously called group EF-9) isolated from clinical sources in the United States, Canada, and Puerto Rico. A total of 68% of these isolates were from sputum specimens. The members are Gram-stain-negative, oxidase-negative, fermentative rods, and grow on MacConkey agar. The distinctive biochemical characteristics of 44 *Ta. ptyseos* isolates were as follows: acid, but no gas from D-glucose, sucrose, and usually (71%) from D-xylose (62% delayed); no acid from lactose, maltose, or D-mannitol; negative tests for indole, urea, methyl red, gelatin, L-lysine decarboxylase, and L-ornithine decarboxylase; L-arginine dihydrolase variable; phenylalanine deaminase positive; Voges–Proskauer test positive by the Coblentz method, but negative by the O'Meara method; non-motile at 36°C, but 66% weakly motile (30% delayed) at 25°C; Simmons citrate positive at 25°C (89%), but negative at 36°C. DNA-DNA relatedness studies on 26 *Ta. ptyseos* strains showed that they were 80%–100% related at 60°C indicating that they comprise a single species. The DNA relatedness to other species within the *Enterobacteriaceae* was 7%–38%. Therefore, it is evident that this species belongs to this family, is distinct from all described species, and needs to be placed in a new genus. *Ta. ptyseos* isolates studied were susceptible to all the antimicrobial agents, namely, amikacin, ampicillin, cephalothin, chloramphenicol, gentamicin, kanamycin, tetracycline, and tobramycin tested by broth dilution. Three striking differences between *Ta. ptyseos* and other members of the *Enterobacteriaceae* were its large zones of inhibition around penicillin (mean diameter 24 mm), its tendency to die on some laboratory media (such as blood agar) within 7 days, and its lesser number (usually one) of flagella. Strain H36 (=ATCC 33301, =CDC D6168, =CDC 9591–78) is the type strain of this new species.

The below given emended description of genus *Tatumella* is based on the data of Hollis et al. (1981), Kageyama et al. (1992), and Brady et al. (2010b).

Gram-stain-negative, non-encapsulated, non-sporing, small rods that are 0.6–1.2 × 0.9–3.0 μm in size. Cells are motile by means of polar, sub polar, or lateral flagella or non-motile at 36°C. Facultatively anaerobic, fermentative, catalase positive (weak and slow), and oxidase-negative. Glucose dehydrogenase, gluconate dehydrogenase, and 2-ketogluconate dehydrogenase are produced. Nitrates are reduced to nitrites. Tests for indole, urease, gelatin, H_2S (TSI [triple sugar iron, agar media]), lysine decarboxylase, ornithine decarboxylase, tryptophan deaminase, potassium cyanide (KCN), lipase, and DNase are negative, but positive for Voges–Proskauer (Coblentz), methyl red, and citrate (Simmons) and variable for phenylalanine, arginine dihydrolase, and ortho-nitrophenyl-β-galactose (ONPG). The growth is inhibited by many antibiotics. The bacterium was isolated from human clinical samples, fruit, and soil. The G + C content of the DNA ranges from 49.8 to 53 mol%.

Brady et al. (2010b) transferred *Pa. citrea*, *Pa. punctata*, and *Pa. terrea* to the genus *Tatumella* emend. as *Ta. citrea* comb. nov., *Ta. punctata* comb. nov., and *Ta. terrea* comb. nov., and described *Ta. morbirosei* sp. nov.

Type species: *Tatumella ptyseos*

Plant pathogenic species: *Ta. citrea*, *Ta. morbirosei*, *Ta. punctata*, *Ta. terrea*.

c. Family: *Pectobacteriaceae*

Genera: *Brenerria*, *Dickeya*, *Lonsdalea*, and *Pectobacterium*

4.3.1.3.23 Genus Brenneria *Hauben et al. 1999*

Hauben et al. (1999) created this genus by transferring the species of cluster III of the then genus *Erwinia*. These species were transferred on the basis of complete 16S rDNA sequence analysis. One newly described species, *Er. alni* was also transferred. Later on, Samson et al. (2005) transferred one species, *Br. paradisiaca* to a newly created genus *Dickeya*.

The genus was named after Don J. Brenner, an American bacteriologist. The below given description of the genus is based on the data of Hauben et al. (1998) and of Wilson et al. (1957), Dye (1968), Lelliott and Dickey (1984), Verdonck (1992), Rijckaert (1994), and Surico et al. (1996). Cells are Gram-stain-negative, non-sporing rods having rounded ends, and measure 0.5–1.0 × 1.3–3.0 μm. Cells occur singly, rarely in pairs, and are motile by means of peritrichous flagella. Facultatively anaerobic, catalase positive, and oxidase negative. Optimal temperature for growth is 27°C–30°C. The bacterium does not produce arginine dicarboxylases, arginine dihydrolase, lysine dicarboxylases, and ornithine dicarboxylases. Members do not hydrolyze starch. Metabolism is fermentative type.

The major fatty acids include 12:0, 14:0, unknown 14.503, 16:0, and 17:0 cyclopropane. The species of the genus comprise a distinct phylogenetic group, as determined by 16S rRNA gene sequence comparisons and are characterized of having 12 signature nucleotides. The G + C contents of the members range from 50.1 to 56.1 mol%. The species of the genus cause tree diseases like wilts, necrosis, rots, blights, and cankers. The entry of the bacteria into the host tissue generally occurs through wounds and natural openings.

The emended description of the genus based on the data from Hauben and Swings (2005) and Brady et al. (2010a) is given below.

Cells measure 0.5–1.0 × 1.3–3.0 μm, have rounded ends, occur singly or rarely in pairs, Gram-stain-negative, and motile by means of peritrichous flagella. Facultatively anaerobic, but anaerobic growth of some species is weak. Optimum temperature for growth is 27°C–30°C and maximum is 40°C. Oxidase is negative and catalase positive. The tests for arginine dihydrolase, lysine decarboxylase, ornithine decarboxylase, tryptophan deaminase, and starch hydrolysis are negative. Do not utilize citrate and do not produce H_2S, urease, indole, or gelatinase. Nitrates are not reduced to nitrites. Major fatty acids are $C_{12:0}$, $C_{14:0}$, $C_{16:0}$, and $C_{17:0}$. Members of the genus *Brenneria* form a phylogenetic clade as determined by both 16S rRNA gene sequence analysis and multi-locus sequence analysis (MLSA) based on four housekeeping genes.

Ten strains of Gram-stain-negative, facultatively anaerobic bacteria isolated from symptomatic oak tissue in the United Kingdom and United States were placed in the genus *Brenneria* on the basis of partial *gyrB* sequencing. Further investigations using a polyphasic approach including MLSA [based on partial *gyrB*, RNA polymerase β (*rpoB*), *infB*, and *atpD* gene sequences], 16S rRNA gene sequencing, DNA-DNA relatedness studies, and both phenotypic and chemotaxonomic assays were carried out. DNA-DNA relatedness values revealed a closer relationship between two groups, indicating the same species, but their differentiation into two groups phenotypically and by enterobacterial repetitive intergenic consensus (ERIC) PCR fingerprints resulted in classifying them into two sub species. The name *Brenneria roseae* sp. nov. was proposed, with *Brenneria roseae* subsp. *roseae* ssp. nov. for the strains from the UK and *Brenneria roseae* subsp. *americana* ssp. nov. for the strains from the US (Brady et al. 2014).

On the basis of phylogenetic and phenotypic data, Denman et al. (2012) concluded that a group of nine Gram-stain-negative, facultatively anaerobic bacterial strains isolated from native oak trees displaying symptoms of acute oak decline in the UK represent a novel species in the genus *Brenneria*, for which the name *Brenneria goodwinii* sp. nov. is proposed.

Li et al. (2015b) isolated five Gram-stain-negative, facultatively anaerobic, motile, bacterial strains from symptomatic bark tissue of *Populus* × *euramericana* canker and on the basis of analysis of 16S rRNA gene sequences and multi-locus sequence analysis classified these strains as a novel species of the genus *Brenneria* with the proposed name *Brenneria populi* sp. nov.

Type species: *Brenneria salicis*

Plant pathogenic species: *Br. alni*, *Br. goodwinii*, *Br. nigrifluens*, *Br. populi*, ***Br. quercina***: pv. *lupnicola*, pv. *quercina*; ***Br. roseae***: subsp. *americana*, subsp. *roseae*; *Br. rubrifaciens*, *Br. salicis.*

4.3.1.3.24 Genus *Dickeya* Samson et al. 2005

The genus was created by Samson et al. (2005). The bacterium causing hollow stalk and blight of *Chrysanthemum morifolium* was described and identified as *Erwinia chrysanthemi*. Subsequently, similar bacteria were isolated from several plants causing typical soft rot and wilt symptoms. After extensive biochemical testes, all these pathogens were finally grouped into a single species, *Er. chrysanthemi* and designated as *Er. chrysanthemi* complex. However, plant bacteriologists for their convenience, divided the members of this complex into six pathovars, namely, pv. *chrysanthemi*, pv. *dianthicola*, pv. *dieffenbachiae*, pv. *parthenii*, pv. *zeae*, and pv. *paradisiaca*. Later on, Hauben et al. (1998) renamed *Er. chrysanthemi* pv. *paradisiaca* as *Brenneria paradisiaca* and transferred the other members to *Pectobacterium chrysanthemi*. Although these pathogens were grouped together as pathovars of *Pe. chrysanthemi*, lot of diversity existed among these pathogens. Samson et al. (2005) resolved this problem by using DNA-DNA hybridization and analysis of 16S rRNA gene sequences. They found that on the basis of analysis of 16S rRNA gene sequences, *Pe. chrysanthemi* strains and *Br. paradisiaca* formed a clade distinct from the genera *Pectobacterium* and *Brenneria*; therefore, they transferred all these pathogens to a newly created genus *Dickeya*. On the basis of DNA-DNA hybridization, all the strains of *Pe. chrysanthemi* formed six genomic species and hence were divided into five species, one species having two biovars. Parkinson et al. (2009), using sequences from *recA* locus, developed a phylogenetic method to differentiate species of *Dickeya* genus. The method provides a simple means for identification to species and intraspecies level.

The genus was named in honor of an American plant pathologist, Robert S. Dickey, for his contributions to the research on the *Er. chrysanthemi* complex. The below given description of the genus is taken from Samson et al. (2005), which is based on their data and those of Waldee (1945) and Hauben et al. (1998). Cells are Gram-stain-negative rods, measuring 0.5–1.0 × 1.0–3.0 μm having rounded ends. They occur mostly singly or in pairs, but sometimes in chains, and are usually motile by means of peritrichous flagella. Facultatively aerobic/anaerobic bacteria that catabolize glucose by a fermentative pathway and reduce nitrates to nitrites. The bacterium produces pectolytic enzymes and indole and grows at 36°C. Members of the genus form a distinct clade based on the sequence analyses of 16S rRNA gene. The G + C contents of its members range from 56.4 to 59.5 mol%. The members cause vascular wilts or soft rots on a range of host plants.

Eight isolates, seven from potato (plants exhibiting blackleg and slow wilt symptoms) and one from hyacinth, of Gram-stain-negative, motile, rod-shaped bacteria were isolated. Their characterization by whole-cell matrix-assisted laser desorption ionization time-of-flight (MALDI-TOF) mass spectrometry, pulsed field gel electrophoresis after digestion of whole-genome DNA with rare-cutting restriction enzymes, average nucleotide identity analysis, and DNA-DNA hybridization studies showed that although related to *D. dadantii*, these isolates represent a novel species within the genus *Dickeya*, for which the name *Dickeya solani* sp. nov. is proposed (van der Wolf et al. 2014).

Type species: *Dickeya chrysanthemi*

Tian et al. (2016) reported that three strains of Gram-stain-negative, pectinolytic bacteria, isolated from pear trees displaying symptoms of bleeding canker in China, had identical 16S rRNA gene sequences that shared 99% similarity to the type strain of *D. dadantii*. Phylogenetic analysis of these strains with isolates representing all species of the genus *Dickeya* and related *Pectobacterium* species supported their affiliation to *Dickeya*. Multi-locus sequence typing employing concatenated sequences encoding *recA*, *fusA*, *gapA*, *purA*, *rplB*, *dnaX*, and the intergenic spacer illustrated a phylogeny which placed these strains as a distinct clade, separate from all other species of the genus *Dickeya*. The name *Dickeya fangzhongdai* sp. nov. was proposed to accommodate pear strains.

Plant pathogenic species: *D. chrysanthemi*, ***D. chrysanthemi***: pv. *chrysanthemi*, pv. *parthenii*; *D. dadantii*, ***D. dadantii***: subsp. *dadantii*, subsp. *dieffenbachiae*; *D. dianthicola*, *D. fangzhongdai*, *D. paradisiaca*, *D. solani*, *D. zeae*.

4.3.1.3.25 Genus Lonsdalea *Brady et al. 2012 gen. nov.*

The genus was named in honor of David Lonsdale for his contributions to British forest pathology. The below given description is based on the data of Hildebrand and Schroth (1967) and Brady et al. (2012). Cells are Gram-stain-negative, short rods (0.5–1.0 × 1.0–2.0 μm) occurring singly, in pairs, or in groups, and motile by peritrichous flagella. Facultatively anaerobic, oxidase negative, but catalase positive. Optimum temperature for growth is 28°C–30°C. The colonies on tryptone soya agar are white to cream, round, convex, and smooth with entire margins. The tests for beta-galactosidase, arginine dihydrolase, lysine decarboxylase, ornithine decarboxylase, and tryptophan deaminase activities are negative. Citrate is utilized, but no production of H_2S, urease, indole, or gelatinase. Nitrates are not reduced to nitrites. Major fatty acids are $C_{14:0}$, $C_{16:0}$, $C_{18:1}$v7*c*, $C_{17:0}$ cyclo and summed features 2 (iso-$C_{16:1}$ and/or $C_{14:0}$ 3-OH) and 3 ($C_{16:1}$v7*c* and/or iso-$C_{15:0}$ 2-OH). The members of the genus *Lonsdalea* form a phylogenetic clade as determined by both 16S rRNA gene sequence analysis and MLSA based on four housekeeping genes. Members of the genus may cause disease on oak trees including drippy nut disease and bark canker and may be associated with acute oak decline. The DNA G + C contents range from 54.5 to 55.1 mol%.

Type species: *Lonsdalea quercina*

Brady et al. (2012) carried out taxonomic studies of bacterial isolates from oak trees in Spain and Britain, showing symptoms of bark canker and acute oak decline, respectively, by a polyphasic approach. Both 16S rRNA gene sequencing and MLSA, based on partial sequences of *gyrB*, *rpoB*, *infB*, and *atpD* genes, separated the isolates into two genetic groups according to their origin. *Br. quercina*, the causal agent of drippy nut disease of oak, was their closest phylogenetic relative, which clustered distant to the other species of the genus *Brenneria*. MLSA data for species of the genera *Brenneria*, *Pectobacterium*, *Dickeya*, *Erwinia*, *Pantoea*, and *Samsonia* confirmed the polyphyletic nature of the genus *Brenneria* and indicated synonymy of *D. dadantii* and *D. dieffenbachiae*. DNA-DNA hybridization experiments confirmed this synonymy and also revealed DNA-DNA relatedness values of 58%–73% between the new oak isolates and *Br. quercina*. Phenotypic and/or chemotaxonomic methods allowed *Br. quercina* and the two genetic groups of new oak isolates to be discriminated from other recognized species of the genus *Brenneria* and from members of the closely related genera *Dickeya*, *Pectobacterium*, and *Samsonia*. Based on the data obtained, they made the following taxonomic proposals.

1. Reclassification of *Br. quercina* as the type species of a novel genus, *Lonsdalea* gen. nov., as *Lonsdalea quercina* comb. nov.
2. Classification of the oak isolates as *Lonsdalea quercina* subsp. *iberica* subsp. nov. and *Lonsdalea quercina* subsp. *britannica* subsp. nov. and leading to the automatic creation of *Lonsdalea quercina* subsp. *quercina* subsp. nov.
3. Emendation of the description of the genus *Brenneria*
4. Reclassification of *D. dieffenbachiae* as *Dickeya dadantii* subsp. *dieffenbachiae* comb. nov. with the automatic creation of *Dickeya dadantii* subsp. *dadantii* subsp. nov.

Toth et al. (2013) reported that seven Gram-stain-negative bacterial strains isolated from oozing bark canker of poplar (*Populus* × *euramericana*) trees in Hungary showed high (>98.3%) 16S rRNA gene sequence similarity to *Lo. quercina*; but they differed from this species in several phenotypic characteristics. MLSA based on three housekeeping genes (*gyrB*, *atpD*, and *infB*) revealed, and DNA-DNA hybridization analysis confirmed, that this group of bacterial strains formed a distinct lineage within the species *Lo. quercina*. A detailed study of phenotypic and physiological characteristics confirmed the separation of poplars' isolates from other sub species of *Lo. quercina*. Therefore, a novel sub species, *Lonsdalea quercina* subsp. *populi* was proposed.

Li et al. (2017) studied four sub species of *Lo. quercina*, namely, *Lo. quercina* subsp. *quercina*, *Lo. quercina* subsp. *britannica*, *Lo. quercina* subsp. *iberica*, and *Lo. quercina* subsp. *populi* by genome sequence-derived average nucleotide identity, phylogenetic analysis based on 16S rRNA gene sequences, multi-locus sequence analysis, and phenotypic characteristics and elevated three sub species to the species level as *Lo. britannica* sp. nov., *Lo. iberica* sp. nov., and *Lo. populi* sp. nov.

Plant pathogenic species: *Lo. britannica*, *Lo. iberica*, *Lo. populi*, ***Lo. quercina***: subsp. *populi*.

4.3.1.3.26 Genus Pectobacterium *Waldee 1945 emend. Hauben et al. 1999*

Waldee (1945) proposed the emendment of the genus *Erwinia* and creation of a new genus *Pectobacterium* to include biochemically more active and soft rotting bacteria like *Er. carotovora* and *Er. chrysanthemi*. Graham (1964), while reviewing the taxonomy of soft rot coliform bacteria supported Waldee's suggestion, followed by many other workers. However, the genus *Pectobacterium* was not recognized in the 8th edition of *Bergey's Manual of Determinative Bacteriology,* but it appeared in the *"Approved Lists of Bacterial Names"* (Skerman et al. 1980). Later in 1998, the species of genus *Erwinia* were further divided into different phylogenetic groups based on complete 16S rDNA sequences. Hauben et al. (1998) united the species constituting the cluster II group and transferred them to this genus. Gardan et al. (2003a) elevated three sub species of *Pe. carotovorum* to the species level.

The below given description of the genus is taken from the emended description of the genus given by Hauben et al. (1999). Cells are Gram-stain-negative, straight rods with rounded ends, and measure 0.5–1.0 × 1.0–3.0 μm. Occur singly or in pairs and motile by peritrichous flagella. Optimum temperature for growth is between 27°C and 30°C. Facultatively anaerobic. Metabolism fermentative. Catalase positive and oxidase negative. Esculin is hydrolyzed, but not starch. No decarboxylases are formed for arginine, lysine, or ornithine. Strains do not possess tryptophan deaminase or urease. Utilize L-alanine, arginine, asparagine, citrulline, glucosamine, glutamine, glycine, histidine, leucine, L-serine, tyrosine, L-tryptophan, ammonium chloride, asparaginic acid, urea, L-methionine, and phenylalanine, but not betaine, choline, quinolinic acid, spermine, spermidine, trigonelline, trimethyl-ammonium, anthranilic acid, hydroxyproline, and sarcosine as nitrogen sources.

The major fatty acids of the type strains include 12:0, 14:0, 15:0, 16:0, 17:1ω8C, and 17:0. The species of the genus comprise a distinct phylogenetic group, as determined by 16S rRNA gene sequence comparisons, and are characterized by 17 signature nucleotides. The G + C content of the DNA ranges from 50.5 to 56.1 mol%. *Pectobacterium* species produce pectolytic enzymes and cause plant diseases, mainly the soft rots and also blights, cankers, necroses, and wilts. The entry of the bacteria into the host plant is mainly through wounds and natural openings.

Several pectolytic bacterial strains, mainly isolated from monocotyledonous plants and previously identified as *Pe. carotovorum*, were thought to belong to a novel species after several taxonomic analyses including DNA-DNA hybridization. In 16S rRNA gene sequence analyses, these strains had a similarity of >97.9% to the 16S rRNA gene sequence of strains representing six other pectobacterial species and sub species. These strains also showed some unique chemotaxonomic features and quantitative differences in polar lipids, lipoquinones, and fatty acids. As these strains represented a novel species, Nabhan et al. (2013) classified these strains as *Pectobacterium aroidearum* sp. nov.

Khayi et al. (2016) reported that on the basis of multi-locus sequence analysis, *Pe. wasabiae* strains RNS 08–42–1A, SCC3193, CFIA1002, and WPP163, collected from potato plant environment, constituted a separate clade from the original Japanese horseradish *Pe. wasabiae*. Besides, these strains were phenotypically distinct from *Pe. wasabiae* strains by producing acids from (+)-raffinose, α-D(+)-α-lactose, D(+)-galactose, and (+)-melibiose, but not from methyl α-D-glycopyranoside, (+)-maltose, or malonic acid. Hence, the name *Pectobacterium parmentieri* sp. nov. was proposed for this taxon. Wu et al. (2017) reported *Pe. carotovorum* subsp. *actinidiae* on kiwifruit in China's Sichuan province. Sarfraz et al. (2018) identified *Pectobacterium punjabense* sp. nov. as the causal agent of blackleg of potato from Punjab province, Pakistan.

Type species: *Pectobacterium carotovorum*

Plant pathogenic species: *Pe. aroidearum*, *Pe. atrosepticum*, *Pe. betavasculorum*, *Pe. cacticida*, *Pe. carotovorum*, ***Pe. carotovorum***: subsp. *actinidiae*, subsp. *brasiliense*, subsp. *carotovorum*, subsp. *odoriferum*; *Pe. parmentieri*, *Pe. punjabense*, *Pe. wasabiae*.

d. Family: *Yresiniaceae*
   Genera: *Samsonia* and *Serratia*

4.3.1.3.27 Genus Samsonia *Sutra et al. 2001*

The genus was established in 2001 and named in honor of R. Samson, a French phytobacteriologist, who did a pioneering work on pectolytic species of genus *Erwinia*. At present, the genus contains only one plant pathogenic species *Sam. erythrinae*, which causes bark necrotic lesions on *Erythrina* species. The strains of this species isolated from *Erythrina* species formed a homogenous phenon among plant pathogenic species of *Enterobacteriaceae* on the basis of phenotypic characters, 16S rDNA sequence analyses, and DNA-DNA hybridization.

The below given description of the genus is based on the data of Sutra et al. (2001) and Sutra et al. (1999). Cells are Gram-stain-negative, non-sporing, catalase positive, and oxidase negative rods having rounded ends. Motile by means of peritrichous flagella. The strains are β-galactosidase positive, produce acetoin, but not hydrogen sulphide, and do not possess arginine dihydrolase, lysine decarboxylase, ornithine decarboxylase, urease, or tryptophan deaminase. Glucose is fermented without gas production. Nitrate is reduced. Strains do not utilize citrate and malonate as carbon sources and do not produce indole from tryptophan. They hydrolyze esculin, but not gelatin, casein, lecithin, or Tween 80.

Type species: *Samsonia erythrinae*

Plant pathogenic species: *Sa. erythrinae*

4.3.1.3.28 Genus Serratia *Bizio 1823*

The genus was established by Bizio in 1823. The genus is characterized by Gram-stain-negative straight rods (0.5–0.8 × 0.9–2.0 μm) with rounded ends and generally motile by peritrichous flagella. Facultatively anaerobic. Growth factors are generally not required. Temperature range for growth lies between 10°C and 36°C. Colonies on nutrient agar are generally opaque, somewhat iridescent, and appear white, pink, or red in color. Red color of the colonies is due to the production of a characteristic red color, prodigiosin. Members of the genus can be distinguished from other members of the family *Enterobacteriaceae* by their unique production of three enzymes, namely, DNase, lipase, and gelatinase. Glucose is fermented and gluconate is oxidized to 2-ketogluconate. Esculin is hydrolyzed by most strains. Tween 80 is hydrolyzed by all species except one. Acetoin is produced from pyruvate by all species except *Se. fonticola*. Nitrate and chlorate are reduced anaerobically. Generally, the members of this genus occur on plant surfaces, in soil, and water, and as opportunistic human pathogens. *Se. marcescens* is mainly the human pathogen and causes nosocomical infections. The genus did not contain any plant pathogen till 1919, when Paine and Stansfield (1919) suggested the role of *Se. proteamaculans* in causing leaf spot disease of king protea (*Protea cynaroides*), which was later on confirmed by Grimont et al. (1978). The mol% G + C of the DNA is 52.0–60.0.

Type species: *Serratia marcescens*

Plant pathogenic species: *Se. marcescens*, *Se. proteamaculans*.

ii. Order: "*Alteromonadales*"
Family: *Alteromonadaceae*
Genus: *Pseudoalteromonas*

4.3.1.3.29 Genus *Pseudoalteromonas* Gauthier et al. 1995 emend. Beurmann et al. 2017

The genus was created by Gauthier et al. (1995). They determined the sequences of small-sub unit rDNA of 17 strains of genera, *Alteromonas*, *Shewanella*, *Vibrio*, and *Pseudomonas*, and these sequences were analyzed by phylogenetic methods. On the basis of data obtained, they emended the existing genus *Alteromonas* and created a new genus *Pseudoalteromonas* to accommodate 11 species that were previously included in *Alteromonas* and transferred one species to genus *Pseudomonas*. Later on, an aerobic, Gram-negative, marine plant pathogenic bacterium, the causal agent of red spot disease of "makonbu" (*Laminaria japonica*), identified as *Pseudoalteromonas bacteriolytica* sp. nov. was transferred to this genus (Sawabe et al. 1998).

The following description of the genus is taken from Gauthier et al. (1995). Cells are Gram-stain-negative, nonsporing, straight rods, which measure 0.2–1.5 × 1.8–3.0 μm and occur singly or in pairs. Motile by a single polar flagellum, either unsheathed or sheathed. Strains do not accumulate poly-β-hydroxybutyrate. Non-bioluminescent. Several species produce pigments. Chemoorganotrophs. Aerobic, possessing a strictly respiratory type of metabolism with oxygen as the terminal electron acceptor. Oxidase is positive, but catalase is generally weak and irregular. None of the strains has a constitutive arginine dihydrolase system. One species, namely, *Psa. denitrificans* is capable of denitrification. Mostly psychrotolerant, with growth occurring between 0°C and 40°C, but all species grow at 20°C. Many strains require amino acids for growth. All species require seawater base for growth. The major quinone is ubiquinone-8 (Q-8). All the abovementioned 12 species of the genus are positive for gelatinase, lipase, lecithinase, and DNase activities and utilization of D-glucose as a sole source of carbon; and negative for utilization of D-ribose, L-rhamnose, salicin, erythritol, sorbitol, *meso*-inositol, and adonitol. The mol% G + C of the DNA is 38.0–48.0 (T_m).

Type species: *Pseudoalteromonas haloplanktis*

Plant pathogenic species: *Psa. bacteriolytica*.

iii. Order: "*Pseudomonadales*"
a. Family: *Moraxellaceae*
Genus: *Acinetobacter*

4.3.1.3.30 Genus *Acinetobacter* Brisou and Prévot (1954)

Cells are rods (0.9–1.6 × 1.5–2.5 μm) becoming spherical in the stationary phase of growth. Cells are Gram-stain-negative, but occasionally difficult to destain, generally occur in pairs and in chains of variable length, and nonsporing. Colonies are generally non-pigmented and are mucoid when the cells are encapsulated. Swimming motility is not observed, but the cells display twitching motility presumably due to the presence of fimbriae. Grow aerobically, having a strictly respiratory type of metabolism with oxygen as the terminal electron acceptor. Most strains do not reduce nitrate to nitrite. Oxidase test is negative, while catalase is positive. Most strains grow between 20°C and 37°C with optimum temperature of 33°C–35°C. However, some strains cannot grow at 37°C. Grow well in most complex media. Most strains can grow in defined media having a single carbon or energy source, such as acetate or lactate, utilizing ammonium or nitrate salts, or one of several common amino acids, as a source of nitrogen. For majority of strains, glutamic acid or aspartic acid can serve as a single source of carbon, energy, and nitrogen in a defined mineral medium. Growth factor requirements are displayed rarely. Most members are saprophytes, occurring naturally in soil, water, sewage, and foods such as raw vegetables. Can also occur, possibly indigenously, on the human skin and in the human respiratory tract. Can be the cause of nosocomial infections such as bacteremia, secondary meningitis, pneumonia, and urinary tract infections in humans. The mol% G + C of the DNA is 38–47.

Type species: *Acinetobacter calcoaceticus*

Patro et al. (2006) reported top-rot phase of red stripe of sugarcane from Srikakulam district of Andhra Pradesh, India and identified the bacterium associated with the phase as *Acin. baumannii*.

On the basis of phenotypic, genotypic, and phylogenetic characteristics, Li et al. (2015a) identified five Gram-stain-negative, non-motile, rod-shaped bacterial strains isolated from cankers of *Populus × euramericana* collected from different locations in Puyang city, Henan province, China as *Acinetobacter populi* sp. nov. These strains could be distinguished from members of most species of the genus *Acinetobacter* by their inability to assimilate L-arginine and benzoate. Based on phenotypic, genotypic, and phylogenetic characteristics, two Gram-stain-negative, non-motile, rod-shaped bacterial strains, isolated from canker bark of *Populus* × *euramericana* collected from different locations in Puyang City, Henan province, China, were identified as *Acinetobacter qingfengensis* sp. nov. (Li et al. 2014).

Plant pathogenic species: *Acin. baumannii*, *Acin. populi*, *Acin. qingfengensis*.

b. Family: *Pseudomondaceae*
Genera: *Pseudomonas* and *Rhizobacter*

4.3.1.3.31 Genus Pseudomonas *Migula 1894 emend. Palleroni 2005 emend. Yang et al. 2013*

The genus was established by Migula in 1894. It contains different types of bacteria, i.e., epiphytic and saprophytic species, animal, human, and plant pathogens. The genus is characterized by straight or slightly curved rods, but not helical, measuring 0.5–1.0 × 1.5–5.0 µm. The cells are motile by one to several polar flagella. Xanthomonadins are not produced. No resting stages are known, Gram-stain-negative. Aerobic, having a strictly respiratory type of metabolism with oxygen as the terminal electron acceptor. However, in some cases, nitrate can be used as an alternate electron acceptor, allowing anaerobic growth. Most species do not accumulate polyhydroxybutyrate granules. Catalase positive, oxidase positive or negative. Chemoorganotrophs. Organic growth factors are not required for most species. The mol% G + C of the DNA is 58.0–69.0.

Yang et al. (2013) have emended the description of the genus as given below. The emended description is as given previously (Migula 1894; Palleroni 2005) with the modification that most species bear one or several polar flagella, but some species have peritrichously flagellated cells.

Type species: *Pseudomonas aeruginosa*

In 1st edition of *Bergey's Manual of Systematic Bacteriology*, Palleroni (1984) enlisted 92 species in this genus, out of which 23 were plant pathogens. Taxonomy of this genus has been revised many times. Several phytopathogenic species of this genus have been transferred to other genera, namely, *Acidovorax* (Willems et al. 1992), *Herbaspirillum* (Baldani et al. 1996), *Burkholderia* and *Ralstonia* (Yabuuchi et al. 1993, 1995; Urakami et al. 1994), and *Xanthomonas*. Many phytopathogenic species of the genus have been designated as pathovars of *Psm. savastanoi* and *Psm. syringae*. Some new species, namely, *Psm. salomonii* and *Psm. palleroniana* (Gardan et al. 2002), *Psm. syringae* pv. *broussonetiae* (Takahashi et al. 1996), and *Psm. costantinii* (Munsch et al. 2002) have been added to this genus.

Cunty et al. (2015) performed a multi-locus sequence analysis based on four housekeeping genes (*gapA*, *gltA*, *gyrB*, and *rpoD*) on 72 strains representative of the French outbreak that occurred since the beginning of 2010, and found that all the strains fell into two phylogenetic groups; one clonal corresponding to biovar 3 and the other corresponding to biovar 4. Strains of biovar 4 were found substantially different from those of the other biovars as they were less aggressive and caused only leaf spots, whereas *Psm. syringae* pv. *actinidiae* (*Psa*) biovars 1, 2, and 3 also caused canker and shoot dieback. Based on these pathogenic differences, supported by phenotypic, genetic, and phylogenetic differences, the authors proposed that *Psa* biovar 4 be renamed as *Psm. syringae* pv. *actinidifoliorum* pv. nov.

Kałużna et al. (2016) reported that based on an analysis of 16S rDNA sequences, cherry isolates shared the highest over 99.9% similarity with *Psm. ficuserectae* and *Psm. congelans*. Phylogenetic analysis using housekeeping genes *gyrB*, *rpoD*, and *rpoB* revealed that they formed a separate cluster and confirmed their closest relation to *Psm. syringae* and *Psm. congelans*. On the basis of this polyphasic taxonomic study, the eight strains from cherry were considered to represent a novel species of the genus *Pseudomonas* for which the name *Psm. cerasi* sp. nov. (Non Griffin 1911) was proposed. Plasencia-Márquez et al. (2017) reported *Psm. aeruginosa* causing leaf infection in soybean (*Glycine max*) from Cuba.

Timilsina et al. (2018) isolated an unusual fluorescent pseudomonad from tomato exhibiting leaf spot symptoms similar to bacterial speck. The bacterium elicited a hypersensitive reaction on tobacco and caused soft rot of potato slices. On the basis of several tests coupled with Biolog GEN III tests, fatty acid methyl ester profiles, and phylogenetic analysis using 16S rRNA and multiple housekeeping gene sequences, the authors identified the bacterium as *Psm. floridensis* sp. nov. Zhang et al. (2018) reported *Psm. qessardii* causing ginseng bacterial root rot in China. Harmon et al. (2018), based on molecular tests, identified the causal agent of stem galls of *Loropetalum chinense* as *Psm. amygdali* pv. *loropetali* pv. nov.

Vinatzer et al. (2017) proposed to use genome similarity-based codes, called life identification numbers (LINs), to describe and name bacterial taxa. Using 93 genomes of *Psm. syringae sensu lato*, LINs were compared with a *Psm. syringae* genome tree, whereby the assigned LINs were found to be informative of a majority of phylogenetic relationships. LINs also reflected host range and outbreak association for strains of *Psm. syringae* pv. *actinidiae*, a pathovar for which many genome sequences are available. They concluded that LINs could provide the basis for a new taxonomic framework to address the shortcomings of the current pathovar system and to complement the current taxonomic system of bacteria in general.

Plant pathogenic species: *Psm. aeruginosa*, *Psm. agarici*, *Psm. amygdali*, ***Psm. amygdali***: pv. *loropetali*; *Psm. asplenii*, *Psm. avellanae*, *Psm. beteli*, *Psm. cannabina*, ***Psm. cannabina***: pv. *alisalensis*, pv. *cannabina*; *Psm. caricapapayae*, *Psm. cerasi*, *Psm. cichorii*, *Psm. cissicola*, *Psm. corrugata*, *Psm. costantinii*, *Psm. ficuserectae*, *Psm. flectens*, *Psm. floridensis*, *Psm. fuscovaginae*, *Psm. hibiscicola*,

Psm. marginalis, ***Psm. marginalis***: pv. *alfalfae*, pv. *marginalis*, pv. *pastinacae*; *Psm. mediterranea*, *Psm. meliae*, *Psm. palleroniana*, *Psm. qessardii*, *Psm. rubrilineans*, *Psm. rubrisubalbicans*, *Psm. salomonii*, *Psm. savastanoi*, ***Psm. savastanoi***: pv. *fraxini*, pv. *glycinea*, pv. *nerii*, pv. *phaseolicola*, pv. *retacarpa*, pv. *savastanoi*; *Psm. syringae*, ***Psm. syringae***: pv. *aceris*, pv. *actinidiae*, pv. *actinidifoliorum*, pv. *aesculi*, pv. *allii*, pv. *antirrhini*, pv. *apii*, pv. *aptata*, pv. *atrofaciens*, pv. *atropurpurea*, pv. *avii*, pv. *berberidis*, pv. *broussonetiae*, pv. *castaneae*, pv. *cerasicola*, pv. *ciccaronei*, pv. *coriandricola*, pv. *coronafaciens*, pv. *coryli*, pv. *cunninghamiae*, pv. *daphniphylli*, pv. *delphinii*, pv. *dendropanacis*, pv. *eriobotryae*, pv. *garcae*, pv. *helianthi*, pv. *hibisci*, pv. *japonica*, pv. *lachrymans*, pv. *lapsa*, pv. *maculicola*, pv. *mellea*, pv. *mori*, pv. *morsprunorum*, pv. *myricae*, pv. *oryzae*, pv. *panici*, pv. *papulans*, pv. *passiflorae*, pv. *persicae*, pv. *philadelphi*, pv. *photiniae*, pv. *pisi*, pv. *porri*, pv. *primulae*, pv. *rhaphiolepidis*, pv. *ribicola*, pv. *sesami*, pv. *solidagae*, pv. *spinaceae*, pv. *striafaciens*, pv. *syringae*, pv. *tabaci*, pv. *tagetis*, pv. *theae*, pv. *tomato*, pv. *ulmi*, pv. *viburni*, pv. *zizaniae*; *Psm. tolaasii*, *Psm. tremae*, *Psm. viridiflava*.

4.3.1.3.32 Genus *Rhizobacter* Goto and Kuwata 1988

The genus was created by Goto and Kuwata (1988) to classify a bacterium causing galls on carrots. The following description of the genus is based on the publication of Goto and Kuwata (1988). Cells are Gram-stain-negative, nonsporing, encapsulated, straight to slightly curved rods, measuring 0.9–1.3 × 2.1–2.5 µm. Cells are motile by polar or lateral flagella or by both, but motile cells are rare in the populations. Poly-β-hydroxybutyrate granules are formed. Chemoorganotrophs. Aerobic, having respiratory type of metabolism of glucose, and D-glucose is utilized as a sole source of carbon and energy. On agar plates, the colonies are white or yellowish-white, plicated, tough, or viscid; if the color is yellow, the pigment differs from xanthomonadins. In liquid media, floccular growth consisting of globular units occurs without finger-like projections. Oxidase and catalase tests are positive. Nitrates are reduced to nitrites and H_2S is produced from cysteine. Growth is inhibited by KCN, and *o*-nitrophenyl-β-D-galactopyranoside test is positive. Arginine dihydrolase, nitrogenase, fluorescent pigment, indole, and acetoin are not produced; and the tests for methyl red and denitrification are also negative. Starch, dextrin, and glycogen are hydrolyzed. The bacterium is susceptible to 10 µg vibriostatic agent O/129 phosphate. Does not require NaCl or growth factors. Contains ubiquinone Q-8. Benzene derivatives are not utilized as carbon sources. The G + C content of the DNA is 67–71 mol%. The type species is a plant pathogen that causes galls on carrot roots and is found in soil.

Type species: *Rhizobacter dauci*

Plant pathogenic species: *Rhba. dauci.*

i. Order: "*Xanthomonadales*"
   Family: "*Xanthomonadaceae*"
   Genera: *Xanthomonas* and *Xylella*

4.3.1.3.33 Genus *Xanthomonas* Dowson 1939 emend. van den Mooter and Swings 1990 emend. Vauterin et al. 1995

The genus was established by Dowson in 1939 to include plant pathogens which cause leaf spots, cankers, blights, etc. on various crop plants. The cells are Gram-stain-negative straight rods, which are 0.4–0.6 × 0.8–2.0 µm; occur mostly singly or in pairs, occasionally in short chains, and rarely as filaments. Cells are motile by means of a single polar flagellum. Usually produce yellow colonies, which are smooth and butyrous, mucoid, or viscid. The exopolysaccharide xanthan, which is responsible for the mucoid or viscous growth, is characteristic of the genus. No growth occurs at pH 4.5 or 4.0, and at 37°C. Growth is inhibited by 6% NaCl, 30% glucose, 0.01% methyl green, 0.01% lead acetate, or 0.1% triphenyl tetrazolium chloride. Chemoorganotrophs and may require one or more growth factors. Glycine, L-glutamine, and L-asparagine are not utilized as sole sources of both carbon and nitrogen. Obligately aerobic and metabolism is strictly respiratory type with oxygen as the terminal electron acceptor. Catalase is produced, oxidase production negative or weak, but urease and indole or acetoin are not produced. Nitrate not reduced, and litmus milk is not acidified.

In Biolog GN microplate tests, D-fructose, α-D-glucose, D-mannose, methylpyruvate, and α-ketoglutaric acid are oxidized, and these reactions are the characteristic of the genus, while α-cyclodextrin, adonitol, D-arabitol, *meso*-erythritol, *meso*-inositol, xylitol, D-glucosaminic acid, γ-hydroxybutyric acid, itaconic acid, sebacic acid, L-ornithine, L-pyroglutamic acid, D-serine, DL-carnitine, γ-aminobutyric acid, phenyl ethylamine, putrescine, 2-aminoethanol, and 2, 3 butanediol are not oxidized. Among the nine fatty acids that predominate in whole-cell preparations, three acids, namely, 9-methyl decanoic acid, 3-hydroxy-9-methyl decanoic acid, and 3-hydroxy-11-methyl dodecanoic acid are highly characteristic of the genus and serve as useful criteria for differentiating *Xanthomonas* strains from other bacteria. The ubiquinone that is present has eight isoprene units. Spermidine is the major polyamine present in the *Xanthomonas* strains. Spermine is usually present in detectable amounts, while 2-hydroxyputrescine and 1, 3-diaminopropane are never present. According to Saddler and Bradbury (2015), species so far described are plant pathogens or are plant associated.

The mol% G + C of the DNA is 63.3–69.7 (T_m).

In 8th edition of *Bergey's Manual of Determinative Bacteriology*, only five species, namely, *Xa. albilineans*, *Xa. ampelina*, *Xa. axonopodis*, *Xa. campestris*, and *Xa. fragariae* were included in this genus. The number of species increased with the passage of time, but with the introduction of pathovar system (Young et al. 1978), many species were downgraded and designated as pathovars of *Xa. campestris* and *Xa. axonopodis*.

Vauterin et al. (1995), on the basis of DNA-DNA hybridization study of 183 strains of the genus, divided it into 20 DNA homology groups and proposed a comprehensive reclassification of the genus. They found that the previously described species *Xa. campestris* was heterogeneous, and it comprised of 16 DNA homology groups and one of these groups exhibited a high level of DNA homology with *Xa. axonopodis*. They emended *Xa. campestris* to include only the pathovars infecting crucifers and transferred its 34 pathovars to *Xa. axonopodis*. They recognized a total of 16 species in the genus leaving 66 former pathovars of *Xa. campestris* unassigned to any species of the genus. The other changes in the taxonomy of the genus include reclassification of *Psm. cissicola* as *Xa. cissicola*, transfer of the bacterium isolated from passion fruit plants (*Passiflora* spp.), originally classified as *Xa. campestris* pv. *passiflorae*, to *Xa. axonopodis* pv. *passiflorae* on the basis of DNA-DNA hybridization data (Goncalves and Rosato 2000), and inclusion of *Xa. cynarae*, the causal agent of bacterial bract spot of artichoke (*Cynara scolymus*) as a new species (Trebaol et al. 2000). Roumagnac et al. (2004), on the basis of DNA-DNA hybridization, thermal stability of DNA reassociation, and fluorescent amplified fragment length polymorphism analysis, identified the causal agent of bacterial blight of onion, garlic, and Welsh onion (*Allium spp.*) as *Xa. axonopodis* pv. *allii*. The pathogens causing bacterial spot of tomato, pepper, or both have been divided into four species, namely, *Xa. vesicatoria*, *Xa. euvesicatoria*, *Xa. perforans*, and *Xa. gardneri* (Jones et al. 2004). Parkinson et al. (2007) showed that phylogenetic analysis of *Xanthomonas* species based on comparison of partial gyrase B gene sequences was a useful method to determine inter-species relatedness and for rapid and accurate species-level identification. Ah-You et al. (2009) used amplified fragment length polymorphism, multi-locus sequence analysis, and DNA-DNA hybridization for genotypic classification of *Xanthomonas* pathovars associated with the plant family Anacardiaceae and concluded that the causal agents of mango bacterial canker and cashew bacterial spot should be classified as pathovars of *Xa. citri*, namely, *Xa. citri* pv. *mangiferaeindicae* and *Xa. citri* pv. *anacardii*, respectively. J.P. Euzeby listed 30 species under this genus (List of prokaryotes with standing in nomenclature: http://www.bacterio.cict.fr/). An addition to the genus is the inclusion of *Xa. dyei* with its three pathovars, namely, *eucalypti*, *laureliae*, and *dysoxyli* classified on the basis of multi-locus sequence analysis (Young et al. 2010). The strains of this species have a unique fatty acid methyl ester profile.

On the basis of the MLSA and virulence phenotypes, Huang et al. (2013) classified a xanthomonad causing a bacterial spot of rose in Florida and Texas as *Xa. alfalfae* subsp. *citrumelonis* pv. *rosa*.

Vicente et al. (2017) reported *Xa. nasturtii* sp. nov. causing a disease in leaves of watercress (*Nasturtium officinale*) from Florida, US.

López et al. (2018), on the basis of polyphasic approach that included both phenotypic and genomic methods, identified three isolates obtained from symptomatic nectarine trees (*Prunus persica* var. *nectarina*) cultivated in Murcia, Spain as *Xa. prunicola*.

Type species: *Xanthomonas campestris*

Plant pathogenic species: *Xa. albilineans*, *Xa. alfalfae*, ***Xa. alfalfae***: subsp. *alfalfae*, subsp. *citrumelonis*, subsp. *citrumelonis* pv. *rosa*; *Xa. arboricola*, ***Xa. arboricola***: pv. *celebensis*, pv. *corylina*, pv. *fragariae*, pv. *juglandis*, pv. *populi*, pv. *pruni*; *Xa. axonopodis*, ***Xa. axonopodis***: pv. *allii*, pv. *anacardii*, pv. *axonopodis*, pv. *bauhiniae*, pv. *begoniae*, pv. *betlicola*, pv. *biophyti*, pv. *cajani*, pv. *cassiae*, pv. *clitoriae*, pv. *coracanae*, pv. *cyamopsidis*, pv. *desmodii*, pv. *desmodiigangetici*, pv. *desmodiilaxiflori*, pv. *desmodiirotundifolii*, pv. *dieffenbachiae*, pv. *erythrinae*, pv. *fascicularis*, pv. *glycines*, pv. *khayae*, pv. *lespedezae*, pv. *maculifoliigardeniae*, pv. *mangiferaeindicae*, pv. *manihotis*, pv. *martyniicola*, pv. *melhusii*, pv. *nakataecorchori*, pv. *patelii*, pv. *pedalii*, pv. *phaseoli*, pv. *phyllanthi*, pv. *physalidicola*, pv. *poinsettiicola*, pv. *punicae*, pv. *rhynchosiae*, pv. *ricini*, pv. *rosa*, pv. *sesbaniae*, pv. *spondiae*, pv. *tamarindi*, pv. *vasculorum*, pv. *vignaeradiatae*, pv. *vignicola*, pv. *vitians*; *Xa. bromi*, *Xa. campestris*, ***Xa. campestris***: pv. *aberrans*, pv. *alangii**, pv. *amaranthicola**, pv. *amorphophalli**, pv. *aracearum**, pv. *arecae**, pv. *argemones**, pv. *armoraciae*, pv. *arracaciae**, pv. *asclepiadis**, pv. *azadirachtae**, pv. *badrii**, pv. *barbareae*, pv. *betae**, pv. *bilvae**, pv. *blepharidis**, pv. *boerhaaviae**, pv. *brunneivaginae**, pv. *campestris*, pv. *cannabis**, pv. *cannae**, pv. *carissae**, pv. *centellae**, pv. *citrumelo*, pv. *clerodendri**, pv. *convolvuli**, pv. *coriandri**, pv. *daturae**, pv. *durantae**, pv. *esculenti**, pv. *euphorbiae**, pv. *fici**, pv. *guizotiae**, pv. *gummisudans**, pv. *heliotropii**, pv. *incanae*, pv. *ionidii**, pv. *lantanae**, pv. *lawsoniae**, pv. *leeana**, pv. *leersiae**, pv. *malloti**, pv. *merremiae**, pv. *mirabilis**, pv. *mori**, pv. *musacearum**, pv. *nigromaculans**, pv. *obscurae**, pv. *olitorii**, pv. *papavericola**, pv. *parthenii**, pv. *passiflorae*, pv. *paulliniae**, pv. *pennamericanum**, pv. *phormiicola**, pv. *physalidis**, pv. *plantaginis*, pv. *raphani*, pv. *sesami**, pv. *spermacoces**, pv. *syngonii**, pv. *tardicrescens**, pv. *thespesiae**, pv. *thirumalacharii**, pv. *tribuli**, pv. *trichodesmae**, pv. *uppalii**, pv. *vernoniae**, pv. *viegasii**, pv. *vitians*, pv. *viticola**, pv. *vitiscarnosae**, pv. *vitistrifoliae**, pv. *vitiswoodrowii**, pv. *zantedeschiae**, pv. *zingibericola**, pv. *zinniae**; *Xa. cassavae*, *Xa. citri*,

* These pathovars were not examined by Vauterin et al. (1995) nor reclassified by others since 1995 and therefore are *Xa. campestris sensu lato* and may not belong to the species as emended by Vauterin et al. (1995) and Bull et al. (2010).

Xa. citri: subsp. *citri*, subsp. *malvacearum*; *Xa. codiaei*, *Xa. cucurbitae*, *Xa. cynarae*, *Xa. dyei*, ***Xa. dyei***: pv. *dysoxyli*, pv. *eucalypti*, pv. *laureliae*; *Xa. euvesicatoria*, *Xa. fragariae*, *Xa. fuscans*, ***Xa. fuscans***: subsp. *aurantifolii*, subsp. *fuscans*; *Xa. gardneri*, *Xa. hortorum*, ***Xa. hortorum***: pv. *carotae*, pv. *hederae*, pv. *pelargonii*, pv. *taraxaci*; *Xa. hyacinthi*, *Xa. maltophilia*, *Xa. melonis*, *Xa. nasturtii*, *Xa. oryzae*, ***Xa. oryzae***: pv. *oryzae*, pv. *oryzicola*; *Xa. perforans*, *Xa. pisi*, *Xa. populi*, *Xa. prunicola*, *Xa. sacchari*, *Xa. theicola*, *Xa. translucens*, ***Xa. translucens***: pv. *arrhenatheri*, pv. *cerealis*, pv. *graminis*, pv. *phlei*, pv. *phleipratensis*, pv. *pistaciae*, pv. *poae*, pv. *secalis*, pv. *translucens*, pv. *undulosa*; *Xa. vasicola*, ***Xa. vasicola***: pv. *holcicola*, pv. *vasculorum*; *Xa. vesicatoria*.

4.3.1.3.34 Genus Xylella *Wells et al. 1987*

The genus was established by Wells et al. (1987) to accommodate xylem-limited procaryotes responsible for causing Pierce's disease of grapevines, phony peach disease, leaf scorches of oak, mulberry, sycamore, elm, almond, and plum, and periwinkle wilt. The below given description of the genus is taken from Wells et al. (1987).

Cells are Gram-stain-negative; predominantly single straight rods measuring 0.25–0.35 × 0.9–3.5 μm. Under certain cultural conditions, long filamentous strands are formed. Non-motile and non-sporing. Colonies are creamy to white and are generally of two types: convex to pulvinate, smooth, opalescent with entire margins and umbonate, rough with finely undulated margins. The bacteria are strictly aerobic and growth is inhibited by 2.5% carbon dioxide. Metabolism is respiratory in which the oxygen is used as terminal electron acceptor. Non-fermentative, non-halophilic, and non-pigmented; oxidase test negative and catalase test positive. Optimal temperature for growth is 26°C–28°C and optimum pH 6.5–6.9. The bacteria are nutritionally fastidious, requiring a specialized medium such as buffered cysteine-yeast extract agar containing charcoal or glutamine-peptone medium containing serum albumin. The G + C content of the DNA is 51.0–52.4 mol% (T_m) or 52.0–53.1 mol% (Bd). Members are found mainly in the xylem of plant tissue.

Type species: *Xylella fastidiosa*

Although Wells et al. (1987) recognized only one species to include all the strains, there are definitely different pathotypes, most of which have not been characterized and compared with each other (Hopkins and Purcell 2002). Schaad et al. (2004) proposed three new sub species, namely, *Xyl. fastidiosa* subsp. *piercei*, *Xyl. fastidiosa* subsp. *multiplex*, and *Xyl. fastidiosa* subsp. *pauca*. *Xyl. fastidiosa* subsp. *pauca* has not been validated due to the lack of deposition of a culture of the type strain. As the type strain designated for subsp. *piercei* is the type strain of the species, the name *Xyl. fastidiosa* subsp. *fastidiosa* takes priority over *Xyl. fastidiosa* subsp. *piercei*.

Su et al. (2016) proposed a novel species, *Xylella taiwanensis* sp. nov. for a Gram-stain negative fastidious bacterium causing pear leaf scorch in Taiwan.

Plant pathogenic species: *Xyl. fastidiosa*, ***Xyl. fastidiosa***: subsp. *fastidiosa*, subsp. *multiplex*; *Xyl. taiwanensis*.

II. Phylum: "*Actinobacteria*"

Class: *Actinobacteria*

i. Order: *Corynebacteriales* ord. nov.

a. Family: *Corynebacteriaceae*

Genus: *Corynebacterium*

4.3.1.3.35 Genus Corynebacterium *Lehmann and Neumann 1896*

The genus was created by Lehmann and Neumann in 1896 with *Cor. diphtheriae* as type species. It contained human, animal, and plant pathogens. All the plant pathogenic species of this genus except *Cor. ilicis* have been transferred to different genera, namely, *Clavibacter*, *Curtobacterium*, *Rathayibacter*, and *Rhodococcus*. The Judicial Commission of the International Committee on Systematics of Prokaryotes ruled (opinion 87) that the name *Cor. ilicis* (Mandel et al. 1961) is represented by the type strain ICMP 2608 = ICPB CI144 and is reported to be a plant pathogenic species (*Int. J. Syst. Evol. Microbiol.*, 58: 1976–1978).

Type species: *Corynebacterium diphtheriae*

Plant pathogenic species: *Cor. ilicis*.

b. Family: *Nocardiaceae*

Genera: *Nocardia* and *Rhodococcus*

4.3.1.3.36 Genus Nocardia *Trevisan 1889*

The genus was created by Trevisan in 1889. In 1952, one phytopathogenic bacterium causing galls, abundant proliferation of buds, and witches' broom effect on blueberry was included in this genus.

The genus was named after Edmond Nocard, a French veterinarian, who discovered the first member of the genus in 1889 as the cause of bovine infection. *Nocardia* are strict aerobes and catalase positive bacteria, which stain from Gram-stain-positive to Gram-stain variable. Non-sporing, non-motile, chemoorganotrophs, having an oxidative type of metabolism. Rudimentary to extensively branched vegetative hyphae measuring 0.5–1.2 μm in diameter are produced. These hyphae often fragment into rod-shaped and coccoid forms. Conidia are occasionally produced in chains on the aerial hyphae. Colonies have a variable shape and may be pigmented or white. Colonies are also of variable appearance and most species appear to have aerial hyphae when viewed with a dissecting microscope. Some strains are partially acid-fast, i.e., they require less concentration of sulphuric or hydrochloric acid during the staining procedure due to the presence of mycolic acids of intermediate length in their cell walls. *Nocardia* species have a wide range of temperature for growth, but grow slowly on non-selective culture media. Colonies become evident in 3–5 days on most commonly used media. However, sometimes a prolonged incubation period of 2–3 weeks may be needed.

Members are distributed world widely in soils that are rich in organic matter and are also found in aquatic habitats.

They are involved in the degradation of hydrocarbons and waxes and can contribute to the biodeterioration of rubber joints in water and sewage pipes. Although most are free-living saprophytes, some species such as *No. asteroides* are opportunistic pathogens, which cause nocardiosis in humans and animals. Persons with a weak immune system like children, the elderly, and the immunocompromised are mostly at risk. The most commonly infected organs are lungs, but the central nervous system and other organs may also be involved. Some *Nocardia* species also cause a variety of cutaneous infections like actinomycetoma. The virulence factors are the catalase and superoxide dismutase, which inactivate reactive oxygen species that would otherwise prove toxic to the bacteria, and a cord factor (trehalose 6–6′ dimycolate). The cord factor interferes with fagocytosis by macrophages by preventing the fusion of the phagosome with the lysosome. The G + C content of the DNA varies from 64.0 to 72.0 mol%.

Type species: *Nocardia asteroides*

Plant pathogenic species: *No. vaccinii.*

4.3.1.3.37 Genus Rhodococcus *Zopf 1891*

The genus was described in 1891 to include a saprophytic Gram-stain-positive bacterium producing red pigment and forming irregular masses of cells. Later on, Goodfellow (1984) transferred *Cor. fascians*, the causal agent of fasciation of sweet pea, and leafy galls and shoot proliferations on many plants to this genus.

The genus is characterized by Gram-stain-positive, non-sporing, non-motile, aerobic bacteria. The cells are nocardioform, meaning they may have mycelial growth with fragmentation into rod-shaped or coccoid forms. The cocci may germinate into short rods, form filaments with side projections, show elementary branching, or in some cases produce extensively branched substrate hyphae. The fragmentation of rods, filaments, and hyphae produces next generation of short rods or cocci. In some strains, feeble, aerial hyphae are produced. No conidia are produced. Colony appearance may be rough, smooth, or mucoid, and pigmented, but rarely colorless. Catalase positive. Chemoorganotrophs, having an oxidative type of metabolism. A wide range of organic compounds are utilized as sole sources of carbon for growth and energy. The growth is generally good at 30°C on laboratory media. The cell walls contain mycolic acids.

The genus *Rhodococcus* is genetically diverse and well adapted to the soil environment. As a group, the rhodococci have tolerance for starvation, which may aid in their persistence in soil. The other habitats are aquatic habitat and herbivore dung, and a few reside in eucaryotic cells. The members of the genus are important due to their ability to metabolize/degrade harmful environmental pollutants such as toluene, naphthalene, polychlorinated biphenyls, petroleum hydrocarbons, detergents, benzene, and various herbicides/pesticides. Strains of *Rhodococcus* are also important due to their ability to catabolize a wide range of compounds and produce bioactive steroids, acrylamide, and acrylic acid, and their involvement in fossil fuel biodesulfurization. *Rho. qingshengii* is capable of degrading carbendazim, a fungicide widely used for plant disease control. The genome of *Rhodococcus* sp. strain RHA 1 was fully sequenced in October 2006 (McLeod et al. 2006). The genome is 9.7 mega base pairs long and is one of the largest bacterial genomes completely sequenced to date. The genetic and catabolic diversity of the genus is not only due to the large bacterial chromosome, but the presence of three large linear plasmids. *Rho. equi* contains a circular plasmid, while *Rho. fascians* harbors a linear plasmid (about 200 kb). Bacterial isolates identified as *Rho. fascians* have been found in diverse habitats, including polar seas and ice sediments in Greenland. Two isolates have been reported as fixing nitrogen. The G + C content of the DNA ranges from 63.0 to 72.0 mol% (T_m).

Type species: *Rhodococcus rhodochrous*

Plant pathogenic species: *Rho. fascians.*

ii. Order: *Micrococcales*
  a. Family: *Micrococcaceae*
  Genus: *Arthrobacter*

4.3.1.3.38 Genus Arthrobacter *Conn and Dimmick 1947*

The genus was created by Conn and Dimmick in 1947. The bacterium causing bacterial blight of leaves and twigs of American hollyhock, and earlier classified as *Cor. ilicis*, was transferred to this genus as *Arthrobacter ilicis*. However, according to Young et al. (2004), *Ar. ilicis* is a validly named species, but probably is not a plant pathogen. The Judicial Commission of the International Committee on Systematics of Prokaryotes ruled (opinion 87) that the name *Ar. ilicis* is represented by the type strain DSM 20138 = ATCC 14264 = NCPPB 1228 and is not a homotypic synonym of *Cor. ilicis* (Mandel et al. 1961), and is reported not to be a plant pathogen (*Int. J. Syst. Evol. Microbiol.*, 58: 1976–78).

Arthrobacter (from the Greek, "**jointed small stick**") are coryneform bacteria characterized by pleomorphism and Gram-stain variable reaction. In young cultures, slender rods are Gram-stain-negative and in about 30-hour-old cultures, very short rods and coccoids are Gram-stain-positive. Coccoid cells measure from ~0.6–1.0 μm in diameter. Colonies on blood agar are small, measuring 2 mm in diameter and range in color from yellow to white. The bacterium can be readily isolated on sheep blood agar. Moderate to good growth occurs on yeast extract-peptone medium. Optimal temperature for growth is between 25°C and 30°C. They have a complex life cycle composed of two distinct stages when grown in complex media. It is also called a **rod-coccus growth cycle** and is a most distinctive feature of the genus. In exponential phase, the bacteria are irregular, branched rods and may reproduce by a process called snapping division. As they enter stationary phase, the cells change to coccoid form. Upon transfer to fresh complex medium, the coccoid cells swell and produce one or more overgrowths and again form actively growing irregular rods.

When viewed under the microscope, some cells are seen forming V-shaped arrangements and also more complex angular arrangements.

Most species of arthrobacteria are obligate aerobes, but all exhibit a pure respiratory, never fermentative, metabolism. However, two species, namely, *Ar. globiformis* and *Ar. nicotianae* exhibit anaerobic metabolism. Both the species utilize nitrate as an electron acceptor at the end of their respiratory chain, reducing it to ammonia via nitrite. Little or no acid is produced from glucose and other sugars in peptone media. Non-sporing. Many species exhibit a week motility that often goes unnoticed. Are nutritionally versatile and use a variety of substrates in their oxidative metabolism including nicotine, nucleic acids, and various herbicides and pesticides. Pyridone is used as a sole source of carbon. Chemoorganotrophs. Catalase positive.

Although arthrobacters are often isolated from sewage, fish, and plant surfaces, soil is their most important habitat, where they form a significant component of the microbial flora. They are well adapted to this habitat because the cocci are resistant to desiccation and starvation. Due to their metabolic diversity, they degrade a variety of chemicals including agricultural pesticides. Hexavalent chromium is 100 times more toxic than trivalent chromium due to its oxidation state and is also much more soluble in water, thus allowing it to seep into ground water very easily. Arthrobacteria cannot only grow in its presence, but can also reduce it to trivalent chromium, a much less toxic form. *Arthrobacter* species in association with several strains of *Streptomyces* can completely degrade an organophosphate insecticide, diazinon. They can use this as the only source of carbon and energy, but individually none can grow on this compound. The genome of *Arthrobacter* sp. FB 24 has been sequenced, and it contains 5,011,598 base pairs. Currently TIGR is sequencing another species, *Ar. aurescens* TC 1 (http://microbewicki.kenyon.edu/index.php/Arthrobacter). The G + C content of the DNA ranges from ~59.0–70.0 mol% (T_m).

Busse et al. (2016) reviewed the taxonomy of the genus *Arthrobacter* and emended the genus *Arthrobacter sensu lato*. The emphasis was also given to intrageneric phylogeny and chemotaxonomic characteristics, concentrating on quinone systems, peptidoglycan compositions, and polar lipid profiles.

Type species: *Arthrobacter globiformis*

Plant pathogenic species: *Ar. ilicis.*

b. Family: *Intrasporangiacae*
Genus: *Janibacter*

4.3.1.3.39 *Genus* Janibacter *Martin et al. 1997*

Originally, the genus was created to accommodate a Gram-stain-positive actinomycete, isolated from environmentally polluted samples and whose cell walls contained *meso*-diaminopimelic acid. Yoon et al. (2004) described a bacterium isolated from abnormally spoiled oriental melons and transferred it to this genus on the basis of 16S rRNA gene sequence analyses and genomic relatedness.

The below given description of the genus is as proposed by Martin et al. (1997). The name of the genus is derived from a Latin word, *Janus*, a God in Roman mythology, who is said to have had two faces. Cells are Gram-stain-positive, coccoid to rod-shaped, and non-motile. Cells occur singly, in pairs, or occasionally in irregular clumps. Non-sporing and not acid-fast. Aerobic, chemoorganotrophs. Catalase is produced, while oxidase test is negative. The cell wall type is A1γ (directly cross-linked *meso*-A_2 pm). Mycolic acids are absent; the menaquinone is M2K-8 (H_4). The complex fatty acid profile contains straight-chain saturated and monounsaturated fatty acids as the major components and the polar lipids are diphosphatidylglycerol, phosphatidylglycerol, and phosphatidylinositol. The strains of the genus are inhabitants of sludge and sewage waste. The G + C content of the DNA is 70.0 mol%.

Type species: *Janibacter limosus*

Plant pathogenic species: *Jani. melonis.*

c. Family: *Microbacteriaceae*
Genera: *Clavibacter, Curtobacterium, Leifsonia, Leucobacter,* and *Rathayibacter*

4.3.1.3.40 *Genus* Clavibacter *Davis et al. 1984*

The genus was created by Davis et al. (1984) to accommodate some phytopathogenic coryneform bacteria whose cell walls contain diaminobutyric acid. Two plant pathogens, *Cla. xyli* subsp. *xyli* and *Cla. xyli* subsp. *cynodontis*, causal agents of ratoon stunting disease of sugarcane and Bermuda grass stunting disease, respectively, were classified in this genus. The creation of this genus was considered necessary, because no other genus as circumscribed for coryneform bacteria was suitable for these phytopathogens. Later on, both *Cla. xyli* subsp. *xyli* and *Cla. xyli* subsp. *cynodontis* were transferred to *Leifsonia*. Three other species of this genus, namely, *Cla. toxicus, Cla. rathayi* and *Cla. tritici* have been transferred to genus *Rathayibacter.*

González and Trapiello (2014) identified a yellow Gram-stain-positive bacterium (causal agent of a newly identified bacterial disease named as bacterial bean leaf yellowing) isolated from bean seeds (*Phaseolus vulgaris* L.) as *Cla. michiganensis* by 16S rRNA gene sequencing. The results of molecular methods employed such as amplification of specific sequences by PCR, 16S amplified rDNA restriction analysis, restriction fragment length polymorphism (RFLP), and multi-locus sequence analysis as well as the analysis of biochemical and phenotypic traits including API 50CH and API ZYM results showed that the bacterium did not represent any known sub species of *Cla. michiganensis*. Therefore, the name *Clavibacter michiganensis* subsp. *phaseoli* subsp. nov. was proposed.

The below given description of the genus is based on Davis et al. (1984). Cells are Gram-stain-positive, not acid-fast, pleomorphic rods, which are often arranged at an angle to give V-shaped arrangements; a marked rod-coccus cycle is not observed. Cells are non-motile and endospores are not formed. Obligately aerobic. Optimal temperature for growth is ca. 20°C–29°C and maximum is ca. 29°C–35°C. The bacterium has exact nutritional requirements and growth occurs on suitable solid media. Slow and weak oxidative production of acids occurs from some carbohydrates. Some organic acids are utilized. Neither nitrate is reduced nor utilized. Oxidase, urease, lipase, and tyrosinase are not generally produced.

Cell wall peptidoglycan contains 2, 4-diamino butyric acid. Rhamnose is the major component of the cell wall, while arabinose is absent. Mycolic acids are also absent. Non-hydroxylated fatty acids consist predominantly of *anteiso*- and *iso*-methyl branched chains; small amounts of straight-chain saturated acids are found. Respiratory quinones are menaquinones. Polar lipids comprise diphosphatidylglycerol, phosphatidylglycerol, and lipids. The G + C content of the DNA is about 70 ± 5 mol%.

Type species: *Clavibacter michiganensis*

Based on multi-locus sequence typing and other biochemical and physiological features including colony color, utilization of carbon sources, and enzyme activities, Oh et al. (2016) found that strain PF008^T isolated from pepper was categorically different from eight known sub species of *Cla. michiganensis* and proposed a novel sub species, *Clavibacter michiganensis* subsp. *capsici* for the bacterium causing bacterial canker of pepper.

Li et al. (2018) reported that whole-genome analysis based on average nucleotide identity and digital DNA-DNA hybridization as well as MLSA of seven housekeeping genes support raising each of the *Cla. michiganensis* sub species to species status. On the basis of whole-genome and MLSA data, the authors proposed the establishment of two new species and three new combinations, namely, *Clavibacter capsici* sp. nov., comb. nov., *Clavibacter tessellarius* sp. nov., comb. nov., *Clavibacter insidiosus* comb. nov., *Clavibacter nebraskensis* comb. nov., and *Clavibacter sepedonicus* comb. nov.

Plant pathogenic species: *Cla. capsici*, *Cla. insidiosus*, *Cla. michiganensis*, *Cla. michiganensis* subsp. *phaseoli*, *Cla. nebraskensis*, *Cla. sepedonicus*, *Cla. tessellarius*.

4.3.1.3.41 Genus Curtobacterium *Yamada and Komagata 1972*

The genus was proposed by Yamada and Komagata in 1972. In 1983, Collins and Jones proposed the transfer of *Corynebacterium flaccumfaciens* to *Curtobacterium* as *Cur. flaccumfaciens* and transfer of *Cor. betae, Cor. oortii* and *Cor. poinsettiae* as pathovars of *Cur. flaccumfaciens*.

The characteristics of the genus as described by Yamada and Komagata (1972) are given below. Cells are weakly Gram-stain-positive and small short rods. Coccoid cells are found in old cultures and old cells frequently loose the character of Gram-stain-positive reaction. Generally motile by means of lateral flagella. Cells multiply by bending type of cell division. Pleomorphism is slightly recognized. Metachromatic granules are not recognized. Ornithine is a principal amino acid found in the cell walls. Acid production is slow and weak from several sugars. Chemoheterotrophs. Various kinds of organic acids are assimilated in addition to acetate, pyruvate, and lactate. Gelatin is slowly hydrolyzed. The G + C content of the DNA is 66.0–71.0 mol%. The bacterium is widely distributed in soil and plant material.

Type species: *Curtobacterium citreum*

Plant pathogenic species: *Cur. flaccumfaciens*, ***Cur. flaccumfaciens***: pv. *betae*, pv. *beticola*, pv. *flaccumfaciens*, pv. *ilicis*, pv. *oortii*, pv. *poinsettiae*.

4.3.1.3.42 Genus Leifsonia *Evtushenko et al. 2000 emend. Dastager et al. 2009*

The genus was created by Evtushenko and associates in 2000 to accommodate two new species, namely, *Leifsonia poae* and *Leifsonia aquatica* found in *Poa annua* root galls induced by the nematode *Subanguina radicicola* and *Cor. aquaticum. Cor. aquaticum* was reclassified as *L. aquatica. Cla. xyli* was reclassified with two sub species, namely, *L. xyli* subsp. *xyli* and *L. xyli* subsp. *cynodontis* in this genus. Members of the genus are characterized by coryneform morphology. The species clearly differ from each other phylogenetically and phenotypically. The species can be distinguished on the bases of their morphologies, composition of cell wall sugars, the requirement of complex media for growth, and many physiological characteristics, including the oxidase reaction. Currently, the genus contains only one phytopathogenic species having two sub species.

The below given description of the genus is taken from Evtushenko et al. (2000). The colonies are yellow or white, circular, somewhat convex, glistening, opaque, and butyrous. The cells are Gram-stain-positive, non-sporing, irregular rods or filaments, which usually fragment into shorter rods or coccoid elements. Primary branching occurs in young cultures of some species. Cells are usually motile, not acid-fast, obligately aerobic, and mesophilic. Catalase test is negative, while oxidase test is variable among species. Cell wall peptidoglycan of B-type contains glycine, glutamic acid, diaminobutyric acid, (DAB) and alanine in a molar ratio close to 1:1:2:1; both isomers, L-DAB and D-DAB, are usually present in almost equal proportions. The major menaquinone is MK-11; MK-10 is also present. Rhamnose is a major constituent of cell wall sugars with small amounts of glucose, galactose, and mannose, while some species also contain fucose. Cell wall teichoic acids and mycolic acids are not present. The principal phospholipids are phosphatidylglycerol and diphosphatidylglycerol. Among fatty acids, *anteiso*-15:0, *anteiso*-17:0, and *iso*-16:0 predominate. The concentrations of polyamines are low, while putrescine is the predominant compound. The G + C content of the DNA ranges from 66.0 to 73.0 mol%. Some species can be distinguished by their tolerance to antibiotics. The genus forms a coherent phylogenetic cluster attached to the branch of *Agromyces* spp.

Type species: *Leifsonia aquatica*

Plant pathogenic species: *L. xyli*, ***L. xyli***: subsp. *cynodontis*, subsp. *xyli*.

4.3.1.3.43 Genus *Leucobacter* Takeuchi et al. 1996

The below given description of the genus is based primarily on the publication of Takeuchi et al. (1996). Cells are Gram-stain-positive, non-motile rods, and non-sporulating. No mycelium is produced. Catalase is produced, but oxidase, arginine dihydrolase, hydrogen sulphide, and urease are not produced. The cell wall peptidoglycan contains 2, 4-diaminobutyric acid as the characteristic diamino acid. The peptidoglycan acyl type is acetyl. The quinone system contains either menaquinone MK-10 and MK-11 or MK-11 only as major isoprenoid quinine. Mycolic acids are not present. The major cellular fatty acids are $C_{15:0}$ anteiso, $C_{16:0}$ iso, and $C_{17:0}$ anteiso. Diphosphatidylglycerol, phosphatidylglycerol, and one unknown glycolipid are present. The G + C content of the DNA is 62.8–69.5 mol%.

Type species: *Leucobacter komagatae*

Fang et al. (2016) isolated a Gram-stain-positive, aerobic, non-motile, rod-shaped bacterium from the symptomatic bark of a *Populus* × *euramericana* canker. Based on the molecular data and physiological and biochemical characteristics, the isolated bacterium was considered to represent a novel species of the genus *Leucobacter*, for which they proposed the name *Leucobacter populi* sp. nov.

Plant pathogenic species: *Leb. populi.*

4.3.1.3.44 Genus *Rathayibacter* Zgurskaya et al. 1993

The genus was created by Zgurskaya et al. (1993) to accommodate three species of Gram-stain-positive, aerobic, coryneform bacteria, namely, *Cla. rathayi*, *Cla. tritici.* and *Cla. iranicus*, previously classified in the genus *Clavibacter*, as well as six strains that were isolated from annual cereal grasses, responsible for ryegrass toxicity and also very similar to *Cla. toxicus*. Later on, Sasaki et al. (1998) transferred *Cla. toxicus* to this genus.

The below given description of the genus is taken from Zgurskaya et al. (1993). The genus was named after E. Rathay, an Australian plant pathologist who first isolated members of this genus. Cells are irregular, pleomorphic, short rods, and measure 0.4–0.8 × 0.5–1.8 µm. V-forms and sometimes short cell chains are formed, but a clear rod-coccus cycle is not observed. Gram-stain-positive, non-sporing, and not acid-fast. Obligately aerobic. Chemoorganotrophic, having an oxidative type of catabolism. Nutritionally exacting. Optimal growth occurs between 24°C and 28°C; no growth occurs at 10°C or 37°C. Cell wall peptidoglycan contains alanine, glutamic acid, glycine, and diaminobutyric acid (1:1:1:2). Cell wall sugars include glucose, mannose, and rhamnose. Cell walls of some strains also contain xylose and galactose. Menaquinones are of the MK-10 type. Mycolic acids are absent. Branching saturated fatty acids of the *anteiso* type (*anteiso*-15:0 and *anteiso*-17:0) are predominant. Diagnostic phospholipids are represented by phosphatidylglycerol, diphosphatidylglycerol, and unknown glycosyldiacylglycerols; several strains contain lyso-phosphatidylglycerol.

Phenylalanine and methionine are assimilated, but most amino acids are not used as sole sources of nitrogen. Casein, guanine, Tween 60, Tween 80, starch, and urea are not hydrolyzed or decomposed. Nitrate is not reduced to nitrite. Methyl red and Voges–Proskauer tests are variable. The G + C content of the DNA ranges from 63.0 to 72.0 mol%. The bacterium is isolated mainly from cereal crops and annual cereal grasses.

Type species: *Rathayibacter rathayi*

Plant pathogenic species: *Rat. iranicus*, *Rat. rathayi*, *Rat. toxicus*, *Rat. tritici.*

iii. Order: *Streptomycetales* ord. nov.
Family: *Streptomycetaceae*
Genus: *Streptomyces*

4.3.1.3.45 Genus *Streptomyces* Waksman & Henrici (1943)

The genus was described in 1943 by Waksman and Henrici. The members of the genus are characterized by the production of **coenocytic mycelium** and Gram-stain-positive reaction. The aerial mycelium produces chains of three to many spores, which may be round, oval, or cylindrical in shape. More than 500 species of *Streptomyces* have been described, but most of them are saprophytes. However, some species attack tuber and root crops such as potato, sweet potato, beet, radish, turnip, and carrot and cause scabbing. Till now, 21 phytopathogenic species of this genus have been validly published (Goyer et al. 1996; Miyajima et al. 1998; Park et al. 2003; Bouchek-Mechiche et al. 2000, 2006). A few species are also pathogens of animals including humans. Guo et al. (2008) reported the results of a multi-locus sequence analysis scheme developed to address the phylogeny of *St. griseus* 16S rRNA gene clade. Sequence fragments of six housekeeping genes, *atpD*, *gyrB*, *recA*, *rpoB*, *trpB*, and 16S rRNA, were obtained for 53 reference strains representing 45 valid species and sub species. Analysis of each individual locus confirmed the suitability of loci and the congruence of single-gene trees for concatenation. Concatenated trees of three, four, five, and all six genes were constructed, and the stability of the topology and discriminatory power of each tree were analyzed. It was concluded from the results that phylogenetic analysis based on multi-locus sequences is more accurate and robust for species delineation within *Streptomyces*. A multi-locus phylogeny of six genes proved to be optimal for elucidating the inter-species relationships within the *St. griseus* 16S rRNA gene clade. Therefore, this multi-locus sequence analysis scheme provides a valuable tool that can be applied to other *Streptomyces* clades for refining the systematic framework of this genus.

The characteristic feature of the genus is the production of an extensively branched mycelium resembling fungi. The diameter of vegetative hyphae ranges from 0.5 to 2.0 µm. The chains of three to many spores, called **conidia**, are formed on mature aerial hyphae. In a few species, substrate mycelium produces conidia in short chains. The conidia are often pigmented having smooth, hairy, or spiny outer surface. Conidia are non-motile and responsible for the reproduction and dispersal of *Streptomyces*.

The colonies are discrete and may be leathery or butyrous. Initially, the colonies have a smooth surface, but later on they appear powdery, granular, or velvet-like due to the formation of a weft of aerial mycelium. Produce a wide variety of pigments responsible for the color of colonies. Gram-stain-positive, strict aerobes, and not acid-fast. Chemoorganotrophs having an oxidative type of metabolism. Use many organic compounds as sole source of carbon for energy and growth. Catalase positive. Nitrates are generally reduced to nitrites and gelatin, casein and starch are hydrolyzed. Members of the genus are widely distributed and found predominantly in soil and decaying vegetation. They play an important role in the degradation of organic matter, most commonly noted in compost piles. In fact, the odor of the moist earth is largely the result of production of volatile substances such as **geosmin** by *Streptomyces* spp. Members are a great source for the production of many antibacterial pharmaceutical agents. Some of these are erythromycin, neomycin, streptomycin, vancomycin, tetracycline, rifamycin, lincomycin, and chloramphenicol. The complete genome sequence of *St. coelicolor* strain A 3 was published in 2002. The genome sequences of *St. scabiei*, the cause of potato common scab, and of *St. avermetilis* have also been completed. In all these species, the chromosome is a linear structure, in contrast to most bacterial chromosomes, which are circular in shape. The G + C content of the DNA is 69.0–78.0 mol% (T_m).

Type species: *Streptomyces albus*

Labeda (2011) reported that phylogenetic analysis of single gene alignments and a concatenated four-gene alignment demonstrated that the phytopathogenic species are taxonomically distinct from each other in spite of high 16S rRNA gene sequence similarities and also provided a tool for the identification of unknown putative phytopathogenic *Streptomyces* strains at the species level.

Huguet-Tapia et al. (2016) reported that genome sequencing, comparative genomics, and phylogenetic analysis of genomes of *Streptomyces* species, including four plant pathogenic species, enabled to discriminate pathogenic strains from saprophytic *Streptomyces* strains. The pathogen-specific genome contained 4,662 orthologs. Phylogenetic reconstruction suggested that *St. scabies* and *St. ipomoeae* share an ancestor, but that their biosynthetic clusters encoding the required virulence factor thaxtomin have diverged. In contrast, *St. turgidiscabies* and *St. acidiscabies*, two relatively unrelated pathogens, possess highly similar thaxtomin biosynthesis clusters, which suggests that the acquisition of these genes was through lateral gene transfer.

Plant pathogenic species: *St. acidiscabies*, *St. albidoflavus*, *St. aureofaciens*, *St. bottropensis*, *St. candidus*, *St. caviscabies*, *St. collinus*, *St. diastatochromogenes*, *St. europaeiscabiei*, *St. galilaeus*, *St. intermedius*, *St. ipomoeae*, *St. luridiscabiei*, *St. niveiscabiei*, *St. puniciscabiei*, *St. reticuliscabiei*, *St. scabiei*, *St. setonii*, *St. steliiscabiei*, *St. turgidiscabies*, *St. wedmorensis*.

III. Phylum: "*Firmicutes*"
1. Class: "*Bacilli*"
   i. Order: *Bacillales*
      a. Family: *Bacillaceae*
         Genus: *Bacillus*

4.3.1.3.46 Genus Bacillus Cohn 1872

It is a very old genus and was created by Cohn in 1872. The important characteristics of the genus are given below. Cells are Gram-stain-positive, endospore forming rods that are usually motile and peritrichously flagellate. The presence of spores makes the cells (rods) to bulge at one end. The spores may be thin- or thick-walled. The bacterium is heterotrophic, aerobic or sometimes facultative, and catalase positive. Ubiquitous in nature, the genus includes both free-living and pathogenic species. The type species, *Ba. subtilis* is one of the best understood procaryotes, in terms of molecular biology and cell biology. The cell wall is composed of teichoic and teichuronic acids. *Ba. subtilis* and *Ba. megaterium* are strict aerobes and do not require growth factors. The former is a powerful producer of extracellular amylases and proteases and serves as a source of commercial production of these enzymes.

Many species of *Bacillus* are of considerable importance. Some species produce antibiotics, namely, bacitracin, polymyxin, gramicidin, etc. *Ba. cereus* causes some forms of food poisoning and can infect humans. *Ba. anthracis* is the causal agent of anthrax disease that afflicts both farm animals and humans. *Ba. thuringiensis* and *Ba. sphaericus* have insecticidal values. Both the species form a solid protein crystal, the parasporal body, next to their spores during endospore formation. The parasporal body of *Ba. thuringiensis* contains protein toxins that can kill larvae of more than 100 lepidopterous insects. Most of these *Ba. thuringiensis* toxin genes are carried on large plasmids. The genes of the bacterium which encode toxin production (BT genes) have been incorporated into cotton, corn, and brinjal. These genetically modified crops are called BT cotton, BT corn, and BT brinjal and possess the insecticidal properties. The parasporal body of *Ba. sphaericus* contains proteins toxic for mosquito larvae and may prove useful in controlling mosquitoes that carry the malaria parasite, *Plasmodium*. The 4.2 million base pairs genome of *Ba. subtilis* has been sequenced. The genome contains genes for the catabolism of many diverse carbon sources, and for protein secretion and antibiotic synthesis. The G + C content of the DNA ranges from 32.0 to 69.0 mol% (T_m).

Type species: *Bacillus subtilis*

Generally, the members of this genus are not active plant pathogens. However, some of its members are found associated with plant material and cause its breakdown particularly at or after the crop harvest. This breakdown also occurs at relatively higher temperatures in comparison to the temperature required by other plant pathogenic bacteria for their pathogenicity. In 1982, a pathogen of cereals, namely, *Ba. megaterium* pv. *cerealis* was transferred to this genus. Peng et al. (2013)

identified a Gram-positive, spore-forming, rod-shaped bacterium, causing ginger rhizome rot in Shandong province, China, as *Bacillus pumilus*. Elbanna et al. (2014) reported for the first time, *Ba. altitudinis* causing soft rot of apple and pear fruits.

Plant pathogenic species: *Ba. altitudinis*, *Ba. megaterium*, *Ba. megaterium* pv. *cerealis*, *Ba. pumilus*

b. Family: *Paenibacillaceae*
Genus: *Paenibacillus*

4.3.1.3.47 Genus Paenibacillus *Ash et al. 1994 emend. Shida et al. 1997 emend. Behrendt et al. 2010*

Ash et al. (1993) proposed that members of "group 3" within the genus *Bacillus* should be transferred to the genus *Paenibacillus*, for which they proposed *Paenibacillus polymyxa* as the type species. Latin *paene* means almost, so the paenibacilli are literally "**almost bacilli.**"

Cells are rod shaped and motile by means of peritrichous flagella. Gram-stain reaction is positive, but usually stain variable or negative in the laboratory. Oval endospores are formed that distend the sporangium. Facultatively anaerobic or strictly aerobic. Most species are catalase positive. Colonies are generally smooth and translucent, light brown, white, or sometimes light pink in color. Optimal temperature for growth is 28°C–40°C. Optimal pH is 7.0, although strains of some species are alkaliphilic. Growth is inhibited by 10% sodium chloride. Major fatty acid is C15:0 anteiso. Additional important fatty acid fractions often contain C16:0, C15:0 iso, and C17:0 anteiso. The DNA G + C content (mol%) ranges from 40 to 59.

Bacteria belonging to this genus have been detected in a variety of environments, such as soil, water, rhizosphere, vegetable matter, forage, and insect larvae, as well as clinical samples. The genus includes *Pae. larvae*, which causes American foulbrood in honeybees, and *Pae. polymyxa*, which fixes nitrogen in soil and is thus used in agriculture and horticulture. Various *Paenibacillus* spp. also produce antimicrobial substances that affect a wide spectrum of microorganisms such as fungi, soil bacteria, plant pathogenic bacteria, and even important anaerobic pathogens such as *Clostridium botulinum*. *Pae. vortex* and *Pae. dendritiformis* form complex colonies with intricate architectures.

Type species: *Paenibacillus polymyxa*.

Zhang et al. (2017) reported *Pae. polymyxa* as the causal agent of bacterial stem rot of *Hylocereus undulates* from China.

Plant pathogenic species: *Pae. polymyxa*.

ii. Order: *Lactobacillales*
a. Family: *Lactobacillaceae*
Genus: *Lactobacillus*

4.3.1.3.48 Genus Lactobacillus *Beijerinck 1901*

Cells are Gram-stain-positive and non-sporing. Cells vary from long and slender, sometimes bent rods, to short often coryneform coccobacilli; chain formation is common. Motility is uncommon, but when present, it is by peritrichous flagella.

Metabolism is fermentative; obligately saccharoclastic, at least half of end product carbon is lactate. Lactate is usually not fermented. Microaerophilic; surface growth on solid media is generally enhanced by anaerobiosis or reduced oxygen pressure and 5%–10% CO_2. Some are anaerobes on isolation. Pigment production is rare; if present, yellow or orange-to-rust or brick red. Nitrate reduction is highly unusual; if occurs, only when terminal pH is above 6.0. Gelatin is not liquefied. Casein is not digested. Indole and H_2S are not produced. Catalase and cytochrome tests and benzidine reaction are negative.

Growth temperature ranges from 2°C to 53°C, while the optimal lies between 30°C and 40°C. Optimal pH ranges from 5.5 to 6.2; growth generally occurs at 5.0 or lower pH. Cell multiplication is often reduced at neutral or initially alkaline pH. The members are found in dairy, meat, fish, and grain products, sewage, water, wine, beer, fruits, and fruit products. They are also found in vegetable pickles, silage, sour dough, and mash. They are a part of the normal flora of the mouth, intestinal tract, and vagina of many homothermic animals including humans. Pathogenic action is rare. The mole% G + C of the DNA ranges from 32 to 53 (Bd, T_m).

Type species: *Lactobacillus delbrueckii*

Based on growth characteristics and sequence analysis of 16S rRNA and *rpoA* genes, Bonasera et al. (2017) identified lactic acid bacteria isolated from diseased onion plants and bulbs as *Lactobacillus plantarum*. *Lactob. plantarum* caused symptoms in both leaves and bulbs. This is the first report of a Gram-stain-positive bacterium to cause a disease on onions.

Plant pathogenic species: *Lactob. plantarum*

b. Family: *Streptococcaceae*
Genus: *Lactococcus*

4.3.1.3.49 Genus Lactococcus *Schleifer et al. 1986 gen. nov*

Lactococcus is a genus of lactic acid bacteria that were formerly included in the genus *Streptococcus* Group N1. They are known as homofermentors because they produce a single product, i.e., lactic acid in this case, as the major or only product of glucose fermentation. However, this homofermentative character can be altered by adjusting environmental conditions such as pH, glucose concentration, and nutrient limitation. Cells are Gram-stain-positive, cocci, found singly, in pairs, or in chains, non-motile, and catalase-negative. The genus contains strains known to grow at or below 7°C.

These organisms are commonly used in the dairy industry for manufacturing fermented dairy products such as cheeses. They can be used as single-strain starter cultures, or as mixed-strain cultures with other lactic acid bacteria such as *Lactobacillus* and *Streptococcus*. Special emphasis is placed in the study of *Lactoc. lactis* subsp. *lactis* and *Lactoc. lactis* subsp. *cremoris*, as these are the strains used as starter cultures in industrial dairy fermentations. Their main role in dairy fermentation is the rapid acidification of milk, which causes a drop in the pH of the fermented product and prevents the growth of spoilage bacteria. These bacteria also play a role in the flavor of the final product. Lactococci are also currently used in the biotechnology industry. They are easily grown at industrial scale on whey-based media. As food-grade bacteria, they are used in the production of foreign proteins that are used in the food industry.

Type species: *Lactococcus lactis*

On the basis of growth characteristics and sequence analysis of 16S rRNA and *rpoA* genes, Bonasera et al. (2017) identified lactic acid bacteria isolated from diseased onion bulbs as *Lactococcus lactis*. *Lactoc. lactis* caused scale discoloration in bulbs, but failed to cause lesions on leaves. This is the first report of *Lactoc. lactis* to cause a disease in onion bulbs.

Plant pathogenic species: *Lactoc. lactis*

c. Family: *Leuconostocaceae*
Genus: *Leuconostoc*

4.3.1.3.50 Genus Leuconostoc van Tieghem 1878

Cells are Gram-stain-positive, generally ovoid cocci, and often form chains. *Leuconostoc* spp. are intrinsically resistant to vancomycin and are catalase-negative (which distinguishes them from staphylococci). All species within this genus are heterofermentative and are able to produce dextrin from sucrose. They are generally slime-forming. Blamed for causing the "stink" when creating a sourdough starter, some species are also capable of causing human infection. Because they are an uncommon cause of disease in humans, standard commercial identification kits are often unable to identify the organism.

Leuconostoc, along with other lactic acid bacteria such as *Pediococcus* and *Lactobacillus*, is responsible for the fermentation of cabbage, making it sauerkraut. In this process, fresh cabbage is fermented in light brine, where the sugars in the cabbage are transformed by lacto-fermentation to lactic acid which gives the cabbage a sour flavor and good keeping qualities. *Leuconostoc* is similarly part of the symbiotic colonies of bacteria and yeast or symbiotic culture of bacteria and yeast (SCOBY) involved in the fermentation of kefir, a fermented milk beverage.

Type species: *Leuconostoc mesenteroides*

Bonasera et al. (2017), on the basis of growth characteristics and sequence analysis of 16S rRNA and *rpoA* genes, identified the strains of lactic acid bacteria isolated from diseased onion plants and bulbs as *Leuconostoc citreum*, *Len. mesenteroides*, and *Len. pseudomesenteroides*. All the three *Leuconostoc* species caused symptoms in both leaves and bulbs. *Len. citreum* caused bulb decay in 7 days at 18°C as well as 37°C. This is the first report of *Leuconostoc* species to cause a disease in onion leaves and bulbs.

Plant pathogenic species: *Len. citreum*, *Len. mesenteroides*, *Len. pseudomesenteroides*.

2. Class: *Mollicutes*
Order: *Entomoplasmatales*
Family: *Spiroplasmataceae*
Genus: *Spiroplasma*

4.3.1.3.51 Genus Spiroplasma Saglio et al. 1973

The genus was first described by Saglio et al. (1973) to classify a mycoplasma-like organism associated with citrus stubborn disease. Later on, two more species, namely, *Sp. kunkelii* and *Sp. phoeniceum*, causal agents of corn stunt and stubborn of periwinkles, respectively, were added to this genus.

The salient characteristics of the genus are given below. Cells are pleomorphic and vary in shape and size. Different shapes found are helical and branched non-helical filaments and spherical and ovoid forms. The helical forms are generally 100–200 nm in width and 3–5 μm long. Spherical cells are approximately 300 nm in diameter. The colonies have a typical **fried-egg appearance**. The colony margin is frequently diffuse due to the movement of growing cells. Colony size varies from 0.1 to 4.0 mm. Cells divide by binary fission. Facultatively anaerobic. Flagella, periplasmic fibrils, or other organs of locomotion are not found. However, helical filaments show motility due to flexional and twitching movements. Chemoorganotrophs. Acid is produced from glucose, but fermentation of other carbohydrates is variable. Arginine is generally hydrolyzed, but urea hydrolysis, esculin fermentation, and serum digestion are negative. Cholesterol or possibly other sterols are required for growth. Sensitive to tetracycline, erythromycin, and amphotericin B, but resistant to penicillin. The genome size is very small and G + C content of the DNA varies from 25.0 to 31.0 mol%. Members can be isolated from phloem fluid of vascular plants and insects that feed on this fluid and from surfaces of flowers and other plant parts. *Sp. citri* causes stubborn disease of citrus and corn stunt.

Type species: *Spiroplasma citri*

Plant pathogenic species: *Sp. citri*, *Sp. kunkelii*, *Sp. phoeniceum*.

Genus: '*Candidatus* Phytoplasma'- The order and family are not yet established.

4.3.1.3.52 Genus 'Candidatus Phytoplasma'

Several wall-less procaryotes are associated with plants and cause yellow-type diseases. These were, earlier called mycoplasma-like organisms due to their morphological similarities to mycoplasmas. Some of these microorganisms, which are spiral-shaped and can grow on artificial culture media; have been assigned to the genus *Spiroplasma*. The remaining wall-less, non-helical, and not yet cultured procaryotes that colonize plant phloem and sap feeding insects are called phytoplasmas. In 1994, the trivial name "Phytoplasma" was adopted by the Phytoplasma Working Team at the 10th Congress of International Organization of Mycoplasmology, replacing the term "mycoplasma-like organisms." Zreik et al. (1995) proposed the genus '*Candidatus* Phytoplasma' to include '*Candidatus* Phytoplasma aurantifolia,' the causal agent of witches' broom disease of lime. Since then, many other *Candidatus* phytoplasmas causing different plant diseases have been added to this genus (Jung et al. 2002, 2003a, 2003b; Verdin et al. 2003; Hiruki and Wang 2004; Lee et al. 2004a, 2004b, 2006, 2011; Marcone et al. 2004a, 2004b; Seemuller and Schneider 2004; Arocha et al. 2005, 2007; Schneider et al. 2005; Valiunas et al. 2006; Al-Saady et al. 2008; Zhao et al. 2009a; Malembic-Maher et al. 2011; Davis et al. 2012, 2013, 2017; Martini et al. 2012; Nejat et al. 2013; Quaglino et al. 2013; Win et al. 2013; Harrison et al. 2014; Šafářová et al. 2016; Naderali et al. 2017; Miyazaki et al. 2018). Most plant pathogenic *Candidatus* species are members of this genus.

'*Ca.* Phytoplasma' cells lack rigid cell walls, are surrounded by a single unit membrane, and are pleomorphic in shape. When observed by transmission electron microscopy, they appear as round to filamentous pleomorphic bodies with a mean diameter of 200–800 nm. Some of the filamentous forms are branched. Phytoplasmas are sensitive to tetracyclines, but non-sensitive to penicillin and other cell wall inhibiting antibiotics. They are resistant to digitonin and sensitive to hypotonic salt solutions, and are therefore, similar to those mollicutes that do not require sterols. Members use UGA as a stop codon and not as a tryptophan codon, as in several other mycoplasmas. Phytoplasmas are closest relatives of the members of the genus *Acholeplasma*. The genome size of different phytoplasmas varies considerably, ranging from 530 to 1,350 kilo base pairs The Bermuda grass white leaf phytoplasma has the smallest genome size of 530 kbp found in phytoplasma to date and may represent the smallest chromosome known for any living cell. Like other members of the mollicutes, phytoplasmas contain one circular double-stranded chromosomal DNA molecule. Short, circular extra chromosomal DNAs or plasmids were found in all the members of the aster yellows group and stolbur group and in some members of the X-disease and clover proliferation groups. Large plasmids may also be present in some phytoplasmas. The G + C contents of phytoplasma chromosomal DNA are estimated to be between 23 and 29 mol% (Bd) (Lee et al. 2000). Members inhabit the phloem sieve elements of vascular plants, and the gut, hemolymph, salivary glands, and other organs of sap-sucking insects (leafhoppers and planthoppers). In insect hosts, they may cause premature mortality. In plant hosts, they cause a wide range of symptoms, such as virescence, phyllody, sterility of flowers, proliferation of auxiliary shoots, abnormal elongation of internodes, and many others, often less specific symptoms.

Classification and taxonomic assignment of phytoplasmas have been based primarily on molecular analysis of 16S rRNA gene sequences because of the inaccessibility of measurable phenotypic characters suitable for conventional microbial characterization. In the recent past, many molecular techniques have been developed for the identification and classification of phytoplasmas. Wei et al. (2007), on the basis of distinctive virtual RFLP patterns and calculated similarity coefficients (obtained by using computer-simulated RFLP analysis of 16S rRNA genes), classified phytoplasma strains into 28 groups. They were able to classify hundreds of previously unclassified phytoplasmas and delineate ten new phytoplasma groups representing three formerly described and seven novel putative '*Candidatus* Phytoplasma' taxa. Wei et al. (2008), extending their abovementioned efforts in the exploitation of computer-simulated 16S rRNA gene RFLP analysis and virtual gel plotting for rapid classification of phytoplasmas, developed a **Perl program** for automated RFLP pattern comparison and similarity coefficient calculation. By streamlining virtual RFLP pattern analysis, the program has led to the establishment of a criterion for phytoplasma 16Sr sub group classification and to the delineation of new and distinct sub group lineages in the clover proliferation phytoplasma group (16SrVI).

Martini et al. (2007) showed that ribosomal protein (rp) genes-, rplV (rpl22)-, and rpsC (rps3)-based phylogeny allowed for finer resolution of distinct lineages within the phytoplasma 16Sr groups. RFLP analysis of rp gene sequences permitted finer differentiation of phytoplasma strains in a given 16Sr group. Hodgetts et al. (2008) have highlighted the utility of virtual RFLP analysis of secA gene sequences to differentiate 16Sr groups and sub groups suggesting that this gene may provide an informative alternative molecular marker for pathogen identification and diagnosis of phytoplasma diseases. Zhao et al. (2009b) developed an interactive online tool, **iPhyClassifier**, to expand the efficacy and capacity of the current 16S rRNA gene sequence-based phytoplasma classification system. iPhyClassifier performs sequence similarity analysis, simulates laboratory restriction enzyme digestions and subsequent gel electrophoresis, and generates virtual RFLP profiles. iPhyClassifier makes instant suggestions on tentative phytoplasma 16Sr group/sub group classification status and '*Candidatus* Phytoplasma' species assignment based on calculated RFLP pattern similarity coefficients and overall sequence similarity scores. Using iPhyClassifier, the authors have revised and updated the classification of strains affiliated with the peach X-disease phytoplasma group. The online tool is available at http://www.ba.ars.usda.gov/data/mppl/iPhyClassifier.html.

As the DNA-dependent RNA polymerase (DpRp) β-subunit gene (*rpoB*) exists as a single copy in the phytoplasma genome, Valiunas et al. (2013) explored the use of *rpoB* for phytoplasma classification and phylogenetic analysis. They sequenced a clover phyllody phytoplasma genetic locus containing ribosomal protein genes, a complete *rpoB* gene, and a partial *rpoC* gene encoding the β′-subunit of DpRp. The *rpoB* gene sequences from phytoplasmas classified in groups 16SrI, 16SrII, 16SrIII, 16SrX, and 16SrXII were subjected to sequence similarity and phylogenetic analyses. The *rpoB* gene sequences were more variable than 16S rRNA gene sequences, more clearly distinguishing among phytoplasma lineages. Phylogenetic trees based on 16S rRNA and *rpoB* gene sequences had similar topologies, and branch lengths in the *rpoB* tree facilitated distinctions among closely related phytoplasmas. Virtual RFLP analysis of *rpoB* gene sequences also improved distinctions among closely related lineages. The results indicated that as the *rpoB* gene provides a useful additional marker for phytoplasma classification, it should facilitate studies of disease etiology and epidemiology.

Lee et al. (2010) emphasized the importance of *secY* gene sequence for delineating phytoplasma strains within each 16Sr group. Phylogenetic analysis based on the *secY* gene permitted finer differentiation of phytoplasma strains. The *secY* gene-based phylogeny not only readily resolved 16Sr sub groups within a given 16Sr group, but also delineated distinct lineages irresolvable by 16S rRNA gene-based phylogeny. Such high resolving power makes the *secY* gene a more useful genetic marker than the 16S rRNA gene for finer differentiation of closely related phytoplasma strains based on RFLP analysis with selected restriction enzymes.

Wei et al. (2015) have emphasized the importance of occurrence, distribution, and possible functional roles of simple sequence repeats (SSRs) in phytoplasma genomes. SSRs of microbes are mutation-prone, phase-variable short DNA tracts that function as "evolutionary rheostats" and enhance rapid adaptations to vastly different environments. The overall density of SSRs in phytoplasma genomes was higher than in representative strains of other prokaryotes. While mono- and tri-nucleotide SSRs were significantly over represented in the phytoplasma genomes, dinucleotide SSRs and other higher-order SSRs were under represented. The occurrence and distribution of long SSRs in the prophage islands and phytoplasma-unique genetic loci indicated that SSRs played a role in compounding the complexity of sequence mosaics in individual genomes and in increasing allelic diversity among genomes.

Type species: '*Candidatus* Phytoplasma aurantifolia'

Plant pathogenic species: '*Ca.* Phy. allocasuarinae', '*Ca.* Phy. americanum', '*Ca.* Phy. asteris', '*Ca.* Phy. aurantifolia', '*Ca.* Phy. australasia', '*Ca.* Phy. australiense', '*Ca.* Phy. balanitae', '*Ca.* Phy. brasiliense', '*Ca.* Phy. caricae', '*Ca.* Phy. castaneae', '*Ca.* Phy. cirsii', '*Ca.* Phy. convolvuli', '*Ca.* Phy. costaricanum', '*Ca.* Phy. cynodontis', '*Ca.* Phy. fragariae', '*Ca.* Phy. fraxini', '*Ca.* Phy. graminis', '*Ca.* Phy. hispanicum', '*Ca.* Phy. japonicum', '*Ca.* Phy. luffae', '*Ca.* Phy. lycopersici', '*Ca.* Phy. malaysianum', '*Ca.* Phy. mali', '*Ca.* Phy. meliae', '*Ca.* Phy. noviguineense', '*Ca.* Phy. omanense', '*Ca.* Phy. oryzae', '*Ca.* Phy. palmicola', '*Ca.* Phy. phoenicium', '*Ca.* Phy. pini', '*Ca.* Phy. pruni', '*Ca.* Phy. prunorum', '*Ca.* Phy. pyri', '*Ca.* Phy. rhamni', '*Ca.* Phy. rubi', '*Ca.* Phy. solani', '*Ca.* Phy. spartii', '*Ca.* Phy. sudamericanum', '*Ca.* Phy. tamaricis', '*Ca.* Phy. trifolii', '*Ca.* Phy. ulmi', '*Ca.* Phy. wodyetiae', '*Ca.* Phy. ziziphi'.

3. Class: '*Clostridia*'
   Order: *Clostridiales*
   Family: *Clostridiaceae*
   Genus: *Clostridium*

4.3.1.3.53 Genus Clostridium *Prazmowski 1880*

It is a very old genus established way back in 1880. At present, it contains only one plant pathogenic species associated with spoilage of potatoes and carrots.

The genus has the following salient characteristics. Cells are Gram-stain-positive, spore-forming rods. Obligate anaerobes. Most species are obligate fermenters, and they ferment by pathways that generate organic solvents such as butantol. No doubt, clostridia are rod-shaped, but when producing spores, they appear more like drum sticks with a bulge at one end. Do not carry out dissimilatory sulphate reduction. An exposure of vegetative cells to oxygen for 5–10 minutes may cause massive death of the population. However, spores are not killed by the oxygen. In many cases, the death results from the production of hydrogen peroxide by the bacteria, which is highly toxic. Most species completely lack heme proteins, including catalase, and are incapable of decomposing hydrogen peroxide.

Clostridium species are commonly found in the environment. They inhabit soil, sewage, marine sediments, and the intestines of both animals and humans. Some species produce toxins and cause diseases. *Clo. tetani* causes tetanus. The bacterium releases an exotoxin called tetanus toxin, which blocks the release of neurotransmitters from the presynaptic membrane of inhibitory interneurons of spinal cord and brainstem of mammals that regulate muscle contraction. *Clo. botulinum* is an etiological agent of botulism, and it is contracted usually through the ingestion of botulinum toxin. *Clo. perfringens* is one of the several species of the genus that causes gas gangrene and is the causative agent in 95% gas gangrene cases. *Clo. sordellii* has been linked to the death of a few women after child birth.

Apart from causing diseases and harmful effects, *Clostriduim* species have commercial uses and industrial value. *Clo. thermocellum* can utilize lignocellulosic waste and generate ethanol. *Clo. ljungdahlii* can produce ethanol from single-carbon sources including synthesis gas, a mixture of carbon monoxide and hydrogen that can be generated from the partial combustion of either fossil fuels or biomass. *Clo. butyricum* and *Clo. pasteurianum* fix nitrogen. The genomes of *Clo. tetani*, *Clo. perfringens*, and *Clo. acetobutylicum* have been sequenced. *Clo. tetani* also contains a plasmid that measures 74082 bp long with 61 open reading frames and encodes the tetanus toxin. The genome of *Clo. perfringens* encodes the typical anaerobic fermentation enzymes leading to gas production, but no enzymes for the tricarboxylic acid cycle of respiratory chain (*Clostridium*—Microbe Wiki). The G + C content of the DNA ranges from 22.0 to 55.0 mol% (T_m).

Type species: *Clostridium butyricum*

Plant pathogenic species: *Clo. puniceum*.

REFERENCES

Adeolu, M., Alnajar, S., Naushad, S., and Gupta, R. S. 2016. Genome-based phylogeny and taxonomy of the "Enterobacteriales": proposal for Enterobacterales ord. nov. divided into the families Enterobacteriaceae, Erwiniaceae fam. nov., Pectobacteriaceae fam. nov., Yersiniaceae fam. nov., Hafniaceae fam. nov., Morganellaceae fam. nov., and Budviciaceae fam. nov. *Int. J. Syst. Evol. Microbiol.* 66: 5575–5599.

Ah-You, N., Gagnevin, L., Grimont, P. A. D., Brisse, S., Nesme, X., et al. 2009. Polyphasic characterization of xanthomonads pathogenic to members of the Anacardiaceae and their relatedness to species of *Xanthomonas*. *Int. J. Syst. Evol. Microbiol.* 59: 306–318.

Allahverdi, T., Rahimian, H., and Ravanlou, A. 2016. First report of bacterial canker in mulberry caused by *Citrobacter freundii* in Iran. *Plant Dis.* 100: 1774.

Al-Saady, N. A., Khan, A. J., Calari, A., Al-Subhi, A. M., and Bertaccini, A. 2008. "*Candidatus* Phytoplasma omanense," associated with witches'-broom of *Cassia italica* (Mill.) Spreng. in Oman. *Int. J. Syst. Evol. Microbiol.* 58: 461–466.

Arocha, Y., Antesana, O., Montellano, E., Franco, P., Plata, G., et al. 2007. "*Candidatus* Phytoplasma lycopersici," a phytoplasma associated with "hoja de perejil" disease in Bolivia. *Int. J. Syst. Evol. Microbiol.* 57: 1704–1710.

Arocha, Y., Lopez, M., Pinol, B., Fernández, M., Picornil, B., et al. 2005. "*Candidatus* Phytoplasma graminis" and "*Candidatus* Phytoplasma caricae," two novel phytoplasmas associated with diseases of sugarcane, weeds and papaya in Cuba. *Int. J. Syst. Evol. Microbiol.* 55: 2451–2463.

Asai, T. 1935. Taxonomic studies on acetic acid bacteria and allied oxidative bacteria isolated from fruits. A new classification of the oxidative bacteria. *J. Agri. Chem. Soc. Japan.* 11: 674–708.

Ash, C., Priest, F. G., and Collins, M. D. 1994. List of new names and new combinations previously effectively, but not validly, published. *Int. J. Syst. Bacteriol.* 44: 852. Effective publication: Ash, C., Priest, F. G., and Collins, M. D. 1993. Molecular identification of rRNA group 3 bacilli (Ash, Farrow, Wallbanks and Collins) using a PCR probe test. *Antonie van Leeuwenhoek* 64: 253–260.

Baldani, J. I., Baldani, V. L. D., Seldin, L., and Döbereiner, J. 1986. Characterization of *Herbaspirillum seropedicae* gen. nov., sp. nov., a root-associated nitrogen-fixing bacterium. *Int. J. Syst. Bacteriol.* 36: 86–93.

Baldani J. I., Pot, B., Kirchhof, G., Falsen, E., Baldani, V. L. D., et al. 1996. Emended description of *Herbaspirillum*; inclusion of [*Pseudomonas*] *rubrisubalbicans*, a mild plant pathogen, as *Herbaspirillum rubrisubalbicans* comb. nov.; and classification of a group of clinical isolates (EF Group 1) as *Herbaspirillum* species 3. *Int. J. Syst. Bacteriol.* 46: 802–810.

Behrendt, U., Schumann, P., Stieglmeier, M., Pukall, R., Augustin, J., et al. 2010. Characterization of heterotrophic nitrifying bacteria with respiratory ammonification and denitrification activity–Description of *Paenibacillus uliginis* sp. nov., an inhabitant of fen peat soil and *Paenibacillus purispatii* sp. nov., isolated from a spacecraft assembly clean room. *Syst. Appl. Mirobiol.* 33: 328–336.

Beijerinck, M. W. 1898. Über die Arten der Essigbakterien. *Zentralbl Bakteriol Parasitenkd Infektionskr Hyg.* Abt II 4: 209–216.

Beijerinck, M. W. 1901. Sur les ferments lactiques de l'industrie. *Archives Néerlandaisesdes Sciences Exactes et Naturelles (Section 2)*, 6: 212–243.

Beijerinck, M. W., and Folpmers, T. 1916. Verslagen van de gewone vergadering der wis-en naturkundige afdeeling. *Konenklijke Akademie van Westenchappen te Amsterdam*, 18: 1198–1200.

Beurmann, S., Ushijima, B., Svoboda, C. M., Videau, P., Smith, A. M., et al. 2017. *Pseudoalteromonas piratica* sp. nov., a budding, prosthecate bacterium from diseased *Montipora capitata*, and emended description of the genus *Pseudoalteromonas*. *Int. J. Syst. Evol. Microbio*l. 67: 2683–2688.

Bizio, B. 1823. Lettera di Bartolomeo Bizio al chiarissimo canonico Angelo Bellani sopra il fenomeno della polenta porporina Biblioteca Italiana o sia Giornale di letteratura, *Scienze e Arti (Anno VIII)* 30: 275–295.

Bonasera, J. M., Asselin, J. A. E., and Beer, S. V. 2017. Lactic acid bacteria cause a leaf blight and bulb decay of onion (*Allium cepa*). *Plant Dis.* 101: 29–33.

Bouchek-Mechiche, K., Gardan, L., Andrivon, D., and Normand, P. 2006. *Streptomyces turgidiscabies* and *Streptomyces reticulsiscabiei*: One genomic species, two pathogenic groups. *Int. J. Syst. Evol. Microbiol.* 56: 2771–2776.

Bouchek-Mechiche, K., Gardan, L., Normand, P., and Jouan, B. 2000. DNA relatedness among strains of *Streptomyces* pathogenic to potato in France: Description of three new species, *S. europaeiscabiei* sp. nov. and *S. steliiscabiei* sp. nov. associated with common scab, and *S. reticulsiscabiei* sp. nov. associated with netted scab. *Int. J. Syst. Evol. Microbiol.* 50: 91–99.

Bouzar, H., and Jones, J. B. 2001. *Agrobacterium larrymoorei* sp. nov., a pathogen isolated from aerial tumours of *Ficus benjamina*. *Int. J. Syst. Evol. Microbiol.* 51: 1023–1026.

Bradbury, J. F. 1982. The new bacterial nomenclature—What to do? *Trop. Pest Manag.* 28: 42–44.

Brady, C., Denman, S., Kirk, S., Venter, S., Rodríguez-Palenzuela, P., et al. 2011. Validation list no. 139. *Int. J. Syst. Evol. Microbiol.* 61: 1011–13. Effective publication: Brady, C., Denman, S., Kirk, S., Venter, S., Rodríguez-Palenzuela, P., et al. 2010a. Description of *Gibbsiella quercinecans* gen. nov., sp. nov., associated with acute oak decline. *Syst. Appl. Microbiol.* 33: 444–450.

Brady, C., Hunter, G., Kirk, S., Arnold, D., and Denman, S. 2014. Description of *Brenneria roseae* sp. nov. and two subspecies, *Brenneria roseae* subspecies *roseae* ssp. nov. and *Brenneria roseae* subspecies *americana* ssp. nov. isolated from symptomatic oak. *Syst. Appl. Microbiol.* 37: 396–401.

Brady, C. L., Cleenwerck, I., Denman, S., Venter, S. N., Rodriguez-Palenzuela, P., et al. 2012. Proposal to reclassify *Brenneria quercina* (Hildebrand and Schroth 1967) Hauben et al. 1999 into a new genus, *Lonsdalea* gen. nov., as *Lonsdalea quercina* comb. nov., descriptions of *Lonsdalea quercina* subsp. *quercina* comb. nov., *Lonsdalea quercina* subsp. *iberica* subsp. nov. and *Lonsdalea quercina* subsp. *britannica* subsp. nov., emendation of the description of the genus *Brenneria*, reclassification of *Dickeya dieffenbachiae* as *Dickeya dadantii* subsp. *dieffenbachiae* comb. nov., and emendation of the description of *Dickeya dadantii*. *Int. J. Syst. Evol. Microbiol.* 62: 1592–1602.

Brady, C. L., Venter, S. N., Cleenwerck, I., Engelbeen, K., Vancanneyt, M., et al. 2009. *Pantoea vagans* sp. nov., *Pantoea eucalypti* sp. nov., *Pantoea deleyi* sp. nov. and *Pantoea anthophila* sp. nov. *Int. J. Syst. Evol. Microbiol.* 59: 2339–2345.

Brady, C. L., Venter, S. N., Cleenwerck, I., Vandemeulebroecke, K., De Vos, P., et al. 2010b. Transfer of *Pantoea citrea*, *Pantoea punctata* and *Pantoea terrea* to the genus *Tatumella* emend. as *Tatumella citrea* comb. nov., *Tatumella punctata* comb. nov. and *Tatumella terrea* comb. nov. and description of *Tatumella morbirosei* sp. nov. *Int. J. Syst. Evol. Microbiol.* 60: 484–494.

Breed, R. S., Murray E. G. D., and Smith, N. R. eds. 1957. *Bergey's Manual of Determinative Bacteriology*. 7th Ed. Williams and Wilkins Co. Baltimore, MD, 1094 pp.

Brenner, D. J., Staley, J. T., and Krieg, N. R. 2001. Classification of prokaryotic organisms and the concept of bacterial speciation. pp. 27–31 in: *Bergey's Manual of Systematic Bacteriology*. 2nd ed. Vol. 1, *The Archaea and the Deeply Branching and Phototrophic Bacteria*. D. R. Boone, and R. W. Castenholz (volume editors), G. M. Garrity (editor-in-chief). Springer-Verlag, New York.

Brisou, J., and Prévot, A. R. 1954. Études de systématique bactérienne. X. Révision des espèces réunies dans le genre *Achromobacter*. *Annales de l'Institut Pasteur* (*Paris*), 86: 722–728.

Buchanan, R. E., and Gibbons, N. E. eds. 1974. *Bergey's Manual of Determinative Bacteriology*, 8th ed. Williams and Wilkins Co., Baltimore, MD, 1268 p.

Bull, C. T., Coutinho, T. A., Denny, T. P., Firrao, G., and Fischer-Le Saux, M. 2014. List of new names of plant pathogenic bacteria (2011–2012). *J. Plant Pathol.* 96: 223–226.

Bull, C. T., De Boer, S. H., Denny, T. P., Firrao, G., Fischer-Le Saux, M., et al. 2008. Demystifying the nomenclature of bacterial plant pathogens. *J. Plant Pathol.* 90: 403–417.

Bull, C. T., De Boer, S. H., Denny, T. P., Firraro, G., Fischer-Le Saux, M., et al. 2010. Comprehensive list of names of plant pathogenic bacteria, 1980–2007. *J. Plant Pathol.* 92: 551–592.

Bull, C. T., De Boer, S. H., Denny, T. P., Firrao, G., Fischer-Le Saux, M., et al. 2012. List of new names of plant pathogenic bacteria (2008–2010). *J. Plant Pathol.* 94: 21–27.

Buonaurio, R., Stravato, V. M., Kosako, Y., Fujiwara, N., Naka. T., et al. 2002. *Sphingomonas melonis* sp. nov., a novel pathogen that causes brown spots on yellow Spanish melon fruits. *Int. J. Syst. Evol. Microbiol.* 52: 2081–2087.

Busse, H.-J. 2016. Review of the taxonomy of the genus *Arthrobacter*, emendation of the genus *Arthrobacter sensu lato*, proposal to reclassify selected species of the genus *Arthrobacter* in the novel genera *Glutamicibacter* gen. nov., *Paeniglutamicibacter* gen. nov., *Pseudoglutamicibacter* gen. nov., *Paenarthrobacter* gen. nov. and *Pseudarthrobacter* gen. nov., and emended description of *Arthrobacter roseus*. *Int. J. Syst. Evol. Microbiol.* 66: 9–37.

Chatton, E. 1937. *Titres et Travaux Scientifiques.* Sottano, Sète, France.

Chun, J., and Rainey, F. A. 2014. Integrating genomics into the taxonomy and systematics of the *Bacteria* and *Archaea*. *Int. J. Syst. Evol. Microbiol.* 64: 316–324.

Cohn, F. 1872/1875. Untersuchungen über Bakterien. *Beitr. Biol. Pflanz.* 1(Heft 2): 127–224.

Conn, H. J. 1942. Validity of the genus *Acaligenes*. *J. Bacteriol.* 44: 353–360.

Conn, H. J., and Dimmick, I. 1947. Soil bacteria similar in morphology to mycobacterium and *Corynebacterium*. *J. Bacteriol.* 54: 291–303.

Cunty, A., Poliakoff, F., Rivoal, C., Cesbron, S., Fischer-Le Saux, M., et al. 2015. Characterization of *Pseudomonas syringae* pv. *actinidiae* (Psa) isolated from France and assignment of Psa biovar 4 to a *de novo* pathovar: *Pseudomonas syringae* pv. *actinidifoliorum* pv. nov. *Plant Pathol.* 64: 582–596.

Dastager, S. G., Lee, J. C., Ju, Y. J., Park, D.-J., and Kim, C.-J. 2009. *Leifsonia kribbensis* sp. nov., isolated from soil. *Int. J. Syst. Evol. Microbiol.* 59: 18–21.

Davis, M. J., Gillaspie, A. G., Jr., Vidaver, A. K., and Harris, R. W. 1984. *Clavibacter*: A new genus containing some phytopathogenic coryneform bacteria, including *Clavibacter xyli* subsp. *xyli* sp. nov., subsp. nov. and *Clavibacter xyli* subsp. *cynodontis* subsp. nov., pathogens that cause ratoon stunting disease of sugar-cane and Bermuda grass stunting disease. *Int. J. Syst. Bacteriol.* 34: 107–117.

Davis, R. E., Zhao, Y., Dally, E. L., Jomantiene, R., Lee, I.-M., et al. 2012. "*Candidatus* Phytoplasma sudamericanum," a novel taxon, and strain PassWB-Br4, a new subgroup 16SrIII-V phytoplasma, from diseased passion fruit (*Passiflora edulis* f. *flavicarpa* Deg.). *Int. J. Syst. Evol. Microbiol.* 62: 984–989.

Davis, R. E., Zhao, Y., Dally, E. L., Lee, I.-M., Jomantiene, R., et al. 2013. "*Candidatus* Phytoplasma pruni," a novel taxon associated with X-disease of stone fruits, *Prunus* spp.: Multilocus characterization based on 16S rRNA, *secY*, and ribosomal protein genes. *Int. J. Syst. Evol. Microbiol.* 63: 2766–2776.

Davis, R. E., Zhao, Y., Wei, W., Dally, E. L., and Lee, I.-M. 2017. "*Candidatus* Phytoplasma luffae," a novel taxon associated with witches' broom disease of loofah, *Luffa aegyptica* Mill. *Int. J. Syst. Evol. Microbiol.* 67: 3127–3133.

De Ley, J., Segers, P., and Gillis, M. 1978. Intra- and intergeneric similarities of *Chromobacterium* and *Janthinobacterium* ribosomal ribonucleic acid cistrons. *Int. J. Syst. Bacteriol.* 28: 154–168.

Denman, S., Brady, C., Kirk, S., Cleenwerck, I., Venter, S., et al. 2012. *Brenneria goodwinii* sp. nov., associated with acute oak decline in the UK. *Int. J. Syst. Evol. Microbiol.* 62: 2451–2456.

Dhanvantari, B. N., Dye, D. W., and Young, J. M. 1978. *Pseudomonas pomi* Cole 1959 is a latter subject synonym of *Acetobacter pasteurianus* (Hansen 1879) Beijerinck 1898 and *Pseudomonas melophthora* Allen and Riker 1932 is a nomen dubium. *Int. J. Syst.Bacteriol.* 28: 532–537.

Dobritsa, A. P., and Samadpour, M. 2016. Transfer of eleven species of the genus *Burkholderia* to the genus *Paraburkholderia* and proposal of *Caballeronia* gen. nov. to accommodate twelve species of the genera *Burkholderia* and *Paraburkholderia*. *Int. J. Syst. Evol. Microbiol.* 66: 2836–2846.

Doi, R. H., and Igarashi, R. T. 1965. Conservation of ribosomal and messenger ribonucleic acid citrons and *Bacillus* species. *J. Bacteriol.* 90: 384–390.

Dowson, W. J. 1939. On the systematic position and generic names of the gram negative bacterial plant pathogens. *Zentralbl. Bakteriol. Parasitenk. Infektionskr. Hyg. Abt. II* 100: 177–193.

Dubnau, D., Smith, I., Morel, P., and Marmur, J. 1965. Gene conservation in *Bacillus* species. 1. Conserved genetic and nucleic acid base sequence homologies. *Proc. Natl. Acad. Sci.USA* 54: 491–498.

Dye, D. W. 1968. A study of the genus *Erwinia*. I. The'*amylovora*' group. *N. Z. J. Sci.* 11: 590–607.

Dye, D. W., Bradbury, J. F., Goto, M., Hayward, A. C., Lelliott, R. A., et al. 1980. International standards for naming pathovars of phytopathogenic bacteria and a list of pathovar names and pathotype strains. *Rev. Plant. Pathol.* 59: 153–168.

Elbanna, K., Elnaggar, S., and Bakeer, A. 2014. Characterization of *Bacillus altitudinis* as a new causative agent of bacterial soft rot. *J. Phytopath.* 162: 712–722.

Evtushenko, L. I., Dorofeeva, L. V., Subbotin, S. A., Cole, J. R., and Tiedje, J. M. 2000. *Leifsonia poae* gen. nov., sp. nov., isolated from nematode galls on *Poa annua*, and reclassification of "*Corynebacterium aquaticum*" Leifson 1962 as *Leifsonia aquatica* (ex Leifson 1962) gen. nov., nom. rev., comb. nov. and *Clavibacter xyli* Davis et al. 1984 with two subspecies as *Leifsonia xyli* (Davis et al. 1984) gen. nov., comb. nov. *Int. J. Syst. Evol. Microbiol.* 50: 371–380.

Fang, W., Li, X., Tan, X.-m., Wang, L.-f., Piao, C.-g., et al. 2016. *Leucobacter populi* sp. nov. isolated from a symptomatic bark of *Populus* × *euramericana* canker. *Int. J. Syst. Evol. Microbiol.* 66: 2254–2258.

Feng, G.-D., Yang, S.-Z., Xiong, X., Li, H.-P., and Zhu, H.-H. 2017. *Sphingomonas spermidinifaciens* sp. nov., a novel bacterium containing spermidine as the major polyamine, isolated from an abandoned lead-zinc mine and emended descriptions of the genus *Sphingomonas* and the species *Sphingomonas yantingensis* and *Sphingomonas japonica*. *Int. J. Syst. Evol. Microbiol.* 67: 2160–2165.

Francis, I. M., Jochimsen, K. N., De Vos, P., and van Bruggen, A. H. C. 2014. Reclassification of rhizosphere bacteria including strains causing corky root of lettuce and proposal of *Rhizorhapis suberifaciens* gen. nov., comb. nov., *Sphingobium mellinum* sp. nov., *Sphingobium xanthum* sp. nov. and *Rhizorhabdus argentea* gen. nov., sp. nov. *Int. J. Syst. Evol. Microbiol.* 64: 1340–1350.

Frank, B. 1889. Über die Pilzsymbiose der Leguminosen. *Ber. Deutsch. Bot. Ges.* 7: 332–346.

Furtado, G. Q., Guimarães, L. M. S., Lisboa, D. O., Cavalcante, G. P., Arriel, D. A. A., et al. 2012. First report of *Enterobacter cowanii* causing bacterial spot on *Mabea fistulifera*, a native forest species in Brazil. *Plant Dis.* 96: 1576.1.

Gardan, L., Bella, P., Meyer, J. M., Christen, R., Rott, P., et al. 2002. *Pseudomonas salomonii* sp. nov., pathogenic on garlic, and *Pseudomonas palleroniana* sp. nov., isolated from rice. *Int. J. Syst. Evol. Microbiol.* 52: 2065–2074.

Gardan, L., Christen, R., Achouak, W., and Prior, P. 2004. *Erwinia papayae* sp. nov., a pathogen of papaya (*Carica papaya*). *Int. J. Syst. Evol. Microbiol.* 54: 107–113.

Gardan, L., Dauga, C., Prior, P., Gillis, M., and Saddler, G. S. 2000. *Acidovorax anthurii* sp. nov., a new phytopathogenic bacterium which causes bacterial leaf-spot of anthurium. *Int. J. Syst. Evol. Microbiol.* 50: 235–246.

Gardan, L., Gouy, C., Christen, R., and Samson, R. 2003a. Elevation of three subspecies of *Pectobacterium carotovorum* to species level: *Pectobacterium atrosepticum* sp. nov., *Pectobacterium betavasculorum* sp. nov. and *Pectobacterium wasabiae* sp. nov. *Int. J. Syst. Evol. Microbiol.* 53: 381–391.

Gardan, L., Stead, D. E., Dauga, C., and Gillis, M. 2003b. *Acidovorax valerianellae* sp. nov., a novel pathogen of Lamb's lettuce [*Valerianella locusta* (L.) Laterr.]. *Int. J. Syst. Evol. Microbiol.* 53: 795–800.

Garnier, M. 2015. *Candidatus* Phlomobacter. pp. 1–3 in: *Bergey's Manual of Systematics of Archaea and Bacteria*, John Wiley and Sons, New Jersey.

Garrity, G. M., Bell, J. A., and Lilburn, T. 2005. The revised road map to the manual. pp. 159–200 in: *Bergey's Manual of Systematic Bacteriology.* 2nd ed. Vol. 2, The Proteobacteria, Part A, Introductory Essays. D. J. Brenner, N. R. Krieg, and J. T. Staley (volume editors), G. M. Garrity (editor-in-chief). Springer-Verlag, New York.

Gauthier, G., Gauthier, M., and Christen, R. 1995. Phylogenetic analysis of the genera *Alteromonas*, *Shewanella*, & *Moritella* using genes coding for small-subunit rRNA sequences and division of the genus *Alteromonas* into two genera, *Alteromonas* (emended) and *Pseudoalteromonas* gen. nov., and proposal of twelve new species combinations. *Int. J. Syst. Bacteriol.* 45: 755–7561.

Gavini, F., Mergaert, J., Beji, A., Mielcarek, C., Izard, D., et al. 1989. Transfer of *Enterobacter agglomerans* (Beijerinck 1888) Ewing and Fife 1972 to *Pantoea* gen. nov. as *Pantoea agglomerans* comb. nov. and description of *Pantoea dispersa* sp. nov. *Int. J. Syst. Bacteriol.* 39: 337–345.

Gillis, M., Van Van, T., Bardin, R., Goor, M., Hebbar, P., et al. 1995. Polyphasic taxonomy in the genus *Burkholderia* leading to an emended description of the genus and proposition of *Burkholderia vietnamiensis* sp. nov. for N_2-fixing isolates from rice in Vietnam. *Int. J. Syst. Bacteriol.* 45: 274–289.

Goncalves, E. R., and Rosato, Y. B. 2000. Genotypic characterization of xanthomonad strains isolated from passion fruit plants (*Passiflora* spp.) and their relatedness to different *Xanthomonas* species. *Int. J. Syst. Evol. Microbiol.* 50: 811–821.

González, A. J., and Trapiello, E. 2014. *Clavibacter michiganensis* subsp. *phaseoli* subsp. nov., pathogenic in bean. *Int. J. Syst. Evol. Microbiol.* 64: 1752–1755.

Goodfellow, M. 1984. Reclassification of *Corynebacteirum fascians* (Tilford) Dowson in the genus *Rhodococcus*, as *Rhodococcus fascians* comb. nov. *Syst. Appl. Microbiol.* 5: 225–229.

Goto, M., and Kuwata, H. 1988. *Rhizobacter daucus* gen. nov., sp. nov., the causal agent of carrot bacterial gall. *Int. J. Syst. Bacteirol.* 38: 233–239.

Goyer, C., Faucher, E., and Beaulieu, C. 1996. *Streptomyces caviscabies* sp. nov., from deep-pitted lesions in potatoes in Quebec, Canada. *Int. J. Syst. Bacteriol.* 46: 635–639.

Graham, D. C. 1964. Taxonomy of the soft rot coliform bacteria. *Annu. Rev. Phytopathol.* 2: 13–42.

Grimont, P. A. D., Grimont, F., and Starr, M. P. 1978. *Serratia proteamaculans* (Paine and Stansfield) comb. nov., a senior subjective synonym of *Serratia liquefaciens* (Grimes and Hennerty) Bascomb et al. *Int. J. Syst. Bacteriol.* 28: 503–510.

Guo, Y., Zheng, W., Rong, X., and Huang, Y. 2008. A multilocus phylogeny of the *Streptomyces griseus* 16S rRNA gene clade: Use of multilocus sequence analysis for streptomycete systematics. *Int. J. Syst. Evol. Microbiol.* 58: 149–159.

Harmon, C. L., Timilsina, S., Bonkowski, J., Jones, D. D., Sun, X., et al. 2018. Bacterial gall of *Loropetalum chinense* caused by *Pseudomonas amygdali* pv. *loropetali* pv. nov. *Plant Dis.* 102: 799–806.

Harrison, N. A., Davis, R. E., Oropeza, C., Helmick, E. E., Narváez, M., et al. 2014. "*Candidatus* Phytoplasma palmicola," associated with a lethal yellowing-type disease of coconut (*Cocos nucifera* L.) in Mozambique. *Int. J. Syst. Evol. Microbiol.* 64: 1890–1899.

Hauben, L., Moore, E. R. B., Vauterin, L., Steenackers, M., Mergaert, J., et al. 1999. Validation of publication of new names and new combinations previously effectively published outside the IJSB. *Int. J. Syst. Bacteriol.* 49: 1–3. Effective publication: Hauben, L., Moore, E. R. B., Vauterin, L., Steenackers, M., Mergaert, J., et al. 1998. Phylogenetic position of phytopathogens within the Enterobacteriaceae. *Syst. Appl. Microbiol.* 21: 384–397.

Hauben, L., and Swings, J. 2005. Genus *Brenneria*. pp. 628–33 in: *Bergey's Manual of Systematic Bacteriology.* 2nd ed. Vol. 2, The Proteobacteria, Part B The Gammaproteobacteria, D. J. Brenner, N. R. Krieg, and J. T. Staley eds. Springer, New York.

Hildebrand, D. C., and Schroth, M. N. 1967. A new species of *Erwinia* causing the drippy nut disease of live oaks. *Phytopathology* 57: 250–253.

Hiruki, C., and Wang, K. 2004. Clover proliferation phytoplasma: "*Candidatus* Phytoplasma trifolii." *Int. J. Syst. Evol. Microbiol.* 54: 1349–1353.

Hodgetts, J., Boonham, N., Mumford, R., Harrison, N., and Dickinson, M. 2008. Phytoplasma phylogenetics based on analysis of *secA* and 23S rRNA gene sequences for improved resolution of candidate species of "*Candidatus* Phytoplasma." *Int. J. Syst. Evol. Microbiol.* 58: 1826–1837.

Hollis, D. G., Hickman, F. W., Fanning, G. R., Farmer, III, J. J., Weaver, R. E., et al. 1982. Validation List No. 8. *Int. J. Syst. Bacteriol.* 32: 266–68. Effective publication: Hollis, D. G., Hickman, F. W., Fanning, G. R., Farmer, III, J. J., Weaver, R. E., et al. 1981. *Tatumella ptyseos* gen. nov., sp. nov., a member of the family Enterobacteriaceae found in clinical specimens. *J. Clin. Microbiol.* 14: 79–88.

Holt, J. G., Krieg, N. R., Sneath P. H. A., Staley, J. T., and Williams, S. T. eds. 1994. *Bergey's Manual of Determinative Bacteriology.* 9th ed. Williams and Wilkins, Baltimore, MD, 787 pp.

Hopkins, D. L., and Purcell, A. H. 2002. *Xylella fastidiosa*: Cause of Pierce's disease of grapevine and other emergent diseases. *Plant Dis.* 86: 1056–1066.

Hormaeche, E., and Edwards, P. R. 1960. A proposed genus *Enterobacter. Int. Bull. Bacteriol. Nomen. Taxon.* 10: 71–74.

Huang, C.-H., Vallad, G. E., Adkison, H., Summers, C., Margenthaler, E., et al. 2013. A novel *Xanthomonas* sp. causes bacterial spot of rose (*Rosa* spp.). *Plant Dis.* 97: 1301–1307.

Huguet-Tapia, J. C., Lefebure, T., Badger, J. H., Guan, D., Pettis, G. S., et al. 2016. Genome content and phylogenomics reveal both ancestral and lateral evolutionary pathways in plant-pathogenic *Streptomyces* species. *Appl. Environ. Microbiol.* 82: 2146–2155.

Hulse, M., Johnson, S., and Ferrieri, P. 1993. Agrobacterium infections in humans: Experience at one hospital and review. *Clin. Infect. Dis.* 16: 112–117.

IRPCM Phytoplasma/Spiroplasma Working Team—Phytoplasma Taxonomy Group. 2004. "*Candidatus* Phytoplasma," a taxon for the wall-less, non-helical prokaryotes that colonize plant phloem and insects. *Int. J. Syst. Evol. Microbiol.* 54: 1243–1255.

Jagoueix-Eveillard, S., Bove, J. M., and Garnier, M. 1994. The phloem-limited bacterium of greening disease of citrus is a member of the ∞ subdivision of the *Proteobacteria. Int. J. Syst. Bacteriol.* 44: 379–386.

Jones, J. B., Lacy, G. H., Bouzar, H., Stall, R. E., Schaad, N. W., et al. 2004. Reclassification of the *xanthomonads* associated with bacterial spot disease of tomato and pepper. *Syst. Appl. Microbiol.* 27: 755–762.

Judicial Commission of the International Committee on Systematics of Prokaryotes. 2008. *Corynebacterium ilicis* is typified by ICMP 2608 =ICPB CI144, *Arthrobacter ilicis* is typified by DSM 20138 =ATCC 14264 =NCPPB 1228 and the two are not homotypic synonyms, and clarification of the authorship of these two species. Opinion 87. *Int. J. Syst. Evol. Microbiol.* 58: 1976–1978.

Jung, H. Y., Sawayanagi, T., Kakizawa, S., Nishigawa, H., Miyata, S.-i., et al. 2002. "*Candidatus* Phytoplalsma castaneae," a novel phytoplasma taxon associated with chestnut witches' broom disease. *Int. J. Syst. Evol. Microbiol.* 52: 1543–1549.

Jung, H. Y., Sawayanagi, T., Kakizawa, S., Nishigawa, H., Wei, W., et al. 2003a. "*Candidatus* Phytoplasma ziziphi," a novel phytoplasma taxon associated with jujube witches'-broom disease. *Int. J. Syst. Evol. Microbiol.* 53: 1037–1041.

Jung, H. Y., Sawayanagi, T., Wongkaew, P., Kakizawa, S., Nishigawa, H., et al. 2003b. "*Candidatus* Phytoplasma oryzae," a novel phytoplasma taxon associated with rice yellow dwarf disease. *Int. J. Syst. Evol. Microbiol.* 53: 1925–1929.

Kado, C. I. 2010. *Plant Bacteriology.* The American Phytopathological Society, St. Paul, MN, 336 pp.

Kageyama, B., Nakae, M., Yagi, S., and Sonoyama, T. 1992. *Pantoea punctata* sp. nov., *Pantoea citrea* sp. nov., and *Pantoea terrea* sp. nov. isolated from fruit and soil samples. *Int. J. Syst. Bacteriol.* 42: 203–210.

Kałużna, M., Willems, A., Pothier, J. F., Ruinelli, M., Sobiczewski, P., et al. 2017. Validation list no. 173. *Int. J. Syst. Evol. Microbiol.* 67: 1–3. Effective publication: Kałużna, M., Willems, A., Pothier, J. F., Ruinelli, M., Sobiczewski, P., et al. 2016. *Pseudomonas cerasi* sp. nov. (non Griffin, 1911) isolated from diseased tissue of cherry. *Syst. Appl. Microbiol.* 39: 370–377.

Khayi, S., Cigna, J., Chong, T. M., Quêtu-Laurent, A., Chan, K.-G., et al. 2016. Transfer of the potato plant isolates of *Pectobacterium wasabiae* to *Pectobacterium parmentieri* sp. nov. *Int. J. Syst. Evol. Microbiol.* 66: 5379–5383.

Kim, P. S., Shin, N.-R., Kim, J. Y., Yun, J.-H., Hyun, D.-W., et al. 2013. *Gibbsiella papilionis* sp. nov., isolated from the intestinal tract of the butterfly *Mycalesis gotama*, and emended description of the genus *Gibbsiella. Int. J. Syst. Evol. Microbiol.* 63: 2607–2611.

Kim, W. S., Gardan, L., Rhim, S. L., and Geider, K. 1999. *Erwinia pyrifoliae* sp. nov., a novel pathogen that affects Asian pear trees (*Pyrus pyrifolia* Nakai). *Int. J. Syst. Bacteriol.* 49: 899–906.

Klerks, M. M., van Gent-Pelzer, M., Franz, E., Zijlstra, C., and van Bruggen, A. H. C. 2007. Physiological and molecular responses of *Lactuca sativa* to colonization by *Salmonella enterica* serovar Dublin. *Appl. Environ. Microbiol.* 73: 4905–4914.

Kontaxis, D. G., and Hayward, A. C. 1978. The pathogen and symptomatology of pink disease of pineapple fruit in the Philippines. *Plant Dis. Reptr* 62: 446–450.

Labeda, D. P. 2011. Multilocus sequence analysis of phytopathogenic species of the genus *Streptomyces. Int. J. Syst. Evol. Microbiol.* 61: 2525–2531.

Lapage, S. P., Sneath, P. H. A., Lessel, E. F., Skerman, V. B. D., Seeliger, H. P. R., and Clark, W. A. 1975. *International Code of Nomenclature of Bacteria: 1976 Revision.* American Society for Microbiology, Washington, DC.

Lee, I. M., Bottner, K. D., Secor, G., and Rivera-Vara, V. 2006. "*Candidatus* Phytoplasma americanum," a phytoplasma associated with a potato purple top wilt disease complex. *Int. J. Syst. Evol. Microbiol.* 56: 1593–1597.

Lee, I. M., Davis, R. E., and Gundersen-Rindal, D. E. 2000. Phytoplasma: Phytopathogenic mollicutes. *Annu. Rev. Microbiol.* 54: 221–255.

Lee, I.-M., Bottner-Parker, K. D., Zhao, Y., Davis, R. E., and Harrison, N. A. 2010. Phylogenetic analysis and delineation of phytoplasmas based on *secY* gene sequences. *Int. J. Syst. Evol. Microbiol.* 60: 2887–2897.

Lee, I.-M., Bottner-Parker, K. D., Zhao, Y., Villalobos, W., and Moreira, L. 2011. "*Candidatus* Phytoplasma costaricanum" a novel phytoplasma associated with an emerging disease in soybean (*Glycine max*). *Int. J. Syst. Evol. Microbiol.* 61: 2822–2826.

Lee, I.-M., Gundersen-Rindal, D. E., Davis, R. E., Bottner, K. D., Marcone, C., et al. 2004a. "*Candidatus* Phytoplasma asteris," a novel phytoplalsma taxon associated with aster yellows and related diseases. *Int. J. Syst. Evol. Microbiol.* 54: 1037–1048.

Lee, I.-M., Martini, M., Marcone, C., and Zhu, S. F. 2004b. Classification of phytoplasma strains in the elm yellows group (16SrV) and proposal of "*Candidatus* Phytoplalsma ulmi" for the phytoplasma associated with elm yellows. *Int. J. Syst. Evol. Microbiol.* 54: 337–347.

Lehmann, K. B., and Neumann, R. 1896. *Atlas und Grundriss der Bakteriologie und Lehrbuch der speciellen bakteriologischen Diagnostik*, 1st ed., J. F. Lehmann, München, Germany.

Lelliott, R. A., and Dickey, R. S. 1984. Genus VII. *Erwinia* Winslow, Broadhurst, Buchanan, Krumwiede, Rogers, and Smith 1920. pp. 469–476 in: *Bergey's Manual of Systematic Bacteriology.* Vol. I, N. R. Krieg, and J. G. Holt eds. Williams and Wilkins, Baltimore, MD.

Li, X., Tambong, J., Yuan, K. (X.), Chen, W., Xu, H., et al. 2018. Reclassification of *Clavibacter michiganensis* subspecies on the basis of whole-genome and multilocus sequence analyses. *Int. J. Syst. Evol. Microbiol.* 68: 234–240.

Li, Y., Chang, J., Guo, L.-m., Wang, H.-M., Xie, S.-j., et al. 2015a. Description of *Acinetobacter populi* sp. nov. isolated from symptomatic bark of *Populus × euramericana* canker. *Int. J. Syst. Evol. Microbiol.* 65: 4461–4468.

Li, Y., Fang, W., Xue, H., Liang, W-x., Wang, L.-f., et al. 2015b. *Brenneria populi* sp. nov., isolated from symptomatic bark of *Populus × euramericana* canker. *Int. J. Syst. Evol. Microbiol.* 65: 432–437.

Li, Y., He, W., Wang, T., Piao, C.-g., Guo, L.-m., et al. 2014. *Acinetobacter qingfengensis* sp. nov., isolated from canker bark of *Populus × euramericana. Int. J. Syst. Evol. Microbiol.* 64: 1043–1050.

Li, Y., Xue, H., Guo, L.-m., Koltay, A., Palacio-Bielsa, A., et al. 2017. Elevation of three subspecies of *Lonsdalea quercina* to species level: *Lonsdalea britannica* sp. nov., *Lonsdalea iberica* sp. nov. and *Lonsdalea populi* sp. nov. *Intr. J. Syst. Evol. Microbiol.* 67: 4680–4684.

Lignieres, J. 1900. Maladies du porc. *Bulletin of the Society for Central Medical Veterinarians ns*, 18: 389–431.

Lincoln, S. P., Fermor, T. R., and Tindall, B. J. 1999. *Janthinobacterium agaricidamnosum* sp. nov., a soft rot pathogen of *Agaricus bisporus*. *Int. J. Syst. Bacteriol.* 49: 1577–1589.

López, M. M., Lopez-Soriano, P., Garita-Cambronero, J., Beltrán, C., Taghouti, G., et al. 2018. *Xanthomonas prunicola* sp. nov., a novel pathogen that affects nectarine (*Prunus persica* var. *nectarina*) trees. *Int. J. Syst. Evol. Microbiol.* 68: 1857–1866.

López, M. M., Roselló, M., Llop, P., Ferrer, S., Christen, R., et al. 2011. *Erwinia piriflorinigrans* sp. nov., a novel pathogen that causes necrosis of pear blossoms. *Int. J. Syst. Evol. Microbiol.* 61: 561–567.

Luo, M., Sheng, Q., Wang, C. L., and Zhang, X. L. 2018. First report of fruit spot on pepper caused by *Erwinia aphidicola* in China. *Plant Dis.* 102: 1445.

Ma, Y., Rong, C., Wang, S., Liu, Y., Wang, J., et al. 2016. Development of a species-specific real-time PCR assay for diagnosing *Pantoea beijingensis*. *J. Plant Pathol.* 98: 365–368.

Malembic-Maher, S., Salar, P., Filippin, L., Carle, P., Angelini, E., et al. 2011. Genetic diversity of European phytoplasmas of the 16SrV taxonomic group and proposal of "*Candidatus* Phytoplasma rubi." *Int. J. Syst. Evol. Microbiol.* 61: 2129–2134.

Marcone, C., Gibb, K. S., Streten, C., and Schneider, B. 2004a. "*Candidatus* Phytoplasma spartii," "*Candidatus* Phytoplasma rhamni" and "*Candidatus* Phytoplasma allocasuarinae," respectively associated with spartium witches'-broom, buckthorn witches'-broom and allocasuarina yellows diseases. *Int. J. Syst. Evol. Microbiol.* 54: 1025–1029.

Marcone, C., Schneider, B., and Seemuller, E. 2004b. "*Candidatus* Phytoplasma cynodontis," the phytoplasma associated with Bermuda grass white leaf disease. *Int. J. Syst. Evol. Microbiol.* 54: 1077–1082.

Martin, K., Schumann, P., Rainey, F. A., Schuetze, B., and Groth, I. 1997. *Janibacter limosus* gen. nov., sp. nov., a new actinomycete with *meso*-diaminopimelic acid in the cell wall. *Int. J. Syst. Bacteriol.* 47: 529–534.

Martini, M., Lee, I. M., Bottner, K. D., Zhao, Y., Botti, S., et al. 2007. Ribosomal protein gene-based phylogeny for finer differentiation and classification of phytoplasmas. *Int. J. Syst. Evol. Microbiol.* 57: 2037–2051.

Martini, M., Marcone, C., Mitrović. J., Maixner, M., Delić, D., et al. 2012. "*Candidatus* Phytoplasma convolvuli," a new phytoplasma taxon associated with bindweed yellows in four European countries. *Int. J. Syst. Evol. Microbiol.* 62: 2910–2915.

Matsuura, T., Mizuno, A., Tsukamoto, T., Shimizu, Y., Saito, N., et al. 2012. *Erwinia uzenensis* sp. nov., a novel pathogen that affects European pear trees (*Pyrus communis* L.). *Int. J. Syst. Evol. Microbiol.* 62: 1799–1803.

McLeod, M. P., Warren, R. L., Hsiao, W. W. L., Araki, N., Myhre, M., et al. 2006. The complete genome of *Rhodococcus* sp. RHA1 provides insights into a catabolic powerhouse. *Proc. Natl. Acad. Sci. USA* 103: 15582–15587.

McNeill, J., Barrie, F. R., Buck, W. R., Demoulin, V., Greuter, W., et al. eds. 2012. *International Code of Nomenclature for algae, fungi, and plants (Melbourne Code) adopted by the Eighteenth International Botanical Congress Melbourne, Australia, July 2011. Regnum Vegetabile 154.* Koeltz Scientific Books, Koenigstein, Germany.

Mergaert, J., Verdonck, L., and Kersters, K. 1993. Transfer of *Erwinia ananas* (synonym, *Erwinia uredovora*) and *Erwinia stewartii* to the genus *Pantoea* emend. as *Pantoea ananas* (Serrano 1928) comb. nov. and *Pantoea stewartii* (Smith 1898) comb. nov., respectively, and description of *Pantoea stewartii* subsp. *indologenes* subsp. nov. *Int. J. Syst. Bacteriol.* 43: 162–173.

Migula, W. 1894. Über ein neues System der Bakterien. Arb Bakteriol Inst Karlsruhe 1, 235–238 (in German).

Miyajima, K., Tanaka, F., Takeuchi, T., and Kuninaga, S. 1998. *Streptomyces turgidiscabies* sp. nov. *Int. J. Syst. Bacteriol.* 48: 495–502.

Miyazaki, A., Shigaki, T., Koinuma, H., Iwabuchi, N., Rauka, G. B., et al. 2018. "*Candidatus* Phytoplasma noviguineense," a novel taxon associated with Bogia coconut syndrome and banana wilt disease on the island of New Guinea. *Int. J. Syst. Evol. Microbiol.* 68: 170–175.

Munsch, P., Alatossava, T., Marttinen, N., Meyer, J.-M., Christen, R., et al. 2002. *Pseudomonas costantinii* sp. nov. another causal agent of brown blotch disease, isolated from cultivated mushroom sporophores in Finland. *Int. J. Syst. Evol. Microbiol.* 52: 1973–1983.

Murray, R. G. E. 1968. Microbial structure as an aid to microbial classification and taxonomy. Spisy (Faculte des Sciences de l' Univesité J. E. Purkyne Brno) 43: 249–252.

Murray, R. G. E., and Schleifer, K. H. 1994. Taxonomic notes: A proposal for recording the properties of putative taxa of procaryotes. *Int. J. Syst. Bacteriol.* 44: 174–176.

Murray, R. G. E., and Stackebrandt, E. 1995. Taxonomic note: Implementation of the provisional status *Candidatus* for incompletely described procaryotes. *Int. J. Syst. Bacteriol.* 45: 186–187.

Nabhan, S., De Boer, S. H., Maiss, E., and Wydra, K. 2013. *Pectobacterium aroidearum* sp. nov., a soft rot pathogen with preference for monocotyledonous plants. *Int. J. Syst. Evol. Microbiol.* 63: 2520–2525.

Naderali, N., Nejat, N., Vadamalai, G., Davis, R. E., Wei, W., et al. 2017. "*Candidatus* Phytoplasma wodyetiae," a new taxon associated with yellow decline disease of foxtail palm (*Wodyetia bifurcata*) in Malays. *Int. J. Syst. Evol. Microbiol.* 67: 3765–3772.

Nejat, N., Vadamalai, G., Davis, R. E., Harrison, N. A., Sijam, K., et al. 2013. "*Candidatus* Phytoplasma malaysianum," a novel taxon associated with virescence and phyllody of Madagascar periwinkle (*Catharanthus roseus*). *Int. J. Syst. Evol. Microbiol.* 63: 540–548.

Oh, E.-J., Bae, C., Lee, H.-B., Hwang, I. S., Lee, H.-I., et al. 2016. *Clavibacter michiganensis* subsp. *capsici* subsp. nov., causing bacterial canker disease in pepper. *Int. J. Syst. Evol. Microbiol.* 66: 4065–4070.

Ohl, M. E., and Miller, S. I. 2001. Salmonella: A model for bacterial pathogenesis. *Annu. Rev. Med.* 52: 259–274.

Oliveira, W. J., Silva, W. A., Silva, A. M. F., Candeia, J. A., Souza, E. B., et al. 2017. First report of *Burkholderia cenocepacia* causing sour skin of onion (*Allium cepa*) in Brazil. *Plant Dis.* 101: 1950.

Oren, A., da Costa, M. S., Garrity, G. M., Rainey, F. A., Rosselló-Móra, R. 2015. Proposal to include the rank of phylum in the International Code of Nomenclature of Prokaryotes. *Int. J. Syst. Evol. Microbiol.* 65: 4284–4287.

Paine, S. G., and Stansfield, H. 1919. Studies in bacteriosis III. A bacterial leaf-spot disease of *Protea cynaroides* exhibiting a host reaction of possibly bacteriolytic nature. *Ann. Appl. Boil.* 6: 27–39.

Palleroni, N. J. 1984. Family I, Pseudomonadaceae. pp. 144–99 in: *Bergey's Manual of Systematic Bacteriology*, 1st ed. Vol. I, N. R. Krieg, and J. G. Holt eds. Williams and Wilkins, Baltimore, MD.

Palleroni, N. J. 2005. Genus I. *Pseudomonas* Migula 1894, 237AL (nom. cons., Opin. 5 of the Jud. Comm. 1952, 121). pp. 323–79 in: *Bergey's Manual of Systematic Bacteriology.* 2nd ed. Vol. 2B, D. J. Brenner, N. R. Krieg, J. T. Staley, and G. M. Garrity eds. Springer, New York.

Park, D. H., Kim J. S., Kwon S. W., Wilson, C., Yu, Y. M., et al. 2003. *Streptomyces luridiscabiei* sp. nov., *Streptomyces puniciscabiei* sp. nov. and *Streptomyces niveiscabiei* sp. nov., which cause potato common scab disease in Korea. *Int. J. Syst. Evol. Microbiol.* 53: 2049–2054.

Parker, C. T., Tindal, B. J., and Garrity, G. M., eds. 2015. *International code of Nomenclature of Prokaryotes. Prokaryotic Code. 2008 Revision.* Published for the International Committee on Systematics of Prokaryotes. *Int. J. Syst. Evol. Microbiol.* doi:10.1099/ijsem.0.000778.

Parkinson, N., Aritua, V., Heeney, J., Cowie, C., Bew, J., et al. 2007. Phylogenetic analysis of *Xanthomonas* species by comparison of partial gyrase B gene sequences. *Int. J. Syst. Evol. Microbiol.* 57: 2881–2887.

Parkinson, N., Stead, D., Bew, J., Heeney, J., Tsror Lahkim, L., et al. 2009. *Dickeya* species relatedness and clade structure determined by comparison of *recA* sequences. *Int. J. Syst. Evol. Microbiol.* 59: 2388–2393.

Patro, T. S. S. K., Nageswara Rao, G. V., and Gopalakrishnan, J. 2006. Association of *Acinetobacter baumannii* with a top rot phase of sugarcane red stripe disease in India. *Indian Phytopathol.* 59: 501–502.

Peng, Q., Yuan, Y., and Gao, M. 2013. *Bacillus pumilus,* a novel ginger rhizome rot pathogen in China. *Plant Dis.* 97: 1308–15.

Plasencia-Márquez, O., Corzo, M., Martínez-Zubiaur, Y., Rivero, D., Devescovi, G., et al. 2017. First report of soybean (*Glycine max*) disease caused by *Pseudomonas aeruginosa* in Cuba. *Plant Dis.* 101: 1950.

Prazmowski, A. 1880. *Untersuchungen über die Entwickelungsgeschichte und Firment-wirking einiger Bacterien-Arten.* Inagural Dissertation, Hugo Voigt, Leipzig, Germany.

Pulawska, J., Willems, A., and Sobiczewski, P. 2012. *Rhizobium skierniewicense* sp. nov., isolated from tumours on chrysanthemum and cherry plum. *Int. J. Syst. Evol. Microbiol.* 62: 895–899.

Quaglino, F., Zhao, Y., Casati, P., Bulgari, D., Bianco, P. A., et al. 2013. "*Candidatus* Phytoplasma solani," a novel taxon associated with stolbur- and bois noir-related diseases of plants. *Int. J. Syst. Evol. Microbiol.* 63: 2879–2894.

Retana, K., Castro, O., Quesada, A., and Blanco, M. 2017. Etiology of stem rot on *Hylocereus* spp. caused by *Enterobacter hormaechei* in Costa Rica (Abstr.) *Phytopathology* 107: S4.18–19.

Rijckaert, C. 1994. Taxonomische studie van de *Erwinia amylovora* groep (Dye 1968). PhD thesis, University Gent, Belgium.

Roberts, R., Steenkamp, E. T., and Pietersen, G. 2015. Three novel lineages of "*Candidatus* Liberibacter africanus" associated with native rutaceous hosts of *Trioza erytreae* in South Africa. *Int. J. Syst. Evol. Microbiol.* 65: 723–731.

Roumagnac, P., Gagnevin, L., Gardan, L., Sutra, L., Manceau, C., et al. 2004. Polyphasic characterization of xanthomonads isolated from onion, garlic and Welsh onion (*Allium* spp.) and their relatedness to different *Xanthomonas* species. *Int. J. Syst. Evol. Microbiol.* 54: 15–24.

Saddler, G. S., and Bradbury, J. F. 2015. *Xanthomonas. Bergey's Manual of Systematics of Archaea and Bacteria.* 1–53.

Šafářová, D., Zemánek, T., Válová, P., and Navrátil, M. 2016. "*Candidatus* Phytoplasma cirsii," a novel taxon from creeping thistle [*Cirsium arvense* (L.) Scop.]. *Int. J. Syst. Evol. Microbiol.* 66: 1745–1753.

Safni, I., Cleenwerck, I., De Vos, P., Fegan, M., Sly, L., et al. 2014. Polyphasic taxonomic revision of the *Ralstonia solanacearum* species complex: Proposal to emend the descriptions of *Ralstonia solanacearum* and *Ralstonia syzygii* and reclassify current *R. syzygii* strains as *Ralstonia syzygii* subsp. *syzygii* subsp. nov., *R. solanacearum* phylotype IV strains as *Ralstonia syzygii* subsp. *indonesiensis* subsp. nov., banana blood disease bacterium strains as *Ralstonia syzygii* subsp. *celebesensis* subsp. nov. and *R. solanacearum* phylotype I and III strains as *Ralstonia pseudosolanacearum* sp. nov. *Int. J. Syst. Evol. Microbiol.* 64: 3087–3103.

Saglio, P., L'hospital, M., Lafleche, D., Dupont, G., Bové, J. M., et al. 1973. *Spiroplasma citri* gen. and sp. n.: A mycoplalsma-like organism associated with "Stubborn" disease of citrus. *Int. J. Syst. Bacteriol.* 23: 191–204.

Sakazaki, R. 1984. Genus IV. *Citrobacter* Werkman and Gillen 1932. pp. 458–61 in: *Bergey's Manual of Systematic Bacteriology.* I ed. Vol. 1, N. R. Krieg, and J. G. Holt eds. The Williams and Wilkins Co, Baltimore, MD.

Samson, R., Legendre, J. B., Christen, R., Saux MF-L., Achouak, W., et al. 2005. Transfer of *Pectobacterium chrysanthemi* (Burkholder et al. 1953) Brenner et al. 1973 and *Brenneria paradisiaca* to the genus *Dickeya* gen. nov. as *Dickeya chrysanthemi* comb. nov. and *Dickeya paradisiaca* comb. nov. and delineation of four novel species, *Dickeya dadantii* sp. nov., *Dickeya dianthicola* sp. nov., *Dickeya dieffenbachiae* sp. nov. and *Dickeya zeae* sp. nov. *Int. J. Syst. Evol. Microbiol.* 55: 1415–1427.

Sarfraz, S., Riaz, K., Oulghazi, S., Cigna, J., Talib, S., et al. 2018. *Pectobacterium punjabense* sp. nov., isolated from blackleg symptoms of potato plants in Pakistan. *Int. J. Syst. Evol. Microbiol.* 68: 3551–3556.

Sasaki, J., Chijimatsu, M., and Suzuki, K.-I. 1998. Taxonomic significance of 2, 4-diaminobutyric acid isomers in the cell wall peptidoglycan of actinomycetes and reclassification of *Clavibacter toxicus* as *Rathayibacter toxicus* comb. nov. *Int. J. Syst. Bacteriol.* 48: 403–410.

Sawabe, T., Makino, H., Tatsumi, M., Nakano, K., Tajima, K., et al. 1998. *Pseudoalteromonas bacteriolytica* sp. nov., a marine bacterium that is the causative agent of red spot disease of *Laminaria japonica. Int. J. Syst. Bacteriol.* 48: 769–774.

Sawada, H., Ieki, H., Oyaizu, H., and Matsumoto, S. 1993. Proposal for rejection of *Agrobacterium tumefaciens* and revised descriptions for the genus *Agrobacterium* and for *Agrobacterium radiobacter* and *Agrobacterium rhizogenes. Int. J. Syst. Bacteriol.* 43: 694–702.

Sawana, A., Adeolu, M., and Gupta, R. S. 2015. List of new names and new combinations previously effectively, but not validly, published. *Int. J. Syst. Evol. Microbiol.* 65: 2017–2025. Effective publication: Sawana, A., Adeolu, M., and Gupta, R. S. 2014. Molecular signatures and phylogenomic analysis of the genus *Burkholderia*: Proposal for division of this genus into the emended genus *Burkholderia* containing pathogenic organisms and a new genus *Paraburkholderia* gen. nov. harboring environmental species. *Front. Genet.* 5: 429.

Schaad, N. W., Postnikova, E., Lacy, G., M'Barck, F., Chang, C.-J., et al. 2004. *Xylella fastidiosa* subspecies: *X. fastidiosa* subsp. *piercei,* subsp. nov., *X. fastidiosa* subsp. *multiplex*, subsp. nov., and *X. fastidiosa* subsp. *pauca* subsp. nov. *Syst. Appl. Microbiol.* 27: 290–300.

Schaad, N. W., Postnikova, E., Sechler, A., Claflin, L. E., Vidaver, A. K., et al. 2009. List of new names and new combinations previously effectively, but not validly, published. *Int. J. Syst. Evol. Microbiol.* 59: 923–925. Effective publication: Schaad, N. W., Postnikova, E., Sechler, A., Claflin, L. E., Vidaver, A. K., et al. 2008. Reclassification of subspecies of *Acidovorax avenae* as *A. avenae* (Manns 1905) emend., *A. cattleyae* (Pavarino 1911) comb. nov., *A. citrulli* (Schaad et al. 1978) comb. nov., and proposal of *A. oryzae* sp. nov. *Syst. Appl. Microbiol.* 31: 434–446.

Schikora, A., Carreri, A., Charpentier, E., and Hirt, H. 2008. The dark side of the salad: *Salmonella typhimurium* overcomes the innate immune response of *Arabidopsis thaliana* and shows an endopathogenic lifestyle. *PLoS One* 3:e2279.

Schleifer, K. H., Kraus, J., Dvorak, C., Kilpper-Bälz, R., Collins M. D., et al. 1986. Validation list no. 20. *Int. J. Syst. Bacteriol.* 36: 354–356. Effective publication: Schleifer, K. H., Kraus, J., Dvorak, C., Kilpper-Bälz, R., Collins M. D. et al. 1985. Transfer of *Streptococcus lactis* and related streptococci to the genus *Lactococcus* gen. nov. *Syst. Appl. Microbiol.* 6: 183–195.

Schneider, B., Torres, E., Martin, M. P., Schröder, M., Behnke, H. D., et al. 2005. "*Candidatus* phytoplasma pini," a novel taxon from *Pinus silvestris* and *Pinus halepensis. Int. J. Syst. Evol. Microbiol.* 55: 303–307.

Sechler, A., Schuenzel, E. L., Cooke, P., Donnua, S., Thaveechai, N., et al. 2009. Cultivation of "*Candidatus* Liberibacter asiaticus," "*Ca.* L. africanus" and "*Ca.* L. americanus" associated with huanglongbing. *Phytopathology* 99: 480–486.

Seemuller, E., and Schneider, B. 2004. "*Candidatus* Phytoplasma mali," "*Candidatus* Phytoplasma pyri" and "*Candidatus* Phytoplasma prunorum," the causal agents of apple proliferation, pear decline and European stone fruit yellows, respectively. *Int. J. Syst. Evol. Microbiol.* 54: 1217–1226.

Shida, O., Takagi, H., Kadowaki, K., Nakamura, L. K., and Komagata, K. 1997. Transfer of *Bacillus alginolyticus, Bacillus chondroitinus, Bacillus curdlanolyticus, Bacillus glucanolyticus, Bacillus kobensis,* and *Bacillus thiaminolyticus* to the genus *Paenibacillus* and emended description of the genus *Paenibacillus. Int. J. Syst. Bacteriol.* 47: 289–298.

Skerman, V. B. D., McGowan, V., and Sneath, P. H. A. eds. 1980. Approved lists of bacterial names. *Int. J. Syst. Bacteriol.* 30: 225–420.

Su, C.-C., Deng, W.-L., Jan, F.-J., Chang, C.-J., Huang, H., et al. 2016. *Xylella taiwanensis* sp. nov., causing pear leaf scorch disease. *Int. J. Syst. Evol. Microbiol.* 66: 4766–4771.

Sugiyama, L. S., Bushe, B. C., Heller, W. P., and Keith, L. M. 2017. First report of *Paraburkholderia andropogonis* causing bacterial leaf streak of *Strelitzia reginae* in Hawaii detected from imported plants. *Plant Dis.* 101: 1030.

Surico, G., Mugnai, L., Pastorelli, R., Giovannetti, L., and Stead, D. E. 1996. *Erwinia alni*, a new species causing bark canker of alder (*Alnus* Miller) species. *Int. J. Syst. Bacteriol.* 46: 720–726.

Sutra, L., Christen, R., Bollet, C., Simoneau, P., and Gardan, L. 2001. *Samsonia erythrinae* gen. nov., sp. nov., isolated from bark necrotic lesions of *Erythrina* sp., and discrimination of plant-pathogenic Enterobacteriaceae by phenotypic features. *Int. J. Syst. Evol. Microbiol.* 51: 1291–1304.

Sutra, L., Prior, P., Perlemoine, K., Rise'de, J. M., Cao-Van, P., et al. 1999. Description of a new disease on *Erythrina* sp. in Martinique (French West Indies) and preliminary characterization of the causal agent as a novel *Erwinia* species. *Plant Pathol.* 48: 253–259.

Takahashi, K., Nishiyama, K., and Sato, M. 1996. *Pseudomonas syringae* pv. *broussonetiae* pv. nov., the causal agent of bacterial blight of paper mulberry (*Broussonetia kazinoki* × *B. papyrifera*). *Ann. Phytopathol. Soc. Japan* 62: 17–22.

Takeuchi, M., Hamana, K., and Hiraishi, A. 2001. Proposal of the genus Sphingomonas sensu stricto and three new genera, *Sphingobium*, *Novosphingobium* and *Sphingopyxis*, on the basis of phylogenetic and chemotaxonomic analyses. *Int. J. Sys. Evol. Microbiol.* 51: 1405–1417.

Takeuchi, M., Sakane, T., Yanagi, M., Yamasato, K., Hamana, K., et al. 1995. Taxonomic study of bacteria isolated from plants: Proposal of *Sphingomonas rosa* sp. nov., *Sphingomonas pruni* sp. nov., *Sphingomonas asaccharolytica* sp. nov. and *Sphingomonas mali* sp. nov. *Int. J. Syst. Bacteriol.* 45: 334–341.

Takeuchi, M., Weiss, N., Schumann, P., and Yokota, A. 1996. *Leucobacter komagatae* gen. nov., sp. nov., a new aerobic gram-positive, nonsporulating rod with 2,4-diaminobutyric acid in the cell wall. *Int. J. Syst. Bacteriol.* 46: 967–971.

Taormina, P. J., Beuchat, L. R., and Slutsker, L. 1999. Infections associated with eating seed sprouts: An international concern. *Emerg. Infect. Dis.* 5: 626–634.

Teixeira, D. C., Saillard, C., Eveillard, S. Danet, J. L., da Costa, P. I., et al. 2005. "*Candidatus* Liberibacter americanus," associated with citrus huanglongbing (greening disease) in São State, Brazil. *Int. J. Syst. Evol. Microbiol.* 55: 1857–1862.

Tian, Y., Zhao, Y., Yuan, X., Yi, J., Fan, J., et al. 2016. *Dickeya fangzhongdai* sp. nov., a plant-pathogenic bacterium isolated from pear trees (*Pyrus pyrifolia*). *Int. J. Syst. Evol. Microbiol.* 66: 2831–2835.

Timilsina, S., Minsavage, G. V., Preston, J., Newberry, E. A., Paret, M. L., et al. 2018. *Pseudomonas floridensis* sp. nov., a bacterial pathogen isolated from tomato. *Int. J. Syst. Evol. Microbiol.* 68: 64–70.

Tóth, T., Lakatos, T., and Koltay, A., 2013. *Lonsdalea quercina* subsp. *populi* subsp. nov., isolated from bark canker of poplar trees. *Int. J. Syst. Evol. Microbiol.* 63: 2309–2313.

Trebaol, G., Gardan, L., Manceau, C., Tanguy, J. L., Tirilly, Y., et al. 2000. Genomic and phenotypic characterization of *Xanthomonas cynarae* sp. nov., a new species that causes bacterial bract spot of artichoke (*Cynara scolymus* L.) *Int. J. Syst. Evo. Microbiol.* 50: 1471–1478.

Trevisan, V. 1889. *I generie le specie delle bacteriacee.* Zanaboni and Gabuzzi, Milan, Italy, pp. 1–33.

Urakami, T., Ito-Yoshida, C., Araki, H., Kijima, T., Suzuki, K.-I., et al. 1994. Transfer of *Pseudomonas plantarii* and *Pseudomonas glumae* to *Burkholderia* as *Burkholderia* spp. and description of *Burkholderia vandii* sp. nov. *Int. J. Syst. Bacteriol.* 44: 235–245.

Uribe-Lorío, L., and Wang, W. A. 2017. Identification of *Erwinia billingiae* as a causal agent of bacterial canker on mango (*Mangifera indica*) trees in Costa Rica (Abstr.) *Phytopathology* 107: S4.12–13.

Valiunas, D., Jomantiene, R., and Davis, R. E. 2013. Evaluation of the DNA-dependent RNA polymerase β-subunit gene (*rpoB*) for phytoplasma classification and phylogeny. *Int. J. Syst. Evol. Microbiol.* 63: 3904–3914.

Valiunas, D., Staniulis, J., and Davis, R. E. 2006. "*Candidatus* Phytoplasma fragariae," a novel phytoplasma taxon discovered in yellows diseased strawberry, *Fragaria* × *ananassa*. *Int. J. Syst. Evol. Microbiol.* 56: 277–281.

Van Bruggen, A. H. C., Jochimsen, K. N., and Brown, P. R., 1990. *Rhizomonas suberifaciens* gen. nov., sp. nov., the causal agent of corky root of lettuce. *Int. J. Syst. Bacteriol.* 40: 175–188.

van den Mooter, M., and Swings J. 1990. Numerical analysis of 295 phenotypic features of 266 *Xanthomonas* strains and related strains and an improved taxonomy of the genus. *Int. J. Syst. Bacteriol.* 40: 348–369.

van der Wolf, J. M., Nijhuis, E. H., Kowalewska, M. J., Saddler, G. S., Parkinson, N., et al. 2014. *Dickeya solani* sp. nov., a pectinolytic plant-pathogenic bacterium isolated from potato (*Solanum tuberosum*). *Int. J. Syst. Evol. Microbiol.* 64: 768–774.

Van Tieghem, P. 1878. Sur la gomme de sucrerie (*Leuconostoc mesenteroides*). *Annales des Sciences Naturelles Botanique* 7: 180–203.

Vaneechoutte, M., Kampfer, P., Baere, T. D., Falsen, E., and Verschraegen, G. 2004. *Wautersia* gen. nov., a novel genus accommodating the phylogenetic lineage including *Ralstonia eutropha* and related species, and proposal of *Ralstonia* [*Pseudomonas*] *syzygii* (Roberts et al. 1990) comb. nov. *Int. J. Syst. Evol. Microbiol.* 54: 317–327.

Vauterin, L., Hoste, B., Kersters, K., and Swings, J. 1995. Reclassification of *Xanthomonas*. *Int. J. Syst. Bacteriol.* 45: 472–489.

Verdin, E., Salar, P., Danet, J.-L., Choueiri, E., Jreijiri, F. et al. 2003. "*Candidatus* Phytoplasma phoenicium" sp. nov., a novel phytoplasma associated with an emerging lethal disease of almond trees in Lebanon and Iran. *Int. J. Syst. Evol. Microbiol.* 53: 833–838.

Verdonck, L. 1992. Taxonomische studie van de "soft rot" verwekkende soorten van het bacteriëngeslacht *Erwinia*. PhD. thesis, University Gent, Belgium.

Vicente, J. G., Rothwell, S., Holub, E. B., and Studholme, D. J. 2017. Pathogenic, phenotypic and molecular characterisation of *Xanthomonas nasturtii* sp. nov. and *Xanthomonas floridensis* sp. nov., new species of *Xanthomonas* associated with watercress production in Florida. *Int. J. Syst. Evol. Microbiol.* 67: 3645–3654.

Vinatzer, B. A., Weisberg, A. J., Monteil, C. L., Elmarakeby, H. A., Sheppard, S. K., et al. 2017. A proposal for a genome similarity-based taxonomy for plant-pathogenic bacteria that is sufficiently precise to reflect phylogeny, host range, and outbreak affiliation applied to *Pseudomonas syringae sensu lato* as a proof of concept. *Phytopathology* 107: 18–28.

Von Naegeli, C. 1857. In Caspary, Bericht über die Ver handlungen der 33, Ver sammlung Deutscher Naturforsches und Aertzte gehalten in Bonn von 18 bis 24 September 1857. *Artz Bot Ztg* 15: 749–792.

Waksman, S. A., and Henrici, A. T. 1943. The nomenclature and classification of actinomycetes. *J. Bacteriol.* 46: 337–341.

Waldee, E. L. 1945. Comparative studies of some peritrichous phytopathogenic bacteria. *Iowa State Coll. J. Sci.* 19: 435–484.

Wei, W., Davis, R. E., Lee, I.-M., and Zhao, Y. 2007. Computer-simulated RFLP analysis of 16S rRNA genes: Identification of ten new phytoplasma groups. *Int. J. Syst. Evol. Microbiol.* 57: 1855–1867.

Wei, W., Davis, R. E., Suo, X., and Zhao, Y. 2015. Occurrence, distribution and possible functional roles of simple sequence repeats in phytoplasma genomes. *Int. J. Syst. Evol. Microbiol.* 65: 2748–2760.

Wei, W., Lee, I.-M., Davis, R. E., Suo, X., and Zhao, Y. 2008. Automated RFLP pattern comparison and similarity coefficient calculation for rapid delineation of new and distinct phytoplasma 16Sr subgroup lineages. *Int. J. Syst. Evol. Microbiol.* 58: 2368–2377.

Wells, J. M., Raju, B. C., Hung, H.-Y., Weisburg, W. G., Mandelco-Paul, L., et al. 1987. *Xylella fastidiosa* gen. nov., sp. nov.: Gram-negative, xylem-limited, fastidious plant bacteria related to *Xanthomonas* spp. *Int. J. Syst. Bacteriol.* 37: 136–143.

Werkman, C. H., and Gillen, G. F. 1932. Bacteria producing trimethylene glycol. *J. Bacteriol.* 23: 167–82.

Whitman, W. B., Oren, A., Chuvochina, M., da Costa, M. S., Garrity, G. M., et al. 2018. Proposal of the suffix *–ota* to denote phyla. Addendum to "Proposal to include the rank of phylum in the International Code of Nomenclature of Prokaryotes." *Int. J. Syst. Evol. Microbiol.* 68: 967–969.

Willems, A., Falsen, E., Pot, B., Jantzen, E., Hoste, B., et al. 1990. *Acidovorax*, a new genus for *Pseudomonas facilis, Pseudomonas delafieldii,* E. Falsen (EF) group 13, EF group 16, and several clinical isolates, with the species *Acidovorax facilis* comb. nov., *Acidovorax delafieldii* comb. nov., and *Acidovorax temperans* sp. nov. *Int. J. Syst. Bacteriol.* 40: 384–398.

Willems, A., Gillis, M., Kersters, K., Van Den Broecke, L., and De Ley, J. 1987. Transfer of *Xanthomonas ampelina* Panagopoulos 1969 to a new genus, *Xylophilus* gen. nov., as *Xylophilus ampelinus* (Panagopoulos 1969) comb. nov. *Int. J. Syst. Bacteriol.* 37: 422–430.

Willems, A., Goor, M., Thielemans, S., Gillis, M., Kerster, K., et al. 1992. Transfer of several phytopathogenic *Pseudomonas* species to *Acidovorax* as *Acidovorax avenae* subsp. *avenae* subsp. nov., comb. nov., *Acidovorax avenae* subsp. *citrulli*, *Acidovorax avenae* subsp. *cattleyae*, and *Acidovorax konjaci*. *Int. J. Syst. Bacteriol.* 42: 107–119.

Wilson, E. E., Starr, M. P., and Berger, J. A. 1957. Bark canker, a bacterial disease of the Persian walnut tree. *Phytopathology* 47: 669–673.

Win, N. K. K., Lee, S.-Y., Bertaccini, A., Namba, S., and Jung, H.-Y. 2013. "*Candidatus* Phytoplasma balanitae" associated with witches' broom disease of *Balanites triflora*. *Int. J. Syst. Evol. Microbiol.* 63: 636–646.

Winslow, C. E. A., Broadhurst, J., Buchanan, R. E., Krumwiede, C. J., Rogers, L. A., et al. 1920. The families and genera of the bacteria. Final report of the Committee of the Society of American Bacteriologists on characterization and classification of bacterial types. *J. Bacteriol.* 5: 191–229.

Woese, C. R., Kandler, O., and Wheelis, M. L., 1990. Towards a natural system of organisms: Proposal for the domains *Archaea*, *Bacteria* and *Eubacteria*. *Proc. Natl. Acad. Sci. USA* 87: 4576–4579.

Wu, W. X., Liu, Y., Huang, X. Q., and Zhang, L. 2017. First report of summer canker caused by *Pectobacterium carotovorum* subsp. *actinidiae* on kiwifruit in China's Sichuan province. *Plant Dis.* 101: 1540.

Yabuuchi, E., Kosako, Y., Fujiwara, N., Naka, T., Matsunaga, I., et al. 2002. Emendation of the genus *Sphingomonas* Yabuuchi et al. 1990 and junior objective synonymy of the species of three genera, *Sphingobium, Novosphingobium* and *Sphingopyxis*, in conjuction with *Blastomonas ursincola.Int. J. Syst. Evol. Microbiol.* 52: 1485–1496.

Yabuuchi, E., Kosako, Y., Naka, T., Suzuki, S., and Yano, I. 1999. Validation of publication of new names and new combinations previously effectively published outside the IJSB. *Int. J. Syst. Bacteriol.* 49: 935–936. Effective publication: Yabuuchi, E., Kosako, Y., Naka, T., Suzuki, S., and Yano, I. 1999. Proposal of *Sphingomonas suberifaciens* (van Bruggen, Jochimsen and Brown 1990) comb. nov., *Sphingomonas natatoria* (Sly 1985) comb. nov., *Sphingomonas ursincola* (Yurkov et al. 1997) comb. nov., and emendation of the genus *Sphingomonas*. *Microbiol. Immunol.* 43: 339–349.

Yabuuchi, E., Kosako, Y., Oyaizu, H., Yano, I., Hotta, H., et al. 1993. Validation of the publication of new names and new combinations previously effectively published outside the IJSB. *Int. J. Syst. Bacteriol.* 43: 398–99. Effective publication: Yabuuchi, E., Kosako, Y., Oyaizu, H., Yano, I., Hotta, H., et al. 1992. Proposal of *Burkholderia* gen. nov. and transfer of seven species of the genus *Pseudomonas* homology group II to the new genus, with the type species *Burkholderia cepacia* (Palleroni and Holmes 1981) comb. nov. *Microbiol. Immunol.* 36: 1251–1275.

Yabuuchi, E., Kosako, Y., Yano, I., Hotta, H., and Nishiuchi, Y. 1996. Validation of the publication of new names and new combinations previously effectively published outside the IJSB. *Int. J. Syst. Bacteriol.* 46: 625–26. Effective publication: Yabuuchi, E., Kosako, Y., Yano, I., Hotta, H., and Nishiuchi, Y. 1995. Transfer of two *Burkholderia* and an *Alcaligenes* species to

Ralstonia gen. nov.: Proposal of *Ralstonia pickettii* (Ralston, Palleroni and Doudoroff 1973) comb. nov., *Ralstonia solanacearum* (Smith 1896) comb. nov. and *Ralstonia eutropha* (Davis 1969) comb. nov. *Microbiol. Immunol.* 39: 897–904.

Yabuuchi, E., Yano, I., Oyaizu, H., Hashimoto, Y., Ezaki, T., et al. 1990. Validation of the publication of new names and new combinations previously effectively published outside the IJSB. *Int. J. Syst. Bacteriol.* 40: 320–321. Effective publication: Yabuuchi, E., Yano, I., Oyaizu, H., Hashimoto, Y., Ezaki, T., et al. 1990. Proposals of *Sphingomonas paucimobilis* gen. nov. and comb. nov., *Sphingomonas parapaucimobilis* sp. nov., *Sphingomonas yanoikuyae* sp. nov., *Sphingomonas adhaesiva* sp. nov., *Sphingomonas capsulata* comb. nov., and two genospecies of the genus *Sphingomonas. Microbiol. Immunol.* 34: 99–119.

Yamada, K., and Komagata, K. 1972. Taxonomic studies on coryneform bacteria V. Classification of coryneform bacteria. *J. Gen. Appl. Microbiol.* 18: 417–431.

Yang, G., Han, L., Wen, J., and Zhou, S. 2013. *Pseudomonas guangdongensis* sp. nov., isolated from an electroactive biofilm, and emended description of the genus *Pseudomonas* Migula. 1894. *Int. J. Syst. Evol. Microbiol.* 63: 4599–4605.

Yoon, J.-H., Lee, H.-B., Yeo, S.-H., and Choi, J. E. 2004. *Janibacter melonis* sp. nov., isolated from abnormally spoiled oriental melon in Korea. *Int. J. Syst. Evol. Microbiol.* 54: 1975–1980.

Young, J. M. 2008. An overview of bacterial nomenclature with special reference to plant pathogenic bacteria. *Syst. Appl. Microbiol.* 31: 405–424.

Young, J. M., Bull, C. T., De Boer S. H., Firrao, G., Saddler, G. E., et al. 2004. Names of plant pathogenic bacteria, 1864–2004. Online [http://www.isppweb.org/names_bacterial_ revised.asp]

Young, J. M., Dye, D. W., Bradbury, J. F., Panagopoulos, C. G., and Robbs, C. F. 1978. A proposed nomenclature and classification for plant pathogenic bacteria. *N. Z. J. Agric. Res.* 21: 153–177.

Young, J. M., Kuykendall, L. D., Martinez-Romero, E., Kerr, A., and Swada, H. 2001. A revision of *Rhizobium* Frank 1889, with an emended description of the genus, and the inclusion of all species of *Agrobacteriun* Conn 1942 and *Allorhizobium undicola* de Laudie et al. 1998 as new combinations: *Rhizobium radiobacter, R. rhizogenes, R. rubi, R. undicola*, and *R. vitis. Int. J. Syst. Evol. Microbiol.* 51: 89–103.

Young, J. M., Saddler, G. S., Takikawa, Y., De Boer, S. H., Vauterin, L., et al. 1996. Names of plant pathogenic bacteria 1864–1995. *Rev. Plant Pathol.* 75: 721–763.

Young, J. M., Wilkie, J. P., Park, D. C., and Watson, D. R. W. 2010. List of new names and new combinations previously effectively, but not validly, published. *Int. J. Syst. Evol. Microbiol.* 60: 1009–10. Effective Publication: Young, J. M., Wilkie, J. P., Park, D. C. and Watson, D. R. W. 2010. New Zealand strains of plant pathogenic bacteria classified by multilocus sequence analysis; proposal of *Xanthomonas dyei* sp. nov. *Plant Pathol.* 59: 270–281.

Zgurskaya, H. I., Evtushenko, L. I., Akimov, V. N., and Kalakoutskii, L. V. 1993. *Rathayibacter* gen. nov., including the species *Rathayibacter rathayi* comb. nov., *Rathayibacter tritici* comb. nov., *Rathayibacter iranicus* comb. nov., and six strains from annual grasses. *Int. J. Syst. Bacteriol.* 43: 143–149.

Zhang, A. H., Zhang, X. X., Lei, F. J., and Zhang, L. X. 2018. First report of bacterial soft rot of ginseng caused by *Pseudomonas qessardii* in Jilin Province of China. *Plant Dis.* 102: 437.

Zhang, R. Y., Zhao, S. X., Tan, Z. Q., and Zhu, C. H. 2017. First report of bacterial stem rot disease caused by *Paenibacillus polymyxa* on *Hylocereus undulatus* in China. *Plant Dis.* 101: 1031.

Zhao, Y., Sun, Q., Wei, W., Davis, R. E., Wu, W., et al. 2009a. "*Candidatus* Phytoplasma tamaricis," a novel taxon discovered in witches'-broom-diseased salt cedar (*Tamarix chinensis* Lour.). *Int. J. Syst. Evol. Microbiol.* 59: 2496–2504.

Zhao, Y., Wei, W., Lee, I.-M., Shao, J., Suo, X., et al. 2009b. Construction of an interactive online phytoplasma classification tool, *i*PhyClassifier, and its application in analysis of the peach X-disease phytoplasma group (16SrIII). *Int. J. Syst. Evol. Microbiol.* 59: 2582–2593.

Zhu, B., Lou, M.-M., Xie, G.-L., Wang, G.-F., Zhou, Q., et al. 2011. *Enterobacter mori* sp. nov., associated with bacterial wilt on *Morus alba* L. *Int. J. Syst. Evol. Microbiol.* 61: 2769–74.

Zopf, W. 1891. Über Ausscheidung von Fellfarbstoffen (Lipochromen) seitens gewisser Spattpilze. *Berichte der Deut Schen Botanischen Gesellschaft.* 9: 22–28.

Zreik, L., Bove, J. M., and Garnier, M. 1998. Phylogenetic characterization of the bacterium-like organism associated with marginal chlorosis of strawberry and proposition of a "*Candidatus*" taxon for the organism, "*Candidatus* Phlomobacter fragariae." *Int. J. Syst. Bacteriol.* 48: 257–261.

Zreik, L., Carle, P., Bové, J. M., and Garnier, M. 1995. Characterization of the mycoplasma like organism associated with witches'-broom disease of lime and proposition of a "*Candidatus*" taxon for the organism, "*Candidatus* Phytoplama aurantifolia." *Int. J. Syst. Bacteriol.* 45: 449–453.

5 Molecular Mechanisms of Virulence and Pathogenesis

Gene activation is essential for plant pathogens to sense, swim towards, colonize, and infect their hosts. Many plant pathogenic bacteria recognize the presence of their plant hosts through the perception of plant generated chemical signals. The signals perceived by the bacterial cells are transduced through the bacterial membranes to initiate specific motility events. The pathogen reaches its host plant following the chemical gradient of the signal. The movement of the bacterial cells through the gradient occurs either through flagella or pili. Various chemicals serve as specific signals and are recognized by the bacteria. *Agrobacterium tumefaciens* recognizes methoxyphenols, namely, sinapinic and cinnamic acids, acetosyringone and coniferyl alcohol, derived from the lignin biosynthetic pathway produced by the plant. A transmembrane-docked histidine kinase called VirA assists in perceiving the methoxyphenols potentiated by simple sugars and low pH and in activating the response regulator VirG in this bacterium. Yuan et al. (2008) reported that acidic environment around 5.5 pH have been shown to induce as many as 78 genes and to repress 74 genes in *Ag. tumefaciens*. Venturi and Fuqua (2013) reviewed the knowledge in chemical signaling that takes place between plants and plant pathogenic bacteria.

5.1 QUORUM SENSING

Plant pathogens not only recognize plant generated chemical signals, but they also recognize self-generated signals. Quorum sensing (dependent on cell population size), an ability to communicate and coordinate behavior via signaling molecules is widespread among both Gram-stain-positive and Gram-stain-negative bacteria. A given species may possess more than one quorum sensing (QS) system. In this process, as the bacteria grow, they regulate gene expression by sensing the concentration of their own diffusible signal molecule. Each bacterial cell produces a small amount of a QS signal molecule that becomes sufficiently concentrated as the cell population increases. There are certain traits that are only expressed when bacteria are crowded together. Traits regulated in phytopathogenic bacteria by QS include the production of extracellular polysaccharides, degradative enzymes, antibiotics, siderophores, and pigments, as well as hypersensitive reaction and pathogenicity protein secretion, Ti-plasmid transfer, motility, biofilm formation, and epiphytic fitness. As QS regulatory systems are often required for pathogenesis, interference with QS signaling may offer a means of controlling bacterial diseases of plants. A brief account of the QS mechanisms operating in some important bacterial plant pathogenic species is given below.

In Gram-stain-negative bacteria, the best characterized signal molecules are the *N*-acyl-homoserine lactones (AHLs). QS systems are tuned to preferentially synthesize, and respond to, specific AHLs having an acyl side group that may range in length from four to six carbons, may contain a degree of unsaturation, and may be modified with a carbonyl or hydroxyl functional group at carbon position 3. AHL is used and produced by many Gram-stain-negative plant pathogenic bacteria and endosymbionts, including *Pseudomonas syringae* pathovars, *Xanthomonas campestris* pathovars, *Ag. tumefaciens*, *Pantoea stewartii* subsp. *stewartii*, *Pectobacterium carotovorum*, *Erwinia amylovora*, and *Tatumella citrea*.

Considerable information is available on the molecular and structural basis of QS control of Ti-plasmid conjugal transfer in *Ag. tumefaciens*. In the nopaline strain C 58, ApoTraR appears to be plasma membrane-associated, but disassociates from the membrane when complexed with AHL. AHL-TraR dimerizes and positions the C-terminal helix-turn-helix domain to attach to the major groove of the 18-bp *tra*-box recognition site; *tra*-boxes are located immediately upstream of the *traAFB*, *traCDG*, *traI-trb*, and *rep* operons. The activation of *traI-trb* operon by TraR results in creation of a positive feedback loop for *traI*. TraR activity also stimulates *traM* gene expression, which encodes a small antiactivator with affinity for the C-terminal domain of TraR. It is assumed that TraM titrates TraR to prevent Ti-plasmid transfer until TraR levels exceed than those of TraM.

In young tumors, sufficient nutrients are present, which favor rapid growth of colonizing agrobacteria over Ti-plasmid maintenance. At this stage, the galls contain a significant population of avirulent agrobacteria due to plasmid loss, and these bacteria appear to have a definite growth advantage. At later stages of gall formation, nutrients may be limiting and the bacterial cell density may be sufficiently high to trigger QS to result in Ti-plasmid transfer to more cells in the galls. These cells then would be capable of causing new infections. The higher plasmid copy number would also increase the expression of the opine catabolic genes resulting in more efficient use of the available opines.

QS in *Pe. carotovorum* subsp. *carotovorum* controls the population density-dependent expression of pathogenicity factors such as extracellular enzymes and the hypersensitive response and pathogenicity (Hrp) secretion system, as well as carbapenem antibiotic production. The primary AHL, 3-oxoC6HL, is produced by the LuxI-like signal synthase CarI. Mutants defective in *carI* do not produce carbapenem, pectolytic enzymes, endoglucanases, and proteases, and fail to secrete harpin. Therefore, they are completely nonpathogenic. The Car QS system directly regulates the genes for the production of carbapenem, which constitute the *carA-H* biosynthetic operon. The QS mechanism of *Pectobacterium* species causing soft rot has been called the "**mob attack**" mechanism of pathogenesis. The bacteria avoid production of pectic enzymes in early stages of infection possibly to avoid triggering a defense response. Early production of cell wall degrading enzymes can release pectic wall fragments that can elicit host defense responses prematurely. The pathogen thus requires building a sufficient population before risking detection. In an experiment, *carI* was introduced into transgenic tobacco plants. These plants produced 3-oxoC6HL and presumably activated the *Pe. carotovorum* subsp. *carotovorum* to produce pectic enzymes early in infection leading to triggering of a defense response. This increased resistance of transgenic plants to the pathogen suggests that such an early enzyme production might be harmful to the pathogen. The production of carbapenem late in infection avoids the tissue colonization by other bacteria and thus protects the food supply of the pathogen.

The QS signaling molecule *N*-(3-oxo-hexanoyl)-L-homoserine lactone, produced by *Pe. carotovorum* subsp. *carotovorum* is required for the production of a β-lactam antibiotic called carbapenem and for the production of pectin lyases, cellulases, and proteases that are required for efficacious survival and for optimal virulence (McGowan et al. 2005). Cui et al. (2005) reported that *Pe. carotovorum* subsp. *carotovorum* encodes ExpR, a homolog of LuxR that activates the expression of a global negative regulator that binds RNA.

Ralstonia solanacearum causes vascular wilt of many solanaceous and a few non-solanaceous plants. After entering the vascular system, the bacterium colonizes the entire plant, and its population exceeds 10^{10} cells cm^{-1} length of stem tissue. Most of its traits related to infection and virulence are regulated by the Phc (*ph*enotype *c*onversion) regulatory system in a population density-dependent manner. PhcA, a LysR-type transcriptional regulator is the most important component of a complex regulatory system, and its activity is regulated by a unique, volatile signal molecule, 3-OH palmitic acid methyl ester (3-OH PAME). *phcA* mutants do not produce extracellular polysaccharide (EPS), pectin methyl esterase, and endoglucanase, but they have an increased motility and higher production of polygalacturonase and siderophore. 3-OH PAME acts as a QS signal and is used to sense population density and to regulate the production of EPS and degradative enzymes such as pectin methyl esterase, endoglucanase, polygalacturonase, etc. Low levels of 3-OH PAME lead to reduced EPS and exoenzyme synthesis, but increased levels of 3-OH PAME promote PhcA activity, resulting in enhanced production of EPS and exoenzymes, but decreased motility and siderophore synthesis. Thus, the Phc regulatory system serves as a master control switch to turn off behaviors suited to free-living survival of the bacterium and to turn on these when needed for initial host contact, microcolony/biofilm formation, and pathogenesis.

Pa. stewartii subsp. *stewartii* causes Stewart's wilt of sweet corn and maize. After entering the plant, the bacterium grows in the apoplast (intercellular spaces) of young leaves causing "water-soaked" lesions and colonizes the xylem vessels leading to subsequent wilting. The accumulation of large amounts of EPS on pit membranes causes vascular occlusion. EPS synthesis in *Pa. stewartii* subsp. *stewartii* strain DC283 is regulated in part by the EsaI/EsaR QS system. The AHL synthase, EsaI, catalyzes the production of 3-oxoC6HL and small amounts of 3-oxoC8HL. The *esaI* gene is constitutively expressed and is not subjected to EsaR-mediated autoregulation. Mutations in the *esaI* gene cause pleiotropic effects, eliminating ALH production, EPS synthesis, and virulence, while the mutations in *esaR* gene result in constitutive growth-independent hypermucoidy. In contrast, the wild-type strain produces EPS in a population density-dependent manner, with measurable levels detected primarily at population densities $>10^8$ cells mL^{-1}. The inoculation of *esaI* mutant strain on sweet corn seedlings fails to induce wilting and causes only slight water-soaking as this strain has a functional copy of EsaR that represses EPS synthesis and Hrp-mediated water-soaking. In contrast, the inoculation of wild-type strains results in the death of seedlings. The *esaR* mutant and *esaI* and *esaR* double-deletion mutants are significantly less virulent than the wild-type strains in spite of the fact that they produce large amounts of EPS and express *hrp* genes. These findings clearly show that QS control of pathogenicity genes is important. The reduced virulence of the mutants indicates that early production of some virulence factors such as EPS may be counterproductive unless they are produced at the proper time and location of infection. *In vitro* studies with the wild-type and QS mutants showed that EPS synthesis interferes with the ability of the pathogen to attach to plastic surfaces. The same attachment deficiency could be artificially induced in the *esaI* mutant strain through exogenous addition of AHL, the degree of attachment being inversely proportional to the amount of AHL supplied. In plant essays, the similar behavior for bacterial attachment to plants was observed; the *esaI* mutants remained localized at the site of infection, whereas the wild-type strains moved at an impressive rate through the vascular system. Another species of *Pantoea*, *Pa. agglomerans* pv. *gypsophilae*, the causal agent of gypsophyla galls, produces C4HL.

Er. amylovora harbors *luxI* and *lusR* homologs of *expI* and *expR*, *hslI* and *hslR* genes of *Pec. carotovorum*, and *expI* and *expR* genes of *Dickeya chrysanthemi*. These genes are used for AHL synthesis and activation in these plant pathogens (Molina et al. 2005). AHL in *Er. amylovora* regulates the expression of virulence effectors and subsequent symptom development.

Cheng et al. (2018) studied the interactions between AHL producing transgenic tobacco plants and *Psm. syringae* pv. *tabaci* 11528. Both a reduction in disease incidence and decrease in the growth of bacterium were observed in AHL producing plants compared to wild-type plants, indicating that plant produced AHLs enhance disease resistance against this pathogen. Subsequent RNA sequencing analysis showed that the exogenous addition of AHLs up-regulated the expression of *Psm. syringae* pv. *tabaci* 11528 genes for flagella production. It is evident that plant produced AHLs activated a wide spectrum of defense responses in plants following inoculation, including the oxidative burst, hypersensitive response, cell wall strengthening, and the production of certain metabolites. Exogenous AHLs alter the gene expression patterns of pathogens, while the plant produced AHLs, either directly or indirectly, enhance plant local immunity during the early stage of plant infection.

Xa. campestris requires the QS gene cluster *rpf* that encodes sensors of the diffusible signal factor and is required for virulence. The HD-GYP domain regulator RpfG in this organism regulates the synthesis of extracellular virulence factors and negatively regulates biofilm formation (Dow et al. 2003). The recognition of the diffusible signal factor is somehow coordinated with the degradation of cyclic di-GMP via a two-component RpfC (histidine kinase) and the RpfG response regulator (Ryan et al. 2006).

Xa. oryzae pv. *oryzae* does not produce AHL, but it produces a modular protein called OryR that has an AHL domain and a helix-turn-helix domain, both of which are typical of the LuxR-family subgroup of QS regulators (Fuqua et al. 2001; Ferluga et al. 2007). Rice signal molecule that is present in high concentration in the xylem sap of rice plants solubilizes OryR protein (Ferluga and Venturi 2009). The chemical structure of rice signal molecule is unknown, but the molecular size is less than 1 kDa.

Counteracting of QS signals: Plant pathogenic bacteria depend greatly on QS regulation to coordinate their colonization and infection of plant hosts. A significant percentage of soil bacteria do produce enzymes for degrading or inactivating AHLs. The best studied of these enzymes is AiiA, a lactonase from *Bacillus cereus* that opens the lactone ring of all AHL tested, reducing their potency as signals by about 1000-fold. Interestingly, when AiiA was expressed in transgenic tobacco and potato plants, the plants showed a high degree of resistance to *Pe. carotovorum* subsp. *carotovorum* infection, which depends heavily on 3-oxo-C6-HSL-mediated QS for pathogenicity. Therefore, destruction or inactivation of AHL signals *in situ* could be a useful strategy for engineering disease resistant plants.

The information available so far suggests that AHLs are not degraded substantially by plant enzymes, at least in the rhizosphere of soil-grown plants. However, plants and animals have innate defense mechanisms for disrupting bacterial QS by producing compounds that mimic AHL signals (Givskov et al. 1996; Gao et al. 2003). Available information shows that some higher plants, namely, pea, rice, tomato, soybean, and *Medicago truncatula* secrete various compounds that act like (mimic) bacterial QS signals. Such "signal-mimic" compounds may provide plants with important tools to disrupt or manipulate QS regulation in associated bacteria. First of all, QS signal-mimic compounds were found in marine red alga, *Delisera pulchra*. The active compounds were a set of halogenated furanones that are structurally similar to AHLs and interact specifically with AHL receptors. These compounds also inhibit AHL-mediated QS in many bacteria, interfere with biofilm formation, and substantially alter the structure of natural bacterial communities on the algal surface in marine environments. Higher plants also produce signal-mimic compounds that can inhibit bacterial responses to an added AHL signal. However, in contrast to the strictly inhibitory activities of the furanone mimics of the *D. pulchra*, many of the signal mimics from higher plants stimulate AHL-induced QS behaviors (Von Bodman et al. 2003).

Williams et al. (2004) reported that the AHL signals of *Psm. aeruginosa* enter the eucaryotic cells and activate transcription factors. The chemical cross talk between plants and their associated bacteria appears to reflect a highly evolved and complicated biological system that seems to benefit both the parties of the interaction. Therefore, it is clear that QS is used to synchronize the production of virulence factors in pathogens utilizing this signaling system. More than 350 genes have been observed to be regulated in *Psm. aeruginosa* by QS (Dong et al. 2005).

5.2 ROLE OF PATHOGENICITY FACTORS IN DISEASE DEVELOPMENT

5.2.1 Bacterial Secretion Systems

Most determinants of pathogenicity need to be secreted to interact with host cells. The successful infection by plant pathogenic bacteria utilizes various influx and efflux systems as well as enzyme- and effector-protein translocation pathways. The influx systems are used for the uptake of essential components needed for survival, which are usually produced by the enzymatic breakdown of complex substrates into smaller compounds accessible by the influx system. The efflux systems protect the bacterial cells from toxic compounds and antimetabolites, including pH differences and induced innate host defense compounds, such as hydrogen peroxide, nitric oxide, phytoalexins, and phenolics.

Six secretion systems, i.e., I to VI are found in Gram-stain-negative bacteria, which deliver various chemicals, including proteins and nucleic acids, to perform a multitude of biochemical and physiological functions. The secretion systems of plant pathogenic bacteria play a major role in their pathogenicity as they transport bacterial proteins and other molecules, required for pathogenicity, into the host cells. As the secretion machineries of these secretion systems utilize the inner and outer membranes of the bacterial envelop, therefore, only the Gram-stain-negative bacteria can be equipped with these secretion pathways. *Pe. atrosepticum* strain 1043 contains genes encoding all the six secretion systems, suggesting a complex secretome and an impressive ability to interact with its environment(s). All the six secretion systems are found in plant pathogenic bacteria.

5.2.1.1 Type I Secretion System

Type I secretion involves protein translocation from the cytoplasm across the inner and outer membranes without any interaction in the periplasm. The type I export mechanism requires only three proteins and does not generate periplasmic intermediates. The type I secretion apparatus is composed of membrane ATPase (the ATP-binding cassette [ABC] protein), a membrane fusion protein that belongs to the membrane fusion protein family, and an outer membrane peptide. The ABC transporter containing these membrane proteins can translocate a wide variety of substrates across extra- and intra-cellular membranes. Polypeptides that are secreted via the type I secretion apparatus are transferred from the cytosol, without a periplasmic intermediate, through the inner and outer membranes out into the extracellular environment. The secreted polypeptides can be toxin or enzymes, such as proteases of *D. chrysanthemi* and *Psm. aeruginosa*. Based on genome sequence analysis, *Ag. tumefaciens*, *Leifsonia xyli*, *Pe. carotovorum*, *Ral. solanacearum*, *Xylella fastidiosa*, *Xa. oryzae* pv. *oryzae*, *Xa. campestris* and pathovars of *Xa. Axonopodis,* and *Psm. syringae* are equipped with type I secretion systems (TISSs).

It is present nearly in all plant pathogenic bacteria. It secretes proteins of various sizes, often with calcium-binding glycine rich repeats, from the cytoplasm to the extracellular matrix in a single step. Various toxins like hemolysins, rhizobiocin, and cyclolysin are also secreted. They consist of ATP-binding cassette proteins and carry out in and out movement of a variety of compounds for which the energy is provided by the hydrolysis of ATP. In the tribe erwiniae, the TISS is required for secretion of proteases, which apparently play only a minor role in pathogenesis. The genome sequence of *Pe. atrosepticum* strain 1043 has revealed additional TISSs. On the whole, TISS plays a minor role in pathogenicity.

5.2.1.2 Type II Secretion System

The type II secretion system (TIISS), conserved among Gram-stain-negative bacteria, is characterized by a two-step protein translocation process. The first step is protein processing through the inner (cytoplasmic) membrane followed by protein modifications to exit the outer membrane. The major route of protein translocation across the inner membrane is secretion pathway, to which substrate are targeted by amino-terminal signal peptides. For more details of the system, the readers may refer to Sandkvist (2001) and Kado (2010).

It is common among Gram-stain-negative bacteria and shares many features with the type IV pilus system. TIISSs are found in several phytopathogens, including xanthomonads, pseudomonads, *Xyl. fastidiosa*, *Ral. solanacearum,* and the tribe erwiniae. In *Xa. campestris* pv. *campestris* and *Xa. oryzae* pv. *oryzae,* it is involved in the secretion of plant cell wall-degrading enzymes (PCWDEs), polysaccharide, and proteases, and hence plays an important role in virulence. TIISS has been studied in detail in soft rot erwiniae (SRE). It is responsible for the secretion of PCWDEs and is, therefore, essential for soft rot disease development. TIISS is involved in the translocation of various virulence factors including toxins, enzymes, and proteins. The translocation of proteins involves a two-step process. In the first step, unfolded proteins move to the periplasm via the secretion (Sec) pathway across the inner membrane. In the second step, processed and folded proteins move through the periplasm and across the outer membrane via an apparatus consisting of 12–14 proteins encoded by a cluster of genes. *Xanthomonas* and *Ralstonia* species possess two TIISS per cell and use them for secretion of virulence factors such as pectinolytic and cellulolytic enzymes. *Xylella* and *Agrobacterium* species possess only one TIISS per cell. *Agrobacterium* species have the genes only for the first step of protein transport across the inner membrane, and these use the type IV secretion system (TIVSS) for the remaining process. TIISS is regulated in part by a quorum-sensing mechanism (Agrios 2004).

5.2.1.3 Type III Secretion System

In plant pathogens, type III secretion apparatus is encoded by *hrp* genes. As the type III secretion system (TIIISS) in phytopathogens was first recognized through its association with the HR, the genes encoding it have been named as *hrp* genes. The *hrp* clusters from several phytopathogenic bacteria have been thoroughly studied. The first *hrp* cluster to be characterized was that of *Psm. syringae* pv. *phaseolicola* (Lindgren et al. 1986), while for *Xanthomonas* species, the first one was of *Xa. campestris* pv. *campestris* (Arlat et al. 1991). Later on, the *hrp* clusters from *Xa. campestris* pv. *vesicatoria, Xa. oryzae* pv. *oryzae,* and *Xa. axonopodis* pv. *glycines* were also sequenced and characterized (Zhu et al. 2000; Buttner & Bonas 2002; Kim et al. 2003). The widely studied *hrp* cluster from *Xa. campestris* pv. *vesicatoria* has a region of 23 kb and contains six operons from hrpA to hrpF. Nine out of the twenty-two proteins encoded by the cluster, conserved among plant and animal pathogens, are named *hrc* (*hrp* conserved), and constitute the core of the type III translocon. Structurally, type III protein secretion system in phytopathogens is composed of a flagellum—the Hrp pilus and a basal body that crosses the two bacterial membranes and ensures the stabilization of the whole structure. This complex structure facilitates the bacterial attachment to the host cell membrane to ensure the translocation of effector proteins to the interior of the plant cell (Hueck 1998; Tampakaki et al. 2004).

There are a number of genes in the *hrp* cluster, which show a high level of similarity to those in the animal pathogens, and these have been termed *hrp* conserved (*hrc*) genes. The *hrp* cluster in the soft rot erwiniae has been demonstrated to play an important role in pathogenesis. *Er. amylovora* (*Eam*), a close relative of soft rot erwiniae, has been best studied of the erwiniae in terms of TIIISS and its effectors. Three TIIISS secreted proteins, namely, $DspE/A_{Eam}$, $HrpN_{Eam}$ (harpin), and $HrpW_{Eam}$ have been characterized in *Er. amylovora.* $DspE/A_{Eam}$ appears to be widely conserved in many plant pathogens and is currently the only demonstrated effector protein in the erwiniae. $HrpW_{Eam}$ and $HrpN_{Eam}$, on the other hand, have been termed helper proteins that may assist translocation of effectors (such as DspE/A), but are believed to remain outside the host cell (Alfano and Collmer 2004).

A number of plant pathogenic bacteria use one or more TIIISSs to alter host cell machinery and cause disease. TIIISSs are found in a variety of bacterial pathogens (Desvaux et al. 2006), including major human pathogens, such as *Salmonella*, *Shigella*, *Yersinia*, and *Pseudomonas* species and enteropathogenic *Escherichia coli*; and are composed of at least 40 genes, making their temporal and spatial regulation very complex. The optimum temperature and pH are critical for the functioning of TIIISS. It plays a most important role in the pathogenicity of bacteria belonging to the genera *Xanthomonas*, *Pseudomonas*, and *Ralstonia*. Its primary function is the transport of effector proteins across the bacterial membrane and further into the plant cell. In addition to the above-mentioned three genera, TIIISS also occurs in most Gram-stain-negative bacteria (*Erwinia* and *Pantoea* species), including those causing diseases in humans and animals. The translocation of the bacterial effector proteins into the host cells interferes with host cell signal transduction and other cellular processes, leading to the enhancement of bacterial virulence in the susceptible hosts. Without the TIIISS, phytopathogens cannot defeat basal defenses of the host, grow in plants, produce disease symptoms on host, or elicit hypersensitive response in non-hosts. TIIISS mutants fail to suppress the basal plant defenses. Some effectors also act as virulence proteins that are recognized by corresponding host resistance (R) proteins, triggering the HR that restricts pathogen spread. Alfano and Collmer (2004) have presented a conceptual overview of TIIISS effectors functioning as both elicitors and suppressors of plant defenses.

The type III secretion apparatus is usually encoded by a set of genes often on a plasmid or within a chromosomal pathogenicity island. The secretion apparatus is composed of a macromolecular complex spanning both the inner and outer membranes. All TIIISSs mediate injection of virulence proteins, called type III effectors as well as nodulation outer proteins.

Psm. syringae utilizes the TIIISS to inject effector proteins into plant host cells. The bacterium has a wide host range, and this wide host range is caused, in part, due to the diversity of effectors employed by this phytopathogen. More than 60 different effector families exist in *Psm. syringae,* and one such family is HopF, which contains over 100 distinct alleles. However, despite this diversity, researchers have focused on only two members of this family, i.e., HopF1 from *Psm. syringae* pv. *phaseolicola* 1449B and HopF2 from *Psm. syringae* pv. *tomato* DC3000. Lo et al. (2017) have reviewed the work done on HopF family members, including their host targets and molecular mechanisms of immunity suppression, and their enzymatic function; and have also provided a phylogenetic analysis of this expanding effector family which provides a basis for a proposed nomenclature to guide the future research. The extensive genetic diversity within the HopF family presents a great opportunity to study how functional diversification of an effector family contributes to host specialization.

Chakravarthy et al. (2018) constructed *Psm. syringae* pv. *tomato* DC3000 derivatives, differentially producing coronatine (COR), the TIIISS machinery, and subsets of key effectors, and assayed in leaves of *Nicotiana benthamiana*. The inoculations were made by dipping the whole plants in bacterial cell suspension and the plants assayed for bacterial population and the production of chlorotic spots on leaves. The strains fell into three classes. Class I strains were TIIISS$^+$, but functionally effectorless, grew poorly *in planta,* and produced faint chlorotic spots only if COR$^+$. Class II strains were TIIISS$^-$, or if TIIISS$^+$, produced effectors avirulence (Avr) PtoB and HopM1. However, class II strains grew better than class I strains *in planta* and, if COR$^+$, also produced robust chlorotic spots. Class III strains were TIIISS$^+$ and minimally produced AvrPtoB, HopM1, and three other effectors encoded in the *Psm. syringae* conserved effector locus. These strains grew better than class II strains *in planta* and produced chlorotic spots without COR if the precursor coronafacic acid was produced. COR made no significant contribution to the bacterial colonization of the apoplast, but enabled a gratuitous, semi-quantitative, surface indication of bacterial growth, which is determined by the strain's effector composition.

Fan et al. (2017) screened 56 plant phenolic compounds and derivatives for their effects on the *Xa. oryzae* pv. *oryzae* TIIISS. Ten of these compounds significantly inhibited the promoter activity of a harpin gene, *hpa1*. Further testing of these inhibitors for their impact on the HR caused by *Xa. oryzae* pv. *oryzae* on non-host tobacco plants revealed that pre-treatment of *Xa. oryzae* pv. *oryzae* with TS006 (*o*-coumaric acid, OCA), TS010, TS015, and TS018 resulted in significantly attenuated HR without affecting bacterial growth or survival. Additionally, Cya translocation assays demonstrated that the translocation of two TIII effectors was suppressed by the four inhibitors. Quantitative reverse transcription-polymerase chain reaction analysis showed that mRNA levels of representative genes in the *hrp* cluster, as well as the regulatory genes *hrpG* and *hrpX*, were reduced by treatment with the four inhibitors, suggesting that expression of the *Xa. oryzae* pv. *oryzae* TIIISS was suppressed. The expression of other virulence factors was not suppressed, which indicated possible TIIISS-specific inhibition. These inhibitors reduced the disease symptoms of *Xa. oryzae* pv. *oryzae* and *Xa. oryzae* pv. *oryzicola* on the rice cultivar IR24 to varying extents.

Bismerthiazol provides an excellent control of bacterial blight of rice caused by *Xa. oryzae* pv. *oryzae*, but does not greatly inhibit *in vitro* growth of the bacterium. According to RNA-sequencing analysis, the transcription of the histidine utilization (Hut) pathway genes of *Xa. oryzae* pv. *oryzae* ZJ173 was inhibited after 4.5 and 9.0 h of bismerthiazol treatment. Functional studies of *hutG* and *hutU* indicated that the Hut pathway had little effect on the growth and bismerthiazol sensitivity of the bacterium *in vitro*, but significantly reduced the aggregation of the cells. Deletion mutants of *hutG* or *hutU* were more motile, produced less biofilm, and were less virulent than the wild-type, indicating that the Hut pathway is involved in QS and contributes to virulence. The over expression of the *hutG-U* operons in ZJ173 reduced inhibition of the bacterium by bismerthiazol. Bismerthiazol did not inhibit the transcription of Hut pathway genes, QS, or virulence of the bismerthiazol-resistant strain 2–1–1. It is evident that bismerthiazol reduces virulence of the bacterium by inhibiting the Hut pathway and QS (Liang et al. 2018).

Gottig et al. (2010) have given a detailed account of type III protein secretion system involved in citrus canker caused by *Xa. axonopodis* pv. *citri* (recent name *Xa. citri* subsp. *citri*).

Hajri et al. (2009) predicted several effector proteins to be secreted by this system in *Xa. axonopodis* pv. *citri,* and among them is the well characterized PthA, a member of the *avrBs3* gene family which has been first identified as the elicitor of the symptoms of citrus canker (Swarup et al. 1992). *Hrp* cluster is critical for the development of the disease. Dunger et al. (2005) reported that *Xa. axonopodis* pv. *citri* mutants in the *hrp* cluster, one in the gene that codifies for HrcN, the ATPase that may be providing the energy for the transport system or for the assembly of the secretion apparatus; another in an operon bearing genes, which codify for proteins of the core of the translocon (HrcQ, HrcR, and HrcS); and the last one in a gene that codifies for a pore-forming protein that mediates translocation of effector proteins into the plant cell (HrpF), showed no virulence when infiltrated in citrus leaves, showing consistency with the crucial role of this system in the pathogenesis. In contrast to the compatible interactions produced in citrus canker, *Xa. axonopodis* pv. *citri* induces non-host resistance with HR in cotton, tobacco, tomato, bean, and pepper, and a functional *hrp* cluster is required for the induction of HR on these non-host plants, since the above-mentioned mutants produced no HR lesions in these plants.

Xa. citri subsp. *citri* uses transcription activator-like effectors (TALEs) as major pathogenicity factors. TALEs, which are delivered into plant cells through TIIISS, interact with effector binding elements (*EBE*s) in host genomes to activate the expression of downstream susceptibility genes to promote disease. Predictably, TALEs bind *EBE*s in host promoters via known combinations of TALE amino acids to DNA bases, known as the TALE code.

Shantharaj et al. (2017) fused $ProBs3_{1EBE}$ and $ProBs3_{14EBE}$ to the *Xanthomonas* gene, *avrGf1*, which encodes a bacterial effector that elicits an HR in grapefruit and sweet orange and demonstrated, in transient assays, that activation of $ProBs3_{14EBE}$ by *Xa. citri* subsp. *citri* TALEs is TIIISS-dependent, and that the expression of AvrGf1 triggers HR and correlates with reduced bacterial growth. All tested virulent *Xa. citri* subsp. *citri* strains from diverse geographical locations activated $ProBs3_{14EBE}$. TALEs are essential for the virulence of *Xa. citri* subsp. *citri* strains and, because the engineered promoter traps are activated by multiple TALEs, this concept has the potential to confer broad-spectrum, durable resistance to citrus canker in stably transformed plants.

Xa. fuscans subsp. *aurantifolii* group C strains possess a type III effector, AvrGf2, belonging to the *Xanthomonas* outer protein (Xop)AG effector gene family, which restricts the host range among citrus species. Gochez et al. (2017) dissected the modular nature and mode of action of AvrGf2 in grapefruit resistance. Mutation of GPLL to AASL in AvrGf2 abolished the elicitation of the HR, while the mutation in only the first amino acid to SPLL delayed the HR in grapefruit. Yeast two-hybrid experiments showed strong interaction of AvrGf2 with grapefruit cyclophilin (GfCyp), while the AvrGf2-SPLL and AvrGf2-AASL mutants exhibited weak and no interaction, respectively. Molecular modeling and *in silico* docking studies for the cyclophilin-AvrGf2 interaction predicted the binding of citrus cyclophilins (CsCyp, GfCyp) to hexameric peptides spanning the cyclophilin-binding domain of AvrGf2 and AvrGf2 mutants (VAGPLL, VASPLL, and VAAASL) with affinities equivalent to or better than a positive control peptide (YSPSA) previously demonstrated to bind CsCyp. Additionally, the C-terminal domain of XopAG family effectors contains a highly conserved motif, CLNA × YD, which was identified to be crucial for the induction of HR based on site-directed mutagenesis (CLNA × YD to CASA × YD). These results suggest a model in which grapefruit cyclophilin promotes a conformational change in AvrGf2, thereby triggering the resistance response.

Xa. axonopodis pv. *manihotis* strain CIO151 possesses 17 predicted type III effectors (TIIIEs) belonging to the Xop class. Medina et al. (2018) characterized nine Xop effectors present in strain CIO151 for their role in virulence and modulation of plant immunity and demonstrated the importance of XopZ, XopX, XopAO1, and AvrBs2 for full virulence, as well as a redundant function in virulence between XopN and XopQ in susceptible cassava plants. Their role in pathogen-associated molecular pattern (PAMP)-triggered immunity (PTI) and effector-triggered immunity (ETI) using heterologous systems revealed that AvrBs2, XopR, and XopAO1 were capable of suppressing PTI. ETI suppression activity was only detected for XopE4 and XopAO1. The results show the overall importance and diversity in functions of major virulence effectors AvrBs2 and XopAO1 in *Xa. axonopodis* pv. *manihotis* during cassava infection.

Nissan et al. (2018) generated draft genome sequences of *Pa. agglomerans* pv. *betae* and pv. *gypsophilae* and employed these sequences in combination with a machine-learning approach and a translocation assay into beet roots to identify the pools of TIIIEs in these pathovars. Mutational analysis revealed that in pv. *betae* 4188, eight TIIIEs (HsvB, HsvG, PseB, DspA/E, HopAY1, HopX2, HopAF1, and HrpK) contribute to pathogenicity on beet and gypsophila. In pv. *gypsophilae* 824-1, nine TIIIEs (HsvG, HsvB, PthG, DspA/E, HopAY1, HopD1, HopX2, HopAF1, and HrpK) contribute to pathogenicity on gypsophila, while the PthG effector triggers HR on beet. HsvB, HsvG, PthG, and PseB appear to endow pathovar specificities to pv. *betae* and pv. *gypsophilae,* and no homologous TIIIEs were identified for these proteins in other phytopathogenic bacteria. Conversely, the remaining TIIIEs contribute to the virulence of both the pathovars, and homologous TIIIEs were found in other phytopathogenic bacteria. Remarkably, HsvG and HsvB, which act as host-specific transcription factors, contributed largely to the disease development.

Sang et al. (2018) found that the *Ral. solanacearum* type III effector RipAY suppresses plant immune responses triggered by bacterial elicitors and by the phytohormone salicylic acid. The biochemical analysis indicated that RipAY associates *in planta* with thioredoxins from *N. benthamiana* and Arabidopsis. RipAY displays γ-glutamyl cyclotransferase activity to degrade glutathione in plant cells, which is required for the reported suppression of immune responses. Seeing the importance of thioredoxins and glutathione as major redox regulators in eukaryotic cells, RipAY activity may constitute a novel and powerful virulence strategy employed by *Ral. solanacearum* to suppress immune responses and potentially alter general redox signaling in host cells.

Duan et al. (2018) obtained gene expression profile of *Ral. solanacearum* PO41 under the root secretions environment of *Solanum tuberosum* at 8, 16, and 24 h after infection, with RNA microarray technology. It was found that the virulence factors of *Ral. solanacearum* mainly focused on the output pathways of toxic protein [Sec pathway, twin-arginine translocation (Tat) pathway, and TIIISS], the aggregation and transfer of exopolysaccharides and the chemotactic movement, and adhesion of flagellum in the potato root secretion ecological niche, whereas the virulence factors in the atypical output pathway mainly distributed in Sec (*secB*, *secDF*, *yidc*) and Tat (*tatA*, *tatC*) pathways to promote the output of folded and unfolded toxic proteins. The *fliI* ATPase was obviously upregulated 8 h after inoculation, showing that type III secretion system was only active at the early stage of *Ral. solanacearum* infection. The upregulated expression of phosphoglucomutase and epimerase showed that the virulence factor of exopolysaccharides was synthesized at the early stage of bacterial infection. Chemotactic receptor and motor protein were obviously upregulated within 24 h after inoculation. It was evident that *Ral. solanacearum* PO41 had already colonized to the roots within 24 h due to the stimulation of root secretion. Some pathogenic genes were upregulated during this period.

5.2.1.4 Type IV Secretion System

This secretion system transports macromolecules between bacteria and from bacteria into their eucaryotic hosts. These transferred proteins are very similar to those which mobilize plasmids among bacteria. The *virB* operon of *Ag. tumefaciens* encodes 11 proteins that form an organized structure. These proteins are involved in the transfer of T-DNA strand from the bacterium to the cytoplasm of the plant cell. The transporting structure originating from bacterial inner membrane stretches through the outer membrane and terminates into a pilus-like structure that protrudes from the bacterial cell (Agrios 2004). Three other phytopathogenic bacteria, namely, *Pe. atrosepticum* strain 1043, *Xa. citri* subsp. *citri,* and *Xyl. fastidiosa* possess gene clusters homologous to *VirB*. The presence of *virB* cluster of *Pe. atrosepticum* strain 1043 within horizontally acquired island 7 suggests that it was acquired through horizontal gene transfer.

Type IV secretion system (TIVSS) transports proteins, nucleoproteins, and nucleic acid substrates through specific channeling complexes incorporated into the outer membrane and cell envelop. The highly conserved TIVSSs have been identified in both plant and mammalian pathogens. This secretion system can also include pathogen-based effectors that are injected into eucaryotic cells. Like the TIIISSs, TIVSSs are used by various bacterial pathogens to transfer bacterial proteins and DNA into eucaryotic cells. Among the bacterial plant pathogens, *Ag. tumefaciens* harbors the well-characterized prototypical TIVSS used to transfer the T-DNA into plant cell to cause crown gall. Genome and plasmid sequences of *Ag. tumefaciens*, *Ag. rhizogenes*, *Ag. vitis*, *Xyl. fastidiosa,* and *Xanthomonas* species have revealed the presence of type IV secretion components. Their functional presence and the role of the Ti-plasmid-based TIVSS in *Ag. tumefaciens* are relatively well characterized. *Xa. citri* pv. *citri* contains two TIVSS loci; one on the chromosome and the other on a 64-kb pXAC64 plasmid (da Silva et al. 2002). Alegria et al. (2005) reported that these loci are VirB orthologs, including aVirB7 protein discovered in *Xa. citri* pv. *citri*.

More than one type IV secretion pathway can occur in a given pathogen, and a pathogen can be also equipped with type II and type III as well as type IV secretion systems, e.g., *Burkholderia cepacia*. Genes encoding the TIVSS component can be found on plasmids and on chromosomes such as that of *Bu. cepacia*. The available information indicates that TIVSS loci are highly conserved and that more than one TIVSS has been widely dispersed among Gram-stain-negative bacteria via lateral gene transfer. *Spiroplasma kunkelii* harbors a 14.6-kb plasmid, pSKU146, which apparently encodes a TIVSS. Plasmid pSKU146 contains open reading frames that encode proteins having similarity to conjugative elements and components of the bacterial type IV secretion apparatus (Davis et al. 2005). Basically, TIVSS are used to either transfer protein effectors into host cells or to take up protein and DNA (retrotransfer) for environmental adaptation (Kado 2002). Interestingly, TIVSSs operate to move substrates in a bidirectional manner. That is, certain TIVSSs function to deliver or take up protein or nucleic acid substrates.

5.2.1.5 Type V Secretion System

Type V secretion system (TVSS), known as the autotransported protein system, is composed of proteins that are sub classified as either an autotransporter or a two-partner secretion pathway family member. Characteristic of this family is the capacity of each member to mediate its own translocation across the bacterial outer membrane. The Sec system is used to mediate translocation across the inner membrane. Autotransporter proteins use a *C*-terminal translocater domain to mediate secretion of an *N*-terminal effector domain across the outer membrane to the bacterial surface in a process requiring no accessory proteins. Proteins of two-partner secretion pathway differ from autotransporters in that the effector and translocater domains are expressed as separate proteins (Henderson et al. 2004). In contrast to type I to IV secretion systems, only a single gene encodes the secreted protein, and the complete machinery is required for secretion. TVSS is found in a wide range of Gram-stain-negative bacteria. *Xanthomonas* and *Xylella* species contain genes for TVSS (autotransporters), which encode surface associated adhesins. Two-partner system also occurs in *Pe. atrosepticum* strain 1043, *Er. amylovora,* and *D. chrysanthemi* (Toth et al. 2006).

5.2.1.6 Type VI Secretion System

In *Vibrio cholerae*, the virulence-associated secretion (VAS) system transports proteins to the exterior of the bacterial cell by a mechanism that does not require hydrophobic *N*-terminal signal sequences. Several *vas* genes products constitute the structural components of the VAS apparatus that delivers effector proteins via the type VI secretion system (TVISS) (Pukatzki et al. 2006). Several Gram-stain-negative bacterial plant pathogens and endosymbionts, such as *Ag. tumefaciens* and *Rhizobium leguminosarum,* respectively, harbor genes homologous to *vas* genes. Wu et al. (2008) reported that the TVISS is used by *Ag. tumefaciens* to secrete a hemolysin-regulated protein that is required for efficient tumorigenesis. TVISS has been found associated with secretion of hemolysins and cytolysins in *Serratia marcescens.* A homologous system is present in *Xyl. fastidiosa* and *Ral. solancearum.* It is hoped that TVISS will be discovered in other Gram-stain-negative bacterial plant pathogens also.

5.2.1.7 Twin-Arginine Translocation Protein Secretion System

Besides the six secretion pathways (types I–VI) described above, there is another pathway used by bacteria called the Tat secretion system, which is found in the cytoplasmic membranes of most bacteria and in the energy transducing-membranes of plant organelles (Berks et al. 2005; Müller and Klösgen 2005). This pathway transports proteins in a Sec-independent manner. Such proteins are synthesized with amino-terminal signal peptides that contain a distinctive invariant peptide sequence of SRR × FLK, which contains the twin arginine amino acid motif. Berks et al. (2000) reported that proteins bearing twin-arginine signal peptides are directed post-translationally to the membrane-embedded Tat secretion system that transports fully folded proteins across the inner membrane using energy provided by the transmembrane pH gradient. The core of the Tat transporter is a large oligomeric complex of TatA, TatB, and TatC proteins. TatA protein provides the protein-conducting channel within the complex through which folded proteins are transported.

Studies on a Tat secretion system in *Ag. tumefaciens* showed that inactivation of this system abolishes anaerobic respiration and chemotaxis, and attenuates virulence (Ding and Christie 2003). Ferreira et al. (2006) found substrates of Tat secretion system in *Psm. syringae* pv. *tomato.* Phospholipases were found to be substrates that are translocated across the outer membrane by a *Psm. syringae* TIISS encoded by the *gsp* operon (Bronstein et al. 2005). It indicates that the Tat secretion system and TIISS coordinate the secretion of these substrates. With intensification of work on this aspect in the future, plant pathogenic bacteria like *Psm. syringae* (Caldelari et al. 2006) and *Ral. solanacearum* (González et al. 2007) will be found to be using this pathway to deliver effectors used in virulence and pathogenesis.

5.3 CIRCUMVENTING HOST RESISTANCE BY PLANT PATHOGENIC BACTERIA

The successful plant pathogenic bacteria are equipped with an arsenal of products that help evade innate and induced host defense mechanisms. Innate immunity depends on germline-encoded receptors that have evolved to recognize highly conserved PAMPs. These receptors are called pattern recognition receptors (PRRs). Two distinct types of elicitors, namely, general elicitors and specific elicitors activate host defenses. General elicitors comprise of bacterial lipopolysaccharides, phospholipids, flagellins, peptidoglycans, proteins such as elongation factors, double stranded nucleic acids, cell wall pieces, and methylated DNA that can bind to host PRRs and trigger host defenses (Kunze et al. 2004; Ingle et al. 2006). Very specific receptors that are often host cultivar specific, perceive specific elicitors.

During the infection process, the pathogenic bacteria inject several proteins into the host tissue, via the secretion systems described above. The effector proteins, which include proteins that suppress defense reactions and promote bacterial survival, include race specific elicitor proteins encoded by *avr* genes. If the host plant carries the corresponding resistant (*R*) gene, such Avr proteins usually induce defense responses such as HR. This is known as an incompatible host-pathogen interaction, in which the pathogen produces an Avr effector protein that is recognized by a cognate *R* gene product in the resistant cultivar (Keen 1990; Martin et al. 2003). R proteins are thought to recognize the modification of the virulence target by the Avr protein. R proteins elicit HR in response to cell death suppression activity, which is only observed in the absence of the cognate R protein, is a simple model. It has been replaced by a more complex model called the **trump model** proposed by Abramovitch and Martin (2005). The trump model includes a trump (T) factor that would act on the cognate R protein to generate an HR-based immunity in response to a cell death suppressor. Certain effector proteins, such as HopPtoN, suppress cell death in both compatible and incompatible interactions (López-Solanilla et al. 2004).

Morel and Dangl (1997) reported that HR is observed in resistant plants challenged with a pathogen as well as a consequence of non-host resistance in non-host plants. This response is associated with cell death at the initial stage of infection that results in rapid localized necrosis that initially involves the activation of pre-existing compounds, followed by induction of plant defense mechanisms. This is accompanied by the production of reactive oxygen species leading to plant cell death, and consequently, elimination of the pathogen (Mysore and Ryu 2004).

Chisholm et al. (2006) reported that activation of R protein-mediated resistance also suppresses microbial growth, but after the limited proliferation of the pathogen. These results challenge the gene-for-gene hypothesis of Flor (1971) in which disease resistance occurs only with the simultaneous presence of *R* gene resistance in the plant and corresponding avirulence (*Avr*) gene in the pathogen, suggesting that a direct Avr-R interaction activates the resistance. However, despite many efforts, the physical interaction between Avr and R proteins has been observed only in a few cases. As many Avr are known to increase the virulence of the pathogen in susceptible plants, an alternative model has been proposed in which the Avr protein is able to interact and modify a target specific to the plant to promote disease in the absence of the corresponding R protein. The function of R protein would be to guard or protect these targets and activate the resistance signaling pathways when these plant targets are modified by a specific Avr (Chisholm et al. 2006).

PAMPs trigger innate immunity defenses in the host (Abramovitch et al. 2006; Jones and Dangl 2006). These defenses are regulated by interconnected chemical signaling networks that utilize ethylene, jasmonic acid, and salicylic acid. PAMPs activate mitogen-activated protein kinases (MAPKs) and the production of reactive oxygen species (ROSs) by the host. Superoxide dismutases produced by the pathogen help reduce the effects of ROSs. Espinosa and Alfano (2004) reported that such virulence factors can inhibit PAMP-elicited basal defenses. Recruitment of host factors by the pathogen is a highly evolved means of circumventing innate defenses. Djamei et al. (2007) reported that the *Agrobacterium* VirE2-binding protein VIPI, produced by the plant host, appears to be a substrate of MAPK3 that phosphorilates VIPI at serine position 79 of the VIPI peptide in order to translocate the T-DNA strand-VirE2 complex into the plant cell nucleus. Thus, *Ag. tumefaciens* has recruited MAPK3 to facilitate nuclear localization and eventual integration of the T-DNA strand in the host chromosome.

Cysteine proteases play an important role in plant pathogenesis. Hotson and Mudgett (2004) reported that the effectors AvrXv4, AvrPphb, AvrRpt2, and XopD have cysteine protease functions. As cysteine proteases have been identified as effectors injected into plant cells by the TIIISS, their role is likely to be targeting defense related proteins.

5.4 NEUTRALIZATION OF REACTIVE OXYGEN SPECIES

When perturbed, signaled, or damaged, lysosomes in the host produce ROSs as a host defense response. The pathogens equipped with a ROS detoxification pathway can effectively dodge this defense. Among plant pathogenic bacteria, *Ag. tumefaciens* harbors the *katG* gene that encodes catalase to hydrolyze toxic peroxides, and as reported by Yuan et al. (2005), it is induced by low phosphate concentration. A catalase peroxidase encoded by *katG* in *Xa. campestris* pv. *campestris* appears to be required for virulence by conferring resistance to the detrimental effects of the hydrogen peroxide defense system (Jittawuttipoka et al. 2009). Sukchawalit et al. (2001) reported that ROS is also detoxified by an alkyl-hydroperoxide reductase encoded by the *ahpC* gene in *Xa. campestris* pv. *phaseoli*. The hydrogen peroxide-induced transcription factor OxyR regulates the expression of both *katG* and *ahpC* genes (Zheng et al. 2001). Hydrogen peroxide activates OxyR by the formation of an intramolecular disulphide bond. Oxidized OxyR in turn activates transcription of *katG* and *ahpC* among other reductases. Saenkham et al. (2007) reported that *Ag. tumefaciens* possesses three iron-containing superoxide dismutases called SodBI, SodBII, and SodBIII, all of which play a role in virulence.

5.5 ROLE OF PHYTOALEXINS AND THEIR DETOXIFICATION

Phytoalexins are antimicrobial compounds produced by the plants in response to pathogen invasion. Some of these compounds restrict or inhibit the growth of bacteria by hydrogen bonding interactions with vital proteins, such as enzymes. Virulent pathogens trigger plant defense responses at later stages of infection. They are able to detoxify phytoalexins using various mechanisms that include modification of these compounds followed by their metabolism as well as efficient efflux systems, which prevent intracellular accumulation of toxic compounds. Due to their broad substrate specificity, efflux pumps expel a wide range of harmful compounds from the cell directly into the extracellular environment, bypassing the periplasm. Palumbo et al. (1998) reported that an efflux pump induced by isoflavonoids was found in *Ag. tumefaciens*. The pump greatly decreases the cellular accumulation of isoflavonoids produced by alfalfa roots as exudates. The efflux system is encoded by *ifeABR* locus and is induced by various isoflavonoids and a chalcone. The mutations in *ifeABR* reduced the ability of *Ag. tumefaciens* to colonize the root system. Burse et al. (2004) reported the presence of an efflux pump, encoded by a putative *acrB* operon, in *Er. amylovora*. Disruption of the efflux pump by mutation in *acrB* dramatically reduces the virulence of this bacterium.

Plants have the ability to counteract the bacterial efflux system by producing inhibitors of the multidrug resistance pumps. Glazebrook and Ausubel (1994) found the accumulation of phytoalexin camalexin in *Arabidopsis thaliana* on inoculation with *Psm. syringae* pv. *maculicola*. *Arabidopsis* mutants deficient in camalexin biosynthesis enhanced the sensitivity of host to *Psm. syringae* pv. *maculicola*. The authors further reported that camalexin is not required for resistance to avirulent *Psm. syringae* strains in *A. thaliana*.

Bacterial pathogens that incorporate parts of their genome into plant cells are equipped with mechanisms to suppress the small interfering RNAs (siRNAs) produced by the plant's defense reaction. Dunoyer et al. (2006) reported that *Ag. tumefaciens* suppressed siRNAs whose production was induced by the incorporation of multiple copies of the T-DNA into the plant genome. The tumor-inducing pathogen triggers the formation of siRNAs 3 days after inoculation, but these siRNAs are not found in mature crown gall tumors. The formation of microRNAs that are required for tumor formation, presumably by a degradation pathway of target mRNAs, induces formation of siRNAs in the tumor that are subsequently degraded by uncharacterized *Agrobacterium*-mediated factor(s).

5.6 OTHER VIRULENCE FACTORS

There is considerable information on virulence and avirulence genes and the contribution of their protein/peptide products in pathogenicity. However, there are other chemicals produced by bacterial plant pathogens that are injurious and lethal to the host plants. Many bacterial plant pathogens produce peptide- or protein-containing toxins, such as tabtoxin, phaseolotoxin, coronatine, syringomycin, tagetitoxin, and rhizobitoxine. Tagetitoxin specifically inhibits the action of RNA polymerase III in eucaryotic 5S and tRNA gene transcription and of chloroplast RNA polymerase.

Foliar infection of plants by bacteria such as *Psm. syringae* pv. *tomato* occurs through stomata (Underwood et al. 2007), which serve as critical entry sites and allow bacteria to transition from epiphytic to endophytic lifestyle. Earlier, it was assumed that the entry of bacteria into leaf tissues through natural openings was a passive process, where the plant lacked mechanisms for preventing bacterial entry, and the bacterium lacked active virulence mechanisms to promote entry. Later on, it was found that entry of bacteria into leaf tissue through stomata is more complex and dynamic than the simple act of swimming into the leaf through passive openings (Gudesblat et al. 2009). Several lines of evidence suggest that stomata actively close in response to plant pathogenic and human pathogenic bacteria or when exposed to conserved molecules found on the surface of bacterial cells known as pathogen/microbe-associated molecular patterns (Melotto et al. 2006). Bacterium-induced stomatal closure (also known as stomatal defense) is a part of plant immune defenses (Figure 5.1). Thus, stomatal closure is an integral basal plant defense mechanism to restrict the invasion of pathogenic bacteria into plant tissues.

Production of phytotoxins and type III secretion system have emerged as important factors to overcome stomatal defense by bacterial pathogens. COR is produced by several pathovars of *Psm. syringae* including *tomato*, *maculicola*, *glycinea*, and *atropurpurea*, where it is known to function as a virulence factor promoting chlorosis in several host plants (Mittal et al. 1995, Brooks et al. 2004). Bender et al. (1999) reported that COR induces modifications in the plant's physiology such as anthocyanin production, alkaloid accumulation, ethylene emission, tendril coiling, and root inhibition. This toxin acts as a virulence factor and contributes to disease development. Ishiga et al. (2009) reported that COR induced effects on photosynthetic machinery and on ROS in modulating necrotic cell death in bacterial speck of tomato.

COR is one of the most extensively studied bacterial phytotoxins and is a non-host-specific phytotoxin. The available evidence suggests that COR plays multiple roles in bacterial pathogenesis including promoting entry of bacteria through stomata at the initial stages of infection (Melotto et al. 2006) and suppression of defenses mediated by the plant hormone salicylic acid later in the infection process (Uppalapati et al. 2007).

The possibility that COR could suppress early defense responses during *Psm. syringae* pv. *tomato* DC3000 infection of Arabidopsis and tomato was suggested by Mittal and Davis (1995) and later on confirmed by Melotto et al. (2006). The discovery that COR is required for re-opening stomata by *Psm. syringae* pv. *tomato* DC3000 represents the first identification of a bacterial virulence factor that suppresses stomatal closure. COR defective mutants have reduced multiplication and symptom production *in planta* compared to the wild-type (Mittal and Davis 1995; Brooks et al. 2004). Investigations of host pathogen interactions, involving both *Psm. syringae* pv. *tomato* and *Psm. syringae* pv. *glycinea*, have suggested that COR may be important for bacterial invasion of plant tissue.

The discovery of this virulence mechanism of *Psm. syringae* pv. *tomato* DC3000 has generated a lot of interest in elucidating the mode of action of COR at the molecular level. Mino et al. (1987) showed that 10 μM COR promoted opening of dark-closed stomata of broad bean and Italian ryegrass. The effect of COR on the stomatal aperture was more pronounced on Italian ryegrass. The authors also reported that COR activates membrane-bound ATPase activity inducing stomatal opening. Liu et al. (2009) reported that the plant protein RIN4 (RPM1 interacting protein 4) and activation of H^+-ATPase have been found to be necessary for COR to re-open stomata. Coronatine insensitive 1 (COI1) is another plant protein necessary for COR functioning in the guard cell (Melotto et al. 2006). In fact, COI1 has been shown to be a receptor for COR (Yan et al. 2009).

Figure 5.2, given below, illustrates signaling components and interactions between molecules to be involved in

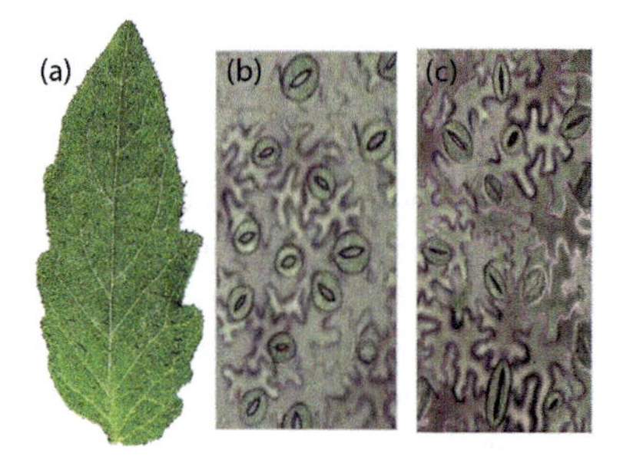

FIGURE 5.1 Light-conditioned tomato leaf (a) and leaf surface under the microscope prior to exposure to *Psm. syringae* pv. *tomato* (*Pst*) DC3000 showing mostly open stomata (b). The same leaf was exposed to *Pst* DC3000, and after 1 h of exposure most stomata were closed (c). (Reproduced with thanks from Baker, C.M. et al., *Braz. J. Med. Biol. Res.*, 43, 698–704, 2010, CCAL.)

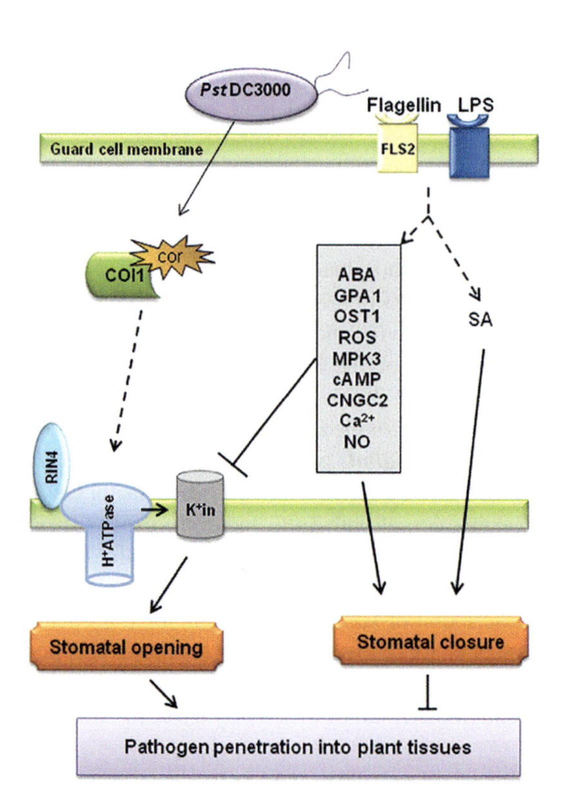

FIGURE 5.2 Model illustrating the signaling components and interactions between molecules that have been experimentally demonstrated to be involved in stomatal defense and bacterial counter defense. Dashed arrows indicate possibly indirect pathways. (Reproduced with thanks from Baker, C.M. et al., *Braz. J. Med. Biol. Res.*, 43, 698–704, 2010, CCAL.)

stomatal defense and bacterial counter defense. In the stomatal guard cell, microbe-associated molecular patterns (MAMPs), e.g., flagellin and lipopolysaccharide (LPS) are perceived by cognate immune receptors (e.g., flagellin receptor FLS2). Perception of MAMPs triggers stomatal closure, which requires the phytohormones salicylic acid (SA) and abscisic acid (ABA), as well as ABA signaling components listed in the grey rectangle (components are listed in order where the top one is the most upstream, their localization in the cell is not shown). Flagellin also prevents stomatal opening by inhibiting inwardly rectifying K^+ channels (K^+ in) through ABA signaling components (GPA1 and possibly others). COR-mediated inhibition of MAMP-triggered stomatal closure requires the plant proteins COI1 (a COR receptor) and RIN4. While COI1 physically binds to COR, RIN4 binds and activates the proton pump (H^+ ATPase) causing membrane hyperpolarization and activation of K^+ influx (K^+ in), a condition that promotes stomatal opening.

Molecular action of coronatine in plant cells: The molecular mechanisms through which COR may operate in the plant cell to cause disease is given below.

Staswick and Tiryaki (2004) reported that coronatine is a structural and functional mimic of the plant hormone jasmonate (JA) conjugated to the amino acid isoleucine (JA-Ile). Biological concentrations of COR activate the JA signaling pathway in the plant. cDNA microarray analysis indicated that the induction of JA-responsive genes in the tomato-*Psm. syringae* pv. *tomato* DC3000 interaction depends on the bacterial production of COR (Zhao et al. 2003). For more details, the readers may refer to Baker et al. (2010).

Yan et al. (2009) demonstrated COI1 to be a receptor for JA-Ile and COR. Based on this, Baker et al. (2010) developed a plausible model for the entire set of interactions where COR produced by the bacterium binds to COI1 and leads to the degradation of jasmonate ZIM-domain (JAZ) proteins through the SCF^{COI1} complex. In this model, JAZ proteins and other adaptor proteins act as repressors of JA signaling. Degradation of JAZ proteins allows for the expression of JA-responsive genes in the plant cell (Chini et al. 2007; Thines et al. 2007; Chung and Howe 2009; Pauwels et al. 2010) blocking plant innate immune responses including stomatal defense (Figure 5.3).

In their update on stomatal defense, Melotto et al. (2017) have given a perspective of the basic mechanisms underlying the process, and their implications in the understanding of plant-microbe interactions. The authors have also focused on significant advances towards a mechanistic understanding of stomatal defense.

Bacterium-triggered stomatal closure is a fast response (<1 h), and the basic mechanism underlying this process includes the following. ABA has long been recognized to induce stomatal closure under drought stress, thereby minimizing water loss through the leaves. However, the role of this hormone in Arabidopsis defense against *Psm. syringae* differs depending on the stage of infection. The current experimental evidence suggests a prominent role of ABA signaling in stomatal defense (Eisenach and de Angeli 2017; Inoue and Kinoshita 2017; Vialet-Chabrand et al. 2017).

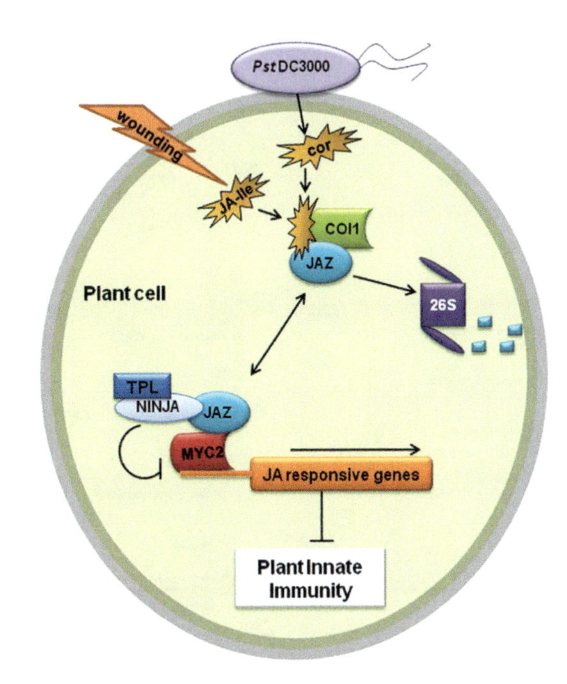

FIGURE 5.3 A model depicting the molecular action of coronatine in plant cells (possibly for all cell types). COR is secreted by *Psm. syringae* pv. *tomato* DC3000 into the plant cell and increases the affinity of the COI1 protein (as part of the SCFCOI1 ubiquitin-ligase complex, not shown here) towards the JAZ repressor. The SCFCOI1 complex catalyzes ubiquitination of JAZ, which is then degraded through the 26S proteasome (denoted as "26S"). JAZ protein is part of a repressor complex that also contains NINJA and TPL, and physically binds to transcriptional activators (such as MYC2) of jasmonate response genes. Upon degradation of JAZ, JA response genes are activated blocking plant innate immune responses including stomatal closure. COR = coronatine; JA = jasmonate; JAZ = jasmonate ZIM-domain; NINJA = novel interactor of JAZ; TPL = TOPLESS. (Reproduced with thanks from Baker, C.M. et al., *Braz. J. Med. Biol. Res.*, 43, 698–704, 2010, CCAL.)

Melotto et al. (2017) have elucidated a putative signaling cascade by which COR prevents bacterium triggered stomatal closure (Figure 5.4). COR, a molecular mimic of JA-isoleucine, promotes physical interaction between the F-box protein COI1 and the transcriptional repressor JAZ, leading to ubiquitination and proteasome-mediated degradation of JAZ proteins (Chini et al. 2007; Thines et al. 2007; Yan et al. 2007). Degradation of JAZ proteins derepresses bHLH transcription factors, such as MYC2, MYC3, and MYC4, leading to activation of downstream transcriptional responses (Zhang et al. 2017). Zheng et al. (2012) reported that COR activation of JA signaling induced the expression of three homologous *NAC* transcription factor genes, *ANAC019*, *ANAC055*, and *ANAC072* that are the direct targets of MYC2. Genetic characterization of *nac* null mutants reveals that NACs mediate COR-induced stomatal reopening and bacterial multiplication in plant tissues by inhibiting the accumulation of SA. Specifically, these NACs exert an inhibitory effect by repressing the expression of genes involved in SA biosynthesis and activating the expression of genes involved in SA metabolism, resulting in overall reduction of

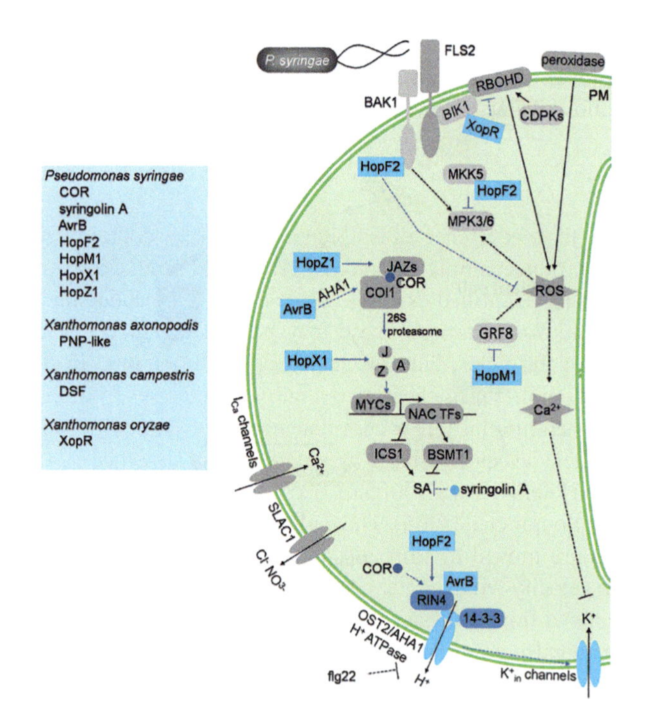

FIGURE 5.4 A simplified diagram of microbial virulence factors that manipulate MAMP-induced stomatal closure. NADPH oxidase RBOHD mediates flg22-induced ROS production and stomatal defense through BIK1-/CDPKs-regulated phosphorylation. Type III peroxidase also contributes to flg22-induced ROS production. MAMP-induced stomatal closure includes accumulation of ROS and NO, cytosolic calcium oscillations, activation of S-type anion channels, and inhibition of K^+_{in} channels. COR induces COI1-JAZ interaction, mediates JAZ degradation, and activates the expression of NAC TFs, which inhibit the expression of *ICS1* and induce expression of *BSMT1*, thereby leading to decreased SA level. COR also may induce stomatal opening through RIN4, which interacts with H^+-ATPase AHA1 and AHA2, resulting in induction of K^+_{in} channels. Syringolin A inhibits SA signaling via its proteasome inhibitor activity. Multiple bacterial effectors HopM1, HopF2, HopX1, AvrB, HopZ1, and XopR could disrupt stomatal defense. HopM1 disrupts the function of GRF8 (a 14–3–3 protein), resulting in reduction of MAMP-induced ROS production. HopF2 directly targets BAK1, MKK5, and RIN4, resulting in inhibition of ROS production. AvrB enhances the interaction of RIN4 and AHA1, which could promote the interaction between COI1 and JAZs, leading to stomatal opening. HopX1 and HopZ1 induce stomatal opening after PAMP-triggered immunity-mediated stomatal closure, through a COI1-independent and COI1-dependent manner, respectively. XopR interacts with BIK1 and suppresses stomatal closure possibly by inhibition of RBOHD. (Reproduced with thanks from Melotto, M. et al., *Plant Physiol.*, 174, 561–71, 2017. With permission from American Society of Plant Biologists.)

SA in infected plants (Zheng et al. 2012; Gimenez-Ibanez et al. 2017). Du et al. (2014) showed that JA and COR activate the expression of the NAC tomato homolog, *JASMONIC ACID2-LIKE*. This transcription factor binds to and activates the expression of *SAMT1* and *SAMT2* that encode enzymes that deactivate SA by methylation, thereby suppressing the accumulation of SA and promoting stomatal opening.

ANAC032, another member of the NAC family of transcription factors, acts as both a positive regulator of SA signaling and a negative regulator of JA signaling. ANAC032 directly binds to the promoter of *MYC2* (positive regulator of JA signaling) and *NIMIN1* (negative regulator of SA signaling) and concomitantly suppresses their transcription within 6 h of *Pst* DC3000 infection. Accordingly, over expression of ANAC032 in Arabidopsis inhibits COR-dependent reopening of stomata (Allu et al. 2016).

COR has also been shown to trigger cellular responses that do not depend on the canonical COI1-JAZ signaling pathway to manipulate guard cell movement. For example, RIN4, a negative regulator of plant innate immunity is required by COR to open the stomatal pore as evidenced by the facts that neither *Pst* DC3000 nor COR was able to reopen stomata of *rpm1/rps2/rin4* mutants (Liu et al. 2009; Lee et al. 2015; Zhou et al. 2015). The perception of flg22 via the FLS2 receptor leads to phosphorylation of the plasma membrane H^+-ATPases and subsequent alkalization of the apoplast that along with the induction of ROS via NADPH oxidase RbohD, triggers stomatal closure (Liu et al. 2009; Li et al. 2014). RIN4 interacts with the plasma membrane H^+-ATPases AHA1 and AHA2, resulting in their inhibition and acidification of the apoplast through hyperpolarization of the plasma membrane and subsequent induction of inward K^+ channels, promoting stomatal opening (Zhang et al. 2008; Liu et al. 2009). Zhang et al. (2008) showed that COR reversed the inhibitory effects of flg22 on K^+ currents to promote stomatal opening.

Syringolin A, a peptide toxin produced by *Psm. syringae* pv. *syringae* is another virulence factor that has a proteasome inhibitory function and impacts stomatal defense (Groll et al. 2008). Syringolin A promotes stomatal opening (Schellenberg et al. 2010). In contrast to *Pst* DC3000, *Psm. syringae* pv. *syringae* does not induce an initial stomatal closure on either its bean host or Arabidopsis (Schellenberg et al. 2010; Panchal et al. 2016). These finding show that this bacterium constitutively produces syringolin A, which is a stronger signal than the MAMPs produced by *Psm. syringae* pv. *syringae* and/or that these plants do not recognize *Psm. syringae* pv. *syringae* MAMPs efficiently. Misas-Villamil et al. (2013) demonstrated that syringolin A diffuses from the site of infection and suppresses SA signaling, thereby decreasing immune responses in adjacent tissues. Schellenberg et al. (2010) reported that this might be the mechanism involved in syringolin A inhibition of MAMP-induced stomatal closure because syringolin A inhibits the proteasome-mediated turnover of NPR1, an important component of SA signaling to induce stomatal defense.

Xa. campestris pv. *campestris* is also capable of interfering with stomatal closure induced by MAMP or ABA signaling. The bacterium does so by inducing the production of diffusible signal factor that is involved in bacterium-to-bacterium signaling (Gudesblat et al. 2009). The mechanism of stomatal reopening by diffusible signal factor upon bacterial invasion is still unknown. *Xa. axonopodis* pv. *citri* produces a compound, known as plant natriuretic peptide-like, that can open stomata during plant infection that correlates with enhanced disease symptoms. Plant natriuretic peptide-like controls stomatal aperture in a cGMP-dependent manner (Gottig et al. 2008).

In addition to phytotoxins, plant pathogenic bacteria also produce type-III-secretion-system effectors (T3SSEs) that can counter stomatal defense by either inhibiting MAMP-triggered stomatal closure or actively inducing stomatal opening. Lozano-Durán et al. (2014) reported that the T3SSE HopM1 produced by *Psm. syringae* disrupts the function of a 14–3–3 protein, GRF8, leading to reduction in MAMP-triggered ROS burst and stomatal defense. However, it is not clear in this case whether stomatal opening is a consequence of HopM1-mediated ROS suppression. Similarly, the *Psm. syringae* effector HopF2 inhibits flg22-induced ROS and stomatal defense (Hurley et al. 2014). The XopR effector from *Xa. oryzae* pv. *oryzae* strain PXO99A also inhibits flg22-induced stomatal closure; however, the mechanism remains elusive (Wang et al. 2016).

Benali et al. (2014) have reviewed the development in virulence strategies of phytopathogenic bacteria and their role in plant disease pathogenesis. They have dealt with secretion systems (type I–VI) employed by phytopathogenic bacteria, and the role of pectin degrading enzymes, adhesins, and extracellular polysaccharides, siderophores, COR, phaseolotoxins, syringomycins, syringopeptins, QS, and biofilm production in causing different plant diseases.

REFERENCES

Abramovitch, R. B., Anderson, J. C., and Martin, G. B. 2006. Bacterial elicitation and evasion of plant innate immunity. *Nat. Rev. Mol. Cell. Biol.* 7: 601–611.

Abramovitch, R. B., and Martin, G. B. 2005. AvrPtoB: A bacterial type III effector that both elicits and suppresses programmed cell death associated with plant immunity. *FEMS Microbiol. Lett.* 245: 1–8.

Agrios, G. N. 2004. *Plant Pathology*. 5th ed. Elsevier Academic Press, San Diego, CA, 922 pp.

Alegria, M. C., Souza, D. P., Andrade, M. O., Docena, C., Khater, L., et al. 2005. Identification of new protein-protein interactions involving the products of the chromosome- and plasmid-encoded type IV secretion loci of the phytopathogen *Xanthomonas axonopodis* pv. *citri*. *J. Bacteriol.* 187: 2315–2325.

Alfano, J. R., and Collmer, A. 2004. Type III secretion system effector proteins; Double agents in bacterial disease and plant defense. *Annu. Rev. Phytopathol.* 42: 385–414.

Allu, A. D., Brotman, Y., Xue, G. P., and Balazadeh, S. 2016. Transcription factor ANAC032 modulates JA/SA signalling in response to *Pseudomonas syringae* infection. *EMBO Rep.* 17: 1578–1589.

Arlat, M., Gough, C. L., Barber, C. E., Boucher, C., and Daniels, M. J. 1991. *Xanthomonas campestris* contains a cluster of hrp genes related to the larger hrp cluster of *Pseudomonas solanacearum*. *Mol. Plant-Microbe Interact.* 4: 593–601.

Baker, C. M., Chitrakar, R., Obulareddy, N., Panchal, S. Williams, P., et al. 2010. Molecular battles between plant and pathogenic bacteria in the phyllosphere. *Braz. J. Med. Biol. Res.* 43: 698–704.

Benali, S., Mohamed, B., and Eddine, H. J. 2014. Virulence strategies of phytopathogenic bacteria and their role in plant disease pathogenesis. *Afr. J. Microbiol. Res.* 8: 2809–2815.

Bender, C. L., Alarcon-Chaidez, F., and Gross, D. C. 1999. *Pseudomonas syringae* phytotoxins: Mode of action, regulation, and biosynthesis by peptide and polyketide synthetases. *Microbiol. Mol. Biol. Rev.* 63: 266–292.

Berks, B. C., Palmer, T., and Sargent, F. 2005. Protein targeting by the bacterial twin-arginine translocation (Tat) pathway. *Curr. Opin. Microbiol.* 8: 174–181.

Berks, B. C., Sargent, F., and Palmer, T. 2000. The Tat protein export pathway. *Mol. Microbiol.* 35: 260–274.

Bronstein, P. A., Marrichi, M., Cartinhour, S., Schneider, D. J., and DeLisa, M. P. 2005. Identification of a twin-arginine translocation system in *Pseudomonas syringae* pv. *tomato* DC3000 and its contribution to pathogenicity and fitness. *J. Bacteriol.* 187: 8450–8461.

Brooks, D. M., Hernandez-Guzman, G., Kloek, A. P., Alarcon-Chaidez, F., Sreedharan, A., et al. 2004. Identification and characterization of a well-defined series of coronatine biosynthetic mutants of *Pseudomonas syringae* pv. *tomato* DC3000. *Mol. Plant-Microbe Interact.* 17: 162–174.

Burse, A., Weingart, H., and Ullrich, M. S. 2004. The phytoalexin-inducible multidrug efflux pump AcrAB contributes to virulence in the fire blight pathogen, *Erwinia amylovora*. *Mol. Plant-Microbe Interact.* 17: 43–54.

Buttner, D., and Bonas, U. 2002. Getting across-bacterial type III effector proteins on their way to the plant cell. *EMBO J.* 21: 5313–5322.

Caldelari, I., Mann, S., Crooks, C., and Palmer, T. 2006. The Tat pathway of the plant pathogen *Pseudomonas syringae* is required for optimal virulence. *Mol. Plant-Microbe Interact.* 19: 200–212.

Chakravarthy, S., Worley, J. N., Montes-Rodriguez, A., and Collmer, A. 2018. *Pseudomonas syringae* pv. *tomato* DC3000 polymutants deploying coronatine and two type III effectors produce quantifiable chlorotic spots from individual bacterial colonies in *Nicotiana benthamiana* leaves. *Mol. Plant Pathol.* 19: 935–947.

Cheng, F., Ma, A., Zhuang, G., and Fray, R. G. 2018. Exogenous *N*-acyl-homoserine lactones enhance the expression of flagella of *Pseudomonas syringae* and activate defence responses in plants. *Mol. Plant Pathol.* 19: 104–115.

Chini, A., Fonseca, S., Fernandez, G., Adie, B., Chico, J. M., et al. 2007. The JAZ family of repressors is the missing link in jasmonate signalling. *Nature* 448: 666–671.

Chisholm, S. T., Coaker, G., Day, B., and Staskawicz, B. J. 2006. Host-microbe interactions: Shaping the evolution of the plant immune response. *Cell* 124: 803–814.

Chung, H. S., and Howe, G. A. 2009. A critical role for the TIFY motif in repression of jasmonate signaling by a stabilized splice variant of the jasmonate ZIM-domain protein JAZ10 in Arabidopsis. *Plant Cell* 21: 131–145.

Cui, Y., Chatterjee, A., Hasegawa, H., Dixit, V., Leigh, N., et al. 2005. ExpR, a LuxR homolog of *Erwinia carotovora* subsp. *carotovora*, activates transcription of *rsmA*, which specifies a global regulatory RNA-binding protein. *J. Bacteriol.* 187: 4792–4803.

da Silva, A. C., Ferro, J. A., Reinach, F. C., Farah, C. S., Furlan, L. R., et al. 2002. Comparison of the genomes of two *Xanthomonas* pathogens with differing host specificities. *Nature* 417: 459–463.

Davis, R. E., Dally, E. L., Jomantiene, R., Zhao, Y., Roe, B., et al. 2005. Cryptic plasmid pSKU146 from the wall-less plant pathogen *Spiroplasma kunkelii* encodes an adhesion and components of a type IV translocation-related conjugation system. *Plasmid* 3: 179–190.

Desvaux, M., Hébraud, M., Henderson, I. R., and Pallen, M. J. 2006. Type III secretion: What's in a name? *Trends Microbiol.* 14: 157–160.

Ding, Z., and Christie, P. J. 2003. *Agrobacterium tumefaciens* twin-arginine-dependent translocation is important for virulence, flagellation, and chemotaxis but not type IV secretion. *J. Bacteriol.* 185: 760–771.

Djamei, A., Pitzschke, A., Nakagami, H., Rajh, I., and Heribert, H. 2007. Trojan horse strategy in *Agrobacterium* transformation: Abusing MAPK defense signaling. *Science* 318: 453–456.

Dong, Y. H., Zhang, X.-F., Xu, J.-L., Tan, A.-T., and Zhang, L.-H. 2005. VqsM, a novel AraC-type global regulator of quorum-sensing signaling and virulence in *Pseudomonas aeruginosa*. *Mol. Microbiol.* 58: 552–564.

Dow, J. M., Crossman, L. C., Findlay, K., He, Y. Q., Feng, J. X., et al. 2003. Biofilm dispersal in *Xanthomonas campestris* is controlled by cell-cell signaling and is required for full virulence to plants. *Proc. Natl. Acad. Sci. USA* 100: 10995–11000.

Du, M., Zhai, Q., Deng, L., Li, S., Li. H., et al. 2014. Closely related NAC transcription factors of tomato differentially regulate stomatal closure and reopening during pathogen attack. *Plant Cell* 26: 3167–3184.

Duan, T. T., Luo, Y. P., Kong, C. Y., Gao, X. D., Zhao, X., et al. 2018. Gene expression profile of *Ralstonia solanacearum* for the rhizosphere ecological niche of *Solanum tuberosum*. *J. Phytopathol.* 166: 151–160.

Dunger, G., Arabolaza, L. N., Gottig, N., Orellano, E. G., and Ottado, J. 2005. Participation of *Xanthomonas axonopodis* pv. *citri hrp* cluster in citrus canker and in non-host plants responses. *Plant Pathol.* 54: 781–788.

Dunoyer, P., Himber, C., and Voinnet, O. 2006. Induction, suppression and requirement of RNA silencing pathways in virulent *Agrobacterium* infections. *Nat. Genet.* 38: 258–263.

Eisenach, C., and de Angeli, A. 2017. Ion transport at the vacuole during stomatal movements. *Plant Physiol.* 174: 520–530.

Espinosa, A., and Alfano, J. R. 2004. Disabling surveillance: Bacterial type III secretion system effectors that suppress innate immunity. *Cell Microbiol.* 6: 1027–1040.

Fan, S., Tian, F., Li, J., Hutchins, W., Chen, H., et al. 2017. Identification of phenolic compounds that suppress the virulence of *Xanthomonas oryzae* on rice via the type III secretion system. *Mol. Plant Pathol.* 18: 555–568.

Ferluga, S., Bigirimana, J., Hőfte, M., and Venturi, V. 2007. A LuxR homologue of *Xanthomonas oryzae* pv. *oryzae* is required for optimal rice virulence. *Mol. Plant Pathol.* 8: 529–538.

Ferluga, S., and Venturi, V. 2009. OryR is a LuxR-family protein involved in interkingdom signaling between pathogenic *Xanthomonas oryzae* and rice. *J. Bacteriol.* 191: 890–897.

Ferreira, A. O., Myers, C. R., Gordon, J. S., Martin, G. B., Vencato, M., et al. 2006. Whole-genome expression profiling defines the HrpL regulon of *Pseudomonas syringae* pv. *tomato* DC3000, allows de novo reconstruction of the Hrp *cis* element and identifies novel coregulated genes. *Mol. Plant-Microbe Interact.* 19: 1167–1179.

Flor, H. H. 1971. Current status of the gene-for-gene concept. *Annu Rev Phytopathol.* 9: 275–298.

Fuqua, C., Parsek, M. R., and Greenberg, E. P. 2001. Regulation of gene expression by cell-to-cell communication: Acyl-homoserine lactone quorum sensing. *Annu. Rev. Genet.* 35: 439–468.

Gao, M., Teplitski, M., Robinson, J. B., and Bauer, W. D. 2003. Production of substances by *Medicago truncatula* that affect bacterial quorum sensing. *Mol. Plant-Microbe Interact.* 16: 827–834.

Gimenez-Ibanez, S., Boter, M., Ortigosa, A., García-Casado, G., Chini, A., et al., 2017. JAZ2 controls stomata dynamics during bacterial invasion. *New Phytol.* 213: 137892.

Givskov, M., De Nys, R., Manefield, M., Gram, L., Maximilien, R., et al. 1996. Eukaryotic interference with homoserine lactone-mediated prokaryotic signaling. *J. Bacteriol.* 178: 6618–6622.

Glazebrook, J., and Ausubel, F. M. 1994. Isolation of phytoalexin-deficient mutants of *Arabidopsis thaliana* and characterisation of their interactions with bacterial pathogens. *Proc. Natl. Acad. Sci. USA* 91: 8955–8959.

Gochez, A. M., Shantharaj, D., Potnis, N., Zhou, X., Minsavage, G. V., et al. 2017. Molecular characterization of XopAG effector AvrGf2 from *Xanthomonas fuscans* ssp. *aurantifolii* in grapefruit. *Mol. Plant Pathol.* 18: 405–419.

González, E. T., Brown, D. G., Swanson, J. K., and Allen, C. 2007. Using the *Ralstonia solanacearum* Tat secretome to identify bacterial wilt virulence factors. *Appl. Environ. Microbiol.* 73: 3779–3786.

Gottig, N., Garavaglia, B. S., Daurelio, L. D., Valentine, A., Gehring, C., et al. 2008. *Xanthomonas axonopodis* pv. *citri* uses a plant natriuretic peptide-like protein to modify host homeostasis. *Proc. Natl. Acad. Sci. USA* 105: 18631–18636.

Gottig, N., Garavaglia, B. S., Garofalo, C. G., Zimaro, T., Sgro, G. G., et al. 2010. Mechanisms of infection used by *Xanthomonas axonopodis* pv. *citri* in citrus canker disease. pp. 196–204 in: *Current Research, Technology and Education Topics in Applied Microbiology and Microbial Biotechnology.* A. Méndez-Vilas ed. Formatex Research Center, Badajoz, Spain.

Groll, M., Schellenberg, B., Bachmann, A. S., Archer, C. R., Huber, R., et al. 2008. A plant pathogen virulence factor inhibits the eukaryotic proteasome by a novel mechanism. *Nature* 452: 755–758.

Gudesblat, G. E., Torres, P. S., and Vojnov, A. A. 2009. *Xanthomonas campestris* overcomes Arabidopsis stomatal innate immunity through a DSF cell-to-cell signal-regulated virulence factor. *Plant Physiol.* 149: 1017–1027.

Hajri, A., Brin, C., Hunault, G., Lardeux, F., Lemaire, C., et al. 2009. A "repertoire for repertoire" hypothesis: Repertoires of type three effectors are candidate determinants of host specificity in *Xanthomonas*. *PLoS ONE*. 4: e6632.

Henderson, I. R., Navarro-Garcia, F., Desvaux, M., Fernandez, R. C., and Ala'Aldeen, D. 2004. Type V protein secretion pathway: The autotransporter story. *Microbiol. Mol. Biol. Rev.* 68: 692–744.

Hotson, A., and Mudgett, M. B. 2004. Cysteine proteases in phytopathogenic bacteria: Identification of plant targets and activation of innate immunity. *Curr. Opin. Plant Biol.* 7: 384–390.

Hueck, C. J. 1998. Type III protein secretion systems in bacterial pathogens of animals and plants. *Microbiol. Mol. Biol. Rev.* 62: 379–433.

Hurley, B., Lee, D., Mott, A., Wilton, M., Liu, J., et al. 2014. The *Pseudomonas syringae* type III effector HopF2 suppresses Arabidopsis stomatal immunity. *PLoS One* 9: e114921.

Ingle, R. A., Carstens, M., and Denby, K. J. 2006. PAMP recognition and the plant-pathogen arms race. *Bioessays* 28: 880–89.

Inoue, S., and Kinoshita, T. 2017. Blue light regulation of stomatal opening and the plasma membrane H+-ATPase. *Plant Physiol.* 174: 531–538.

Ishiga, Y., Uppalapati, S. R., Ishiga, T., Elavarthi, S., Martin, B., et al. 2009. The phytotoxin coronatine induces light-dependent oxygen species in tomato seedlings. *New Phytol.* 181: 147–160.

Jittawuttipoka, T., Buranajitpakorn, S., Vattanaviboon, P., and Mongkolsuk, S. 2009. The catalase-peroxidase KatG is required for virulence of *Xanthomonas campestris* pv. *campestris* in a host plant by providing protection against low levels of H_2O_2. *J. Bacteriol.* 191: 7372–7377.

Jones, J. D. G., and Dangl, J. L. 2006. The plant immune system. *Nature* 444: 323–329.
Kado, C. I. 2002. Horizontal transmission of genes by *Agrobacterium* species. pp. 45–50 in: *Horizontal Gene Transfer.* 2nd ed. M. Syvanen, and C. I. Kado eds. Academic Press, San Diego, CA.
Kado, C. I. 2010. *Plant Bacteriology.* American Phytopathological Society, St. Paul, MN, 336 pp.
Keen, N. T. 1990. Gene-for-gene complementarity in plant-pathogen interactions. *Annu. Rev. Genet.* 24: 447–463.
Kim, J. G., Park, B. K., Yoo, C. H., Jeon, E., Oh, J., et al. 2003. Characterization of the *Xanthomonas axonopodis* pv. *glycines* Hrp pathogenicity island. *J. Bacteriol.* 185: 3155–3166.
Kunze, G., Zipfel, C., Robatzek, S., Niehaus, K., Boller, T., et al. 2004. The N terminus of bacterial elongation factor TU elicits innate immunity in *Arabidopsis* plants. *Plant Cell* 16: 3496–3507.
Lee, D., Bourdais, G., Yu, G., Robatzek, S., and Coaker, G. 2015. Phosphorylation of the plant immune regulator RPM1-INTERACTING PROTEIN4 enhances plant plasma membrane H+-ATPase activity and inhibits flagellin triggered immune responses in Arabidopsis. *Plant Cell* 27: 2042–2056.
Li, L., Li, M., Yu, L., Zhou, Z., Liang, X., et al. 2014. The FLS2-associated kinase BIK1 directly phosphorylates the NADPH oxidase RbohD to control plant immunity. *Cell Host Microbe* 15: 329–338.
Liang, X., Yu, X., Pan, X., Wu, J., Duan, Y., et al. 2018. A thiadiazole reduces the virulence of *Xanthomonas oryzae* pv. *oryzae* by inhibiting the histidine utilization pathway and quorum sensing. *Mol. Plant Pathol.* 19: 116–128.
Lindgren, P. B., Peet, R. C., and Panopoulos, N. J. 1986. Gene cluster of *Pseudomonas syringae* pv. "*phaseolicola*" controls pathogenicity of bean plants and hypersensitivity of nonhost plants. *J. Bacteriol.* 168: 512–522.
Liu, J., Elmore, J. M., Fuglsang, A. T., Palmgren, M. G., Staskawicz, B. J., et al. 2009. RIN4 functions with plasma membrane H+-ATPases to regulate stomatal apertures during pathogen attack. *PLoS Biol.* 7: e1000139.
Lo, T., Koulena, N., Seto, D., Guttman, D. S., and Desveaux, D. 2017. The HopF family of *Pseudomonas syringae* type III secreted effectors. *Mol. Plant Pathol.* 18: 457–468.
López-Solanilla, E., Bronstein, P. A., Schneider, A. R., and Collmer, A. 2004. HopPtoN is a *Pseudomonas syringae* Hrp (type III secretion system) cysteine protease effector that suppresses pathogen-induced necrosis associated with both compatible and incompatible plant interactions. *Mol. Microbiol.* 54: 353–365.
Lozano-Durán, R., Bourdais, G., He, S. Y., and Robatzek, S. 2014. The bacterial effector HopM1 suppresses PAMP-triggered oxidative burst and stomatal immunity. *New Phytol.* 202: 259–269.
Martin, G. B., Bogdanova, A. J., and Sessa, G. 2003. Understanding the functions of plant disease resistance proteins. *Annu. Rev. Plant Biol.* 54: 23–61.
McGowan, S. J., Barnard, A. M. L., Bosgelmez, G., Sebalhia, M., Simpson, N. J. L., et al. 2005. Carbapenem antibiotic biosynthesis in *Erwinia carotovora* is regulated by physiological and genetic factors modulating the quorum sensing-dependent control pathway. *Mol. Microbiol.* 55: 526–545.
Medina, C. A., Reyes, P. A., Trujillo, C. A., Gonzalez, J. L., Bejarano, D. A., et al. 2018. The role of type III effectors from *Xanthomonas axonopodis* pv. *manihotis* in virulence and suppression of plant immunity. *Mol. Plant Pathol.* 19: 593–606.
Melotto, M., Underwood, W., Koczan, J., Nomura, K., and He, S. Y. 2006. Plant stomata function in innate immunity against bacterial invasion. *Cell* 126: 969–980.
Melotto, M., Zhang, L., Oblessuc, P. R., and He, S. Y. 2017. Stomatal defense a decade later. *Plant Physiol.* 174: 561–571.
Mino, Y., Matsushita, Y., and Sakai, R. 1987. Effect of coronatine on stomatal opening in leaves of broadbean and Italian ryegrass. *Ann. Phytopathol. Soc. Jpn.* 53: 53–55.
Misas-Villamil, J. C., Kolodziejek, I., Crabill, E., Kaschani, F., Niessen, S., et al. 2013. *Pseudomonas syringae* pv. *syringae* uses proteasome inhibitor syringolin A to colonize from wound infection sites. *PLoS Pathog* 9: e1003281.
Mittal, S., and Davis, K. R. 1995. Role of the phytotoxin coronatine in the infection of *Arabidopsis thaliana* by *Pseudomonas syringae* pv. *tomato. Mol. Plant-Microbe Interact.* 8: 165–171.
Molina, L., Rezzonico, F., Défago, G., and Duffy, B. 2005. Autoinduction in *Erwinia amylovora*: Evidence of an acyl-homoserine lactone signal in the fire blight pathogen. *J. Bacteriol.* 187: 3206–3213.
Morel, J. B., and Dangl, J. L. 1997. The hypersensitive response and the induction of cell death in plants. *Cell Death Differ.* 4: 671–683.
Müller, M., and Klösgen, R. B. 2005. The Tat pathway in bacteria and chloroplasts. *Mol. Membr. Biol.* 22: 113–121.
Mysore, K. S., and Ryu, C. M. 2004. Nonhost resistance: How much do we know? *Trends Plant Sci.* 9: 97–104.
Nissan, G., Gershovits, M., Morozov, M., Chalupowicz, L., Sessa, G., et al. 2018. Revealing the inventory of type III effectors in *Pantoea agglomerans* gall-forming pathovars using draft genome sequences and a machine-learning approach. *Mol. Plant Pathol.* 19: 381–392.
Palumbo, J. D., Kado, C. I., and Phillips, D. A. 1998. An isoflavonoid-inducible efflux pump in *Agrobacterium tumefaciens* is involved in competitive colonization of roots. *J. Bacteriol.* 180: 3107–3113.
Panchal, S., Chitrakar, R., Thompson, B. K., Obulareddy, N., Roy, D., et al. 2016. Regulation of stomatal defense by air relative humidity. *Plant Physiol.* 172: 2021–2032.
Pauwels, L., Barbero, G. F., Geerinck, J., Tilleman, S., Grunewald, W., et al. 2010. NINJA connects the co-repressor TOPLESS to jasmonate signalling. *Nature* 464: 788–791.
Pukatzki, S., Ma, A. T., Sturtevant, D., Krastins, B., Sarracino, D., et al. 2006. Identification of a conserved bacterial protein secretion system in *Vibrio cholerae* using the *Dictyostelium* host model system. *Proc. Nat. Acad. Sci. USA* 103: 1528–1533.
Ryan, R. P., Fouhy, Y., Lucey, J. F., and Dow, J. M. 2006. Cyclic di-GMP signaling in bacteria: Recent advances and new puzzles. *J. Bacteriol.* 188: 8327–8334.
Saenkham, P., Eiamphungporn, W., Farrand, S. K., Vattanaviboon, P., and Mongkolsuk, S. 2007. Multiple superoxide dismutases in *Agrobacterium tumefaciens*: Functional analysis, gene regulation, and influence on tumorigenesis. *J. Bacteriol.* 189: 8807–8817.
Sandkvist, M. 2001. Biology of type II secretion. *Mol. Microbiol.* 40: 271–83.
Sang, Y., Wang, Y., Ni, H., Cazalé, A.-C., She, Y.-M., et al. 2018. The *Ralstonia solanacearum* type III effector RipAY targets plant redox regulators to suppress immune responses. *Mol. Plant Pathol.* 19: 129–142.
Schellenberg, B., Ramel, C., and Dudler, R. 2010. *Pseudomonas syringae* virulence factor syringolin A counteracts stomatal immunity by proteasome inhibition. *Mol. Plant-Microbe Interact.* 23: 1287–1293.
Shantharaj, D., Römer, P., Figueiredo, J. F. L., Minsavage, G. V., Krönauer, C., et al. 2017. An engineered promoter driving expression of a microbial avirulence gene confers recognition of TAL effectors and reduces growth of diverse *Xanthomonas* strains in citrus. *Mol. Plant Pathol.* 18: 976–989.

Staswick, P. E., and Tiryaki, I. 2004. The oxylipin signal jasmonic acid is activated by an enzyme that conjugates it to isoleucine in Arabidopsis. *Plant Cell* 16: 2117–2127.

Sukchawalit, R., Loprasert, S., Atichartpongkul, S., and Mongkolsuk, S. 2001. Complex regulation of the organic hydroperoxide resistance gene (*ohr*) from *Xanthomonas* involves Ohr, a novel organic peroxide-inducible negative regulator, and posttranscriptional modifications. *J. Bacteriol.* 183: 4405–4412.

Swarup, S., Yang, Y., Kingsley, M. T., and Gabriel, D. W. 1992. An *Xanthomonas citri* pathogenicity gene, *pthA*, pleiotropically encodes gratuitous avirulence on nonhosts. *Mol. Plant-Microbe Interact.* 5: 204–213.

Tampakaki, A. P., Fadouloglou, V. E., Gazi, A. D., Panopoulos, N. J., and Kokkinidis, M. 2004. Conserved features of type III secretion. *Cell Microbiol.* 6: 805–816.

Thines, B., Katsir, L., Melotto, M., Niu, Y., Mandaokar, A., et al. 2007. JAZ repressor proteins are targets of the SCF(COI1) complex during jasmonate signalling. *Nature* 448: 661–665.

Toth, I. K., Pritchard, L., and Birch, P. R. J. 2006. Comparative genomics reveals what makes an enterobacterial plant pathogen. *Annu. Rev. Phytopathol.* 44: 305–336.

Underwood, W., Melotto, M., and He, S. Y. 2007. Role of plant stomata in bacterial invasion. *Cell Microbiol.* 9: 1621–1629.

Uppalapati, S. R., Ishiga, Y., Wangdi, T., Kunkel, B. N., Anand, A., et al. 2007. The phytotoxin coronatine contributes to pathogen fitness and is required for suppression of salicylic acid accumulation in tomato inoculated with *Pseudomonas syringae* pv. *tomato* DC3000. *Mol. Plant-Microbe Interact.* 20: 955–965.

Venturi, V., and Fuqua, C. 2013. Chemical signaling between plants and plant-pathogenic bacteria. *Annu. Rev. Phytopathol.* 51: 17–37.

Vialet-Chabrand, S., Hills, A., Wang, Y., Griffiths, H., Lew, V. L., et al. 2017. Global sensitivity analysis of OnGuard models identifies key hubs for transport interaction in stomatal dynamics. *Plant Physiol.* 174: 680–688.

Von Bodman, S. B., Bauer, W. D., and Coplin, D. L. 2003. Quorum sensing in plant-pathogenic bacteria. *Annu. Rev. Phytopathol.* 41: 455–482.

Wang, S., Sun, J., Fan, F., Tan, Z., Zou, Y., et al. 2016. A *Xanthomonas oryzae* pv. *oryzae* effector, XopR, associates with receptor-like cytoplasmic kinases and suppresses PAMP-triggered stomatal closure. *Sci. China Life Sci.* 59: 897–905.

Williams, S. C., Patterson, E. K., Carty, N. L., Grisold, J. A., Hamood, A. N., et al. 2004. *Pseudomonas aeruginosa* autoinducer enters and functions in mammalian cells. *J. Bacteriol.* 186: 2281–2287.

Wu, H.-Y., Chung, P.-C., Shih, H.-W., Wen, S.-R., and Lai, E.-M. 2008. Secretome analysis uncovers an Hcp-family protein secreted via a type VI secretion system in *Agrobacterium tumefaciens. J. Bacteriol.* 190: 2841–2850.

Yan, Y., Stolz, S., Chételat, A., Reymond, P., Pagni, M., et al. 2007. A downstream mediator in the growth repression limb of the jasmonate pathway. *Plant Cell* 19: 2470–2483.

Yan, J., Zhang, C., Gu, M., Bai, Z., Zhang, W., et al. 2009. The Arabidopsis CORONATINE INSENSITIVE 1 protein is a jasmonate receptor. *Plant Cell* 21: 2220–2236.

Yuan, Z.-C., Liu, P., Saenkham, P., Kerr, K., and Nester, E. W. 2008. Transcriptome profiling and functional analysis of *Agrobacterium tumefaciens* reveals a general conserved response to acidic conditions (pH 5.5) and a complex acid-mediated signaling involved in *Agrobacterium*-plant interactions. *J. Bacteriol.* 190: 494–507.

Yuan, Z.-C., Zaheer, R., and Finan, T. M. 2005. Phosphate limitation induces catalase expression in *Sinorhizobium meliloti, Pseudomonas aeruginosa* and *Agrobacterium tumefaciens. Mol. Microbiol.* 58: 877–894.

Zhang, L., Zhang, F., Melotto, M, Yao, J, and He, S. Y. 2017. Jasmonate signaling and manipulation by pathogens and insects. *J. Exp. Bot.* 68: 1371–1385.

Zhang, W., He, S. Y., and Assmann, S. M. 2008. The plant innate immunity response in stomatal guard cells invokes G-protein-dependent ion channel regulation. *Plant J.* 56: 984–996.

Zhao, Y., Thilmony, R., Bender, C. L., Schaller, A., He, S. Y., et al. 2003. Virulence systems of *Pseudomonas syringae* pv. *tomato* promote bacterial speck disease in tomato by targeting the jasmonate signaling pathway. *Plant J.* 36: 485–499.

Zheng, M., Wang, X., Doan, B., Lewis, K. A., Schneider, T. D., et al. 2001. Computation-directed identification of OxyR DNA binding sites in *Escherichia coli. J. Bacteriol.* 183: 4571–4579.

Zheng, X.-Y., Spivey, N. W., Zeng, W., Liu, P.-P., Fu, Z. Q., et al. 2012. Coronatine promotes *Pseudomonas syringae* virulence in plants by activating a signaling cascade that inhibits salicylic acid accumulation. *Cell Host Microbe* 11: 587–596.

Zhou, Z., Wu, Y., Yang, Y., Du, M., Zhang, X., et al. 2015. An Arabidopsis plasma membrane proton ATPase modulates JA signaling and is exploited by the *Pseudomonas syringae* effector protein AvrB for stomatal invasion. *Plant Cell* 27: 2032–41.

Zhu, W., MaGbanua, M. M., and White, F. F. 2000. Identification of two novel hrp-associated genes in the hrp gene cluster of *Xanthomonas oryzae* pv. *oryzae. J. Bacteriol.* 182: 1844–1853.

6 Plants as Carriers of Human Enteric Bacterial Pathogens

The cases of human illness associated with foodborne diseases caused by fresh produce have been noticed since early 1970s and reported outbreaks associated with this commodity have grown steadily. Earlier it was thought that most cases of illness linked to fresh produce were the result of post-harvest microbial contamination rather than contamination in the field. With the recognition of pre-harvest crop contamination, being the cause of foodborne illness, new ecological concepts regarding the fitness of enteric pathogens on crop plants need to be answered. Various aspects, such as bacterial epiphytic fitness, endophytic growth and survival, resource utilization in the phyllosphere and the rhizosphere, plant-microbe interactions, and microbe-microbe interactions will provide information for a better understanding of the ecology of enteric pathogens on plants. The review article by Brandl (2006) presents an overview of issues related to the microbial contamination of fresh produce with a focus on the behavior of enteric pathogens on plant surfaces in the context of concepts in food safety and plant microbiology. Tyler and Triplett (2008), in their review article, have stated that enteric bacteria are common inhabitants of the interior of plants and *Salmonella* and *Escherichia coli* O157:H7 have been shown to enter into plants. The extent of endophytic colonization by strains is regulated by plant defenses and several genetic determinants necessary for this interior colonization in endophytic bacteria have been identified. Common virulence factors in plant and animal pathogens have also been described in bacteria that can infect both plant and animal models.

6.1 ASSOCIATION OF HUMAN ENTERIC BACTERIAL PATHOGENS WITH PLANT ENVIRONMENT

Several bacterial pathogens, including *Sal. enterica*, pathogenic *Es. coli*, *Shigella* spp., *Campylobacter* spp., *Listeria monocytogenes*, *Staphylococcus aureus*, *Yersinia* spp., and *Bacillus cereus* have caused fresh produce-associated epidemics of enteric illness. The association of human enteric viruses with foodborne illness has also been reported by Hirneisen and Kniel (2013).

Sal. enterica and pathogenic *Es. coli* are not plant pathogens. Schikora et al. (2008) and Klerks et al. (2007b) reported that plants showed chlorosis or reduced vigor following inoculation with *Sal. enterica*, but Koch's postulates were not completed in either case. Therefore, *Sal. enterica* and pathogenic *Es. coli* remain characterized as plant-associated bacteria. Certain bacterial species that commonly inhabit plant surfaces, including *Pseudomonas aeruginosa*, *Burkholderia cepacia* complex members, *Enterobacter cloacae*, *Enterococcus faecalis,* and *Erwinia* spp., have been shown to infect both plant and human tissue. The traditional view that human enteric bacterial pathogens have a defined ecological niche has been challenged by several investigators reporting outbreaks of foodborne illness associated with these pathogens. The available evidence that enteric bacterial pathogens have a high incidence on fresh fruits and vegetables, and that their presence on produce can lead to epidemics of foodborne illness, has caused a further change in the concept regarding the niche specificity of these pathogens. Furthermore, the unexpected increase in produce-associated bacterial infections is indicative of a much more important role of plants as a secondary habitat for enteric pathogens than previously thought. The role of plants, as a vector of these cross domain pathogens, to their human hosts has remained overlooked in the past. In their review article, Barak and Schroeder (2012) have highlighted the significant steps throughout the life cycle of human enteric bacterial pathogens to focus the role of plants as their vectors.

Epidemics of foodborne illness linked to fresh produce in the United States increased from 1973 through 1997, both in absolute number and in the proportion of total food-linked outbreaks with a known etiologic agent (Sivapalasingam et al. 2004). During this period, lettuce, melon, seed sprouts, and fruit juice were the four most important single produce items implicated in epidemics of foodborne illness in the United States. A survey by Center for Science in the Public Interest (2005) revealed that fresh produce caused 28,315 cases of foodborne illness and was the second most common single-food vehicle linked to outbreaks in the United States during 1990–2003. In England and Wales, salad, vegetables, and fruit caused 6.4% and 10.1% of all outbreaks with a known food vehicle during the periods of 1993–1998 (WHO 2000) and 1999–2000 (WHO 2003), respectively. Other fresh produces commonly associated with outbreaks of bacterial enteric disease include tomato, cilantro, parsley, spinach, green onions, carrot, and cabbage.

Earlier outbreaks of foodborne disease epidemics linked to fresh produce were initially suspected to result from cross contamination of fruit or vegetables during food preparation with other food items, particularly meat products. The surveys conducted by the United States (US) Food and Drug Administration (2001a, 2001b) of samples obtained from major distributors indicated that 1.6% of the domestic and 4.4% of the imported produce were contaminated with human enteric pathogens. A study by the Microbial Data Program of the US Department of Agriculture found that in 2002, 0.62% of samples of domestic or imported cantaloupe, celery, lettuce, and tomatoes were contaminated with *Es. coli* strains harboring virulence factors (US Department of Agriculture 2004). In other surveys, Harris et al. (2003) reported the presence of pathogenic enteric bacteria on produce and in unpasteurized fruit or vegetable juices sampled during production or at retail markets. The above-mentioned surveillance studies support the hypotheses of possible contamination of fresh produce with human pathogens in the pre-harvest environment, or in a post-harvest during washing and minimal processing, and consequently, of outbreaks linked to fresh produce not necessarily being the outcome of cross-contamination events with other foods. The occurrence of foodborne illness from contaminated fresh produce has changed the belief that such disease was linked only to the consumption of foods of animal origin, including meat, poultry, eggs, and milk.

Many outbreaks of foodborne illness of humans mainly due to pre-harvest contamination of food have been mentioned above, but the major outbreak caused by *Es. coli* O157:H7 in Japan in 1996 is of great significance. It caused sickening of more than 10,000 people, the majority being school going children. Moreover, the food was not of animal origin, but radish sprouts contaminated with Shiga toxin-producing *Es. coli* (STEC) (Watanabe et al. 1999). This Japanese outbreak began to highlight the fact that plant and animal pathogens are similar in many ways, which was not appreciated by many people before 1990. The numerous produce-implicated outbreaks are testament to *Sal. enterica* and pathogenic *Es. coli*'s ability to persist and use plants as vectors to humans (Taormina et al. 1999). Some important outbreaks associated with foodborne illness are listed in Table 6.1.

In US, there are more chances of *Sal. enterica* contamination from fresh plant produce than animal products. Sivapalasingam et al. (2004) reported that *Sal. enterica*, the most frequent etiological agent of outbreaks of fresh produce caused 48% of such outbreaks in the US between 1973 and 1997. However, of special concern was the occurrence of outbreaks caused by multidrug-resistant strains of *Sal. enterica* serovar Typhimurium DT104 and linked to the consumption of lettuce (Horby et al. 2003; Takkinen et al. 2005). It also commonly occurs on tomato and cantaloupe. Yearly salmonellosis outbreaks have occurred in the United States resulting from contaminated tomato fruits produced on the eastern shore of Virginia since 2002 (Bidol et al. 2007). Pathogenic *Es. coli* is the second most important causal agent of outbreaks from fresh produce, and the pathogenic strains include the enterotoxigenic serotype O157:H7 that causes hemolytic uremic syndrome leading to death, particularly in children, the elderly, and the immunocompromised. *Es. coli* O157:H7 caused 21% of all produce-linked outbreaks during 1982–2002 in the US (Rangel et al. 2005). In the US during 1990–2004, 76% of the epidemics from contaminated fruits were caused by *Sal. enterica* and 48% of the epidemics from contaminated fresh leafy vegetables were caused by *Es. coli*. Thus, America's salad bowl contains *Es. coli* O157:H7 and *Sal. enterica* that can lead to leafy green contamination and sometimes to human illness outbreaks.

The data are also available for various plant parts linked to human illness caused by *Sal. enterica*, pathogenic *Es. coli*, *Shigella* spp., and *Campylobacter* spp. for United States for the period from 1990 to 2004 (Table 6.2). During this period of 14 years, fruit contaminated with *Sal. enterica* caused 76% of the epidemics, whereas pathogenic *Es. coli* strains were the causal agent of the highest proportion of 48% foodborne illness outbreaks associated with fresh leafy vegetables. These leafy vegetables consisted mostly of lettuce, while the pathogenic *Es. coli* was predominantly serotype O157:H7.

Besides enteric bacteria, agricultural produce can also become contaminated with human viral pathogens in the field through soil, feces, or water used for irrigation; through application of manure, biosolids, pesticides, and fertilizers; and through animals, insects, and dust. The Norovirus (also known as winter vomiting bug in the United Kingdom) was originally named the 'Norwalk agent' after Norwalk, Ohio in the United States, where an outbreak of acute gastroenteritis occurred among children at Bronson Elementary School in November 1968. The genus name *Norovirus* is derived from Norwalk

TABLE 6.1
Foodborne Outbreaks Caused by Human Enteric Pathogens Linked to Contaminated Seed

Seed	Outbreak Year	Pathogen	Clinical Cases	References
Mung	1988	*Salmonella enterica* Saint Paul	143	O'Mahony et al. (1990)
Alfalfa	1994	*Sal. enterica* Bovismorbificans	210	Puohiniemi et al. (1997)
Alfalfa	1995	*Sal. enterica* Stanley	242	Mahon et al. (1997)
Alfalfa	1996	*Sal. enterica* Meleagridis	75	Mohle-Boetani et al. (2001)
Alfalfa	1997	*Escherichia coli* O157:H7	85	Como-Sabetti et al. (1997)
Alfalfa	1998	*Sal. enterica* Cubana	22	Mohle-Boetani et al. (2001)
Alfalfa	1999	*Sal. enterica* Muenchen	157	Proctor et al. (2001)
Alfalfa	2001	*Sal. enterica* Kottbus	31	Winthrop et al. (2003)
Alfalfa	2009	*Sal. enterica* Bovismorbificans	42	Rimhanen-Finne et al. (2011)
Radish	1996	*Es. coli* O157:H7	10,000	Nathan (1997)
Clover	1999	*Sal. enterica* Typhimurium	112	Brenner (1984)
Fenugreek	2011	*Es. coli* O104:H4	4,321	Buchholz et al. (2011)

TABLE 6.2
Number of Outbreaks[a] Linked to Single Items of Fresh Produce in the United States During 1990–2004

Produce Type	*Salmonella enterica*	Pathogenic *Escherichia coli*	*Shigella* spp.	*Campylobacter* spp.
Fruits and fruit juices	32 (76)	8 (19)	1 (2)	1 (2)
Leafy vegetables	8 (30)	13 (48)	3 (11)	3 (11)
Seed sprouts	9 (60)	6 (40)	0	0
Total	49	27	4	4

Source: Brandl, M.T., *Annu. Rev. Phytopathol.*, 44, 367–392, 2006.
Note: Figures in parentheses indicate percentage of outbreaks caused by a given pathogen within a given produce category.
[a] Data are compiled from the publications: Center for Science in the Public Interest (2002); Harris et al. (2003); The US Centers for Disease Control and Prevention (2003, 2004, 2005). The authors of these publications did not report outbreaks from single items of fresh produce caused by *Sta. aureus*, *Li. monocytogenes*, or *Y. enterocolitica* in the United States during the given time period.

virus, the only species of the genus. The species causes approximately 90% of epidemic outbreaks of non-bacterial gastroenteritis around the world (Lindersmith et al. 2003) and may be responsible for 50% of all foodborne outbreaks of gastroenteritis in the United States (Widdowson et al. 2005; Anonymous 2007). The virus affects around 267 million people and causes over 200,000 deaths each year; these deaths are usually in less developed countries and in the very young, elderly, and immunosuppressed persons (Debbink et al. 2012).

Hirneisen and Kniel (2013) studied the survival and stability of human noroviruses and norovirus surrogates, namely, *Murine norovirus* (MNV) and *Tulane virus* (TV) on foliar surfaces of spinach plants in pre-harvest growth stages. The site of virus inoculation showed the greatest impact on virus survival as viruses inoculated on adaxial leaf surfaces had lower decimal reduction time (D values) than viruses inoculated on abaxial leaf surfaces. Under certain conditions, spinach type also influenced virus survival, with greater D values observed for survival on semi-savoy spinach leaves compared to those on smooth spinach leaves. Additional UVA and UVB exposure to mimic sunlight influenced virus survival on adaxial surfaces of both semi-savoy and smooth spinach plants for both the viruses. The average D value of human genogroup II norovirus inoculated onto semi-savoy spinach was not statistically significant from those of MNV and TV, suggesting that these surrogates may have similar survival on spinach leaves compared with human noroviruses. The information on the behavior of enteric viruses on spinach leaves can help the growers for a better risk assessment under certain growing conditions. The following factors are responsible for the frequent occurrence of produce-linked outbreaks:

1. Introduction of minimally processed produce
2. Increased importation of fresh produce
3. Increased consumption of meals outside the home
4. Increased popularity of salad bars
5. Increased consumption of fresh fruits and vegetables, and fresh fruit juices
6. Improved methods to identify and track pathogens.

6.2 SOURCES OF CONTAMINATION

The common sources of contamination of fresh plant produce with these pathogens are contaminated manure (enteric pathogens can survive for prolonged periods in animal feces), contaminated water used for irrigation, pesticide and fertilizer application, and soil. In general, enteric bacterial pathogens enter agricultural environments via animal feces. The verified routes of crop produce contamination from feces are soil, compost, water, and seed.

6.2.1 Animals

After the occurrence of outbreaks caused by contaminated fresh produce, animals found in agricultural production areas have been identified as carriers of Shiga toxin-producing *Es. coli* and *Sal. enterica*. These may be freely wandering animals in the fields, but also include those contained in pastures or high-density feeding operations. The feces from these contained animals are a common source of human bacterial pathogens as substantiated by many reports (Brashears et al. 2003; Hutchison et al. 2005a, 2005b; Thurston-Enriquez et al. 2005; Soupir et al. 2006). The pathogens associated with these feces are moved by rain and flood waters or aerosolized by high velocity winds. After mixing with water, these pathogens can move along with surface water, commonly used for irrigation and also as pesticide and fertilizer diluents. In addition to direct flow into field crops, surface water can also percolate through the soil column into groundwater.

6.2.2 Soil

The survival of human enteric bacterial pathogens in agricultural fields depends on many characteristics of the soil, such as soil structure, clay content, organic matter, and resident microflora. In temperate climates, the populations of both *Sal. enterica* and pathogenic *Es. coli* in soil decline with time following entry of pathogen either via manure or water (Natvig et al. 2002; Semenov et al. 2008).

The survival of *Salmonella* and/or pathogenic *Es. coli* in manure-emended soil ranges from several weeks to more than 8 months (Islam et al. 2004a, 2004b, 2004c; Johannessen et al. 2005; Nicholson et al. 2005; Holley et al. 2006; You et al. 2006; Semenov et al. 2008). The survival of these pathogens in manure, compost, and ultimately in soil, leading to contamination of fresh produce, is influenced by the prevailing environment. The form of manure applied to fields as fertilizer amendments can influence the survival of *Sal. enterica* and *Es. coli* O157:H7 in soil and on plants. Semenov et al. (2009) reported that in general, slurry application led to more pathogen cells percolating to greater depths than manure. After mixing farmyard manure or injecting slurry into the top 10 cm of soil, *Sal. enterica* survived approximately 25 days in soil treated with manure or slurry, and *Es. coli* O157:H7 survived longer (~30 days) in soil when injected as slurry compared to mixing of manure (~18 days). More pathogens are found in the rhizosphere in comparison to bulk soil after a slurry application, but no difference is observed after addition of manure.

Sal. enterica survived for more than 8 months in soil of a radish or carrot crop following an October application of contaminated manure or irrigation water and was isolated from radishes and carrots up to 84 and 203 days, respectively, after the seeds were sown in the contaminated manure or irrigated soil (Islam et al. 2004b). *Es. coli* O157:H7 persisted for 217 days in soils emended with contaminated compost and was detected on lettuce and parsley up to 77 and 177 days, respectively, after planting (Islam et al. 2004a). *Es. coli* O157:H7 persisted longer in soil in association with plants, similar to leaf spot pathogens such as *Xanthomonas campestris* pv. *vitians* (Barak et al. 2001) and *Psm. syringae* (Hirano and Upper 1983). Plant pathogenic bacteria that lack the ability to survive in bulk soil can utilize plant debris to persist between crops or overwinter/oversummer. Although *Es. coli* O157:H7 can survive in the bulk soil for more than 8 months under certain conditions, its populations do decline. However, its survival in soil is enhanced by the presence of plant roots (Gagliardi and Karns 2002) and crop debris.

6.2.3 Water

The most common route of human enteric bacterial pathogens' entry into agricultural environments is via contaminated fecal matter deposited directly into surface water as reported by Jay et al. (2007) or animal feces transported into surface water following rain or flood events. On the basis of a survey conducted in Salinas Valley, California, Cooley et al. (2007) found that frequency of isolation of *Es. coli* O157:H7 from surface water, e.g., creeks and streams, was more likely in areas where cattle were present. Isolation frequency was also higher following rain or flooding events.

Besides *Es. coli* O157:H7, *Sal. enterica* is often isolated from surface waters commonly used for irrigation water in some agricultural production areas (Winfield and Groisman 2003; Haley et al. 2009). In a survey for *Sal. enterica* in and around Monterey County, California, Gorski et al. (2011) isolated several strains from surface water over a period of several months, indicating their persistence in the environment. However, the occurrence of these human pathogens in the fresh produce production environment is not limited to California only, but could be anywhere where animals, including humans, deposit feces.

6.2.4 Seed

Contaminated seed is a confirmed route of contamination in foodborne illness outbreaks. Cooley et al. (2003) reported that both *Sal. enterica* and *Es. coli* O157:H7 moved from sites of inoculation on roots to leaves and flowers. This extensive movement was along the outside of the plant, and the flagella-mediated motility leads to seed contamination following *Sal. enterica* introduction to the rhizosphere. The US Food and Drug Administration has recommended soaking of seed for 20 min in a 20,000 mg L^{-1} of calcium hypochlorite solution. This treatment does reduce, but does not eliminate the contaminating bacteria from the seeds.

Sprouts are a common fresh produce vehicle in causing foodborne illness (CDC 2009). Alfalfa seeds are the most common contaminated seed identified, but it is unknown whether this is due to the popularity of this type of sprout with consumers, to seed production practices, or to sprout production practices. As only a small quantity of alfalfa seed is used for human consumption, therefore, no specific production practices are followed to check its contamination with human enteric pathogens. Mohle-Boetani et al. (2001) reported that alfalfa seeds that tested positive for an outbreak strain were fertilized with un-composted chicken manure and irrigated with non-potable canal water. As seed treatment recommended by US Food and Drug Administration is not fully effective against human enteric pathogens, their use leads to occurrence of severe outbreaks. Brooks et al. (2001) have given an interesting account of clover sprouts responsible for a 1999 salmonellosis outbreak, which were grown by two independent producers using a common seed lot. Producer X did not treat the clover seeds prior to sprouting, while the producer Y soaked seeds for 20 min in a 20,000 mg L^{-1} calcium hypochlorite solution (based on the recommendations of US Food and Drug Administration). The seed produced by producer X showed four times more illnesses than those of producer Y. It is evident that seed treatment reduces, but does not eliminate the bacterium responsible for illness from the sprouts. Subsequently, Proctor et al. (2001) confirmed these findings in multistate outbreaks resulting from alfalfa sprouts.

6.3 COLONIZATION OF PLANTS BY HUMAN ENTERIC BACTERIAL PATHOGENS

6.3.1 Attraction by Plants

The rhizosphere is a reservoir of plant-associated bacteria including human enteric pathogens. Human enteric pathogenic bacteria are capable of sensing available plant nutrients and use these signals to actively move towards plants for subsequent attachment and colonization. The rhizosphere of plants attracts *Salmonella* spp. This active movement of *Sal. enterica* to the plant roots is different from the previously reported passive contamination due to soil splashing, irrigation, insect transmission, or even passive uptake through roots (Solomon et al. 2002). *Sal. enterica* is attracted to the lettuce rhizosphere by a sugar-like carbon source in root exudates (Klerks et al. 2007b). Many bacteria colonize the roots of plants or persist in the rhizosphere by using these exudates as a nutritional source. *Salmonella*'s behavior is similar to that of *Ralstonia solanacearum*, which demonstrates a preferential chemotaxis towards exudates of host plants (Yao and Allen 2006). Whether the numerous human pathogens associated with plants or different serovars of specific pathogens differentially chemotax to various plants remains to be determined.

Due to chemotaxis, *Sal. enterica* in the soil can find the rhizosphere and subsequently colonizes the phyllosphere. *Sal. enterica* differentially colonizes agricultural plants growing in the contaminated soil, which suggests a case of differential chemotaxis towards specific plants (Barak et al. 2008). It preferred radish, turnip, and broccoli for colonization over carrot, lettuce, and tomato. Although, the bacterium showed no preference among lettuce cultivars, it preferred the specific leafy greens radicchio and endive in comparison to lettuce. This preference for specific crop plants may reflect differences in the root exudates or a more favorable rhizosphere microflora, which is also influenced by root exudates. The steep gradient of root exudates in soil is continually modulated by the indigenous soil microflora.

After attraction by the root exudates, *Sal. enterica* actively moves towards the roots and colonizes them. After colonization of roots, *Sal. enterica* and *Es. coli* O157:H7 can move into the phyllosphere by flagella-mediated motility. It is not known whether human pathogens, such as *Sal. enterica*, require directed motility to locate roots, although niche exploration on the root has been shown to be flagella-mediated (Cooley et al. 2003). Root exudates, therefore, have a dual role, i.e., the induction of chemotaxis of *Sal. enterica* to the roots and the simultaneous conditioning of its cells for plant attachment and subsequent colonization (Klerks et al. 2007a; Barak et al. 2008).

Like rhizosphere, the bacteria are also attracted to phyllosphere. Depending on the plant species, *Sal. enterica* preferentially colonizes specific areas of leaves, i.e., cell junctions (Cooley et al. 2003), along veins (Brandl and Mandrell 2002), stomata (Kroupitski et al. 2009), and/or trichomes (Barak et al. 2011). It is unknown whether any attractant, other than a preferred microclimate, shelter from UV exposure, or a place of free water accumulation, exists for bacteria at cell junctions or along veins. The attractant produced by trichomes is also unknown, although it is suspected that the nutrient-rich secretions common of glandular trichomes (Wagner 1991) could be alluring. Nutrients produced by actively photosynthesizing guard cells of lettuce attract *Sal. enterica* to stomata and addition of exogenous sugars disrupts the attraction (Kroupitski et al. 2009). Tropism towards stomata is shared by *Sal. enterica* (Kroupitski et al. 2009). Bacteria aggregate near open stomata for subsequent entry to the leaf apoplast. For plant-associated bacteria, entry to the leaf apoplast serves two purposes; (1) bacteria escape the harsh climate of the leaf surface and (2) apoplast entry is the first step in leaf infection.

6.3.2 Attachment and Persistence

Colonization of plants by bacteria requires their adhesion to the plant surface and subsequent multiplication. Flagella, pili, and extracellular polysaccharides are commonly employed by bacteria for plant colonization. Flagella are best known for their role in bacterial motility; however, other uses of the flagella that could be useful during plant colonization exist. Wang et al. (2005) reported sensing of external wetness by flagella of *Sal. enterica*. Sensing the external environment by flagella leads to regulation of its own biogenesis as well as a type third secretion system (TIIISS) in the *Salmonella* pathogenicity island-1. In addition to chemotaxis to roots or movement along plants, *Sal. enterica* and *Es. coli* O157:H7 use flagella for adhesion to leaves as reported by Kroupitski et al. (2009) and Saldana et al. (2011), respectively.

The enteric plant-associated bacteria, both animal and plant pathogens, require extracellular polysaccharide, cellulose, and protein filaments for biofilm formation. Biofilm formation by plant-associated bacteria is a multicellular and aggregative behavior, and it is used by them to successfully colonize plants (Danhorn and Fuqua 2007). AgfD is the master regulator of aggregative behaviour in *Sal. enterica* (Gibson et al. 2006), while the CsgD is the *Es. coli* homolog of AgfD (Römling et al. 1998). AgfD-regulated aggregative components, namely, thin aggregative fimbriae (but curli in *Es. coli*), cellulose, and O-antigen capsules, are involved in adhesion and colonization to alfalfa sprouts (Barak et al. 2005, 2007; Jeter and Matthysse et al. 2005). *Sal. enterica* and *Es. coli* O157:H7 differ in colonization of plants. *Es. coli* O157:H7 utilizes only curli fimbriae, and not cellulose, during leaf colonization (Saldana et al. 2011), whereas *Sal. enterica* requires both type of fimbriae, although curli to a lesser extent. These AgfD-regulated aggregative components have no role in pathogenicity studies carried out with animals (van der Velden et al. 1998). Hence, AgfD is the first enteric bacterial regulator identified outside an animal host. The fact that AgfD and those yet uncharacterized regulators that govern plant fitness factors are necessary for these enteric pathogens to complete their life cycle on plants gives support to the better understanding that plants have a crucial role in the life cycle of these pathogens. Brandl et al. (2005) reported that 54% of the cell population of *Sal. enterica* was present in aggregates composed of at least 128 cells, while some comprised of as many as 4096 cells, 3 days after inoculation onto moist cilantro leaves. These bacterial cells, while in the exopolymer matrix of aggregates, may remain protected from the detrimental effect of rapid and frequent fluxes in water activity on plants.

Although thin aggregative fimbriae (curli), cellulose, O-antigen, and flagella are required for aggregative behavior in plant-associated bacteria, they have distinct roles in their persistence on plant tissue, such as bacterial cell-cell or bacterial cell-plant cell interactions. Thin aggregative fimbriae and cellulose protect *Sal. enterica* from desiccation and the capsule delays the oxidative degradation (Gibson et al. 2006; Jonas et al. 2007). For persistence on plants, it is paramount to ward off desiccation due to the rapid and daily fluctuations in free water on leaf surfaces (Lindow and Leveau 2002). *Sal. enterica* strains with stronger biofilm-forming ability *in vitro* have stronger adhesion and persistence on lettuce leaves (Kroupitski et al. 2009). Biofilm formation is equally important for colonization of plant roots (Barak et al. 2007). *Sal. enterica*'s tolerance to desiccation stress coupled with its ability to multiply under subsequent wet conditions helps in its survival on plant surface (Kroupitski et al. 2009).

Kroupitski et al. (2013), using recombinase-based *in vivo* expression technology, identified *Sal. enterica* genes involved in the persistence of this organism on post-harvest lettuce during cold storage. Knockout mutations in eight upregulated genes revealed that four of them have a role in persistence of the pathogen in this system. These genes were *stfC*, *bcsA*, *misL*, and *yidR*, encoding a fimbrial outer membrane usher, a cellulose synthase catalytic subunit, an adhesin of the autotransporter family expressed from the *Salmonella* pathogenicity island-3, and a putative ATP/GTP-binding protein, respectively. *bcsA*, *misL*, and *yidR*, but not *stfC* mutants were impaired also in attachment and biofilm formation, suggesting that these functions are required for survival of *Sal. enterica* on post-harvest lettuce. This being the first report that gene *misL*, which has a role in *Sal. enterica* binding to fibronectin in animal hosts, is involved also in adhesin to post-harvested lettuce.

In addition to biofilm formation, *Sal. enterica* has a quorum-sensing system and the cells can swarm. Swarming is a coordinated, flagella-mediated motility carried out by both Gram-stain-positive and Gram-stain-negative bacteria across semi-solid surfaces (Kearns 2010). The swarming cells of *Sal. enterica* are morphologically and physiologically distinct from swimming or stationary cells. During swarm state induction, *Sal. enterica* cells elongate, produce more flagella (Harshey and Matsuyama 1994), produce either a surfactant (Mireles et al. 2001) or osmolite (Chen et al. 2007), exhibit adaptive resistance to multiple antibiotics (Butler et al. 2009), and initiate de novo biosynthesis (Kim et al. 2004). Root exudates induced swarming in plant-associated bacteria such as *Serratia liquefaciens* (Teplitski et al. 2000) as well as *Sal. enterica* (Barak et al. 2009). Bacteria preferentially colonize the root zone of elongation and sites of lateral root expansion (Walker et al. 2003), a nutrient rich environment due to root exudates and hence, colonized by large, diverse bacterial communities. For survival in this niche, bacteria capable of inducing swarming may avoid antimicrobial compounds produced by the plant to defend against pathogens or compete with rival microbes for nutrients, and propel themselves through the semi-solid sloughing of root cellular debris and mucilage to ideal niches. Inability of *Sal. enterica* to swarm results in its failure to colonize the roots (Barak et al. 2009). Non-flagellated and non-motile strains neither move beyond the point of inoculation on roots nor invade lateral roots with a small initial inoculum (10 cfu mL^{-1}) (Cooley et al. 2003). Swarming is the only one type of flagella-mediated motility used by *Sal. enterica* to locate preferential niches.

Plant-associated bacteria also use swimming to colonize the plants. Cooley et al. (2003) reported that *Sal. enterica* travelled from the roots into the phyllosphere by swimming. However, the ability to swim requires coordinated behavior. A hypermotile strain of *Ral. solanacearum* poorly colonized roots (Meng et al. 2011). Therefore, colonization is a balanced act, and it requires coordination of cells for flagella-mediated motility.

Continuous association of a bacterium with a single host or an environment may reduce its fitness for other hosts. Perpetual passage from animal to soil and/or water/to plants/ to animal may push pathogens to retain and gain fitness for each environment. Analysis of *Es. coli* O157:H7 spinach outbreak that occurred in 2006 revealed that disease severity increased compared with the 1996 sprout outbreak (Manning et al. 2008). Genomic sequencing of the strain causing spinach outbreak showed substantial genomic differences that contributed to more severe disease. Fitness factors important for survival on plants, or at least outside the animal host, may have selected for such a deadly strain.

6.3.3 Growth and Multiplication

Human enteric bacterial pathogens grow and multiply on plants before their entry into the intestinal tract of humans along with seed and other plant products. Their multiplication on plants in the phyllosphere, rhizosphere, and spermosphere is influenced by many biological and environmental factors. The phyllosphere is a comparatively harsh environment for the multiplication and survival of microorganisms compared to rhizosphere. A brief account of multiplication and survival of human enteric bacterial pathogens on plants is given below.

6.3.3.1 In Phyllosphere

For an effective colonization of plants, the bacteria need to grow and multiply on the plants. The plant phyllosphere and rhizosphere are highly heterogeneous environments and vary in nutrient availability. Depending on the plant species, physiological state, and plant growth conditions; sugars, organic acids, and amino acids are available to bacteria in the phyllosphere. *Es. coli* O157:H7 and *Sal. enterica* persist on tomato, lettuce, and cilantro plants with minimal growth (Islam et al. 2004a, 2004b).

Koukkidis et al. (2017) reported that traces of juices released from damaged salad leaves can significantly enhance colonization of salad leaves by *Sal. enterica*. Salad juices in water increased *Salmonella* growth by 110% and in host-like serum-based media by more than 2,400-fold compared to control. Other aspects relevant to salad leaf colonization and retention, such as motility and biofilm formation, increased over control levels by >220% and 250%, respectively; direct attachment to salad leaves increased by >350% when the salad leaf juice was present. Salad leaf juices also enhanced pathogen attachment to the salad bag plastic. Over 5 days of refrigeration (a standard storage time for bagged salad leaves), even traces of juice within the salad bag fluids, when mixed in water, increased *Salmonella* growth

280-times higher than the control, as well as enhanced salad bag colonization, which could lead to an infection of the consumers.

Mishra et al. (2017) attempted to develop a mathematical system model to understand the pathway of *Es. coli* O157:H7 in the production of leafy greens. The results indicate that the seasonality of outbreaks of *Es. coli* O157:H7-associated contamination of leafy greens was in good agreement with the prevalence of this pathogen in cattle and wild pig feces in a major leafy greens producing region in California. On the basis of results obtained in different scenarios, it is evident that the concentration of *Es. coli* O157:H7 in leafy greens can be reduced considerably, if contamination of soil with wild pig and cattle feces is mitigated.

The leaf surface is a harsh site for bacterial growth due to rapid fluctuations in temperature and availability of free water, effect of UV radiation, and the heterogeneity of available nutrients (Hirano and Upper 1983; Lindow and Brandl 2003). Many authors have reported the persistence of enteric bacteria, namely, *Sal. enterica*, *En. cloacae,* and *Es. coli* O157:H7 in the phyllosphere for extended periods, although their populations may decline with time (Pederson and Leser 1992; Takikawa et al. 2002; Cooley et al. 2003; Islam et al. 2004a, 2004b, 2004c). The colonization of leaf tissue by *Sal. enterica* led to its colonization of fruit and seed (Cooley et al. 2003; Barak et al. 2011). A free film of water enhances bacterial colonization and survival rates. Relative humidity of 85%–95% sustained *Psm. aeruginosa* (Green et al. 1974) and *Sal. enterica* (Cooley et al. 2003; Barak et al. 2011) populations on leaves. However, survival on leaves at lower humidity requires fitness factors like UV-mediated DNA repair systems (Fernandez et al. 2006), pigmentation (Poplawsky et al. 2000), and extracellular polysaccharide production (Kemp et al. 2004) commonly expressed by phytopathogenic bacteria. Escape from such an environment by bacteria lacking these fitness factors increases their survival in the phyllosphere.

Es. coli O157:H7 occurs to a greater extent on the external tissue on plants, but it can also occur in the internal tissue of plants. Wright et al. (2013) using high-resolution microscopy determined the extent of internalization of *Es. coli* O157:H7 in live spinach and lettuce plants. The bacterium was found in internal tissue of both the plant species. Colonization occurred within the apoplast between the plant cells. Furthermore, colonies were detected inside the cell wall of epidermal and cortical cells of spinach and *Nicotiana benthamiana* roots. Internal colonization of epidermal cells resembled that of *Pe. atrosepticum* on potato. In contrast, only sporadic cells of *Es. coli* strain K-12 were found on spinach, but there was no occurrence of bacteria inside the cells. It appears that internal colonization of plants by bacteria is limited to a specific group of plant-interacting bacteria, including *Es. coli* O157:H7.

O'Brien and Lindow (1989) demonstrated that population sizes of *Sal. enterica* Typhimurium and *Es. coli* reached equal to those of *Psm. syringae* in the bean and corn phyllosphere under constant wet conditions, while their population sizes were significantly less than those of the epiphytic bacteria after incubation under dry conditions. The colonization studies of a strain of *Sal. enterica* serovar Thompson, isolated during an outbreak linked to cilantro, on the cilantro phyllosphere revealed that the pathogen was less fit than the common epiphytic bacteria *Psm. chlororaphis* and *Pantoea agglomerans* under wet conditions at 22°C (Brandl and Mandrell 2002). However, the competitiveness of *Sal. enterica* serovar Thompson on wet cilantro leaves significantly increased when the plants were incubated at 30°C and 37°C, enabling it to achieve rapidly higher population sizes than at 24°C. In growth chamber experiments, Natvig et al. (2002) detected *Sal. enterica* at harvest from the leaves of arugula and the roots of radish plants (raised from seeds planted in soil cores amended with contaminated manure) grown under conditions simulating a wet and warm summer environment, but not under those simulating a comparatively drier and cooler spring environment. The higher multiplication rates of the enteric pathogen at warm temperatures likely allow it to use a larger share of the nutrient resources that it can assimilate on the plant surface. Hence, it is probable that free water from rain fall, dew, and irrigation, and accompanying warm temperatures enable *Sal. enterica* to undergo bursts of growth on plants in the field, thereby making the required population sizes to cause foodborne illness. However, *Sal. enterica* and other enteric pathogens are likely not as much adapted as plant-colonizing bacteria to utilize the full spectrum of nutrients present on plant surfaces and, therefore, are limited by the amount and range of available nutrients that they can assimilate to thrive on plants. Lin (1996) reported that unlike many plant-associated bacteria and *Es. coli*, *Sal. enterica* was typically unable to assimilate sucrose, one of the main sugars present in leaf and root exudates. This fact may explain the relatively lower number of outbreaks caused by *Sal. enterica* on leaves in comparison to fruit as is evident in Table 6.2. Gu et al. (2013) investigated the effects of soil management on movement of *Sal. enterica* Typhimurium in tomato plants and found that more *Salmonella* spp. survived in inoculated leaves on plants grown in conventional than in organic soil. Only tomato plants grown in conventional soil produced contaminated fruit (1%) and approximately 5% of the seed from infested fruit were internally contaminated. No *Salmonella* sp. was detected in plants grown from contaminated seed.

Low water availability is considered to be one of the main limitations of bacterial survival on plant surfaces. However, *Sal. enterica* has a high tolerance to long-term desiccation stress in non-host environments. This is the reason why *Sal. enterica* is the most common etiologic agent of foodborne illness linked to spices (Vibha et al. 2006) and also responsible for the occurrence of two salmonellosis outbreaks associated with dry raw almonds (The Centers for Disease Control and Prevention 2004; Isaacs et al. 2005). It was hypothesized that at least in one of these almond-linked outbreaks, *Sal. enterica* multiplied on the moist almond fruits on the orchard floor at harvest and survived on the dry kernels to which it had migrated. It shows the pathogen's ability to cope with low water availability as the moisture content of dried nuts in storage is less than 6%. Brandl and Mandrell (2002) reported that the growth of *Sal. enterica* declined at rates similar to those of epiphytic bacteria on the dry leaf surface of cilantro plants under conditions of low relative humidity imposed for 24 h, but it resumed at significant rates upon reoccurrence of wet conditions on the leaves. Therefore, the tolerance of *Sal.*

enterica to desiccation stress, combined with its potential to multiply under subsequent wet conditions may confer on this pathogen the ability to persist despite the high fluctuations in water availability on plant surfaces in the field. It indicates that even small populations of the pathogen capable of survival on crop plants before harvest may increase to infectious dose levels during growth-conducive conditions while the produce is stored or processed for consumption.

Generally, *Sal. enterica* has comparatively low epiphytic fitness on plants. Poza-Carrion et al. (2013) reported that viable population sizes of *Sal. enterica* applied to plants pre-inoculated with *Psm. syringae* or either of two *Er. herbicola* strains were ≥10-fold higher than that on control plants that were not pre-colonized by such indigenous bacteria when assessed 24–72 h after the imposition of desiccation stress. Hence, the survival of immigrant cells of *Sal. enterica* on plants appears to be strongly influenced by the epiphytic bacteria, and the presence of common epiphytic bacteria on plants can protect such immigrants from a key stress like desiccation encountered on leaf surfaces.

The occurrence of large number of cases of human illness linked to fresh produce, spices, and nuts contaminated with non-typhoidal *Salmonella* supports the hypothesis that colonization of plants is an important part of the life cycle of this human pathogen. Although plant responses to human enteric pathogens are distinct from the more specific responses to phytopathogenic bacteria, plants appear to recognize *Salmonella*, likely by detecting conserved microbial patterns, which subsequently activate basal defenses. Numerous *Salmonella* genes have been identified as playing a role in its colonization of plant surfaces and tissues and in its various interactions with other members of the plant-microbial community. Importantly, *Salmonella* employs diverse and overlapping strategies to interact with plants and their microflora and to successfully colonize its vertebrate hosts (Brandl et al. 2013).

The phyllosphere contamination on agricultural plants is influenced by the traits unique to the phyllosphere of different plants. Barak et al. (2008) reported that *Sal. enterica* failed to colonize some cultivars of tomato following their sowing in contaminated soil. This differential persistence in the phyllosphere was also observed following dip inoculation of the plants in a *Sal. enterica* suspension as reported by Barak et al. (2011). Leaf characteristics that varied among the tomato cultivars might have contributed to the differential survival of *Sal. enterica*. Physical structures, such as leaf waxes, are known to influence bacterial colonization of leaves. Marcell and Beattie (2002) showed that the reduction in the density of crystalline waxes of corn increased the epiphytic populations of *Clavibacter michiganensis* and *Pa. agglomerans*. The changes in cuticular wax biosynthesis that produce glossy corn exhibit differential effects on populations of *Cla. michiganensis* and *Pa. agglomerans*. Those leaf characteristics that encourage colonization of non-pathogen may also benefit plant-associated human pathogens, such as *Sal. enterica* and pathogenic *Es. coli*. The trichome, a known preferential site of bacterial colonization on the leaf surface identified by Beattie and Lindow (1995), also influences persistence of *Sal. enterica*. Barak et al. (2011) identified glandular trichomes as the preferential site of *Sal. enterica* on tomato leaves.

Bacteria, including the phytopathogenic bacteria, lack the ability to directly penetrate the intact plant surface. However, they rely entirely on natural openings or accidental wounds to enter the internal plant tissues. High humidity or wet weather results in opening of stomata to avoid leaf water-soaking. The bacteria that fail to open stomata, namely, *Psm. aeruginosa* (Starkey and Rahme 2009), *Sal. enterica* (Kroupitski et al. 2009), and *Bu. cepacia* (Kawamoto and Lorbeer 1974) can enter the leaf apoplast. Entry in the substomatal chamber places the bacteria in a protected niche that facilitates their persistence and is the first step towards plant infection by phytopathogenic bacteria.

The route of human enteric bacterial pathogens internalization into plant tissue has been a subject for extensive discussion (Warriner and Namvar 2010). Human bacterial pathogens' internalization through plant stomata has been studied for *Es. coli* O157:H7 (Melotto et al. 2006) and *Sal. enterica* serovar Typhimurium (Kroupitski et al. 2009). Thilmony et al. (2006) reported that *Es. coli* O157:H7 triggers stomatal closure, but it is not able to overcome this plant immune response when inoculated as pure cultures under laboratory conditions. However, Kroupitski et al. (2009) documented a remarkable ability of *Sal. enterica* serovar Typhimurium to migrate towards stomata and enter plant tissues without triggering stomatal immune response. This finding suggests the possibility that not only plant pathogens, but also some human enteric bacterial pathogens have evolved mechanisms to subvert plant stomata-based defence to enter plant tissues. The underlying mechanism of this observation is not understood and is a topic of active research (Baker et al. 2010)

Several investigators have examined the role of plant diseases in the contamination of produce with foodborne pathogens. Although *Salmonella* and *Es. coli* O157:H7 fail to infect plants or grow in the phyllosphere unaided, they grow in plant cell lysates, root exudates, and nutrients liberated by plant pathogens (Aruscavage et al. 2006; Kyle et al. 2010) and persist as infective doses, documented by the numerous foodborne illness outbreaks. Wells and Butterfield (1997) observed twice the incidence of *Salmonella* species on fruit and vegetables affected by soft rot in comparison to healthy produce. They found that the population of *Salmonella* species on potato, carrot, and bell pepper discs co-inoculated with the soft rot pathogen, *Er. carotovora* or *Psm. viridiflava* was 10- and 3-fold higher, respectively, than when inoculated with the enteric pathogen alone. The hypothesis that *Salmonella* spp. exchange AI-2 signals with *Pe. carotovorum* to increase its competitive fitness was tested by Cox et al. (2013), using mutants involved in AI-2 production (*luxS*) or perception (*lsrACDBF* or *lsrG*). Co-infections of a wild-type *Salmonella* sp. and its AI-2 mutants (at ≈$3–10^4$ cfu mL^{-1}) were established in green or red tomato ("FL 47" or "Campari") for 3 or 5 days, as well as tomato co-infected with *Pectobacterium* (at 10^9 cfu mL^{-1}) or its *luxS* mutant. The results of this study showed that AI-2 signalling offers no significant benefit to *Salmonella* spp. in this model of colonization of tomato or soft rots.

Carlin et al. (1995) reported that soft rot also increased the population of *Li. monocytogenes* on endive leaves. It is presumed that the damaged cells in rotten plant tissue leak abundant nutrients, which otherwise are not available to enteric pathogens on healthy plant surfaces. This assumption is supported by the fact that the growth of enteric pathogens is also enhanced in plant tissue that is simply mechanically injured. The degradation of plant tissue by plant pathogens may offer enteric pathogens a broad spectrum of nutrients for growth, in addition to nutrient leakage from the damaged plant cells due to disease. However, not all plant pathogens affect the colonization of plant tissue by enteric pathogens to the same extent. Brandl (2004) reported that co-inoculation of *Sal. enterica* with *Er. chrysanthemi* or *Psm. viridiflava* revealed the better correlation in population increase of *Sal. enterica* on the diseased leaves of cilantro plants with those of *Er. chrysanthemi* than with those of *Psm. viridiflava*. Both the growth rate and final population sizes of *Sal. enterica* were greater in *Er. chrysanthemi* infected leaves than in leaves infected with *Psm. viridiflava*. In another study, Liao and Sapers (1999) revealed that the presence of *Psm. fluorescens* and *Psm. viridiflava* strongly inhibited the growth of *Li. monocytogenes*, while the presence of *Er. carotovora* or *Xa. campestris* did not affect its growth, either negatively or positively, after co-inoculation with one of these four pathogens onto potato tuber slices. They suggested the competition for iron via production of siderophores as a possible mechanism for the antagonism by the pseudomonads in this study. Conversely, in other situations, the growth of enteric pathogens may be enhanced by their heterologous utilization of siderophores produced by plant-associated microorganisms or by the plants themselves. Some plant pathogenic bacteria promote the growth of enteric pathogens even without causing disease. *Xa. vesicatoria* promotes *Sal. enterica* growth in the phyllosphere in the absence of plant disease (Barak and Liang 2008). The populations of *Sal. enterica* grew several hundred times higher in phyllosphere in the presence of *Xa. vesicatoria* than *Sal. enterica* populations on leaves without xanthomonads. It shows that apart from benefiting the enteric pathogens via nutrient liberation, plant pathogens benefit the enteric pathogens also in other ways.

6.3.3.2 Inhabitation of Spermosphere and Rhizosphere

Both spermosphere and rhizosphere are rich in plant exudates that can be utilized by the surrounding microbial community. Bacterial multiplication varies in spermospheres of different plants and is higher when seed exudates include essential nutrients and lack microbial inhibitors. Roberts et al. (2009) reported that *En. cloacae* exhibited a higher metabolic activity in the spermosphere of pea than that of cucumber, and the difference might be due to sugar concentration or composition, as replication could be restored in the cucumber spermosphere by addition of xylose and fructose. Colonization of the corn, radish, and wheat spermospheres was not related to the total peptide catabolic activity of *En. cloacae* as reported by Roberts et al. (1996).

The rhizosphere is a reservoir for plant-associated bacteria, including human enteric pathogens. Nutrient availability in the rhizosphere is also active and dynamic. Germinating seeds and subsequent roots release soluble and insoluble products into the rhizosphere via root exudates. After seed inoculation, *Sal. enterica* preferred to colonize the roots over the phyllosphere, most likely owing to root exudates as reported by Charkowski et al. (2002). Composition and distribution of root exudates vary with many factors, including plant age, different root zones, nutritional status, and soil types. These factors influence the microbial community associated with the root system, as bacterial species differ in their ability to utilize and compete for various nutrient sources (Petra et al. 2004; Haichar et al. 2008). Populations of *Bur. cepacia* complex members and *Es. coli* O157:H7 in the rhizosphere were generally up to 100 times more than those in bulk soil (Pallud et al. 2001; Gagliardi and Karns 2002; Jacobs et al. 2008).

6.4 MULTIPLE SEROVAR INFECTIONS

In a trace back investigation of the 1988 salmonellosis outbreak from seeds or sprouts in the United Kingdom, *Sal. enterica* ser. SaintPaul, the outbreak strain; ser. Virchow, a second outbreak strain; ser. Muenchen, ser. Lancing, ser. Litchfield, and ser. Arizona were isolated by O'Mahony et al. (1990). In another salmonellosis outbreak from alfalfa seed lot occurring in 1995, *Sal. enterica* ser. Newport, the outbreak strain; ser. Albany and ser. Schwarzengrund were isolated. In these outbreaks, different serovars were associated with the plants involved. When the produce is contaminated with multiple serovars, it is not known why one or two particular serovars cause human illness and not the others. The following possibilities may be there: (1) the outbreak strain had a higher population on the seed and thus had more chances to infect; (2) some *Sal. enterica* serovars are more virulent or infectious than others; and (3) the outbreak strain was more easily isolated in the clinical examination than those isolated in the research laboratory.

6.5 PLANT RESPONSE/REACTION

Two of the human enteric pathogens, namely, *Sal. enterica* and *Es. coli* have generally been reported to cause outbreaks on different fresh produce commodities. *Sal. enterica* was found on tomato and cantaloupe, pathogenic *Es. coli* on lettuce, and *Sal. enterica* and pathogenic *Es. coli* on sprouted seeds. Different *Sal. enterica* serovars were found to colonize the plants in these outbreaks. It is not known whether this differential serovar colonization was due to the pathogen involved or the host. Some *Sal. enterica* serovars have caused repeated outbreaks in different years on the same fresh produce commodity, e.g., Poona serovar on cantaloupe and Newport and Javiana serovars on tomato. The reports of outbreaks of foodborne illness occurred till date, have not identified a single dominant serovar on all the plants tested.

6.6 DEFENSE MECHANISMS OF PLANTS

The plants rely on the innate immunity of each cell, which acts as a first line of defence. FLS2 is one of the best described molecular pattern receptors, and it recognizes flagellin

(Zipfel et al. 2004). FLS2 acts as a first line of defence by recognizing the most conserved amino acid sequence in flagellin, which is common to all plant-associated bacteria, including *Sal. enterica* (Felix et al. 1999). On their arrival on leaves, the bacteria are detected by plants and are normally prevented from entering the apoplast by immediate closure of stomata and left on the leaf surface to survive the harsh conditions.

Roy et al. (2013) reported that *Sal. enterica* serovar Typhimurium SL1344, like *Psm. syringae* pv. *tabaci* DC3000 causes a transient stomatal closure in lettuce. The mechanisms underlying stomatal closure and reopening mediated by SL1344 are not understood, but it appears that not only plant pathogens, but also some human pathogens may have evolved mechanisms to modulate plant stomatal movements as part of their colonization strategy of the phyllosphere (Roy et al. 2013; Melotto et al. 2014). As such, the study of stomatal defence may have implications beyond plant diseases.

The role played by plant defence mechanisms in restricting colonization and/or persistence of human pathogens has been elucidated by bacterial mutational analysis and plant response assays. Iniguez et al. (2005) reported that elimination of the microbe-associated molecular pattern, flagella, increased endophytic colonization of *Sal. enterica*. This finding suggests that *Medicago truncatula* recognizes *Sal. enterica* on arrival via flagellin and mounts a defence response leading to a reduction in the bacterial population. *Sal. enterica* induced both salicylic acid (SA)-dependent and -independent plant defenses as reported by Iniguez et al. (2005). Ethylene, a signal molecule for induced systemic resistance in plants, decreases endophytic colonization of *Sal. enterica*. It shows that the plants use at least two different defence pathways to recognize the human pathogens and mount a defence to keep their population in check.

Phytopathogenic bacteria utilize a TIIISS and inject effectors to overcome plant innate immunity and facilitate pathogenic action. Plant and human pathogens share TIIISSs and their effectors, as well as transcriptional regulators. Two TIIISSs required for animal infection by *Salmonella* occur within *Salmonella* pathogenicity island (SPI) 1 or 2. *SPI*-1 promotes invasion of mammalian cells. Iniguez et al. (2005) reported that mutations that eliminate the structural component of the *SPI*-1 TIIISS needle apparatus are required for translocation of other effectors and increase endophytic populations of *Sal. enterica* in the roots. TIIISS mutants of *Sal. enterica* had even higher populations on plants that fail to accumulate SA or are disrupted in both SA-mediated and SA-independent defence responses (Iniguez et al. 2005). These results show that both SA-dependent and -independent pathways restrict populations of *Sal. enterica* in wild-type plants, and that the TIIISSs, including flagella, trigger either innate immunity and/or effector-triggered immunity. Hirano et al. (1999) reported that TIIISS mutants of *Psm. syringae* had no fitness defect in the spermosphere, although their populations in the phyllosphere were small. The role of TIIISS in plant pathogenic bacteria appears to facilitate bacterial replication because without growth the disease does not occur.

6.7 FUTURE STRATEGIES

Roy et al. (2013) investigated plant defence responses induced by the fully pathogenic bacteria *Es. coli* O157:H7 and *Sal. enterica* serovar Typhimurium SL1344 in both *Arabidopsis thaliana* and lettuce (*Lactuca sativa*). Unlike SL1344, O157:H7 induced strong plant immunity at both pre-invasion and post-invasion steps of infection. O157:H7 triggered stomatal closure even under high relative humidity. It indicates that plants may recognize and respond to some human pathogens more effectively than others. It is very important to understand how plant responses can diminish contamination of human bacterial pathogens for preventing disease outbreaks and improving the safety of food supplies.

Several investigators have demonstrated that the outcome of interactions of *Salmonella* with plants depends on the plant species and genotype. Internalization of *Salmonella* into plant tissue varies greatly among plant species and plant colonization also varies among cultivars of a species. Barak et al. (2011) reported that there was an approximately 100-fold variation in the phyllosphere populations of *Salmonella* on four tomato cultivars, and *Solanum pimpinellifolium* variety WVa700 supported the lowest number of bacteria. Interestingly, WVa700 was also significantly less susceptible to *Psm. syringae* pv. *tomato*, causal agent of bacterial speck of tomato. Quilliam et al. (2012) showed an important role of plant genotype also in the proliferation of *Es. coli* O157:H7 in the lettuce phyllosphere. These findings suggest that specific genetic factors governing the plant response to microbial colonization reflect the outcome of the interaction of enteric pathogens with plants, or that the differences in physicochemical properties, such as availability of certain nutrients or surface morphology, associated with different crop genotypes impact proliferation of *Salmonella* and *Es. coli* on and inside plants. These findings give a clue to the potential of breeding for resistance to colonization by enteric pathogens, keeping in view the economic feasibility of such breeding programmes. However, it is unknown whether there is any correlation between plant basal immune responses to phytopathogens and to human pathogens. A positive correlation would provide an opportunity to integrate breeding for increased basal resistance of crops to both plant and human enteric pathogens.

In addition to competition for nutrients on plant surfaces, enteric pathogens may be exposed to phages and antibacterial compounds producing bacteria. Janisiewicz et al. (1999) reported that a strain of *Psm. syringae*, previously shown having fungicidal properties, reduced growth of *Es. coli* O157:H7 on wounded apples by 10- to 1,000-fold. This and other such findings prompted further experiments on biological control of human enteric pathogens associated with fresh produce. Fett (2006) showed that a well-characterized biocontrol strain of *Psm. fluorescens* 2–79 effectively reduced

Salmonella populations on alfalfa sprouts. Some phytobacteria, including those involved in biocontrol of plant diseases, significantly inhibit attachment and plant colonization by non-typhoidal *Salmonella* and enterovirulent *Es. coli* by producing antibiotics or competing for nutrients in the phyllosphere (Teplitski et al. 2011). Similarly, bacteriophages can also be used to reduce the contamination of enteric pathogens on the produce. With the development of host range mutant (h-mutants) phages by Jackson (1989), the efficacy of phages for controlling bacteria has increased many-fold, and they have been used successfully for controlling some bacterial diseases of plants in the field. It is evident that phages and certain bacteria have a proven antibacterial activity and can be used as potential tools for controlling enteric pathogens throughout the produce production cycle. However, it is important to consider that zero tolerance for most human pathogens on fresh fruit and vegetables implies that even the most effective biocontrol agent would need to be integrated into a general control strategy.

It is evident from the above given account that the plants play a critical role in the life cycle of human enteric bacterial pathogens. These enteric pathogens use plants as vectors between animal hosts, all the while following the life cycle script of plant-associated bacteria (Barak and Schroeder 2012). Also, animal/human bacterial pathogens and bacterial plant pathogens have some common mechanisms such as type III secretion systems and their effectors, and transcriptional regulators, which function in both animal and plant hosts. It is just possible that during the process of evolution, *Sal. enterica* and pathogenic *Es. coli*, presently characterized as plant-associated bacteria, may acquire pathogenic capabilities and become plant pathogens. It will be a matter of great concern for plant bacteriologists and plant pathologists.

Fletcher et al. (2013) reported that the frequent occurrences of outbreaks of foodborne illness associated with fresh fruits and vegetables have revived interest among public health agencies and sparked a new wave of research on food safety issues related to microbial contamination of fresh produce. Recent efforts to address concerns about microbial contamination of food plants and resulting foodborne human illness have prompted new collaborations and interactions between scientific communities of plant pathology and food safety. The repeated outbreaks of human illness resulting from pre-harvest contamination of plant products by human enteric pathogens such as Shiga toxin-producing *Es. coli* and *Salmonella* spp. have led some plant pathologists to broaden the application of their science in the past two decades to address problems of human enteric pathogens on plants. New collaborations between members of plant pathology and food safety communities have led to enhanced research activity and understanding of the issues for which research is needed. Therefore, continued interactions and communication among plant pathologists and food safety scientists is essential and can be achieved by the creation of an interdisciplinary research coordination network. Brandl and Sundin (2013) have emphasized that the critical aspect of produce safety is relevant to research in plant microbial ecology and intersects with numerous concepts that are explored in plant pathology.

REFERENCES

Anonymous 2007. Norovirus: Technical Fact Sheet. National Center for Infectious Diseases, CDC. http://www.cdc.gov/Ncidod/dvrd/revb/gastro/norovirus-factsheet.htm.

Aruscavage, D., Lee, K., Miller, S. A., and LeJeune, J. T. 2006. Interactions affecting the proliferation and control of human pathogens on edible plants. *J. Food Sci.* 71: 89–98.

Baker, C. M., Chitrakar, R., Obulareddy, N., Panchal, S., Williams, P., et al. 2010. Molecular battles between plant and pathogenic bacteria in the phyllosphere. *Braz. J. Med. Biol. Res.* 43: 698–704.

Barak, J. D., Gorski, L., Liang, A. S., and Narm, K.-E. 2009. Previously uncharacterized *Salmonella enterica* genes required for swarming play a role in seedling colonization. *Microbiology* 155: 3701–3709.

Barak, J. D., Gorski, L., Naraghi-Arani, P., and Charkowski, A. O. 2005. *Salmonella enterica* virulence genes are required for bacterial attachment to plant tissue. *Appl. Environ. Microbiol.* 71: 5685–5691.

Barak, J. D., Jahn, C. E., Gibson, D. L., and Charkowski, A. O. 2007. The role of cellulose and O-antigen capsule in the colonization of plants by *Salmonella enterica*. *Mol. Plant-Microbe Interact.* 20: 1083–1091.

Barak, J. D., Koike, S. T., and Gilbertson, R. L. 2001. Role of crop debris and weeds in the epidemiology of bacterial leaf spot of lettuce in California. *Plant Dis.* 85: 169–178.

Barak, J. D., Kramer, L. C., and Hao, L.-Y. 2011. Colonization of tomato plants by *Salmonella enterica* is cultivar dependent and type I trichomes are preferred colonization sites. *Appl. Environ. Microbiol.* 77: 498–504.

Barak, J. D., and Liang, A. S. 2008. Role of soil, crop debris, and a plant pathogen in *Salmonella enterica* contamination of tomato plants. *PLoS One* 3: e1657.

Barak, J. D., Liang, A., and Narm, K.-E. 2008. Differential attachment to and subsequent contamination of agricultural crops by *Salmonella enterica*. *Appl. Environ. Microbiol.* 74: 5568–5570.

Barak, J. D., and Schroeder, B. K. 2012. Interrelationships of food safety and plant pathology: The life cycle of human pathogens on plants. *Annu. Rev. Phytopathol.* 50: 241–266.

Beattie, G. A., and Lindow, S. E. 1995. The secret life of foliar bacterial pathogens on leaves. *Annu. Rev. Phytopathol.* 33: 145–172.

Bidol, S. A., Daly, E. R., Ricker, R. E., Hill, T. A., Taylor, T. H., et al. 2007. Multistate outbreaks of *Salmonella* infections associated with raw tomatoes eaten in restaurants: United States, 2005–2006. *CDC Morb. Mortal. Wkly. Rep.* 56: 909–911.

Brandl, M. T. 2004. Fitness of *Salmonella enterica* in the phyllosphere. *Phytopathology* 94(Suppl.): 128.

Brandl, M. T. 2006. Fitness of human enteric pathogens on plants and implications for food safety. *Annu. Rev. Phytopathol.* 44: 367–392.

Brandl, M. T., and Mandrell, R. E. 2002. Fitness of *Salmonella enterica* serovar Thompson in the cilantro phyllosphere. *Appl. Environ. Microbiol.* 68: 3614–3621.

Brandl, M. T., and Sundin, G. W. 2013. Focus on food safety: Human pathogens on plants. *Phytopathology*, 103: 304–305.

Brandl, M. T., Cox, C. E., and Teplitski, M. 2013. *Salmonella* interactions with plants and their associated microbiota. *Phytopatholgy* 103: 316–325.

Brandl, M. T., Miller, W. G., Bates, A. H., and Mandrell, R. E. 2005. Production of autoinducer 2 in *Salmonella enterica* serovar Thompson contributes to its fitness in chickens but not on cilantro leaf surfaces. *Appl. Environ. Microbiol.* 71: 2653–2662.

Brashears, M. M., Galyean, M. L., Loneragan, G. H, Mann, J. E., and Killinger-Mann, K. M. 2003. Prevalance of *Escherichia coli* O157 and performance of beef feedlot cattle given *Lactobacillus* direct-fed antimicrobials. *J. Food. Prot.* 66: 748–754.

Brenner, D. J. ed. 1984. Family 1. Enterobacteriaceae Rahn 1937, Nom. fam. cons. pp. 408–20 in: *Bergey's Manual of Systematic Bacteriology*. 1st ed. Vol. 1. Williams & Wilkins Co., Baltimore, MD.

Brooks, J. T., Rowe, S. Y., Shillam, P., Heltzel, D. M., Hunter, S. B., et al. 2001. *Salmonella typhimurium* infections transmitted by chlorine-pretreated clover sprout seeds. *Am. J. Epidemiol.* 154: 1020–1028.

Buchholz, U., Bernard, H., Werber, D., Böhmer, M. M., Remschmidt, C., et al. 2011. German outbreak of *Escherichia coli* O104:H4 associated with sprouts. *N. Engl. J. Med.* 365: 1763–1770.

Butler, M. T., Wang, Q., and Harshey, R. M. 2009. Cell density and mobility protect swarming bacteria against antibiotics. *Proc. Natl. Acad. Sci. USA* 107: 3776–3781.

Carlin, F., Nguyen-The, C., and Abreu da Silva, A. 1995. Factors affecting the growth of *Listeria monocytogenes* on minimally processed fresh endive. *J. Appl. Bacteriol.* 78: 636–646.

The Centers for Disease Control and Prevention (CDC). 2003. 2002-Foodborne disease outbreaks due to bacterial etiologies. http://www.cdc.gov/foodborneoutbreaks/us_outb/fbo2002./bacterial02.htm.

The Centers for Disease Control and Prevention (CDC). 2004. Outbreaks of *Salmonella* serotype Enteritidis infections associated with raw almonds-United States and Canada, 2003–2004. *MMWR* 53: 1–3.

The Centers for Disease Control and Prevention (CDC). 2005. Foodborne outbreaks due to bacterial etiologies. 2004. http://www.cdc.gov/foodborneoutbreaks/us_outb/fbo2004./summary04.htm.

The Centers for Disease Control and Prevention (CDC). 2009. Surveillance for food borne disease outbreaks: United States, 2006. *Morb. Mortal.Wkly. Rep.* 58: 609–615.

Center for Science in the Public Interest. 2002. Outbreak Alert! Closing the Gaps in Our Federal Food Safety Net. Washington, DC. http://www.cspinet.org/Report.

Center for Science in the Public Interest. 2005. Outbreak Alert! Closing the Gaps in Our Federal Food Safety Net. Washington, DC. http://www.cspinet.org/Report.

Charkowski, A. O., Barak, J. D., Sarreal, C. Z., and Mandrell, R. E. 2002. Differences in growth of *Salmonella enterica* and *Escherichia coli* O157:H7 on alfalfa sprouts. *Appl. Environ. Microbiol.* 68: 3114–3120.

Chen, B. G., Turner, L., and Berg, H. C. 2007. The wetting agent required for swarming in *Salmonella enterica* serovar Typhimurium is not a surfactant. *J. Bacteriol.* 189: 8750–8753.

Como-Sabetti, K., Reagan, S., Allaire, S., Parrott, K., Simonds, C. M., et al. 1997. Outbreaks of *Escherichia coli* O157:H7 infection associated with eating alfalfa sprouts: Michigan and Virginia, June–July 1997. *Morb. Mortal. Wkly. Rep.* 46: 741–744.

Cooley, M. B., Miller, W. G., and Mandrell, R. E. 2003. Colonization of A*rabidopsis thaliana* with *Salmonella enterica* and enterohemorrhagic *Escherichia coli* O157:H7 and competition by *Enterobacter asburiae*. *Appl. Environ. Microbiol.* 69: 4915–4926.

Cooley, M., Carychao, D., Crawford-Miksza, L., Jay, M. T., Myers, C., et al. 2007. Incidence and tracking of *Escherichia coli* O157:H7 in a major produce production region in California. *PLoS One* 2: e1159.

Cox, C. E., McClelland, M., and Teplitski, M. 2013. Consequences of disrupting *Salmonella* AI-2 signaling on interactions within soft rots. *Phytopathology* 103: 352–361.

Danhorn, T., and Fuqua, C. 2007. Biofilm formation by plant-associated bacteria. *Annu. Rev. Microbiol.* 61: 401–422.

Debbink, K., Lindesmith, L. C., Donaldson, E. F., and Baric, R. S. 2012. Norovirus immunity and the great escape. *PLoS Pathog* 8 (10): e1002921.

Felix, G., Duran, J. D., Volko, S., and Boller, T. 1999. Plants have a sensitive perception system for the most conserved domain of bacterial flagellin. *Plant J.* 18: 265–276.

Fernandez Zenoff, V., Sineriz, F., and Farias, M. E. 2006. Diverse responses to UV-B radiation and repair mechanisms of bacteria isolated from high-altitude aquatic environments. *Appl. Environ. Microbiol.* 72: 7857–7863.

Fett, W. F. 2006. Inhibition of *Salmonella enterica* by plant-associated pseudomonads *in vitro* and on sprouting alfalfa seed. *J. Food Prot.* 69: 719–728.

Fletcher, J., Leach, J. E., Eversole, K., and Tauxe, R. 2013. Human pathogens on plants: Designing a multidisciplinary strategy for research. *Phytopathology* 103: 306–315.

Gagliardi, J. V., and Karns, J. S. 2002. Persistence of *Escherichia coli* O157:H7 in soil and on plant roots. *Environ. Microbiol.* 4: 89–96.

Gibson, D. L., White, A. P., Snyder, S. D., Martin, S., Heiss, C., et al. 2006. *Salmonella* produces an O-antigen capsule regulated by AgfD and important for environmental persistence. *J. Bacteriol.* 188: 7722–7730.

Gorski, L., Parker, C. T., Liang, A., Cooley, M. B., Jay-Russell, M. T., et al. 2011. Prevalence, distribution, and diversity of *Salmonella enterica* in a major produce region of California. *Appl. Environ. Microbiol.* 77: 2734–2748.

Green, S. K., Schroth, M. N., Cho, J. J., Kominos, S. D., and Vitanza-Jack, V. B. 1974. Agricultural plants and soil as a reservoir for *Pseudomonas aeruginosa*. *Appl. Microbiol.* 28: 987–991.

Gu, G., Cevallos-Cevallos, J. M., Vallad, G. E., and van Bruggen, A. H. C. 2013. Organically managed soils reduce internal colonization of tomato plants by *Salmonella enterica* serovar Typhimurium. *Phytopathology* 103: 381–388.

Haichar, F. Z., Marol, C., Berge, O., Rangel-Castro, J. I., Prosser, J. I., et al. 2008. Plant host habitat and root exudates shape soil bacterial community structure. *Int. Soc. Microbiol. Ecol. J.* 2: 1221–1230.

Haley, B. J., Cole, D. J., and Lipp, E. K. 2009. Distribution, diversity and seasonality of waterborne *Salmonella* in a rural watershed. *Appl. Environ. Microbiol.* 75: 1248–1255.

Harris, L. J., Farber, J. N., Beuchat, L. R., Parish, M. E., Suslow, T. V., et al. 2003. Outbreaks associated with fresh produce: Incidence, growth, and survival of pathogens in fresh and fresh-cut produce. *Comp. Rev. Food Sci. Food Saf.* 2 (Suppl.): 78–141.

Harshey, R. M., and Matsuyama, T. 1994. Dimorphic transition in *Escherichia coli* and *Salmonella typhimurium*: Surface-induced differentiation into hyperflagellate swarmer cells. *Proc. Natl. Acad. Sci. USA* 91: 8631–8635.

Hirano, S. S., and Upper, C. D. 1983. Ecology and epidemiology of foliar bacterial plant pathogens. *Annu. Rev. Phytopathol.* 21: 243–270.

Hirano, S. S., Charkowski, A. O., Collmer, A., Willis, D. K., and Upper, C. D. 1999. Role of the Hrp type III protein secretion system in growth of *Pseudomonas syringae* pv. *syringae* B728a on host plants in the field. *Proc. Natl. Acad. Sci. USA* 96: 9851–956.

Hirneisen, K. A., and Kniel, K. E. 2013. Norovirus surrogate survival on spinach during preharvest growth. *Phytopathology* 103: 389–94.

Holley, R. A., Arrus, K. M., Ominski, K. H., Tenuta, M., and Blank, G. 2006. *Salmonella* survival in manure-treated soils during simulated seasonal temperature exposure. *J. Environ. Qual.* 35: 1170–1180.

Horby, P. W., O'Brien, S. J., Adak, G. K., Graham, C., Hawker, J. I., et al. 2003. A national outbreak of multi-resistant *Salmonella enterica* serovar Typhimurium definitive phage type (DT) 104 associated with consumption of lettuce. *Epidemiol. Infect.* 130: 169–178.

Hutchison, M. L., Walters, L. D., Avery, S. M., Munro, F., and Moore, A. 2005a. Analyses of livestock production, waste storage, and pathogen levels and prevalences in farm manures. *Appl. Environ. Microbiol.* 71: 1231–1236.

Hutchison, M. L., Walters, L. D., Moore, T., Thomas, D. J. I., and Avery, S. M. 2005b. Fate of pathogens present in livestock wastes spread onto fescue plots. *Appl. Environ. Microbiol.* 71: 691–696.

Iniguez, A. L., Dong, Y., Carter, H. D., Ahmer, B. M., Stone, J. M., et al. 2005. Regulation of enteric endophytic bacterial colonization by plant defenses. *Mol. Plant-Microbe Interact.* 18: 169–178.

Isaacs, S., Aramini, J., Ciebin, B., Farrar, J. A., Ahmed, R., et al. 2005. An international outbreak of salmonellosis associated with raw almonds contaminated with a rare phage type of *Salmonella enteritidis*. *J. Food Prot.* 68: 191–198.

Islam, M., Doyle, M. P., Phatak, S. C., Millner, P., and Jiang, X. 2004a. Persistence of enterohemmorrhagic *Escherichia coli* O157:H7 in soil and on leaf lettuce and parsley grown in fields treated with contaminated manure composts or irrigation water. *J. Food Prot.* 67: 1365–1370.

Islam, M., Morgan, J., Doyle, M. P., Phatak, S. C., Millner, P., et al. 2004b. Fate of *Salmonella enterica* serovar Typhimurium on carrots and radishes grown in fields treated with contaminated manure composts or irrigation water. *Appl. Environ. Microbiol.* 70: 2497–2502.

Islam, M., Morgan, J., Doyle, M. P., Phatak, S. C., Millner, P., et al. 2004c. Persistence of *Salmonella enterica* serovar Typhimurium on lettuce and parsley and in soils on which they were grown in fields treated with contaminated manure composts or irrigation water. *Foodborne Pathog. Dis.* 1: 27–35.

Jackson, L. E. 1989. Bacteriophage prevention and control of harmful plant bacteria. U. S. Patent No. 4828999.

Jacobs, J. L., Fasi, A. C., Ramette, A., Smith, J. J., Hammerschmidt, R., et al. 2008. Identification and onion pathogenicity of *Burkholderia cepacia* complex isolates from the onion rhizosphere and onion field soil. *Appl. Environ. Microbiol.* 74: 3121–3129.

Janisiewicz, W. J., Conway, W. S., and Leverentz, B. 1999. Biological control of postharvest decays of apple can prevent growth of *Escherichia coli* O157: H7 in apple wounds. *J. Food Prot.* 62: 1372–1375.

Jay, M. T., Cooley, M., Carychao, D., Wiscomb, G. W., Sweitzer, R. A., et al. 2007. *Escherichia coli* O157:H7 in feral swine near spinach fields and cattle, central California coast. *Emerg. Infect. Dis.* 13: 1908–1911.

Jeter, C., and Matthysse, A. G. 2005. Characterization of the binding of diarrheagenic strains of *E. coli* to plant surfaces and the role of curli in the interaction of the bacteria with alfalfa sprouts. *Mol. Plant-Microbe Interact.* 18: 1235–1242.

Johannessen, G. S., Bengtsson, G. B., Heier, B. T., Bredholt, S., Wasteson, Y., et al. 2005. Potential uptake of *Escherichia coli* O157:H7 from organic manure into crisphead lettuce. *Appl. Environ. Microbiol.* 71: 2221–2225.

Jonas, K., Tomenius, H., Kader, A., Normark, S., Römling, U., et al. 2007. Roles of curli, cellulose and BapA in *Salmonella* biofilm morphology studied by atomic force microscopy. *BMC Microbiol.* 7: 70.

Kawamoto, S. O., and Lorbeer, J. W. 1974. Infection of onion leaves by *Pseudomonas cepacia*. *Phytopathology* 64: 1440–45.

Kearns, D. B. 2010. A field guide to bacterial swarming motility. *Nat. Rev. Microbiol.* 8: 634–644.

Kemp, B. P, Horne, J., Bryant, A., and Cooper, R. M. 2004. *Xanthomonas axonopodis* pv. *manihotis gumD* gene is essential for EPS production and pathogenicity and enhances epiphytic survival on cassava (*Manihot esculenta*). *Physiol. Mol. Plant Pathol.* 64: 209–218.

Kim, J. W., Kim, Y. S., and Kyung, K. H. 2004. Inhibitory activity of essential oils of garlic and onion against bacteria and yeasts. *J. Food Prot.* 67: 499–504.

Klerks, M. M., Franz, E., van Gent-Pelzer, M., Zijlstra, C., and van Bruggen, A. H. C. 2007a. Differential interaction of *Salmonella enterica* serovars with lettuce cultivars and plant-microbe factors influencing the colonization efficiency. *Int. Soc. Microbiol. Ecol. J.* 1: 620–631.

Klerks, M. M., van Gent-Pelzer, M., Franz, E., Zijlstra, C., and van Bruggen, A. H. C. 2007b. Physiological and molecular responses of *Lactuca sativa* to colonization by *Salmonella enterica* serovar Dublin. *Appl. Environ. Microbiol.* 73: 4905–4914.

Koukkidis, G., Haigh, R., Allcock, N., Jordan, S., and Freestone, P. 2017. Salad leaf juices enhance *Salmonella* growth, colonization of fresh produce, and virulence. *Appl. Environ. Microbiol.* 83: e02416–e02516.

Kroupitski, Y., Brandl, M. T., Pinto, R., Belausov, E., Tamir-Ariel, D., et al. 2013. Identification of *Salmonella enterica* genes with a role in persistence on lettuce leaves during cold storage by recombinase-based *in vivo* expression technology. *Phytopathology* 103: 362–372.

Kroupitski, Y., Golberg, D., Belausov, E., Pinto, R., Swartzberg, D., et al. 2009. Internalization of *Salmonella enterica* in leaves is induced by light and involves chemotaxis and penetration through open stomata. *Appl. Environ. Microbiol.* 75: 6076–6086.

Kyle, J. L., Parker, C. T., Goudeau, D., and Brandl, M. T. 2010. Transcriptome analysis of *Escherichia coli* O157:H7 exposed to lysates of lettuce leaves. *Appl. Environ. Microbiol.* 76: 1375–1387.

Liao, C. H., and Sapers, G. M. 1999. Influence of soft rot bacteria on growth of *Listeria monocytogenes* on potato tuber slices. *J. Food Prot.* 62: 343–348.

Lin, E. 1996. Dissimilatory pathways for sugars, polyols, and carboxylates. pp. 307–342 in: *Escherichia coli and Salmonella: Cellular and Molecular Biology*. F. Neidhardt, ed. ASM Press, Washington, DC.

Lindesmith, L., Moe, C., Marionneau, S., Ruvoen, N., Jiang, X., et al. 2003. Human susceptibility and resistance to Norwalk virus infection. *Nat. Med.* 9: 548–553.

Lindow, S. E., and Brandl, M. T. 2003. Microbiology of the phyllosphere. *Appl. Environ. Microbiol.* 69: 1875–83.

Lindow, S. E., and Leveau, J. H. J. 2002. Phyllosphere microbiology. *Curr. Opin. Biotechnol.* 13: 238–243.

Mahon, B. E., Ponka, A., Hall, W. N., Komatsu, K., Dietrich, S. E., et al. 1997. An international outbreak of *Salmonella* infections caused by alfalfa sprouts grown from contaminated seeds. *J. Infect. Dis.* 175: 876–882.

Manning, S. D., Motiwala, A. S., Springman, A. C., Qi, W., Lacher, D. W., et al. 2008. Variation in virulence among clades of *Escherichia coli* O157:H7 associated with disease outbreaks. *Proc. Nat. Acad. Sci. USA* 105: 4868–4873.

Marcell, L. M., and Beattie, G. A. 2002. Effect of leaf surface waxes on leaf colonization by *Pantoea agglomerans* and *Clavibacter michiganensis. Mol. Plant-Microbe Interact.* 15: 1236–1244.

Melotto, M., Panchal, S., and Roy, D. 2014. Plant innate immunity against human bacterial pathogens. *Front. Microbiol.* 5: 411.

Melotto, M., Underwood, W., Koczan, J., Nomura, K., and He, S. Y. 2006. Plant stomata function in innate immunity against bacterial invasion. *Cell* 126: 969–980.

Meng, F., Yao, J., and Allen, C. 2011. A MotN mutant of *Ralstonia solanacearum* is hypermotile and has reduced virulence. *J. Bacteriol.* 193: 2477–2486.

Mireles, J. R., Toguchi, A., and Harshey, R. M. 2001. *Salmonella enterica* serovar Typhimurium swarming mutants with altered biofilm-forming abilities: Surfactin inhibits biofilm formation. *J. Bacteriol.* 183: 5848–5854.

Mishra, A., Pang, H., Buchanan, R. L., Schaffner, D. W., and Pradhan, A. K. 2017. A system model for understanding the role of animal feces as a route of contamination of leafy greens before harvest. *Appl. Environ. Microbiol.* 83: e02775–e02816.

Mohle-Boetani, J. C., Farrar, J. A., Werner, S. B., Minassian, D., Bryant, R., et al. 2001. *Escherichia coli* O157 and *Salmonella* infections associated with sprouts in California, 1996–1998. *Ann. Intern. Med.* 135: 239–247.

Nathan, R. 1997. American seeds suspected in Japanese food poisoning epidemic. *Nat. Med.* 3: 705–706.

Natvig, E. E., Ingham, S. C., Ingham, B. H., Cooperband, L. R., and Roper, T. R. 2002. *Salmonella enterica* serovar Typhimurium and *Escherichia coli* contamination of root and leaf vegetables grown in soils with incorporated bovine manure. *Appl. Environ. Microbiol.* 68: 2737–2744.

Nicholson, F. A., Groves, S. J., and Chambers, B. J. 2005. Pathogen survival during livestock manure storage and following land application. *Bioresour. Technol.* 96: 135–43.

O'Brien, R. D., and Lindow, S. E. 1989. Effect of plant species and environmental conditions on epiphytic population sizes of *Pseudomonas syringae* and other bacteria. *Phytopathology* 79: 619–627.

O'Mahony, M., Cowden, J., Smyth, B., Lynch, D., Hall, M., et al. 1990. An outbreak of *Salmonella* saint-paul infection associated with bean sprouts. *Epidemiol. Infect.* 104: 229–235.

Pallud, C., Viallard, V., Balandreau, J., Normand, P., and Grundmann, G. 2001. Combined use of a specific probe and PCAT medium to study *Burkholderia* in soil. *J. Microbiol. Methods* 47: 25–34.

Pederson, J. C., and Leser, T. D. 1992. Survival of *Enterobacter cloacae* on leaves and in soil detected by immunofluorescence microscopy in comparison with selective plating. *Microbiol. Releases* 1: 95–102.

Petra, M., David, C., and Ching Hong, Y. 2004. Development of specific rhizosphere bacterial communities in relation to plant species, nutrition and soil type. *Plant Soil* 261: 199–208.

Poplawsky, A. R., Urban, S. C., and Chun, W. 2000. Biological role of xanthomonadin pigments in *Xanthomonas campestris* pv. *campestris. Appl. Environ. Microbiol.* 66: 5123–5127.

Poza-Carrion, C., Suslow, T., and Lindow, S. 2013. Resident bacteria on leaves enhance survival of immigrant cells of *Salmonella enterica. Phytopathology* 103: 341–351.

Proctor, M. E., Hamacher, M., Tortorello, M. L., Archer, J. R., and Davis, J. P. 2001. Multistate outbreak of *Salmonella* serovar Muenchen infections associated with alfalfa sprouts grown from seeds pretreated with calcium hypochlorite. *J. Clin. Microbiol.* 39: 3461–3465.

Puohiniemi, R., Heiskanen, T., and Siitonen, A. 1997. Molecular epidemiology of two international sproutborne *Salmonella* outbreaks. *J. Clin. Microbiol.* 35: 2487–2491.

Quilliam, R. S., Williams, A. P., and Jones, D. L. 2012. Lettuce cultivar mediates both phyllosphere and rhizosphere activity of *Escherichia coli* O157:H7. *PLoS One* 7(3): e33842.

Rangel, J. M., Sparling, P. H., Crowe, C., Griffin, P. M., and Swerdlow, D. L. 2005. Epidemiology of *Escherichia coli* O157:H7 outbreaks, United States, 1982–2002. *Emerg. Infect. Dis.* 11: 603–609.

Rimhanen-Finne, R., Niskanen, T., Lienemann, T., Johansson, T., Sjöman, M., et al. 2011. A nationwide outbreak of *Salmonella* Bovismorbificans associated with sprouted alfalfa seeds in Finland, 2009. *Zoonoses Public Health* 58: 589–596.

Roberts, D. P, Baker, C. J., McKenna, L., Liu, S., Buyer, J. S., et al. 2009. Influence of host seed on metabolic activity of *Enterobacter cloacae* in the spermosphere. *Soil Biol. Biochem.* 41: 754–761.

Roberts, D. P., Dery, P. D., and Hartung, J. S. 1996. Peptide utilization and colonization of corn, radish and wheat spermospheres by *Enterobacter cloacae. Soil Biol. Biochem.* 28: 1109–1111.

Römling, U., Sierralta, W. D., Eriksson, K., and Normark, S. 1998. Multicellular and aggregative behaviour of *Salmonella typhimurium* strains is controlled by mutations in the *agf*D promoter. *Mol. Microbiol.* 28: 249–264.

Roy, D., Panchal, S., Rosa, B. A., and Melotto, M. 2013. *Escherichia coli* O157:H7 induces stronger plant immunity than *Salmonella enterica* Typhimurium SL1344. *Phytopathology* 103: 326–332.

Saldana, Z., Sánchez, E., Xicohtencatl-Cortes, J., Puente, J. L., and Giron, J. A. 2011. Surface structures involved in plant stomata and leaf colonization by shiga-toxigenic *Escherichia coli* O157:H7. *Front. Microbiol.* 2: 119.

Schikora, A., Carreri, A., Charpentier, E., and Hirt, H. 2008. The dark side of the salad: *Salmonella typhimurium* overcomes the innate immune response of *Arabidopsis thaliana* and shows an endopathogenic lifestyle. *PLoS One* 3: e2279.

Semenov, A. V., Franz, E., van Overbeek, L., Termorshuizen, A. J., and van Bruggen, A. H. C. 2008. Estimating the stability of *Escherichia coli* O157:H7 survival in manure-amended soils with different management histories. *Environ. Microbiol.* 10: 1450–1459.

Semenov, A. V., van Overbeek, L., and van Bruggen, A. H. C. 2009. Percolation and survival of *Escherichia coli* O157:H7 and *Salmonella enterica* serovar Typhimurium in soil amended with contaminated dairy manure or slurry. *Appl. Environ. Microbiol.* 75: 3206–3215.

Sivapalasingam, S., Friedman, C. R., Cohen, L., and Tauxe, R. V. 2004. Fresh produce: A growing cause of outbreaks of foodborne illness in the United States, 1973 through 1997. *J. Food Prot.* 67: 2342–2353.

Solomon, E. B., Yaron, S., and Matthews, K. R. 2002. Transmission of *Escherichia coli* O157:H7 from contaminated manure and irrigation water to lettuce plant tissue and its subsequent internalization. *Appl. Environ. Microbiol.* 68: 397–400.

Soupir, M., Mostaghimi, S., Yagow, E., Hagedorn, C., and Vaughan, D. 2006. Transport of fecal bacteria from poultry litter and cattle manures applied to pastureland. *Water Air Soil Pollut.* 169: 125–136.

Starkey, M., and Rahme, L. G. 2009. Modeling *Pseudomonas aeruginosa* pathogenesis in plant hosts. *Nat. Protoc.* 4: 117–124.

Takikawa, Y., Mori, H., Otsu, Y., Matsuda, Y., Nonomura, T., et al. 2002. Rapid detection of phylloplane bacterium *Enterobacter cloacae* based on chitinase gene transformation and lytic infection by specific bacteriophages. *J. Appl. Microbiol.* 93: 1042–1050.

Takkinen, J., Nakari, U. M., Johansson, T., Niskanen, T., Siitonen, A., et al. 2005. A nationwide outbreak of multiresistant *Salmonella* Typhimurium va. Copenhagen DT104B infection in Finland due to contaminated lettuce from Spain. *Eurosurveill. Wkly.* http://www.eurosurveillance.org. Accessed January 4, 2019.

Taormina, P. J., Beuchat, L. R., and Slutsker, L. 1999. Infections associated with eating seed sprouts: An international concern. *Emerg. Infect. Dis.* 5: 626–634.

Teplitski, M., Robinson, J. B., and Bauer, W. D. 2000. Plants secrete substances that mimic bacterial n-acyl homoserine lactone signal activities and affect population density-dependent behaviours in associated bacteria. *Mol. Plant-Microbe Interact.* 13: 637–648.

Teplitski, M., Warriner, K., Bartz, J., and Schneider, K. R. 2011. Untangling metabolic and communication networks: Interactions of enterics with phytobacteria and their implications in produce safety. *Trends Microbiol.* 19: 121–127.

Thilmony, R., Underwood, W., and He, S. Y. 2006. Genome-wide transcriptional analysis of the *Arabidopsis thaliana* interaction with the plant pathogen *Pseudomonas syringae* pv. *tomato* DC3000 and the human pathogen *Escherichia coli* O157:H7. *Plant J.* 46: 34–53.

Thurston-Enriquez, J. A., Gilley, J. E., and Eghball, B. 2005. Microbial quality of runoff following land application of cattle manure and swine slurry. *J. Water Health* 3: 157–171.

Tyler, H. L., and Triplett, E. W. 2008. Plants as a habitat for beneficial and/or human pathogenic bacteria. *Annu. Rev. Phytopathol.* 46: 53–73.

US Department of Agriculture. 2004. Progress update and 2002 data summary. Microbial data program. http://www.ams.usda.gov/ (Cited by Brandl, M.T. 2006. Fitness of human enteric pathogens on plants and implications for food safety. *Ann. Rev. Phytopath.* 44: 367–392).

US Food and Drug Administration (FDA). 2001a. FDA survey of imported fresh produce. FY 1999 Field Assignment. US Food and Drug Administration. Center for Food Safety and Applied Nutrition. Office of Plant and Dairy Foods and Beverages. January 30, 2001. http://www.cfsan.gov/~dms/prodsur6.html. Accessed January 25, 2002.

US Food and Drug Administration (FDA). 2001b. FDA Survey of domestic fresh produce: interim results. US Food and Drug Administration. Center for Food Safety and Applied Nutrition. June 30, 2001. http://vm.cfsan.gov/~dms/prodsur8.html. Accessed January 25, 2002.

van der Velden, A. W., Baumler, A. J., Tsolis, R. M., and Heffron, F. 1998. Multiple fimbrial adhesins are required for full virulence of *Salmonella typhimurium* in mice. *Infect. Immun.* 66: 2803–2808.

Vibha, V., Ailes, E., Wolyniak, C., Angulo, F., and Klontz, K. 2006. Recalls of spices due to bacterial contamination monitored by the U.S. Food and Drug Administration: The predominance of Salmonellae. *J. Food Prot.* 69: 233–237.

Wagner, G. J. 1991. Secreting glandular trichomes: More than just hairs. *Plant Physiol.* 96: 675–679.

Walker, T. S., Bais, H. P., Grotewold, E., and Vivanco, J. M. 2003. Root exudation and rhizosphere biology. *Plant Physiol.* 132: 44–51.

Wang, Q., Suzuki, A., Mariconda, S., Porwollik, S., and Harshey, R. M. 2005. Sensing wetness: A new role for the bacterial flagellum. *EMBO J.* 24: 2034–2042.

Warriner, K., and Namvar, A. 2010. The tricks learnt by human enteric pathogens from phytopathogens to persist within the plant environment. *Curr. Opin. Biotechnol.* 21: 131–136.

Watanabe, Y., Ozasa, K., Mermin, J. H., Griffin, P. M., Masuda, K., et al. 1999. Factory outbreak of *Escherichia coli* O157:H7 infection in Japan. *Emerg. Infect. Dis.* 5: 424–428.

Wells, J. M., and Butterfield, J. E. 1997. *Salmonella* contamination associated with bacterial soft rot of fresh fruits and vegetables in the market place. *Plant Dis.* 81: 867–872.

WHO. 2000. Seventh report of WHO surveillance program for control of foodborne infections and intoxications in Europe. *FAO/WHO Collab. Cent. Res. Train. Food Hygiene Zoonoses.*

WHO. 2003. Eighth report of WHO surveillance program for control of foodborne infections and intoxications in Europe. *Collab. Cent. Res. Train. Food Hygiene Zoonoses.*

Widdowson, M-A., Sulka, A., Bulens, S. N., Beard, R. S., Chaves, S. S., et al. 2005. Norovirus and foodborne disease, United States, 1991–2000. *Emerg. Infect. Dis.* 11: 95–102.

Winfield, M. D., and Groisman, E. A. 2003. Role of nonhost environments in the lifestyles of *Salmonella* and *Escherichia coli*. *Appl. Environ. Microbiol.* 69: 3687–3694.

Winthrop, K. L., Palumbo, M. S., Farrar, J. A., Mohle-Boetanim, J. C., Abbott, S., et al. 2003. Alfalfa sprouts and *Salmonella* Kottbus infection: A multistate outbreak following inadequate seed disinfection with heat and chlorine. *J. Food Prot.* 66: 13–17.

Wright, K. M., Chapman, S., McGeachy, K., Humphris, S., Campbell, E., et al. 2013. The endophytic lifestyle of *Escherichia coli* O157:H7: Quantification and internal localization in roots. *Phytopathology* 103: 333–340.

Yao, J., and Allen, C. 2006. Chemotaxis is required for virulence and competitive fitness of the bacterial wilt pathogen *Ralstonia solanacearum*. *J. Bacteriol.* 188: 3697–3708.

You, Y., Rankin, S. C., Aceto, H. W., Benson, C. E., Toth, J. D., et al. 2006. Survival of *Salmonella enterica* serovar Newport in manure and manure-amended soils. *Appl. Environ. Microbiol.* 72: 5777–5783.

Zipfel, C., Robatzek, S., Navarro, L., Oakeley, E. J., Jones, J. D. G., et al. 2004. Bacterial disease resistance in *Arabidopsis* through flagellin perception. *Nature* 428: 764–767.

Section II

Specific Plant Diseases

7 Description of Plant Diseases

7.1 BACTERIAL BLIGHT OF RICE

Bacterial blight (hereafter referred to as BB) is one of the most devastating diseases of rice (*Oryza sativa* L.) due to its very high epiphytotic potential and its destructiveness to nitrogen-responsive high yielding cultivars in both temperate and tropical climates. Its occurrence in 1970 in Africa has added a new dimension to concerns about its transmission and dissemination. It has also been a serious problem on hybrid rice. In the 1980s when hybrid rice was first released in China, epiphytotics occurred widely because of lack of resistance in the hybrid rice, resulting in severe yield losses.

It is probably one of the most extensively studied diseases and a large number of research publications, dealing with different aspects of the pathogen and the disease, have appeared in the literature. Mew et al. (1993), in their paper "Focus on bacterial blight of rice," have provided information on various aspects of the pathogen and the disease. Several review papers containing valuable information on the disease have also been published by workers from different countries. Two such review papers have also appeared from India (Srivastava 1972; Thind 2002). The review paper by Nino-Liu et al. (2006) provides very useful information on different aspects of the casual agent of the disease. Khan et al. (2014), in a mini-review, have consolidated the existing information on the disease and the progress made both in conventional as well as in molecular dimensions of breeding together with potential findings and constraints. For additional information, the readers may also see annual reports of International Rice Research Institute, Philippines and Central Rice Research Institute, Cuttack, India.

The disease was first seen by farmers in the Fukuoka Prefecture of Japan in 1884 and was called **white withering disease**. Originally, it was believed to be caused by acidic soil. In 1908, Takaishi observed bacterial masses in dewdrops of infected rice leaves, successfully isolated a bacterium and proved its pathogenicity, but did not name the organism. Bokura, in 1911, also isolated a bacterium, proved its pathogenicity, and named it *Bacillus oryzae* Hori & Bokura. Reitsma and Schure (1950) studied a rice disease called **Kresek** in Indonesia and considering it different from bacterial blight, named the bacterium as *Xanthomonas kresek*. Later, Goto (1964) reported the **pale yellow leaf phase** of the disease, which occurs at a later stage of growth. Both kresek phase and pale yellow leaf phase are different phases of bacterial blight.

The disease is widespread and occurs in the Philippines, Thailand, Indonesia, China, Taiwan, Malaysia, Cambodia, Vietnam, Myanmar, Laos, Bangladesh, Mexico, Southeast Russia, Australia, Pakistan, West Africa, Latin America, the Caribbean, and India. However, the presence of the disease has not been confirmed in the United States and not yet found in Europe. Quarantines for BB pathogen are in place in the United States and other rice growing countries where the disease is not endemic, but also in places where it is present, to prevent the introduction of new virulent strains. It is also an European and Mediterranean Plant Protection Organization (EPPO) A1 quarantine pest.

In India, Sreenivasan et al. (1959) reported the disease for the first time from Maharashtra, but the symptoms described by them were not exactly of bacterial blight. Bhapkar et al. (1960) reported correct symptoms of the disease, but they did not isolate the bacterium. Later, Srivastava and Rao (1963) reported its epiphytotic on cv. BR 34 from Shahabad district of Bihar. During this period, a Formosian rice variety, Taichung Native 1, a high nitrogen-responsive, but resistant to lodging was introduced into India to step up the rice production. As this variety was highly susceptible to the disease, the disease that was unknown in all but two states of India earlier, became pandemic within a couple of years with the introduction of this variety.

Although the disease occurs in many countries, it is most destructive in Southeast Asia. The yield losses vary with host cultivar, age, and nutrition of the host and environmental conditions. They are usually high in early infected crop, and the occurrence of kresek phase may lead to total failure of the crop in some cases. Losses due to BB increased significantly following the widespread cultivation of nitrogen-responsive, high yielding, semi-dwarf rice cultivars derived from Taichung Native 1 and its derivatives in the 1960s. Prior to the introduction of resistant varieties and implementation of strict quarantine measures in Japan, losses generally ranged from 20% to 30%, but in some cases went as high as 50% (Ou 1972). Reports from India, the Philippines, and Indonesia estimate that losses due to kresek phase have reached 60%–75% depending on weather, location, and rice variety (Reddy et al. 1979a; Ou 1985). Ray and Sengupta (1970) also reported losses up to 50% from semi-dwarf high yielding varieties. Singh et al. (1977) reported that four rice cultivars with a similar leaf score of 9 showed variable loss in grain yield. Reddy et al. (1979b) established a relationship between severity of BB at different growth stages of host plant and rice yield and found that early epiphytotics reduced grain yield, panicle fertility, and kernel weight, while late epiphytotics had no measurable effect on crop yield. They also devised a scale for calculating BB-associated losses. In addition to reduction in yield, the disease may also affect grain quality by interfering with maturation.

7.1.1 Symptoms

It is a typical vascular disease. The below given three different types of symptoms appear.

7.1.1.1 Leaf Blight Phase

These are the most common symptoms of the disease and generally appear 4–6 weeks after transplanting. Initially, the symptoms start as small, green, water-soaked spots at the tips and margins of fully developed leaves. The spots extend forming linear, yellow- to straw-colored stripes having wavy inner margins. The stripes appear generally on both the margins and rarely on one margin of the leaves. The stripes progress rapidly, extending lengthwise and crosswise. The drying and twisting of the leaf tips occur (Figure 7.1a). Occasionally, linear stripes may develop anywhere on the leaf lamina or midrib with or without marginal stripes. The blightening extends to leaf sheaths, culms, killing the tillers. In humid weather, yellow- or amber-colored drops of bacterial ooze appear on young lesions at the margins of the leaves. On drying, these drops form yellowish hard beads. The symptoms on panicles appear as grey to light brown lesions on glumes, resulting in poor fertility and low quality grains. The symptoms of BB often occur simultaneously with those of bacterial leaf streak, and individual leaves may show symptoms of both the diseases.

7.1.1.2 Kresek (Wilt) Phase

It is the most destructive phase of the disease and results from systemic infection of plants in the nursery or early infection in the main field. In an infected plant, the bacteria spread through xylem of the infected leaf to the culm and infect the crown (base of other leaves). Subsequently, all the tillers of the plants are infected resulting in quick wilting of the entire plant. The leaves of the affected plants roll completely, droop, and turn yellow or grey. Ultimately, the diseased tillers wither away and in severely infected plants, stools are killed. The plants that survive, exhibit arrested growth of tillers and stunted appearance (Figure 7.1b). Kresek phase is common in tropical countries. Temperature range of 28°C–34°C favors kresek development. Kresek phase is generally confused with stem borer attack. However, in the latter case, affected tillers/panicles can be easily pulled out and detached part of the tiller shows the signs of insect feeding at the broken end.

7.1.1.3 Pale Yellow Leaf Phase

The youngest leaves in a clump turn pale yellow or whitish. The diseased leaves later turn yellowish brown and tillers do not grow fully.

The disease can be easily diagnosed in the field on the basis of symptoms described above. Additionally, the field diagnosis of the disease can be supplemented by placing infected leaf bits in a glass of clean water. If the water turns turbid after some time, presence of the disease is confirmed. However, further confirmation of the diagnosis can be done by observing the bacterial ooze in the laboratory.

Symptoms of bacterial blight on rice leaves, characteristics to those caused by *Xa. oryzae* pv. *oryzae*, have been reported to be caused by *Pantoea ananatis* and *Pa. stewartii* in Togo (Kini et al. 2017).

7.1.2 Causal Organism

Xanthomonas oryzae pv. *oryzae* (Ishiyama 1922) Swings et al. 1990
Syn. *Xa. oryzae*, *Xa. campestris* pv. *oryzae*

The bacterium was first named as *Bacillus oryzae* Hori & Bokura. Ishiyama (1922) renamed it as *Pseudomonas oryzae* Uyeda & Ishiyama according to Migula's system of classification. In 1927, it was renamed as *Bacterium oryzae* (Uyeda & Ishiyama) Nakata according to E.F. Smith's concept and subsequently to *Xanthomonas oryzae* (Uyeda & Ishiyama) Dowson with the creation of *Xanthomonas* genus. With the introduction of pathovar concept, the name *Xa. campestris* pv. *oryzae* (Ishiyama) Dye was adopted (Dye 1978). Swings et al. (1990), on the basis of phenotypic, genotypic, and chemotaxonomic data, proposed *Xanthomonas oryzae* (*ex* Ishiyama 1922) sp. nov., nom. rev. and renamed the blight pathogen as *Xanthomonas oryzae* pv. *oryzae* (Ishiyama) Swings et al. (hereafter referred to as *Xoo*). The following description of *Xa. oryzae* (*ex* Ishiyama 1922) sp. nov., nom. rev. is taken primarily from Swings et al. (1990).

Cells are straight rods (0.4–0.8 × 1.5–2.9 μm) having round ends, Gram-stain-negative, non-sporulating, and motile with a single polar flagellum. Cells occur singly, in pairs,

(a)

(b)

FIGURE 7.1 Bacterial blight of rice: (a) Leaf blight phase. (b) Kresek phase.

or sometimes in chains and filaments may also occur. The bacterium is obligately aerobic and catalase positive. Colonies on solid media containing glucose are round, convex, mucoid, and yellow in color due to production of xanthomonadin. Xanthomonadin, though not required for virulence, protects the bacterium form UV light and probably contributes to its survival in the field. The cells produce copious, capsular extracellular polysaccharide (EPS). EPS plays a role in virulence, and *Xoo* mutants lacking this ability are severely attenuated in virulence. The tests for indole and 2-ketogluconate formation, urease, egg yolk hydrolysis, nitrate reduction, and oxidase are negative. Starch is hydrolyzed and carbon sources are used only oxidatively. Litmus milk is not acidified, but H_2S is produced. Esculin, Tween 40, and 80 are hydrolyzed. No growth occurs at 4°C or 35°C or in the presence of 3% NaCl, but poor growth occurs at 10°C and 32°C. Optimum temperature for growth is between 25°C and 30°C. The G + C contents of the DNA range from 64.6 to 65.0 mol%.

Xoo differs from *Xa. oryzae* pv. *oryzicola* in causing different type of symptoms on rice. *Xoo* also differs from pathovar *oryzicola* in not producing acetoin, inability to use L-alanine as sole source of carbon, and unable to grow on 0.2% vitamin-free casamino acids. However, *Xa. oryzae* pv. *oryzicola* is positive for all the above given three characteristics. *Xoo* is able to grow in the presence of 0.001% cupric nitrate, while *Xa. oryzae* pv. *oryzicola* is not.

Xu et al. (2010) obtained 534 single-colony isolates of *Xoo from* rice leaves showing bacterial blight symptoms, collected from southern China in 2007 and 2008 and tested these isolates on plates for sensitivity to streptomycin. Four strains (0.75%) isolated from the same county of province Yunnan were resistant to streptomycin, and the resistance factor (the ratio of the mean median effective concentration inhibiting growth of resistant isolates to that of sensitive isolates) was approximately 226. The resistant isolate also showed streptomycin resistance *in vivo*. Besides resistant isolates, isolates of less sensitivity were also present in the population of *Xoo* from province Yunnan. Mutations in the *rpsL* (encoding S12 protein) and *rrs* genes (encoding 16S rRNA) and the presence of the *strA* gene accounting for streptomycin resistance in other phytopathogenic or animal and human pathogenic bacteria, were examined on sensitive and resistant strains of *Xoo* by polymerase chain reaction amplification and sequencing. Neither the presence of *strA* gene nor mutations in *rpsL* or *rrs* genes were found, suggesting that different resistance mechanisms are involved in the resistant isolates of *Xoo*.

Elshakh et al. (2016) reported that the culture filtrates of three *Bacillus* strains, namely, rice seed-associated strain *Ba. subtilis* A15, rhizobacterial strains *Ba. amyloliquefaciens* D29 and *Ba. methylotrophicus* H8 significantly suppressed the growth and biofilm formation of *Xoo* along with changes in bacterial cell morphology such as swelling of cells and severe cell wall alterations. Polymerase chain reaction (PCR) analysis revealed that all the three strains harbor the antimicrobial-associated genes that are responsible for biosynthesis of bacillomycin, fengycin, iturin and surfactin. Subsequent real-time quantitative (q)PCR analysis revealed the upregulated expression of *fenD* and *srfAA* genes in D29 and H8, and *fenD* and *ituC* genes in A15 during their *in vitro* interaction with *Xoo*. It indicates that the antibacterial mechanisms of these strains may be at least partially associated with their ability to secrete corresponding lipopeptides. Further, the application of these strains under greenhouse conditions was found to be effective in controlling bacterial blight, which was achieved through the activation of inducing systemic resistance, resulted from the enhanced activities of defense-related enzymes. This is the first report of demonstration of the mode of antibacterial effect of *Bacillus* strains against *Xoo*.

The genomes of two *Xoo* strains, namely, Japanese strain MAFF 311018 (also called T 7174) and Korean strain KACC 10331 (also called KXO 85) have been sequenced by Lee et al. (2005) and Ochiai et al. (2005), respectively. Salzberg et al. (2008) reported the genome sequencing of a third strain, PXO 99^A of the bacterium. The genome of PXO 99^A is a single circular chromosome of 5,240,075 bp and considerably longer than the genomes of other two strains (4,941,439 bp for KACC 10331 and 4,940,217 bp for MAFF 311018). The genome of PXO 99^A contains 5,083 protein coding genes, including 87 not found in the genomes of other two strains. The findings of Salzberg et al. (2008) provide striking evidence of **genome plasticity** and **rapid evolution** within *Xoo*. The comparisons point to sources of genomic variation and candidates for strain-specific adaptation of this pathogen that help to explain the extraordinary diversity of *Xoo* genotypes and races that have been isolated throughout the world.

Zhao et al. (2007) developed a novel and highly sensitive real-time method for the identification of *Xoo* based on a TaqMan probe, which is designed to recognize the sequence of a putative siderophore receptor gene coding sequence (cds) specific to *Xoo* and can be identified from either a bacterial culture or naturally infected rice seeds and leaves in about 2 h. The sensitivity of the method is 100 times higher than that of the currently used polymerase chain reaction gel electrophoresis method for diagnosis.

Song et al. (2012) developed a triplex PCR method for the race-specific detection of *Xoo*, designing three primer sets: for specific internal regions of two genes (*hpaA* and *XorII* very-short-patch-repair endonuclease) and for a genomic locus derived from an amplified fragment length polymorphism fragment specific for the K3 and K5 races. The pathogen was quickly detected in the asymptomatic rice leaf 3 days after inoculation and at a distance of 6 cm from the lesion site. This simple and rapid assay will be useful for the detection and identification of *Xoo* as well as for disease forecasting in paddy fields.

Pathogenic variation: A high degree of race-cultivar specificity occurs in the host-pathogen interactions. The first indication of host specialization in *Xoo* came when Asakaze, a resistant rice cultivar was unexpectedly found to be severely attacked in Kyushu Island areas of Japan. This prompted active research on this aspect. Sato et al. (1976) analyzed the distribution of bacterial population in Kyushu area and divided it into two races: one attacked all differential cultivars including Wase Aikoku 3 and the other attacked Kinmaze and Wase Aikoku 3, but not the other cultivars. In 1977, Yamamoto and associates tested 71 isolates from Indonesia on Kozaka's differential cultivars (including Rantaj-emas along with others) and found no isolates belonging to Kozaka's groups I or II (Kozaka 1969), but 46 isolates with virulence similar to group III in Japan, i.e., virulent to Rantaj-emas. They also identified two more groups, IV and V, and the last two groups showed pathogenicity similar to the two races identified by Sato et al. (1976). These results not only demonstrated pathogenic specialization/variation, but also showed a strong interaction between rice cultivars and strains of the pathogen. At present, there are more than 30 races of the pathogen reported from different countries (Nino-Liu et al. 2006). A set of races identified in the Philippines using five differential rice cultivars (Mew 1987) has been used widely for identifying and classifying resistance to the pathogen in other cultivars.

Lore et al. (2011) found that the current *Xoo* population from the Punjab state, India is highly variable and can be classified into seven distinct pathotypes (PbXo-1 to PbXo-7) on the basis of differential reactions on a set of near-isogenic lines in the background of IR24 and some international, national, and regional cultivars. PbXo-7, the most dominant pathotype, was found to be virulent and induced susceptible/moderately susceptible reaction on 22 of the 40 test genotypes followed by PbXo-1, PbXo-5, and PbXo-6; PbXo-2 was the least virulent pathotype. Chen et al. (2016) reported a new race that was highly virulent on a near-isogenic line CBB23 (*Xa23*) derived from a cross between a wild rice *O. rufipogon* accession (RBB16) and a susceptible *indica* rice variety Jingang 30. Wang et al. (2014) reported this line resistant to earlier races of the pathogen in the province.

7.1.3 Disease Cycle

7.1.3.1 Perpetuation

It has been carefully studied in temperate regions, particularly in Japan (Tagami and Mizukami 1962; Mizukami 1966). The following sources help in the survival of the bacterium in different regions of the world.

7.1.3.1.1 Weed Hosts

In Japan, natural infection is found on weeds such as *Leersia* spp. and *Zizania latifolia* and among them, *L. sayanuka* is the most important source of perpetuation. The bacterium survives in rhizosphere of this weed and multiplies in early spring when new shoots of the weed begin to emerge. From the weed, the bacterium is dispersed through irrigation water to the nursery or transplanted fields of rice. Gonzalez et al. (1991) reported that *L. hexandra* acts as a symptomless carrier of the bacterium. In India, Brar and Thind (1994) reported *Echinochloa colonum* as a natural host of the bacterium and also proved that seeds from infected plants transmit the bacterium to the seedlings. In the tropics, *Leptochloa* spp., *Cyperus difformis* and *C. rotundus* also act as natural hosts of the bacterium. Other hosts of the bacterium include several species of wild rice (*O. rufipogon*, *O. australiensis*) and a number of weeds such as, *Cynodon dactylon*, *Cenchrus ciliaris*, *Brachiaria mutica*, *E. crusgalli*, *Panicum maxicum,* and *Paspalum scrobiculatum.*

7.1.3.1.2 Rice Straw

The bacterium can survive in piles of rice straw kept outdoors in temperate regions (Isaka 1962). In tropical areas like Punjab, India, the bacterium has been found to survive in rice straw kept in big heaps protected from weathering agencies, but it does not play any role in the perpetuation of the bacterium (personal observation).

7.1.3.1.3 Rice Stubbles

The bacterium survives in the stubbles (base of stems and roots) of the plants left behind in the field after harvest until the following spring in Japan. It is the most important source of infection in Thailand (Tabei and Eamchit 1974).

7.1.3.1.4 Volunteer Plants

Infected seeds from the previous crop shed in the field give rise to diseased seedlings, and these serve as a source of infection in double-cropped areas. In north India, off-type tall plants usually infected with bacterial blight are generally found in rice fields. As these plants generally mature before harvesting of the crop, they shed their seeds, and these infected seeds initiate the disease in the subsequent year (Durgapal 1985).

7.1.3.1.5 Rhizosphere of Succeeding Crops

Thind and Brar (1989), on the basis of two-year field trials, demonstrated that the bacterium survived in the rhizosphere of berseem, wheat, and winter maize (sown in fields after harvesting bacterial blight infected rice) until the sowing of next rice crop. The survival in the rhizosphere of wheat and rapeseed has also been reported by Wakimoto (1956) and Mohiuddin et al. (1977). The survival of the bacterium in the rhizosphere of wheat and berseem in Punjab and other north Indian states, where rice is generally sown after these crops, has an important bearing because major rice acreage in these areas is sown after harvesting wheat and berseem.

7.1.3.1.6 Seed

The seed-borne nature of the bacterium was first reported by Fang et al. (1956), and according to them, the bacterium was present mainly in glumes and occasionally in the endosperm. Since then, the seed-borne nature of the bacterium has remained controversial, some workers reporting positive results, while others reporting negative findings. Different methods like isolation of the bacterium from seed, inoculation of seed macerate on young seedlings, bacterial streaming from seed parts, and seedling symptom test have been used to prove the seed-borne nature. Among them, the seedling symptom test is considered best, being natural and reliable.

Srivastava and Rao (1963, 1964) and Singh (1972) reported positive seed-plant transmission of the bacterium. Reddy (1983) using 9-month-old seeds observed typical blight symptoms 60 days after sowing. Brar and Thind (1995) developed a technique called **Nitrogen-Supplemented Seedling Symptom Test** (NSSST) using irrigation water containing 300 μg of nitrogen mL^{-1}, to prove seed-plant transmission of the bacterium. The typical symptoms of the disease appear in 10–12 days after sowing in NSSST. Thind and Brar (1998) reported that the bacterium survived in all the seed samples, until the sowing of next rice crop. These seed samples were collected from naturally infected crop from different places in Punjab, India and tested at monthly intervals. However, the number of seeds showing positive seed-plant transmission decreased with storage period. The seed-borne inoculum is a major source of perpetuation in Punjab and possibly in other northwestern states of India where only one crop of rice is taken in a year.

Tian et al. (2014), based on 16S–23S internal transcribed spacer ribosomal DNA sequence data, designed padlock probe (PLP), P-Xoo and tested to detect *Xoo* in rice seed by combining with dot-blot hybridization. With this technique, a detection sensitivity of 1 pg of *Xoo* genomic DNA was observed. The technique facilitated the detection of *Xoo* in rice seed lots with 2% artificially infested seed. PLP assay detected *Xoo* in 39.3% (13 of 33) of naturally infested commercial rice seed lots. In contrast, conventional polymerase chain reaction using OSF1/OSR1 primers sets detected *Xoo* in 9.1% (3 of 33) of the same rice seed lots. The combining of both the padlock probes (P-Xoo for pv. *oryzae* and P-Xoc for pv. *oryzicola*) with reverse dot-blot hybridization enabled to simultaneously detect *Xoo* and *X. oryzae* pv. *oryzicola* in naturally infested commercial rice seed lots. This technique can serve as an important tool to detect multiple pathogens in seed and thereby prevent the spread of important pathogens.

7.1.3.2 Disease Development

The bacterium enters the host through hydathodes, wounds, and growth cracks formed due to the development of new roots. The new wounds are more conducive than old wounds for infection. Bacteria multiply in the intercellular spaces of the underlying epitheme, then enter and spread into the plant through the xylem. The pathogen moves vertically through the leaf through primary veins, but also progresses laterally through commisural veins. Within a few days, bacterial cells and EPS produced by the bacterium fill the xylem vessel and bacteria ooze out from hydathodes, forming beads or strands of exudate on the leaf surface which acts as a source of secondary inoculum. The EPS protects the bacterial cells from desiccation.

The temperature range of 25°C–31°C is highly favorable for disease development. The kresek phase appears in 20 days when rice plants are kept at 31°C, while it takes 40 days or more at 21°C. Younger plants, i.e., less than 21 days old, are the most susceptible, and the susceptibility decreases with the increase in age. The common practice of cutting leaf tips of seedlings before transplanting, particularly when the seedlings are quite tall, plays an important role in the development of kresek symptoms because the cut ends allow the entry of the bacterium directly into the xylem vessels. In addition, broken roots resulting from pulling the seedlings off the seed-bed serve as entry points for bacteria present in flood-irrigated fields.

The disease is favored by high relative humidity. The epiphytotics of the disease are common in rainy or monsoon season in India. The same varieties, which show high disease severity in rainy season, either escape infection or show traces to moderate infection in the *Rabi* season. Secondary spread of the disease occurs through wind-splashed rain, irrigation water, and contact of healthy leaves with diseased leaves. Wind-driven rains and persistent warm weather are highly conducive for the spread of pathogen and disease development. Severe epiphytotics often occur following typhoons, strong and violent winds, wind-blown rain, and hail, as all these wound rice plants and disperse bacteria.

Heavy doses of nitrogenous fertilizers influence the disease incidence and even the source of nitrogen also influences disease severity. However, increase in disease intensity with increased levels of fertilizers differs with different cultivars. In general, nitrogen-induced susceptibility is expressed by susceptible cultivars, while it is not so with resistant cultivars. Reddy et al. (1979a) showed that high levels of nitrogen (>100 kg ha^{-1}) caused reduction in yield due to increased bacterial blight incidence in susceptible variety IET 2895, but resistant variety IET 4141 was least affected by all the nitrogen levels. The increase in disease severity due to application of excess doses of nitrogen results from enhanced vegetative growth leading to high humidity.

The disease severity also increases under shade. In Punjab, India, it is a common sight to see high disease severity on plants growing under tree shade in transplanted rice (personal observation). The high disease severity results from high relative humidity and favorable temperature maintained for longer period under shade.

Attempts have been made to forecast the disease by different methods based on climatic factors and counting bacterial and bacteriophage populations. The unusual changes in climate affect forecasting by all these methods. In temperate regions, monitoring of bacteriophage strains specific to *Xoo* has been used for forecasting since 1960s. The time when phage population in irrigation water exceeds 200 particles mL^{-1} corresponds to time 10 days before the initial disease development (Yoshimura et al. 1960). For an accurate forecasting, the water sample should be taken in the morning at the outflow. However, the bacteriophage forecasting system is not practiced extensively in tropical Asia, because rice cultivation is mostly rain fed, limiting the use of phage detection in paddy fields. Moreover, the disease forecasting has a limited utility because no satisfactory chemical control of the disease is available.

Mondal et al. (2016) studied the role of XopF, one of the conserved effectors in *Xoo*, using a null mutant developed through a PCR-based homologous recombination strategy. XopF regulated the growth of *Xoo* in plants and suppressed pathogen-associated molecular patterns-triggered immune (PTI) response in rice. *Xoo* wild, but not mutant, *Xoo* Δ*xopF*, produced intense blight lesions after inoculation by leaf clipping method, while *Xoo* Δ*xopF* showed significant reduction in colonization of host in comparison to the wild strain. The relative expression analysis of PTI marker genes *PR10*, *OsWRKY13*, *OsRLCK16*, and *OsFLS2* showed that these genes were up-regulated 1.5- to 5-fold upon challenge-inoculation with *Xoo*Δ*xopF* indicating the role of XopF in suppressing PTI in rice. The findings show that XopF suppresses basal PTI response in plants and favors growth and pathogenic action of *Xoo* in rice.

Wu et al. (2017) identified 15 potential Tat (twin-arginine translocation)-dependent translocation proteins (TDTPs) by using comparative proteomics and bioinformatics analyses. Combining systematic mutagenesis, phenotypic characterization, and gene expression, they further found that multiple TDTPs play key roles in *Xoo* adaption or virulence. In particular, four TDTPs were involved in virulence, three contributed to colonization in planta, one had a key role in attachment to leaf surface, four were involved in tolerance to multiple stresses, and two were required for full swarming motility. These findings suggest that multiple TDTPs may have differential contributions to involvement of the Tat pathway in *Xoo* adaption, physiology, and pathogenicity.

Talreja and Nerurkar (2018) reported that cinnamic acid and its derivatives like ferulic acid, caffeic acid, and p-coumaric acid showed significant reduction in promoter activity of xopQ, an effector secreted through type third secretion system by *Xoo* BXO43. They also found significant reduction in disease symptoms confirming cinnamic acid, ferulic acid, caffeic acid, and p-coumaric acid as effective virulence attenuating agents of *Xoo*.

7.1.3.3 Host Resistance

The development of resistant cultivars carrying major resistance (*R*) genes has been the most effective approach for controlling BB. To date, 42 genes imparting resistance to BB pathogen have been identified (Busungu et al. 2016). Most of these genes have been found in cultivars of *O. sativa* sub species *indica,* but some are also from *japonica* cultivars and from wild species of rice such as, *O. longistaminata*, *O. rufipogon*, *O. nivara*, *O. minuta,* and *O. officinalis*. Some of these *R* genes like *Xa21* are effective only in adult plants, while most of them do not seem to be related with plant growth stage. However, *Xa3* is typically effective only in adult plants bearing at least one race against which it is effective at all stages of growth. Some genes like *Xa21* and *Xa31* impart resistance against a wide range of *Xoo* races, while others are effective against only one or a few races of the pathogen. Most *R* genes are dominant, while some are recessive (*xa5*, *xa8*, *xa13*, *xa19*, *xa20*, *xa24,* and *xa28*), while still others are semi-dominant. The effectiveness of disease resistance genes is sometimes influenced by plant developmental stage. Webb et al. (2010) reported that gene *Xa7* was not fully functional in very young seedlings, but was completely effective by 21 days after sowing. However, if a partial or fully functional *avrXa7* allele was present, *Xa7* resistance was effective at all growth stages tested after the transplant stage, i.e., more than 21 days after sowing.

Near isogenic lines (NILs) have been developed by introducing single *R* genes into the background of susceptible *indica* cultivar IR 24 to serve as differential lines. These differential lines are used for the identification of races of the bacterium. Rice differentials recommended by the scientists of All India Coordinated Rice Improvement Programme of Directorate of Rice Research, Hyderabad are given in Table 7.1. Rice workers also include commercially grown rice varieties in their regions, as additional differentials, to have a better assessment of the pathogenic variation in the region.

Resistance of rice to specific *Xoo* races is governed by major *R* genes with a qualitative effect that impart complete resistance and polygenes with a quantitative effect, which impart partial resistance. It was believed that gene *Xa21* derived from *O. longistaminata* conferred resistance to all the pathotypes/races of *Xoo* from India and the Philippines. However, Saini and Goel (1998) reported that even this growth stage specific

TABLE 7.1
Rice Differentials Recommended by the Scientists of All India Coordinated Rice Improvement Programme of Directorate of Rice Research, Hyderabad for Field Monitoring Virulences in *Xanthomonas oryzae* pv. *oryzae*

S. No.	Line/Differential	Gene Combination
1	IRBB 1	*Xa1*
2	IRBB 3	*Xa3*
3	IRBB 4	*Xa4*
4	IRBB 5	*xa5*
5	IRBB 7	*Xa7*
6	IRBB 8	*xa8*
7	IRBB 10	*Xa10*
8	IRBB 11	*Xa11*
9	IRBB 13	*xa13*
10	IRBB 14	*Xa14*
11	IRBB 21	*Xa21*
12	IRBB 50	*Xa4+xa5*
13	IRBB 51	*Xa4+xa13*
14	IRBB 52	*Xa4+Xa21*
15	IRBB 53	*xa5+xa13*
16	IRBB 54	*xa5+Xa21*
17	IRBB 55	*xa13+Xa21*
18	IRBB 56	*Xa4+xa5+xa13*
19	IRBB 57	*Xa4+xa5+Xa21*
20	IRBB 58	*Xa4+xa13+Xa21*
21	IRBB 59	*xa5+xa13+Xa21*
22	IRBB 60	*Xa4+xa5+xa13+Xa21*
23	DV 85	*xa5+Xa7*
24	Ajaya (IET 4141/ CR 98-7216)	
25	RP B10-226	
26	TN 1	*Xa14*

Note: 1–9 scale of Standard Evaluation System for Rice (IRRI 2014) is used to record the disease reaction.

gene was ineffective in 70-day-old plants against some isolates of the bacterium from Punjab, India. Priyadarisini and Gnanamanickam (1999) also found that out of 140 isolates of *Xoo* isolated from epiphytotic affected plants that occurred in Kerala, south India in 1998, 20 isolates were virulent on rice line IRBB 21 (carrying *Xa21* gene) but the rice line NH 56 carrying four resistance genes (*Xa4*, *xa5*, *xa13*, and *Xa21*) was resistant to all the isolates. Sirisha et al. (2004) again reported breakdown of resistance of *Xa21* against *Xoo* due to possible gain of virulence by the pathogen in the process of co-evolution. Swamy et al. (2006) also reported that Pusa Basmati 1 transgenic lines expressing *Xa21* gene were susceptible to all the six isolates collected from different regions of India, while another line MH 2R into which three resistant genes, *xa5*, *xa13*, and *Xa21* were introduced through marker-assisted selection (MAS) showed resistant reaction to all the isolates. Sanchez et al. (2000) have already emphasized the usefulness of MAS for pyramiding of bacterial blight resistance genes, particularly for recessive genes like *xa5* and *xa13* that are difficult to select through conventional breeding in the presence of dominant genes, such as *Xa21*. They also reported that MAS showed an accuracy of 95% and 96% for identifying homozygous resistant plants containing *xa5* and *xa13* genes, respectively. Host resistance conferred by single genes against *Xoo* is generally not stable due to considerable variation in the pathogen population. Therefore, pyramiding of BB resistance genes in rice is based on pathogen population response to these genes. Thus, the selected resistance genes are combined in a cultivar through MAS. This helps in developing varieties with broader-spectrum and longer-lasting resistance. Shanti et al. (2001) also found that three gene combinations, i.e., *Xa4* + *xa5*, *xa5* + *Xa21*, and *Xa4* + *xa5* + *Xa21* conferred a broad-spectrum of resistance to all the strains evaluated (collected from Orissa, Madhya Pradesh, and Uttar Pradesh of India), thus supporting the strategy of pyramiding appropriate resistance genes. Pradhan et al. (2016) developed pyramided lines carrying two BB resistance gene combinations (*Xa21* + *xa13* and *Xa21* + *xa5*) in a lowland cultivar Jalmagna background through backcross breeding by integrating molecular markers. The genotypes with combination of two BB resistance genes conferred high levels of resistance to the predominant *Xoo* isolates prevalent in Odisha, India. The pyramided lines showed similarity with the recipient parent with respect to major agromorphological traits.

Yugander et al. (2017) collected four hundred isolates of *Xoo* from diverse rice growing regions of India and carried out their virulence profiling on a set of differentials consisting of 22 near isogenic lines of IR24 possessing different BB resistance genes and their combinations along with the checks. They found that none of the single BB resistance genes can provide broad spectrum resistance in India. However, two-gene combinations like *xa5* + *xa13* and different three- or four-gene combinations like *Xa4* + *xa5* + *xa13*, *Xa4* + *xa13* + *Xa21*, *xa5* + *xa13* + *Xa21*, and *Xa4* + *xa5* + *xa13* + *Xa21* are broadly effective throughout India.

Xa21 is the first resistance gene cloned in the monocots. It is a member of multigene family in rice and of a larger family of genes in plants and animals involved in recognition of pathogens and innate immunity. In addition to *Xa21*, other genes like *Xa1*, *xa5*, *xa13*, and *Xa27* have been cloned and represent gene families. The *avrXa7* and *avrXa10* corresponding to *R* genes *Xa7* and *Xa10*, respectively, are the first avirulence genes of *Xoo* cloned from race 2 strain PXO 86. These avirulence genes are members of the large gene family of *avrBs3* and share an extensive sequence homology with *avrBs3* gene cloned from the pepper and tomato pathogen, *Xa vesicatoria*. This gene family is widely distributed in *Xanthomonas* species, and *Xoo* possesses 12–14 copies. Type III secretion system-mediated secretion and delivery of *avrXa10* into rice cells by *Xoo* have been demonstrated using a reporter protein fusion.

Dossa et al. (2016) reported that African rice, *O. glaberrima* accessions contain bacterial blight resistance different from the *Xa* gene type and high temperature enhances the resistance of this species. Dossa et al. (2017) found that drought stress affected rice response to *Xoo,* and the response varied according to the rice genotype. They concluded that evaluation of rice varieties under combined abiotic and biotic stresses will be the best strategy to determine biotic stress resistance durability under climate change.

Zhang et al. (2015) reported that a rice introgression line, FF329, derived from the cross between donor PSBRC66 and recipient Huang-Hua-Zhan, exhibited a typical hypersensitive response when inoculated with all 21 representative *Xoo* strains and found that BB resistance in FF329 was controlled by a single dominant gene. As FF329 showed broad-spectrum BB resistance, different from both the parents, the gene identified was novel and was designated *Xa39*. This gene can provide rice breeders with a new option to incorporate BB resistance into newly developed cultivars.

7.1.4 Management

The measures adopted for the management of BB vary depending on the location and disease incidence. The following measures help in reducing the disease severity.

1. Destruction of diseased plant debris including rice straw and stubbles
2. Removal of weeds from canals and ridges
3. Seed treatment before sowing is practiced in tropical Asia. Seed can be treated with Streptocycline (100 μg mL^{-1}) + Agallol (0.1%) or Aretan/Tafasan (0.05%) followed by hot water treatment at 52°C–54°C for 30 min. Alternately, use the seed from disease-free crop
4. Nursery should be raised away from diseased fields and avoid its flooding with contaminated water. It should also not be sown under shade. Raising the nursery under upland dry conditions has been found effective in Sri Lanka
5. Keep proper plant spacing and avoid high doses of nitrogenous fertilizers

6. Chemical control of BB began in 1950s with the prophylactic sprays of Bordeaux mixture. Since then many fungicides including copper and mercurial compounds and several antibiotics have been tried singly and in combination with limited success. In temperate regions, probenazole is applied in the irrigation water in nursery and in the transplanted rice to prevent the bacterial multiplication. The chemicals used for spraying include tecloftalam, phenazine oxide, and nickel dimethyl dithiocarbamate. In the tropical monsoon climate of Asia, chemical sprays are not generally given due to the non-availability of effective commercial chemicals and difficulty in spraying due to uncertain weather. However, the following three reports suggest the possibility of managing the disease with chemical sprays. Foliar sprays of Blitox-50WP (0.25%) + Streptocycline (100 μg mL^{-1}) have been found to give significant disease control along with significant increase in grain yield (Thind and Mehra 1992; Sinha and Sinha 2000). Tilak et al. (2001) obtained significant disease control along with significant increase in grain yield with two sprays of Bacterimycin-2000 (0.04%) in a multi-location trial
7. Bacterial antagonists of *Xoo* have received special attention as biocontrol candidates, and as a result of it, some success has been achieved in this respect. In India, more than 40 bacterial isolates antagonistic to the bacterium have been identified through *in vitro* and *in vivo* tests. Among them, rice-associated rhizobacteria, *Psm. fluorescens* and *Psm. putida* strain V14i significantly suppressed the disease severity when sprayed on leaves (Sivamani et al. 1987). The application of *Bacillus* species, as seed treatment before sowing, root-dip prior to transplanting, and foliar sprays in the field, has also given encouraging results. In at least one study, the antagonists suppressed the disease by almost 60% and increased the plant height and grain yield by two-fold (Vasudevan et al. 2002).

   Thind and Ahmad (1994) reported that out of 32 microorganisms, isolated from phylloplane and spermosphere of four rice cultivars and from phylloplane of four rice weeds, nine proved antagonistic to the bacterium in *in vitro* evaluation, and these antagonists also gave significant control of the disease under field conditions for 2 consecutive years along with an increase in grain yield. Three antagonists, namely, *Ba. subtilis*, *Erwinia herbicola,* and *Penicillium oxalicum* gave significantly higher grain yield in addition to significantly better disease control in comparison to Streptocycline (100 μg mL^{-1}). *Ba. subtilis*, the most promising antagonist, gave 35.4% disease control and 10.55% increase in grain yield. Vidhyasekaran et al. (1996) reported that treatment of rice seeds with talc-based formulation of *Psm. fluorescens* at 10 g kg^{-1} of seed induced systemic resistance and gave protection up to 60 days against bacterial blight pathogen, while the foliar sprays imparted resistance up to 15 days. Kamala and Vidhyasekaran (1998) tested talc-based formulation of *Psm. fluorescens* str. Pf 1 as seed treatment, root-dip application, and foliar sprays individually and in combination, and found that the combination treatment induced resistance and gave 60%–70% disease control in field along with an increase in grain yield. Despite these encouraging results, biocontrol agents have not attained a widespread use for the control of BB
8. Use of resistant cultivars is the most effective and economical method to control BB of rice. However, the development of resistant varieties and durability of the resistance is complicated by the high diversity in the pathogen population. Resistant varieties, based on pathogen population structure of different regions, have been developed and released for cultivation. Rice cultivars, namely, PR 113, PR 114, PR 121, PR 122, PR 123, PR 124, and PR 127 developed by Punjab Agricultural University, Ludhiana, Punjab have been released for cultivation in Punjab, India. PR 127 is resistant to all the ten pathotypes of the bacterium prevalent in Punjab (Anonymous 2018). Pusa 1460, a variety developed by pyramiding resistance genes *xa5* + *xa13* + *Xa21* in the background of Pusa Basmati 1 through marker-assisted backcross breeding has been released for cultivation in Punjab, Haryana, Jammu and Kashmir, Delhi, and Uttarakhand of India. Transgenic rice developed from cultivar IR 72 showed high level of resistance to existing races of the pathogen in the Philippines (Mew and Vera Cruz 2001). The growing of cultivar mixtures containing appropriate single resistance genes has also been tried for the management of the disease.

REFERENCES

Anonymous. 2018. *Package of Practices for Kharif Crops*. Punjab Agricultural University, Ludhiana, India, pp. 2–3.

Bhapkar, D. G., Kulkarni, N. B., and Chavan, V. M. 1960. Bacterial blight of paddy. *Poona Agri. Coll. Mag.* 51: 36–46.

Brar, J. S., and Thind, B. S. 1994. A new weed host of *Xanthomonas oryzae* pv. *oryzae*, the causal agent of bacterial leaf blight of rice. *Ann. Plant Prot. Sci.* 2: 79–80.

Brar, J. S., and Thind, B. S. 1995. Nitrogen supplemented seedling symptom test—A new method for proving seed transmission of *Xanthomonas oryzae* pv. *oryzae*. *Oryza* 32: 42–45.

Busungu, C., Taura, S., Sakagami, J.-I., and Ichitani, K. 2016. Identification and linkage analysis of a new rice bacterial blight resistance gene from XM14, a mutant line from IR24. *Breed. Sci.* 66: 636–645.

Chen, X. L., Yan, Q., Li, R. F., Li, K. H., and Gao, L. J. 2016. First report of a new and highly virulent race of *Xanthomonas oryzae* pv. *oryzae*, the causal agent of bacterial leaf blight of rice in Guangxi province, China. *Plant Dis.* 100: 1492.

Dossa, G. S., Oliva, R., Maiss, E., Cruz, C. V., and Wydra, K. 2016. High temperature enhances the resistance of cultivated African rice, *Oryza glaberrima*, to bacterial blight. *Plant Dis.* 100: 380–387.

Dossa, G. S., Torres, R., Henry, A., Oliva, R., Maiss, E., et al. 2017. Rice response to simultaneous bacterial blight and drought stress during compatible and incompatible interactions. *Eur. J. Plant Pathol.* 147: 115–127.

Durgapal, J. C. 1985. Self sown plants from bacterial blight infected rice seeds—A possible source of primary infection in North West India. *Curr. Sci.* 54: 1283–1284.

Dye, D. W. 1978. Genus IX *Xanthomonas* Dowson 1939. In: Young, J. M., Dye, D. W., Bradbury, J. F., Panagopoulos, C. G., Robbs, C. F. A proposed nomenclature and classification for plant pathogenic bacteria. *N. Z. J. Agric. Res.* 21: 162–166.

Elshakh, A. S. A., Anjum, S. I., Qiu, W., Almoneafy, A. A., Li, W., et al. 2016. Controlling and defence-related mechanisms of *Bacillus* strains against bacterial leaf blight of rice. *J. Phytopathol.* 164: 534–546.

Fang, C. T., Lin, C. F., and Chu, C. L. 1956. A preliminary study on the disease cycle of the bacterial blight of rice. *Acta Phytopathol. Sin.* 2: 173–185 (Ch. en).

Gonzalez, C. F., Xu, D. W., and Li, H. L. 1991. *Leersia hexandra*, an alternate host for *Xanthomonas campestris* pv. *oryzae* in Texas. *Plant Dis.* 75: 159–162.

Goto, M. 1964. "Kresek" and pale yellow leaf systemic symptoms of bacterial leaf blight of rice caused by *Xanthomonas oryzae* (Uyeda et Ishiyama) Dowson. *Plant Dis. Reptr* 48: 858–861.

IRRI 2014. *Standard Evaluation System for Rice (SES)*. 5th ed. International Rice Research Institute, Los Banos, Philippines, p. 22.

Isaka, M. 1962. Overwintering of bacterial leaf blight organism in damaged leaf in paddy field. *Proc. Assoc. Plant Prot.* Hokuriku, 10, 90.

Ishiyama, S. 1922. Studies on the white leaf disease of rice plants. *Rept. Agri. Expt. Sta. Tokyo* 45: 233–251 [Ja.].

Kamala, N., and Vidhyasekaran, P. 1998. Evaluation of new biopesticide formulation for the management of some foliar diseases of rice. pp. 382–388 in: *Ecological Agricultural and Sustainable Development*. Vol. 2, G. S. Dhaliwal, N. S. Randhawa, R. Arora, and A. K. Dhawan eds. Indian Ecological Society, Punjab Agricultural University, Ludhiana and Centre for Research in Rural and Industries Development (CRRID), Chandigarh, India.

Khan, M. A., Naeem, M., and Iqbal, M. 2014. Breeding approaches for bacterial leaf blight resistance in rice (*Oryza sativa* L.), current status and future directions. *Eur. J. Plant Pathol.* 139: 27–37.

Kini, K., Agnimonhan, R., Afolabi, O., Soglonou, B., Silué, D., et al. 2017. First report of a new bacterial leaf blight of rice caused by *Pantoea ananatis* and *Pantoea stewartii* in Togo. *Plant Dis.* 101: 241.

Kozaka, T. 1969. Control of rice diseases with resistant varieties. *Agric. Hortic.* 44: 208–212.

Lee, B.-M., Park, Y.-J., Park, D.-S., Kang, H.-W., Kim, J.-G., et al. 2005. The genome sequence of *Xanthomonas oryzae* pathovar *oryzae* KACC 10331, the bacterial blight pathogen of rice. *Nucleic Acids Res.* 33: 577–586.

Lore, J. S., Vikal, Y., Hunjan, M. S., Goel, R. K., Bharaj, T. S., et al. 2011. Genotypic and pathotypic diversity of *Xanthomonas oryzae* pv. *oryzae*, the cause of bacterial blight of rice in Punjab state of India. *J. Phytopathol.* 159: 479–487.

Mew, T. W. 1987. Current status and future prospects of research on bacterial blight of rice. *Annu. Rev. Phytopathol.* 25: 359–382.

Mew, T. W., Alvarez, A. M., Leach, J. E., and Swings, J. 1993. Focus on bacterial blight of rice. *Plant Dis.* 77: 5–12.

Mew, T. W. and Vera Cruz, C. M. 2001. Bacterial blight of rice. pp. 71–74 in: *Encyclopedia of Plant Pathology*. Vol. 1, O. C. Maloy, and T. D. Murray eds. John Wiley & Sons, New York.

Mizukami, T. 1966. Epidemiology of bacterial leaf blight of rice and use of phages for forecasting. *Eleventh Pacific Science Congress, Symposium on Plant Diseases in the Pacific*, pp. 15–32.

Mohiuddin, M. S., Rao, Y. P., Mohan, S. K., and Verma, J. P. 1977. Survival of *Xanthomonas oryzae*, the casual incitant of bacterial blight of rice in the rhizosphere of wheat. *Sci. Cult.* 43: 124–125.

Mondal, K. K., Verma, G., Junaid, M. A., and Mani, C. 2016. Rice pathogen *Xanthomonas oryzae* pv. *oryzae* employs inducible *hrp*-dependent XopF type III effector protein for its growth, pathogenicity and for suppression of PTI response to induce blight disease. *Eur. J. Plant Pathol.* 144: 311–323.

Nino-Liu, D. O., Ronald, P. C., and Bogdanove, A. J. 2006. *Xanthomonas oryzae* pathovars: Model pathogens of a model crop. *Mol. Plant Pathol.* 7: 303–324.

Ochiai, H., Inoue, Y., Takeya, M., Sasaki, A., and Kaku, H. 2005. Genome sequence of *Xanthomonas oryzae* pv. *oryzae* suggests contribution of large numbers of effector genes and insertion sequences to its race diversity. *Jpn. Agr. Res. Q.* 39: 275–287.

Ou, S. H. 1972. *Rice Diseases*. 1st ed. Commonwealth Mycological Institute, Kew, UK, 368 pp.

Ou, S. H. 1985. *Rice Diseases*. 2nd ed. C. A. B. International, Slough, UK, 380 pp.

Pradhan, S. K., Nayak, D. K., Pandit, E., Behera, L., Anandan, A., et al. 2016. Incorporation of bacterial blight resistance genes into lowland rice cultivar through marker-assisted backcross breeding. *Phytopathology* 106: 710–718.

Priyadarisini, V. B., and Gnanamanickam, S. S. 1999. Occurrence of a subpopulation of *Xanthomonas oryzae* pv. *oryzae* with virulence to rice cv. IRBB 21 (*Xa21*) in Southern India (Abstr.). *Plant Dis.* 83: 781.

Ray, P. R., and Sengupta, T. K. 1970. A study on the extent of loss of yield in rice due to bacterial blight of rice. *Indian Phytopathol.* 23: 713–714.

Reddy, A. P. K., Katyal, J. C., Rouse, D. I., and MacKenzie, D. R. 1979a. Relationship between nitrogen fertilization, bacterial leaf blight severity and yield of rice. *Phytopathology* 69: 970–973.

Reddy, A. P. K., MacKenzie, D. R., Rouse, D. I., and Rao, A. V. 1979b. Relationship of bacterial leaf blight severity to grain yield of rice. *Phytopathology* 69: 967–969.

Reddy, P. R. 1983. Evidence for seed transmission of *Xanthomonas campestris* pv. *oryzae*. *Curr. Sci.* 52: 265–266.

Reitsma, J., and Schure, P. S. J. 1950. "Kresek," a bacterial disease of rice. Contributions of the Central Agricultural Research Station, Bogor No. 117, 17 pp.

Saini, R. G., and Goel, R. K. 1998. Disease resistance in rice (*Oryza sativa* L.)—An appraisal of the Indian work. *Crop Improv.* 25: 141–154.

Salzberg, S. I., Sommer, D. D., Schatz, M. C., Phillippy, A. M., Rabinowicz, P. D., et al. 2008. Genome sequence and rapid evolution of the rice pathogen *Xanthomonas oryzae* pv. *oryzae* PXO 99^A. *BMC Genomics* 9: 204–219.

Sanchez, A. C., Brar, D. S., Huang, N., Li, Z., and Kush, G. S. 2000. Sequence tagged site marker-assisted selection for three bacterial blight resistance genes in rice. *Crop Sci.* 40: 792–797.

Sato, T., Choi, Y. S., Iwasaki, M., and Watanabe, B. 1976. Distribution of races of *Xanthomonas oryzae* in Kyushu (Abstr.). *Ann. Phytopathol. Soc. Jpn.* 17: 42.

Shanti, M. L., George, M. L. C., Vera Cruz, C. M., Bernardo, M. A., Nelson, R. J., et al. 2001. Identification of resistance genes effective against rice bacterial blight pathogen in Eastern India. *Plant Dis.* 85: 506–512.

Singh, G. P., Srivastava, M. K., Singh, R. V., and Singh, R. M. 1977. Variation in quantitative and qualitative losses caused by bacterial blight in different rice varieties. *Indian Phytopathol.* 30: 180–185.

Singh, R. N. 1972. Perpetuation of the bacterial blight disease of rice in North India. *Indian Phytopathol.* 25: 148–150.

Sinha, B. B. P., and Sinha, R. K. P. 2000. Integrated disease management of bacterial blight of rice with special reference to neem product. *Ann. Plant Prot. Sci.* 8: 260–261.

Sirisha, C., Reddy, J. N., Mishra, D., Das, K. M., Bernardo, M. A., et al. 2004. Susceptibility of IRBB 21 carrying the resistance gene *Xa21* to bacterial blight. *Rice Genet. Newslet.* 21: 75–77.

Sivamani, E., Anuratha, C. S., and Gnanamanickam, S. S. 1987. Toxicity of *Pseudomonas fluorescens* towards bacterial plant pathogens of banana (*Pseudomonas solanacearum*) and rice (*Xanthomonas campestris* pv. *oryzae*). *Curr. Sci.* 56: 547–548.

Song, E.-S., Lee, B.-M., Lee, C.-S., and Park, Y.-J. 2012. PCR-based rapid assay for discriminative detection of latent infections of rice bacterial blight. *J. Phytopathol.* 160: 195–200.

Sreenivasan, M. C., Thirumalachar, M. J., and Patel, M. K. 1959. Bacterial blight of rice. *Curr. Sci.* 28: 469–470.

Srivastava, D. N. 1972. Bacterial blight of rice. *Indian Phytopathol.* 25: 1–16.

Srivastava, D. N., and Rao, Y. P. 1963. Epidemic of bacterial blight disease of rice in North India. *Indian Phytopathol.* 16: 393–94.

Srivastava, D. N., and Rao, Y. P. 1964. Seed transmission and epidemiology of bacterial blight of rice in North India. *Indian Phytopathol.* 17: 77–78.

Swamy, P., Panchbhai, A. N., Dodiya, P., Naik, V., Panchbhai, S. D., et al. 2006. Evaluation of bacterial blight resistance in rice lines carrying multiple resistance genes and *Xa21* transgenic lines. *Curr. Sci.* 90: 818–824.

Swings, J., Van den Mooter, M., Vauterin, L., Hoste, B., Gillis, M., et al. 1990. Reclassification of the casual agents of bacterial blight (*Xanthomonas campestris* pv. *oryzae*) and bacterial leaf streak (*Xanthomonas campestris* pv. *oryzicola*) of rice as pathovars of *Xanthomonas oryzae* (ex Ishiyama 1922) sp. nov., nom. rev. *Int. J. Syst. Bacteriol.* 40: 309–311.

Tabei, H., and Eamchit, S. 1974. Infection source of bacterial leaf blight of rice in Thailand. *Jpn. Agric. Res. Q.* 8, 123–125.

Tagami, Y., and Mizukami, T. 1962. Historical review of the researches on bacterial leaf blight of rice caused by *Xanthomonas oryzae* (Uyeda *et* Ishiyama) Dowson. Special Report of the Plant Disease and Insect Pests Forecasting Service No. 10, 112 pp.

Talreja, S. S., and Nerurkar, A. S. 2018. Small molecules cause virulence attenuation of *Xanthomonas oryzae* pv. *oryzae*, the pathogen causing bacterial blight of rice. *Eur. J. Plant Pathol.* 151: 229–241.

Tian, Y., Zhao, Y., Xu, R., Liu, F., Hu, B., et al. 2014. Simultaneous detection of *Xanthomonas oryzae* pv. *oryzae* and *X. oryzae* pv. *oryzicola* in rice seed using a padlock probe-based assay. *Phytopathology* 104: 1130–1137.

Thind, B. S. 2002. Bacterial blight of rice—An overview. *Plant Dis. Res.* 17: 227–251.

Thind, B. S. and Ahmad, M. 1994. Biological control of *Xanthomonas oryzae* pv. *oryzae*. pp. 867–73 in: *Proceedings of the 8th International Conference on Plant Pathogenic Bacteria.* M. Lemattre, S. Freigoun, K. Rudolph, and J. G. Swings eds. Versailles, France.

Thind, B. S., and Brar, J. S. 1989. Perpetuation of *Xanthomonas campestris* pv. *oryzae* (Ishiyama) Dye in the rhizosphere of succeeding crops. pp. 1021–27 in: *Proceedings of the 7th International Conference on Plant Pathogenic Bacteria.* Part A, Z. Klement ed. Budapest, Hungary.

Thind, B. S., and Brar, J. S. 1998. Perpetuation of *Xanthomonas oryzae* pv. *oryzae* under North Indian conditions. pp. 409–19 in: *Proceedings of the 9th International Conference on Plant Pathogenic Bacteria.* A. Mahadevan ed. Chennai, India.

Thind, B. S., and Mehra, R. K. 1992. Chemical control of bacterial blight of rice. *Plant Dis. Res.* 7: 226–234.

Tilak, K. V. B. R., Udaiyan, K., and Ramarethinam, S. 2001. Bioefficacy of bacterimycin—2000, a plant immunomodulator in the control of bacterial leaf blight (BLB) in rice. *Pestology* 25: 1–10.

Vasudevan, P., Kavitha, S., Priyadarisini, V. B., Babujee, L., and Gnanamanickam, S. 2002. Biological control of rice diseases. pp. 11–32 in: *Biological Control of Crop Diseases.* S. Gnanamanickam ed. Marcel Dekker, New York.

Vidhyasekaran, P., Kamala, N., and Ramanathan, A. 1996. Induced systemic resistance against rice bacterial blight. Paper presented in *9th International Conference on Plant Pathogenic Bacteria.* August 26–29, 1996, Chennai, India, p. 98.

Wakimoto, S. 1956. Considerations of the overwintering and the infection mechanisms of *Xanthomonas oryzae. Plant Prot. Jpn.* 10: 421–424.

Wang, C.-L., Qin, T.-F., Yu, H.-M., Zhang, X.-P., Che, J.-Y., et al. 2014. The broad bacterial blight resistance of rice line CBB23 is triggered by a novel transcription activator-like (TAL) effector of *Xanthomonas oryzae* pv. *oryzae*. *Mol. Plant Pathol.* 15: 333–341.

Webb, K. M., Garcia, E., Vera Cruz, C. M., and Leach, J. E. 2010. Influence of rice development on the function of bacterial blight resistance genes. *Eur. J. Plant Pathol.* 128: 399–407.

Wu, G., Su, P., Wang, B., Zhang, Y., Qian, G., et al. 2017. Novel insights into Tat pathway in *Xanthomonas oryzae* pv. *oryzae* stress adaption and virulence: Identification and characterization of Tat-dependent translocation proteins. *Phytopathology* 107: 1011–1021.

Xu, Y., Zhu, X.-F., Zhou, M.-G., Kuang, J., Zhang, Y., et al. 2010. Status of streptomycin resistance development in *Xanthomonas oryzae* pv. *oryzae* and *Xanthomonas oryzae* pv. *oryzicola* in China and their resistance characters. *J. Phytopathol.* 158: 601–608.

Yoshimura, S., Aoyagi, K., Morihasni, T., Yoshino, M., Nishimura, H., et al. 1960. *Proc. Assoc. Plant Prot.* Hokuriku, 8: 31–41.

Yugander, A., Sundaram, R. M., Ladhalakshmi, D., Hajira, S. K., Prakasam, V., et al. 2017. Virulence profiling of *Xanthomonas oryzae* pv. *oryzae* isolates, causing bacterial blight of rice in India. *Eur. J. Plant Pathol.* 149: 171–191.

Zhang, F., Zhuo, D.-L., Zhang, F., Huang, L.-Y., Wang, W.-S., et al. 2015. *Xa39*, a novel dominant gene conferring broad-spectrum resistance to *Xanthomonas oryzae* pv. *oryzae* in rice. *Plant Pathol.* 64: 568–575.

Zhao, W. J., Zhu, S. F., Liao X. L., Chen, H. Y., and Tan, T. W. 2007. Detection of *Xanthomonas oryzae* pv. *oryzae* in seeds using a specific TaqMan probe. *Mol. Biotechnol.* 35: 119–127.

7.2 BACTERIAL LEAF STREAK OF RICE

This is another important disease of rice (*Oryza sativa* L.), but comparatively less destructive and less widespread than bacterial blight (BB). Reinking (1918) was the first to report the disease from the Philippines, but he named it as bacterial stripe. However, the disease was erroneously referred to as bacterial blight (BB) for several years. Fang et al. (1957) studied the disease in southern China, distinguished it from BB, and renamed it as bacterial leaf streak (hereafter referred to as BLS). They also found its causal agent different from that of BB and named it as *Xanthomonas oryzicola*.

The occurrence of the disease is restricted largely to tropical and sub tropical Asia including Thailand, Indonesia, Bangladesh, Malaysia, Vietnam, and India in addition to southern China and the Philippines. It is one of the major rice diseases in southern China. It is also present in regions of northern Australia and has become a significant problem in parts of West Africa. Most recently, it has been reported from Kenya, East Africa (Onaga et al. 2018). Quarantines for the causal pathogen are in place in the United States and other rice growing countries where the disease is not present, but also in places where the disease occurs, to guard against the introduction of new virulent strains. The pathogen is also listed in the 2002 Agricultural Bioterrorism Protection Act of the USA as a potential bioterrorism agent, necessitating strict biosecurity measures in addition to standard biosafety measures. The available reports suggest that the yield losses range from 0% to 20% depending upon the rice variety, climatic conditions, and severity of the disease (Ou 1985). Under conditions favorable for disease development, reduction in grain weight up to 32% has been reported. The incidence of BLS is increasing in parts of Asia, where hybrid rice is grown, such as China, because hybrid rice cultivars are susceptible to the pathogen.

7.2.1 Symptoms

Although it is mainly a foliage disease infecting the parenchymatous tissue exclusively, infection of the seed coat is common. The symptoms start as minute, water-soaked, translucent, interveinal pinheads that can be better seen when the leaf is held against light. These pinheads enlarge gradually into 1–10 cm long, water-soaked, dark green, translucent, discrete, vein-limited streaks, which are typically yellow in color. The width of the streaks varies from 0.5 to 1.00 mm depending on the rice cultivar, but is generally more in broad-leaf varieties. At later stage, infected leaves turn grayish-white and die. Under moist conditions, yellow- to amber-colored bacterial exudate is deposited in the form of beads in linear rows all along the streak on both the leaf surfaces. This bacterial ooze is a typical sign of the disease. Irregular, diffused, cholorotic areas appear around the lesions in highly susceptible cultivars (Figure 7.2). In severe cases, streaks may coalesce resulting in blotchy lesions and blightening of the leaves. Sheath infection is rare, but symptoms are similar to that of leaf lamina. Glume infection occurs in the form of water-soaked lesions.

FIGURE 7.2 Bacterial leaf streak of rice. (Courtesy of Donald Groth, Louisiana State University AgCenter, Bugwood.org.)

In resistant varieties, water-soaked pinheads turn brown without elongation or enlarge into 5–10 mm long reddish-brown streaks. Under conditions favorable for the spread of pathogen, the disease may affect the entire field.

When infection results from entry of the bacterium through breaks in the leaf, which might occur due to high wind velocity, symptoms may expand length-wise killing most or whole of the leaf. The symptoms of BB and BLS are quite distinct in the initial stage, but they become less differentiable in the advanced stage. Moreover, the severe infection of BLS may also result in partial or total leaf blightening resembling BB symptoms. However, even in such cases, both the diseases can be correctly and conveniently diagnosed on the basis of diagnostic hints given below (Shekhawat and Rao 1972).

BB infected leaves, when viewed against light, appear uniformly opaque, but in case of BLS, numerous interveinal, translucent streaks are seen. In case of BB, bacterial exudate appears in the form of isolated and infrequently formed beads near the leaf margins on the upper surface of the leaves, while in case of BLS, numerous beads of yellow to amber color, arranged closely in a linear fashion, appear all along the streak length on both the leaf surfaces.

The symptoms of BB and BLS often occur simultaneously in rice fields, and the symptoms of both the diseases may be present on the same leaves.

7.2.2 Causal Organism

Xanthomonas oryzae pv. *oryzicola* (Fang et al. 1957) Swings et al. 1990
Syn. *Xa. oryzicola*, *Xa. campestris* pv. *oryzicola*

Fang et al. (1957) first named the bacterium as *Xa. oryzicola*. Later on, it was changed to *Xa. translucens* f. sp. *oryzae* (Fang et al.) Pordesimo (1958) and then to *Xa. translucens* f. sp. *oryzicola* (Fang et al.) Bradbury (1971). With the introduction of pathovar concept, it was renamed as *Xa. campestris* pv.

oryzicola (Fang et al.) Dye (Dye 1978). Swings et al. (1990) reclassified it as *Xa. oryzae* pv. *oryzicola*. *Xa. oryzae* pv. *oryzicola* (hereafter referred to as *Xoc*), being a pathovar of *Xa. oryzae*, shares the characteristics of *Xa. oryzae*. The description of *Xanthomonas oryzae* (Ishiyama 1922) sp. nov., nom. rev. is given under BB of rice. *Xoc* differs from *Xa. oryzae* pv. *oryzae* (hereafter referred to as *Xoo*) in its ability to produce acetoin, to utilize L-alanine as sole source of carbon, to grow on 0.2% vitamin-free casamino acids, and sensitivity to 0.001% cupric nitrate. The growth of *Xoc* on solid media is fast in comparison to *Xoo*. Variation in *Xoc* has not been studied as extensively as in *Xoo*. However, the analysis of strains isolated from different locations in the Philippines indicates the diversity in the pathogen (Raymundo et al. 1999). Clusters of isolates were not robust, and there was no correlation between clusters and the reactions of a set of diverse rice varieties. Like *Xoo*, *Xoc* also serves as a model for understanding fundamental aspects of bacterial interaction with plants (Nino-Liu et al. 2006).

The genome of *Xoc* strain BLS 256 has been sequenced (available in the Comprehensive Microbial Resource, http://www.tigr.org). The genome has overall similarity to *Xoo* genomes sequenced with respect to size, G + C content, abundance of IS elements and TAL effector genes, and the presence of inversions and rearrangements. However, the genome also has some notable unique features like a gene cluster with predicted functions in synthesis and transport of lipopolysaccharides and bordered by two highly conserved genes and is distinct both in gene order and gene content from the corresponding locus in the *Xoo* genomes. Half of the cluster is orthologous with *Xa. citri* subsp. *citri* cluster present at this locus, while half is apparently unique.

The genome of *Xoc* annotates one uncharacterized gene, *XOC_3841*, only one open reading frame (ORF) in this strain is annotated to encode Phosphohexosemutase (XanA), which reversibly converts glucose 1-phosphate to glucose 6-phosphate that implicates in the carbon metabolism pathways. Guo et al. (2014) reported that *xanA* is required for extracellular polysaccharides production, cell motility, and the full virulence of *Xoc*.

Xu et al. (2010) obtained 827 single colony isolates of *Xoc from* rice leaves showing bacterial leaf streak symptoms, collected from southern China in 2007 and 2008 and tested these isolates on plates for sensitivity to streptomycin. All the isolates examined were sensitive to streptomycin and no isolates showed decreased sensitivity to the antibiotic.

7.2.3 Disease Cycle

Infected seed is the main source of perpetuation of the bacterium. The bacterium can also survive on wild rice species. In tropical areas, where only one crop of rice is taken in a year, plant debris, stubbles, and soil do not play any role in the survival. In India, where two or three crops of rice are taken in a year, the bacterium can survive on the crop itself. The problem of survival is only in areas like north or northwestern India, where only one crop of rice is taken in a year. Tian et al. (2014), based on 16S–23S internal transcribed spacer ribosomal DNA sequence data, designed padlock probe, P-Xoc to detect *Xoc* in rice seed by combining it with dot-blot hybridization. With this technique, a detection sensitivity of 1 pg of *Xoc* genomic DNA was observed. The technique facilitated the detection of *Xoc* in rice seed lots with 0.2% artificially infested seed. The padlock probe assay detected *Xoc* in 21.3% (10 of 47) of naturally infested commercial rice seed lots. In contrast, conventional polymerase chain reaction using XoocF/XoocR primers sets detected *Xoc* in 8.5% (4 of 47) of the same rice seed lots. The combining of both the padlock probes (P-Xoo for pv. *oryzae* and P-Xoc for pv. *oryzicola*) with reverse dot-blot hybridization enabled to simultaneously detect *Xoo* and *Xoc* in naturally infested commercial rice seed lots. This technique can serve as an important tool to detect multiple pathogens in seed and thereby prevent the spread of important pathogens.

During germination of seed, the bacterium present in the inner surface of the glumes infects the developing plumule. The first leaf carries the bacterium, and the subsequent leaves are infected successively as they emerge. The bacterium penetrates the leaf mainly through stomata, multiplies in the substomatal cavity and then colonizes the intercellular spaces of parenchyma. Leaf veins act as a barrier to the lateral spread of the pathogen. The penetration can also occur through wounds, but it remains restricted to the apoplast of the mesophyll tissue and does not invade the xylem. In humid weather, masses of bacterial cells ooze out in the form of bacterial exudate from stomata and act as inoculum for secondary spread of the pathogen through wind, rain, irrigation water, insects, and contact. Wind-driven rain is the most important agent for the dissemination of the bacterium. Bacterial exudate on the flag leaf provides abundant inoculum for the infection of seeds in the panicle. Without strong winds and rain, secondary spread of the disease is limited, and the effect of disease dwindles rapidly as the growth of new leaves compensates for damage to infected leaves.

The disease is more severe in rainy season, but disappears during short periods of dry spell or bright sunshine. Relative humidity of 90% or above or dew during morning hours for 2–3 days is necessary for rapid infection. Temperature is more important than relative humidity for the enlargement of lesions. Optimal temperature for disease development is 26°C–30°C and below 22°C, enlargement of lesions is checked. Younger plants are more susceptible and after 60 days of age, only top young leaves remain susceptible.

Different methods, like spraying of bacterial suspension without injury, spraying the cell suspension containing carborundum powder, and smearing the cell suspension on the underside of the leaves with a piece of cloth, are used for artificial inoculation of plants to evaluate the germplasm. Out of these methods, smear method gives better and consistent results even under dry conditions, provided water is sprayed twice a day on inoculated plants. The best results are obtained up to 1 month age of plants.

Qian et al. (2012) detected six putative hypothetical secreted proteins from *Xoc* strain BLS 256, but these were absent in *Xoo*. Disruption-based mutagenesis study revealed that one of them, *Xoc_15235*, named as extracellular polysaccharide and virulence related gene (*epv*), were required for the optimal virulence in rice host, but not for the induction of a hypersensitive reaction in non-host tobacco. Sequence analysis revealed that *epv* was highly conserved in *Xanthomonas* spp., except *Xoo*. In-frame deletion of *epv* in *Xoc* dramatically impaired pathogen virulence and extracellular polysaccharide production, one of the important known virulence-associated functions in *Xanthomonas* spp. Quantitative real-time reverse-transcription polymerase chain reaction revealed that expressions of both encoding exopolysaccharide xanthan biosynthesis export protein (*gumB*) gene and a known virulence related encoding phosphoglycerate kinase (*pgk*) gene were obviously reduced in the *epv*-deletion mutant compared with the wild-type strain Rs 105. Additionally, *epv* was positively regulated by both diffusible signal factor and global regulator Clp in *Xoc*. This is the first investigation concerning the role and genetics of *epv* in the extracellular polysaccharide production and virulence of *Xoc*.

Zhao et al. (2012) characterized a novel pathogenicity-related gene, *Xoryp_010100018570* (named *hshB*), of *Xoc*. *hshB* encodes a hydrolase with a putative signal peptide, which is a homolog of imidazolonepropionase. Reverse-transcription polymerase chain reaction analysis showed that *hshB* and its upstream gene, *Xoryp_010100018565* (named *hshA*), are co-transcribed in *Xoc*. Subsequent experimental results indicated that mutation of *hshB* remarkably impaired the virulence, extracellular protease activity, extracellular polysaccharide production, growth in minimal medium, and resistance to oxidative stress and bismerthiazol of *Xoc*. Mutation of *clp*, encoding a global regulator, resulted in similar phenotypes. Real-time PCR assays showed that *hshB* transcription is positively regulated by *clp* and diffusible signal factor (DSF) and induced by poor nutrition. The finding not only discovered a novel gene *hshB* regulated by DSF-dependent quorum-sensing system and *clp,* but also showed that *hshB* was required for virulence of *Xoc*.

Su et al. (2018) reported that the *his* operon was involved in histidine biosynthesis and two genes within this operon, *trpR* and *hisB*, were required for virulence and bacterial growth *in planta*. Hence, these two genes play an important role in the pathogenicity of *Xoc* Rs105.

Host resistance: No major genes resistant to *Xoc* have been identified in rice. BLS resistance in rice is quantitatively inherited and is controlled by multiple quantitative trait loci (QTLs). Several QTLs conditioning resistance to BLS have been identified in *indica* rice. A QTL with relatively large effect, *qBlsr5a*, has been mapped in a ~380 kb region on chromosome 5. To fine map *qBlsr5a* further, Xie et al. (2014) developed overlapping sub chromosome segment substitution lines (sub-CSSLs) from a secondary F_2 population of 7000 plants, in which only the chromosomal region harboring *qBlsr5a* segregated. *qBlsr5a* was delimited to a 30-kb interval, in which only three genes were predicted. These three putative genes did not show significant response to the infection of BLS pathogen in both resistant and susceptible parental lines. Two nucleotide substitutions in the coding sequence of gene LOC_Os05g01710, which encodes the gamma chain of transcription initiation factor IIA (TFIIAγ), resulted in a change of the 39th amino acid from valine (in the susceptible parent) to glutamic acid (in the resistant parent). The resistant parent allele of LOC_Os05g01710 is identical to *xa5*, a major gene resistant to BB of rice. LOC_Os05g01710 is very possibly the candidate gene of *qBlsr5a*.

Song et al. (2015) identified two genes, *hshA* and *hshC* within *nodB-rghB* locus that were unique to *Xoc* Rs 105 compared with *Xoo* PXO99^A, and found that the transcription of these genes was regulated by DSF signaling in *Xoc*. When the low inoculation density of *Xoc* was used, the mutation of these genes impaired the virulence of the wild-type Rs 105. However, the mutation of *hshA* or *hshC* significantly reduced the *in planta* growth ability and level of epiphytic survival of *Xoc* cells, which were the probable mechanisms of involvement of these two genes in virulence. The identification of these two novel DSF-controlled virulence associated genes, *hshA* and *hshC* will add to our knowledge of the regulatory mechanisms of conserved DSF virulence signaling in *Xanthomonas* species.

Zhao et al. (2005) reported that *Rxo1* gene cloned from maize provided resistance against several strains of *Xoc*, as a transgene in rice. As a transgene, *Rox1* holds a great promise for rice breeding as a simply inherited source of resistance to BLS. An avirulent gene *avrRxo1* has also been cloned and it is present in all the strains of *Xoc* tested. Hence, the indication is there that it may play a critical role in fitness including the virulence of the bacterium. The activity of *avrRxo1* is dependent on type third secretion system, but its product is not TAL effector and lacks similarity to any known protein.

7.2.4 Management

Not many studies have been conducted for the management of the disease, but many of the measures used for controlling BB can be expected to work for this disease also. As in case of BB, host genetic resistance is also important for the control of BLS. Brief information on this aspect is given above under host resistance. The other management measures include:

1. Use of pathogen-free seed or treatment of seed by soaking in Streptocycline (250 μg mL^{-1}) solution for 12 h or in hot water at 50°C for 30 min
2. As the disease is exclusively parenchymatous, it can be effectively controlled by spraying Streptocycline (100 μg mL^{-1}). The number of sprays will vary with disease severity, but in general, 2–3 sprays at 10–15 days intervals should give satisfactory results. Yuan et al. (2017) identified two compounds, namely, berberine and jatrorrhizine extracted from stems of *Mahonia fortunei* having a strong inhibitory effect on *Xoc*. The antibacterial activity of berberine was stronger, with a half-maximal inhibitory concentration (IC_{50}) of 2.9008 mg L^{-1}. At the concentration of 0.5 g L^{-1}, its control efficacy on BLS of

rice was more than 84%. Furthermore, berberine was absorbed by rice leaves, it translocated up and down in the rice plant, and the effective period was also long, but its lateral translocation inside the blade was poor.

REFERENCES

Bradbury, J. F. 1971. Nomenclature of the bacterial leaf streak pathogen of rice. *Int. J. Syst.Bacteriol.* 21: 72.

Dye, D. W. 1978. Genus IX *Xanthomonas* Dowson 1939. In: Young, J. M., Dye, D. W., Bradbury, J. F., Panagopoulos, C. G., and Robbs, C. F. A proposed nomenclature and classification for plant pathogenic bacteria. *N. Z. J. Agric. Res.* 21: 162–166.

Fang, C. T., Ren, H. C., Chen. T. Y., Chu, Y. K., Faan, H. C., et al. 1957. A comparison of the rice bacterial leaf blight organism with the bacterial leaf streak organism of rice and *Leersia hexandra* Swartz. *Acta Phytopathol. Sin.* 3: 99–124 (Ch. en).

Guo, W., Chu, C., Yang, X.-X., Fang, Y., Liu, X., et al. 2014. Phosphohexosemutase of *Xanthomonas oryzae* pv. *oryzicola* is negatively regulated by HrpG and HrpX, and required for the full virulence in rice. *Eur. J. Plant Pathol.* 140: 353–364.

Ishiyama, S. 1922. Studies of bacterial leaf blight of rice. *Rep. Imp. Agric. Stn. Konosu* 45: 233–261.

Nino-Liu, D. O., Ronald, P. C., and Bogdanove, A. J. 2006. *Xanthomonas oryzae* pathovars: Model pathogens of a model crop. *Mol. Plant Pathol.* 7: 303–324.

Onaga, G., Murori, R., Habarugira, G., Nyongesa, O., Bigirimana, J., et al. 2018. First report of *Xanthomonas oryzae* pv. *oryzicola* causing bacterial leaf streak of rice in Kenya. *Plant Dis.* 102: 1025.

Ou, S. H. 1985. *Rice Diseases*. 2nd ed. C.A.B. International, Slough, UK, 380 pp.

Pordesimo, A. N. 1958. Bacterial blight of rice. *Philippine Agri.* 42: 115–128.

Qian, G., Zhang, Y., Zhou, Y., Liu, C., Zhao, Y., et al. 2012. *epv*, encoding a hypothetical protein, is regulated by DSF-mediating quorum sensing as well as global regulator Clp and is required for optimal virulence in *Xanthomonas oryzae* pv. *oryzicola*. *Phytopathology* 102: 841–847.

Raymundo, A. K., Briones, A. M. Jr., Ardales, E. Y., Perez, M. T., Fernandez, L. C., et al. 1999. Analysis of DNA polymorphism and virulence in Philippine strains of *Xanthomonas oryzae* pv. *oryzicola*. *Plant Dis.* 83: 434–440.

Reinking, O. A. 1918. Philippine economic plant diseases. *Oryza sativa* L. Rice bacterial leaf stripe. *Philippine J. Sci.* Sec. A, 13: 225–226.

Shekhawat, G. S., and Rao, Y. P. 1972. Bacterial leaf streak of rice. pp. 13–21 in: *Plant Bacteriology*. Vol. I, *Bacterial Diseases of Plants in India*. P. N. Patel ed. Indian Agricultural Research Institute, New Delhi, India.

Song, Z., Zhao, Y., Zhou, X., Wu, G., Zhang, Y., et al. 2015. Identification and characterization of two novel DSF-controlled virulence-associated genes within the *nodB-rhgB* locus of *Xanthomonas oryzae* pv. *oryzicola* Rs105. *Phytopathology* 105: 588–596.

Su, P., Song, Z., Wu, G., Zhao, Y., Zhang, Y., et al. 2018. Insights into the roles of two genes of the histidine biosynthesis operon in pathogenicity of *Xanthomonas oryzae* pv. *oryzicola*. *Phytopathology* 108: 542–551.

Swings, J., Van den Mooter, M., Vauterin, L., Hoste, B., Gillis, M., et al. 1990. Reclassification of the causal agents of bacterial leaf blight (*Xanthomonas campestris* pv. *oryzae*) and bacterial leaf streak (*Xanthomonas campestris* pv. *oryzicola*) of rice as pathovars of *Xanthomonas oryzae* (ex Ishiyama 1922) sp. nov., nom. rev. *Int. J. Syst. Bacteriol.* 40: 309–311.

Tian, Y., Zhao, Y., Xu, R., Liu, F., Hu, B., et al. 2014. Simultaneous detection of *Xanthomonas oryzae* pv. *oryzae* and *X. oryzae* pv. *oryzicola* in rice seed using a padlock probe-based assay. *Phytopathology* 104: 1130–1137.

Xie, X., Chen, Z., Cao, J., Guan, H., Lin, D., et al. 2014. Toward the positional cloning of *qBlsr5a*, a QTL underlying resistance to bacterial leaf streak, using overlapping sub-CSSLs in rice. *PLoS ONE* 9: e95751.

Xu, Y., Zhu, X.-F., Zhou, M.-G., Kuang, J., Zhang, Y., et al. 2010. Status of streptomycin resistance development in *Xanthomonas oryzae* pv. *oryzae* and *Xanthomonas oryzae* pv. *oryzicola* in China and their resistance characters. *J. Phytopathol.* 158: 601–608.

Yuan, G., Chen, Y., Li, F., Zhou, R., Li, Q., et al. 2017. Isolation of an antibacterial substance from *Mahonia fortunei* and its biological activity against *Xanthomonas oryzae* pv. *oryzicola*. *J. Phytopathol.* 165: 289–296.

Zhao, B., Lin. X., Poland, J., Trick, H., Leach, J., et al. 2005. A maize resistance gene functions against bacterial streak disease in rice. *Proc. Natl. Acad. Sci. USA* 102: 15383–15388.

Zhao, Y., Qian, G., Fan, J., Yin, F., Zhou, Y., et al. 2012. Identification and characterization of a novel gene, *hshB*, in *Xanthomonas oryzae* pv. *oryzicola* co-regulated by quorum sensing and *clp*. *Phytopathology* 102: 252–259.

7.3 BACTERIAL SHEATH BROWN ROT OF RICE

Bacterial sheath brown rot of rice (*Oryza sativa* L.) was first reported from Hokkaido, Japan, where it was widespread (Tanii et al. 1976). Earlier, the causal organism was referred to as *Pseudomonas marginalis*, but the re-examination of its characteristics showed that the rice bacterium differs from that isolated from Chinese cabbage and garlic in levan and 2-keto-gluconate production, β-glucosidase and pectinase activities, acid production from sucrose, sorbitol, inositol, and other sugars, host range, and serological relationships. Therefore, Tanii et al. (1976) named it as *Pseudomonas fuscovaginae*, a new sp. Besides causing sheath brown rot of rice, *Psm. fuscovaginae* also causes grain rot of rice and is worldwide in occurrence (Zeigler and Alavarez 1990).

The disease occurs in areas of high altitude (1200–1700 m above sea level) having low temperature (20°C–22°C) and high humidity, under both temperate and tropical conditions. The occurrence of the disease has been reported from Mexico, Cuba, Venezuela, Argentina, Guatemala, Panama, Suriname, Colombia, Chile, Ecuador, Peru, Romania, Kenya, Rwanda, Tanzania, Mozambique, Madagascar, China, Pakistan, Nepal, Russia, Republic of Korea, the Philippines, Malaysia, Indonesia, Australia, and Brazil (Xie 2003; Sharma et al. 1997; Plantwise Technical Factsheet 2015). It has been found to cause substantial yield losses in South America (Webster and Gunnell 1992). In Madagascar, *Psm. fuscovaginae* is the principal limiting factor in irrigated rice cultivation at altitudes higher than 1500 m, with losses reaching 100%, especially above 1800 m (Rott 1987). Razak et al. (2009) reported that in Peninsular Malaysia, the highest disease incidence of 62% was recorded in the states of Pahang and Selangor. Although not common, yield losses as high as 72.2% have also been reported in Indonesia.

7.3.1 Host Range

In addition to rice, the bacterium infects Bentgrasses (*Agrostis* spp.), oats (*Avena sativa*), Mountain brome (grass) (*Bromus marginatus*), barley (*Hordeum vulgare*), perennial ryegrass (*Lolium perenne*), smooth meadow-grass (*Poa pratensis*), rye (*Secale cereale*), sorghum (*Sorghum bicolor*), Triticale, wheat (*Triticum aestivum*), and maize (*Zea mays*).

7.3.2 Symptoms

The symptoms typically occur on the flag leaf sheath at booting stage and on the panicle. Initially, infected seedlings show yellow to brown discoloration on the lower leaf sheath. The discoloration later turns greyish-brown to dark-brown. Ultimately, the infected seedlings may rot and die. The leaves of infected seedlings may show a systemic discoloration of the midrib and veins.

The symptoms on mature plants may be observed on the flag leaf sheath during booting to heading, on other leaf sheaths, and the panicle. Oblong to irregular, dark green, water-soaked lesions occur, which later turn greyish-brown or brown and may be surrounded by an effuse dark brown margin. The leaf sheath may also display general water-soaking and necrosis without distinct lesions. In severe infections, the entire leaf sheath may become necrotic and dry out, and the panicle withers.

The spikelets of emerging panicles may be discolored, sterile, or symptomless except for small brown spots. Glumes of panicles may show water-soaked lesions, later turning light-brown. Grains may be discolored, deformed or empty. Severely affected panicles often fail to emerge properly from the boot, producing severely discolored and poorly filled grains. In acute infections, the sheaths turn completely greyish brown or dark brown and panicles shrivel and dry.

7.3.3 Causal Organism

Pseudomonas fuscovaginae Miyajima et al. 1983

The name *Psm. fuscovaginae* Tanii et al. (1976) was omitted from the *Approved Lists of Bacterial Names*. Therefore, this name is here revived for the organism to which it originally referred (Miyajima et al. 1983).

Cells are aerobic, Gram-stain-negative, and rod-shaped having polar flagella. They oxidize glucose in oxidation-fermentation medium and produce a green fluorescent pigment, oxidase, and arginine dihydrolase. Characteristics that distinguish this species from other fluorescent pseudomonads, which are positive for arginine dihydrolase and oxidase, are its ability to produce a hypersensitive reaction in tobacco plants, and its inability to utilize 2-ketogluconate and inositol. Several characteristics that distinguish this species from *Psm. marginalis* are production of levan, potato soft rot, nitrate reduction, production of 2-ketogluconate, the formation of pits on polypectate gels, and the use of 2-ketogluconate, polygalacturonic acid and various carbohydrates (Rott et al. 1989). Twenty-two *Psm. fuscovaginae* strains from Madagascar were similar to reference strains from Burundi and Japan in basic biochemical and serological tests, and several showed a higher aggressiveness in pathogenicity tests (Duveiller et al. 1990).

Patel et al. (2012) determined the draft genome sequence of *Psm. fuscovaginae* UPB0736; the strain isolated on April 26, 1986 from a sheath brown rot lesion of rice at Antsirabe (1550 m elevation) in Madagascar. It is the first genome sequence report of this species. The G + C content was 61.46%, similar to that of other sequenced *Pseudomonas* genomes.

Xie et al. (2012) sequenced and annotated the draft genome of *Psm. fuscovaginae* CB98818, a strain isolated from diseased rice in Zhejiang province, China. The draft genome will provide the basis for a better understanding of the molecular pathogenesis of rice bacterial sheath brown rot and is an important data resource for epidemiological studies, comparative genomics, and quarantine measures of this broad host-range pathogen. The draft genome sequence of *Psm. fuscovaginae* CB98818 comprises 6,541,443 bp and has a G + C content of 61.4%. A total of 6,290 genes were predicted using Glimmer (Salzberg et al. 1998). The draft sequence contains 74 RNAs, including two 23S rRNA genes, 7 5S rRNA genes, and 65 tRNAs. Furthermore, 94.2% of the ORFs have orthologs in the reference strain *Psm. fuscovaginae* CB98818 (BLASTP E value of 1E-20), but 372 ORFs were not found in the released genomes of members of the genus *Pseudomonas*. Furthermore, functional annotation of *Psm. fuscovaginae* strain CB98818 revealed pathogenicity-related genes like type VI secretion system, type III, and IV secretion system, Hcp- and VgR-like protein, hypersensitive reaction and pathogenicity (Hrp) protein, and flagellin. These genes are virulence associated in many phytopathogenic Gram-stain-negative bacteria (Drummond et al. 2011).

7.3.4 Disease Cycle

Psm. fuscovaginae is a seed-borne and seed-transmitted pathogen of rice. The infected rice seeds are an important source of primary inoculum and a means of dissemination of the pathogen to new areas (Zeigler and Alvarez 1987). The bacterium also survives in the field as an epiphyte on rice plants and weeds. The secondary spread occurs through wind-splashed rain and irrigation water.

Adorada et al. (2015) reported the endophytic growth of *Psm. fuscovaginae* in rice seedlings using its rifampicin-marked strain. The bacterium was found in most tested seeds indicating that, even without visible discoloration of seed, seed transmission is possible. Crushed discolored seeds contained more bacterial cells than the non-crushed discolored seeds. The bacterial cells were released during seed soaking, contaminating clean seed, and lowering seed germination. The recovery of rifampicin-resistant cells of the bacterium, which were inoculated onto rice seeds, from different rice tissues at different growth stages, indicated their endophytic colonization. As symptomless seeds do not assure pathogen-free seeds, the presence of seed-borne bacteria results in poor seed germination and poor establishment of seedlings.

Psm. fuscovaginae produces three different types of phytotoxic metabolites, syringotoxin, fuscopeptin A, and fuscopeptin B that have been shown to be involved in generating the disease symptoms (Flamand et al. 1996). Ballio et al. (1996) determined the structures of fuscopeptin A and fuscopeptin B. The major bioactive compound among these three phytotoxic metabolites is syringotoxin (Batoko et al. 1994a), a non-host specific toxin considered as a virulence factor of different pathovars of *Psm. syringae* (Gross 1991). The toxins enhanced seed germination, but drastically inhibited elongation of seedlings, which was correlated to the varietal susceptibility to the disease in the field. After denaturation of the bioactive compounds of the extract, all previously observed effects were lost. These phytotoxins can, therefore, be considered as an integral part of the plant pathogen-interaction (Batoko et al. 1994a).

The treatment of rice callus with the toxins resulted in a high electrolyte leakage. The kinetics of leaching and the percentage of total electrolyte lost showed that the toxins *per se* affected the cells of red tolerant genotype less dramatically than those of susceptible ones (Batoko et al. 1994b). Batoko et al. (2012) summarized that the toxins thus produced the same alteration as those induced by the bacteria in the field.

Mattiuzzo et al. (2011) investigated the role of the quorum sensing regulatory system in causing sheath brown rot by *Psm. fuscovaginae*. The bacterium possesses two conserved *N*-acyl homoserine lactone (AHL) quorum sensing (QS) systems which were designated as PfsI/R and PfvI/R. The PfsI/R system is homologous to the BviI/R system of *Burkholderia vietnamiensis* and produces and responds to C10-HSL and C12-HSL, whereas PfvI/R is homologous to the LasI/R system of *Psm. aeruginosa* and produces several long chain 3-oxo-HSLs and responds to 3-oxo-C10-HSL and 3-oxo-C12-HSL and at high concentrations of AHL can also respond to structurally different long chain AHLs. Both the systems were found to be negatively regulated by a repressor protein which was encoded by a gene located intergenically between the AHL synthase and LuxR-family response regulator. The *pfsI/R* system was regulated by a novel repressor designated RsaM, while the *pfvI/R* system was regulated by both the RsaL repressor and RsaM. The two systems are not transcriptionally hierarchically organized, but share a common AHL response and both are required for plant virulence. Therefore, *Psm. fuscovaginae* has a unique complex regulatory network composed of at least two different repressors which directly regulate the AHL QS systems and pathogenicity.

Patel et al. (2014) characterized some of the genes involved in *Psm. fuscovaginae* virulence through an *in planta* screening of 1,000 Tn5 mutants and identified nine mutants that showed virulence deficiency compared to the wild type. The inactivated loci in these mutants include some metabolic functions and also some known virulence associated functions such as type IV pilus biogenesis protein PilZ, type VI secretion system machinery, and syringopeptin synthetase. These findings show that *Psm. fuscovaginae* might share features of some of its virulence mechanisms with other phytopathogens. New loci, never reported as being involved in virulence and encoding for hypothetical proteins, were also found. Genome mining with future virulence studies will further highlight the mechanisms of virulence of this broad host range emerging phytopathogen.

7.3.5 Management

1. Since the pathogen is seed-borne, preventive measure, such as use of pathogen-free seed is very important. Hot water treatment of seed at 65°C is effective
2. Eradication and burning of sprouted rice plants and plant litter immediately after harvest, and off-season cultivation of non-host crops, such as potatoes or lupins (*Lupinus* sp.) may be used to minimize the disease incidence (Rott 1987)
3. Adjust sowing time to avoid low temperatures during the crop season
4. Transplant seedlings that are 20- to 30-day-old rather than older seedlings (Macapuguay and Mnzaya 1988)
5. Limit the number of irrigations as bacterial sheath brown rot has not yet been found on rainfed rice in Madagascar (Rott 1987)
6. Kakar et al. (2014) reported biocontrol of the disease using rhizo-bacterial isolate *Bacillus amyloliquefaciens* Bk7. *In vitro* laboratory assay revealed that strain Bk7 and its metabolites significantly suppressed the growth of *Psm. fuscovaginae* with 93% efficacy. In glasshouse experiments, strain Bk7 exhibited biocontrol efficacy of 76.6% by reducing the disease incidence to 16.9%, compared to 72.8% observed in control. Additionally, the isolate Bk7 showed the growth promotion efficacy of plant height (GPE, 46.4%) and fresh weight (GPE, 84.3%). Characterization of isolate Bk7 revealed its strong capability for biofilm formation, inorganic phosphate solubilization, and production of high amounts of indole-3 acetic acid, siderophores, and ammonia *in vitro*.

REFERENCES

Adorada, D. L., Stodart, B. J., Pangga, I. B., and Ash, G. J. 2015. Implications of bacterial contaminated seed lots and endophytic colonization by *Pseudomonas fuscovaginae* on rice establishment. *Plant Pathol.* 64: 43–50.

Ballio, A., Bossa, F., Camoni, L., Di Giorgio, D., Flamand, M.-C., et al. 1996. Structure of fuscopeptins, phytotoxic metabolites of *Pseudomonas fuscovaginae*. *FEBS Lett.* 381: 213–216.

Batoko, H., Bouharmont, J., and Maraite, H. 1994a. Inhibition of rice (*Oryza sativa* L.) seedling elongation by a *Pseudomonas fuscovaginae* toxin. *Euphytica*, 76: 139–143.

Batoko, H., Hsissou, D., Bouharmont, J., Maraite, H., and Kinet, J.-M. 1994b. Phytotoxicity of a *Pseudomonas fuscovaginae* toxin on rice cells: Enhancement of electrolyte leakage from treated calli. *Archives Internationales de Physiologic, de biochimie et de biophysique* 102: 1–11.

Batoko, H., Bouharmont, J., Kinet, J.-M., and Maraite, H. 2012. Inhibition of rice (*Oryza sativa* L.) internode elongation by *Pseudomonas fuscovaginae* toxins. pp. 227–229 In: *Pseudomonas Syringae Pathovars and Related Pathogens*. K. Rudolph, T. J. Burr, J. W. Mansfield, D. E. Stead, A. Vivian, and J. von Kietzell eds. Springer Science and Business Media, Dordrecht, the Netherlands, 665 pp.

Drummond, A. J., et al. 2011. Geneious v5.6 [http://www.geneious.com/]

Duveiller, E., Notteghem, J. L., Rott, P., Snacken F., and Maraite, H. 1990. Bacterial sheath brown rot of rice caused by *Pseudomonas fuscovaginae* in Malagasy. *Trop. Pest Manag.* 36: 151–153.

Flamand, M.-C., Pelsser, S., Ewbank, E., and Maraite, H. 1996. Production of syringotoxin and other bioactive peptides by *Pseudomonas fuscovaginae*. *Physiol. Mol. Plant Pathol.* 48: 217–231.

Gross, D. C. 1991. Molecular and genetic analysis of toxin production by pathovars of *Pseudomonas syringae*. *Ann. Rev. Phytopath.* 29: 247–278.

Kakar, K. U., Duan, Y.-P., Nawaz, Z., Sun, G., Almoneafy, A. A., et al. 2014. A novel rhizobacterium Bk7 for biological control of brown sheath rot of rice caused by *Pseudomonas fuscovaginae* and its mode of action. *Eur. J. Plant Pathol.* 138: 819–834.

Macapuguay, F., and Mnzaya, M. 1988. Progress on irrigated rice agronomy at Usangu village, M'Beya zone, Tanzania. *Intl. Rice Commission Newsletter* 37: 40–43.

Mattiuzzo, M., Bertani, I., Ferluga, S. Cabrio, L., Bigirimana, J., et al. 2011. The plant pathogen *Pseudomonas fuscovaginae* contains two conserved quorum sensing systems involved in virulence and negatively regulated by RsaL and the novel regulator RsaM. *Environ. Microbiol.* 13: 145–162.

Miyajima, K., Tanii, A., and Akita, T. 1983. *Pseudomonas fuscovaginae* sp. nov., nom. rev. *Int. J. Syst. Bacteriol.* 33: 656–657.

Patel, H. K., da Silva, D. P., Devescovi, G., Maraite, H., Paszkiewicz, K., et al. 2012. Draft genome sequence of *Pseudomonas fuscovaginae*, a broad-host-range pathogen of plants. *J. Bacteriol.* 194: 2765–2766.

Patel, H. K., Matiuzzo, M., Bertani, I., de Paul Bigirimana, V., Ash, G. J., et al. 2014. Identification of virulence associated loci in the emerging broad host range plant pathogen *Pseudomonas fuscovaginae*. *BMC Microbiol.* 14: 274–277.

Plantwise Technical Factsheet 2015. Sheath brown rot (*Pseudomonas fuscovaginae*).

Razak, A., Zainudin, N., Sidiqe, S., Ismail, N., Mohamad, N., et al. 2009. Sheath brown rot disease of rice caused by *Pseudomonas fuscovaginae* in the Peninsular Malaysia. *J. Plant Prot. Res.* 49: 244–249.

Rott, P. 1987. Brown rot (*Pseudomonas fuscovaginae*) of the leaf sheath of rice in Madagascar. Institute de Recherches Agronomiques Tropicales et des Cultures Vivrieres, Montpellier, France, 22 pp.

Rott, P., Notteghem, J. L., and Frossard, P. 1989. Identification and characterization of *Pseudomonas fuscovaginae*, the causal agent of bacterial sheath brown rot of rice, from Madagascar and other countries. *Plant Dis.* 73: 133–137.

Salzberg, S. L., Delcher, A. L., Kasif, S., and White, O. 1998. Microbial gene identification using interpolated Markov models. *Nucleic Acids Res.* 26: 544–548.

Sharma, S., Sthapit, B. R., Pradhanang, P. M., and Joshi, K. D. 1997. Bacterial sheath brown rot of rice caused by *Pseudomonas fuscovaginae* in Nepal. Rice Cultivation in Highland Areas. pp. 107–112 in: *Proceedings of the CIRAD Conference held at Antananarivo, Madagascar 29 March-5 April 1996*, C. Poisson, and J. Rakotoarisoa eds.CIRAD-CA, Montpellier, France.

Tanii, A., Miyajima, K., and Akita, T. 1976. The sheath brown rot disease of rice plant and its causal bacterium, *Pseudomonas fuscovaginae*. *Ann. Phytopathol. Soc. Japan* 42: 540–548. (in Japanese with English abstract).

Webster, R. K., and Gunnell, P. S. 1992. *Compendium of Rice Diseases*. American Phytopathological Society, St. Paul, MN, 62 pp.

Xie, G. L. 2003. First report of sheath brown rot of rice in China and characterization of the causal organism by phenotypic tests and biolog. *Int. Rice Res. Notes* 28(1): 50–52.

Xie, G., Cui, Z., Tao, Z., Qiu, H. Liu, H., et al. 2012. Genome sequence of the rice pathogen *Pseudomonas fuscovaginae* CB98818. *J. Bacteriol.* 194: 5479–5480.

Zeigler, R. S., and Alvarez, E. 1987. Bacterial sheath brown rot of rice caused by *Pseudomonas fuscovaginae* in Latin America. *Plant Dis.* 71: 592–597.

Zeigler, R. S., and Alvarez, E. 1990. Characteristics of *Pseudomonas spp. causing* grain discoloration and sheath rot of rice, and associated pseudomonad epiphytes. *Plant Dis.* 74: 917–922.

7.4 YELLOW EAR ROT OF WHEAT

The disease is also known by other names such as, **yellow slime rot** and **spike blight** and in India as ***Tundu* disease**. It was first reported by Hutchinson (1917) from Punjab, India, but he did not establish the Koch's postulates. According to him, the disease was present since 1908. The disease is present in the entire wheat belt of north India. In general, the losses in India range from 1% to 2%, but in extreme cases, losses up to 50% have also been reported. Paruthi and Gupta (1987) reported an infection level of 60.8% from India. The disease is quite severe in Pakistan, and the yield losses ranging from 69.6% to 93.5% (depending upon the cultivars) have been reported by Anwar and Inam-ul-Haq (1992). They also reported drastic reductions in other plant parameters, i.e., 15.1%–21.7% in plant height, 20.8%–27.7% in tillering, 10.9%–14.5% in ear length, and 28.6%–84.3% in number of grains. Besides India and Pakistan, the occurrence of disease has been reported from China, Egypt, Australia, Iraq, Iran, Cyprus, and Ethiopia.

Yellow ear rot is actually a **disease complex**, i.e., it is caused by a bacterium, *Rathayibacter tritici* and a nematode, *Anguina tritici*. *A. tritici* alone causes another disease, namely, **ear cockle of wheat**. Milne in 1925 gave the indication that the disease is caused by a bacterium and a nematode when he observed that some ear cockle galls could be obtained from certain ear heads showing yellow ear rot. However, the experimental evidence for this was provided by Fahmy and Mikhail (1925) by showing that bacterium alone could not produce the disease. Later on, Chaudhuri (1935) claimed that the bacterium alone can cause the disease, but his claim has not been confirmed by the subsequent workers. Vasudeva and Hingorani (1952) confirmed the work of Fahmy and Mikhail (1925) and the latest confirmation is by Sahebani et al. (2006).

7.4.1 Symptoms

As *A. tritici* is also a part of the causal complex, initial symptoms of ear cockle and yellow ear rot are the same. The first symptom is the enlargement of the stem near the soil line in 20- to 25-day-old seedlings. The leaves emerging from such seedlings are twisted and wrinkled. Affected plants are dwarfed and spreading type, and many die in case of severe infection.

The symptoms of yellow ear rot become evident at the time of ear emergence and are characterized by the presence of a bright yellow slime or gum on abortive ears and the leaves in contact with such ears in the boot leaf. The affected ears fail to produce any grain. In humid weather, yellow slimy mass in the ears trickles down and may fall on the lower leaves. On drying, the slimy mass becomes hard, brittle, and brown. Affected culms either die in young stage or grow until heading. In the latter case, the emerging spike is narrow and short with grains partially or completely replaced by the bacterial mass. Sometimes, the spike may fail to come out of the boot leaf. The stalk is always distorted when the bacterial mass is produced in the ears. If the conditions favorable for the multiplication of the bacteria continue, the slimy mass completely fills the ear resulting into a **complete *Tundu* ear**. However, if the conditions are not that favorable, then partial ear cockle and partial yellow ear rot symptoms develop. In the latter case, ear cockle galls carry the bacterial inoculum for initiating the disease in the next year. Bright yellow streaks or slimy mass also appear on the leaf surface of young plants.

7.4.2 Causal Organisms

Rathayibacter tritici (*ex* Hutchinson 1917) Zgurskaya et al. 1993
Syn. *Clavibacter tritici*, *Corynebacterium michiganense* pv. *tritici*, *Cor. tritici*
Anguina tritici

Rat. tritici: The bacterium was first named *Pseudomonas tritici* by Hutchinson (1917), and Burkholder (1948) changed it to *Corynebacterium tritici*. With the introduction of pathovar concept for the classification of plant pathogenic bacteria, it was changed to *Cor. michiganense* pv. *tritci*. Later on, Davis et al. (1984) reclassified it as *Clavibacter tritci*. Zgurskaya et al. (1993) created a new genus *Rathayibacter* and renamed it as *Rathayibacter tritici*. The below given description of the bacterium is taken from Zgurskaya et al. (1993).

Cells are irregular, pleomorphic, short rods and measure 0.4–0.8 × 0.5–1.8 µm. V-forms and sometimes short cell chains are formed, but a clear rod-coccus cycle is not observed. Non-motile, non-sporulating, Gram-stain-positive, not acid-fast, and obligately aerobic. Chemoorganotrophs, having an oxidative type of metabolism and catalase test positive. The optimal growth occurs between 24°C and 28°C. The cell walls contain galactose and xylose. The bacterium is capable of utilizing xylose, lactose, L-arabinose, mannitol, sorbitol, inulin, citrate, tartrate, and sebacinate. Acid is produced from glucose, galactose, fructose, mannose, maltose, sucrose, and xylose. Glutamic acid and DL-ornithine are utilized, but not DL-valine. The bacterium is able to hydrolyze Tween 21, 40, and 85 and grow in the presence of 5% NaCl and 0.03% potassium tellurite. Voges–Proskauer test is negative, but methyl red is positive. The bacterium contains a 50 kb plasmid.

Park et al. (2017) reported the first complete genome sequence of *Rat. tritici* NCPPB 1953 with diverse features of this strain. The whole genome consists of one circular chromosome of 3,354,681 bp having a GC content of 69.48%. A total of 2,979 genes were predicted, comprising 2,866 protein coding genes and 49 RNA genes. The comparison of genomic analyses of *Rat. tritici* NCPPB 1953 and *Rat. toxicus* strains identified 1,052 specific genes in *Rat. tritici* NCPPB 1953. Using the BlastKOALA database, the authors found that the flexible genome of *Rat. tritici* NCPPB 1953 is highly enriched in "Environmental Information Processing" system and metabolic processes for diverse substrates. Moreover, many specific genes of *Rat. tritici* NCPPB 1953 are distributed in substrate-binding proteins for extracellular signals including saccharides, lipids, phosphates, amino acids, and metallic cations. These findings provide clues for rapid and stable colonization of *Rat. tritici* for disease mechanism and nematode association.

It is generally believed that both the bacterium and nematode are equally important in causing the disease. However, the opinion is changing in favor of a major role played by the bacterium. Now it is believed that the disease is caused mainly by the bacterium, and the nematode helps in carrying the bacterium to the ears and in its entry into the host tissue by causing wounds in addition to the survival of the bacterium in or the nematode galls.

7.4.3 Disease Cycle

The primary source of inoculum for both the bacterium and the nematode are cockles or nematode galls. These galls may be present in the soil or in the seed. The galls carry the bacterium inside and on their surface. Under dry conditions, the nematode can survive inside the galls for several years, while the survival of the bacterium is at least for 5 years.

The nematode galls present in the soil, or brought to the soil along with the contaminated seed, absorb moisture, swell, and release 2nd stage larvae. These bacteria laden larvae find their way to growing point of the seedlings and feed ectoparasitically. With the natural growth of the plants, the larvae are carried to the inflorescence, where they enter the floral primordia and feed endoparasitically. The bacteria also gain entry into the developing ear heads along with the nematode. Under favorable weather conditions, the bacteria multiply profusely inside the ears producing an abundant slimy mass, which envelops the developing spikelets. Poorly drained soils, early sowing, and more number of irrigations increase the disease incidence.

7.4.4 Management

As the nematode galls of *A. tritici* are responsible for the survival of the bacterium and its larvae initiate the disease, the control of this nematode will automatically eliminate the disease. The following measures are adopted for the control of *A. tritici* and ultimately the yellow ear rot.

1. Use of gall-free seed. If the seed is not gall-free, it should be made gall-free by adopting the following measures
   a. Soak the seed in 20% brine solution (NaCl or KCl). Vigorously stir the seed and remove the galls, which float on the surface of water. Destroy these galls by burning. After soaking in brine solution, wash the seed 2–3 times with water to avoid reduction in germination
   b. Soak the seed in water for 2 h and heat the water to raise its temperature to 51°C and maintain it for 30 min. The heat treatment activates the dormant larvae and then kills them
   c. Galls can also be removed by sieving, fanning, or winnowing. In China, nematode eliminators are used to remove the galls
2. Crop rotation for 2–3 years with non-graminaceous crops
3. Grow the disease resistant or tolerant cultivars
4. Use of chemicals for seed/soil treatment

References

Anwar, M. J., and Inam-ul-Haq, M. 1992. Effect of *Tundu* disease on growth and yield of wheat. *Pakistan J. Phytopathol.* 4: 66–69.

Burkholder, W. H. 1948. Genus I. *Corynebacterium* Lehmann & Neumann. pp. 381–408 in: *Bergey's Manual of Determinative Bacteriology*. 6th ed. R. S. Breed, E. G. D. Murray, and A. P. Hitchens eds. Baillière, Tindall & Cox, London, UK and Williams & Wilkins Co., Baltimore, MD.

Chaudhuri, H. 1935. A bacterial disease of wheat in Punjab. *Proc. Indian Acad. Sci.* 1(B): 579–585.

Davis, M. J., Gillaspie, A. J. Jr., Vidaver, A. K., and Harris, R. W. 1984. *Clavibacter*: A new genus containing some phytopathogenic coryneform bacteria, including *Clavibacter xyli* subsp. *xyli* sp. nov., subsp. nov. and *Clavibacter xyli* subsp. *cynodontis* subsp. nov., pathogens that cause ratoon stunting disease of sugarcane and bermudagrass stunting disease. *Int. J. Syst. Bacteriol.* 34: 107–117.

Fahmy T., and Mikhail, T. 1925. The bacterial disease of wheat caused by *Pseudomonas tritci* Hutchinson. *Agri. J. Egypt* 1: 64–72.

Hutchinson, C. M. 1917. A bacterial disease of wheat in Punjab. *Mem. Dept. Agric. India Bact. Ser.* 1: 169–175.

Park, J., Lee, P. A., Lee, H. H., Choi, K., Lee, S. W., et al. 2017. Comparative genome analysis of *Rathayibacter tritici* NCPPB 1953 with *Rathayibacter toxicus* strains can facilitate studies on mechanisms of nematode association and host infection. *Plant Pathol. J.* 33: 370–381.

Paruthi, I. J., and Gupta, D. C. 1987. Incidence of "tundu" in barley and kanki in wheat field infested with *Anguina tritici*. *Haryana Agric. Univ. J. Res.* 17: 78–79.

Sahebani, N., Kheiri, A., Rahimian, H., Sharifi-Tehrani, A., and Zakeri, Z. 2006. A study on interaction between wheat seed gall nematode *Anguina tritici* and wheat yellow ear rot *Rathayibacter tritici*. *Iranian J. Agric. Sci.* 36: 1355–1360.

Vasudeva, R. S., and Hingorani, M. K. 1952. Bacterial disease of wheat caused by *Corynebacterium tritici* (Hutchinson) Bergey et al. *Phytopathology* 42: 291–293.

Zgurskaya, H. I., Evtushenko, L. I., Akimov, V. N., and Kalakoutskii, L. V. 1993. *Rathayibacter* gen. nov., including the species *Rathayibacter rathayi* comb. nov., *Rathayibacter tritici* comb. nov., *Rathayibacter iranicus* comb. nov., and six strains from annual grasses. *Int. J. Syst. Bacteriol.* 43: 143–149.

7.5 Bacterial Leaf Blight and Basal Glume Rot of Wheat

Bacterial leaf blight (also called **leaf necrosis**) and basal glume rot (also called **basal glume blotch** or **spike symptoms**) of wheat (*Triticum aestivum* L.) are generally considered to be of minor economic importance. Weather conditions strongly influence severity of these diseases and economic losses caused by them. These occur sporadically and only under extremely humid conditions in spring and summer. The high losses have been recorded only in a few cases, for example, in South Dakota, where epiphytotic outbreaks of leaf necrosis were recorded over a period of 7 years, and during this period, the fields having 75% or more necrotic leaves were commonly found (Otta 1974). In Germany, losses due to basal glume rot in areas of marshy soils were estimated to exceed 50% (Mavridis et al. 1991). In addition to yield losses, infection due to basal glume rot can severely reduce the grain quality of bread wheat (Vassilev and Karov 1985; Mavridis et al. 1991).

Since the first description of a wheat disease caused by *Pseudomonas syringae* in 1920 (McCulloch 1920), wheat pathogens of the *Psm. syringae* group have been reported from nearly all temperate and sub tropical wheat growing regions. These pathogens occur in maritime climates (warm summers, mild winters, and constant humidity) and under temperate climatic conditions (warm, rainy summers, and cold winters). Bacterial leaf blight has been reported from Argentina, USA, Canada, Italy, Pakistan, and South Africa. Comparatively, basal glume rot has been reported from more countries, namely, Canada, South Africa, Mexico, New Zealand, Australia, Bulgaria, Germany, Syria, Ukraine, Denmark, and Belgium. However, in several cases, the symptoms present on the spike were not similar to those observed for basal glume rot. It is possible that in such cases, the widespread populations of *Psm. syringae* that live on plant surfaces as epiphytes, but do not produce disease symptoms, might have been mistaken to cause the spike symptoms.

Psm. syringae pv. *syringae* infects many different plant species, including stone fruit trees, wheat, sorghum, lilac, bush bean, and many weeds (Elliott 1951; Bradbury 1986). *Psm. syringae* pv. *syringae* strains from corn, sorghum, foxtail, and peach have been found to be pathogenic on wheat seedlings (Otta 1974). **Bacterial black nod** that produces slightly different symptoms on leaves and spikes is caused by

Psm. syringae pv. *japonica* (Mukoo) Dye et al. The bacterium mainly attacks barley, but it also infects wheat.

7.5.1 Symptoms

7.5.1.1 Bacterial Leaf Blight

Initial symptoms appear at booting to early heading as numerous, tiny, water-soaked spots on the flag leaf and the first and second leaves below the flag leaf. Within 2–3 days, these lesions expand and often coalesce into large, greyish-green, desiccated areas. These areas become necrotic and bleach to a light tan or white within a week. Necrosis often progresses until 75%–100% area of the leaf blade is destroyed. Bacterial ooze has been observed on the lesions in some cases, especially after light dew or rain, on the dorsal side of old water-soaked lesions.

7.5.1.2 Basal Glume Rot

The infected glumes show a dull, brownish black area at the base. Usually, only 1/3 lower part of the glume, or less is darkened. Symptoms on the inner surfaces of the glumes are more conspicuous than those on the outer side. Sometimes, small, water-soaked areas are found on the margins of the lesions. Basal glumes blotch can reduce yields by reducing seed fill (Capparelli et al. 2005).

Due to a publication by Wilkie (1973), atypical symptoms on spikes and stems have often been wrongly attributed to basal glume rot by some authors (Wiese 1987; Smith and Hattingh 1991). In these publications, the entire glume, from the base to the top, shows dark-brown or black streaks. Symptoms may develop on a few or all the glumes of a spike, and longitudinal streaks are also found on the upper part of the stem, especially just above the nodes. In some cases, the described symptoms appeared in addition to the typical symptoms of basal glume rot (Wilkie 1973). It is true that *Psm. syringae* pv. *atrofaciens* was isolated from plants showing atypical symptoms, but it was never convincingly proved by artificial inoculation that those symptoms were caused by this bacterium. The observed **atypical symptoms** have also been described as **brown necrosis**, **melanism**, or **pseudo-black chaff**, and can be induced by abiotic stress. Therefore, it is possible that in the abovementioned cases epiphytic pseudomonads might have been isolated by chance and assumed to be the cause of the observed symptoms. The mere isolation of these pathovars does not prove that the observed disease symptoms are caused by them. These pathogens can be assumed to have a causal role only when their high concentrations (e.g., 10^8 cfu g^{-1} fresh weight) occur in the diseased tissue (von Kietzell and Rudolph 1997a).

In diseased grains, symptoms appear at the basal or germ end, varying in color from a scarcely noticeable brown to charcoal black (McCulloch 1920). No doubt this symptom is typical of *Psm. syringae* pv. *atrofaciens*, it is not specific and has also been attributed to other pathogens, such as *Alternaria alternata* and *Bipolaris sorokiniana* (Huguelet and Kiesling 1973; Statler et al. 1975).

7.5.2 Causal Organisms

Pseudomonas syringae pv. *syringae* van Hall 1902 causes leaf blight

Psm. syringae pv. *atrofaciens* (McCulloch) Young et al. 1978 causes basal glume rot

The cells are straight or slightly curved Gram-stain-negative rods, occur singly or in chains of a few cells, and motile by a tuft of polar flagella. The bacteria are strictly aerobic. The colonies on King's medium B after 24 h of incubation are circular, convex, translucent to opaque, whitish-grey, and emit a blue fluorescence that turns green after 2 days. The KBC medium is highly selective for the isolation of *Psm. syringae* pv. *atrofaciens* and promotes excellent growth of the bacterium if the concentration of boric acid is reduced from the original 1.5 g to 0.5 g L^{-1} (Mohan and Schaad 1987). Tests for arginine dihydrolase, oxidase, and potato rotting are negative. Levan is produced on 5% sucrose and starch is not hydrolyzed. Gelatin, esculin, and arbutin are hydrolyzed, and protocatechinate is cleaved at the *ortho* position (Bradbury 1986).

In BIOLOG-tests (Biolog, Inc.), both the pathovars utilized Tween 40 and 80, L-arabinose, D-arabinose, D-fructose, D-galactose, α-D-glucose, *m*-inositol, D-mannitol, D-mannose, sucrose, methyl pyruvate, cis-aconitic acid, citric acid, formic acid, D-galactonic acid lactone, D-gluconic acid, D-glucosaminic acid, DL lactic acid, malonic acid, D-saccharic acid, bromo succinic acid, D-alanine, L-alanine, L-alanyl-glycine, L-asparagine, L-aspartic acid, L-glutamic acid, L-serine, γ-amino butyric acid, uridine, glycerol, D-sorbitol, and DL-alpha-glycerol phosphate (Bochner 1989).

Psm. syringae pv. *syringae* and pv. *atrofaciens* cannot be differentiated on the basis of morphological, physiological, serological, and genetic characteristics. The only way to distinguish between these pathovars is to conduct a pathogenicity test on their respective hosts. Both the pathovars produce hypersensitive reaction on tobacco.

Kazempour et al. (2010) identified *Psm. syringae* pv. *atrofaciens* using specific primers, SyD1 and SyD2 and *Psm. syringae* pv. *syringae* using specific primers, PSF and PSR.

7.5.3 Disease Cycle

Psm. syringae pvs. *atrofaciens* and *syringae* survive between crops in infected crop debris, pathogenically on collateral hosts, and epiphytically on collateral hosts and weeds. Both the pathogens have important epiphytic phases. In Germany, von Kietzell and Rudolph (1997b) isolated *Psm. syringae* pv. *atrofaciens* from 10% to 48% of symptomless wheat leaves and glumes. Bacterial populations obtained in these studies were higher than 10^5 cfu/cm^2. On barley, it was isolated from 35% to 63% of leaf and glume samples showing no disease symptoms and bacterial concentrations reached higher than 10^7 cfu/cm^2. *Psm. syringae* pv. *syringae* was shown to move from inoculated wheat seed to the seedlings and to survive as an epiphyte on the first leaves (Fryda and Otta 1978). In other field experiments, von Kietzell and Rudolph (1997b) also isolated *Psm. syringae* pv. *atrofaciens* from the first leaves of cereals after seed inoculation. Initially, the subsequent leaves were pathogen-free, but the bacterium successively invaded the upper leaves and finally reached the flag leaf without causing symptoms of basal glume rot, due to unfavorable weather conditions. Fessehaie (1993) isolated *Psm. syringae* pv. *atrofaciens* strains from grassy weeds.

Psm. syringae pvs. *syringae* and *atrofaciens* are also commonly found on wheat seeds. Otta (1977) found *Psm. syringae* pv. *syringae* in all the 21 seed lots representing eight winter wheats from nine locations in the USA and Canada. *Psm. syringae* pv. *atrofaciens* was also found on 42 out of 54 wheat seed samples and on 35 out of 37 barley seed samples collected randomly from different cultivars and locations in Germany (Fessehaie 1993; von Kietzell and Rudolph 1997b). Bacterial concentrations reached up to 10^5 cfu per 50 g seed in these samples.

As *Psm. syringae* pv. *syringae* and *Psm. syringae* pv. *atrofaciens* are abundantly present as epiphytic populations on their hosts; it is evident that weather conditions are more relevant to disease outbreaks than the presence of inoculum. Wind-splashed rain and overhead-irrigation act as dispersal agents of the bacteria. The bacteria are readily moved in irrigation water. Wind-blown dust or residue fragments carry *Psm. syringae* pv. *syringae* to seeds (Capparelli et al. 2005). Infected seed plays an important role in the long distance dissemination of the bacteria. The periods of extraordinarily humid and cool weather favor outbreaks of both the diseases.

Besides other factors, syringomycins and polysaccharides produced by these bacteria play an important role in their virulence and/or pathogenicity. The syringomycins and cyclic lipodepsinona peptides produced by *Psm. syringae* pv. *atrofaciens* and by *Psm. syringae* pv. *syringae* are phytotoxins having broad antimicrobial activity. Polysaccharides excreted by the bacteria include levan, alginate, and lipopolysaccharides. Ovod et al. (1995) characterized lipopolysaccharides by an *O*-specific chain consisting of D-rhamnose tetrasaccharide repeating units with α1-2 and α1-3 bonds, and lateral branching substitutes of D-rhamnose, D-fucose, or *N*-acetyl-D-glucosamine.

7.5.4 Management

Psm. syringae pathovars that cause bacterial leaf blight and basal glume rot of wheat are weak pathogens and outbreaks of these diseases are quite infrequent. Therefore, routine control measures for their management have not been worked out. However, the following measures can prove useful to minimize the disease incidence.

1. Contaminated seed plays an important role in inoculum transmission; therefore, seed lots from heavily infested fields should not be used for sowing. Seed treatment with copper bactericides may provide some disease control
2. Sow resistant or tolerant varieties
3. Deep plough and burry the infected plant residue
4. Avoid overhead-irrigation, especially during cool and wet weather
5. Avoid reuse of irrigation tail water

REFERENCES

Bochner, B. 1989. "Breathprints" at the microbial level. *ASM News* 55(10): 536–539.

Bradbury, J. F. 1986. *Guide to Plant Pathogenic Bacteria*. CAB International Mycological Institute, Kew, UK. 332 pp.

Capparelli, R., Amoroso, M. G., Palumbo, D., Iannaccone, M., Faleri, C., et al. 2005. Two plant puroindolines colocalize in wheat seed and *in vitro* synergistically fight against pathogens. *Plant Microb. Biol.* 58: 857–867.

Elliott, C. 1951. *Manual of Bacterial Plant Pathogens*. 2nd ed. Waltham, MA: Chronica Botanica. 186 pp.

Fessehaie, A. 1993. Über die Isolation des Erregers der basalen Spelzenfäule an Getreide, *Pseudomonas syringae* pv.*atrofaciens* ((McCull.) Young, Dye, Wilkie) aus Gerste, Hafer und Wildgräsern. Diploma thesis, Universität Göttingen, Institut für Pflanzenpathologie und Pflanzenschutz.

Fryda, S. J., and Otta, J. D. 1978. Epiphytic movement and survival of *Pseudomonas syringae* on spring wheat. *Phytopathology* 68: 1064–1067.

Huguelet, J. E., and Kiesling, R. L. 1973. Influence of inoculum composition on the black point disease of durum wheat. *Phytopathology* 63: 1220–1225.

Kazempour, M. N., Kheyrgoo, M., Pedramfar, H., and Rahimian, H. 2010. Isolation and identification of bacterial glume blotch and leaf blight on wheat (*Triticum aestivum* L.) in Iran. *Afr. J. Biotechnol.* 9: 2860–2865.

Mavridis, A., Meyer, D., Mielke, H., and Steinkampf, G. 1991. Zum Auftreten und zur Schadwirkung der basalen Spelzenfäule beim Sommerweizen. *Kali-Briefe (Büntehof)* 20: 469–473.

McCulloch, L. 1920. Basal glume rot of wheat. *J. Agric. Res.* 18: 543–549.

Mohan, S. K., and Schaad, N. W. 1987. An improved agar plating assay for detecting *Pseudomonas syringae* pv. *syringae* and *P. s.* pv. *phaseolicola* in contaminated bean seed. *Phytopathology* 77: 1390–1395.

Otta, J. D. 1974. *Pseudomonas syringae* incites a leaf necrosis on spring and winter wheats in South Dakota. *Plant Dis. Reptr* 58: 1061–1064.

Otta, J. D. 1977. Occurrence and characteristics of isolates of *Pseudomonas syringae* on winter wheat. *Phytopathology* 67: 22–26.

Ovod, V., Ashorn, P., Yakovleva, L., and Krohn, K. 1995. Monoclonal antibodies against the core oligosaccharide and O polysaccharide of *Pseudomonas syringae* lipopolysaccharide: Generation, characterization and use for bacterial classification and partial epitope mapping. *Phytopathology* 85: 226–232.

Smith, J., and Hattingh, M. J. 1991. Fluorescent pseudomonads associated with diseases of wheat in South Africa. *J. Phytopathol.* 133: 36–48.

Statler, G. D., Kiesling, R. L., and Busch, R. H. 1975. Inheritance of black point resistance in durum wheat. *Phytopathology* 65: 627–629.

van Hall, C. J. J. 1902. Bijdragen tot de kennis der Bakterieele Plantenzeikten. Coöperatieve Drukkerij-vereeniging "Plantijn," Inaugural dissertation, Amsterdam, the Netherlands.

Vassilev, V. I., and Karov, S. 1985. Effect of cereal basal bacteriosis on the economic physico-technological and biochemical properties of wheat grain and flour. *Plant Sci.* 22: 13–20 (in Bulgarian).

von Kietzell, J., and Rudolph, K. 1997a. Wheat diseases caused by *Pseudomonas syringae* pathovars. pp. 49–58 in: *The Bacterial Diseases of Wheat: Concepts and Methods of Disease Management*. E. Duveiller, L. Fucikovsky, and K. Rudolph eds. CIMMYT, Mexico City, Mexico.

von Kietzell, J., and Rudolph, K. 1997b. Epiphytic occurrence of *Pseudomonas syringae* pv. *atrofaciens*. pp. 29–34 in: *Pseudomonas syringae Pathovars and Related Pathogens*. K. Rudolph, T. J. Burr, J. W. Mansfield, D. Stead, A. Vivian, and J. von Kietzell eds. Kluwer Academic Publishers, Dordrecht, the Netherlands.

Wiese, M. V. 1987. *Compendium of Wheat Diseases*. 2nd ed. American Phytopathological Society, St. Paul, MN, pp. 5–10.

Wilkie, J. P. 1973. Basal glume rot of wheat in New Zealand. *N. Z. J. Agric. Res.* 16: 155–160.

Young, J. M., Dye, D. W., and Wilkie, J. P. 1978. Genus *Pseudomonas* Migula 1894. In: Young, J. M., Dye, D. W., Bradbury, J. F., Panagopoulos, C. G., and Robbs, C. F. 1978. A proposed nomenclature and classification for plant pathogenic bacteria. *N. Z.J. Agric. Res.* 21: 153–177.

7.6 BACTERIAL LEAF STREAK (BLS) AND BLACK CHAFF OF WHEAT

Bacterial leaf streak (BLS) of cereals, also known as **bacterial leaf stripe**, infects many important cereal crops and grasses. The disease is also called black chaff when the symptoms occur on the glumes. The pathogen was first described on barley (*Hordeum vulgare* L.) (Jones et al. 1917), and later on wheat (*Triticum aestivum* L.) (Smith et al. 1919), rye (*Secale cereale* L.) (Reddy et al. 1924), grasses (Wallin 1946), and lastly on triticale (× *Triticosecale* Wittmack) (Zillinsky and Borlaug 1971).

Bacterial leaf streak is a sporadic, but widespread disease of wheat that can cause significant losses. The seed-borne nature of the pathogen is a major problem with the possibility that primary inoculum may increase during seed multiplication. The seed-borne nature of the pathogen is also a major constraint for international germplasm exchange. Fortunately, a specific succession of events is necessary for the outbreak of an epiphytotic. If one of the events required for disease development does not occur, the epiphytotic may not occur. Black chaff incidence, severity, and distribution may thus vary from year to year, even in disease-prone areas.

Epiphytotics of BLS may occur in various scenarios. This explains why the disease has a global distribution and is sporadic in areas as different as sprinkler-irrigated wheat fields in the United States, Mexican highlands characterized by marked daytime temperature changes, and the Southern Cone countries of South America, where warm and cloudy days may occur alternately. Because of its sporadic occurrence, research on epidemiology and resistance is particularly difficult and, consequently, significant achievements for its management have not been made.

Yield losses due to BLS and black chaff are variable, ranging from negligible to 40%, depending on the stage of infection and severity. During 2008–2009 growing seasons, surveys of wheat in five locations in North Dakota showed an average of 80% disease incidence. Yield losses up to 40% have occurred in the most severely diseased fields in Idaho, although losses are generally 10% or less. Using a modified single tiller approach over three seasons in the USA, Shane et al. (1987) calculated that 50% disease severity on the flag leaf caused an 8%–13% loss in kernel weight and 100% disease severity on the flag leaf resulted in a 13%–34% loss. In Mexico, yield loss in wheat was evaluated in a high rainfall, temperate environment based on infection and yield in single tillers. It was found that on an average, losses below 5% can be expected when the percent infected flag leaf area is less than 10%. However, an average yield reduction up to 20% can be anticipated, if 50% of the flag leaf is diseased (Duveiller and Maraite 1993).

Yield loss is a linear function of the percent infected flag leaf area, and even a small infected leaf area will affect the yield. Although the disease is usually observed late in the growing season, the negative effect of the pathogen on yield can be observed as soon as lesions develop on the flag leaf because even a small percentage of diseased leaf area has an immediate effect. The effect of the disease on yield is probably related to a reduction in photosynthesis resulting from the extent of diseased leaf area because similar effects on yield can be observed when leaves are detached from plants (Duveiller and Maraite 1993).

Bacterial leaf streak has a widespread geographical distribution (Duveiller et al. 1997). In North America, it has been reported from the United States, Canada, and Mexico. In South America, it occurs in Argentina, Bolivia, parts of Brazil, Paraguay, Peru, and Uruguay (Tessi 1949; Mohan and Mehta 1985; Frommel 1986; Mehta 1990; Duveiller et al. 1991). The disease occurs on wheat in China (Sun and He 1986), Pakistan (Akhtar and Aslam 1985), and Iran (Alizadeh et al. 1995), and on triticale in India (Richardson and Waller 1974). The disease affects durum (*T. turgidum* var. *durum* L.) and bread wheat in irrigated areas of Syria (Mamluk et al. 1990), Israel (CIMMYT 1977), Turkey (Demir and Üstün 1992), and Yemen (Bragard et al. 1995). Paul and Smith (1989) reported the absence of disease from Western Europe, probably due to unfavorable environmental conditions, particularly low temperatures. In Africa, it has been reported from Kenya (Burton 1931), Ethiopia (Korobko et al. 1985), South Africa (Smit and Van A. Bredenkamp 1988), Tanzania (Bradbury 1986), Libya and Madagascar (Bragard et al. 1995), Morocco (Sands and Fourest 1989), and Zambia (Bragard et al. 1997). In Australia, its occurrence has been recorded on wheat and rye in New South Wales (Noble 1935).

7.6.1 Symptoms

Typical symptoms on the leaves appear as elongated, light brown lesions, several centimeters long, which are initially distinct, but later coalesce to cover large leaf areas. In the early stage, the symptoms are characterized by translucent stripes that are easily seen under incident light. Initially, lesions are water-soaked and produce honey-like bacterial exudates under humid conditions. On drying, the exudates harden into yellowish, resinous granules on the surface of the lesions and are easily detachable. When there is dew, rain, or guttation water, exudates dissolve to form conspicuous milky drops that may later spread over the leaf surface and dry up as thin, greyish, almost transparent flakes. In dry weather and in wheat fields without sprinkler-irrigation, lesions are usually without exudate that may make it difficult to identify the cause of the disease without isolating the pathogen (Duveiller et al. 1997).

Culms, leaves, rachis, glumes, and awns may be infected, and symptoms on wheat plants vary with the environment, variety, disease severity, and interaction with fungi. However, Duveiller (1994) experimenting extensively with different genotypes in Mexico found that variations of symptoms on wheat leaves are limited. Most of the variations observed are probably due to melanic reactions on ears and nodes caused under abiotic stress and mistaken for disease symptoms. An early infection may occur in the field, but the symptoms often go undetected in seedlings.

On the glumes (black chaff phase Figure 7.3), BLS is characterized by black, longitudinal, more or less parallel stripes that are more numerous and conspicuous on the upper parts. It can be recognized by a greasy appearance or alternating bands of diseased and healthy areas on the awns. Purple-black symptoms may extend to the peduncle between the inflorescence and the flag leaf, and may sometimes show a yellow center. On triticale, the bacterium produces moist grey to whitish lesions on the glumes, and discoloration is rarely seen on the peduncle.

Brown melanism also referred to as **pseudo-black chaff**, a non-parasitic disorder, is often confused with symptoms of bacterial leaf streak on the glumes. Goulden and Neatby (1929) and Waldron (1929) found that susceptibility to melanism on the spike is often inherited from stem rust resistant parents. Johnson and Hagborg (1944) showed that high temperature, especially when combined with high humidity, favored the development of melanic areas on the glumes, lemmas, peduncles, and internodes of rust resistant varieties. As brown melanosis is known to be associated with the *Sr2* gene for stem rust resistance (McIntosh 1988), it is possible that earlier reports of black chaff were in fact of pseudo-black chaff, a non-parasitic disorder not caused by the bacterium.

FIGURE 7.3 Bacterial leaf streak and black chaff of wheat: Black chaff symptoms. (Courtesy of Mary Burrows, Montana State University, Bugwood.org.)

Sharp discolored interveinal streaks on the glumes indicate the presence of *Xa. translucens* pv. *undulosa*, particularly if also irregularly distributed on the spike, and if accompanied by an abundant BLS on the leaves. In contrast, melanosis on the peduncle that occurs on the same side of most culms in a field as a result of exposure to sunshine (UV light), is indicative of **brown melanosis**. Besides this non-parasitic disorder, other very similar symptoms on the spike not caused by *Xa. translucens* pv. *undulosa* may be caused by *Bipolaris sorokiniana*, *Alternaria* spp., *Stagonospora nodorum* (syn. *Septoria nodorum*) and *Pseudomonas syringae* pv. *atrofaciens*.

7.6.2 Causal Organism

Xanthomonas translucens pv. *undulosa* (Smith et al. 1919) Vauterin et al. 1995
Syn: *Xa. campestris* pv. *translucens*

Different names have been proposed, depending on the host plant, for the closely related cereal streak pathogens and often grouped together under the name "**translucens group**." The taxonomy of this group of bacteria and the phytopathological relevance of the classification has been re-examined in detail and clarified (Vauterin et al. 1995; Bragard et al. 1997). The following four names of bacterial leaf streak inducing pathovars of *Xa. translucens* are included in the list of plant pathogenic bacteria of International Society for Plant Pathology (Young et al. 1996):

1. *Xa. translucens* pv. *translucens* (Jones et al. 1917) Vauterin et al. 1995
2. *Xa. translucens* pv. *cerealis* (Hagborg 1942) Vauterin et al. 1995
3. *Xa. translucens* pv. *secalis* (Reddy et al. 1917) Vauterin et al. 1995
4. *Xa. translucens* pv. *undulosa* (Smith et al. 1919) Vauterin et al. 1995

In this publication, the name *Xa. translucens* pv. *undulosa*, as suggested by Duveiller et al. (2002), is used to refer to the pathogen that causes BLS on wheat.

Xa. translucens pv. *undulosa*: The bacterium can be cultured on common media such as nutrient agar, yeast peptone glucose agar, King's medium B, and Wilbrink's medium. Among the non-selective media, nutrient agar and Wilbrink's media are suitable for culturing *Xa. translucens* pv. *undulosa*. Semi-selective media for culturing include KM-1, XTS, and WBC (Wilbrink's boric acid-cephalexin medium). When no selective medium is available, Wilbrink's medium is preferred, because typical yellow mucoid colonies of the bacterium are best distinguished from those of saprophytes on this medium. The optimum temperature for growth *in vitro* is 28°C–30°C.

The cells are Gram-stain-negative, rod-shaped (0.4–0.8 × 1.0–2.5 μm), non-sporing, and motile by a single polar flagellum.

The cells occur singly or in pairs, except in peptonized nutrient broth with 2% sodium chloride in which long non-motile chains are formed. The bacterium is oxidative, i.e., produces acid from glucose aerobically, nitrates are not reduced and the reactions for Kovacs' oxidase and arginine dihydrolase are also negative. There is no production of 2-ketogluconate, esculin is hydrolyzed, but not starch, and lactose is not used. Hypersensitive reaction on tobacco is positive, but this reaction is not always clear, and a cool incubation temperature is required. Very few biochemical and physiological tests are useful for differentiating pathovars of *Xa. translucens*. These pathovars can only be identified on the basis of pathogenicity and host range studies.

Gardiner et al. (2014) reported the characterization and genome sequencing of *Xa. translucens* isolate DAR61454 pathogenic on wheat and barley and found isolate DAR61454 most closely related to other *Xa. translucens* strains and the sugarcane- and banana-infecting *Xanthomonas* strains on the basis of phylogenetic analysis of the ATP synthase beta sub unit (*atpD*) gene. However, it shared a type III secretion system (TIIISS) with *Xa. translucens* pv. *graminis* and more distantly related xanthomonads. Assays with an adenylate cyclase reporter protein demonstrated that TIIISS of DAR61454 was functional in delivering proteins to wheat cells. Comparative analysis of 18 different *Xanthomonas* isolates revealed 84 proteins unique to isolates infecting rice and the wheat/barley isolate DAR61454. Genes encoding 60 of these proteins are found in gene clusters in the *Xa. translucens* DAR61454 genome, suggesting cereal-specific pathogenicity islands.

Langlois et al. (2017) sequenced genomes of 15 *Xa. translucens* strains representing six different pathovars and compared them with additional publicly available *Xa. translucens* genome sequences to obtain a genome-based phylogeny for robust classification of this species. Three main clusters: one consisting of pv. *cerealis*, second consisting of pvs. *undulosa* and *translucens*, and the third consisting of pvs. *arrhenatheri*, *graminis*, *phlei*, and *poae* were obtained. Based on genomic differences, diagnostic loop-mediated isothermal amplification (LAMP) primers were developed that clearly distinguish strains that cause disease on cereals, such as pvs. *undulosa*, *translucens*, *hordei*, and *secalis* from strains of pvs. *arrhenatheri*, *cerealis*, *graminis*, *phlei*, and *poae* that cause disease on non-cereal hosts. Additional LAMP assays were also developed that selectively amplify strains belonging to pvs. *cerealis* and *poae*, distinguishing them from other pathovars. These primers will be instrumental in diagnostics for implementing quarantine regulations to limit further geographic spread of *Xa. translucens* pathovars.

Adhikari et al. (2012) reported pathogenic and genetic diversity in *Xa. translucens* pv. *undulosa* in North Dakota after inoculating 226 strains of the bacterium, collected from five locations in North Dakota, on a set of 12 wheat cultivars and other cereal hosts. Similar haplotypes were detected in more than one location. Although diversity was greatest (≈92%) among individuals within a location, statistically significant ($P \leq 0.001$ or 0.05) genetic differentiation among locations was also found, indicating geographic differentiation between pathogen populations.

7.6.3 Disease Cycle

Seed is the most important source of primary inoculum and is responsible for the large-scale dissemination of the pathogen. Forster and Schaad (1990) estimated that the bacterium in seed can die in 63–81 months depending on the storage conditions. Due to a very low seed-plant transmission rate of the bacterium, low levels of seed contamination generally do not produce a diseased crop (Schaad 1988). In Idaho, more than 60% of spring wheat seed lots were found to be contaminated, and seed lots with less than 1,000 cfu per gram of seed did not result in field epiphytotics. This suggests that methods for detecting the pathogen on the seed do not need to be very sensitive (Schaad 1987; Forster and Schaad 1987). However, the situation may vary from one environment to another, and the pathogen's ability to multiply in the host should not be underestimated.

The survival of bacterium is poor in soil, but it is better in crop debris (Boosalis 1952). However, in warm and humid climates, plant stubbles usually decay very fast, and wheat pathogenic bacteria cannot survive in decomposing debris.

Due to its broad host range, *Xa. translucens* pv. *undulosa* can survive on weeds and grasses. Thompson et al. (1989) detected epiphytic populations of the pathogen in Idaho on *Poa pratensis*, *Festuca arundinacea*, *F. rubra*, *Hordeum leporinum*, and *Medicago sativa* growing near spring wheat fields. The overwintering of *Xa. translucens* on perennial hosts, such as smooth brome (*Bromus inermis* Leyss.) and timothy (*Phleum pratense* L.) enable the pathogen to spread to nearby cereals. The bacterium also seems to overwinter on winter wheat and rye.

The moisture facilitates the pathogen's release from the seed and also contributes to leaf colonization and invasion of leaf tissue by the pathogen. Free water facilitates the dispersal and spread of the pathogen in the field, thereby increasing the number of lesions. Wind-splashed rain and overhead-irrigation greatly influence the short distance spread of the pathogen. In addition, visitors to demonstration plots, particularly in the morning when dew is at its maximum, increase the spread of bacteria. In Brazil, Mehta (1990) showed that the spread of the pathogen from one field to another is limited, and spread through splashing rain is restricted to a distance of 4–5 m. The role of aphids in long distance transmission of the disease is probably limited. Bacteria enter through the stomata and multiply in large masses in the parenchyma. Micro-injuries to awns and leaves caused by hail or wind also contribute to bacterial penetration.

The pathogen causes infection over a wide range of temperatures ranging from 15°C to 30°C, but maximum growth occurs above 26°C. Duveiller and Maraite (1995) have shown that temperature has a major impact on epiphytotics. Pathogen multiplication in leaf tissue is directly dependent on temperature, and dry weather (relative humidity less than 30%) does not limit disease progress. The symptoms only appear when temperature allows the bacterial population to reach an estimated threshold of 10^8 cfu per leaf. Low temperatures retard the multiplication of the pathogen and disease progress.

Kim et al. (1987) showed that strains of *Xa. translucens* pv. *translucens* express ice nucleation activity at temperatures from −2°C to −8°C. Damage caused to plant tissue by the ice provides conditions suitable for pathogen invasion and multiplication. Frost conditions may thus explain the frequent incidence of disease in high elevation environments or in regions, such as southern Brazil, where wheat is grown during the winter season.

Stanton et al. (2016) tested four equipments to create an artificial infection of *Xa. translucens* pv. *undulosa* and found that all the sprayer attachments (except the CO_2 boom) generated significantly higher levels of infection than the non-inoculated and mock-inoculated controls. The SOLO sprayer increased disease severity over the CO_2-powered sprayer in years when conditions for infection were less favorable.

Silva et al. (2010) reported that the application of silicon to wheat plants increased resistance to leaf streak possibly through an increase in tissue lignification and the participation of chitinases and peroxidises.

No genotype is immune to BLS pathogen. As the resistance is incomplete, it is not easily observed under strong disease pressure. Although resistance to *Xa. translucens* pv. *undulosa* in wheat has been identified globally (Boosalis 1952; Akhtar and Aslam 1985; Thompson and Souza 1989; Duveiller 1990; Milus and Mirlohi 1994; El Attari et al. 1996; Milus et al. 1996), the information available on its mode of inheritance is limited. Tillman and Harrison (1996) reported that heritability of resistance to bacterial streak (*Xa. campestris* pv. *translucens*) was relatively low and ranged from 0.12 to 0.70 (average 0.31) with parent-offspring regression and from 0.18 to 0.40 (average 0.36) with variance components. It increased by 0.22, 0.29, and 0.29 for population 1, 2, and 3, respectively, with means of two replications and two environments. The field trials carried out in Mexico showed that five genes condition the resistance in five wheat lines, namely, Turaco, Alondra, Angostura, Mochis, and Pavon. Cultivars Pavon and Mochis showed the highest level of resistance. None of the five genotypes tested contained all the five identified resistance genes, suggesting that there may be cultivars having more resistance than Pavon and Mochis (Duveiller et al. 1993).

Adhikari et al. (2011) screened 605 winter wheat accessions of diverse origin and improvement status, and found 35 accessions resistant, one accession, PI 266860, susceptible, and six accessions showed differential responses. Kandel et al. (2012) evaluated 45 hard red spring wheat genotypes with diverse genetic backgrounds for disease severity and observed clear differences in level of resistance among these genotypes, but none was immune. SD4205 was found resistant to the disease.

Sapkota et al. (2018) evaluated a worldwide collection of triticale accessions and the major North Dakota hard red spring and durum wheat cultivars for reaction to two local *Xa. translucens* pv. *undulosa* strains. All the wheat cultivars showed a susceptible reaction, but a wide range of reactions was observed among triticale accessions. Out of the 502 accessions tested, 45 and 10 accessions were resistant to two virulent strains LB10 and P3, respectively, and five accessions, PI 428736, PI 428854, PI 428913, PI 542545, and PI 587229, being highly resistant to both the strains. Statistical analysis showed significant difference among the accessions, strains, and the accession by strain interaction ($P < 0.001$). Bacterial multiplication in resistant triticale accessions was significantly slower than that in susceptible ones. Molecular cytogenetic characterization in four representative triticale accessions confirmed the hexaploid level of the species and the presence of 12 or 14 rye chromosomes. The identified resistant triticale accessions are valuable materials for developing wheat germplasm with high levels of resistance to *Xa. translucens* pv. *undulosa*.

Curland et al. (2016) evaluated multi-locus sequencing typing (MLST), multi-locus sequence analysis (MLSA), and *in planta* virulence assays for determining the genetic diversity, pathovar designations, and host preferences of a large collection of field isolates of *Xa. translucens* associated with outbreaks of bacterial leaf streak in midwestern United States and reported that MLSA and MLST corroborated host virulence and were effective means for assigning *translucens* and *undulosa* pathovar designations.

Curland et al. (2018) used MLST and MLSA of four common housekeeping genes (*rpoD*, *dnaK*, *fyuA*, and *gyrB*) to evaluate the genetic diversity of 82 strains of *Xa. translucens* isolated between 2006 and 2013 from wheat, barley, rye, and intermediate wheatgrass. *In planta,* the disease assays were also conducted on 75 strains to measure relative virulence in wheat and barley. All strains were determined by MLSA to be related to *Xa. translucens* pv. *undulosa* and *Xa. translucens* pv. *translucens*. The clustering of strains based on Bayesian, network, and minimum spanning trees correlated with relative virulence levels in inoculated wheat and barley. Therefore, phylogeny based on *rpoD*, *dnaK*, *fyuA*, and *gyrB* correlated with host of isolation and was an effective means for predicting virulence of strains belonging to *Xa. translucens* pv. *translucens* and *Xa. translucens* pv. *undulosa*.

Jayawardana et al. (2016) identified a genomic region on the rye chromosome 5R largely associated with resistance to BLS through QTL mapping. This is the first report of chromosomal location of a major resistance gene to BLS, and the identified markers tightly linked to the resistance gene will be useful in transferring the resistance gene into wheat to develop germplasm with high level of resistance to BLS.

7.6.4 Management

Use pathogen-free certified seed because sowing pathogen-free seed is the first logical step in avoiding an outbreak. Foundation seed should be multiplied in disease-free areas where climatic conditions are unfavorable for the development of epiphytotics.

1. Different seed treatments, namely, cupric hydroxide (Kocide SD), non-volatile mercury (Mist-O-Matic) and volatile mercury compound (Panogen 15), Guazatine Plus (syn. Panoctine Plus), acidified cupric

acetate (0.5% dose at 45°C for 20 min), and dry heat treatment at 72°C for 7 days have been evaluated, but none has given satisfactory control. Nevertheless, seed should be disinfected before sowing even if currently available seed treatments are not fully effective. Therefore, seed treatment with dry heat at 72°C for 7 days or a product such as Panoctine Plus is recommended

2. Breeding resistant genotypes appears to be the best way to reduce the risk of yield losses. Screening for resistance should be done through artificial inoculation, and resistance evaluation should be based on adult plant response in the field. Cultivars, namely, Blade, Cromwell, Faller, Howard, or Knudson are resistant to BLS.

REFERENCES

Adhikari, T. B., Gurung, S., Hansen, J. M., and Bonman, J. M. 2012. Pathogenic and genetic diversity of *Xanthomonas translucens* pv. *undulosa* in North Dakota. *Phytopathology* 102: 390–402.

Adhikari, T. B., Hansen, J. M., Gurung, S., and Bonman, J. M. 2011. Identification of new sources of resistance in winter wheat to multiple strains of *Xanthomonas translucens* pv. *undulosa*. *Plant Dis.* 95: 582–588.

Akhtar, M. A., and Aslam, M. 1985. Bacterial stripe of wheat in Pakistan. *Rachis* 4: 49.

Alizadeh, A., Barrault, G., Sarrafi, A., Rahimian, H., and Albertini, L. 1995. Identification of bacterial leaf streak of cereals by their phenotypic characteristics and host range in Iran. *Eur. J. Plant Pathol.* 101: 225–229.

Boosalis, M. G. 1952. The epidemiology of *Xanthomonas translucens* (J. J. and R.) Dowson on cereals and grasses. *Phytopathology* 42: 387–395.

Bradbury, J. F. 1986. *Guide to Plant Pathogenic Bacteria*. Slough, UK, Mycological Institute, CAB International, 332 pp.

Bragard, C., Singer, E., Alizadeh, A., Vauterin, L., Maraite, H., et al. 1997. *Xanthomonas translucens* from small grains: Diversity and phytopathological relevance. *Phytopathology* 87: 1111–1117.

Bragard, C., Verdier, V., and Maraite, H. 1995. Genetic diversity among *Xanthomonas campestris* strains pathogenic for small grains. *Appl. Environ. Microbiol.* 61: 1020–1026.

Burton, G. J. L. 1931. Annual report of the senior plant breeder 1931. *Kenya Dept. Agric. Ann. Rep.* 176–209.

CIMMYT. 1977. Israel. in: *CIMMYT Report on Wheat Improvement*. p. 237. Mexico City, Mexico.

Curland, R., Gao, L., Bull, C., Vinatzer, B., Dill-Macky, R., et al. 2016. Differentiation of *Xanthomonas translucens* pathovars in wheat and barley (Abstr.). *Phytopathology* 106: S4.187.

Curland, R. D., Gao, L., Bull, C. T., Vinatzer, B A., Dill-Macky, R., et al. 2018. Genetic diversity and virulence of wheat and barley strains of *Xanthomonas translucens* from the Upper Midwestern United States. *Phytopathology* 108: 443–453.

Demir, G., and Üstün, N. 1992. Studies on bacterial streak disease (*Xanthomonas campestris* pv. *translucens* (Jones et al.) Dye) of wheat and other gramineae. *J. Turk. Phytopathol.* 21: 33–40.

Duveiller, E. 1990. Screening criteria for bacterial leaf streak in bread wheat, durum wheat and triticale in CIMMYT. pp. 1011–1016 in: *Proceedings of the 7th International Conference on Plant Pathogenic Bacteria*. Budapest, Hungary, Z. Klement ed.

Duveiller, E. 1994. A study of *Xanthomonas campestris* pv. *undulosa* populations associated with symptomless wheat leaves. *Parasitica* 50: 109–117.

Duveiller, E., Bragard, C., and Maraite, H. 1991. Bacterial diseases of wheat in the warmer areas - reality or myth? pp. 189–202 in: *Wheat for the Non-traditional Warm Areas. The Proceedings of the International Conference*. D. Saunders ed. Iguazu Falls, Brazil, CIMMYT, Mexico City, Mexico.

Duveiller, E., Bragard, C., and Maraite, H. 1997. Bacterial leaf streak and black chaff caused by *Xanthomonas translucens*. in: *The Bacterial Diseases of Wheat: Concepts and Methods of Disease Management*. E. Duveiller, L. Fucikovsky, and K. Rudolph eds. CIMMYT, Mexico City, Mexico.

Duveiller, E., Bragard, C., and Maraite, H. 2002. Bacterial leaf streak and black chaff. in: *Bread Wheat: Improvement and Production*. B. C. Curtis, S. Rajaram, and H. G. Macpherson eds. Food and Agriculture Organization of the United Nations, Rome, Italy.

Duveiller, E., and Maraite, H. 1993. Study of yield loss due to *Xanthomonas campestris* pv. *undulosa* in wheat under high rainfall temperate conditions. *J. Plant Dis. Prot.* 100: 453–459.

Duveiller, E., and Maraite, H. 1995. Effect of temperature and air humidity on multiplication of *Xanthomonas campestris* pv. *undulosa* and symptom expression in susceptible and field-tolerant wheat genotypes. *J. Phytopathol.* 143: 227–232.

Duveiller, E., van Ginkel, M., and Tijssen, M. 1993. Genetic analysis of resistance to bacterial leaf streak caused by *Xanthomonas campestris* pv. *undulosa* in bread wheat. *Euphytica* 66: 35–43.

El Attari, H., Sarrafi, A., Garrigues, S., Dechamp-Guillaume, G., and Barrault, G. 1996. Diallel analysis of partial resistance to an Iranian strain of bacterial leaf streak (*Xanthomonas campestris* pv. *cerealis*) in wheat. *Plant Pathol.* 45: 1134–1138.

Forster, R. L., and Schaad, N. W. 1987. Tolerance levels of seed borne *Xanthomonas campestris* pv. *translucens*, the causal agent of black chaff of wheat. pp. 974–975 in: *Proceedings of the 6th International Conference on Plant Pathogenic Bacteria*. E. L. Civerelo, A. Collmer, R. E. Davis, and A. G. Gillaspie eds. Martinus Nijhoff Publishers, Dordrecht, the Netherlands.

Forster, R. L., and Schaad, N. W. 1990. Longevity of *Xanthomonas campestris* pv. *translucens* in wheat seed under two storage conditions. pp. 329–331 in: *Proceedings of the 7th International Conference on Plant Pathogens Bacteria*. Part A, Z. Klement ed. Budapest, Akadémiai Kiadó.

Frommel, M. I. 1986. *Xanthomonas campestris* pv. *translucens*, causal agent of bacterial streak of wheat (*Triticum aestivum*). Montevideo, Uruguay, Dirección de Sanidad Vegetal (in Spanish).

Gardiner, D. M., Upadhyaya, N. M., Stiller, J., Ellis, J. G., Dodds, P. N., et al. 2014. Genomic analysis of *Xanthomonas translucens* pathogenic on wheat and barley reveals cross-kingdom gene transfer events and diverse protein delivery systems. *PLoS ONE* 9(1): e84995.

Goulden, C. H., and Neatby, K. W. 1929. A study of disease resistance and other varietal characters of wheat: Application of the analysis of variance and correlation. *Sci. Agr.* 9: 575–86.

Hagborg, W. A. F. 1942. Classification revision in *Xanthomonas translucens*. *Can. J.Res.* 20: 312–326.

Jayawardana, M., Li, X., Fiedler, J., and Liu, Z. 2016. Genetic mapping of a gene conditioning resistance to bacterial leaf streak in triticale (Abstr.). *Phytopathology* 106: S4.189.

Johnson, T., and Hagborg, W. A. F. 1944. Melanism in wheat induced by high temperature and humidity. *Can. J. Res.* 22(C): 7–10.

Jones, L. R., Johnson, A. G., and Reddy, C. S. 1917. Bacterial blight of barley. *J. Agric. Res.* 11: 625–643.

Kandel, Y. R., Glover, K. D., Tande, C. A., and Osborne, L. E. 2012. Evaluation of spring wheat germplasm for resistance to bacterial leaf streak caused by *Xanthomonas campestris* pv. *translucens*. *Plant Dis.* 96: 1743–1748.

Kim, H. K., Orser, C., Lindow, S. E., and Sands, D. C. 1987. *Xanthomonas campestris* pv. *translucens* strains active in ice nucleation. *Plant Dis.* 71: 994–997.

Korobko, A. P., Wondimagegne, E., and Anisimoff, B. V. 1985. *Bacterial Stripe and Black Chaff of Wheat in Ethiopia.* Scientific Phytopathological Laboratory, Ambo, Ethiopia, 5 pp.

Langlois, P. A., Snelling, J., Hamilton, J. P., Bragard, C., Koebnik, R., et al. 2017. Characterization of the *Xanthomonas translucens* complex using draft genomes, comparative genomics, phylogenetic analysis, and diagnostic LAMP assays. *Phytopathology* 107: 519–527.

Mamluk, O. F., Al-Ahmed, M., and Makki, M. A. 1990. Current status of wheat diseases in Syria. *Phytopath. Medit.* 29: 143–150.

McIntosh, R. A. 1988. The role of specific genes in breeding for durable stem rust resistance in wheat and triticale. pp. 1–9 in: *Breeding Strategies for Resistance to the Rusts of Wheat.* N. W. Simmonds and S. Rajaram eds. CIMMYT, Mexico City, Mexico.

Mehta, Y. R. 1990. Management of *Xanthomonas campestris* pv. *undulosa* and *hordei* through cereal seed testing. *Seed Sci. Tech.* 18: 467–476.

Milus, E. A., Duveiller, E., Kirkpatrick, T. L., and Chalkey, D. B. 1996. Relationships between disease reactions under controlled conditions and severity of wheat bacterial streak in the field. *Plant Dis.* 80: 726–730.

Milus, E. A., and Mirlohi, A. F. 1994. Use of disease reactions to identify resistance in wheat to bacterial streak. *Plant Dis.* 78: 157–161.

Mohan, S. K., and Mehta, Y. R. 1985. Studies on *Xanthomonas campestris* pv. *undulosa* in wheat and triticale in Paraná State. *Fitopatologia Brasileira* 10: 447–453 (in Portuguese).

Noble, R. J. 1935. Australia: Notes on plant diseases recorded in New South Wales for the year ending 30th June 1935. *Int. Bull. Plant Prot.* 12: 270–273 (abstract: RAM 15, 280).

Paul, V. H., and Smith, I. M. 1989. Bacterial pathogens of gramineae: Systematic review and assessment of quarantine status for the EPPO region. *Bull. OEPP/EPPO Bull.* 19: 33–42.

Reddy, C. S., Godkin, J., and Johnson, A. G. 1924. Bacterial blight of rye. *J. Agric. Res.* 28: 1039–1040.

Richardson, M. J., and Waller, J. M. 1974. Triticale diseases in CIMMYT trial locations. pp. 193–99 in: *Triticale: Proceedings of an International Symposium.* El Batan, Mexico, Monograph 024e, International Development Research Center, Ottawa, Canada.

Sands, D. C., and Fourest, E. 1989. *Xanthomonas campestris* pv. *translucens* in North and South America and in the Middle East. *Bull. OEPP/EPPO Bull.* 19: 127–1230.

Sapkota, S., Zhang, Q., Chittem, K., Mergoum, M., Xu, S. S., et al. 2018. Evaluation of triticale accessions for resistance to wheat bacterial leaf streak caused by *Xanthomonas translucens* pv. *undulosa. Plant Pathol.* 67: 595–602.

Schaad, N. W. 1987. The use and limitations of methods to detect seed borne bacteria. pp. 324–32 in: *Seed Pathology*, Vol. 2, *International advanced course*, University of Passo Fundo, Passo Fundo, Brazil.

Schaad, N. W. 1988. Bacteria. In symposium: Inoculum thresholds of seed-borne pathogens. 76th annual meeting of the American Phytopathological Society. *Phytopathology* 78: 872–875.

Shane, W. W., Baumer, J. S., and Teng, P. S. 1987. Crop losses caused by *Xanthomonas* streak on spring wheat and barley. *Plant Dis.* 71: 927–930.

Silva, I. T., Rodrigues, F. Á., Oliveira, J. R., Pereira, S. C., Andrade, C. C. L., et al. 2010. Wheat resistance to bacterial leaf streak mediated by silicon. *J. Phytopathol.* 158: 253–262.

Smit, I. B. J., and Van A. Bredenkamp, T. 1988. Items from South Africa: International nurseries. *Ann. Wh. Newsl.* 34: 84.

Smith, E. F., Jones, L. R., and Reddy, C. S. 1919. The black chaff of wheat. *Science* 50: 48.

Stanton, J., Smith, M., and Dill-Macky, R. 2016. Testing field equipment for the application of *Xanthomonas translucens* pv. *undulosa*, the causal agent of bacterial leaf streak of wheat (Abstr.). *Phytopathology* 106: S4.193.

Sun, F., and He, L. 1986. Studies on determinative techniques for resistance of wheat to black chaff (*Xanthomonas translucens* f. sp. *undulosa*). *Acta Phytophylacica Sinica* 13: 109–115 (in Chinese; English summary).

Tessi, J. L. 1949. Current status of work on *Xanthomonas translucens* var. *cerealis*. *Presented at the 4th Wheat, Oats, Barley and Rye Meeting*, Castellar, Argentina, p. 200.

Thompson, D. C., Schaad, N. W., and Forster, R. L. 1989. New perennial hosts of epiphytic populations of *Xanthomonas campestris* pv. *translucens*. *Phytopathology* 79: 1168 (Abstr.).

Thompson, D. C., and Souza, E. J. 1989. Reaction of spring wheat cultivars to black chaff, 1988. American Phytopathological Society. *Biol. Cult. Tests* 4: 51.

Tillman, B. L., and Harrison, S. A. 1996. Heritability of resistance to bacterial streak in winter wheat. *Crop Sci.* 36: 412–418.

Vauterin, L., Hoste, B., Kersters, K., and Swings, J. 1995. The relationship within genus *Xanthomonas* and a proposal for a new classification. *Int. J. Syst. Bacteriol.* 45: 472–489.

Waldron, L. R. 1929. The relationship of black chaff disease of wheat to certain physical and pathological characters. *Science* 70: 268.

Wallin, J. R. 1946. Parasitism of *Xanthomonas translucens* (J. J. and R.) Dowson on grasses and cereals. *Iowa St. Coll. J. Sci.* 20: 171–193.

Young, J. M., Saddler, G. S., Takikawa, Y., De Boer, S. H., Vauterin, L., et al. 1996. Names of plant pathogenic bacteria 1864–1995. *Rev. Plant Pathol.* 75: 721–763.

Zillinsky, F. J., and Borlaug, N. E. 1971. Progress in developing triticale as an economic crop. *Int. Maize Wheat Improv. Cen. Res. Bull.* 17: 18–21.

7.7 BACTERIAL STRIPE AND BLACK CHAFF OF BARLEY

The disease was first reported by Jones et al. (1917) as bacterial blight of barley (*Hordeum vulgare* L.) from the USA. The currently used name of the disease is bacterial stripe and black chaff of barley. It is prevalent in Mexico, USA, South America, France, Belgium, Spain, and India. In India, the disease was first reported by Patel and Shekhawat (1971) as bacterial blight of barley from Ludhiana, Punjab, India. Shekhawat and Patel (1972) have also given a brief account of the disease. Later on, Maharshi (1996) reported it from Rajasthan in 1989, where it is now more or less endemic (Maharshi and Sharma 2006). The disease incidence may reach nearly 100% in overhead-irrigated fields in contrast to fields irrigated with conventional flood irrigation system.

Sharma and Maharshi (2012) have given a brief account of the occurrence of the disease in Rajasthan along with information on seed-plant transmission of the bacterium and the environmental conditions favoring the disease development. However, there is not much information available on losses caused by the disease.

7.7.1 Symptoms

All the aboveground plant parts are affected. Mainly two types of symptoms, i.e., stripes and black chaff appear. The stripes appear on leaf sheath, leaf lamina, and stems, while the black chaff appears on ears and neck. Initially, water-soaked, dark green flecks/lesions appear on leaf sheaths and leaf blades, which elongate to linear stripes parallel to the veins. The stripes also appear on the stem internodes. Soon the stripes lose their greenish-yellow and water-soaked appearance and become necrotic, brown, and translucent. The length of the stripes varies from 15 to 25 cm or may cover the entire length of the leaf blade (Figure 7.4a and b). Severe infection in susceptible varieties may cause chlorosis of the leaves. Severe infection of flag leaf and its sheath results in sealing of sheath lumen with bacterial mass, resulting in the distortion of emerging spike and its rachis. Often, the emergence of spike is completely prevented. Frequently, the stripes coalesce to form yellowish-brown blotches leading to shriveling and drying of the infected leaves/sheaths. In the field, the symptoms are more pronounced at the tillering stage. In the early morning hours, copious viscous slimy mass of bacterial ooze appears on the surface of stripes, which dries up into numerous beads or crusty translucent scales under dry conditions.

Seed-borne nature of the bacterium leads to the infection of young seedlings. Severely affected plants remain weak and stunted, and often the ears fail to emerge. If the ears emerge, they are dwarfed, twisted, and even sterile having compressed and closed spikelets. Young infected plants at 2–3 leaf stage are sometimes so severely stunted and may be killed or if they grow further, the grain formation is poor accompanied by deformation of spikes. Sometimes, only the awns emerge from the ears followed by the death of plants. The symptoms of the disease can be confused with those of barley stripe caused by *Helminthosporium gramineum,* but it can be conveniently diagnosed on the basis of bacterial ooze/turbidity test in the laboratory and field, respectively.

The initiation of leaf lesions from the apex and their downward extension and starting of the lesions primarily from central part of the blade where the leaf bends have been reported by different workers from different regions. It appears that the symptom development and further progress may vary in different geographical areas.

7.7.2 Causal Organism

Xanthomonas translucens pv. *translucens* (Jones et al. 1917) Vauterin et al. 1995
Syn. *Xa. campestris* pv. *translucens*

Jones et al. (1917) originally described the bacterium as *Bacterium translucens*; and with the establishment of genus *Xanthomonas*, it was transferred to it as *Xa. translucens*. Later on, it was designated as *Xa. translucens* f. sp. *hordei*. With the introduction of pathovar system, it was named as

(a)

(b)

FIGURE 7.4 Bacterial stripe and black chaff of barley: (a) Infected plant. (b) Stripes present on leaves.

Xa. campestris pv. *translucens*. Vauterin et al. (1995), while reclassifying the species of *Xanthomonas*, changed it to *Xa. translucens* pv. *translucens*.

The colonies on nutrient agar are yellow, mucoid, convex, and smooth having entire margins. Cells measure 0.4 × 1.0 µm and are motile by a single polar flagellum. The bacterium produces hydrogen sulphide, but does not reduce nitrate, does not produce indole, does not hydrolyze starch, and does not utilize lactose; catalase test is positive, while oxidase test is negative. The optimum temperature for its growth ranges from 28°C to 30°C.

The following four pathogens, often grouped under the name **translucens group**, cause bacterial stripe and streak disease on cereals:

Xa. translucens pv. *translucens*
Xa. translucens pv. *cerealis*
Xa. translucens pv. *secalis*
Xa. translucens pv. *undulosa*

Jones et al. (1917), who first isolated the bacterium from barley, found that the bacterium was pathogenic only to barley and did not infect wheat, rye, oat, timothy, spelt emmer, and einkorn (all *Triticum* spp.) Maharshi (1996) also reported that the Rajasthan isolate of the bacterium (CABIMIB 12929) was also non-pathogenic to wheat. Following the rules of bacterial nomenclature (International Society for Plant Pathology, Young et al. 1996), the name *Xa. translucens* pv. *translucens* should be reserved for strains pathogenic only to barley as originally reported by Jones et al. (1917). As *Xa. translucens* pv. *undulosa* strains infect wheat and triticale, but can also be isolated from barley and rye; hence, its host range is not only broader than that of *Xa. translucens* pv. *translucens,* but also covers the host range of *Xa. translucens* pv. *cerealis*. Based on amplified fragment length polymorphism and fatty acid methyl ester analysis, Bragard et al. (1997) showed that strains pathogenic on barley and not on wheat, clustered in a genetically different group. They further reported that the pathovars *cerealis*, *translucens* and *undulosa* corresponded to true biological entities.

7.7.3 Disease Cycle

The bacterium survives in infected seed, and it is both externally and internally seed-borne. Maharshi (1996), using paper towel and seedling system test, reported 30% seed-plant transmission at harvesting time, which decreased to 4% after 7 months, i.e., up to the sowing time of next crop. *Xa. translucens* specific medium is better than nutrient glucose agar for the detection of the pathogen from the seeds. No inoculum threshold is known for infected barley seeds, but wheat seeds, the washings of which contained approximately 1,000 cfu of the bacterium mL^{-1} per gram of seed, produced the diseased seedlings (Schaad 1988).

Jones et al. (1917) reported that the bacterium from the seeds enters into lemma and leaf through stomata and further moves into intercellular spaces of thin-walled parenchymatous tissue. Sharma (2003) traced the movement of the bacterium from husk (glumes) and pericarp, but not from endosperm or embryo, to the coleoptile, first true leaf, and then into the aerial parts of the seedlings. The transmission from seed to seedlings is facilitated by the presence of free water, i.e., it is more in high rainfall or under overhead-irrigation.

The survival of the bacterium in infected crop debris and soil is only for a short period and hence of no significance. The bacterium enters the host through stomata and wounds. The dissemination of the bacterium occurs through wind-splashed rain, overhead-irrigation, plant to plant contact, and insects. In Montana (USA), the bacterium has been found to spread from a single infection locus over an area of 28 m^2 in 39 days under sprinkler irrigation. Sprinkler irrigation during the daytime also promotes the penetration of the bacterium, as the stomata are open during the day.

The bacterium grows under a wide range of temperature and moisture conditions, but the disease is more severe in warm and humid conditions. The bacterium multiplies rapidly in the host tissue in tissue creases and intercellular spaces. The incubation period is short and is of approximately 10 days. The bacterium multiplies in all the aerial plant parts including spikes, which contaminate the developing seeds.

7.7.4 Management

1. Use of pathogen-free seed is the best approach for the management of the disease. The seed certification programs and seed assays are necessary for the production of pathogen-free seed
2. Dry heat treatment of barley seeds at 72°C for 5–7 days has been found to eradicate the pathogen. Fourest et al. (1990) found that dry heat treatment of heavily infested barley seed at 71°C, 75°C, or 84°C for 11 days eliminated the pathogen, while treatment at 72°C for 4 days apparently eliminated the pathogen from moderately infested seed. Reduction in seed germination was practically negligible for seed treated at 71°C or 72°C for 7 days or less, while the treatment at 71°C for 11 days caused an average reduction of 8%
3. Avoid the overhead-irrigation or limit its use
4. Use of disease resistant cultivars is the most effective method of disease management, but no disease resistant cultivars are available. However, the available cultivators differ in their level of susceptibility to the pathogen

REFERENCES

Bragard, C., Singer, E., Alizadeh, A. Vauterin, L., Maraite, H., et al. 1997. *Xanthomonas translucens* from small grains: Diversity and phytopathological relevance. *Phytopathology* 87: 1111–1117.

Fourest, E., Rehms, L. D., Sands, D. C., Bjarko, M., and Lund, R. E. 1990. Eradication of *Xanthomonas campestris* pv. *translucens* from barley seed with dry heat treatments. *Plant Dis.* 74: 816–818.

Jones, L. R., Johnson, A. G., and Reddy, C. S. 1917. Bacterial blight of barley. *J. Agric. Res.* 11: 625–643.

Maharshi, R. P. 1996. Bacterial stripe and black chaff disease of barley caused by *Xanthomonas campestris* pv. *translucens*: A report on the natural occurrence in Rajasthan, India. pp. 367–370 in: *Proceedings of the 9th International Conference on Plant Pathogens Bacteria*. A. Mahadevan, ed. Chennai, India.

Maharshi. R. P., and Sharma, H. 2006. Bacterial stripe and black chaff of barley (*Hordeum vulgare* L.) caused by *Xanthomonas campestris* pv. *translucens* in Rajasthan, India. *Ann. Exptl. Agri. Al. Sci.* 1: 103–110.

Patel, P. N., and Shekhawat, G. S. 1971. Occurrence of *Xanthomonas translucens* f. sp. *hordei* in India and its pathogenicity to rice. *Plant Dis. Reptr* 55: 365–368.

Schaad, N. W. 1988. Bacteria. In: Symposium on inoculum thresholds of seed-borne pathogens. *Phytopathology* 78: 872–875.

Sharma, H. 2003. Role of seed transmission and environmental factors on development of bacterial stripe disease of barley (*Hordeum vulgare* L.) caused by *Xanthomonas campestris* pv. *translucens* (Jones, Johnson & Reddy) Dye. PhD thesis, Rajasthan Agricultural University Bikaner, 68 pp.

Sharma, H., and Maharshi, R. P. 2012. *Bacterial Stripe Disease of Barley: A Research Investigation*. Lambert Academic Publishing, Riga, Latvia, 96 pp.

Shekhawat, G. S., and Patel, P. N. 1972. Bacterial blight of barley. pp. 38–42 in: *Plant Bacteriology*. Vol. I, *Bacterial Diseases of Plants in India*. P. N. Patel ed. Indian Agriculture Research Institute, New Delhi, India.

Vauterin, L. Hoste, B., Kerster, K., and Swings, J. 1995. Reclassification of *Xanthomonas*. *Int. J. Syst. Bacteriol.* 45: 472–489.

Young, J. M., Saddler, G. S., Takikawa, Y., De Boer, S. H., Vauterin, L., et al. 1996. Names of plant pathogenic bacteria 1864–1995. *Rev. Plant Pathol.* 75: 721–763.

7.8 BACTERIAL STALK ROT OF MAIZE

Four bacterial plant pathogens, namely, *Dickeya zeae* (syn. *Erwinia chrysanthemi* pv. *zeae*, *Pectobacterium chrysanthemi* pv. *zeae*), *Enterobacter cloacae* subsp. *dissolvens* (syn. *Er. dissolvens*), *Pseudomonas syringae* pv. *lapsa* (syn. *Psm. lapsa*), and *Acidovorax avenae* subsp. *avenae* (syn. *Psm. alboprecipitans*) have been reported to cause stalk rot of maize (*Zea mays* L.). Among the diseases caused by the abovementioned four bacterial pathogens, stalk rot caused by *D. zeae* is the most important and widespread disease of the maize and is described below. It was first reported from United Arab Republic by Sabet (1954), and he identified the bacterium as *Er. carotovora* f. sp. *zeae*. Since then, it has been reported from many countries including the USA, where it is called **top rot**. Martinez-Cisneros et al. (2014) reported it from Mexico and also hinted that the control of causal bacterium's vector, *Chilo partellus*, can aid in the management of the disease.

Hingorani et al. (1959) reported the disease for the first time from India. In India, it is one of the major diseases of maize and is prevalent in Himachal Pradesh, Uttar Pradesh, Delhi, Punjab, Rajasthan, Bihar, West Bengal, and Andhra Pradesh. During *Kharif* 1969, an epiphytotic of the disease occurred in Sundar Nagar area of Mandi district of Himachal Pradesh on hybrid Ganga 5 and the disease incidence was as high as 80%–85% in several fields (Thind 1970). The work done on the disease in India has been reviewed by Thind and Payak (1985). Kumar et al. (2017b), in their review article on bacterial stalk rot of maize, have presented information on nomenclature; biochemical, physiological, and molecular characterization; host range and survival of the pathogen, etiology and epidemiology of the disease, germplasm evaluation, and disease management.

7.8.1 Host Range

The bacterium has a very wide host range. It infects many fleshy fruits and vegetables and solanaceous plants like brinjal, tomato, and tobacco. The bacterium also infects sorghum, pearl millet, and sugarcane.

7.8.2 Symptoms

In nature, the symptoms usually appear at flowering stage. In parts, the seriousness of the disease lies in its often sudden and dramatic appearance and the swiftness with which it can kill the plants within 2–4 days under favorable conditions. The first symptoms are water-soaking and rotting of basal leaves, especially the leaf sheaths. The close examination of infected plants reveals water-soaking, rotting, and discoloration of basal internodes. The rind loses its healthy green color and looks pale straw. The diseased stalks appear as they have been boiled in water. Infected plants wilt.

Infection advances rapidly covering the considerable length or the entire length of the stalk. Two most characteristic symptoms of the disease are: (1) Infected plants emit a sweet fermenting alcoholic smell. (2) Infected plants usually topple down from 2nd or 3rd basal internode. It is mainly a parenchymatous disease. In advanced infection, the pith is completely destroyed, leaving the vascular bundles in a disorganized state, but intact (Figure 7.5).

Occasionally, developing ears, shanks, and stalks get infected at a level higher than the ground level. In majority of

FIGURE 7.5 Bacterial stalk rot of maize: Stalks appear as boiled in water and on the right side, a toppled plant.

these cases, the borer attack is found. In such cases, ears droop down and hang limply. This results due to the softening of shanks. The husk leaves, the grains, if any, as well as the cob become soft and appear water-soaked. In USA, the disease is known as top rot because the symptoms usually appear on apical portion of the plants including the leaves.

The symptoms of stalk rots caused by other bacteria are given below to distinguish them from the one described above.

Psm. syringae pv. *lapsa* causes rapid rotting of stalks and leaves. Affected plants collapse, but do not emit a fermenting smell. *Aci. avenae* subsp. *avenae* causes bacterial leaf blight and stalk rot of maize. The symptoms include leaf blight accompanied by shredding of leaves and rotting of stalks. Infection occurs at the top, leading to rotting and bleaching of the top. Multiple earing is also common. Infection caused by *En. cloacae* subsp. *dissolvens* is generally localized and may not cover the whole circumference or length of the stalk.

7.8.3 Causal Organism

Dickeya zeae (Burkholder et al. 1953) Samson et al. 2005
Syn. *Er. chrysanthemi* pv. *zeae*, *Pe. chrysanthemi* pv. *zeae*

Sabet (1954), on the basis of similarity of maize stalk rot pathogen to *Er. carotovora* in cultural, morphological, and biochemical characters, and its ability to attack maize and other graminaceous hosts in addition to other known hosts of *Er. carotovora*, named it as *Er. carotovora* f. sp. *zeae*. However, assigning the epithet of *forma specialis* to this bacterium by Sabet (1954) was not correct, as there was no specialization or restriction in attacking the hosts by the maize pathogen. Rather, it attacked more hosts than *Er. carotovora*. Graham (1964), while reviewing the taxonomy of soft rot coliform bacteria disfavored the use of *forma specialis* for maize pathogen, as there was no specialization in pathogenicity similar to that found in rust fungi. In the 8th edition of *Bergey's Manual of Determinative Bacteriology*, it was named as *Er. chrysanthemi* Burk. et al. corn pathotype (Buchanan and Gibbons 1974). Victoria et al. (1975) named it as *Er. chrysanthemi* pv. *zeae*. Hauben et al. (1998) renamed it as *Pectobacterium chrysanthemi*. Samson et al. (2005) have renamed it as *Dickeya zeae* sp. nov.

The cells are Gram-stain-negative rods (0.6–0.8 × 1.5–3.0 µm) with rounded ends, motile by peritrichous flagella, facultative anaerobes, and metabolism is fermentative. The bacterium produces indole and H_2S, reduces nitrates, liquefies gelatin, and degrades pectin. Oxidase, urea hydrolase, phenylalanine deaminase, and urease tests are negative. The G + C content of the DNA is 56.4 mol%.

Kumar et al. (2015) reported that 59 isolates of *D. zeae* obtained from maize growing areas of Punjab during *Kharif* 2012 and 2013 showed variation in virulence when tested on susceptible cv. Punjab Sweet Corn-1 and were grouped as highly virulent, moderately virulent, and less virulent. All the isolates produced cavities on crystal violet pectate medium. Characterizations of these isolates on the basis of 27 biochemical tests showed differential reaction for utilization of starch and other carbohydrates, gelatin liquefication, and growth at high salt concentration. Multiple antibiotic resistance was also observed in all the isolates tested. Numerical analysis of phenotypic features revealed two major groups of isolates associated at 60% similarity coefficient.

The bacterium is highly sensitive to 1 µg mL^{-1} of chlorine and 1, 10, 25, and 50 µg mL^{-1} of Streptocycline. Highly motile cultures are also highly virulent, while non-motile cultures are non-infective. Phages specific to the bacterium occur in nature (Thind 1970).

7.8.4 Disease Cycle

Infected plant debris seems to be the only known source of survival of the bacterium. The bacterium can survive for 9 months in soil containing naturally and artificially infected stalks. Maize-potato-maize rotation increases the disease incidence as potato also acts as the host of bacterium. The bacterium enters the host through wounds. Maize borer (*C. partellus*) have been shown to transmit the bacterium (Thind and Singh 1976). Dalmacio et al. (2007) have also reported the association of Asian corn borer (*Ostrinia furnacalis*) with the occurrence of the disease. The optimum temperature for disease development is 28°C–30°C and relative humidity above 90% also favors the disease development. Heavy doses of nitrogenous fertilizers increase the incidence of the disease. Nitrogen doses of 120 and 180 kg per hectare, without any level of potash and phosphorus, significantly increased the disease incidence than 60 kg of nitrogen (Saxena and Lall 1981). They recommended 60, 60, and 40 kg of nitrogen, phosphorus, and potash, respectively, per hectare to keep the disease under check.

A positive correlation exists between early infection/higher disease incidence to loss in grain yield. Early infection leads to heavy losses or death of plants. Screening of 32 maize genotypes including inbred lines, hybrids, composites, and open-pollinated varieties revealed the disease incidence ranging from 42.9% to 100%, yield losses ranging from 21.3% to 98.8%, and high susceptibility of most of the hybrids and composites (Thind and Payak 1978). The high levels of phenols in hybrid Ganga 2 plants after 60 and 75 days of planting, i.e., flowering stage in comparison to 40 and 95 days after planting were implicated with relatively more resistance at flowering.

Kumar et al. (2017a) analyzed activities of phenylalanine ammonia lyase 1 (PAL), peroxidase (POX), and polyphenoloxidase (PPO) from three sets of maize lines showing moderately resistant, moderately susceptible, and highly susceptible reaction to *D. zeae* and found an elevated activity of these enzymes in all the three sets. Moderately resistant lines showed significantly more activity of these enzymes than the highly susceptible lines. The activity of PAL and PPO peaked after 48 h and of POX after 72 h of challenge inoculation with *D. zeae* in all the maize lines. The activity of these enzymes was also negatively correlated with disease development. It is evident that PAL, POX, and PPO play an important role in imparting resistance in maize against *D. zeae*.

7.8.5 Management

1. Removal and destruction of diseased plant debris are of utmost importance for the management of the disease
2. Avoid waterlogging of the field
3. Number of plants should not exceed the recommended population density (55,000/ha)
4. Use of resistant cultivars: Dalmacio et al. (2007) reported that transgenic hybrids showed significantly less bacterial stalk rot mortality and ear rot incidence and higher grain yield in comparison to their conventional counterparts
5. Chlorinated water has been tried and found effective in USA (Thompson 1965) and India (Thind and Payak 1972; Sharma et al. 1982). Thind and Payak (1972), on the basis of 3-year field trials, reported that soil drenching with chlorinated water (100 μg mL^{-1} chlorine) at 2-week intervals, starting from knee-high stage to flowering stage, reduced the disease incidence by 75.92%. One pre-planting application was also given. Lall and Saxena (1978) recommended two applications of bleaching powder at 25 kg ha^{-1} per application. The first application should be given at flowering and the second, 10 days after the first application.

REFERENCES

Buchanan, R. E. and Gibbons, N. E. eds. 1974. *Bergey's Manual of Determinative Bacteriology*. 8th ed. The Williams and Wilkins Co., Baltimore, MD. 1268 pp.

Burkholder, W. H., McFadden, L. A., and Dimock, A. W. 1953. A bacterial blight of chrysanthemums. *Phytopathology* 43: 522–526.

Dalmacio, S. C., Lugod, T. R., Serrano, E. M., and Munkvold, G. P. 2007. Reduced incidence of bacterial stalk rot on transgenic insect-resistant maize in the Philippines. *Plant Dis.* 91: 346–351.

Graham, D. C. 1964. Taxonomy of the soft rot coliform bacteria. *Annu. Rev. Phytopathol.* 2: 13–42.

Hauben, L., Moore, E. R. B., Vauterin, L., Steenackers, M., Mergaert, J., et al. 1998. Phylogenetic position of phytopathogens within the Enterobacteriaceae. *Syst. Appl. Microbiol.* 21: 384–397.

Hingorani, M. K., Grant, U. J., and Singh, N. J. 1959. *Erwinia carotovora* f. sp. *zeae, a* destructive pathogen of maize in India. *Indian Phytopathol.* 12: 151–157.

Kumar, A., Hunjan, M. S., Kaur, H., Dhillon, H. K., and Singh, P. P. 2017a. Biochemical responses associated with resistance to bacterial stalk rot caused by *Dickeya zeae* in maize. *J. Phytopathol.* 165: 822–832.

Kumar, A., Hunjan, M. S., Kaur, H., Rawal, R., Kumar, A., et al. 2017b. A review on bacterial stalk rot disease of maize caused by *Dickeya zeae*. *J. Appl. Natural Sci.* 9: 1214–1225.

Kumar, A., Hunjan, M. S., Kaur, H., and Singh, P. P. 2015. Characterization of *Dickeya zeae* isolates causing stalk rot of maize based on biochemical assays and antibiotic sensitivity. *Indian Phytopathol.* 68: 375–379.

Lall, S. and Saxena, S. C. 1978. Bacterial stalk rot of maize: Loss assessment and possibility of disease control by application of calcium hypochlorite (Abstr.). *Indian Phytopathol.* 31: 120.

Martinez-Cisneros, B. A., Juarez-Lopez, G., Valencia-Torres, N., Duran-Peralta, E., and Mezzalama, M. 2014. First report of bacterial stalk rot of maize caused by *Dickeya zeae* in Mexico. *Plant Dis.* 98: 1267.

Sabet, K. A. 1954. A new bacterial disease of maize in Egypt. *Empire J. Exptl. Agri.* 22: 65–67.

Samson, R., Legendre, J. B., Christen, R., Fischer-Le Saux, M., Achouak, W., et al. 2005. Transfer of *Pectobacterium chrysanthemi* (Burkholder et al. 1953) Brenner et al. 1973 and *Brenneria paradisiaca* to the genus *Dickeya* gen. nov. as *Dickeya chrysanthemi* comb. nov. and *Dickeya paradisiaca* comb. nov. and delineation of four novel species, *Dickeya dadantii* sp. nov., *Dickeya dianthicola* sp. nov., *Dickeya dieffenbachiae* sp. nov. and *Dickeya zeae* sp. nov. *Int. J. Syst. Evol. Microbiol.* 55: 1415–1427.

Saxena, S. C. and Lall, S. 1981. Effect of fertilizer application on the incidence of bacterial stalk rot of maize. *Indian J. Mycol. Plant Pathol.* 11: 164–168.

Sharma, S. C., Randhawa, P. S., Thind, B. S. and Khera, A. S. 1982. Use of Klorocin for the control of bacterial stalk rot and its absorption, translocation and persistence in maize tissue. *Indian J. Mycol. Plant Pathol.* 12: 185–190.

Thind, B. S. 1970. Investigations on Bacterial Stalk Rot of Maize (*Erwinia carotovora* var. *zeae* Sabet). PhD thesis, Postgraduate School, Indian Agricultural Research Institute, New Delhi, India, 113 pp.

Thind, B. S. and Payak, M. M. 1972. Antibiotics and bleaching powder in the control of bacterial stalk rot of maize. *Hindustan Antibiot. Bull.* 15: 9–13.

Thind, B. S. and Payak, M. M. 1978. Evaluation of maize germplasm and estimation of losses due to maize stalk rot. *Plant Dis. Reptr* 62: 319–323.

Thind, B. S., and Payak, M. M. 1985. A review of bacterial stalk rot of maize in India. *Trop. Pest Manag.* 31: 311–316.

Thind, B. S., and Singh, N. J. 1976. Maize borer (*Chilo partellus* Swinhoe) as carrier of *Erwinia carotovora* var. *zeae*, the causal agent of bacterial stalk rot of maize. *Curr. Sci.* 45: 117–118.

Thompson, D. L. 1965. Control of bacterial stalk rot of corn by chlorination of water in sprinkler irrigation. *Crop Sci.* 5: 369–370.

Victoria, J. I., Arboleda, F., and Munoz, S. 1975. La pudricion suave del tallo de maiz (*Zea mays* L.) en Colombia. *Noticas Fitopatologicas* 4: 136–147.

7.9 STEWART'S WILT OF MAIZE

In 1895, F.C. Stewart first observed a bacterial wilt of sweet corn from Long Island, New York (Stewart 1897). He described the symptoms of the disease, isolated the bacterium, and proved Koch's postulates. In 1898, E.F. Smith described and named the causal bacterium, the culture of which was sent to him by Stewart, as *Pseudomonas stewartii* to honor Stewart. Prior to Stewart's report, T.J. Burrill reported a new bacterial disease of field corn from southern Illinois. However, Burrill and an entomologist, S.A. Forbes attributed this disease to dry conditions and large population of clinch bugs in spite of the presence of bacteria in the diseased tissue. Moreover, Burrill could not prove Koch's postulates in spite of isolating the bacterium from diseased plants and insects, and also described many symptoms of the disease presently not attributed to Stewart's wilt. Hence, the credit of reporting Stewart's wilt correctly for the first time goes to F.C. Stewart. The disease was also called **bacterial wilt** or **bacterial leaf blight**.

The disease has been a major threat to corn producers in the eastern corn belt of the United States for over 100 years. Severe outbreaks of the disease occurred in Nebraska in 1999 and 2000. However, the economic losses due to the disease have been reduced due to the development of resistant hybrids, adequate warning systems, and effective control measures. Golden Cross Bantam, a single-cross sweet corn hybrid developed by Glenn Smith and released in 1932 reduced enormous disease losses. It was so widely grown that within a few years of its release, it accounted for 70%–80% of all the sweet corn canned in USA. The disease also occurs in other countries of South and Central America, Europe, and Asia. Some countries including South Africa are believed to be free from the pathogen. Orio et al. (2012) reported the disease from Argentina for the first time.

The disease is generally more severe on sweet corn than on dent corn or popcorn, but some hybrids and inbred lines of popcorn and field corn are also highly susceptible. Economic losses are caused due to reduction in yield and restrictions on import of seed, as the causal bacterium is a quarantined pathogen in several countries (EPPO A2 quarantine pest).

7.9.1 Host Range

Gamma grass (*Tripsacum dactyloides*) and teosinte (*Zea maxicana*) are also natural hosts of the bacterium. Choi and Kim (2013) reported the occurrence of the disease on *Dracaena sanderiana* (common name lucky bamboo) plants in greenhouses from Seongnam, Gyeonggi Province, South Korea, with an incidence of 35%–50%. The artificially infected hosts include sorghum, millet, Sudan grass, and some weed species like yellow foxtail.

7.9.2 Symptoms

There are two phases of the disease, namely, **wilt phase** and **leaf blight phase**. In both the phases, the initial symptoms appear as leaf lesions originating from corn flea beetle feeding wounds.

7.9.2.1 Wilt Phase

In general, this phase is more severe as the plants may be killed outrightly. The symptoms usually occur on young seedlings, but can also appear on older plants of very susceptible sweet corn cultivars. Initially, the leaf tissue surrounding flea beetle feeding wounds becomes water-soaked. Later on, pale-green to white-yellow linear streaks having irregular or wavy margins develop parallel to leaf veins. In susceptible cultivars, these lesions may extend the entire length of the leaves. With age, water-soaked and chlorotic tissues become necrotic and may result into leaf blight, if weather conditions remain favorable.

The wilt phase is associated with systemic infection. After entering the host tissue, the bacteria multiply in the xylem vessels and spread throughout the plant and occasionally infect kernels. The multiplication of bacteria and production of large quantities of extracellular polysaccharides inside the xylem vessels results in plugging of xylem vessels and reduction in flow of water and nutrients. Bacterial slime may exude in the form of droplets from the cob husks. Bacteria penetrate the seed, but do not enter the embryo. The systemically infected plants become stunted, wilt, and die (Figure 7.6a). The infected plants produce dwarfed and bleached tassels, which ultimately shrivel and die. The bacterial multiplication in the stalk tissue causes browning and necrosis of vascular bundles. Yellow bacterial ooze comes out from the cut ends when systemically infected stalks are pressed. In severely infected plants, open cavities may develop in the stalks at the base.

7.9.2.2 Leaf Blight Phase

These symptoms usually appear after tasseling, although these can appear any time during the crop season. Foliar symptoms of this phase are similar to those of wilt phase. The blighted leaves often wither and die, but they are replaced by new leaves in young seedlings. On adult plants, this phase becomes more obvious because no new leaves are produced. In severe cases, the entire leaves may be blighted due to coalescence

(a)

(b)

FIGURE 7.6 Stewart's wilt of corn: (a) Wilted plant. (b) Leaf blight phase. (Courtesy of J.K. Pataky, University of Illinois at Urbana-Champaign, Bugwood.org.)

of lesions (Figure 7.6b). Secondary fungi often grow on the dead tissue.

The leaf symptoms similar to those of Stewart's wilt are also produced by other leaf blights caused by *Acidovorax avenae* subsp. *avenae*, *Burkholderia andropogonis,* and *Aspergillus* species. Therefore, it is advisable to confirm the identity of the disease by isolating the causal agent and proving its pathogenicity. Stewart's wilt lesions on plants after tasseling can also be confused with those of Goss's wilt. However, these two diseases can be distinguished by using a selective growth medium for Goss's wilt bacterium and an enzyme-linked immunosorbent assay (ELISA) antibody technique for Stewart's wilt bacterium.

7.9.3 Causal Organism

Pantoea stewartii subsp. *stewartii* (Smith 1898) Mergaert et al. 1993
Syn. *Erwinia stewartii*

Cells are Gram-stain-negative, non-flagellate, non-sporing, facultatively anaerobic rods and measure approximately 0.4–0.8 × 0.9–2.2 µm. Capsules are formed. Colonies on nutrient glucose agar are cream-yellow to orange-yellow. The tests for methyl red, Voges–Proskauer, H_2S and indole production, gelatin and esculin hydrolysis, and nitrate reduction are negative. The bacterium also does not produce oxidase, urea hydrolase, arginine dihydrolase, and ornithine dicarboxylase, but catalase is produced. The metabolism is fermentative. Pathogenicity is related to the production of extracellular polysaccharides, which require quorum-sensing regulatory proteins. Maltose, D-arabitol, quinate, citrate, betaine, D-saccharate, and 3–*O*-methyl-D-glucose are not used as sole source of carbon and energy. The mol% G + C of the DNA is 54.6–55.1 (T_m, melting temperature).

7.9.4 Disease Cycle

The bacterium primarily overwinters (survives) in the digestive tracts of hibernating adult corn flea beetles (*Chaetocnema pulicaria* Melsheimer, order *Coleoptera*, family *Chrysomelidae*). The beetles hibernate and pass the winter in soil of grassy areas near the corn fields. The beetles come out from soil in spring when soil temperature is around 18°C–21°C and feed on young corn plants and deposit the bacteria in wounds as fecal contamination. The bacteria enter into the leaf veins and cause infection. These beetles are also the primary mode of secondary spread of the pathogen. The beetles feed on the infected plants, acquire the bacterium, and spread it to the healthy plants. An infested beetle can carry and transmit the bacterium throughout its life. Other insects like rootworms and flea beetle species can be vectors of the bacterium, but their role is negligible in comparison to the corn flea beetles. Corn fields near to grassy areas and winter wheat are generally more severely affected by the disease. Menelas et al. (2006) reported that after 48 h of feeding period on diseased plants, corn flea beetles required a minimum period of 3 h to successfully transmit the bacterium and 50% transmission occurred by 7.6 ± 0.87 h. They further concluded that the beetles are highly efficient vectors, both in acquiring and transmitting the bacterium; hence, only those insecticides should be used for seed treatment and foliar sprays, which act quickly to prevent beetles from acquiring and transmitting the bacterium.

The bacterium can also survive in the infected seed, but the risk of seed–plant transmission is extremely low or almost non-existent under commercial conditions. As most of this seed is of poor quality, it would be removed during routine seed conditioning. Moreover, among infected seeds, the frequency of seed–plant transmission is extremely low ranging from 0.022% to 0.038% (Block et al. 1998; Michener et al. 2002). Besides, the seed-borne infection is also influenced by certain disease parameters. It is more common in systemically infected plants. The ratio of infected seeds in systemically infected susceptible plants is 1 in 50,000 seeds, while it is 1 in 20,000,000 seeds in resistant plants showing leaf blight symptoms. The seed–plant transmission of the bacterium is least significant in North America where disease is endemic. More than 60 countries outside North America enforce quarantine restrictions on import of corn seed to prevent the entry of the pathogen (Pataky 2003).

The severity of disease is influenced by many factors like host nutrition, host reaction to the pathogen, temperature and the number of over-wintering corn flea beetles. High levels of ammonium nitrogen, and phosphorus tend to increase the susceptibility, while high levels of calcium and potassium impart resistance. Meyer et al. (2010), on the basis of 741 observations made from 79 field trials conducted at 15 locations throughout Illinois and one each in Kentucky and Delaware, found that host resistance affected the incidence of systemic seedling wilt within ranges of winter temperatures used for Stewart's wilt forecasts. In case of moderate and resistant hybrids, frequency distributions of Stewart's wilt incidence and mean incidence ranging from 0.7% to 1.8% did not differ among three winter temperature ranges above −2.8°C. On the other hand, distributions of Stewart's wilt incidence on susceptible hybrids differed among each of the four ranges of winter temperature from Stevens-Boewe forecast (i.e., >0.6, −1.1 to 0.6, −2.8 to −1.1, and <−2.8), with mean incidence ranging from 0.5% to 8.5%. They also found that levels of host resistance and winter temperature influenced the probability of exceeding economic thresholds of 1% or 5% incidence of systemic Stewart's wilt. Stewart's wilt is unlikely to exceed economic thresholds when the mean winter temperature is below −4.4°C. When mean winter temperature was above −2.8°C, the probability of exceeding 1% incidence of systemic Stewart's wilt was 0.59 for susceptible sweet corn hybrids and 0.28 for moderate and resistant hybrids. With mean winter temperature below −2.8°C, the probability of exceeding 1% incidence of systemic Stewart's wilt was 0.22% for susceptible hybrids and 0.04 for moderate and resistant sweet corn hybrids. The probability of exceeding 5% incidence was less than 0.1, except when the mean winter temperature was above −2.8°C and susceptible hybrids were grown. High temperatures also favor the disease development. The temperature during winter indirectly affects the disease by influencing the survival of corn flea beetles.

Stevens (1934) developed a forecasting system based on winter temperature to predict the occurrence of the disease. Later on, Boewe (1949) modified Stevens's method to predict more accurately the leaf blight phase of the disease. The simplified version of Stevens–Boewe's method uses the average of mean monthly temperatures for December, January, and February. If the mean daily temperature for December, January, and February is above 1°C, the flea beetles will survive and the disease is likely to be severe. If this average daily temperature is less than −3°C, flea beetles are unlikely to survive, and it is unlikely that Stewart's wilt will be severe. Stevens–Boewe's forecasting system appears to be a reliable predictor of the disease, because severe epiphytotics of the disease, which occurred in the early 1930s, early 1950s, and 1990s in Midwest and eastern United States, followed winters that were among the warmest of that century.

Pa. stewartii subsp. *stewartii* is a remarkable bacterium for laboratory studies due to its relative ease of propagation and genetic manipulation, and the fact that it appears to employ a minimal number of pathogenicity mechanisms. Moreover, *Pa. stewartii* subsp. *stewartii* produces copious amounts of its QS signal, AHL, making it an excellent organism for studying QS-controlled gene regulation in a plant-pathogenic bacterium. Therefore, *Pa. stewartii* subsp. *stewartii* has become the microbial paradigm for QS control of gene expression by both repression and activation via a QS regulator that binds DNA in the absence and dissociates in the presence of the signal ligand. As *Pa. stewartii* subsp. *stewartii* is a member of the *Enterobacteriaceae*, the lessons learned from its interaction with plants may be extrapolated to other plant-associated enterobacteria, such as *Erwinia*, *Dickeya*, and *Pectobacterium* spp., or enteric human pathogens associated with plants, such as *Escherichia coli* and *Salmonella* spp. Studies of this bacterium have provided useful insight into how xylem-dwelling bacteria establish themselves and incite disease in their hosts (Roper 2011).

Pa. stewartii subsp. *stewartii* is a serious bacterial pathogen of sweet corn that colonizes both the apoplast and xylem tissues in which reactive oxygen species (ROS) are produced. The genome of *Pa. stewartii* predicts the presence of two redox-sensing transcriptional regulators, OxyR and SoxR, which activate gene expression in response to oxidative stress. ROS exposure in the form of hydrogen peroxide and the superoxide-generating compound paraquat initiates an induced stress response through OxyR and SoxR that includes activation of the ROS-detoxifying enzymes alkyl hydroperoxide reductase and superoxide dismutase. *Pa. stewartii* ΔsoxR was more sensitive to paraquat and was compromised in the ability to form water-soaked lesions, whereas ΔoxyR was more sensitive to hydrogen peroxide treatment and was deficient in exopolysaccharide production and the elicitation of wilting symptoms. It is evident that both SoxR and OxyR play an important role in the virulence of *Pa. stewartii* during the infection process while colonizing different niches (Burbank and Roper 2014).

Yield losses vary with host cultivars, growth stage of the plant at the time of infection, and whether the infection is systemic or non-systemic. The losses may range from 40% to 100% when infection of susceptible sweet corn hybrids occurs before 5-leaf stage. In sweet corn, the yield loss is about 0.8% for every 1% incidence of systemically infected seedlings.

7.9.5 Management

1. Planting resistant hybrids of sweet corn is the best control measure. Many sweet corn hybrids contain sufficient resistance to the wilt pathogen (Michener et al. 2003; Meyer et al. 2010)
2. Another equally effective control measure is the control of corn flea beetles with insecticides. The insecticides can be applied as seed treatment; furrow application before sowing or foliar application, but seed treatment is most effective. Use of insecticide is more viable in sweet corn and seed corn production. Treatment of sweet corn seed of susceptible variety "Sprint" with imidacloprid at of 4 mL kg^{-1} caused 37%–83% reduction in disease incidence (Kuhar et al. 2002)
3. Use of pathogen-free seed. The seed-borne infection can be detected using ELISA

REFERENCES

Block, C. C., Hill, J. H., and McGee, D. C. 1998. Seed transmission of *Pantoea stewartii* in field and sweet corn. *Plant Dis.* 82: 775–780.

Boewe, G. H. 1949. Late season incidence of Stewart's disease on sweet corn and winter temperatures in Illinois, 1944–1948. *Plant Dis. Reptr* 33. 192–194.

Burbank, L., and Roper, M. C. 2014. OxyR and SoxR modulate the inducible oxidative stress response and are implicated during different stages of infection for the bacterial phytopathogen *Pantoea stewartii* subsp. *stewartii*. *Mol. Plant-Microbe Interact.* 27: 479–490.

Choi, O., and Kim, J. 2013. *Pantoea stewartii* causing Stewart's wilt on *Dracaena sanderiana* in Korea. *J. Phytopathol.* 161: 578–581.

Kuhar, T. P., Stivers-Young, L. J., Hoffmann, M. P., and Taylor, A. G. 2002. Control of corn flea beetle and Stewart's wilt in sweet corn with imidacloprid and thiamethoxam seed treatments. *Crop Prot.* 21: 25–31.

Menelas, B., Block, C. C., Esker, P. D., and Nutter, F. W. Jr. 2006. Quantifying the feeding periods required by corn flea beetles to acquire and transmit *Pantoea stewartii*. *Plant Dis.* 90: 319–324.

Mergaert, J. Verdonck, L., and Kersters, K. 1993. Transfer of *Erwinia ananas* (synonym, *Erwinia uredovora*) and *Erwinia stewartii* to the genus *Pantoea* emend. as *Pantoea ananas* (Serrano 1928) comb. nov. and *Pantoea stewartii* (Smith 1898) comb. nov., respectively, and description of *Pantoea stewartii* subsp. *indologenes* subsp. nov. *Int. J. Syst. Bacteriol.* 43: 162–173.

Meyer, M. D., Pataky, J. K., Joos, D. K., Esgar, R.W., and Henry, B. R. 2010. Influence of host resistance on Stewart's wilt forecasts and probability of exceeding thresholds for use of seed-treatment insecticides on sweet corn. *Plant Dis.* 94: 1111–1117.

Michener, P. M., Freeman, N. D., and Pataky, J. K. 2003. Relationship between reactions of sweet corn hybrids to Stewart's wilt and incidence of systemic infection by *Erwinia stewartii*. *Plant Dis.* 87: 223–228.

Michener, P. M., Pataky, J. K., and White, D. G. 2002. Rates of transmitting *Erwinia stewartii* from seed to seedlings of a sweet corn hybrid susceptible to Stewart's wilt. *Plant Dis*. 86: 1031–1035.
Orio, A. G. A., Brücher, E., Plazas, M. C., Sayago, P., Guerra, F., et al. 2012. First report of Stewart's wilt of maize in Argentina caused by *Pantoea stewartii*. *Plant Dis*. 96: 1819.
Pataky, J. K. 2003. Stewart's wilt of corn. APSnet Feature Story, July–August 2003 [https://www.apsnet.org/edcenter/apsnetfeatures/Pages/StewartsWilt.aspx]. Accessed November 8, 2018.
Roper, M. C. 2011. *Pantoea stewartii* subsp. *stewartii*: Lessons learned from a xylem-dwelling pathogen of sweet corn. *Mol. Plant Pathol*. 12: 628–637.
Smith, E. F. 1898. Notes on Stewart's sweet-corn germ. *Pseudomonas stewarti*, n. sp. *Proc. American Assoc*. Advancement of Sci. 47: 422–426.
Stevens, N. E. 1934. Stewart's disease in relation to winter temperatures. *Plant Dis. Reptr* 18: 141–149.
Stewart, F. C. 1897. A bacterial disease of sweet corn. *N.Y. Agric. Exp. Sta. Bull*. 130: 422–439.

7.10 GOSS'S BACTERIAL WILT AND LEAF BLIGHT OF MAIZE

Although the full name of the disease includes both phases, namely, bacterial wilt and leaf blight, it is commonly referred to simply as Goss's wilt. Goss's bacterial wilt and leaf blight of maize is also known by other names such as, **Nebraska leaf freckles and wilt**, **leaf freckles and wilt**, **bacterial freckles and wilt,** and **bacterial leaf blight and wilt of maize**. The disease was named after R.W. Goss, who is credited with pioneering the early development of modern plant pathology in Nebraska. The disease was first observed in Dawson County (near Lexington) of Nebraska in 1969 and has since been confirmed in several states throughout the Midwest, including Illinois, Wisconsin, and Iowa. In 2008, Goss's bacterial wilt and leaf blight caused yield losses up to 60 bushels/acre in northwestern Indiana, showing that the disease can cause substantial yield loss in the state. The disease has also been reported from Minnesota (Malvick et al. 2010), Texas (Korus et al. 2011), and parts of Colorado, Wyoming, Kansas, and South Dakota. Goss's wilt also occurs in Canada. The bacterium can infect food-grade corn, seed corn, and sweet corn, but popcorn is of particular concern in Indiana due to hybrid susceptibility, the high potential of yield loss, and phytosanitary restrictions on exported seed.

Goss's wilt impact on the crop generally depends on the amount of leaf area lost during the grain filling period. Leaf loss may lead to reduced stalk quality and yield, with grain losses ranging from negligible to nearly 50%. Higher levels of corn residue from corn-after-corn production and reduced tillage are likely contributing factors for higher incidence of the disease. In addition, the prevalence of summer storms (hail, wind, and rainstorms) that damage corn leaves has a large impact on the severity of infection and yield loss in a given growing season.

Goss's wilt may reduce corn plant stands and vigour, grain quality, and yield. During the systemic infection phase, the disease may reduce plant stands and weaken surviving plants, both of which are associated with reduced yield. However, in most cases, yield loss is mainly due to the leaf blight phase of the disease, due to reduction in green leaf area and premature death of plants. Timing of leaf blight infection plays a critical role in yield reduction due to disease. Early infections lead to the greatest yield loss, while late infections often have little impact on yield. Yield reductions up to 50% have been documented when susceptible hybrids were infected early in the growing season. Other agronomic issues such as stalk lodging may result from leaf area loss due to disease. This can result in further reductions in yield and also reductions in grain quality, if ears come in contact with the ground.

7.10.1 Host Range

Maize (*Zea mays*) is the main host. Sugarcane (*Saccharum officinarum*), common sorghum (*Sorghum bicolor*), Sudan grass (*S. sudanense*), teosinte (*Zea mexicana*), green foxtail, and shatter cane have also been reported as natural hosts. The experimentally reported hosts are barnyard grass (*Echinochloa crusgalli*), *Triticale* sp., and wheat (*Triticum aestivum*). Ikley et al. (2016) identified three new hosts, namely, annual ryegrass (*Lolium multiflorum*), Johnson grass (*S. halepense*), and large crabgrass (*Digitaria sanguinalis*). Previously cited hosts shattercane, giant foxtail (*Setaria faberi*), green foxtail (*Setaria viridis*), and yellow foxtail (*Setaria pumila*) were also confirmed as hosts. Barnyard grass, cereal rye (*Secale cereale*), and all broadleaf species tested were identified as non-hosts. In a separate greenhouse experiment, they also inoculated annual ryegrass, giant foxtail, and large crabgrass plants with bacterium and 2 weeks after inoculation, treated the plants with seven single mode action herbicides. The bacteria recovered from herbicide treated plants proved pathogenic on reinoculation.

Webster et al. (2016b) reported the grass weed, woolly cupgrass (*Eriochloa villosa*), and the native grasses, big bluestem (*Andropogon gerardi*) and little bluestem (*Schizachyrium scoparium*) as new hosts of the bacterium.

7.10.2 Symptoms

Two types of symptoms are produced, namely, leaf blight phase and wilt phase.

7.10.2.1 Leaf Blight Phase

It is more common and more damaging than the wilt phase. Early leaf symptoms are oblong or elongated lesions of water-soaked, greyish-green tissue that progress into long streaks with wavy, irregular margins. These streaks extend along the leaf veins. Leaf streaks change to grey to light green or yellow to red stripes with wavy or irregular margins (confusion with leaf symptoms of Stewart's wilt and cold damage is possible). As lesions enlarge and coalesce, they form large areas of necrotic tissue on the leaves and eventually, entire leaves may wilt and dry up. Discrete, discontinuous, water-soaked spots (freckles) that develop within the streaks, often near lesion edges, are characteristic of the disease. These spots are dark-green to blackish in appearance and look like **freckles** (hence

FIGURE 7.7 Goss's bacterial wilt and blight of maize: Leaf blight phase. (Courtesy of Larry Osborne, Bugwood.org.)

the name) when infected leaves turn brown (Figure 7.7). In humid weather, bacterial ooze or exudate forms in the streaks, which dries to form a glistening/shiny, varnish-like crust. In dry weather, Goss's wilt symptoms are easily confused with drought stress or leaf scorch from chemical burn.

7.10.2.2 Wilt Phase

Plants may also be infected systemically by Goss's wilt, especially in the seedling stage. Systemically infected plants show discolored, orange to brown vascular bundles, and eventually a slimy stalk rot that can lead to wilting and death of plants. In later stages of infection, plants wilt, show leaf blight, and may be stunted.

Two characteristics used to distinguish Goss's wilt from other diseases are the presence of freckles within the streaks and dried glistening/shiny bacterial exudate on the leaf surface. Development of Goss's wilt is most common after a hailstorm or sand blasting; whereas Stewart's wilt epidemics occur when populations of flea beetles are high. Therefore, Stewart's wilt is accompanied by obvious flea beetle feeding scars on leaves.

7.10.3 Causal Organism

Clavibacter nebraskensis (Vidaver and Mandel 1974) Li et al. 2018

Syn. *Cla. michiganensis* subsp. *nebraskensis*, *Corynebacterium michiganense* subsp. *nebraskense*, *Cor. michiganense* pv. *nebraskense*, *Cor. nebraskense*

The cells are pleomorphic rods, often in angular V-formations with no marked rod-coccus development cycle. Gram-stain-positive, non-acid-fast, non-motile, and non-sporing. Obligate aerobes. Acid production from carbohydrates is slow and weak. Nitrates are not reduced and casein is not hydrolyzed. Catalase is positive, while oxidase, lipase, tyrosinase, and urease are negative. The cell-wall peptidoglycan, based on the presence of 2, 4-diaminobutyric acid, is type B2γ cross-linked at a diaminobutyrate residue. The bacterium does not have a type III secretion system.

The bacterium is a quarantined pest and a phytosanitary certificate is required for export of corn seed.

Gross and Vidaver (1979) developed a selective agar medium (CNS) for isolation of *Cor. nebraskense* from corn tissue and soil. Selectivity depended on a combination of nalidixic acid, polymyxin B sulfate, lithium chloride, cycloheximide, and Bravo 6F®. Recovery of *Cor. nebraskense* on CNS from pure culture was equal to the recovery on nutrient broth-yeast extract medium (NBY). CNS medium is quantitatively and qualitatively superior to other selective media developed for *Corynebacterium* spp. On CNS, *Cor. nebraskense* generally could be isolated from soil having as low as 10^3 cfu per gram of soil. More than 99% of the total recoverable soil bacteria on NBY could not grow on CNS. However, some coryneform and coccoid soil bacteria grew on CNS. In addition to *Cor. nebraskense*, high efficiencies of plating on CNS were noted for some phytopathogenic *Corynebacterium* spp., but *Cor. sepedonicum* and *Cor. insidiosum* did not grow. Soil and airborne fungi were unable to grow on CNS.

Tambong et al. (2016) generated a draft genome sequence of *Cla. michiganensis* subsp. *nebraskensis* strain DOAB 395 and performed genome and proteome analysis of *Cla. michiganensis* subsp. *nebraskensis* strains (DOAB 397 and DOAB 395) isolated in 2014 and compared with the type strain, NCPPB 2581 (isolated over 40 years ago). Further, the genome data were exploited for the development of a multiplex TaqMan real-time PCR tool for rapid detection of *Cla. michiganensis* subsp. *nebraskensis*. The specificity of the assay was validated using 122 strains of *Clavibacter* and non-*Clavibacter* spp. A blind test and naturally infected leaf samples were used to confirm specificity. The sensitivity (0.1–1.0 pg) compared favorably with previously reported real-time PCR assays. This tool should fill the current gap for a reliable diagnostic technique.

Webster et al. (2016a) used MLSA to evaluate genetic variation within a subset containing 50 isolates of *Cla. michiganensis* subsp. *nebraskensis* obtained from infected fields across the midwestern US. Isolates varied by year of isolation (1971–2014), as well as geographic origin and virulence on corn. The use of four genes in MLSA was sufficient to differentiate the bacterium from other *Cla. michiganensis* sub species and revealed it to be a monophyletic clade with small, but discernible genetic differences among isolates.

Agarkova et al. (2011) determined population structure of a wide temporal and geographic collection of 131 strains, originating between 1969 and 2009, using amplified fragment length polymorphism (AFLP) analysis and repetitive

DNA sequence-based BOX-PCR. Analysis of the composite data set of AFLP and BOX-PCR fingerprints revealed two groups with a 60% cut-off similarity: a major group A (118 strains) and a minor group B (13 strains). The clustering in both groups was not correlated with strain pathogenicity. Group A contained two clusters, A1 (78 strains) and A2 (40 strains), with a linkage of 75%. Group A strains did not show any correlation with historical, geographical, morphological, or physiological properties of the strains. Group B was very heterogeneous and eight out of nine clusters were represented by a single strain. The mean similarity between clusters in group B varied from 13% to 63%. All strains in group B were isolated after 1999. The percentage of group B strains among all strains isolated after 1999 (69 strains) was 18.8%.

Yasuhara-Bell et al. (2016) developed a LAMP assay to detect the tripartite ATP-independent periplasmic (TRAP)-type C4-dicarboxylate transport system large permease component. The authors found that out of the strains of *Cla. michiganensis* subsp. *nebraskensis* along with all other *Cla. michiganensis* sub species and several genera of non-target bacteria tested, only strains of *Cla. michiganensis* subsp. *nebraskensis* reacted positively with the LAMP assay. The LAMP assay accurately identified the bacterium from diseased maize samples and has a potential to be used as a field test.

Schlund et al. (2016), based on *in vitro* inhibitory tests of five spray adjuvants, reported that a positive control of copper hydroxide (Kocide® 3000) had the greatest frequency of *Cla. michiganensis* subsp. *nebraskensis* inhibition among treatments at all the three concentrations tested. Non-ionic surfactant (NIS) decreased the growth compared to the rest of the adjuvants. NIS inhibited growth more than the copper hydroxide positive control.

7.10.4 Disease Cycle

The main source of perpetuation of the bacterium appears to be maize crop residue that has been left to overwinter. The bacterium is internally and externally seed-borne and seed-plant transmission occurs, but appears to be of minor importance as compared to residue transmission (Schuster 1975). Seed-plant transmission is only about 2%, but infected seed may introduce the disease into disease-free areas. Grass weeds such as green foxtail, barnyard grass, and shattercane can also serve as perennial hosts for the bacteria, and be a reservoir from which bacteria spread to corn plants.

The bacterium enters the host plant through stomata and trichomes, but can also enter through wounds caused by hails, sand blasts, rainstorms, windstorms, and mechanical injuries. Although the bacteria can enter through wounds created by insect feeding, insects are not known to transmit the bacteria. Mallowa et al. (2016), in greenhouse studies, found that severe wounding is not necessary for infection of maize by the bacterium, and stomata or trichomes may serve as entry points for the bacterium as leaf blight symptoms were observed on non-wounded plants in ambient (37.0% plant incidence) and increased humidity conditions (60.0% plant incidence). Epiphytic populations of the bacterium survived and increased on maize leaves, particularly at increased humidity and was found colonizing maize leaves in localized sites that included epidermal junctions, cuticle depressions, in and around stomata, and at the base of trichomes.

Eggenberger (2016) monitored epiphytic *Cla. michiganensis* subsp. *nebraskensis* population densities and the temporal and spatial spread of Goss's wilt incidence originating from three inoculum point sources, namely, non-inoculated control, inoculum point sources established by wound inoculation, and inoculum point sources consisting of the bacterium-infested corn residue for 2 years. Epiphytic population was detected on asymptomatic corn leaves collected up to 2.5 m away from inoculum sources at 15 days after inoculation and increased until mid-August in both years. Plots infested with bacterial residue had final Goss's wilt incidence of 7.5% and 1.8% in 2012 and 2013, respectively, while those with a wound-inoculated source had final Goss's wilt incidence of 16.6% and 14.0% in 2012 and 2013, respectively.

Botti-Marino et al. (2016) reported that the epiphytic and internal populations of both the *Cla. michiganensis* subs. *nebraskensis* isolates, differing in virulence on corn hybrids, rose over time, but the internal populations of the less virulent strain were significantly lower on the more susceptible hybrid compared to the more virulent strain. This suggests that the virulence status can affect the ability to develop internal populations. The screening of 19 strains of *Cla. michiganensis* subs. *nebraskensis* for biofilm formation via two *in vitro* methods indicated variation in biofilm formation among strains. There was no correlation of biofilm formation to virulence of the pathogen, suggesting that biofilm formation may play other roles in ecology of the bacterium.

Wind-driven rain and overhead irrigation disseminate the bacteria. Harvest and tillage equipment, balers, and wind can also transfer infected residue and soil to pathogen-free fields. To help avoid spreading the pathogen in this way, harvest and till infected fields last of all and clean the equipment of crop residue and soil. Plants may be infected at any stage of development. Wet weather and high relative humidity favor disease development because leaf wetness is required for infection to occur, and the bacteria spread most readily in humid weather. Hot and dry weather will slow disease development, making symptoms of Goss's wilt harder to distinguish from other disorders, such as drought stress. Low levels of the disease may go undetected until environmental conditions favor disease development.

Mbofung et al. (2016) compared colonization of resistant and susceptible hybrids by *Cla. michiganensis* subsp. *nebraskensis* and reported that lesion expansion in the susceptible hybrid was associated with a faster rate of its spread and multiplication in the tissues. In the resistant hybrid, spread and multiplication was reduced ($P < 0.0001$) and 16 days after inoculation, became imperceptible. Initially, the bacterium showed a preference for colonization of the metaxylem vessels in both the hybrids. The spread from cell to cell was accomplished through disruption of cell walls, due to abundance of bacterial cells or enzymatic activity.

Resistance to the pathogen was associated with production of a dense matrix in the xylem that deformed and restricted movement of the bacterial cells.

McNally et al. (2016) reported the identification of CMN_01184 as a novel gene target and its use in conventional PCR (cPCR) and SYBR Green-based quantitative polymerase chain reaction (qPCR) assays for specific detection and quantification of *Cla. michiganensis* subsp. *nebraskensis*. The cPCR and qPCR assays based on primers targeting CMN_01184 specifically amplified only *Cla. michiganensis* subsp. *nebraskensis* from a diverse collection of 129 bacterial and fungal isolates, including many maize bacterial and fungal pathogens, environmental organisms from agricultural fields, and all the known subspecies of *Cla. michiganensis*. The specificity of the assays for detection of only *Cla. michiganensis* subsp. *nebraskensis* was also validated using *Cla. michiganensis* subsp. *nebraskensis* infected field samples and healthy maize leaves, and *Cla. michiganensis* subsp. *nebraskensis* infested and uninfected soil. The detection limits for pure *Cla. michiganensis* subsp. *nebraskensis* DNA were 30 and 3 ng, and 100 and 10 cfu of *Cla. michiganensis* subsp. *nebraskensis* for cPCR and qPCR assays, respectively. These novel, specific, and sensitive PCR assays based on CMN_01184 are effective for the diagnosis of Goss's wilt and also for studies of the epidemiology and host-pathogen interactions of *Cla. michiganensis* subsp. *nebraskensis*.

Treat and Tracy (1990) determined inheritance of resistance to Goss's wilt using ten hybrids from a diallel cross of five sweet corn inbreds in 1987 and 1988. The inbreds used to make the diallel were widely used historically and were chosen on the basis of adaptation and relative maturity. Three hybrids were resistant and seven intermediate, while the field corn controls were extremely susceptible. General combining ability (GCA) and specific combining ability (SCA) sums of squares accounted for 94% and 6% of the variation among crosses, respectively. GCA was highly significant ($P \leq 0.01$), while SCA was non-significant. Year differences were non-significant, but date of rating and hybrid × year interaction effects were significant (P 0.05). Resistance to Goss's wilt is available in sweet corn, and recurrent selection should be effective if improvement in resistance is desired.

7.10.5 Management

As no chemicals are available to control the disease, it is important to prevent the entry and spread of the pathogen. The following measures will help reduce or minimize the disease incidence.

1. Use of pathogen-free seed
2. Crop rotation with non-host crops such as soybeans, dry beans, small grains, or alfalfa allows for an additional year of corn residue decomposition between two maize crops
3. Deep tillage and other practices that encourage residue decomposition are especially effective for incorporating and burying infected residue
4. Grassy weeds, namely, green foxtail, barnyard grass, and shattercane that act, as collateral hosts of the bacterium, should also be controlled to minimize the inoculum
5. Use of resistant cultivars. High resistance in dent maize hybrids and inbreds has been found, but sweet corn varieties appear to be less resistant
6. Follow general hygiene and disinfection of harvesting and tillage equipment and balers. Harvest and till infected fields at the end

REFERENCES

Agarkova, I. V., Lambrecht, P. A., and Vidaver, A. K. 2011. Genetic diversity and population structure of *Clavibacter michiganensis* subsp. *nebraskensis*. *Can. J. Microbiol.* 57: 366–374.

Botti-Marino, M., Jacobs, J., Chilvers, M., and Sundin, G. 2016. Epiphytic and endophytic populations and biofilm formation of *Clavibacter michiganensis* subspecies *nebraskensis*, and the relation to virulence (Abstr.). *Phytopathology* 106: S4.185.

Eggenberger, S., Diaz-Arias, M. M., Gougherty, A. V., Nutter, F. W., Jr., Sernett, J., et al. 2016. Dissemination of Goss's wilt of corn and epiphytic *Clavibacter michiganensis* subsp. *nebraskensis* from inoculum point sources. *Plant Dis.* 100: 686–695.

Gross, D. C., and Vidaver, A. K. 1979. A selective medium for isolation of *Corynebacterium nebraskense* from soil and plant parts. *Phytopathology* 69: 82–87.

Ikley, J., Johnson, W., and Wise, K. 2016. Defining the alternative host range and the effect of herbicides on pathogenicity of *Clavibacter michiganensis* subsp. *nebraskensis*, causal agent of Goss's wilt of corn (Abstr.). *Phytopathology* 106: S1.4.

Korus, K. A., Timmerman, A. D., French-Monar, R. D., and Jackson, T. A. 2011. First report of Goss's bacterial wilt and leaf blight (*Clavibacter michiganensis* subsp. *nebraskensis*) of corn in Texas. *Plant Dis.* 95: 73.

Li, X., Tambong, J., Yuan, K. (X.), Chen, W., Xu, H., et al. 2018. Re-classification of *Clavibacter michiganensis* subspecies on the basis of whole-genome and multi-locus sequence analyses. *Int. J. Syst. Evol. Microbiol.* 68: 234–240.

Mallowa, S. O., Mbofung, G. Y., Eggenberger, S. K., Den Adel, R. L., Scheiding, S. R., et al. 2016. Infection of maize by *Clavibacter michiganensis* subsp. *nebraskensis* does not require severe wounding. *Plant Dis.* 100: 724–731.

Malvick, D., Syverson, R., Mollov, D., and Ishimaru C. A. 2010. Goss's bacterial blight and wilt of corn caused by *Clavibacter michiganensis* subsp. *nebraskensis* occurs in Minnesota. *Plant Dis.* 94: 1064.

Mbofung, G. C. Y., Sernett, J., Horner, H. T., and Robertson, A. E. 2016. Comparison of susceptible and resistant maize hybrids to colonization by *Clavibacter michiganensis* subsp. *nebraskensis*. *Plant Dis.* 100: 711–717.

McNally, R. R., Ishimaru, C. A., and Malvick, D. K. 2016. PCR-mediated detection and quantification of the Goss's wilt pathogen *Clavibacter michiganensis* subsp. *nebraskensis* via a novel gene target. *Phytopathology* 106: 1465–1472.

Schlund, S. A., Jackson-Ziems, T. A., Kruger, G. R., Blankenship, E., and Robertson, A. 2016. Inhibitory effects *in-vitro* by field adjuvants on *Clavibacter michiganensis* subsp. *nebraskensis*, the causal agent of Goss's wilt and leaf blight of corn. *Phytopathology* 106: S1.8.

Schuster, M. L. 1975. Leaf freckles and wilt of corn incited by *Corynebacterium nebraskense* Schuster, Hoff, Mandel and Lazar 1972. *Neb. Agr. Expt. Sta. Res. Bul.* 270, 40 PP.

Tambong, J. T., Xu, R., Daayf, F., Brière, S., Bilodeau, G. J., et al. 2016. Genome analysis and development of a multiplex TaqMan real-time PCR for specific identification and detection of *Clavibacter michiganensis* subsp. *nebraskensis*. *Phytopathology* 106: 1473–1485.
Treat, C. L., and Tracy, W. F. 1990. Inheritance of resistance to Goss's wilt in sweet corn. *J. Amer. Soc. Hort. Sci.* 115: 672–674.
Vidaver, A. K., and Mandel, M. 1974. *Corynebacterium nebraskense*, a new, orange-pigmented phytopathogenic species. *Int. J. Syst. Bacteriol.* 24: 482–485.
Webster, B. T., Curland, R. D., McNally, R. R., Ishimaru, C. A., and Malvick, D. K. 2016a. Comparative phylogenetic analysis and genetic diversity of *Clavibacter michiganensis* subsp. *nebraskensis*. *Phytopathology* 106: S1.9.
Webster, B., Ishimaru, C., and Malvick, D. 2016b. Native grasses and common weeds as hosts for *Clavibacter michiganensis* subsp. *nebraskensis* (Abstr.). *Phytopathology* 106: S4.193–194.
Yasuhara-Bell, J., de Silva, A., Heuchelin, S. A., Chaky, J. L., and Alvarez, A. M. 2016. Detection of Goss's wilt pathogen *Clavibacter michiganensis* subsp. *nebraskensis* in maize by Loop-Mediated Amplification. *Phytopathology* 106: 226–235.

7.11 CORN STUNT DISEASE

Corn stunt is one of the most economically important diseases of maize (*Zea mays* L.) in the USA, Mexico, and Central and South America (Bradfute et al. 1981; Tsai and Falk 1988). Corn stunt was first named by Kunkel (1946). Its incidence has increased in many tropical and sub tropical maize growing areas. The studies carried out in Argentina from 1991 to 2001 showed that the disease has largely spread across the northern part of the country, since its initial discovery in the northeast of Argentina. In the Tucumán province, the disease caused yield reductions ranging from 50% to 90% (with an average of 70%), and the infected plants produced three times less grain yield. Carloni et al. (2013) have reported the prevalence of the disease in the temperate region of Argentina. In the US where the disease was formerly considered sporadic in nature; it has been observed every year in the central valley of California since 1996, and in 2001, its outbreak caused economic losses of more than 5 million US dollars. Earlier, the occurrence of both the leafhopper and corn stunt were sporadic in the southern San Joaquin Valley. However, since 1996, the leafhopper and corn stunt disease have occurred on a yearly basis with significant losses each year. Besides California, the disease occurs in Florida, Louisiana, Michigan, Mississippi, Ohio, and Texas states of US. The disease also occurs in Belize, El Salvador, Guatemala, Honduras, Jamaica, Nicaragua, Panama, Bolivia, Brazil, Colombia, Paraguay, Peru, and Venezuela (CABI 2016).

7.11.1 Host Range

In addition to *Z. mays* L., the other natural hosts of corn stunt spiroplasma are *Z. mays mexicana* (Schrad.) Iltis, *Z. diploperennis* Iltis, Doebley and Guzman, *Z. perennis* (Hitchc.) Reeves and Mangelsd., *Z. mars* × *Tripsacum floridanum*.pn2 Porter exVasey L., and *Z. luxurians* (Durieu and Ascherson) Bird (Nault 1980). *Vicia faba* L., *Catharanthus roseus* (L.) G. Don, *Raphanus sativus* L., *Sinapis alba* L. (*Brassica hirta* Moench), and *Spinacia oleracea* L. are also reported as experimental hosts (Markham and Alivizatos 1983; Markham et al. 1977).

7.11.2 Symptoms

The initial symptoms of corn stunt are characteristic small chlorotic stripes that develop at the leaf bases of young plants after about 25–30 days of sowing. The chlorotic stripes become fused and extend further toward the leaf tips in the older leaves with green spots and stripes on a chlorotic background. Infected maize plants are stunted and show chlorotic stripes on the leaves. They have much shorter internodes with a proliferation of secondary shoots, thus giving the plants a short and bushy appearance. The stalk may have multiple ears, sometimes as many as 6–7 per plant. The ears are small and poorly filled, leaving a large number of blank spaces on the ear (Figure 7.8). The kernels that do develop are frequently loose leading to what is called **loose tooth ears**. Younger leaves near the top of the plant are yellow. With age they take on reddish to reddish-purple color that varies with variety. Reddening on leaves varies depending on the corn genotype and environmental conditions. Symptoms may vary according to climatic conditions, maize cultivar, and the presence of other pathogens. The disease is most severe in corn planted after July 1, but can occur in corn planted as early as March and April.

All current commercial varieties of field and sweet corn appear to be susceptible. Yield loss depends on the growth stage of the corn when it is infected, but can be significant, especially in late sown corn.

Diagnostic techniques: Corn stunt spiroplasma is a motile, helical, cell wall-free procaryote that can readily be seen or detected by phase contrast or dark field microscopy of plant (infected) juice or hemolymph and abdominal smears from

FIGURE 7.8 Corn stunt: Stunted plant. (Courtesy of Clemson University—USDA Cooperative Extension Slide Series, Bugwood.org.)

infected leafhopper vectors (Davis and Worley 1973). It is a phloem-limited organism. It is highly resistant to penicillin in *in vitro* tests (Chang and Chen 1978). Treatment of inoculated plants with tetracycline antibiotic causes remission of symptoms and interferes with leafhopper transmission (Granados 1969). It can readily be cultured and maintained *in vitro* (Chen and Liao 1975; Williamson and Whitcomb 1975; Davis et al. 1984). Infected corn plants and leafhoppers can be tested for the presence of spiroplasma using microscope or by serological assays using spiroplasma deformation tests and ELISA. The ELISA technique developed by Eden-Green (1982) to detect corn stunt spiroplasma from infected corn plants is available. A unique type of cell deformity is found in the stomata of epidermal strips of leaves from the infected plants which can be used as diagnostic feature (Overman et al. 1992), with stomata showing rounded guard cells and subsidiary cells fused into adjoining epidermal cells.

7.11.3 Causal Organism

Spiroplasma kunkelii sp. nov. Whitcomb et al. 1986

The below given description of the species is based largely on the publication of Bradbury (1991). The bacterium is a helical, motile, cell wall-free procaryote, bounded by a single membrane. Cells are approximately 0.15–0.2 μm in diameter, 2.0–15 μm in length, and 0.4–0.6 μm in amplitude of the helical gyre, with a regular gyre length in a given helical filament. The cells are able to pass through a membrane filter of 220 nm pore size, but not through 100 nm pores. In *in vitro* cultivation, cell dimensions vary slightly with medium. Helical cells exhibit flexional and rotational motility, with translational movement in viscous media. The colonies have a **typical "fried-egg"** appearance. The colony margin is frequently diffuse due to the movement of growing cells. Colony size varies from 0.1 to 4.0 mm. Cells divide by binary fission.

Davis et al. (2015) reported the complete nucleotide sequence of the circular chromosome and four plasmids of *Sp. kunkelii* strain CR2-3x, which was isolated in axenic culture from naturally infected field grown plants of corn in Costa Rica. There were 29,904 reads and a total of 174,151,004 nucleotides. The N_{50} read length was 9,508 nucleotides; the mean read length was 5,823 nucleotides; and the average reference consensus concordance was 100.00%. The assembled, circular chromosome of 1,463,926 bp has an overall base composition of 24.97 mol% G + C; the average coverage per base position was 97.35 ×. The four plasmids were 22,558, 14,615, 20,501, and 7,576 bp in size having base compositions of 27.64, 28.14, 24.20, and 21.50 mol% G + C, respectively. The chromosome has 1,646 protein coding regions, multiple insertions of spiroplasma virus sequences, one set of rRNA genes, and 33 tRNA genes.

Carpane et al. (2013) reported that the genome composition of this species is highly conserved among isolates. The degree of polymorphism between isolates of different geographic origins was low, and the level of genomic variability was similar within isolates of different countries. Hence, the instability of maize resistance due to generation of new pathotypes of *Sp. kunkelii* is unlikely. Instead, other components of this complex pathosystem could account for the breakdown of resistance.

7.11.4 Disease Cycle

The spiroplasma is transmitted naturally by *Dalbulus maidis* DeLong and Wolcott, and *D. elimatus* Ball, and experimentally by *Graminella nirifrons* Forbes, *G. sonora* Ball, *Stirellus bicolor* Van Duzee, *Exitianus exitiosus* Uhler, and *Euscelidius variegatus* Kirsch (Granados et al. 1968; Granados 1969; Nault and Knoke 1981). The most important vector throughout Latin America is the corn leafhopper *D. maidis* (*Homoptera: Cicadellidae, Deltocephalinae*). It is also a most common and efficient vector in South Florida (Tsai 1987a, 1987b). When *D. maidis* individuals are given access to the pathogen, nearly 100% are capable of transmitting it (Markham and Alivizatos 1983), but neither the longevity nor reproductivity of *D. maidis* is detectably affected by the infection (Madden et al. 1984). The spiroplasma is transmitted in a persistent manner, requiring a latency period before becoming circulative in either the insect vector or the host plant and is retained for long periods by the vector (Nault 1980). Tsai (1987b) reported a retention period of 45 days for *D. maidis*. It appears that relatively long feeding periods (hours to days) by the vector are required for acquisition and transmission (Anaya Garcia 1975), as is commonly the case for persistent pathogens. An incubation period in *D. maidis* ranges from 17.5 to 21.2 days (Tsai 1987b) and the length of incubation period varies inversely with the length of acquisition access period. There is a specific latency period in the plant before the disease can be transmitted to another vector individual, depending on temperature and the maize variety (Martinez-Lopez 1977; Nault 1980). The spiroplasma overwinters within the adult leafhopper; when the leafhoppers emerge from overwintering in early spring, they can be infective. Disease symptoms appear about 3 weeks after the corn is infected.

Carloni et al. (2013) reported the prevalence of the disease in the temperate region of Argentina. The highest disease prevalence and incidence levels were found in the transition area from the temperate to the sub tropical region, related to the highest *D. maidis* prevalence and insects sampled per location. *D. maidis* adults were found in volunteer corn plants and spontaneous vegetation in autumn and winter months, which were source of inoculum for *Sp. kunkelii*. This overwintering ability was related to detection of *D. maidis* insects in corn crops at early growth stages in the following growing season.

7.11.5 Management

1. Early planting is the best way to minimize damage from the disease
2. Control of vectors
3. Prompt removal and destruction of infected plants
4. Elimination of reservoir hosts such as *Tripsacum* spp. and volunteer corn plants
5. Breeding for resistance to the insect vector and the pathogen

REFERENCES

Anaya Garcia, M. A. 1975. Determinacion del periodo minimo y optimo de inoculacion necesaria para que el vector Dalbulus maidis transmite el patogeno causante del achapar-ramiento del maiz. *Siades (El Salvador)* 4: 9–14.

Bradbury, J. F. 1991. *Spiroplasma kunkelii. IMI Descriptions of Fungi and Bacteria*, No. 1047. Wallingford, UK: CAB Inernational.

Bradfute, O. E., Tsai, J. A., and Gordon, D.T. 1981. Corn stunt spiroplasma and viruses associated with a maize disease epidemic in southern Florida. *Plant Dis.* 65: 837–841.

CABI. 2016. *Invasive Species Compendium*, Datasheet: *Spiroplasma kunkelii* (corn stunt spiroplasma). www.cabi.org/isc/datasheet/50978. Accessed November 11, 2018.

Carloni, E., Carpane, P., Paradell, S., Laguna, I., and Giménez Pecci, M. P. 2013. Presence of *Dalbulus maidis* (*Hemiptera*: *Cicadellidae*) and of *Spiroplasma kunkelii* in the temperate region of Argentina. *J. Econ. Entomol.* 106: 1574–1581.

Carpane, P., Melcher, U., Wayadande, A., de la Paz Gimenez Pecci, M., Laguna, G., et al. 2013. An analysis of the genomic variability of the phytopathogenic mollicute *Spiroplasma kunkelii. Phytopathology* 103: 129–134.

Chang, C. J., and Chen, T.A. 1978. Antibiotic spectrum of pathogenic spiroplasmas of plant, insect and vertebrate animal (Abstr.). *Phytopathol. News* 12: 233.

Chen, T.A., and Liao, C.H. 1975. Corn stunt spiroplasma: Isolation, cultivation, and proof of pathogenicity. *Science* 188: 1015–1017.

Davis, R. E., Shao, J., Dally, E. L., Zhao, Y., Gasparich, G. E., et al. 2015. Complete genome sequence of *Spiroplasma kunkelii* strain CR2-3x, causal agent of corn stunt disease in *Zea mays* L. *Genome Announc.* 3(5): e01216–e01315. doi:10.1128/genomeA.01216-15.

Davis, M. J., Tsai, J. H., and McCoy, R.E. 1984. Isolation of the corn stunt spiroplasma from maize in Florida. *Plant Dis.* 68: 600–604.

Davis, R. E., and Worley, J. F. 1973. Spiroplasma: Motile, helical microorganism associated with corn stunt disease. *Phytopathology* 63: 403–408.

Eden-Green, S. J. 1982. Detection of corn stunt spiroplasma *in vivo* by ELISA using antisera to extracts from infected corn plants (*Zea mays*). *Plant Pathol.* 31: 289–297.

Granados, R. R. 1969. Maize viruses and vectors. pp. 327–359 in: *Viruses, Vectors and Vegetation.* K. Maramorosch ed. Interscience Publishers, New York.

Granados, R. R., Granados, J. S., Maramorosch, K., and Reinitz, J. 1968. Corn stunt virus: Transmission by three cicadellid vectors. *J. Econ. Entomol.* 61: 1282–1287.

Kunkel, L. L. 1946. Leafhopper transmission of corn stunt. *Proc. Nat. Acad. Sci. USA* 32: 246–247.

Madden, L. V., Nault, L. R., Heady, S. E., and Styer, W. E. 1984. Effect of maize stunting mollicutes on survival and fecundity of Dalbulus leafhopper vectors. *Ann. Appl. Biol.* 105: 431–441.

Markham, P. G., and Alivizatos, A. S. 1983. The transmission of corn stunt spiroplasma by natural and experimental vectors. pp. 56–61 in: *Proceedings: International Maize Virus Disease Colloquium and Workshop*, August 2–6, 1982, D. T. Gordon, J. K. Knoke, L. R. Nault, and R. M. Ritter eds. Ohio Agricultural Research and Development Center, Wooster, OH.

Markham, P. G., Townsend, R., Plaskitt, K., and Saglio, P. 1977. Transmission of corn stunt to dicotyledonous plants. *Plant Dis. Reptr* 61: 342–345.

Martinez-Lopez, G. 1977. New maize virus diseases in Colombia. pp. 20–29 in: *Proceedings: International Maize Virus Disease Colloquium and Workshop*, August 1976, L. E. Williams, D. T. Gordon, and L. R. Nault eds. Ohio Agricultural Research and Development Center, Wooster, OH.

Nault, L. R. 1980. Maize bushy stunt and corn stunt: A comparison of disease symptoms, pathogen host ranges, and vectors. *Phytopathology* 70: 659–662.

Nault, L. R., and Knoke, J. K. 1981. Maize vectors. pp. 77–84 in: *Virus and Virus-Like Diseases in Maize in the United States.* D. T. Gordon, J. K. Knoke, and G. E. Scott eds. Southern Cooperative Series Bulletin, 247.

Overman, M. A., Ko, N. J., and Tsai, J. H. 1992. Identification of viruses and mycoplasmas in maize by light microscopy. *Plant Dis.* 76: 318–322.

Tsai, J. H. 1987a. Bionomics of *Dalbulus maidis* (DeLong and Wolcott). A vector of mollicutes and virus (Homoptera: Cicadellidae). pp. 209–221 in: *Mycoplasma Diseases of Crops: Basic and Applied Aspects.* K. Maramorosch, and S. P. Raychaudhuri eds. Springer-Verlag, New York.

Tsai, J. H. 1987b. Mycoplasma diseases of corn in Florida. pp. 317– 325 in: *Mycoplasma Diseases of Crops: Basic and Applied Aspects.* K. Maramorosch, and S. P. Raychaudhuri eds. Springer-Verlag, New York.

Tsai, J. H., and Falk, B. W. 1988. Tropical corn pathogens and their associated vectors. pp. 177–201 in: *Advances in Disease Vector Research.* K. F. Harris ed. Springer-Verlag, New York.

Whitcomb, R. F., Chen, T. A., Williamson, D. L., Liao, C., Tully, J. G., et al. 1986. *Spiroplasma kunkelii* sp. nov.: Characterization of the etiological agent of corn stunt disease. *Int. J. Syst. Bacteriol.* 36: 170–178.

Williamson, D. L., and Whitcomb, R. F. 1975. Plant mycoplasma: A cultivable spiroplasma causes corn stunt disease. *Science* 188: 1018–1020.

7.12 BACTERIAL LEAF STRIPE OF MAIZE

Six bacterial plant pathogens have been reported to attack maize foliage in different parts of the world. Important among them are *Acidovorax avenae* subsp. *avenae* (syn. *Pseudomonas rubrilineans*), *Burkholderia andropogonis* (syn. *Psm. andropogonis*), and *Psm. alboprecipitans.*

Ullasa et al. (1967) observed for the first time a severe leaf stripe disease in some inbred lines of maize during 1965 and 1966 *Kharif* seasons at Delhi, India and identified the bacterium as *Xanthomonas rubrilineans.* Later, Dange (1972) while working on this disease found that it is caused by *Psm. rubrilineans*, the pathogen that causes red stripe of sugarcane and has already been reported on sugarcane from India by Chona and Rao (1963). The first natural occurrence of this pathogen on maize was observed by Orian (1957) in Mauritius. The disease has also been reported from Jiangsu Province (Gao et al. 2007) and Guangdong province, China (Ji et al. 2014). Bacterial leaf spot disease of maize due to *Xanthomonas maydis*, reported by Rangaswami et al. (1961) from India, has not been found subsequently and the identity of the pathogen is also in doubt (Dye 1966).

Since 1965, bacterial leaf stripe has been regularly observed at Delhi and Pantnagar, but has never caused significant losses.

The disease has also been seen at Ludhiana (Punjab) and in Rajasthan and Himachal Pradesh and probably occurs in other maize growing areas of India. As the disease being described here causes symptoms mainly on the leaves, it is appropriate to call it **Bacterial Leaf Stripe of Maize** rather than bacterial stripe disease of maize.

7.12.1 Host Range

In addition to maize, sugarcane, and teosinte, the bacterium attacks sorghum, barley, *Pennisetum purpureum*, *Setaria verticillata*, and *Sorghum halepanse* (Ullasa et al. 1967; Dange 1972).

7.12.2 Symptoms

Initially, water-soaked spots appear on both the leaf surfaces, which develop rapidly into long stripes. The stripes are distributed throughout the leaf lamina, but are generally more concentrated toward the leaf base. The length of stripes varies from 1 to 25 cm or may be more, and they may cover the whole length of the leaf. The stripes often coalesce forming elongated broad bands of necrotic tissue. Later, the stripes turn dark brown and papery. On old papery stripes, many saprophytic fungi grow.

Under humid weather, bacterial exudate appears on the stripes, which on drying forms a dark brown crust. In later stages, stripes split and disintegrate giving the leaves a shredded or tattered appearance. Occasionally, narrow, blackish, longitudinal stripes appear on the midribs. Generally leaf sheaths, stalks, and ears are not infected in nature.

7.12.3 Causal Organism

Acidovorax avenae subsp. *avenae* (Manns 1909) Willems et al. 1992
Syn. *Pseudomonas rubrilineans*

It is primarily a pathogen of sugarcane. Therefore, its taxonomy and cultural, morphological, and biochemical characters are given under red stripe of sugarcane.

Maize, sugarcane, and teosinte isolates have the same cultural, morphological, and biochemical characters. They cross infect the hosts of each other on artificial inoculation. A phage isolated from diseased maize tissues typed all the 12 isolates of maize, but did not type sugarcane isolates in spot testing. It caused lysis of teosinte isolates also, but later on resistant colonies appeared in the lysed areas (Dange 1972).

7.12.4 Disease Cycle

The bacterium is internally and/or externally seed-borne. It enters the host through stomata or wounds. Secondary spread of the pathogen occurs mainly through wind-splashed rain. Young plants are more susceptible than older plants.

Dhkal et al. (2016) investigated the potential role of four antioxidant enzymes in imparting resistance against *Aci. avenae* subsp. *avenae* in six inbreds/hybrids of maize, namely, G-5414, PMH-1, CM-139, LM-13, CM-600, and CML-25. Antioxidant enzyme activity was estimated spectrophotometrically from the leaf samples collected at 0, 24, 48, 72, 96, 120, and 144 h after inoculation. A spike in expression of various antioxidant enzymes following pathogen challenge was recorded, with maximum spike after 48 or 72 h post-inoculation in different maize lines. Inbred line CML-25 showed maximum enzyme activity, while G-5414 showed the least. Significantly high values of Pearson's correlation coefficient obtained between antioxidant enzyme activity and lesion length development indicate a possible involvement of these enzymes in imparting disease resistance against bacterial leaf stripe of maize.

Although the maize, sugarcane, and teosinte isolates artificially cross infect the hosts of each other, it appears that in nature they do not cross infect other hosts. In fields, the pathogen may be present on maize, but absent on sugarcane crop growing adjacent to it, suggesting that maize strain does not infect sugarcane in nature.

Inbred lines, namely, CM 104, CM 105, CM 106, CM 109, CM 112, CM 201, CM 300, and Eto group were found resistant on artificial screening, and most of the commercial grown hybrids and composites are not severely infected in nature (Dange 1972).

7.12.5 Management

1. Use of pathogen-free seed is the most effective method of disease control. In the absence of pathogen-free seed, there is an urgent need to devise the methods for the eradication of seed-borne inoculum
2. As the pathogen attacks only the parenchyma of the host tissue, it can be effectively checked with Streptocycline. Thind (1978) obtained 67% control of the disease with two sprays of Streptocycline (100 μg mL^{-1}) given at an interval of 10 days

REFERENCES

Chona, B. L., and Rao, Y. P. 1963. Association of *Pseudomonas rubrilineans* (Lee et al.) Stapp with the red stripe disease of sugarcane in India. *Indian Phytopathol.* 16: 392–393.

Dange, S. R. S. 1972. Studies on Bacterial Stripe Disease of Maize Caused by *Pseudomonas rubrilineans*. PhD thesis, Postgraduate School, Indian Agricultural Research Institute, New Delhi, India.

Dhkal, M., Hunjan, M. S., Kaur, H., and Kaur, R. 2016. Biochemical basis of bacterial leaf streak resistance in maize. *Indian Phytopathol.* 69: 373–380.

Dye, D. W. 1966. A comparative study of some atypical "Xanthomonads." *N. Z. J. Sci.* 9: 843–854.

Gao, Y., Hu, B-S., Wang, F-X., Liu, F-Q., and Xu, Z-G. 2007. Identification of causal agent of maize bacterial leaf stripe. *Jiangsu J.* Agric. Sci. 23: 22–25.

Ji, C., Ou, J., Xu, D., and Pan, R. 2014. First report of corn bacterial leaf stripe caused by *Acidovorax avenae* in Guangdong Province, China. *Plant Dis.* 98: 1424.

Manns, T. F. 1909. *The Blade Blight of Oats: A Bacterial Disease*. Bulletin of the Ohio Agriculture Experiment Station No. 210, pp. 91–167.

Orian, G. 1957. Plant Pathology Division. Rep. Dep. Agric. Mauritius, 1995 pp. 90–93.

Rangaswami, G., Prasad, N. N., and Easwaran, K. S. S. 1961. A new bacterial leaf spot disease of maize. *Madras Agric. J.* 48: 392–393.

Thind, B. S. 1978. Chemical control of bacterial leaf stripe of maize caused by *Pseudomonas rubrilineans*. *Indian J. Plant Prot.* VI: 70–71.

Ullasa, B. A., Mehta, Y. R., Payak, M. M., and Renfro, B. L. 1967. *Xanthomonas rubrilineans* on *Zea mays* in India. *Indian Phytopathol.* 20: 77–78.

Willems, A., Goor, M., Thielemans, S., Gillis, M., Kersters, K., et al. 1992. Transfer of several phytopathogenic *Pseudomonas* species to *Acidovorax* as *Acidovorax avenae* subsp. *avenae* subsp. nov., comb. nov., *Acidovorax avenae* subsp. *citrulli*, *Acidovorax avenae* subsp. *cattleyae*, and *Acidovorax konjaci*. *Int. J. Syst. Bacteriol.* 42: 107–119.

7.13 RED STRIPE OF SUGARCANE

Lyon (1922) was the first to observe red stripe on tip canes in Hawaii. Lee and Martin (1925) established the bacterial nature of the disease. Another kind of symptom/condition called **top-rot of sugarcane** was reported from Queensland by Tryon (1923). Cottrell-Dormer (1932) showed that red stripe and top-rot are two different phases of the same disease, but not two different diseases as considered earlier, and are caused by the same bacterium. Since these early studies, the disease has been reported from most of the sugarcane growing countries of the world for which the readers can refer to chapter XXII of the book entitled, *Diseases of Sugarcane*: *Major Diseases* edited by Ricaud et al. (1989).

In India, the disease was first observed simultaneously by Mc Rae (1933) and Desai (1933). Since then, several reports of severe infections have appeared infecting varieties like Co 419 in Maharashtra (Albuquerque and Arakeri 1956; Bhide et al. 1956; Summanwar and Bhide 1962), Co 449, Co 527, and Fiji B in Tamil Nadu (Rangaswami 1960a, 1960b), and Co 312 in Delhi and eastern districts of Punjab (Chona and Rao 1963).

The disease was of great economic importance in commercial plantings of tip canes and occasionally on other varieties in Hawaii. The losses can occur if susceptible varieties are grown under environmental conditions favorable for development and dissemination of the pathogen. The losses up to 15% or more have been reported due to top-rot phase (Martin and Wismer 1961). The red stripe phase alone generally does not affect yield, and the symptoms usually disappear with the onset of the new growth.

Girard et al. (2014) reported *Acidovorax avenae* subsp. *avenae* causing sugarcane red stripe for the first time from Gabon. It was also the first description of the occurrence of the top-rot form of the disease in R570, a cultivar that is grown in several locations of Africa, the Mascarene Islands, and the French West Indies. Li et al. (2018) reported the wide spread occurrence of the disease from China.

Patro et al. (2006) reported top-rot phase of the disease from Srikakulam district of Andhra Pradesh, India, but the bacterium associated with the disease was identified as *Acinetobacter baumannii*.

7.13.1 Host Range

In addition to sugarcane, the bacterium naturally infects maize (Orian 1957; Ullasa et al. 1967), maize and teosinte (Dange and Payak 1973), and two grasses, namely, *Paspalum nutans* and *P. paniculatum* (Orian 1956). A few more hosts have been shown to be artificially infected. Bhide et al. (1956) reported successful inoculation of sorghum and Summanwar and Bhide (1962) of *Pennisetum typhoides* (Burm. f.) Stapf and Hubb. with red stripe bacterium. Ullasa et al. (1967) could successfully inoculate only sorghum and barley with *Xa. rubrilineans* isolated from maize, but not several other graminaceous hosts. Dange (1972) reported that maize isolates infected *bajra*, barley, sorghum, elephant grass (*Pennisetum purpureum*), and *Setaria verticillata*. Martin and Wismer (1961), who could infect only several varieties of *Sorghum bicolor* out of many wild and cultivated grasses, stated that the common field grasses were of little or no importance as a source of inoculum for sugarcane crop.

7.13.2 Symptoms

There are two types/phases of symptoms of red stripe, namely, leaf stripe and top-rot. These may occur singly or simultaneously and under field conditions both are favored by high relative atmospheric humidity.

7.13.2.1 Leaf Stripe

In the beginning, the symptoms appear as watery green stripes on the leaf lamina. They usually appear in the middle of the leaf and near the midrib, but sometimes they may concentrate near the leaf base. The stripes spread rapidly upward and downward and develop a reddish color, which later on changes to dark red, hence the name red stripe. They are straight and uniform in width and remain confined to the parenchymatous tissue in the early stages of infection. Two or more stripes often coalesce to form broad bands of diseased leaf tissue. The width of the stripes varies from 0.5 to 4.0 mm and the length from a few centimeters to the entire length of the leaf lamina (Figure 7.9a). However, in some varieties, the length of the stripes may extend to leaf sheath also. The stripes may also appear on the lower surface of the midrib. During the warm moist weather, the bacterial ooze comes out through the stomata on the lower leaf surface of the affected tissue, which often dries to form whitish flakes.

The disease appears mostly on the young and middle-aged leaves and may attack the partially unrolled youngest leaves. In young cane, the entire stool may be killed, while in older

(a)

(b)

FIGURE 7.9 Red stripe of sugarcane: (a) Red stripes on leaf. (b) Top rot phase.

cane, the individual stalks die. Usually the young ratoons are more susceptible to infection than plant cane of the same age.

7.13.2.2 Top-Rot

As the name indicates, it is the rotting of the top portion of the plant. The losses from this phase are much higher than those from leaf stripes. The yellowing and wilting of older leaves occurs and the affected plants may show the typical reddish leaf stripes. Top-rot may result from bud or stem infection without showing leaf symptoms as well as from infection of the leaves. The reddish discoloration on the outside and reddish splashes on the inner surfaces often appear on the leaf sheaths attached to the affected internodes. The sunken areas, initially water-soaked in appearance and later turning reddish-brown, often appear on the affected internodes (Figure 7.9b). As the rotting advances, large cavities are formed inside the internodes. In advanced stage, the leaf spindle is easily pulled out of enclosing sheaths.

A characteristic strong unpleasant odor, an important diagnostic feature of the disease, is emitted by the rotted spindles. Sometimes, the odor is so strong that it may be felt from a distance. Stalks exhibiting top-rot are stunted in growth and usually die; their tops often break and fall down. Sometimes, the uppermost healthy buds of top-rot affected stalks give rise to side shoots, and the leaves of these shoots may exhibit stripe symptoms.

A new disease called **false red stripe**, different from the above-described disease of sugarcane, has been reported from central and southern areas of Brazil (Mantovani et al. 2006). It does not occur in north and northeastern areas of Brazil or any other country. The causal bacterium was identified as a species of *Xanthomonas*. Artificial inoculation, by injecting and spraying of bacterial cell suspension, of 31 host species comprising crops, grasses, and weeds revealed that only sorghum, maize, and oats expressed disease symptoms 15 days after inoculation.

Occasionally, the leaf symptoms of red stripe and those of gumming (*Xa. axonopodis* pv. *vasculorum*) and mottled stripe (*Pseudomonas rubrisubalbicans*) can be confused with each other. Gumming can be distinguished due to its some specific symptoms like formation of gum pockets. There is no top-rot in mottled stripe and in general its symptoms are paler in color.

7.13.3 Causal Organism

Acidovorax avenae subsp. *avenae* (Manns 1909) Willems et al. 1992

Syn. *Pseudomonas avenae* subsp. *avenae*, *Psm. rubrilineans*

The taxonomy of bacterium causing red stripe has undergone many changes. It was first classified as *Phytomonas rubrilineans* by Lee et al. (1925). It was renamed as *Pseudomonas rubrilineans* by Stapp (1928) and as *Bacterium rubrilineans* by Elliott (1930). Following the establishment of genus *Xanthomonas* by Dowson in 1939, Starr and Burkholder (1942) transferred the bacterium to this genus as *Xa. rubrilineans*. However, they pointed out that as far as lipolytic activity is concerned, red stripe pathogen does not agree with a typical *Xanthomonas*. Wernham (1948) also included it in doubtful species of *Xanthomonas*. In spite of these doubts, Breed et al. (1957) listed it as a species of *Xanthomonas* in the 7th edition of *Bergey's Manual of Determinative Bacteriology*.

After a detailed systematic study of bacteria causing sugarcane diseases, Hayward (1962) found that red stripe pathogen has many characteristics of genus *Pseudomonas* and hence proposed that the name, *Pseudomonas rubrilineans* (Lee et al.) Stapp should replace *Xa. rubrilineans* (Lee et al.) Starr and Burkholder. Dye (1963) and Bradbury (1967) also supported the Hayward's proposal. Dange (1972) reported that the red stripe of sugarcane and bacterial stripe disease of maize were caused by the same pathogen. He further reported that earlier reports from India identifying *Xa. rubrilineans* as the causal agent of red stripe of sugarcane (Desai 1933; Albuquerque and Arakeri 1956; Bhide et al. 1956; Rangaswami 1960a, 1960b; Summanwar and Bhide 1962; Chona and Rao 1963) and of bacterial stripe disease of maize (Ullasa et al. 1967) were probably erroneous. Willems et al. (1992), on the basis of DNA-DNA hybridization and other studies, grouped *Psm. avenae*, *Psm. rubrilineans* and *Psm. setariae* into *Acidovorax avenae* subsp. *avenae* subsp. nov., comb. nov. The below given description of *Psm. rubrilineans* (recent name, *Aci. avenae* subsp. *avenae*) is taken from Hayward (1962).

Cells are straight rods and motile by one polar flagellum. Poly-β-hydroxybutyrate is accumulated. Colonies are non-mucoid on 2% glucose peptone agar. No pigments are produced. Oxidase test is positive and H_2S is produced. Nitrates are reduced to nitrites, but denitrification is negative. Tween 80 is hydrolyzed, starch hydrolysis is negative, and gelatin liquefaction is weak. Capable of growth at 40°C. Acid is produced from glucose, fructose, galactose, arabinose, mannitol, glycerol, and sorbitol. No acid is produced from sucrose, raffinose, salicin, lactose, maltose, cellobiose, and *m*-inositol.

Fontana et al. (2013), using randomly amplified polymorphic DNA (RAPD)-PCR, detected the presence of at least four different biotypes among the analyzed isolates of *Aci. avenae* subsp. *avenae* collected from sugarcane producing areas of Tucumán and Salta (northwest of Argentina).

7.13.4 Disease Cycle

As the sugarcane crop is of 10 months duration or is taken as a ratoon crop, undecomposed plant debris or the infected crop may serve as primary source of inoculum. In Java, Bolle (1929) showed that the bacterium isolated from old withered leaves showing red stripes was virulent after 4 months. Cottrell-Dormer (1932) was able to isolate the pathogen from leaf stripe infected plant debris kept in a dry cardboard box for 7 months.

The bacterium is disseminated mainly through wind-blown rain. After multiplying in large numbers in parenchyma of leaves, the bacteria ooze out on the surface of leaf lesions in moist warm weather. Then with the help of wind and rain, these bacteria spread from plant to plant, from one field to another, or fall down on the lower plant parts or may run down the leaves to cause the stem infection. The bacterium is rarely disseminated through cane knives and mechanical equipment.

The bacterium enters the host through stomata and wounds. The leaf wounds are caused due to the scratches made by the marginal spines of the leaves through the action of the wind. The bacterium enters through the stomata present on both the leaf surfaces, but the symptoms from natural infection usually appear first on the lower leaf surfaces. Barnum and Martin (1925) found that the bacterium artificially infected all the plant parts, but the youngest leaves and the youngest internodes showed more susceptibility. Vigorously growing plants were also much more susceptible than the less vigorous plants. Pal et al. (1996) showed that out of four methods tried for screening of sugarcane cultivars, whorl inoculation method was found to be most effective as it gave highest disease incidence and also induced both the phases of the disease.

Johnson et al. (2016) found a negative correlation between yield components and red stripe incidence, with the strongest relationship between sucrose per metric ton and disease incidence. The disease incidence was positively correlated with soil nutrition, i.e., phosphorus, potassium, zinc, and calcium. The incidence also increased with increasing nitrogen rate showing maximum effect in heavy soils. The data also indicated that the use of infected canes as a seed source can significantly decrease shoot emergence, stalk population, and subsequent cane and sugar yields.

7.13.5 Management

The use of resistant varieties is the most effective and economical measure for managing the disease. The replacement of susceptible varieties with resistant varieties has relegated red stripe to be of little economic importance. Before release, the varieties must be screened using artificial inoculation techniques. In North Queensland, top-rot phase of the disease was markedly reduced by altering the planting date. Autumn (March–April) sown crop showed much less disease incidence compared to spring (August–September) sown crop.

Quarantine measures are helpful in restricting the movement of the pathogen to disease-free areas.

REFERENCES

Albuquerque, M. J., and Arakeri, H. R. 1956. Sugarcane red stripe disease on Co 419. *Indian Sugar* 6: 323–324.

Barnum, C. C., and Martin, J. P. 1925. The susceptibility of roots, stalks, leaf sheath and leaf blades to red-stripe disease, and the relationship of maturity of tissues to increasing resistance to red stripe. Red-Stripe Disease Studies. Hawaii. Sugar Plant. Assoc. Exp. Stn., Pathol. Dept., pp. 35–48.

Bhide, V. P., Hedge, P. K., and Desai, M. K. 1956. Bacterial red-stripe disease of sugarcane caused by *Xanthomonas rubrilineans* in Bombay state. *Curr. Sci*. 25: 330.

Bolle, P. C. 1929. De roodestrepenziekte. *Arch. Suikerind*. 37: 1147–1218.

Bradbury, J. F. 1967. *Pseudomonas rubrilineans. C.M.I. Description of Pathogenic Fungi and Bacteria*. Set 13, No. 127.

Breed, R. S., Murray, E. G. D., and Smith, N. R. eds. 1957. *Bergey's Manual of Determinative Bacteriology*. 7th ed. Williams and Wilkins Co., Baltimore, MD, 1094 pp.

Chona, B. L., and Rao, Y. P. 1963. Association of *Pseudomonas rubrilineans* (Lee et al.) Stapp with the red stripe disease of sugarcane in India. *Indian Phytopathol*. 16: 392–393.

Cottrell-Dormer, W. 1932. Red-Stripe disease of sugar-cane in Queensland. *Bull. Bur. Sugar Exp. Stn. Div. Pathol*. 3: 25–59.

Dange, S. R. S. 1972. Studies on bacterial stripe disease of maize caused by *Pseudomonas rubrilineans*. PhD thesis, Postgraduate School, Indian Agricultural Research Institute, New Delhi, India.

Dange, S. R. S., and Payak, M. M. 1973. The taxonomic position of the pathogen of red stripe of sugarcane. *Sugarcane Pathol. Newsl*. 10: 25–28.

Desai, S. V. 1933. Occurrence of the red stripe disease of sugarcane in India. (Abstr.). *Proceedings of the 20th Indian Science Congress*, Patna, India, pp. 61–62.

Dye, D. W. 1963. Comparative study of the biochemical reactions of additional *Xanthomonas* spp. *N. Z. J. Sci*. 6: 483–486.

Elliott, C. 1930. *Manual of Bacterial Plant Pathogens*. Bailliere, Tindall and Cox, London, UK, 349 pp.

Fontana, P. D., Rago, A. M., Fontana, C. A., Vignolo, G. M., Cocconcelli, P. S., et al. 2013. Isolation and genetic characterization of *Acidovorax avenae* from red stripe infected sugarcane in Northwestern Argentina. *Eur. J. Plant Pathol*. 137: 525–534.

Girard, J.-C., Noëll, J., Larbre, F., Roumagnac, P., and Rott, P. 2014. First report of *Acidovorax avenae* subsp. *avenae* causing sugarcane red stripe in Gabon. *Plant Dis*. 98: 684.

Hayward, A. C. 1962. Studies on bacterial pathogens of sugarcane. II. Differentiation, taxonomy and nomenclature of the bacteria causing red stripe and mottled stripe diseases. *Mauritius Sugar Ind. Res. Inst. Occas. Pap*. 13: 13–27.

Johnson, R. M., Grisham, M. P., Warnke, K. Z., and Maggio, J. R. 2016. Relationship of soil properties and sugarcane yields to red stripe in Louisiana. *Phytopathology* 106: 737–744.

Lee, H. A., and Martin, J. P. 1925. The cause of red-stripe disease of sugarcane. Red-Stripe Disease Studies. Hawaiian Sugar Planters' Association, Experiment Station, Pathology Department, pp. 83–93.

Lee, H. A., Purdy, H. A., Barnum, C. C., and Martin, J. P. 1925. A comparison of red-stripe disease with bacterial diseases of sugar cane and other grasses. Red-Stripe Disease Studies. Hawaiian Sugar Planters' Association, Experiment Station, Pathology Department, pp. 64–74.

Li, X.-Y., Sun, H.-D., Rott, P. C., Wang, J.-D., Huang, M.-T., et al. 2018. Molecular identification and prevalence of *Acidovorax avenae* subsp. *avenae* causing red stripe of sugarcane in China. *Plant Pathol.* 67: 929–937.

Lyon, H. L. 1922. A leaf disease of the Tip canes. *Proc. Hawaii. Sugar Plant Assoc.* 246.

Manns, T. F. 1909. *The Blade Blight of Oats: A Bacterial Disease*. Bulletin of the Ohio Agriculture Experiment Station No. 210, pp. 91–167.

Mantovani, E. S., Marini, D. C., and Giglioi, E. A. 2006. Host range of *Xanthomonas* sp., casual agent of the false red stripe of sugarcane, among grasses. *Summa Phytopathol.* 32: 124–130.

Martin, J. P., and Wismer, C. A. 1961. Red stripe. pp. 109–126 in: *Sugar-Cane Diseases of the World*. Vol. I, J. P. Martin, E. V. Abbott, and C. G. Hughes eds. Elsevier, Amsterdam, the Netherlands.

Mc Rae, W. 1933. New diseases reported during the year 1932. *Int. Bull. Pl. Prot.* 7: 79–80.

Orian, G. 1956. Occurrence of a disease similar to red stripe of sugar cane in Mauritius. *Proc. Int. Soc. Sugar Cane Technol.* 9: 1042–1048.

Orian, G. 1957. Plant Pathology Division. Rep. Dep. Agric. Mauritius, 1955, pp. 90–93.

Pal, V., Kumar, Anil, and Vir, S. 1996. Evaluation of inoculation methods for red stripe induction in sugarcane. pp. 430–433 in: *Proceedings of the 9th International Conference on Plant Pathogenic Bacteria*. A. Mahadevan ed. Chennai, India.

Patro, T. S. S. K., Nageswara Rao, G. V., and Gopalakrishnan, J. 2006. Association of *Acinetobacter baumannii* with a top rot phase of sugarcane red stripe disease in India. *Indian Phytopathol.* 59: 501–502.

Rangaswami, G. 1960a. Studies on two bacterial diseases of sugarcane. *Curr. Sci.* 29: 318–319.

Rangaswami, G. 1960b. Further studies on bacterial gummosis and red stripe disease of sugarcane. *J. Annamalai Univ.* 22: 135–150.

Ricaud, C., Egan, B. T., Gillaspie A. G. Jr., and Hughes, C. G. eds. 1989. *Diseases of Sugarcane: Major Diseases*. Elsevier, Amsterdam, the Netherlands, 399 pp.

Stapp, C. 1928. In Sorauer's Handb. *Pflanzenkr.* 2(5): 35.

Starr, M. P., and Burkholder, W. H. 1942. Lipolytic activity of phytopathogenic bacteria determined by means of spirit blue agar and its taxonomic significance. *Phtopathology* 32: 598–604.

Summanwar, A. S., and Bhide, V. P. 1962. Bacterial red stripe disease of sugarcane (*Saccharum officinarum*) caused by *Xanthomonas rubrilineans* var. *indicus* in Maharashtra. *Indian J. Sugarcane Res. Dev.* 6: 65–68.

Tryon, H. 1923. Top rot of the sugar cane. *Bull. Bur. Sugar Exp. Stn., Div. Pathol.* 1.

Ullasa, B. A., Mehta, Y. R., Payak, M. M., and Renfro, B. L. 1967. *Xanthomonas rubrilineans* on *Zea mays* in India. *Indian Phytopathol.* 20: 77–78.

Wernham, C. C. 1948. The species value of pathogenicity in the genus *Xanthomonas*. *Phytopathology* 38: 283–291.

Willems, A., Goor, M., Thielemans, S., Gillis, M., Kersters, K., et al. 1992. Transfer of several phytopathogenic *Pseudomonas* species to *Acidovorax* as *Acidovorax avenae* subsp. *avenae* subsp. nov., comb. nov., *Acidovorax avenae* subsp. *citrulli*, *Acidovorax avenae* subsp. *cattleyae*, and *Acidovorax konjaci*. *Int. J. Syst. Bacteriol.* 42: 107–119.

7.14 RATOON STUNTING DISEASE OF SUGARCANE

Ratoon stunting disease was first observed by Steindl in Queensland, Australia in the summer of 1944–1945 on ratoon crops of sugarcane cultivar Q 28. He also coined the term "ratoon stunting" as the ratoon crops were more severely infected. Due to the seed-borne nature of the pathogen, the disease continued to spread from one country to another country along with the movement of the germplasm, and now it is present in most of the cane growing countries of the world. Fifty-two countries from where the disease has been reported are given in the list compiled in 1986 by the International Society of Sugar Cane Technologists (ISSCT's) Standing Committee on Sugarcane Diseases (Ricaud et al. 1989).

In a review article, Young (2016) traced the origin of ratoon stunting disease to the release of modern commercial hybrids and provided evidence that *Saccharum officinarum*, the major progenitor of modern sugarcane cultivars, is not the natural host of *Leifsonia xyli* subsp. *xyli*. Rather, it is proposed that the wild relative, *S. spontaneum*, is more likely to be the original host, and that *L. xyli* subsp. *xyli* was acquired during interspecific hybridization work undertaken in Java during the 1920s.

In India, the disease was first observed by Prof. S.J.P. Chilton in 1956 in CoS 510 from Golagokarnath, U.P. (Chona 1956). However, the presence of disease was reportedly questioned as most of the experiments conducted subsequently failed to confirm the presence of the disease. Later on, Singh (1966) confirmed the presence of the disease in Co 290 obtained from Jaora, Madhya Pradesh.

The name ratoon stunting disease (hereafter called RSD) is a misnomer as the disease retards the growth of the plant crop as well and causes considerable losses. The cumulative losses due to RSD are generally more than the losses caused by any other disease of sugarcane. The losses caused by the disease are influenced by many factors such as susceptibility of the variety grown, weather conditions, amount of irrigation water available, and whether the crop is freshly planted or a ratoon crop. Overall losses ranging from 5% to 10% have been reported by Hughes (1974). The losses up to 30% or more have also been reported in some years, and in highly susceptible cultivars, these may go as high as 50%. Urashima and Marchetti (2013) reported disease incidence (based on dot blot enzyme immunoassay of sap) of 23.6%, 27.1%, and 25.8% in 2009, 2010, and 2011, respectively, from Brazil. RB867515, the major cultivar in São Paulo, had field incidence up to 70% in 2009, 48% in 2010, and 88% in 2011.

In spite of the reduction in cane yield and sugar per unit area, the sugar content of diseased canes may be increased. This results from the reduced growth and lower water content of the canes. This is the only disease of sugarcane in which the sugar content of the diseased canes increases.

7.14.1 Host Range

No natural hosts other than sugarcane have been reported. Gillaspie and Teakle (1989) have listed 14 hosts, which were artificially infected by the bacterium and from these hosts the bacterium was successfully reinoculated to the sugarcane. These hosts included maize, sorghum, sweet Sudan grass, and many other grasses commonly found in sugarcane fields.

7.14.2 Symptoms

There are two types of symptoms, internal and external symptoms.

7.14.2.1 External Symptoms

The stunting and unthrifty growth of the plants are the only external symptoms of the disease, which can occur not only due to RSD, but due to several other factors like nutritional deficiency, lack of moisture, poor agronomic practices, etc. The stunting and unthriftiness due to RSD are generally more pronounced when the crop grows in reduced soil moisture. The slow growth of the infected crop accompanied with production of small and thinner canes causes reduction in yield in spite of total numbers of canes remaining unchanged. However, in a draught year, the total number of stalks may also be reduced along with the variation in the length of individual canes in a stool. The stunting is not uniform in all the plants even if all the stools in a field are infected. There is a characteristic **up and down appearance**, which means that all the plants are not of uniform height.

The varieties differ in their response to infection and show different degrees of stunting. Ratoon crops usually show more disease severity than plant crops. The losses in clone Q 28 ranged from 12% to 37% in plant crops, while these usually exceeded 60% in ratoon crops. The roots and underground parts of the diseased canes do not show any abnormality, but there is general reduction in root mass.

7.14.2.2 Internal Symptoms

In contrast to generalized external symptoms, there are well-defined and characteristic internal symptoms of the disease. The discoloration of individual vascular bundles in fully differentiated nodes of mature stalks is generally observed in diseased canes. The discoloration occurs in lower part of the node just below the region of leaf sheath attachment and is confined to the nodes.

In the longitudinally splitted diseased stalks, the discoloration appears as dots, commas, or short straight or bent lines of 2–3 mm length. The discoloration varies in type and intensity with severity of disease and the variety involved. Some varieties do not show any discoloration and in those which show discoloration, it may vary in the same variety form time to time. The discoloration may be of yellow, orange, pink-red, and reddish-brown shade (Figure 7.10).

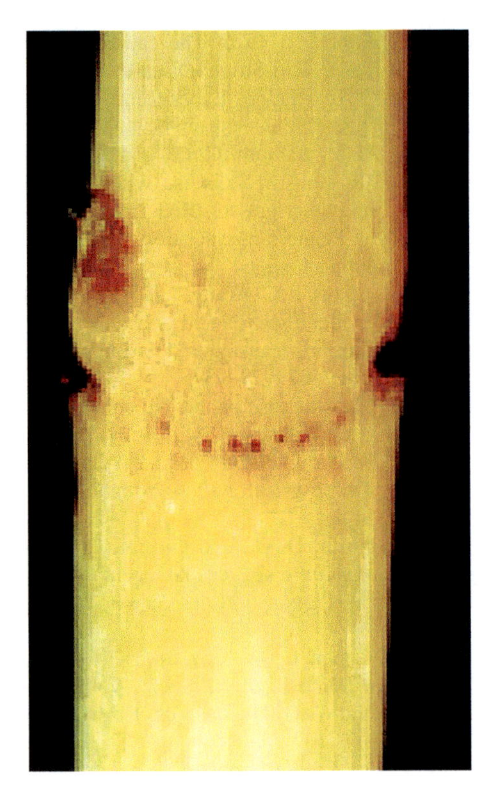

FIGURE 7.10 Ratoon stunting disease of sugarcane: Internal stalk symptoms.

The discoloration is due to the plugging of many large xylem vessels with a colored gummy substance. For a reliable diagnosis, all the nodes in the fully developed stalks should show discoloration. Only the undamaged stalks should be examined for the diagnosis of RSD as the damage to the internal tissues of the sugarcane due to mechanical/pest injuries leads to discoloration of tissues. The diseased stalks should be examined immediately after splitting because the symptoms become inconspicuous as the tissues darken or dry. The nodes of young canes infected with RSD also show pinkish color below the apical meristem, which may extend downward about a centimeter. These symptoms are not as reliable as the mature node symptoms for diagnosing the disease, but they are more helpful in detecting the disease at an early stage in some varieties.

The vascular discoloration also occurs in leaf scald and gumming disease of sugarcane caused by *Xanthomonas albilineans* and *Xa. axonopodis* pv. *vasculorum*, respectively. In these diseases, the intensity of discoloration is more and the discoloration also extends from nodes to internodes in comparison to RSD. Moreover in gumming disease, the formation of gum pockets and exudation of gum form cut ends of the canes are other distinguishing features.

Viswanathan (1996) found that internal/nodal symptoms were often unreliable for diagnosing the disease as they were not produced in all the susceptible varieties and they also varied in varieties in which they were produced. He also found that indirect ELISA and dot blot techniques were more suitable for the detection of the disease, the former being more sensitive than the latter. Duttamajumder (1996) reported that many workers have questioned the reliability of nodal symptoms for diagnosing the disease and at present it is taken as an indication, but not the confirmation of the disease. He further reported that Co 975, a genotype universally susceptible to RSD with 100% nodal symptoms did not yield any RSD bacterium in extracts taken from most of the samples. It shows that Co 975 is not susceptible to the disease and may be a false positive case for RSD. He also tested 25 commercially grown genotypes for the presence of RSD bacterium using dark field, phase contrast, and fluorescence microscopy and found that no genotype was completely free from the bacterium. Dark field microscopy was superior to phase contrast microscopy. The other techniques commonly used to detect RSD bacterium in the infected plants include tissue-blot enzyme immunological assay and fluorescent antibody staining (FAS) and among these techniques, including the abovementioned ones, FAS is considered more effective than others.

In nature, RSD is often found in association with mosaic. Simultaneous occurrence of both the diseases has been more common in certain cultivars. The combined infection of both the diseases causes marked reduction in germination and yield.

7.14.3 Causal Organism

Leifsonia xyli subsp. xy*li* (Davis et al. 1984) Evtushenko et al. 2000
Syn. *Clavibacter xyli* subsp. *xyli*

For nearly about three decades, the causal agent of RSD was believed to be a virus because of its mode of transmission and the failure of the workers to either observe or isolate the pathogen. Gillaspie et al. (1973), Teakle et al. (1973), and Maramorosh et al. (1973) independently reported the association of a microorganism with the disease. Davis et al. (1980) isolated the bacterium and proved its pathogenicity. Colonies on SC medium after 2 weeks of incubation at 28°C are circular with entire margin, non-pigmented, convex, and measure 0.1–0.3 mm in diameter. Bacterial cells measure 0.25–0.35 × 1.0–4.0 μm and are straight or slightly curved rods, some cells swollen in the middle or at the tip. Variations observed in cell size have been attributed to different methods of sample preparation. Cells are non-motile. Gram-stain-positive, non-sporing, aerobic, not acid-fast, oxidase negative, and catalase positive. The cell walls contain fucose and rhamnose as the major sugars and 2, 4-diaminobutyric acid, glutamic acid, glycine and alanine as the major amino acids. All the coryneform plant pathogenic bacteria that contain 2, 4-diaminobutyric acid were placed in the genus, *Clavibacter*. The guanine + cytosine content of the bacterium is 66 mol%, and the whole-cell fatty acid extracts contain 17%–24% 15:0-anteiso, 5%–14% 16:0-iso, and 62%–72% 17:0-anteiso acids. Evtushenko et al. (2000) transferred *Cla. xyli* subsp. *xyli* to a newly created genus *Leifsonia* as *Leifsonia xyli* subsp. *xyli*.

7.14.4 Disease Cycle

The bacterium perpetuates in the diseased canes and infected stubbles and the latter give rise to infected ratoon crop. There is no evidence of its survival in the true seed or soil.

As the bacterium can be readily inoculated mechanically to the healthy plants through the sap of diseased canes, cutting knives/cutter-planter machines and harvesting equipment play an important role in the dissemination of the bacterium. Hughes and Steindl (1955) found that in one case, the cutter machine spread the bacterium from one diseased cane of clone Q 28 to 60 consecutive plants and thereafter, to some scattered plants in Queensland. In Louisiana, the dissemination by harvester equipment increased the infected plants in the plant crop from 16% to 47% and then in the subsequent ratoon crops from 50% to 80%–90% (Steib et al. 1957). The exchange of germplasm from one country to another country has also played a major role in the spread of disease. The field rats, dogs, and foxes play a limited role in spreading the disease.

There seems to be no possibility of the spread of the pathogen through wind-splashed rain as the susceptible varieties grown adjacent to the infected cane fields remained free from infection. In the varietal evaluation trials also, the rows of healthy canes planted adjacent to the rows of inoculated canes, remained free from infection for 3 years when proper sanitation measures were observed (Hughes and Steindl 1955).

The bacterium penetrates through wounds and then it enters the xylem vessels. After penetration, it colonizes the xylem vessels of roots, stalks, and leaves. It seems that the bacterium generally multiplies and spreads in dysfunctional vessels. The enzymatic activity of the bacterium dissolves pit membranes to facilitate its movement through xylem vessels. The extent and nature of anatomical barriers, which restrict the movement of the bacterium in the host, are considered mainly responsible for the susceptible/resistant reaction of the cultivars. The susceptibility of the host is influenced more by the accessibility of vascular tissue to the pathogen than the ability of the pathogen to multiply within the host.

Quecine et al. (2016) studied the colonization of sugarcane tissues using a mutant strain of *L. xyli* subsp. *xyli* that stably expressed the *gfp* gene in sugarcane tissues. In the leaves, *L. xyli* subsp. *xyli*-tagged cells were found in a new niche, i.e., in the mesophyll and in the bundle sheath cells surrounding the vascular system, in addition to, within the xylem vessels as reported earlier. This finding indicates that the bacterium is able to move from the xylem to the parenchyma of the leaf cells and the colonization of sugarcane by this pathogen is not limited to the xylem vessels only as commonly reported.

Zhu et al. (2018) inoculated cuttings of two sugarcane cultivars, Badila and GT11, with *L. xyli* subsp. *xyli* to investigate the effects of infection on their growth, physiological characteristics, and associated gene differential expression. The bacterial population gradually increased with plant growth after inoculation. Plant height, stalk diameter, single stalk weight, and water potential decreased as a result of infection, while membrane permeability and amino acid content increased in infected plants compared to the control. The expression of *PAL*, *ZFP*, and *NBS-LRR* genes also increased in plants subjected to ratoon stunt stress, indicating that these genes are involved in sugarcane responses to *L. xyli* subsp. *xyli* infection.

Farahani et al. (2015) compared conventional PCR, nested PCR, and real-time PCR for the detection of *L. xyli* subsp. *xyli* from infected sugarcane plants and reported that conventional PCR lacked considerable sensitivity, nested PCR had acceptable sensitivity, but real-time PCR was the most efficient technique to be used as a diagnostic tool for the disease.

Young et al. (2016) compared four diagnostic techniques, namely, quantitative PCR on pooled leaf sheath biopsies (LSB-qPCR), conventional PCR on the same templates (LSB-PCR), evaporative-binding enzyme immunoassay (EB-EIA) coupled with phase contrast microscopy (PCM) on expressed xylem sap from the same fields, and conventional PCR on the same xylem sap samples for the detection of causal agent from 100 fields of sugarcane (*Saccharum* interspecific hybrids) of unknown infection status. LSB-qPCR and LSB-PCR detected the causal agent in 27 and 18 fields, respectively, whereas, from samples of expressed xylem sap from the same fields, conventional PCR identified 12 infections and EB-EIA/PCM detected the bacterium from 3 fields. The sensitivities of qPCR and PCR, determined from plate counts of a dilution series, were approximately 10^3 and 10^4 cfu mL^{-1}, respectively. Further tests conducted on 139 LSB samples from across the Australian industry, qPCR and PCR diagnosed the disease in 31 and 25 fields, respectively. Using qPCR and PCR on LSB samples, the disease was diagnosed in a range of cultivars throughout the year, and these methods could detect the pathogen in 3 months to more than 1 year old sugarcane plants.

7.14.5 Management

1. The use of pathogen-free seed and prevention of the spread of the pathogen from the diseased plants to the healthy plants are the most effective control measures. The selection of apparently healthy, thick canes for seed greatly reduces the disease incidence. However, the selection of healthy canes is hampered due to the absence of clearly defined external symptoms of the disease
2. In the absence of pathogen-free seed, the diseased canes/setts should be treated with heat therapy to eliminate the seed-borne inoculum. Four methods that employ heat therapy for treating sugarcane setts are given below

   Hot water treatment: This method is used globally. In India, the hot water treatment is normally carried out at 50°C for ½ hour. The same is done at 50°C for 3 h in Queensland (Australia), Florida, and Hawaii (USA). The seed setts of mature stalks are usually less affected and give better germination after hot water treatment than those from immature or senescing stalks.

   Hot air treatment: In India, it is used to a lesser extent than hot water treatment. It is carried out at 54°C for 8 h. In this treatment, the bud damage is less than that of hot water treatment. It is more suitable for soft, succulent, and immature canes, which cannot withstand hot water treatment. After the treatment, dip the seed cane in water at ambient temperatures to reduce the loss in germination.

   Moist-hot-air treatment: The method was developed in India to check the loss of moisture in hot air treatment. Within 1 h, the required temperature of 54°C is obtained, and it is maintained for 7 h in a sealed chamber (Shukla et al. 1974).

   Aerated-steam method: It is the most recent method in which an aerated-steam, a mixture of air and steam is used. In India, the recommended temperature is 54°C for 4 h, while in the US the treatment is done at 51°C for 4 h. The smooth and uniform distribution of heat in the treatment chamber makes this method more effective. It is also more economical than the other methods. In addition to RSD, the treatment is also effective for grassy shoot, leaf scald, and red rot of sugarcane.

   The major drawbacks with heat therapy are reduction in germination, high cost of equipment, and failure to eradicate the pathogen completely. The percentage of disease-free setts never exceeds more than 90–95, especially in hot water treatment. In the remaining 5%–10% setts, the number of bacteria goes so low, which is difficult to detect with the commonly used methods. Hence, there is an urgent need to develop the more sensitive and accurate methods to detect this low population of the bacterium.
3. Sanitation including disinfection of cutting knives, cultivators, and harvester blades is very helpful in preventing the spread of the pathogen. The commonly used disinfectants include 5%–15% Lysol (neutralized cresylic acid), 1% Dettol, 50% ethanol, or 10% formalin. A minimum contact of 5 min of the disinfectant with the cutting surface is essential for proper disinfection
4. Strict quarantine measures should be followed to check the spread of the pathogen to the disease-free areas
5. Give proper irrigation to the crop to minimize the losses
6. Use of resistant varieties is the most economical method of disease control. Though most commercial sugarcane clones are susceptible to the pathogen, substantial resistance has been found in clones such as CP 29-116, Q 50, and H 60-6909

The Australian sugar industry has never pursued genetic resistance to RSD, despite it being widely considered to be one of the most important diseases of sugarcane. Young (2018), in his review, traces the factors that have led to an industry stance that is apparently without any scientific justification, and which has tended to downplay the significance of RSD on Australian sugarcane productivity, and thus has led to a significant loss in production. Based on the available information, the author calls on the Australian sugar industry to prioritize selection for RSD resistance in the plant improvement program.

REFERENCES

Chona, B. L. 1956. Presidential address for the pathology section. *Proceedings of the International Society of Sugar Cane Technologists 9th Congress*, pp. 975–980.

Davis, M. J., Gillaspie, A. G. Jr., Harris, R. W., and Lawson, R. H. 1980. Ratoon stunting disease of sugarcane: Isolation of the causal bacterium. *Science* 240: 1365–1367.

Davis, M. J., Gillaspie, A. G., Vidaver, A. K., and Harris, R.W. 1984. *Clavibacter*: A new genus containing some phytopathogenic coryneform bacteria, including *Clavibacter xyli* subsp. *xyli* sp. nov., subsp. nov. and *Clavibacter xyli* subsp. *cynodontis* subsp. nov., pathogens that cause ratoon stunting disease of sugarcane and Bermudagrass stunting disease. *Int. J. Syst. Bacteriol.* 34: 107–117.

Duttamajumder, S. K. 1996. Ratoon stunting disease in subtropical India: Prevalence and control. pp. 333–341 in: *Proceedings of the 9th International Conference on Plant Pathogenic Bacteria*. A. Mahadevan ed. Chennai, India.

Evtushenko, L. I., Dorofeeva, L. V., Subbotin, S. A., Cole, J. R., and Tiedje, J. M. 2000. *Leifsonia poae* gen. nov., sp. nov., isolated from nematode galls on *Poa annua* and reclassification of *Corynebacterium aquaticum* Leifson 1962 as *Leifsonia aquatica* (ex Leifson 1962) gen. nov., nom. rev., comb. nov. and *Clavibacter xyli* Davis et al. 1984 with two subspecies as *Leifsonia xyli* (Davis et al. 1984) gen. nov., comb. nov. *Int. J. Syst. Evol. Microbiol.* 50: 371–380.

Farahani, A. S., Taghavi, S. M., and Taher-Khani, K. 2015. Comparison of conventional, nested and real-time PCR for detection of the causal agent of ratoon stunt in Iran. *J. Plant Pathol.* 97: 259–263.

Gillaspie, A. G. Jr., Davis, R. E., and Worley, J. F. 1973. Diagnosis of ratoon stunting disease based on the presence of a specific microorganism. *Plant Dis. Reptr.* 57: 987–990.

Gillaspie, A. G. Jr., and Teakle, D. S. 1989. Ratoon stunting disease. pp. 59–80 in: *Diseases of Sugarcane: Major Diseases*. C. Ricaud, B. T. Egan, A. G. Gillaspie, Jr., and Hughes, C. G. eds. Elsevier, Amsterdam, the Netherlands.

Hughes, C. G. 1974. The economic importance of ratoon stunting disease. *Proc. Int. Soc. Sugar Cane Technol.* 15: 213–217.

Hughes, C. G., and Steindl, D. R. L. 1955. Ratoon stunting disease of sugarcane. *Queensl. Bur. Sugar Exp. Stn. Tech. Comm.* No. 2, 54 pp.

Maramorosh, K., Plavsic-Banjac, B., Bird, J., and Liu, L. J. 1973. Electron microscopy of ratoon stunted sugarcane: Microorganisms in xylem. *Phytopathol. Z.* 77: 270–273.

Quecine, M. C., Silva, T. M., Carvalho, G., Saito, S., Mondin, M., et al. 2016. A stable *Leifsonia xyli* subsp. *xyli* GFP-tagged strain reveals a new colonization niche in sugarcane tissues. *Plant Pathol.* 65: 154–162.

Ricaud, C., Egan B. T., Gillaspie, A. G. Jr., and Hughes, C. G. eds. 1989. *Diseases of Sugarcane: Major Diseases*. Elsevier, Amsterdam, the Netherlands, 399 pp.

Shukla, U. S., Ram R. S., and Tripathi, R. C. 1974. Effect of moist hot air treatments on the control of GSD and RSD. *Annu. Rep. Indian Inst. Sugarcane Res.* Lucknow, India, pp. 72–74.

Singh, K. 1966. Ratoon stunting disease of sugarcane in India. *Indian Sugar* 16: 335–337.

Steib, R. J., Forbes, I. L., and Chilton, S. J. P. 1957. A report on further studies on the ratoon stunting disease of sugarcane in Louisiana. *Sugar J.* 19: 35–37.

Teakle, D. S., Smith, P. M., and Steindl, D. R. L. 1973. Association of small coryneform bacterium with the ratoon stunting disease of sugarcane. *Aust. J. Agric. Res.*, 24: 869–874.

Urashima, A. S., and Marchetti, L. B. L. 2013. Incidence and severity of *Leifsonia xyli* subsp. *xyli* infection of sugarcane in São Paulo State, Brazil. *J. Phytopathol.* 161: 478–484.

Viswanathan, R. 1996. Serological techniques for the detection of a bacterium causing ratoon stunting disease (RSD) in sugarcane. pp. 162–168 in: *Proceedings of the 9th International Conference on Plant Pathogenic Bacteria*. A. Mahadevan ed. Chennai, India.

Young, A. J. 2016. Possible origin of ratoon stunting disease following interspecific hybridization of *Saccharum* species. *Plant Pathol.* 65: 1403–1410.

Young, A. J. 2018. Turning a blind eye to ratoon stunting disease of sugarcane in Australia. *Plant Dis.* 102: 473–482.

Young, A. J., Kawamata, A., Ensbey, M. A., Lambley, E., and Nock, C. J. 2016. Efficient diagnosis of ratoon stunting disease of sugarcane by quantitative PCR on pooled leaf sheath biopsies. *Plant Dis.* 100: 2492–2498.

Zhu, K., Yuan, D., Zhang, X.-Q., Yang, L.-T., and Li, Y.-R. 2018. The physiological characteristics and associated gene expression of sugar cane inoculated with *Leifsonia xyli* subsp. *xyli*. *J. Phytopathol.* 166: 44–52.

7.15 GUMMOSIS OF SUGARCANE

Gummosis, also known as **gumming** or **Cobb's disease**, was the first disease of sugarcane reported in the literature. Dranert (1869) reported this disease for the first time from Brazil, but its bacterial nature was first demonstrated by Cobb (1893) while working in Australia. Immediately, its occurrence was reported from Mauritius in 1894, from Fiji in 1895 and later on, from Reunion, Caribbean, Argentina, Belize, Columbia, Dominican Republic, French Guyana, Jamaica, Ghana, Madagascar, Malawi, Mozambique, Mexico, Panama, Puerto Rico, South Africa, Swaziland, and Zimbabwe (Ricaud and Autrey 1989; Janse 2005). In India, Rangaswami (1960) reported it from Tamil Nadu on Co 449 and Co 527. So far, there is no report of its occurrence from north India.

The disease caused heavy losses until 1930s where highly susceptible noble canes were grown. However, the development of resistant interspecific hybrids brought the disease under control, and it was eradicated from Australia, Brazil, Fiji, and the West Indies (Ricaud and Autrey 1989). At present, it is a minor disease in most countries, except in those where climatic conditions are highly favorable for the development and spread of the disease.

The losses from the disease occur due to reduction in yield and sugar content of the canes. Besides the loss in yield, the presence of gum (polysaccharides) in the cane juice interferes with the milling process leading to low sugar recovery. The race of the pathogen involved and the degree of tolerance of the varieties affected, influence the extent of loss. In the same variety, the loss from systemic infection of the stalks will be more than from stalks showing foliar infection only. The losses as high as 30%–40% in cane yield and 9%–70% in sugar content have been reported under epiphytotic conditions in Australia.

7.15.1 Host Range

Ricaud and Autrey (1989) have listed natural hosts of the bacterium, in addition to sugarcane, reported by different workers. These are maize, three palms, namely, princess palm (*Dictyosperma album*), royal palm (*Roystonia regia*), and betel nut palm (*Areca catechu*); broom bamboo (*Thysanolaena maxima*), and Guatemala grass (*Tripsacum laxum*). The infection was rarely observed on maize, *A. catechu*, and Guatemala grass.

T. maxima often shows a systemic infection, but the bacterium is quite different from the bacterium infecting the sugarcane. Isolates identical to the races infecting sugarcane, obtained from *T. maxima* cause only foliar symptoms on this host. It is evident from the above given account that sugarcane is the only main host of gumming pathogen. The other hosts are only temporary and accidental hosts and do not harbor the pathogen for long periods in the absence of sugarcane.

A more or less similar disease occurs on maize and Guatemala grass. Strains of *Xanthomonas* causing this disease differ in several aspects from the gumming pathogen and probably they belong to a group of strains reclassified as *Xa. vasicola* pv. *vasculorum*.

7.15.2 Symptoms

The symptoms of the disease consist of two phases, namely, **leaf-streaking phase** and **systemic phase**.

In the first phase, yellow to orange, 3–5 mm wide streaks appear on the leaves and progress along the affected veins. With age, their color changes to ashy grey and the affected tissue becomes necrotic. Usually, they originate from the leaf margin and extend toward the base, but they can arise from anywhere on leaves when the infection occurs through the wounds and develop in either direction. The streaks may cover the entire length of lamina and may even extend to leaf sheath, in the later case leading to the systemic infection of the stalks. The streaks of gummosis differ in color and patterns of appearance from those caused by leaf scald and chlorotic streak.

The severity of leaf infection is influenced by the age of leaf, environmental conditions, and varietal reaction of the cane. The intensity of the streaks is maximum on the mature leaves. Under highly favorable conditions of disease development, many streaks may appear on a leaf and several leaves of a stalk may show symptoms.

The characteristic symptoms of the disease are observed, when transversely cut and longitudinally splitted stalks are examined. Longitudinally splitted canes may show reddish discoloration of vascular bundles, which is more prominent at the nodes. The breakdown of the internal stalk tissue results in formation of pockets/cavities, particularly below the apical meristem. These pockets contain gum consisting of bacterial cells and extracellular polysaccharide.

The bacteria ooze out in the form of yellow shiny droplets from infected bundles of the transversally cut surface of stumps or stalks having systemic infection. The bacterial ooze can be seen more clearly by performing a **sweating test**. In this test, cut pieces of affected stalks kept in sealed polythene bags or moist chambers are incubated at 30°C–35°C. Within a few hours, yellowish bacterial ooze can be seen on the cut ends. This test is more helpful in detecting the bacterial ooze from doubtful cases.

Systematic infection may also cause malformation of the stalk making it flat on one side and bulging on the opposite side. Transverse splits appearing as **knife cuts** can be seen near the apex of affected stalk. Systemically infected plants remain stunted. Ratoon crop suffers less than the plant crop.

Systemic infection is usually, but not always associated with the chlorosis of leaves. The leaves of mature stalks may show chlorosis, but it is usually more common in young ratoon shoots. In highly susceptible varieties, the death of plants may occasionally occur after the chlorotic symptoms, while in tolerant varieties, the affected stalks may completely recover from these symptoms, especially under highly favorable growth conditions.

A part or whole of the lamina may show chlorosis and from one to all leaves of a stalk may be affected. In case of partial chlorosis, white patches appear on lamina; starting from the base of the oldest leaf and moving to the tip of the youngest leaf. The chlorotic symptoms of gummosis and leaf scald are indistinguishable from each other. As the pathogen is absent in the affected tissue in both the diseases, the involvement of a toxin is suspected.

Narrow red stripes, indistinguishable from those of red stripe disease, may also appear on the leaves of systemically infected stalks. A film of dried bacterial exudate is often seen on these stripes. The bacterium infects both xylem vessels and mesophyll cells.

Most recently, a new bacterial disease of sugarcane caused by *Pantoea stewartii* subsp. *indologenes* has been reported by Silva-Rojas et al. (2017) from Mexico. Affected plants show chlorotic streaks in the leaves, and these symptoms frequently appear in the middle of the leaves as fine lines with irregular borders. The disease may be confused with gummosis as it has some overlapping symptoms.

7.15.3 Causal Organism

Xanthomonas axonopodis pv. *vasculorum* (*ex* Cobb 1893) Vauterin et al. 1995
Syn. *Xa. vasculorum*, *Xa. campestris* pv. *vasculorum*

Cells are Gram-stain-negative, straight rods (0.4–0.5 × 1.0–1.5 µm); occur predominantly singly, motile by a single polar flagellum, non-sporing, non-capsular, and not acid-fast. Xanthomonadins are produced. Hydrolysis of starch and esculin is positive. Growth on nutrient agar is poor. Amino acids are required for growth. Acid is produced within 21 days from glucose, sucrose, and trehalose in Dye's medium C, but not from arabinose, mannose, galactose, cellobiose, and fructose. Catalase test and H_2S production from peptone are positive, but nitrate reduction and lysine dicarboxylase tests are negative. The PCR-RFLP of the 16S-23S ribosomal (r)DNA spacer region has been used to differentiate *Xanthomonas* species attacking sugarcane. The method also allows differentiation of type A from type B of *Xa. axonopodis* pv. *vasculorum* (Destefano et al. 2003).

7.15.4 Disease Cycle

The bacterium perpetuates in diseased canes and infected stubbles. The diseased setts disseminate the bacterium to short as well as long distances. The diseased setts and contaminated setts give rise to primary infection of the crop. The secondary spread occurs through the cutting knives, harvesting implements, wind-blown rain, and cyclones; the spread of the bacteria is very fast through the last two means. The cutting knives and implements are important agents for spreading the bacterium at harvest time.

During humid weather and periods of heavy dew, bacteria ooze out from wounds of infected stalks and leaves and may be also through stomata of leaves and leaf sheaths. Then they are splashed away/carried with the help of wind/rain and enter the host through wounds and stomatal openings. The bacteria may also enter the stalks via leaf scars and wounds caused due to farm operations. The wounds on the leaves are caused due to the shearing action of the wind and the serrated margins of leaves. For successful infection of the host, high humidity or free moisture on host surface is essential. From vascular infection, leaf streaks develop within 1–3 weeks under favorable weather conditions.

In Mauritius, the severity of epiphytotics was positively correlated with the number and intensity of cyclones in the growing season. Epiphytotics were more severe when cyclones immediately preceded the early beginning of a dry and cool maturing season (Ricaud 1969).

7.15.5 Management

1. Use of resistant varieties is the most effective method of disease management. In Australia and Mauritius, legislative measures were adopted for the replacement of susceptible varieties with resistant varieties. The varieties having high degree of resistance to foliar infection should be recommended in areas where environmental conditions favor frequent epiphytotics of the disease
2. The other management measures like sanitation have a limited utility due to the rapid spread of the disease. However, they can be helpful in areas where conditions for rapid aerial spread occur very infrequently. The solution of an iodophore (250 μg mL^{-1} available iodine) can be used to disinfect cutting knives and blades of harvesters
3. Use of pathogen-free seed. The seed should not be taken form fields showing systemic infection
4. Rouging of diseased plants particularly, those showing systemic infection (chlorosis) will reduce the disease incidence
5. Do not grow the collateral host like maize, sorghum, and pearl millet near the sugarcane fields

REFERENCES

Cobb, N. A. 1893. Plant diseases and their remedies: Diseases of the sugarcane. *Agric. Gaz. N. S. W.*, 4: 777–798.

Destefano, S. A. L., Almeida, I. M. G., Rodrigues, N. J., Ferreira, M., and Balani, D. M. 2003. Differentiation of *Xanthomonas* species pathogenic to sugarcane by PCR-RFLP analysis. *Eur. J. Plant Pathol.* 109: 283–288.

Dranert, F. M. 1869. Bericht uber die Krankheit des Zuckerrohres. *Z. Parasi-tenkd. Bd.* 1: 13–17.

Janse, J. D. 2005. *Phytobacteriology: Principles and Practice.* CABI Publishing, Wallingford, UK, 360 pp.

Rangaswami, G. 1960. Studies on two bacterial diseases of sugarcane. *Curr. Sci.* 29: 318–319.

Ricaud, C. 1969. Investigation on the systemic infection of gumming disease. *Proc. Int. Soc. Sugar Cane Technol.* 13: 1159–1169.

Ricaud, C. and Autrey, L. J. C. 1989. Gumming disease. pp. 21–31 in: *Diseases of Sugarcane: Major Diseases.* C. Ricaud, B. T. Egan, A. G. Gillaspie Jr., and C. G. Hughes eds. Elsevier, Amsterdam, the Netherlands.

Silva-Rojas, H., Aguilar-Granados, A., Rebollar-Alviter, A., Valdes-Balero, A., Sanchez-Pale, J., et al. 2017. *Pantoea stewartii* subsp. *indologenes* is responsible for a new bacterial disease in sugarcane. *Phytopathology* 107: S2.4.

7.16 LEAF SCALD OF SUGARCANE

Leaf scald of sugarcane was first reported by Wilbrink (1920) from Java and later on by North (1926) from Australia and Fiji. Since then, it has been reported from many countries and Ricaud and Ryan (1989) have listed 44 countries from where it has been reported. Later on, Rott and Davis (2000) reported that the disease occurs in at least 66 countries. In India, the disease was first detected by Egan in 1960 from I.A.R.I., New Delhi during his visit to India (Egan 1979), but the first detailed report of the disease was published by Satyanarayana (1974) from Andhra Pradesh. Since then, the incidence of the disease has increased many folds, and now it has become a major disease of sugarcane in the country, occurring in Punjab, Uttar Pradesh, and Bihar. Among the susceptible cultivars, prominent ones include CoJ 64, CoS 767, and CoLK 8001. Mensi et al. (2013) reported the occurrence of leaf scald for the first time from Gabon describing an unusual highly virulent and aggressive strain of *Xanthomonas albilineans*.

Leaf scald was a serious disease of noble canes, which were widely grown in the early years of twentieth century. North (1926) reported substantial losses due to the disease from Australia and in severe outbreaks; the losses were more than

25%. Egan (1971) reported that outbreaks of the disease in early 1900s caused heavy losses in highly susceptible varieties in Queensland. The replacement of susceptible noble canes with varieties containing *Saccharum spontaneum* genes, resistant to the disease, has considerably reduced the incidence of the disease. It is still a major constraint in the successful cultivation of sugarcane in many countries like Mauritius, Fiji, Cuba, Brazil, Taiwan, Indonesia, and the United States.

Apart from reducing cane yield; the disease also lowers the quality of the juice by reducing Brix, Pol, and Purity. The slow growth of affected stalks, reduction in number of millable canes per stool and poor ratooning reduce the yield. The acute stage causes death of stools and reduction in sugar yield. The factors, which influence the growth of sugarcane plants, also influence the disease severity. The onset of adverse conditions soon after the maturity of the canes, give rise to large number of side shoots. The number and size of side shoots are directly related to deterioration of juice quality.

(a)

(b)

FIGURE 7.11 Leaf scald of sugarcane: (a) Pencil-line streaks on leaf lamina. (b) Infected plant bearing adventitious shoots.

7.16.1 Host Range

In addition to sugarcane, the bacterium causes natural infection on *Paspalum dilatatum*, *P. congugatum*, *Brachiaria piligera*, and *Imperata cylindrica* var. *major* (blady grass). The possibility of survival of the bacterium in blady grass for sometime can make this host as a source of inoculum for the sugarcane crop. *Zea mays*, *Rottboellia cochinchinensis*, *I. cylindrica*, *Panicum maximum*, *Pennisetum purpureum*, *Paspalum* sp., and *Sorghum* sp. also act as natural hosts of the bacterium. It seems that *Xa. albilineans* infects sugarcane and some other members of the family Graminae.

7.16.2 Symptoms

There are two phases of symptoms, namely, the chronic phase and the acute phase. The latent infection is also an important characteristic feature of the disease.

7.16.2.1 Chronic Phase

These are the most commonly observed symptoms. The most characteristic symptom of the disease is the appearance of **white pencil-line** streaks, which are narrow, about 1–2 mm wide, and sharply defined on the leaf lamina. The streaks develop parallel to the midrib and often extend to the base of leaf lamina or the leaf sheath (Figure 7.11a). The specific epithet of the causal pathogen has been derived from the description of this streak.

Initially, the streaks are usually well-defined, but they become diffused and broader with the passage of time. However, the close examination of these broader and diffused stripes will show the white pencil-line at the center. As the disease progresses, the stripes start drying causing withering away of the leaves. The drying always starts from the tip and extends downward, giving a scalded appearance to the leaves. The streaks also appear on leaves in gummosis or chlorotic streak of sugarcane, but these streaks are devoid of central white pencil-line. The partial or complete chlorosis of the leaf blade also occurs. The white pencil-line or its reddish necrotic patches can be seen on close examination of these chlorotic blades.

Generally the side shoots, varying in number, develop on the mature infected stalks. Some of the side shoots die when they are still young, while the surviving shoots generally exhibit typical white pencil-line or chlorosis of leaves (Figure 7.11b).

The necrosis of the vascular bundles in the form of bright to dark red streaks can be seen on the longitudinally splitted canes. These streaks are generally more prominent at the nodes than in the internodes and also appear in the side shoots. Inside the severely infected canes, longitudinal lysigenous cavities may be formed in the nodal and internodal areas, particularly near the shoot apex. These cavities are relatively dry compared to those of gumming disease as they contain less amount of bacterial slime or gum.

7.16.2.2 Acute Phase

The sudden wilting and death of mature stalks, often without any previous symptoms, marks the onset of the acute phase. Occasionally, the small shoots having typical streaks appear at the base of stalks. The other symptoms like proliferation of buds, bunchy top appearance, and reductions of the root systems were observed by Agnihotri (1986). Acute phase generally occurs after a stress resulting from a prolonged dry weather or dry weather after a rainy season. To confirm the occurrence of acute phase, one should look for canes showing chronic phase symptoms among the affected plants or wait until the new tillers come out bearing the white pencil-line symptoms.

In India, the symptoms of acute phase have not been observed. However, gradual wilting of plants showing chronic phase symptoms, especially during the summer months and after the first monsoon shower is commonly observed in some susceptible varieties like CoLK 8001, CoJ 81, and Co 62399 (Agnihotri and Duttamajumder 1992).

7.16.2.3 Latent Infection

The disease is characterized by a latent phase/infection in which the infected plants do not show any symptoms. The plants suffering from chronic phase recover from the disease under certain conditions. The recovery is more common in tolerant varieties or when the conditions are suitable for plant growth. The latency may involve a few shoots or stools or an entire field. The symptoms may appear on such plants after ratooning and on the plants growing from setts taken from the infected stools. The latent infection can also lead to the development of acute phase. Various techniques like fluorescent antibody technique, ELISA, and use of indicator plants are used to detect the latent infection. Among them, fluorescent antibody technique is more effective.

7.16.3 Causal Organism

Xanthomonas albilineans (Ashby 1929) Dowson 1943
Syn. *Phytomonas albilineans*, *Bacterium albilineans*

In 1929, Ashby named the bacterium as *Bacterium albilineans*. Later on, it was transferred to genus *Phytomonas* as *Phytomonas albilineans*. With the establishment of genus *Xanthomonas*, it was transferred to it as *Xanthomonas albilineans*. Wilbrink's agar, now known as sucrose peptone agar (SPA), was earlier used for isolation and culturing of this bacterium. Glutamic acid and methionine are required for growth. The optimum temperature for growth on SPA is 25°C and the maximum is 37°C. The isolation of the bacterium can be made by using the macerated infected tissue or juice of infected plants after diluting it 1,000-fold. On SPA, the colonies appear after 5–7 days of incubation, and they are honey-yellow to Naples-yellow, minute, glistening, and viscid having an entire margin.

A selective medium (*Xanthomonas albilineans* selective [XAS] medium) has been devised by supplementing Wilbrink's medium with 5 g of KBr, 100 mg of cycloheximide, 2 mg of benomyl, 25 mg of cephalexin, 30 mg of novobiocin, and 50 mg of kasugamycin per liter. The isolation on this medium gives more than 98% frequency from symptomatic plants. The growth rate, morphology, and pigmentation of colonies on the medium serve as useful differential characteristics to identify the bacterium.

Cells are straight rods (0.25–0.3 × 0.6–1.0 μm), predominantly single, Gram-stain-negative, and motile by a single polar flagellum. Xanthomonadins are produced. The bacterium neither liquefies gelatin nor reduces nitrate and does not produce H_2S, ammonia and indole. Esculin is hydrolyzed, but not starch. Acid is produced within 21 days on Dye's medium C from glucose, sucrose, and mannose. Ammonium salts, nitrates, or asparagine are not used as a sole source of nitrogen.

Variation in genetic, physiological, and serological characteristics, and in pathogenicity of the bacterium, indicating the existence of different pathotypes in the species, has been found, but so far there is no evidence of existence of races. According to Silva et al. (2007), there is evidence for occurrence of pathogenic variation in the bacterium as the same cultivar gives a susceptible reaction to the pathogen in one region, but a different reaction in another region. Repetitive polymerase chain reaction (rep-PCR) analysis of 50 isolates, collected by them from different regions of São Paulo state of Brazil, revealed genetic diversity among these isolates, and they divided them into three groups. Pathogenicity test of these isolates revealed the variation in aggressiveness among the isolates, suggesting a possible relationship between occurrence of bacterial variants and differential reaction of sugarcane cultivars in São Paulo state. It can be inferred from these findings that pathogenic races of the bacterium occur in nature, and there is an urgent need to develop suitable differentials for the identification of these races.

7.16.4 Disease Cycle

The main sources of perpetuation of the bacterium are infected stubbles and infected setts used for sowing. Major dissemination of the bacterium occurs at the time of harvesting and planting, and it occurs mainly through the infected cuttings and implements used for harvesting and cutting of canes. Infected cuttings are an important mode of spread over long distances. Duttamajumder (1996) showed that setts obtained from the upper half of the infected canes were much more efficient (48%) in producing the disease than those from lower half (21%).

The cane knives and base cutters of mechanical harvesters also play a role in the secondary spread of the bacterium within a crop. The harvesting implements carry the inoculum after cutting an infected cane and transmit the same up to the third stool, after which the transmission becomes irregular. Fluff (seed)-plant transmission of the bacterium has also been proved. Aerial dissemination of the pathogen has been reported to occur in several geographical areas of the world and was associated with outbreaks of the disease (Comstock 2001; Daugrois et al. 2003). The role of phyllosphere colonization by the bacterium in the epidemiological cycle of the disease in Guadeloupe was also shown by Daugrois et al. (2003). Role of insects such as grasshoppers, leafhoppers, leaf beetles, and cane borers, and rodents in transmission is suspected, but not yet proved. The disease is favored by wet weather, particularly in combination with cyclones. Usually, the appearance of disease after the first rain shower indicates the involvement of wind-blown rain in disease development. Stress caused by drought and water-logging usually increases the disease severity.

Xa. albilineans is a systemic and xylem-invading pathogen and colonizes only the xylem of the vascular strands of the affected parts. However, occasionally the bacterium may break the xylem vessels and destroy the surrounding cells, producing cavities. Generally, this situation is manifested by the appearance of red areas on the white pencil-line. The bacterium produces gum, a xanthan-like polysaccharide composed of glucose, mannose, and glucuronic acid. The gum occludes phloem and xylem vessels and partially enters mesophyll cells (Solas et al. 2003). The chlorosis-inducing isolates of the bacterium produce toxic compounds, of which the **albicidin** being the most important plays a key role in pathogenicity. Albicidin inhibits DNA replication of protoplastids, thereby blocking plastid development. There is no correlation between toxin production and pathogenicity of *Xa. albilineans*. Hence, it is apparent that some other virulence factors play a key role concurrently with albicidin in pathogenicity (Champoiseau et al. 2006).

Garces et al. (2014) reported that assays with SYBR Green primers and TaqMan probe and primers derived from the albicidin toxin biosynthesis gene cluster efficiently and reproducibly amplified *Xa. albilineans*. Detection was more sensitive with qPCR compared with conventional PCR. Assays were specific for *Xa. albilineans,* and sap extracts did not inhibit the qPCR reaction. Populations of *Xa. albilineans* varied in different tissues and differences were the greatest within tissues in resistant cultivars. The bacterial populations in systemically infected, young, not yet fully emerged leaves exhibited the greatest differences between resistant and susceptible cultivars. As the qPCR is a highly sensitive method for the detection of *Xa. albilineans*, it will provide a reliable method for leaf scald resistance screening. Gutierrez et al. (2016) compared qPCR with the visual symptom rating in 31 clones at different times after inoculation and found the qPCR assay more consistent than the visual symptom rating at 8 weeks after inoculation. The authors also suggested that qPCR can provide an improved method to evaluate resistance to leaf scald in sugarcane.

Daugrois et al. (2014) reported that resistance of sugarcane to leaf scald appears to involve several traits, including limiting size of epiphytic *Xa. albilineans* populations and limiting the capacity of the pathogen to produce leaf necrotic symptoms by invading the leaf vascular system or to move from the leaf into the stalk.

7.16.5 Management

An integrated approach consisting of the following measures can prove very effective for the management of the disease.

1. Resistant varieties: The use of resistant varieties is the most effective, practical, and economical method of control. Before release, the resistant varieties must be tested by artificial inoculation and both plant and ratoon crops should be screened. For assessing the reaction of varieties, take into account only the systemic infection on the non-inoculated leaves and ignore the initial symptoms on the inoculated plants
2. Pathogen-free seed: As the bacterium is transmitted mainly through the diseased setts, the use of pathogen-free seed is of utmost importance for the management of the disease. In areas free from severe epiphytotics, pathogen-free seed can be obtained by regular inspection of seed plots, rouging of diseased plants, and then protecting the crop by adopting phytosanitary measures. Duttamajumder (1996) showed that a directed-selection for an apparently healthy cane progeny yielded symptomless plants in the fourth progeny with no trace of leaf scald bacterium on isolation. From 25% of the apparently healthy canes in the first year, it increased to about 90% at the end of the 4th year
3. Heat therapy: Heat therapy has been tried to eradicate the bacterium from the diseased planting material. Hot air treatment at 54°C for 8 h is not effective. Long hot water (50°C for 3 h) and short hot water (52°C for 20 min) treatments are effective, but they considerably reduce the germination of buds. Nowadays, moist-hot-air treatment (54°C for 4 h) is mostly used. Agnihotri (1984) obtained an effective control of the disease with this method. It is superior to all other heat therapy methods used in India, and it also controls grassy shoot and ratoon stunting diseases of sugarcane
4. Chemotherapy: Dip the single bud setts after desiccating for 12 h, in 100 μg mL^{-1} solution of streptomycin for 1 day. However, its higher concentrations are injurious to plants
5. Phytosanitary measures: The disinfection of cane knives, mechanical harvesters, and planters reduces the spread of the bacterium. Ten percent commercial formalin, 1% Dettol, or 5% Lysol can be used for the disinfection of implements. While cutting setts, the knives should be disinfected at frequent intervals. The destruction of infected debris, stubbles, and of weeds like *I. cylindrica* helps in reducing the inoculum load. *Sorghum* sp. and *Z. mays*, which act as collateral hosts of the bacterium, should not be grown near the sugarcane crop
6. Quarantine measures: Strict quarantines should be enforced to check the spread of the pathogen to the disease-free areas. However, the latent infection of the bacterium is a big hurdle in the effective implementation of the quarantines. All the quarantine stations should be equipped with adequate facilities for the detection of the bacterium in the seed material
7. Cultural practices: High doses of nitrogenous fertilizers, water stress, and water-logging increase the incidence of the disease. The correction or removal of these factors will reduce the incidence of the disease. The early ploughing of heavily infected fields considerably reduces the source of inoculum

REFERENCES

Agnihotri, V. P. 1984. Control of leaf scald of sugarcane, a bacterial disorder, by heat therapy and streptomycin. *Indian Sugar Crops J.* 10: 13–14.

Agnihotri, V. P. 1986. Some new or interesting symptoms of leaf scald of sugarcane. *Indian Sugar Crops J.* 11: 21.

Agnihotri, V. P., and Duttamajumder, S. K. 1992. Leaf scald disease of sugarcane. pp. 27–40 in: *Plant Diseases of International Importance*. Vol. IV, *Diseases of Sugar, Forests and Plantation Crops*. A. N. Mukhopadhyay, J. Kumar, H. S. Chaube, and U. S. Singh eds. Prentice Hall, Englewood Cliffs, NJ.

Ashby, S. F. 1929. The bacterium which causes gumming disease of sugar canes with notes on two other bacterial diseases of the same host. *Trop. Agric.* 6: 135–138.

Champoiseau, P., Daugrois, J. H., Pieretti, I., Cociancich, S., Royer, M., et al. 2006. High variation in pathogenicity of genetically closely related strains of *Xanthomonas albilineans*, the sugarcane leaf scald pathogen, in Guadeloupe. *Phytopathology* 96: 1081–91.

Comstock, J. C. 2001. Foliar symptoms of sugarcane leaf scald. *Sugar J.* 64: 23–32.

Daugrois, J. H., Dumont, V., Champoiseau, P., Costet, L., Boisne, R., et al. 2003. Aerial contamination of sugarcane in Guadeloupe by two strains of *Xanthomonas albilineans*. *Eur. J. Plant Pathol.* 109: 445–458.

Daugrois, J. H., Boisne-Noc, R., and Rott, P. 2014. Leaf surface colonization of sugarcane by *Xanthomonas albilineans* and subsequent disease progress vary according to the host cultivar. *Plant Dis.* 98: 191–196.

de Souza e Silva, M., Bedendo, I. P., and Casagrande, M. V. 2007. Molecular and pathogenic characterization of isolates of *Xanthomonas albilineans* (Ashby) Dowson, causal agent of sugarcane leaf scald. *Summa Phytopathol.* 33: 341–347.

Dowson, W. J. 1943. On the generic names *Pseudomonas*, *Xanthomonas*, and *Bacterium* for certain bacterial plant pathogens. *Trans. Brit. Mycol. Soc.* 26: 1–14.

Duttamajumder, S. K. 1996. Dynamics of field spread of *Xanthomonas albilineans* (Ashby) Dowson causing leaf scald disease in sugarcane. pp. 324–330 in: *Proceedings of the 9th International Conference on Plant Pathogenic Bacteria*. A. Mahadevan ed. Chennai, India.

Egan, B. T. 1971. The decline of leaf scald as a major disease in northern Queensland. *Proc. Queensland Soc. Sugarcane Technol.* 38: 157–161.

Egan, B. T. 1979. Leaf scald disease in India: Letters to the editor. *Sugarcane Pathol. News*. 22: 53.

Garces, F. F., Gutierrez, A., and Hoy, J. W. 2014. Detection and quantification of *Xanthomonas albilineans* by qPCR and potential characterization of sugarcane resistance to leaf scald. *Plant Dis.* 98: 121–126.

Gutierrez, A., Garces, F. F., and Hoy, J. W. 2016. Evaluation of resistance to leaf scald by quantitative PCR of *Xanthomonas albilineans* in sugarcane. *Plant Dis.* 100: 1331–1338.

Mensi, I., Girard, J.-C., Pieretti, I., Larbre, F., Roumagnac, P., et al. 2013. First report of sugarcane leaf scald in Gabon caused by a highly virulent and aggressive strain of *Xanthomonas albilineans*. *Plant Dis.* 97: 988.

North, D. S. 1926. Leaf scald, a bacterial disease of sugarcane. Colonial Sugar Refining Co. Ltd. Sydney Agric. Rep. No. 8, 80 pp.

Ricaud, C., and Ryan, C. C. 1989. Leaf scald. pp. 39–58 in: *Diseases of Sugarcane: Major Diseases*. C. Ricaud, B. T. Egan, A. G. Gillaspie, Jr., and C. G. Hughes eds. Elsevier, Amsterdam, the Netherlands.

Rott, P., and Davis, M. J. 2000. Leaf scald. pp. 38–44 in: *A Guide to Sugarcane Diseases*. P. Rott, R. A. Bailey, J. C. Comstock, B. J. Croft, and A. S. Saumtally eds. La Librairie du Cirad, Montpellier, France.

Satyanarayana, Y. 1974. Leaf scald of sugarcane: A new disease in Andhra Pradesh. *Indian Sugar* 24: 23–26.

Solas, M. T., Piñón, D., Acevedo, R., Fontaniella, B., Legaz, M. E., et al. 2003. Ultrastructural changes and production of xanthan-like polysaccharide associated with scald of sugarcane leaves caused by *Xanthomonas albilineans*. *Eur. J. Plant Pathol.* 109: 351–359.

Wilbrink, G. 1920. De Gomziekte van het Suikerriet, hare Oorzaak en hare Bestrijding. *Arch. Suikerind, Ned. Indie*, 28: 1399–1525.

7.17 BACTERIAL BLIGHT OF COTTON

The disease is also known by other names such as **angular leaf spot** and **black arm of cotton**. It was first reported from Alabama, USA in 1891 by Atkinson, who named it as black rust. During 1900–1905, E.F. Smith proved Koch's postulates, described the causal organism, and also established a relationship between black arm and angular leaf spot phases of the disease, which were earlier considered as different diseases. The disease is worldwide in occurrence and causes heavy losses in some countries. In India, the disease was first observed in 1918 in Madras, and it remained a minor disease until the introduction of tetraploid cottons. It is argued that in India, susceptible cultivars phased out under the selection pressure of the pathogen, and the pathogen and the cultivars reached a balance. On the basis of this co-evolution, Knight and Hutchinson (1950) concluded that the disease originated in India. However, Patel and Kulkarni (1948) are of the view that the disease was introduced in the middle of nineteenth century through exotic material.

In 1940s, a campaign was launched in India to replace the rain fed indigenous diploid cottons with high yielding, but susceptible tetraploid, *Gossypium hirsutum*, and to some extent *G. barbadense* cottons. After the launch of the campaign, the severe epiphytotics of the disease occurred during 1948–1952, and the disease became established in 1960s. Irrigation and high nitrogen applied to achieve the full potential of the introduced cotton cultivars also increased the disease severity. In India, the disease occurs in all the cotton growing areas and the losses range from 5% to 25% (Verma and Singh 1972).

7.17.1 Symptoms

There are four different types or phases of symptoms of the disease. It is not necessary that all the four types of symptoms should appear under all the situations. Their occurrence and predominance varies under different environmental conditions.

7.17.1.1 Seedling Blight

Water-soaked circular spots appear on lower surface of cotyledons. The spots spread forming irregular patches, causing drying and withering of the seedlings. Many plants die. This phase is more common in countries like USA where the temperature is low at the time of sowing of cotton.

7.17.1.2 Leaf Spots

Water-soaked spots first appear on the lower surface of leaves. These spots enlarge, forming dark brown to black, angular dead areas and become visible also on the upper surface of leaves. As the spread of these spots is bounded by veins and veinlets, they are angular in shape. The spots enlarge forming irregular large patches (Figure 7.12a). The infection extends along the veins turning the veins black and infected veins give a **finger-like appearance**. This phase is called **vein blight** or **black vein** (Figure 7.12a). From veins, the infection reaches the petioles through midrib and infects the stem. Petiole infection leads to severe defoliation of plants. Under humid conditions, yellowish bacterial exudate is formed on leaf lesions.

7.17.1.3 Black Arm

Linear, sunken, dark brown to black lesions appear on the stem and branches of the plants and eventually girdle them. In severely affected stems, there is deep cracking and gummosis, which results in breaking off the stem by wind. This phase is more common in Sudan and Uganda.

7.17.1.4 Boll Rot

Water-soaked, circular spots appear on bolls, which later on, turn brown and become sunken (Figure 7.12b). Severely infected bolls drop, but in case of mildly infected bolls, the infection progresses internally. Secondary organisms also enter the infected bolls and cause extensive boll rotting. The infected bolls may fail to open.

Sometimes, atypical symptoms of the disease appear that consist of dull green, flaccid areas extending from periphery of cotyledons and leaves, and elongated, water-soaked, dark brown areas on leaf lamina. Malavolta et al. (2008) have described a new type of symptomatology named **bacterial leaf blight** from São Paulo state of Brazil. The new type of symptomatology is characterized by leaf blight, generally exhibiting chlorotic haloes and sometimes showing V-shaped lesions beginning at the border of the leaves. The bacterium causing the new kind of symptoms was identical to the type strain of *Xanthomonas axonopodis* pv. *malvacearum*.

Earlier, it was considered that the bacterium invades only the parenchyma of the host. Since 1932, information has become available that the bacterium invades the vascular system also. From infected cotyledons, the bacterium moves through the vascular system of the petioles and reaches the stem. In the stem, it moves upward within the vascular system from where it breaks outward causing cortical rotting of young stem tissues and petioles, and infection of the leaf lamina. The movement of the bacterium in the susceptible cultivars indicates the probable systemic or at least spreading nature of the disease. The isolation of the bacterium from embryos and seeds also shows the systemic nature of the pathogen.

7.17.2 Causal Organism

Xanthomonas citri subsp. *malvacearum* (hereafter referred to as *Xcm*) (Smith 1901) Schaad et al. 2007, subsp. nov., nom. rev.

Syn. *Xa. campestris* pv. *malvacearum*, *Xa. axonopodis* pv. *malvacearum*

Cells are rod-shaped, measuring 0.9 × 1.2 μm; occur singly or in pairs, but rarely in chains, monotrichous, non-sporing, not acid-fast, encapsulated, chemoorganotrophic, facultatively aerobic and Gram-stain-negative. The metabolism is respiratory, never fermentative, catalase test is positive, but oxidase negative. Asparagine is not utilized as a sole source of carbon and nitrogen. The bacterium may utilize methionine, glutamic acid, and nicotinic acid in various combinations as minimal growth requirements. Triphenyl tetrazolium chloride at 0.1% to 0.2% in nutrient agar inhibits its growth.

The bacterium liquefies gelatin, hydrolyses starch, digests casein, but not fat. Produces H_2S, but not indole, nitrate is not reduced; methyle red and Voges–Proskauer tests are negative, and can tolerate 3% sodium chloride. Sucrose, maltose, dextrose, levulose, xylose, glycerol, galactose, lactose, and raffinose are utilized with acid production, but without gas. Arabinose, rhamnose, salicin, mannitol, and dulcitol are not utilized. The optimum temperature for growth is 31°C–32°C, while the minimum and maximum are at 6°C and 42°C, respectively. Plasmids varying in number and size are present in the cells (Das and Verma 2003). Schaad et al. (2005), on the basis of DNA-DNA relatedness determined by sequencing the 16S–23S intergenic spacer regions and AFLP analysis,

(a)

(b)

FIGURE 7.12 Bacterial blight of cotton: (a) Angular spots and vein blight symptoms on leaves. (Courtesy of Stephen Allen, www.padil.gov.au, CCAL.) (b) Boll rot. (Courtesy of Stephen Allen, www.csiro.au, CCAL.)

reclassified the bacterium as *Xa. smithii* subsp. *smithii* sp. nov. comb. nov. nom. nov., but later amended the classification and renamed it as *Xa. citri* subsp. *malvacearum* (Schaad et al. 2006, 2007). Schaad et al. (2006) also reported the following characteristics of *Xcm*. The colonies of *Xcm* appear on yeast dextrose carbonate agar (YDC) and Fieldhouse Sasser (FS) agar after 40–44 and 56–60 h of incubation, respectively at 28°C–30°C. The bacterium produces an alkaline reaction in litmus milk without hydrolysis, liquefies gelatin, but does not hydrolyze pectate. *Xcm* does not utilize melizitose. *Xcm* can be differentiated from *Xa. campestris* pv. *campestris* and most other *Xanthomonas* pathovars, sub species and species by DNA-DNA reassociation assays, rep-PCR profiles, intergenic spacer (ITS) sequencing, serological methods, SDS-PAGE patterns of membrane proteins, and phenotypic characters.

Zhai et al. (2013) reported high quality draft sequences of a highly virulent *Xa. axonopodis* pv. *malvacearum* strain (Göttinger Sammlung Phytopathogener Bakterien [GSPB2388]) from Sudan and of a strain of race 18 (GSPB1386) from Nicaragua, using Illumina/Solexa paired-end sequencing. The short sequence reads were assembled into 61 scaffolds for GSPB2388 (N50 of 164 kb) and 127 scaffolds for GSPB1386 (N50 of 100 kb), with draft maps of roughly 5 Mb, which contain 4,665 and 4,520 protein-coding genes, respectively. Based on gene annotation and comparisons with plant pathogen proteins, 181 and 178 potential virulence-related genes, including genes encoding a major group of type III effectors, were identified from GSPB2388 and GSPB1386, respectively. The differential effectors and sequence diversity between the highly virulent strains and race 18 may enable the identification of key factors that contribute to the high virulence of highly virulent strains.

Races: Pathogenic variation in the bacterium is well established and until now 32 races have been reported from different parts of the world (Table 7.2). In 1946, Knight was probably the first to suggest the existence of biological specialization in the bacterium, when Stoneville-20, a line of upland cotton resistant to the pathogen was severely infected in the United States. Bird and Hadley in 1958 identified two races of the pathogen on the basis of their reaction to three lines of upland cotton. Brinkerhoff (1963), on the basis of screening of isolates collected from all over the world on eight differential lines belonging to *G. hirsutum* and *G. barbadense*, identified ten new races. Later on, Hunter et al. (1968) developed a set of differentials (eight lines of upland cotton having different bacterial blight resistance genes) and identified 15 races of *Xcm*. Total number of races was raised to 17 when two more races, i.e., I_1 and I_2 were described from India.

Out of two sets of differentials developed by Nayudu (1964) and Hunter et al. (1968) for the identification of races of *Xcm*, the set developed by Hunter et al. (1968) is more suitable and is widely used. It contains isogenic lines with different bacterial blight resistance genes, which give stable reaction under different environmental conditions. Further, the race identification is based scientifically on the number of genes attacked, and on which, until now, all the isolates have given variable reactions.

Until now, 26 races have been described from India, and these are *Xcm* R-1, 2, 4, 5, 7–10, 13, 14, 16–25, and 27–32. Race 32 was most widely distributed and was present at more than 83% of the locations tested. Out of 133 isolates tested, 59.25% belonged to this race. It was the most virulent race as it attacked five major bacterial blight resistance genes, namely, B_7, B_4, B_2, B_{In}, and B_N. The other prevalent races were *Xcm* R-27, 28, and 29, which attacked three genes. Less virulent races, *Xcm* R-1, 2, 9, and 10, were restricted in their distribution (Verma and Raj 1992).

Xcm R-18H described from Pakistan is capable of attacking five genes. A race from western and central Africa has infected the line 101–102.B, i.e., the combined genes $B_2 + B_3$, as well as other combined gene resistance. A race called post-Barkat race (PBR) is capable of attacking all single and developed combinations of major *B* genes for resistance in Sudan. As some of the presently known races are highly virulent and capable of attacking combined gene resistance, there is an urgent need to develop another set of differentials to properly identify these races.

More than one race may be present on the same plant, on the same leaf, or even in the same lesion. Thirteen races were isolated from four different lesions formed on the same leaf of Acala-44, which is the most susceptible line of differentials, having no blight resistance gene. New races possessing more virulent genotypes could be generated in nature from mixed infections of pathogenic races (in a susceptible cultivar) in a synergistic reaction (Verma et al. 1979).

Delannoy et al. (2005) reviewed the genetic basis of cotton resistance to the pathogen, with reference to resistance genes, resistance gene analogs, and bacterial avirulence genes along with the physiological mechanism involved in the hypersensitive reaction and suggested that this host-pathogen interaction represented the most complex resistance gene/*avr* gene system yet known.

Essenberg et al. (2014) developed a pyramid of lines with all possible combinations of two and three genes using near-isogenic lines of upland cotton (*G. hirsutum*) carrying single, race-specific genes B_4, B_{In}, and b_7 for resistance to bacterial blight to achieve broader and higher resistance approaching that of L.A. Brinkerhoff's exceptional line Im216. Under field conditions in north-central Oklahoma, pyramid lines exhibited broader resistance to individual races and, consequently, higher resistance to a race mixture. Although some enhancement in resistance was observed, the pyramided lines carrying two or three *B* genes did not acquire the level of resistance of Im216. In a growth chamber, bacterial populations of race 1 in and on leaves of the pyramid lines decreased significantly with increasing number of *B* genes in only one of four experiments. The older lines, Im216 and AcHR, exhibited considerably lower bacterial populations than any of the one-, two-, or three-*B*-gene lines.

7.17.3 Disease Cycle

Infected seed is the main source of perpetuation of the bacterium, and it may carry the pathogen both externally (as a slimy

TABLE 7.2
Nomenclature of Races of *Xanthomonas citri* subsp. *malvacearum*

Part A: Primary Differentials for Race Number

Disease Reaction on Differentials: Susceptible = +, Resistant = −							Race Number	
I	II	III	IV	V	VI	VII	USA	India
+	−	−	−	−	−	−	13	1
+	+	−	−	−	−	−	1, 11	2
+	−	+	−	−	−	−	—	3
+	−	−	+	−	−	−	—	4
+	−	−	−	+	−	−	—	5
+	−	−	−	−	+	−	—	6
+	+	+	−	−	−	−	2, 12	7
+	+	−	+	−	−	−	—	8
+	+	−	−	+	−	−	3	9
+	+	−	−	−	+	−	4	10
+	−	+	+	−	−	−	—	11
+	−	+	−	+	−	−	—	12
+	−	+	−	−	+	−	9	13
+	−	−	+	+	−	−	—	14
+	−	−	+	−	+	−	—	15
+	−	−	−	+	+	−	19	16
+	+	+	+	−	−	−	—	17
+	+	+	−	+	−	−	15	18
+	+	+	−	−	+	−	17	19
+	+	−	+	+	−	−	6	20
+	+	−	+	−	+	−	—	21
+	+	−	−	+	+	−	5	22
+	−	+	+	+	−	−	—	23
+	−	+	+	−	+	−	—	24
+	−	+	−	+	+	−	—	25
+	−	−	+	+	+	−	—	26
+	+	+	+	+	−	−	8	27
+	+	+	+	−	+	−	16	28
+	+	+	−	+	+	−	14	29
+	+	−	+	+	+	−	7	30
+	−	+	+	+	+	−	—	31
+	+	+	+	+	+	−	10, 18	32

Source: Verma, J.P. and Raj, S., Blight and wilt of cotton, in: *Plant Diseases of International Importance*, Vol. IV, Diseases of Sugar, Forests and Plantation Crops, Mukhopadhyay, A.N. et al., eds., Prentice Hall, Englewood Cliffs, NJ, pp. 138–160, 1992.
I, Acala-44 (no genes); II, Stoneville 2B-S9 (polygenes); III, Stoneville-20 (B_7+ polygenes); IV, Mebane B-1 (B_2 + polygenes); V, 1–10.B (B_{In} + polygenes); VI, 20–3 (B_N + polygenes); VII, 101–102.B (B_2B_3 + unknown).

Part B: Additional Differentials for Biotypes of Race 32

Disease Reaction on Differentials: Susceptible = +, Resistant = −				
VII	VIII	IX	X	Biotype Number
−	−	−	−	32.0
+	+	−	−	32.1 (probably race 18 of USA)
+	−	+	−	32.2
+	+	+	−	32.3 (Follin's isolate can be placed here)

Source: VII, 101–102.B (B_2B_3 + unknown); VIII, Empire B_4 (B_4 + polygenes); IX, Tamcot CAMD-E ($B_2B_3B_7$ + polygenes); X, a line yet to be developed.
Note: Even when the line X is attacked by a new race of *Xa. citri* subsp. *malvacearum*, the old race number will remain the same and valid. Additional lines can always be added to designate new races/biotypes. Further, additional information can always be given; e.g., race 32.3 can also attack the gene combination of T × Bonham ($B_2B_3B_4B_7B_{Sm}$).

mass in the fuzz) as well as internally. The other sources of perpetuation are infected self-sown seeds and infected plant debris. The bacterium does not survive in the plant debris until the next season of crop when it is buried deep in soil and the soil remains wet. However, it can survive for 7 months in trash on the soil surface. The bacterium also does not survive when wheat crop is taken between two cotton crops. However, the plant debris serves as an important source of infection where cotton is grown as a continuous crop. It seems that under northern Indian conditions, infected seed including the self-sown seed is an important source of primary inoculum.

The bacterium enters the host mainly through stomata. Insect punctures also serve as entry points for boll infection. The disease is favored by high humidity and moderate temperature. Primary infection occurs at relatively low temperature of 30°C at which cotton germinates and grows, but secondary infection occurs at relatively higher temperature, i.e., 35°C. The presence of moisture is essential for secondary spread of the pathogen during the first 48 h of infection. Relative humidity of 85% or above is essential for abundant infection. Dry and hot weather retard the normal development of the disease.

Wind-blown rain is the main disseminating agent of the bacterium. Surface irrigation and overhead-irrigation also spread the bacterium. In some areas, wind-blown detached leaves and wind-blown dust may serve as means of dissemination. The transmission of the bacterium by two cotton pests, namely, *Dysdercus koenigii* and *Earias* spp. is known, but their role in dissemination and disease development is not clear. The bacterium survives in the excreta of larvae of *Earias* spp. for 6 days.

Jatropha curcas, *Lochnera pusilla*, *Thurbaria thespesioides*, *Eriodendron anfractosum*, and *Thespesis* spp. act as collateral host of the pathogen, but their role in the disease epidemiology is not clearly defined.

7.17.4 Management

The eradication of seed-borne inoculum and destruction of infected plant debris and self-sown plants are of utmost importance for the management of the disease.

1. Treat the acid or machine delinted seeds with carboxin, oxycarboxin, or Agrimycin-100 @ of 2 g kg^{-1} of seed
2. Hot water treatment of seed at 56°C for 10 min is also effective
3. Destroy the diseased plant debris and self-sown seedlings. Plant debris should be buried deep in the soil followed by flooding of the field for 24–48 h
4. Protective sprays of phylloplane bacteria, namely, two species of *Flavobacterium* and one species each of *Pseudomonas* and *Aeromonas* are effective. This schedule should be included in the integrated management of the disease
5. To check the secondary spread of the disease, spray the crop with Blitox (500 gm) + Agrimycin-100 (20 g) or Streptocycline (3 g) per acre per spray. Give the first spray on 4- to 6-week-old crop and subsequent sprays at 10–15 days intervals. Three or more sprays should be given depending upon the severity of the disease. These chemicals possessed some curative effect also (Verma and Raj 1992)
6. Use of resistant varieties

REFERENCES

Brinkerhoff, L. A. 1963. Variability of *Xanthomonas malvacearum*: The cotton bacterial blight pathogen. Oklahoma State University Information Technology, 98, 96 pp.

Das, I. K., and Verma, J. P. 2003. Plasmids in bacteria associated with bacterial leaf blight of cotton. *Indian Phytopathol.* 56: 180–182.

Delannoy, E., Lyon, B. R., Marmey, P., Jalloul, J. F., Daniel, J. L., et al. 2005. Resistance of cotton towards *Xanthomonas campestris* pv. *malvacearum*. *Annu. Rev. Phytopathol.* 43: 63–82.

Essenberg, M., Bayles, M. B., Pierce, M. L., and Verhalen, L. M. 2014. Pyramiding *B* genes in cotton achieves broader but not always higher resistance to bacterial blight. *Phytopathology* 104: 1088–1097.

Hunter, R. E., Brinkerhoff, L. A., and Bird, L. S. 1968. The development of a set of upland cotton lines for differentiating races of *Xanthomonas malvacearum*. *Phytopathology* 58: 830–832.

Knight, R. L., and Hutchinson, J. B. 1950. The evolution of black arm resistance in cotton. *J. Genet.* 50: 36–58.

Malavolta, V. A. Jr., Destefano, S. A. L., Beriam, L. O. S., Pizzinatto, M. A., and Cia, E. 2008. Cotton leaf blight, a new symptomatology caused by *Xanthomonas axonopodis* pv. *malvacearum*. *Summa Phytopathol.* 34: 168–171.

Nayudu, M. V. 1964. Variation in *Xanthomonas malvacearum*. *Indian Cott. Grow. Rev.* 18: 350–355.

Patel, M. K., and Kulkarni, Y. S. 1948. *Xanthomonas malvacearum* (Erw. F. Smith) Dowson on exotic cottons in India. *Curr. Sci.* 17: 243–244.

Schaad, N. W., Postnikova, E., Lacy, G., Sechler, A., Agarkova, I., et al. 2005. Reclassification of *Xanthomonas campestris* pv. *citri* (ex Hasse 1915) Dye 1978 forms A, B, C, D and E as *X. smithii* subsp. *citri* (ex Hasse) sp. nov. nom. rev. comb. nov.; *X. fuscans* subsp. *aurantifolii* (ex Gabriel 1989) sp. nov. nom. rev. comb. nov., and *X. alfalfae* subsp. *citrumelo* (ex Riker and Jones) Gabriel et al., 1989 sp. nov. nom. rev. comb. nov.; *X. campestris* pv. *malvacearum* (ex Smith 1901) Dye 1978 as *X. smithii* subsp. *smithii* sp. nov. comb. nov nom. nov.; *X. campestris* pv. *alfalfae* (ex Riker and Jones, 1935) Dye 1978 as *X. alfalfae* subsp. *alfalfae* (ex Riker *et al.*, 1935) sp. nov. nom. rev.; and "var. *fuscans*" of *X. campestris* pv. *phaseoli* (ex Smith, 1897) Dye 1978 as *X. fuscans* subsp. *fuscans* sp. nov. *Syst. Appl. Microbiol.* 28: 494–518.

Schaad, N. W., Postnikova, E., Lacy, G., Sechler, A., Agarkova, I., et al. 2007. List of new names and new combinations previously effectively, but not validly, published. *Int. J. Syst. Evol. Microbiol.* 57: 893–897. Effective Publication: Schaad, N. W., Postnikova, E., Lacy, G., Sechler, A., Agarkova, I., et al. 2006. Emended classification of xanthomonad pathogens on citrus. *Syst. Appl. Microbiol.* 29: 690–695.

Smith, E. F. 1901. The cultural characters of *Pseudomonas hyacinthi, Ps. campestris, Ps. phaseoli,* and *Ps. stewarti,* four one-flagellate yellow bacteria parasitic on plants. *Bull., Division of Vegetable Physiology and Pathology, U.S. Department of Agriculture* 28, 1–153.

Verma, J. P., Chowdhury, H. D., and Singh, R. P. 1979. Interactions between different races of *Xanthomonas malvacearum* in leaves of *Gossypium hirsutum*. *Z. Pflankrankh. Pflschutz.* 86: 460.

Verma, J. P., and Raj, S. 1992. Blight and wilt of cotton. pp. 138–160 in: *Plant Diseases of International Importance*. Vol. IV, *Diseases of Sugar, Forests and Plantation Crops*. A. N. Mukhopadhyay, J. Kumar, H. S. Chaube, and U. S. Singh eds. Prentice Hall, Englewood Cliffs, NJ.

Verma, J. P., and Singh, R. P. 1972. Bacterial blight of cotton. pp. 159–68 in: *Plant Bacteriology*. Vol. I, *Bacterial Diseases of Plants in India*. P. N. Patel ed. Indian Agricultural Research Institute, New Delhi, India.

Zhai, J., Xia, Z., Liu, W., Jiang, X., and Huang, X. 2013. Genomic sequencing globally identifies functional genes and potential virulence-related effectors of *Xanthomonas axonopodis* pv. *malvacearum*. *Eur. J. Plant Pathol.* 136: 657–663.

7.18 BACTERIAL BLIGHT OF SESAME

Bacterial blight of sesame (*Sesamum indicum* L.) was first reported from Sudan by Boughey in 1942. Sabet and Dowson (1960) described the symptoms and identified the bacterium as *Xanthomonas sesami*. The disease occurs in most of the sesame growing regions of the world and causes considerable losses under favorable weather conditions. The complete loss of crop, particularly under rain fed conditions, has been reported from Sudan (Sabet and Dowson 1960). Vijayat and Chakravarti (1977) reported 60% loss in the capsules under field conditions from Rajasthan, India. Approximately 20% loss in yield have been reported from Jabalpur area of Madhya Pradesh, India by Shukla et al. (1972). The disease has also been reported from Pakistan (Fardoos 2009) and Venezuela. Lee et al. (2005) reported the disease from Suwon, Hongchun, and Yeonchun, Korea.

7.18.1 Symptoms

The plants are susceptible at all the growth stages. Small, water-soaked, dark green, marginal spots develop on the lower surface of cotyledons of the seedlings developing from infected seeds. In severe infection, the lesions spread rapidly covering the whole cotyledons, which eventually dry. However, in mild infection, small, scattered, brown, necrotic spots appear on the cotyledons. In highly susceptible varieties, the infection of the growing point may cause the death of seedlings. In surviving seedlings, dark brown, water-soaked spots appear on the true leaves. These spots turn dark olive-green, increase in size (2–3 mm), and turn dark reddish-brown to black. The spots are initially round, but often attain angular shape. The spots may remain small or may coalesce to form large patches/blotches. Later on, the dead tissue of the spots dries out, becomes brittle, and ultimately the leaves fall. The infection may also start from the margin of the leaves and proceeds toward the midrib forming light brown spots. The infected leaves ultimately show severe blightening and finally dry.

On stem, the infection starts as light brown discoloration and spreads along the stem forming slightly raised dark reddish-brown lesions. Honey dew-like bacterial ooze comes out from the infected areas of the stem. On capsules, oval, slightly raised, dark reddish-brown lesions are produced. The infection also occurs on petioles and flowers.

Bashir et al. (2007) found that the combined infection of *Pseudomonas syringae* pv. *sesami* and *Xa. campestris* pv. *sesami*, causal agents of bacterial leaf spot and bacterial blight of sesame, respectively, resulted in more disease severity than the individual infections of these pathogens. The percentages of stem and leaf infection and defoliation were more in combined infection than the individual infections. Under such conditions, it becomes difficult to distinguish the symptoms of the individual diseases.

7.18.2 Causal Organism

Xanthomonas campestris pv. *sesami* (Sabet & Dowson 1960) Dye 1978
Syn. *Xa. sesami*

The bacterium being a pathovar of *Xa. campestris* species has the same morphological, cultural, and biochemical characters as those of *Xa. campestris*. It can be distinguished from *Xa. campestris* and its other pathovars on the basis of host range. The description of *Xa. campestris* is given under black rot of crucifers.

7.18.3 Disease Cycle

Infected seeds carry the bacterium and effectively transmit it to the seedlings. The bacterium survives in the infected seeds up to 16 months and in soil from 4 to 6 months (Habish and Hamad 1971). As the bacterium does not survive in the soil up to the sowing of the next crop, infected seed is the only source of perpetuation of the bacterium. The bacterium enters the host mainly through stomata and quickly invades vascular tissue. Secondary spread is mainly through the wind-splashed rain. The spread to the distant areas occurs through the infected seed (Chand and Singh 1972).

High temperature and high humidity favor the disease development. The disease mainly develops in the rainy season or when the relative humidity is high at night. Maximum disease intensity was observed in September 1993 and 1994 in Kanpur, India when the mean temperature ranged between 29.0°C and 29.4°C, relative humidity was 88.0%–90.5%, and the rainfall was 8.90–9.97 mm. There was considerable reduction in disease severity by the end of September due to the decrease in temperature, relative humidity, and rainfall (Srivastava et al. 1997). The severity of disease is also related to soil moisture (30%–40%) and relative humidity (75%–85%).

Age of leaves influences the severity of disease. The disease index on older leaves is much less than that on fully expanded young leaves. Multiplication of the pathogen in the artificially inoculated leaves reached maximum on 3rd day in both young and old leaves, but the number of bacterial cells in young leaves was 10 times more than those in old leaves (Chand et al. 1970). Nitrogen and photoperiod influence the susceptibility of the varieties. Application of nitrogen under long photoperiod increased the susceptibility of variety Venezuela 51, while its susceptibility decreased under the short photoperiod. However, nitrogen under short photoperiod increased the susceptibility of variety Early Russian with little effect under long photoperiod. Firdous et al. (2007) reported that tomato gave the better hypersensitive reaction than potato to the bacterium, and high relative humidity and 10^9 cfu mL^{-1} were required for the hypersensitive reaction.

Rehman et al. (2013) studied the effect of different concentrations of culture filtrate of *Xa. campestris* pv. *sesami* in Murashige and Skoog medium on seed germination, and height and root length of sesame seedlings, and reported that culture filtrate greatly affected the root length and height of seedlings, and slowed down the process of seed germination. The minimum root length and the smallest seedling height were obtained with 4% culture filtrate compared to normal growth of seedlings in the control.

Inam-ul-Haq et al. (2016) evaluated ELISA with polyclonal antibodies (PAbs) for the detection of *Xa. campestris* pv. *sesami*. PAbs were prepared in rabbits against pure culture of *Xa. campestris* pv. *sesami,* and direct antigen coated-ELISA was used for the analysis of pathogen in sesame germplasm obtained from National Agricultural Research Centre (NARC) (Islamabad), Pakistan. The pathogen from 29 sesame varieties, categorized from resistant to susceptible, was isolated and used as an antigen. After incubation, optical density was calculated at 405 nm. The results revealed that lowest reaction (0.068) was observed in highly resistant sesame genotypes, SG-22 and SG-55, while genotypes, SG-34, SG-33, and SG-72 were categorized as resistant based on reactivity (0.073). The authors concluded that ELISA with PAbs should be preferred for the detection of *Xa. campestris* pv. *sesami* as these are more stable, target multiple epitopes on single antigen and are easy to produce.

Naqvi et al. (2012) screened sesame germplasm against *Xa. campestris* pv. *sesami*. Among the 20 entries screened at Faisalabad, no line was found immune, highly resistant, or resistant, but only four lines were found moderately resistant, 14 were moderately susceptible, and two highly susceptible. Out of 119 lines of germplasm screened at NARC, none was found immune to the pathogen, three lines, SG-34, SG-22, and SG-55 were highly resistant, two lines, SG-72 and SG-33 were resistant, and ten lines were found to be moderately resistant. Forty-nine lines were categorized as moderately susceptible, 40 as susceptible, and 15 as highly susceptible.

7.18.4 Management

The following measures are helpful in reducing the disease incidence.

1. Use of resistant varieties like T 58
2. Sow the crop early, i.e., immediately after the onset of monsoon
3. Use of pathogen-free seed. In the absence of pathogen-free seed, the seed should be treated with hot water at 52°C for 10 min followed by soaking in streptomycin solution (100 μg mL^{-1}) for 6–8 h. Chand et al. (1970) found that seed treatment with Ceresan Wet (0.1%), streptomycin (100 μg mL^{-1}) and their combination for 80 min reduced seedling infection, but did not completely eliminate it, and the secondary spread was so rapid that there were no differences in disease incidence among the treated and untreated seed plots
4. Foliar sprays of Streptocycline solution (100 μg mL^{-1}) are recommended for the control of the disease, but the best control is achieved with the combination of seed treatment and foliar sprays. Rai and Srivastava (2003) obtained an effective control of the disease along with an increase in yield by seed treatment with Streptocycline (100 μg mL^{-1}) followed by three sprays of Streptocycline (100 μg mL^{-1}) + Blue Copper 50 (copper oxychloride at 500 μg mL^{-1}) at 10 days intervals

REFERENCES

Bashir, S., Irfan ul-Haque, M., Mukhtar, T., Irshad, G., and Hussain, M. A. 2007. Pathogenic variation in *Pseudomonas syringae* and *Xanthomonas campestris* pv. *sesami* associated with blight of sesame. *Pak. J. Bot.* 39: 939–943.

Chand, J. N., Mandloi, S. C., and Kulkarni, S. N. 1970. Studies on bacterial blight of sesamum caused by *Xanthomonas sesami*. Symposium on approaches to control important bacterial diseases of plants in India. Indian Phytopathol. Soc. Bull. No. 6, 19–23.

Chand, J. N., and Singh, D. 1972. Bacterial blight of sesamum. pp. 169–72 in: *Plant Bacteriology*. Vol. I, *Bacterial Diseases of Plants in India*. P. N. Patel ed. Indian Agricultural Research Institute, New Delhi, India.

Dye, D. W. 1978. Genus IX. *Xanthomonas* Dowson 1939. In: Young, J. M., Dye, D. W., Bradbury, J. F., Panagopoulos, C. G., and Robbs, C. F. 1978. A proposed nomenclature and classification for plant pathogenic bacteria. *N. Z. J. Agric. Res.* 21: 162–166.

Fardoos, S. 2009. Virulence Analysis of *Xanthomonas campestris* pv. *sesami* and *Pseudomonas syringae* pv. *sesami* The Causal Organisms of Sesame (*Sesamum indicum* L.) Bacterial Blight. PhD thesis, University of Arid Agriculture, Rawalpindi, Pakistan.

Firdous, S. S., Asghar, R., Haque, M. I., and Afzal, S. N. 2007. Development of hypersensitive response by *Xanthomonas campestris* pv. *sesami* on *Lycopersicon esculentum* L. and *Solanum tuberosum* L. leaves. *Pak. J. Bot.* 39: 2135–2139.

Habish, H. A., and Hamad, A. H. 1971. Survival and chemical control of *Xanthomonas sesami*. *F. A. O. Plant Prot. Bull.* No. 19: 36–40.

Inam-ul-Haq, M., Naqvi, S. F., Tahir, M. I., Rehman, H. M., Ahmed, R., et al. 2016. Serodiagnosis of *Xanthomonas campestris* pv. *sesami* causing bacterial leaf blight disease in sesame. *Pak. J. Agri. Sci.* 53: 535–539.

Lee, S.-D., Lee, J.-H., Kim, Y.-K., Heu, S.-G., and Ra, D.-S. 2005. Bacterial blight of sesame caused by *Xanthomonas campestris* pv. *sesami*. *Res. Plant Dis.* 11: 146–151.

Naqvi, S. F., Inam-ul-Haq, M., Tahir, M. I., and Mughal, S. M. 2012. Screening of sesame germplasm for resistance against the bacterial blight caused by *Xanthomonas campestris* pv. *sesami*. *Pak. J. Agri. Sci.* 49: 131–134.

Rai, M., and Srivastava, S. S. L. 2003. Field evaluation of some agrochemicals for management of bacterial blight of sesame. *Farm Sci. J.* 12: 81–82.

Rehman, I., Deeba, F., ul Haque, M. I., and Saqlan Naqvi, S. M. 2013. Inhibition of sesame seedling growth by *Xanthomonas campestris* pv. *sesami* culture secretions. *J. Animal Plant Sci.* 23: 1207–1210.

Sabet, K. A., and Dowson, W. J. 1960. Bacterial leaf spot of sesame (*Sesamum orientale* L.). *Phytopathol. Z.* 37: 252–258.

Shukla, B. N., Chand, J. N., and Kulkarni, S. N. 1972. Changes in sugar contents of sesame leaves infected with *Xanthomonas sesami*. *Nature* (London) 213 (5078): 813.

Srivastava, S. S. L., Rai, M., and Singh, D. V. 1997. Epidemiological study on bacterial blight of sesame. *Indian Phytopathol.* 50: 289–290.

Vijayat, R., and Chakravarti, B. P. 1977. Yield losses due to bacterial leaf spot of *Sesamum orientale* in Rajasthan. *Indian J. Plant Pathol.* 7: 97.

7.19 BACTERIAL LEAF SPOT OF SESAME

The disease was first observed by Malkoff in Bulgaria in 1903, and he named it as **Black rot**. Later on in 1906, Malkoff attributed it to *Bacterium sesami,* but the Koch's postulates of the bacterium were proved by Kovacevski in 1930. Since then, the disease has been reported from Sofia, the USA, Japan, Greece, Yugoslavia, Egypt, and India. In India, the disease was first reported from Delhi by Durgapal and Rao (1967). The disease has now spread to most states of India through infected seed. Isakeit et al. (2012) reported leaf spot of sesame caused by a *Xanthomonas* sp. from Texas, the United States.

7.19.1 Host Range

Sesame appears to be the only natural host although some other hosts are mildly infected artificially (Sutic and Dowson 1962). However, these hosts do not play any role in the disease cycle.

7.19.2 Symptoms

The symptoms develop on all the aboveground plant parts as blackish spots varying in size. On leaves, the spots appear oil-soaked and have a star-like external border, or they are delimited by the veins, giving them a distinct angular shape. From the spots, infection moves along the veins, turning these veins sooty in appearance. In severe infection, the spots may cover the entire leaf area. Finally, the infected leaves curl, die, and defoliate. The infection on the stem appears in the form of lesions, which generally encircle the stem forming rings. The lesions appear on the whole length of the stem and frequently coalesce forming large elongated areas. The capsules, infected at an early stage, turn black and do not bear seeds. Infection on older capsules results in the formation of minute spots (about 2 mm diameter). Severely infected plants may die.

7.19.3 Causal Organism

Pseudomonas syringae pv. *sesami* (Malkoff 1906) Young et al. 1978
Syn. *Psm. sesami*

The major characteristics of the bacterium are given below.

Bacterial cells are Gram-stain-negative, non-sporing, non-encapsulated rods measuring 0.6–0.8 by 1.2–3.8 μm, and motile by 2–5 polar flagella. The cells may occur singly or in pairs. Colonies on beef agar are circular, flat, striate, smooth, and white having entire margins. Green fluorescent pigment is produced in culture. The bacterium liquefies gelatin rapidly and gives an alkaline reaction in litmus milk without coagulation. The tests for starch hydrolysis, nitrate reduction, and production of indole and hydrogen sulphide are negative. Acid, but no gas is produced from glucose, however, from sucrose, lactose, and glycerol, even acid is not produced. Temperature for optimum growth is 30°C. The bacterium is facultatively aerobic.

Pathogenic variability exists in the bacterium. Strains differing in virulence and races giving differential reactions on sesame varieties have been reported. Thomas and Orellana (1962) identified two races of the pathogen, namely, race 1 and race 2. Race 2 infects variety Margo, while race 1 does not. Varieties Delco and Venezuela 51 are susceptible to both the races, while variety Early Russian is resistant. Two races are similar morphologically and also react alike in several standard physiological tests, but differ in glucose-asparagine ratio requirements. At high sugar levels, race 2 requires less asparagine than the race 1. Race 2 also liquefies gelatin and peptonizes milk faster than race 1.

Thomas and Orellana (1962, 1963) observed that the reactions of Margo, Early Russian, and Venezuela 51 varieties to two races of the pathogen were related to differences in concentration of aspartic acid, leucine, histidine, cysteine, and certain other amino acids and were associated closely with apparent differences in the ratio of reducing sugars to these amino acids. Delco and Venezuela 51, the most susceptible varieties, had the highest concentration of amino acids. Their findings substantiate that the two races differ in the glucose asparagine ratios at which they grow.

7.19.4 Disease Cycle

The bacterium is transmitted through infected seed, and this probably is the only source of perpetuation of the bacterium. Fawzi et al. (1991) showed that the bacterium is internally seed-borne. After germination, the bacteria from the infected seeds infect the cotyledonary leaves, and from there the bacteria spread to all the aboveground plant parts. The pathogen enters the host through stomata and wounds and the infection remains confined to the parenchymatous and vascular tissues. The infection of the capsules results in the infection of the seeds. Warm temperature and wind-splashed frequent rains are the most favorable factors for the development of epiphytotics (Chand and Singh 1972).

Firdous et al. (2009) conducted histological studies to elucidate the mode of infection of *Psm. syringae* pv. *sesami* and reported that periodical examination of cleared discs taken from inoculated leaves showed the presence and multiplication of bacterial cells in depressions and around trichome bases for 24 h before penetration through stomata and trichome basal cells. They further reported that appearance of pin-point-sized spots 3 days after inoculation coincided with chloroplast damage; and early appearance of small necrotic spots 4 days after inoculation revealed the presence of the bacterium in parenchymatous tissues, and from there it moved to transverse vascular systems.

Firdous et al. (2014) conducted further histopathological studies, using toluidine blue O, to elucidate the process of infection of *Psm. syringae* pv. *sesami* in a susceptible sesame genotype and reported that the bacterium was found as dark blue masses in infected tissues. The bacterium colonized substomatal and intercellular spaces of the spongy parenchyma cells, when initial water-soaking symptoms developed 2–3 days after inoculation. As water-soaking progressed, disruption of mesophyll cells occurred; mesophyll tissues were surrounded by bacterium followed by thinning and disruption of the cell walls. Later on, when bacterial cells increased in space previously occupied by mesophyll cells, there were empty spaces without any differentiation of host tissues. However, the bacterium was not found in vascular bundles (tracheary elements) of leaf and stem, but some phloem tissues of stem sections were found infected. The damage to the chloroplasts might have occurred due to the production of chlorosis or necrosis inducing toxins, and these toxins played a crucial role in the pathogenesis of *Psm. syringae* pv. *sesami* in sesame plants.

Application of nitrogen and photoperiod considerably influence the reaction of sesame varieties to the pathogen. Thomas (1965) studied the effect of 12 and 16 h photoperiod per day along with and without supplemental nitrogen on the reaction of resistant variety Early Russian and susceptible variety Venezuela 51. The most favorable combination for disease development was short light period coupled with supplemental nitrogen and with this combination both the varieties gave susceptible reaction. However, with either light period and without supplemental nitrogen, the original reactions of the varieties remained unchanged.

7.19.5 Management

1. Grow resistant varieties. White-seeded early varieties perform better than other varieties
2. Seed treatments and foliar sprays recommended for bacterial blight of sesame are also effective against this disease

REFERENCES

Chand, J. N., and Singh, D. 1972. Bacterial leaf spot of sesame. pp. 173–177 in: *Plant Bacteriology*. Vol. I, *Bacterial Diseases of Plants in India*. P. N. Patel ed. Indian Agricultural Research Institute, New Delhi, India.

Durgapal, J. C., and Rao, Y. P. 1967. Bacterial leaf spot of sesamum (*Sesamum orientale* L.) in India. *Indian Phytopathol.* 20: 178–179.

Fawzi, F. G., El Wakil, A. A., and El Deeb, A. A. 1991. Association of *Pseudomonas syringae* pv. *sesami* with some sesame seed lots in Egypt. *Egyptian J. Agric. Res.* 69: 685–693.

Firdous, S. S., Asghar, R., and Murtaza, G. 2014. Histopathology of leaf spot of sesame (*Sesamum indicum* L.) caused by *Pseudomonas syringae* pv. *sesami*. *J. Animal & Plant Sci.* 24: 814–819.

Firdous, S. S., Asghar, R., Ul-Haque, M. I., Waheed, A., Afzal, S. N., et al. 2009. Pathogenesis of *Pseudomonas syringae* pv. *sesami* associated with sesame (*Sesamum indicum* L.) bacterial leaf spot. *Pak. J. Bot.* 41: 927–934.

Isakeit, T., Hassett, B. T., and Ong, K. L. 2012. First report of leaf spot of sesame caused by *Xanthomonas* sp. in the United States. *Plant Dis.* 96: 1222.

Malkoff, K. 1906. Weitere Untersuchungen über die Bakterienkrankheit auf *Sesamum orientale*. *Centralblatt für Bakteriologie, Parasitenkunde und Infektionskrankheiten* 2, 16: 664–666.

Sutic, D., and Dowson, W. J. 1962. Bacterial leaf spot of sesamum in Yugoslavia. *Phytopathol. Z.* 45: 57–65.

Thomas, C. A. 1965. Effect of photoperiod and nitrogen on reaction of sesame to *Pseudomonas sesami* and *Xanthomonas sesami*. *Plant Dis. Rptr.* 49: 119–120.

Thomas, C. A., and Orellana, R. G. 1962. Resistance of sesame varieties and pathogenicity of strains of *Pseudomonas sesami* in relation to amino acids and reduced sugars (Abstr.). *Phytopathology* 52: 30.

Thomas, C. A., and Orellana, R. G. 1963. Amino compounds and reducing sugar in relation to resistance of sesame to *Pseudomonas sesami*. *Phytopathol. Z.* 46: 101–104.

Young, J. M., Dye, D. W., and Wilkie, J. P. 1978. Genus *Pseudomonas* Migula 1894. in: Young, J. M., Dye, D. W., Bradbury, J. F., Panagopoulos, C. G., and Robbs, C. F. 1978. A proposed nomenclature and classification for plant pathogenic bacteria. *N. Z. J. Agric. Res.* 21: 153–177.

7.20 SOYBEAN PUSTULE

Soybean [*Glycine max* (L.) Merril.] pustule was first described by Nakano (1919) from Japan. Hedges in 1922 reported it from USA and named the causal bacterium as *Bacterium phaseoli* var. *sojense*. The disease occurs in most soybean growing areas of the world where warm weather and frequent rains occur during the growing season. Janse (2005) has listed 54 countries, including USA, where the disease occurs. Severe infection causes premature defoliation that decreases yield by reducing size and number of seeds. In a study reporting 11% loss in yield, 63% of it was due to the reduction in seed size.

In India, the disease was first reported by Uppal et al. in 1938 from Jalgaon, Maharashtra. Now, it is present in most parts of the country (Patel and Jindal 1972).

7.20.1 Host Range

Besides soybean, the bacterium infects common bean (*Phaseolus vulgaris*), horse gram (*Macrotyloma uniflorum* = *Dolichos biflorus*), cowpea (*Vigna unguiculata*), lima bean (*P. lunatus*), redvine, buckwheat-vine (*Brunnichia cirrhosa*), and many other *Phaseolus* and *Vigna* spp.

7.20.2 Symptoms

The symptoms are confined mostly to the leaves and are most prevalent during the later growing season of the crop. The symptoms start as small, pale, yellowish-green spots (lesions) having dark reddish-brown centers, formed mostly on the upper leaf surface. As the disease progresses, the central part of each lesion develops into a minute, raised, light-colored pustule usually on the lower surface of the leaf. Sometimes, leaf spots are formed without developing pustules. The lesions merge with each other forming large, irregular areas of dead leaf tissue surrounded by yellowish margins. Dead areas may fall out due to the action of wind, giving the leaves a ragged appearance (Figure 7.13). Severe infection often results in partial defoliation. On the pods, small, reddish brown, slightly raised spots may develop on the susceptible cultivars.

The symptoms of soybean pustule can be confused with those of soybean rust. Mature rust pustules have a round opening on the upper leaf surface for the release of spores. In soybean pustule, if any opening is present, it is typically a linear crack across the surface of pustule. These features can only be seen with magnification of 20X. The absence of water-soaking of bacterial pustules and their raised centers distinguish this disease from bacterial blight of soybean. Moreover, in bacterial blight, the spots are more angular in shape (Giesler 2006).

7.20.3 Causal Organism

Xanthomonas axonopodis pv. *glycines* (Nakano 1919) Vauterin et al. 1995
Syn. *Xa. glycines*, *Xa. campestris* pv. *glycines*

The bacterium being a pathovar of *Xa. axonopodis* species has the same morphological, cultural, and biochemical characters as those of *Xa. axonopodis*. It can be distinguished from *Xa. axonopodis* and its other pathovars on the basis of host range. The description of *Xa. axonopodis* is given under bacterial leaf spot of green gram. Some additional characteristics of the bacterium are given below. Colonies on leaf infusion agar are small, circular, smooth having entire margin; initially pale yellow, but become deep yellow with age. The bacterium grows slowly in culture. The optimum growth occurs at 30°C–33°C. The maximum and minimum temperatures at which growth can take place are 38°C and 10°C, respectively.

FIGURE 7.13 Bacterial pustule of soybean: Leaf pustules. (Courtesy of Daren Mueller, Iowa State University, Bugwood.org.)

Virulence of *Xa. axonopodis* pv. *glycines* is associated with different genetic determinants, including a pathogenicity island that includes nine *hrp*, nine *hrc*, and eight *hpa* genes that are regulated by HrpG and HrpX. This cluster includes genes encoding a type III secretion system that is essential for pathogenesis and for induction of hypersensitive reaction (HR) on some plant species. The HrpG seems to be essential for induction of a non-specific HR on different non-host plants. Mutation of HrpG shows that it plays critical roles in HR on tobacco. Another gene that encodes a pectate lyase in *Xa. axonopodis* pv. *glycines* has also been shown to be necessary for induction of HR on tobacco and pepper, but not on tomato, sesame, and cucumber (Kaewnum et al. 2006). The above given information shows that *Xa. axonopodis* pv. *glycines* contains several genes that are required for disease development and induction of HR.

Athinuwat et al. (2009) identified three races of the bacterium on pustule disease resistant and susceptible soybean cultivars based on virulence phenotype. For race 3, an *avrBs3* homolog, *avrXg1* was identified that conferred resistance expressed as an HR on resistant cultivar Williams 82. Mutations in two predicted functional domains of *avrXg1* led to an increase in virulence on Williams 82 and an increase in bacterial population on susceptible cultivars. Expression of *avrXg1* in race 1 that is predicted to confer an unspecific HR, led to virulence on susceptible cultivars Spencer and PI 520733. Expression of *avrXg1* in race 2 that is predicted of carrying *avrBs3*-like genes, resulted in gained virulence and fitness of pathogen on both resistant and susceptible cultivars. These results demonstrate multi-functions of *avrXg1* depending on pathogen and plant genetic backgrounds.

7.20.4 Disease Cycle

The bacterium survives in the infected crop residue, infected seed, and in the rhizosphere of wheat roots. The bacterium can survive in the seed for 2½ years. The chances of survival in infected crop residue are more when the residue remains on the soil surface. The entry of the bacterium occurs through stomata and wounds. After entry into the host, the bacterium moves into intercellular spaces and multiplies. In response to infection, leaf cells increase in size and also divide at a faster rate. Increase in host growth and in bacterial cell mass in the localized area causes bulging of epidermal cells on both leaf surfaces. These raised areas rupture and become pustules.

The dissemination of the bacterium occurs through wind-splashed rain, overhead-irrigation, and cultural operations during wet foliage. Disease outbreaks occur after several days of rainstorms or hailstorms. The disease is favored by warm and wet weather. The optimum temperature for disease development is 30°C–33°C.

Sain and Gour (2013) reported that fraction of partially purified toxin(s) number 4 and 7, collected using sephadex gel chromatography, was found to be highly toxic to the host plants as well as other plants on detached leaf/twig and seedlings (more than 50% wilting). The partially purified toxin caused chlorosis in leaves of turnip, radish, castor, soybean, and mustard (>50%). Toxin treatment also reduced 100% seed germination of clusterbean, gram, coriander, turnip, and urad bean.

Athinuwat et al. (2018) reported that targeted mutations in *flgK* and *pilD* genes in strain KU-P-SW005 of *Xa. axonopodis* pv. *glycines* led to altered motility phenotypes. The *flgK* mutants lacked a monopolar flagellum and lost swimming motility, whereas the *pilD* mutant lacked type IV pili and was unable to move via twitching. The *flgK* and *pilD* mutants were also altered in biofilm production and caused reduced disease in susceptible soybean cultivar Spencer when compared to KU-P-SW005.

The typical symptoms of the disease are pustules surrounded by small yellow haloes. However, PI 96188 exhibits pustules without chlorotic haloes, which indicates a resistant response. Kim et al. (2011) studied the mode of inheritance of this novel symptom in PI 96188 and investigated whether the gene controlling disease resistance in PI 96188 is identical to the *rxp* gene or not. PI 96188 was crossed with the resistant cultivar SS2-2. All F_1 plants showed the same phenotype as SS2-2, and the F_2 population segregated into 75 typical symptoms (haloes present): 28 novel symptoms (haloes absent) indicating the presence of a single recessive gene. To map the novel symptom in PI 96188, a population of 88 F_7 recombinant inbred lines was developed from a cross between PI 96188 and the susceptible cultivar Jinjoo1. The disease resistance gene from PI 96188 was mapped on chromosome 10 (LG O) rather than chromosome 17 (LG D2). This gene was linked with the simple sequence repeat marker, Sat_108 at the distal end of chromosome 10. Thus, the disease resistance gene from PI 96188 was determined to be a new gene.

Narvel et al. (2001) mapped the *Rxp* locus in soybean that conditions reaction to bacterial pustule by simple sequence repeat marker analysis using a population of 116 F_4-derived lines from a cross between the resistant parent Young and the susceptible parent PI 416937. The *Rxp* locus was mapped 3.9 cM from Satt372 and 12.4 cM from Satt014 on linkage group D2. Linkage associations were confirmed by identifying a close association between the simple sequence repeat genotype at each locus identified as flanking *Rxp* and the bacterial pustule reaction of individual lines derived from a population different from the one used for mapping. An analysis of molecular pedigree showed that bacterial pustule-resistant cultivars inherited the resistance gene *rxp* from the ancestral cultivar CNS based on their consistent genotypic pattern at flanking marker loci. The results show that marker-assisted selection for *rxp* would be very effective.

Michalski et al. (1983) have developed a computer-based advisory system for diagnosing soybean diseases in Illinois.

7.20.5 Management

1. Use of resistant varieties is the most effective method for the management of this disease. The disease has been relegated to minor importance in the USA mainly through host resistance. The *Rxp* gene, originally present in cultivar CNS, a variety immune to soybean pustule, has been used to develop resistant varieties. *Rxp* gene confers resistance by requiring an increased number of bacterial cells for disease development. The *avrRxv* gene inhibits disease development by inducing a hypersensitive reaction. *Xa. axonopodis* pv. *glycines* does not require some other genes, required by other phytopathogenic bacteria, for elicitation of hypersensitive reaction
2. Crop rotation for 2–3 years
3. Destruction of infected crop residue by deep ploughing
4. Avoid or limit the use of overhead-irrigation
5. Avoid cultural operations when foliage is wet
6. Give preventive sprays of copper fungicides

REFERENCES

Athinuwat, D., Brooks, S., Burr, T. J., and Prathuangwong, S. 2018. Flagella and pili of *Xanthomonas axonopodis* pv. *glycines* are associated with motility, biofilm formation and virulence on soybean. *J. Phytopathol.* 166: 590–600.

Athinuwat, D., Prathuangwong, S., Cursino, L., and Burr, T. 2009. *Xanthomonas axonopodis* pv. *glycines* soybean cultivar virulence specificity is determined by *avrBs3* homolog *avrXg1*. *Phytopathology* 99: 996–1004.

Giesler, L. J. 2006. Bacterial diseases of soybean. University of Nebraska, Lincoln [http://extensionpublications.unl.edu/assets/pdf/g2058.pdf]. Accessed December 10, 2018.

Janse, J. D. 2005. *Phytobacteriology: Principles and Practice*. CABI Publishing, Wallingford, UK, 360 pp.

Kaewnum, S., Prathuangwong, S., and Burr, T. J. 2006. A pectate lyase homolog, *xagP*, in *Xanthomonas axonopodis* pv. *glycines* is associated with hypersensitive response induction on tobacco. *Phytopathology* 96: 1230–136.

Kim, K. H., Park, J.-H., Kim, M. Y., Heu, S., and Lee, S.-H. 2011. Genetic mapping of novel symptom in response to soybean bacterial leaf pustule in PI 96188. *J. Crop Sci. Biotechnol.* 14: 119–123.

Michalski, R. S., Davis, J. H., Bisht, V. S., and Sinclair, J. B. 1983. A computer-based advisory system for diagnosing soybean diseases in Illinois. *Plant Dis.* 67: 459–463.

Nakano, K. 1919. Soybean leaf spot. *J. Plant Prot., Tokyo* 6: 217–221.

Narvel, J. M., Jakkula, L. R., Phillips, D. V., Wang, T., Lee, S.-H., et al. 2001. Molecular mapping of *Rxp* conditioning reaction to bacterial pustule in soybean. *J. Heredity* 92: 267–270.

Patel, P. N., and Jindal, J. K. 1972. Bacterial pustule of soybean. pp. 59–65 in: *Plant Bacteriology*. Vol. I, *Bacterial Diseases of Plants in India*. P. N. Patel ed. Indian Agricultural Research Institute, New Delhi, India.

Sain, S. K., and Gour, H. N. 2013. Pathological and physio-biochemical characterization of *Xanthomonas axonopodis* pv. *glycines*, incitent of *Glycine max* leaf pustules. *Indian Phytopathol.* 66: 20–27.

Vauterin, L., Hoste, B., Kersters, K., and Swings, J. 1995. Reclassification of *Xanthomonas*. *Int. J. Syst. Bacteriol.* 45: 472–489.

7.21 BACTERIAL BLIGHT OF SOYBEAN

Bacterial blight of soybean [*Glycine max* (L.) Merril.] was first noticed by Heald in Nebraska, USA in 1905, who described the disease and suggested its bacterial nature. Johnson and Coerper in 1917 isolated the bacterium, proved its pathogenicity, and suggested it to be a member of genus *Pseudomonas*. It is the most common bacterial disease of soybean in areas where cool and wet weather prevails. The disease is most conspicuous in fields during midseason as the bacterium remains active under moist conditions, until checked by hot and dry weather. It occurs throughout the United States, where many cultivars are susceptible. Frequent stormy conditions and rains accompanied with cooler weather have made this disease quite common in Nebraska soybean fields. Williams and Nyvall (1980) reported yield reduction of 17.9%. The bacterium also considerably inhibits germination of infected seeds and inhibition up to 68% has been reported (Nicholson and Sinclair 1971). However, the disease has not caused significant economic losses in most areas of the world due to host resistance or plant tolerance. A survey undertaken in eight districts of Marathwada region of Maharashtra, India during *Kharif* seasons of 2009 and 2010 for recording the severity and incidence of the disease revealed the average disease incidence of 14.5% and maximum pod blight severity on variety JS-335 (Jagtap et al. 2012a).

The disease is widespread and occurs in Europe–Austria, Bulgaria, Czechoslovakia (former), Denmark, France, Germany, Hungary, Italy, Mainland Italy, Moldova, Poland, Romania, Russia Far East, southern Russia, Sweden, UK, Ukraine, and Yugoslavia (Fed. Rep.); Asia–Brunei Darussalam, China, Hebei, Heilongjiang, Henan, Jiangsu, Jilin, Liaoning, Shaanxi, Shandong, Zhejiang, India (Madhya Pradesh, Maharashtra), Indonesia, Japan, Kazakhstan, Korea Republic, Mongolia, Pakistan, Philippines, and Taiwan; Africa–Egypt; South Africa–Zambia and Zimbabwe; North America–Canada (Manitoba, Ontario), Mexico, USA (Alabama, Delaware, Florida, Hawaii, Illinois, Iowa, Minnesota, Missouri, Nebraska, North Carolina, North Dakota, Ohio, and Wisconsin); South America–Argentina, Brazil, Bahia, Goias, Parana, Rio Grande do Sul, Santa Catarina, Colombia, and Venezuela; and Oceania–Australia, New South Wales, Northern Territory, Queensland, South Australia, Victoria, and New Zealand (CABI 1999).

7.21.1 Host Range

Besides soybean, strains of the bacterium infect common bean, cowpea, lima bean, snap bean, and tepary bean.

7.21.2 Symptoms

The disease is most prevalent in early season in contrast to soybean pustule, which appears late. The symptoms appear on all the aboveground plant parts, but are most evident on leaves in the mid to upper canopy. Generally, the lesions first appear on cotyledons as brown spots on margins, which enlarge, and turn dark brown as the tissue collapses. Young plants may show stunting and may die if the infection reaches the growing point.

Symptoms are most conspicuous on leaves although they also appear on stems, petioles, and pods. Young leaves are most susceptible. Symptoms on the leaves start as small, angular, yellow to brown lesions. The lesions are usually water-soaked in the center and surrounded by yellowish-green haloes. Older lesions turn reddish-brown to black as the tissue dies. Angular lesions enlarge in cool and wet weather and may coalesce to produce large, irregular dead areas. Eventually the dried lesions fall out giving the leaves a ragged appearance (Figure 7.14). Often, the leaves are badly shredded after strong winds and stormy rains. Infected young leaves become stunted, distorted, and chlorotic. Severe infection results in defoliation of lower leaves. Large, black lesions may also develop on stem and petioles.

The symptoms on seed appear rarely. However, when lesions appear on pods, developing seeds may show some kind of symptoms. Small, water-soaked lesions appear on pods, which enlarge and merge giving rise to dark brown to black areas. The slimy bacterial ooze may cover the developing seeds. During storage, infected/infested seeds may become shriveled, discolored, and develop sunken or raised lesions. Infected seeds lack vigor and germinate poorly.

Chlorosis, also called **systemic toxemia**, develops throughout the infected plants. It results from the production of a toxin, coronatine, by the bacterium in the older infected leaves and translocated throughout the plant. Strains of the bacterium carrying the *avrB* gene elicit a hypersensitive reaction in cultivars carrying the resistant gene *Rpg1*.

The symptoms of bacterial blight differ from those of soybean pustule that in the former the lesions are water-soaked and flat in contrast to raised centers of lesions of the latter.

FIGURE 7.14 Bacterial blight of soybean. (Courtesy of Clemson University—USDA Cooperative Extension Slide Series, Bugwood.org.)

Bacterial blight can be confused with brown spot/Septoria leaf spot caused by *Septoria glycines*. The presence of yellowish-green halos around the bacterial blight lesions distinguishes it from the latter. Both the diseases can occur together on the same plant, but bacterial blight is more common on young upper leaves, while the brown spot is usually found on older, lower leaves of the plants.

7.21.3 Causal Organism

Pseudomonas savastanoi pv. *glycinea* (Coerper 1919) Gardan et al. 1992
Syn. *Psm. syringae* pv. *glycinea*, *Psm. glycinea*

Gardan et al. (1992) found that *Psm. syringae* subsp. *savastanoi* strains belong to a DNA relatedness group that includes strains of *Psm. syringae* pv. *glycinea* and *Psm. syringae* pv. *phaseolicola*. This DNA group was distinct from *Psm. syringae* pv. *syringae* (including the type strain of *Psm. syringae*). The results of a numerical analysis were in accordance with DNA hybridization data. Therefore, they elevated *Psm. syringae* subsp. *savastanoi* (Janse 1982) to species level as *Pseudomonas savastanoi* sp. nov., which includes *Psm. savastanoi* pv. *savastanoi*, *Psm. savastanoi* pv. *glycinea*, and *Psm. savastanoi* pv. *phaseolicola*.

Psm. savastanoi (*ex* Smith 1908) Gardan et al. 1992: The below given description of the species is based mainly on the publication of Gardan et al. (1992). Cells are Gram-stain-negative rods, measure 0.4–0.8 by 1.0–3.0 µm and are motile by means of one to four polar flagella. Rather slow growing. Colonies are white or creamy, smooth, flat, and glistening with entire or erose margins on Yeast extract Bacto peptone glucose agar (YBGA). Blue fluorescent pigment is produced under UV light on King's medium B. The optimal temperature for growth is 24°C–26°C. Metabolism is respiratory. Oxidase and arginine (Thornley) tests are negative and nitrates are not reduced. Esculin, gelatin, and starch are not hydrolyzed. Assimilates sucrose, L-arabinose, gluconate, caprylate, fumarate, DL-glycerate, L-malate, pyruvate, citrate, D-α-alanine, and L-proline, but does not assimilate lactose, L-xylose, adonitol, 2-aminobutyrate, DL-lactate, DL-3-hydroxybutyrate, D-(-)-tartrate, L-cysteine, L-methionine, and L-valine. Strains isolated from members of various genera of the *Oleaceae* and *Nerium oleander* do not produce levan, whereas strains isolated from *Fraxinus excelsior* (21%), *Phaseolus vulgaris*, and *G. max* produce levan. The G + C content of the DNA is 60 mol% (De Ley 1968). Produces a hypersensitive reaction on tobacco leaves.

Psm. savastanoi pv. *savastanoi* causes knots, galls, and cankers on members of the various genera of the *Oleaceae* and *N. oleander*, *Psm. savastanoi* pv. *glycinea* causes bacterial blight of soybean, and *Psm. savastanoi* pv. *phaseolicola* causes halo blight of bean.

Qi et al. (2011) sequenced genomes of two *Psm. savastanoi* pv. *glycinea* (*Psg*) strains, i.e., race 4 and B076, using 454 pyrosequencing. The genomes of both *Psg* strains share more than 4,900 highly conserved genes, indicating very low genetic diversity between *Psg* genomes. Though conserved, genome rearrangements and recombination events occur commonly within the two *Psg* genomes. When compared to each other, 437 and 163 specific genes were identified in B076 and race 4, respectively. Most specific genes are plasmid-borne, indicating that acquisition and maintenance of plasmids may represent a major mechanism to change the genetic composition of the genome and even acquire new virulence factors. Type three secretion gene clusters of *Psg* strains are near identical with that of *Psm. savastanoi* pv. *phaseolicola* (*Psp*) strain 1448A, and they shared 20 common effector genes. Moreover, the coronatine biosynthetic cluster is present on a large plasmid in strain B076, but not in race 4. *In silico* subtractive hybridization-based comparative genomic analyses with nine sequenced phytopathogenic pseudomonads; identified dozens of specific islands, and revealed that the genomes of *Psg* strains are more similar to those belonging to the same genomospecies such as *Psp* 1448A than to other phytopathogenic pseudomonads. The number of highly conserved genes (core genome) among them decreased dramatically when more genomes were included in the subtraction, suggesting the diversification of pseudomonads, and further indicating the genome heterogeneity among pseudomonads. However, the number of specific genes did not change significantly, suggesting these genes are indeed specific in *Psg* genomes. These results strengthen the idea of a species complex of *Psm. syringae* and support the reclassification of *Psm. syringae* into different species.

Gnanamanickam et al. (1982) reported a correlation between coronatine production by strains of *Psm. glycinea* in culture and induction of systemic symptoms (chlorosis and stunting) in inoculated soybean plants. Application of purified preparations of coronatine to unifoliate leaves of soybean plants resulted in localized chlorosis, development of chlorosis in subsequently developing trifoliate leaves, and stunting of plant growth, similar to symptoms induced by infection. Coronatine was demonstrated in soybean leaves infected with the bacterium, but was not detected in healthy leaves. However, some pathogenic strains do not produce coronatine, which indicates that it may not be essential for pathogenicity.

The phytotoxin coronatine, produced by the bacterium in a temperature-dependent manner, consists of a polyketide, coronafacic acid (CFA) and an amino acid derivative, coronamic acid. It is produced optimally at 18°C, whereas no detectable synthesis occurs at 28°C. Budde and Ullrich (2000) investigated the impact of temperature on coronatine production by *Psm. syringae* (= *Psm. savastanoi*) pv. *glycinea* PG4180 by inoculating soybean and tobacco plants. After spray inoculation, PG4180 caused typical bacterial blight symptoms on soybean plants when the bacteria were grown at 18°C prior to inoculation, but not when derived from cultures grown at 28°C. The disease outcome was quantified by determining the bacterial populations *in planta*. The temperature effect was not observed when PG4180 was artificially infiltrated into soybean leaves, indicating that bacterial invasion via natural plant openings and pre-inoculation temperature were important for phytotoxin synthesis. The inoculation of PG4180 on tobacco plants (incompatible host) elicited the HR irrespective of the bacterial pre-inoculation temperature. However, the HR was significantly delayed when tobacco plants were treated with cells of the CFA-overproducing derivative, PG4180.N9, which were derived from cultures grown at 18°C, compared with parallels incubated at 28°C. CFA biosynthesis by PG4180.N9 was optimal at 18°C and negligible at 28°C. The impact of CFA synthesis on the HR studied with different growth media, mutants, and transconjugants of PG4180, indicated that the amount of synthesized CFA, but not that of coronatine influenced the outcome of the HR. Experiments involving the application of purified coronafacoyl compounds suggested that the observed delay of the HR was mediated by CFA, providing further information on CFA's putative role as a molecular mimic of the plant signaling molecule, jasmonic acid.

7.21.4 Disease Cycle

The bacterium survives in infected/infested seed and the first evidence of its seed-plant transmission was provided by Kendrick and Gardner (1921). Yao et al. (2007) reported that the bacterium could be transmitted through infected seeds and the average incidence of infected seeds and seed-plant transmission rate were 7.6% and 0.48%, respectively. Infection/infestation of seed occurs through the pods during the growing season or at the time of threshing and cleaning.

Plasencia-Márquez et al. (2017) designed generic primers and evaluated the sensitivity of a BIO-PCR system to detect *Pseudomonas* spp. in soybean seeds. *Pseudomonas* strain isolated from soybean was diluted in seed extracts and sensitivity of the BIO-PCR system using generic primers was evaluated. The PCR using PgB1 and PgB2 generic primers was highly specific to detect members of *Pseudomonas* genus. Using BIO-PCR, it was possible to detect bacterial strain until 10^2 cfu mL^{-1}.

The pathogen also survives in infected crop residue, but the survival in crop residue is variable and is influenced by several factors. The survival is better under cold and dry conditions compared to warm and wet conditions. Fields with reduced tillage will tend to have increased bacterial blight as the bacterium resides in the residue. The continuous soybean cultivation, zero-tillage, or fields planted with seeds from an infected crop also show more disease severity. In Brazil, the bacterium overwinters on volunteer bean plants in winter wheat fields.

The bacterium grows epiphytically on leaves, buds, and stems of soybean plants without causing any symptoms. The epiphytic growth is influenced by the reaction of the host cultivar to the bacterium. On the leaves of susceptible host cultivars, the population increases 1,000-fold, while on resistant cultivars, it remains either unchanged or declines. This epiphytic phase may serve as a continuous source of inoculum.

It is typically an early season disease favored by cool and wet weather, which prevails during the early crop season. The optimum temperature for disease development is 21°C–27°C. Warm and dry weather after the midseason of crop prevents the disease development. Disease epiphytotics are commonly seen after windy rainstorms coupled with cool weather. The outbreaks that occur later in the season often follow windy rainstorms. Outbreaks typically develop several days after a rainstorm or hailstorm. Symptoms are most evident on new growth that is expanding at the time of the rain event.

Primary infection usually occurs on the cotyledons, and these lesions serve as a major source of secondary inoculum. The spread of the bacterium occurs through wind-splashed rain, overhead-irrigation, and cultural operations especially during wet foliage and rubbing of wet leaves with each other. After entering through stomata and wounds, bacteria multiply in the intercellular spaces of mesophyll cells. Typical water-soaked lesions develop 5–7 days after infection. During multiplication in the host, pathogen produces toxin called coronatine, which is translocated throughout the plant, mainly through the phloem. Toxin interferes with photosynthesis and chlorophyll production leading to systemic chlorosis and stunting of plants. Fully expanded trifoliate leaves do not show systemic chlorosis, while it is maximum on young trifoliate leaves. Infected plants recover after some time and perform normal photosynthetic activity and chlorophyll production, but stunting remains permanently. Toxin produced by the bacterium produces chlorotic lesions also on tobacco, bean, and tomato, indicating its non-host-specificity.

Four resistance genes, *Rpg1*, *Rpg2*, *Rpg3*, and *Rpg4*, have been identified, using a cross between cultivars, Flambeau × Merit. Line LL 489–605 is resistant. Soybean cultivars resistant to *Psm. syringae* pv. *glycinea* (*Psg*) exhibit a HR (necrotic) to infection. *Psg* strains carrying the *avrB* gene elicit the HR in soybean cultivars carrying the resistance gene *Rpg1*. *Psg* expressing *avrB* at a high level and capable of eliciting the HR in the absence of *de novo* bacterial RNA synthesis have been obtained in *in vitro* culture. Nutritional signals and regions within the *Psg* hrp gene cluster, an approximately 20-kilobase genomic region, also necessary for pathogenicity, control *avrB* transcription (Huynh et al. 1989).

Resistant varieties have been developed in various countries, but most cultivars have moderate levels of resistance. At least nine races of the pathogen have been identified and the occurrence of these races has complicated the breeding of resistant cultivars. Demonstration of a single dominant resistant gene, *Rpg1* in the host, together with molecular cloning of avirulence genes in the pathogen, suggested a gene-for-gene system (McGee 1992).

7.21.5 Management

The following measures will help in reducing the disease incidence.

1. Use of resistant varieties is the best method to manage the disease. Highly susceptible cultivars should not be planted in areas where the disease is a potential problem
2. Use of pathogen-free seed

3. Follow crop rotation with non-host crops for 2–3 years
4. Avoid or limit the cultural operations when the foliage is wet
5. Avoid or limit the overhead-irrigation
6. Practice deep tillage in fields where the disease is frequently severe. Cover soybean residues after harvest by tillage, if possible, or shred residue for quick decomposition. Differences in disease ratings and infestation occurred in one study after no-tillage and conventional tillage systems with 83.3% and 70.3% of plants infected, respectively (Anaele et al. 1990)
7. Biological control: *Bdellovibrio bacteriovorus*, a small comma-shaped bacterium isolated from the rhizosphere of soybean roots, inhibited development of local and systemic symptoms of bacterial blight when inoculated onto soybean with *Psm. glycinea* at ratios of 9:1 and 99:1, respectively (Scherff 1973a). A bacterium, characteristic of a group of yellow bacteria, isolated from soybean leaves infected with *Psm. glycinea* and designated as YB-3, was inoculated onto soybean leaves after mixing with *Psm. glycinea* cells. The mixture of *Psm. glycinea* and YB-3 in the ratio of 1:9 inhibited development of bacterial blight, while at 1:4 ratio, there was ca. 50% reduction in symptoms, and at 1:1 ratio there was no reduction in lesion development. There was complete inhibition of bacterial blight symptoms when 1:4 mixture of *Psm. glycinea* and YB-3 were incubated for 24–48 h before inoculation (Scherff 1973b). It has been shown that naturally occurring epiphytes isolated from soybean leaves have a high antagonistic efficacy against *Psm. syringae* pv. *glycinea* (Völksch et al. 1993; May et al. 1997). One *Erwinia herbicola* (= *Pantoea agglomerans*) and one *Psm. syringae* strain possessed the ability to suppress the growth of the pathogen and the ability to establish and maintain epiphytic populations on soybeans plants under field conditions (Völksch et al. 1996). This suggests that the spread of disease can be curtailed by these bacterial strains through competitive exclusion.

   The epiphyte *Psm. syringae* pv. *syringae* 22d/93 (Pss22d), isolated from soybean leaves, had been characterized as a promising and species-specific biocontrol strain *in vitro* and *in planta* against *Psm. syringae* pv. *glycinea* (Psg). Pss22d produces three toxins, namely, syringomycin, syringopeptin, and 3-methylarginine. In contrast to syringopeptin and syringomycin, 3-methylarginine inhibited the growth of Psg *in vitro*. Braun et al. (2010) reported that the wild-type Pss22d and its toxin-deficient mutants prevented development of disease symptoms normally caused by Psg. It is evident that none of the syringopeptin, syringomycin, and 3-methylarginine are required for Pss22d's antagonistic activity *in planta*. Consequently, factors other than the three toxins may contribute to the intra-species antagonism *in planta*
8. Use of chemicals: Jagtap et al. (2012b) reported that spray applications of Streptocycline (100 µg mL^{-1}) + copper oxychloride (0.25%) gave the lowest disease incidence (12.74%) along with highest seed yield (2,605 kg/ha) and test weight (14.33 g)

REFERENCES

Anaele, A. O., Pacumbaba, R. P., and Bishnoi, U. R. 1990. The effects of conventional and no-tillage systems on incidence of bacterial blight and with interaction of row spacings and weed control methods on the performance of other agronomic characters of soybean. *J. Agron. Crop Sci.* 165: 312–318.

Braun, S. D., Hofmann, J., Wensing, A., Weingart, H., and Ullrich, M. S. 2010. *In vitro* antibiosis by *Pseudomonas syringae* Pss22d, acting against the bacterial blight pathogen of soybean plants, does not influence *in planta* biocontrol. *J. Phytopathol.* 158: 288–295.

Budde, I. P., and Ullrich, M. S. 2000. Interactions of *Pseudomonas syringae* pv. *glycinea* with host and non-host plants in relation to temperature and phytotoxin synthesis. *Mol. Plant-Microbe Interact.* 13: 951–961.

CABI; EPPO. Distribution maps of plant diseases 1999 No. April (Edition 1) pp. Map 782.

Coerper, F. M. 1919. Bacterial blight of soybean. *J. Agr. Res.* 18: 179–194.

De Ley, J. 1968. DNA base composition and hybridization in the taxonomy of phytopathogenic bacteria. *Annu. Rev. Phytopathol.* 6: 63–90.

Gardan, L., Bollet, C., Ghorrah, M. A., Grimont, F., and Grimont P. A. D. 1992. DNA relatedness among the pathovar strains of *Pseudomonas syringae* subsp. *savastanoi* Janse 1982 and proposal of *Pseudomonas savastanoi* sp. nov. *Int. J. Syst. Bacteriol.* 42: 606–612.

Gnanamanickam, S. S., Starratt, A. N., and Ward, E. W. B. 1982. Coronatine production *in vitro* and *in vivo* and its relation to symptom development in bacterial blight of soybean. *Can. J. Bot.* 60: 645–650.

Huynh, T. V., Dahlbeck, D., and Staskawicz, B. J. 1989. Bacterial blight of soybean: Regulation of a pathogen gene determining host cultivar specificity. *Science* 245(4924): 1374–1377.

Jagtap, G. P., Dhopte, S. B., and Dey, U. 2012a. Survey, surveillance and cultural characteristics of bacterial blight of soybean. *Afr. J. Agric. Res.* 7: 4559–4563.

Jagtap, G. P., Dhopte, S. B., and Dey, U. 2012b. Bio-efficacy of different antibacterial antibiotic, plant extracts and bioagents against bacterial blight of soybean caused by *Pseudomonas syringae* pv. *glycinea*. *Scient. J. Microbiol.* 1: 1–9.

Janse, J. D. 1982. *Pseudomonas syringae* subsp. *savastanoi* (ex Smith) subsp. nov., nom. rev., the bacterium causing excrescences on *Oleaceae* and *Nerium oleander* L. *Int. J. Syst. Bacteriol.* 32:166–169.

Kendrick, J. B., and Gardner, M. W. 1921. Seed transmission of soybean bacterial blight. *Phytopathology* 11: 339.

May, R., Volksch, B., and Kampmann, G. 1997. Antagonistic activities of epiphytic bacteria from soybean leaves against *Pseudomonas syringae* pv. *glycinea in vitro* and *in planta*. *Microb. Ecol.* 34: 118–124.

McGee, D. C. 1992. Soybean diseases: A reference source for seed technologists. American Phytopathological Society, St. Paul, MN, 151 pp.
Nicholson, J. F., and Sinclair, J. B. 1971. Amsoy soybean seed germination inhibited by *Pseudomonas glycinea*. *Phytopathology* 61: 1390–1393.
Plasencia-Márquez, O., Corzo, M., Rivero, D., and Martínez-Zubiaur, Y. 2017. BIO-PCR for *Pseudomonas* spp. detection in soybean (*Glycine max*) seeds (Abstr.). *Phytopathology* 107: S4.12.
Qi, M., Wang, D., Bradley, C. A., and Zhao, Y. 2011. Genome sequence analyses of *Pseudomonas savastanoi* pv. *glycinea* and subtractive hybridization-based comparative genomics with nine pseudomonads. *PLoS One*. 6(1): e16451.
Scherff, R. H. 1973a. Control of bacterial blight of soybean by *Bdellovibrio bacteriovorus*. *Phytopathology* 63: 400–402.
Scherff, R. H. 1973b. Bacterial blight of soybeans as influenced by populations of yellow bacteria on leaves and buds. *Phytopathology* 63: 752–755.
Smith, E. F. 1908. Recent studies of the olive-tubercle organism. Bulletin, Bureau of Plant Industry, US Department of Agriculture 131, 25–43.
Völksch, B., Nüske, J., and May, R. 1996. Characterization of two epiphytic bacteria from soybean leaves with antagonistic activities against *Pseudomonas syringae* pv. *glycinea*. *J. Basic Microbiol.* 36: 453–462.
Völksch, B., Ullrich, M. S., and Fritsche, W. 1993. Occurrence of anti-microbial activities of bacteria from soybean leaf spots. *J. Basic Microbiol.* 33: 349–355.
Williams, D. J., and Nyvall, R. F. 1980. Leaf infection and yield losses caused by brown spot and bacterial blight diseases of soybean. *Phytopathology* 70: 900–902.
Yao, H., Xie, G., and Jin, Y. 2007. Study on the causal agent of bacterial leaf spot of soybean in Shanghai. *Acta Agric. Shanghai* 23: 41–45.

7.22 OLIVE KNOT AND OLEANDER KNOT DISEASES

Olive knot disease of olive trees (*Olea europaea* L.) is of worldwide distribution in olive-growing regions. It has been reported from Algeria, Argentina, Australia, Austria, Brazil, Colombia, Cyprus, France, Greece, Iran, Iraq, Israel, Libya, Mexico, Morocco, the Netherlands, New Zealand, Peru, Poland, South Africa, Spain, Sweden, Switzerland, Tanzania, Tunisia, Turkey, UK, Uruguay, USA, former USSR, and former Yugoslavia. Hijazin and Khlaif (2005) reported the occurrence of the disease on oleander, jasmine, and ziziphus from Jordan. The reports of its occurrence have also been made from Egypt (Ahmad et al. 2009) and Nepal (Balestra et al. 2009). Godena et al. (2012) reported it from Istria county of Croatia. Tsuji et al. (2015) reported the occurrence of the disease from Shizuoka Prefecture, Japan.

The increased cultivation of the Manzanillo olive, a highly susceptible cultivar, and the introduction of super high-density olive plantings for oil production, a system where mechanized cultural practices can promote disease development, have increased the incidence and severity of the disease. The reduced growth of the trees adversely affects the fruit yield, and probably quality and flavor of the olive oil. Moreover, severe symptoms of olive knot detract from the aesthetics of olive trees used in commercial and private landscapes.

The effect of the disease on the olive crop is not well understood, but reductions in yield can be expected where heavy infection has affected tree structure (Teviotdale 1994). There are indications that heavy infection affects olive quality and yield (Schroth et al. 1968, 1973). Schroth et al. (1973) reported that olive trees having moderate infection (0.51–1.0 knot/0.3 m of fruitwood) yielded 94.6 kg olives per tree, while those showing light infection (0.1–0.3 knot/0.3 m of fruitwood) yielded 121.3 kg/tree. The average income per tree was $27.30 for moderately infected trees and $36.77 for lightly infected trees. On an acre basis, the loss is around $403.38.

The adverse effects of the disease on oil yield and quality have not received much attention, although several authors have suggested this. Schroth et al. (1968) reported green table olives from infected trees to have an off-flavor (bitter, salty, sour, or rancid). In another study, the samples of olives of cultivar Frantoio were collected from two areas of Umbria, Algeria from healthy trees and trees having intermediate and severe disease severity, and the oil was extracted. It was found that the bacterial infection did not cause any adverse effect on yield, fruit weight, or oil quality (acidity, peroxide value). In fact, all the oil samples were extra virgin grade. It was also demonstrated that total polyphenols content was not related to the disease. The conclusions of the sensory analysis were that it is not possible for consumers to detect defects in oil quality linked to the disease. More detailed studies are needed to have a better understanding of the effects of disease on the yield and quality of fruits and oil.

Pseudomonas savastanoi pv. *nerii* causes oleander (*Nerium oleander*, family *Apocynaceae*) knot. Although *Psm. savastanoi* pv. *nerii* is more commonly associated with oleander knot, it can infect olive trees also and can cause olive knot. While *Psm. savastanoi* pv. *nerii* infects both oleander and olive, *Psm. savastanoi* pv. *savastanoi* infects only olive.

7.22.1 Host Range

Olive tree is the main host of *Psm. savastanoi* pv. *savastanoi*. Japanese privet (*Ligustrum japonicum*), jasmine (*Jasminum* spp.), *Phillyrea* sp., and Spanish broom (*Retama sphaerocarpa*), all belonging to family *Oleaceae*, have also been reported as its hosts (Alvarez et al. 1998). Bozkurt et al. (2014) reported pomegranate (*Punica granatum* L.) a new host of the bacterium from Turkey. Besenyei and Hevesi (2003) described another variety of *Psm. savastanoi* as the cause of galls on golden bells (*Forsythia intermedia*).

7.22.2 Symptoms

On olive, irregular, smooth to warty knots or galls are formed mainly on shoots and branches, but can also occur on roots, trunks, leaves, leaf petioles, and fruit stems (Figure 7.15). Initially, the knots are small (a few millimeters in diameter), soft, pale-green excrescences that expand, sometimes reaching several centimeters in diameter, and gradually turn greenish-brown or brown. Internally, the knots are

FIGURE 7.15 Olive knot. (Courtesy of Tyler Rood, www.oliveoil-commission.org.)

compact spongy tissues in which lysogenic cavities form that are filled with bacterial cells. With age, the knots crack, become partly necrotized, decay, and die within 6–8 months after their formation. Under certain circumstances, they can last for a longer period and increase in size. In case of heavy infections, when knots partially or completely girdle branches, the affected limbs are dwarfed, defoliated, or killed. On young plants in the field, the knots may cause weakness in the plant by partially girdling branches that form the structure of the trees. Although rare, infection of fruit may occur as neoplastic alteration of the mesocarp at the peduncle, preventing fruit development, and causing deformation or formation of numerous roughly circular, brown spots, 0.5–2.5 mm in diameter, initially raised and then depressed at the lenticels.

Symptoms often appear at leaf scars resulting from leaf drop, and wounds/cracks caused due to frost, hailstorms, insects, and careless pruning. The wounds caused during harvest of olives when branches are beaten with a stick, also aid in host infection.

Psm. savastanoi pv. *nerii* infects inflorescences and flowers in oleander. The bacterium can also spread internally via laticiferous ducts and produces a yellow halo around galls present on leaves. The inflorescences, fruits, and leaves of oleander often become distorted.

7.22.3 Causal Organisms

Olive knot

Pseudomonas savastanoi pv. *savastanoi* (*ex* Smith 1908) Young et al. 1996

Syn. *Psm. savastanoi*, *Psm. syringae* subsp. *savastanoi* pv. *oleae*, *Psm. syringae* pv. *savastanoi*

Oleander knot

Pseudomonas savastanoi pv. *nerii* (Janse 1982) Young et al. 1996

Syn. *Psm. nerii*, *Psm. syringae* subsp. *savastanoi* pv. *nerii*, *Psm. syringae* pv. *savastanoi*

Savastano (1889) in Italy described the olive knot bacterium for the first time as *Bacillus oleaetuberculosis*. Later, E.F. Smith (1908) in the USA named it as *Bacterium savastanoi* and completed the description. Gardan et al. (1992) found that *Psm. syringae* subsp. *savastanoi* strains belong to a DNA relatedness group that includes strains of *Psm. syringae* pv. *glycinea* and *Psm. syringae* pv. *phaseolicola,* and this DNA group was distinct from *Psm. syringae* pv. *syringae* (including the type strain of *Psm. syringae*). The results of a numerical analysis were also in accord with DNA hybridization data. Therefore, the authors elevated *Psm. syringae* subsp. *savastanoi* (Janse) 1982 to species level as *Pseudomonas savastanoi* sp. nov., which includes *Psm. savastanoi* pv. *savastanoi*, *Psm. savastanoi* pv. *glycinea*, and *Psm. savastanoi* pv. *phaseolicola*.

Psm. savastanoi pv. *savastanoi* is included in the certification procedure of olive in European countries (Martelli et al. 1995).

The below given description of the bacterium is based on Tsuji et al. (2015). The cells are Gram-stain-negative and aerobic. The colonies are smooth, circular, convex having entire margin, 1–2 mm in diameter on YP agar plates after 3 days of incubation at 27°C. The biochemical tests, for levan production, oxidase activity, potato soft rot, and arginine dihydrolase are negative, but positive for tobacco hypersensitive reaction and production of fluorescent pigment on King's medium B. All the strains did not hydrolyze casein, gelatin, and esculin.

Aragon et al. (2014), on the basis of a phylogenetic analysis of the *iaaM*/*iaaH* operon of *Psm. savastanoi* pv. *savastanoi*, which is involved in the biosynthesis of Indole-3-acetic acid (IAA), suggested that biosynthesis of the IAA, virulence and full fitness of the pathogen depended only on the functionality of the *iaaM-1*/*iaaH-1* operon. In contrast, the *iaaM-2*/*iaaH-2* operon, which carries a 22-nt insertion in the *iaaM-2* gene, did not contribute to the production of IAA by this bacterium. Hijazin and Khlaif (2005) found that PCR was more efficient in detecting olive knot bacteria by *iaaL* gene amplification than by isolation on KB medium.

7.22.4 Disease Cycle

Psm. savastanoi pv. *savastanoi* survives in gall tissues and also as an epiphyte on twigs, leaves, and fruit. The bacterium can occur in an epiphytic phase, especially in spring and autumn (Ercolani 1978). As the pathogen survives better on rough bark surfaces than on foliage, pathogen populations are higher on twigs than on leaves. The populations of the bacterium on plant surfaces vary throughout the year, maximum being during the rainy season. Quesada et al. (2007) monitored the seasonal dynamics of *Psm. savastanoi* pv. *savastanoi* on stems and leaves of symptomless shoots of naturally infected olive trees in Spanish orchards over 3 years and found significant differences between summer and the rest of seasons. The highest population of the bacterium occurred in warm, rainy months, while low numbers were generally found in hot and dry months. Epiphytic populations of bacteria move from inoculated to non-inoculated trees and disease severity, i.e., the number of galls per tree is directly related to the magnitude of the epiphytic pathogen population (Quesada et al. 2010).

The epiphytic pathogen populations and bacterial ooze emitted from galls serve as primary inoculum for new infections. The pathogen can be disseminated both within a plant and to neighboring plants by rain splashes and insects. In the latter case, the spread may be accidental by mining insects, but a more specific relationship exists for *Psm. syringae* subsp. *savastanoi* pv. *oleae* and the olive fly (*Bactrocera* = *Dacus oleae*), where the bacterium was found in the intestinal tract and salivary glands of the fly (Petri 1910). The long distance dissemination occurs mainly through infected nursery stock and to a lesser extent through contaminated pruning tools.

The pathogen can enter the plant through natural openings that occur when the tree drops its leaves, flowers, or fruit, through wounds resulting from natural events such as frost injury or hail damage, or wounds caused by cultural practices such as pruning and harvesting. The leaf scars are the most common points of pathogen entry. Leaf scar abscissions are most susceptible to infection during the first 2 days after leaf fall and remain susceptible for 7 more days (Teviotdale and Krueger 2004). Pruning wounds remain susceptible to infection for at least 14 days. Although olives are evergreen and leaves drop throughout the year, the tree susceptibility increases during rains resulting from increased leaf fall and availability of fresh leaf scars serving as infection courts. Other factors, such as frost damage or olive leaf spot disease, also enhance leaf drop and increase vulnerability to infection by *Psm. savastanoi* pv. *savastanoi*.

Generally, infections by *Psm. savastanoi* pv. *savastanoi* on olive remain localized, resulting in gall formation at the infection sites. Secondary galls, although formed rarely, can be initiated by bacterial movement within the xylem vessels of the olive. The secondary galls typically form in close proximity to the primary gall, and the potential of a plant to support secondary gall development can vary with cultivars. Although all olive cultivars are susceptible to the pathogen, the disease severity can vary with cultivars and age of plants. One-year-old plants are more susceptible to infection than 3-year-old plants. Penyalver et al. (2006) evaluated the effects of pathogen virulence, plant age, the dose/response relationship, and the induction of secondary tumors in olive inoculation assays. The severity of the disease in a given cultivar was strongly related to the pathogen dose applied at the wound sites. Secondary tumors developed in non-inoculated wounds away from the inoculation sites on the stem, suggesting the migration of the pathogen within olive plants. The inoculation of 29 olive cultivars with two pathogen strains at two inoculum doses showed that none of the cultivars were immune to the pathogen.

As the galls are formed only during the active tree growth, infections initiated during the winter do not develop into galls until spring. This latent period between infection and symptom development offers another undetected mode of pathogen dissemination, as movement of asymptomatic nursery stock can result in long-distance pathogen spread and introduction of the disease in disease-free landscapes or orchards. These asymptomatic plants may evade plant health inspections and allow for international movement of the pathogen. Lamichhane and Varvaro (2014) did not find the occurrence of olive knot pathogen on the phylloplane of 28 Italian olive cultivars, recently introduced to Nepal suggesting that the introduction of a new plant species in a given area does not necessarily introduce the pathogen when the propagation materials are rigorously supervised for pathogen exclusion.

Bertolini et al. (2003) designed a nested-PCR, to be performed in a single closed tube, for the detection of *Psm. savastanoi* pv. *savastanoi*. Nested-PCR coupled with dot-blot hybridization was able to detect up to one cell of the target per mL of olive extract, showing the greatest sensitivity compared with all previously reported detection assays. The newly developed nested-PCR assay detected the bacterium in 82 samples compared to the detection of the bacterium in 50 samples by the earlier reported techniques like bacterial isolation and single PCR.

Moretti et al. (2017) investigated genetic and phenotypic diversity of *Psm. savastanoi* pv. *savastanoi* by analyzing its 124 isolates collected from 15 countries of Mediterranean region and suggested that the present population of the bacterium is the result of clonally expansion of a single strain, that moderate migration of the pathogen occurred between countries, and that changes in virulence arose during its evolution.

7.22.5 Management

The best management of both olive and oleander knot diseases relies on exclusion of their pathogens. However, the reduction in their populations on the plant surface can also result into their effective management. The following measures should be adopted.

1. Use of disease-free nursery stock. Plants should be purchased after the spring because plants having latent infections do not show symptoms during winter
2. Remove the galls and infected branches and burn them. Severely infected trees should also be removed and burned. Do not remove the galls and prune during winter and spring rainy season. It should be done during dry summer months and all the tools should be sterilized with 10% bleach solution to prevent disease transmission both within and between trees
3. Minimize wounding of trees and do not apply overdose of fertilizers to avoid excessive vegetation
4. Commercial olive growers can use Bordeaux mixture, copper oxychloride, or copper hydroxide (to reduce epiphytic population of the pathogen) as a component of an integrated disease management program. Sprays of copper compounds are also recommended in the nursery provided there is no leaf injury under local conditions (Young 2004). Copper sprays should be applied in autumn and spring, before rain and after pruning and leaf fall, to protect wound sites.

Addition of an adjuvant may assist in penetration and effectiveness of the copper. Teviotdale and Krueger (2004) reported that the common practice of growers, i.e., of one post-harvest application of copper bactericide, provided only minimal protection and additional sprays in spring were needed to substantially improve disease control. However, no products are available for use on backyard trees. For the homeowners, a combination of cultural practices and sanitation is the most appropriate method of disease management

5. Nguyen et al. (2017) assessed quaternary ammonium compounds (QACs) as sanitizing agents for contaminated equipment as a disease management strategy. *In vitro* tests, QACs exhibited high toxicity against the bacterium over a broad pH range (6–9) using short exposure periods (15–60 seconds) and low concentrations (5 $\mu g\ mL^{-1}$). QACs applied to contaminated hard surfaces in the presence of an organic load reduced bacterial recovery by $\geq 3.6\ \log_{10}$ cfu mL^{-1}. Under field conditions, sanitation of hedging equipment that was contaminated with the pathogen (2×10^7 cfu mL^{-1}) and used to prune olives was successful and sometimes completely prevented new infections from occurring. Application of additional foliar spray treatments of copper or copper-kasugamycin mixtures after hedging significantly improved disease control. A non-phenolic QAC formulation, however, was ineffective as a preventative treatment when applied prior to inoculation of olive wounds, whereas a copper hydroxide application was highly effective. Based on these findings, a QAC formulation has been registered for field use as a sanitizer for olive equipment in California in 2015.

   Nguyen et al. (2018) found that in field trials, the application of copper (metallic copper equivalent of 1,260 $\mu g\ mL^{-1}$), kasugamycin (200 $\mu g\ mL^{-1}$), and copper-kasugamycin mixtures to inoculated wounds within 24 h of inoculation resulted in higher olive knot control than applications at later times. In greenhouse trials, copper or copper-kasugamycin applied to wounds 7 days before inoculation persisted and reduced knot incidence by >50%
6. Lavermicocca et al. (2002) reported that the application of 6,000 arbitrary units mL^{-1} of crude bacteriocin (dialyzed ammonium sulfate precipitate of culture supernatant) of *Psm. syringae* pv. *ciccaronei* at the inoculated V-shaped slits and leaf scars resulted in the formation of knots with weight values reduced by 81% and 51%, respectively, depending on the strains and inoculation method used, compared to the control. The bacteriocin application also reduced the multiplication of epiphytic populations of the pathogen, which after 30 days were at least 350 and 20 times lower than the control populations on twigs and on leaves, respectively
7. Use of resistant varieties: There is some resistance present in varieties of olive tree (Varvaro and Surico 1978). Cultivars, namely, Carolea, Koroneiki, Leccino, and Pendolino were found most tolerant in New Zealand (Young et al. 2004). Sisto et al. (2001) reported that cultivars Carolea, Bella di Spagna, Cerasella, Cima di Melfi, Coratina, Corniola, Dolce Agogia, Leucocarpa, Maiatica di Ferrandina, Nolca, and San Felice exhibited a moderate response to artificial infection
8. Do not harvest olives during rains. Harvesting by beating should be avoided or minimized when the disease is present, as beating causes wounds. Protecting the wounds with copper, either by spraying immediately after harvest or by wrapping beating sticks with cloth soaked in Kocide, will help to reduce the infection

REFERENCES

Ahmad, A. A., Moretti, C., Valentini, F., Hosni, T., Farag, N. S., et al. 2009. Olive knot caused by *Pseudomonas savastanoi* pv. *savastanoi* in Egypt. *J. Plant Pathol.* 91: 235.

Alvarez, F., Garcia de los Rios, J. E., Jimenez, P., Rojas, A., Riche, P., et al. 1998. Phenotypic variability in different strains of *Pseudomonas syringae* subsp. *savastanoi* isolated from different hosts. *Eur. J. Plant Pathol.* 104: 603–609.

Aragón, I. M., Pérez-Martínez, I., Moreno-Pérez, A., Cerezo, M., and Ramos, C. 2014. New insights into the role of indole-3-acetic acid in the virulence of *Pseudomonas savastanoi* pv. *savastanoi*. *FEMS Microbiol. Lett.* 356: 184–192.

Balestra, G. M., Lamichhane, J. R., Kshetri, M. B., Mazzaglia, A., and Varvaro, L. 2009. First report of olive knot caused by *Pseudomonas savastanoi* pv. *savastanoi* in Nepal. *Plant Pathol.* 58: 393.

Bertolini, E., Penyalver, R., García, A., Olmos, A., Quesada, J. M., et al. 2003. Highly sensitive detection of *Pseudomonas savastanoi* pv. *savastanoi* in asymptomatic olive plants by nested-PCR in a single closed tube. *J. Microbiol. Methods* 52: 261–266.

Besenyei, E., and Hevesi, M. 2003. Characterization of the *Pseudomonas savastanoi* pv. *forsythiae* pv. nov.- a novel pathovar of the knot disease bacterium. *Noveny Vedelem* 39: 123–128.

Bozkurt, I. A., Soylu, S. Mİrİk, M., Serce, C. U., Baysal, Ö. 2014. Characterization of bacterial knot disease caused by *Pseudomonas savastanoi* pv. *savastanoi* on pomegranate (*Punica granatum* L.) trees: A new host of the pathogen. *Lett. Appl. Microbiol.* 59: 520–527.

Ercolani, G. L. 1978. *Pseudomonas savastanoi* and other bacteria colonizing the surface of olive leaves in the field. *J. General Microbiol.* 109: 245–257.

Gardan, L., Bollet, C., Ghorrah, M. A., Grimont, F., and Grimont, P. A. D. 1992. DNA relatedness among the pathovar strains of *Pseudomonas sryingae* subsp. *savastanoi* Janse (1982) and proposal of *Pseudomonas savastanoi* sp. nov. *Int. J. Syst. Bacteriol.* 42: 606–612.

Godena, S., Dminić, I., and Đermić, E. 2012. Differential susceptibility of olive varieties to olive knot disease in Istria. *J. Central Eur. Agri.* 13: 85–94.

Hijazin, R., and Khlaif, H. 2005. Detection of *Pseudomonas savastanoi* pv. *savastanoi* from olive and other hosts by polymerase chain reaction (PCR). *Dirasat, Agric. Sci.* 32(1): 2005.

Janse, J. D. 1982. *Pseudomonas sryingae* subsp. *savastanoi* (ex Smith) subsp. nov., nom. rev., the bacterium causing excrescences on *Oleaceae* and *Nerium oleander* L. *Int. J. Syst. Bacteriol.* 32: 166–169.
Lamichhane, J. R., and Varvaro, L. 2014. Olive knot pathogen with pronounced epiphytic lifestyle is not present in association to leaf surface of European olive across the Himalayas in Nepal. *J. Phytopathol.* 162: 170–179.
Lavermicocca, P., Lonigro, S. L., Valerio, F., Evidente, A., and Visconti, A. 2002. Reduction of olive knot disease by a bacteriocin from *Pseudomonas syringae* pv. *ciccaronei*. *Appl. Environ. Microbiol.* 68: 1403–1407.
Martelli, G. P., Savino, V., Di Terlizzi, B., Catalano, L., Sabanadzovic, S., et al. 1995. Viruses and certification of olive in Apulia (Southern Italy). *Acta Hortic.* (ISHS) No. 386, 569–573.
Moretti, C., Vinatzer, B. A., Onofri, A., Valentini, F., and Buonaurio, R. 2017. Genetic and phenotypic diversity of Mediterranean populations of the olive knot pathogen, *Pseudomonas savastanoi* pv. *savastanoi*. *Plant Pathol.* 66: 595–605.
Nguyen, K. A., Förster, H., and Adaskaveg, J. E. 2017. Quaternary ammonium compounds as new sanitizers for reducing the spread of the olive knot pathogen on orchard equipment. *Plant Dis.* 101: 1188–1193.
Nguyen, K. A., Förster, H., and Adaskaveg, J. E. 2018. Efficacy of copper and new bactericides for managing olive knot in California. *Plant Dis.* 102: 892–898.
Penyalver, R., García, A., Ferrer, A., Bertolini, E., Quesada, J. M., et al. 2006. Factors affecting *Pseudomonas savastanoi* pv. *savastanoi* plant inoculations and their use for evaluation of olive cultivar susceptibility. *Phytopathology* 96: 313–319.
Petri, L. 1910. Untersuchungen über die Darmbakterien der Olivenfliege. *Zentralblatt für Bakteriologie und Parasitenkunde*. II. Abteilung 26, 357–367.
Quesada, J. M., García, A., Bertolini, E., López, M. M., and Penyalver, R. 2007. Recovery of *Pseudomonas savastanoi* pv. *savastanoi* from symptomless shoots of naturally infected olive trees. *Inter. Microbiol.* 10: 77–84.
Quesada, J. M., Penyalver, R., Pérez-Panadés, J., Salcedo, C. I., Carbonell, E. A., et al. 2010. Dissemination of *Pseudomonas savastanoi* pv. *savastanoi* populations and subsequent appearance of olive knot disease. *Plant Pathol.* 59: 262–269.
Savastano, L. 1889. Il bacillo della tuberculosi dell'olivo. Atti della Reale Accademia Nazionale dei Lincei. Rendiconti. Classe di scienze fisiche, matematiche e naturali 5, 92–94.
Schroth, M. N., Hildebrand, D. C., and O'Reilly, H. J. 1968. Off-flavor of olives from trees with olive knot tumors. *Phytopathology* 58: 524–525.
Schroth, M. N., Osgood, J. W., and Miller, T. D. 1973. Quantitative assessment of the effect of the olive knot disease on olive yield and quality. *Phytopathology* 63: 1064–1065.
Sisto, A., Lo Cantore, P., and Iacobellis, N. S. 2001. Preliminary results on the response of olive cultivars to artificial inoculation with *Pseudomonas syringae* subsp. *savastanoi*. *Proceedings of 11th Congress of the Mediterranean Phytopathological Union*, pp. 240–242.
Smith, E. F. 1908. Recent studies on the olive-tubercle organism. Bulletin of the Bureau of Plant Industry, US Department of Agriculture 131, 25–43.
Teviotdale, B. L. 1994. Diseases of olive. p. 107 in: *Olive Production Manual*. L. Ferguson, G. S. Sibbett, and G. C. Martin eds. University of California Publication No. 3353.
Teviotdale, B. L., and Krueger, W. H. 2004. Effects of timing of copper sprays, defoliation, rainfall, and inoculum concentration on incidence of olive knot disease. *Plant Dis.* 88 131–135.
Tsuji, M., Ohta, K., Tanaka, K., and Takikawa, Y. 2015. First report of knot disease on olive (*Olea europaea*) in Japan caused by *Pseudomonas savastanoi* pv. *savastanoi*. *Plant Dis.* 99: 1445.
Varvaro, L., and Surico, G. 1978. Behaviour of different olive (*Olea europaea* L.) cultivars to inoculation with *Pseudomonas savastanoi* (E. F. Smith) Stevens. *Phytopathol. Mediterr.* 17: 174–177.
Young, J. M. 2004. Olive knot disease and its pathogens. *Aust. Plant Pathol.* 33: 33–39.
Young, J. M., Saddler, G. E., Takikawa, Y., De Boer, S. H., Vauterin, L., et al. 1996. Names of plant pathogenic bacteria 1864–1995. *Rev. Plant Pathol.* 75: 721–763.
Young, J. M., Wilkie, J. P., Fletcher, M. J., Park, D.-C., Pennycook, S. R., et al. 2004. Relative tolerance of nine olive cultivars to *Pseudomonas savastanoi* causing bacterial knot disease. *Phytopathol. Mediterr.* 43: 395–402.

7.23 COMMON AND FUSCOUS BLIGHTS OF BEAN

Common bacterial blight or more frequently referred to, as common blight is an important disease of common bean (*Phaseolus vulgaris* L.). The disease causes significant yield losses in temperate and tropical areas and is one of the major constraints in dry bean production. Common blight was first reported in 1892 by Beach from New York and by Halsted from New Jersey. In 1897, E.F. Smith described the causal bacterium and named it as *Bacillus phaseoli*. Since its first report, the disease has been reported from many states of the USA and Canada; France, Hungary, Japan, Australia, Bulgaria, Madagascar, Norway, the Philippines, Spain, Turkey, New Zealand, Paraguay, Uruguay, Venezuela, Guatemala, Nicaragua, Honduras, Mexico, Argentina, Brazil, Colombia, Uganda, Burundi, Zaire, Zambia, Rwanda, and South Africa. It is not established in a number of bean growing countries of the EPPO region. Common blight pathogen is an A2 quarantine pest for EPPO. In India, Patel and Jindal first observed the disease in 1971 at Pune, Maharashtra, where it has been occurring for a long time. Osdaghi and Zademohamad (2016) reported the disease on lima bean (*P. lunatus* cv. Christmas) plants from East Azerbaijan and West Azerbaijan provinces, northwestern Iran, while it has already been reported on common bean and *mung* bean from several regions of Iran, including the central plain and southwestern provinces. Although the disease is named common blight, it is not necessarily the most common one in any given situation.

Singh and Miklas (2015), in their review article, have stated that considerable information is available on many aspects of the disease. However, Andean and Middle American common bean cultivars with high levels of combined resistance to less-aggressive and aggressive bacterial strains in all aerial plant parts are not available. They have reviewed the progress achieved in breeding for resistance, briefly describe problems faced, and discuss strategies for integrated genetic improvement to develop cultivars resistant to common blight.

The disease occurs in temperate, sub tropical, and tropical regions, but the maximum losses are caused in tropical areas because the most severe outbreaks of the disease occur in warm humid weather. The losses result from the reduction in photosynthetic area of plants and loss in quality and number of pods and seeds. In 1918, 75% of the fields in New York State of USA were affected with common blight and the disease caused heavy losses. In the subsequent years, losses ranged from 20% to 50%. In 1953, the disease was widespread in Nebraska (USA), and it caused an estimated loss of over 1 million US dollars. In 1976 in USA, common blight was economically the most important disease among the bacterial diseases of beans, and it caused an estimated loss of 4 million US dollars. The losses due to this disease in field bean crop in Ontario (Canada) were over 1,252 tons in 1970 and 218 tons in 1972. A model for the assessment of yield losses from bacterial blight has been developed in Ontario, using aerial infrared photographic surveys to assess disease levels. In 1967, yield losses up to 38% in dry bean in Ontario (Canada) and from 10% to 20% in navy bean in Michigan (USA) during 1976 were reported (Gilbertson and Maxwell 1992). Yield losses ranging from 17% to 22% were also reported from Colombia. In addition to loss in yield, the disease causes reduction in quality of seed through discoloration of infected seeds.

In 1930, Burkholder described two bacterial diseases of bean, namely, fuscous blight and bacterial brown spot. The causal bacterium of the fuscous blight, isolated by him from bean plants collected from Switzerland, was very close to the pathogen of common blight, and it also produced symptoms identical to common blight with some minor differences. Fuscous blight was first found in Switzerland in 1924. Subsequently it has been reported from Australia, Yugoslavia, Canada, New Zealand, South America, Germany, Hungary, and many other countries.

7.23.1 Host Range

In addition to common bean, the host range of *Xanthomonas axonopodis* pv. *phaseoli* (*Xap*) includes field bean, snap bean, lima bean, scarlet runner bean, hyacinth bean, soybean, *mung* bean, *urad* bean, *moth* bean, tepary bean, white flowering lupine, and fenugreek. *Xa. fuscans* subsp. *fuscans*, the causal agent of fuscous blight, infects field bean, snap bean, civet bean, and scarlet runner bean.

7.23.2 Symptoms

The symptoms of common and fuscous blights are similar and generally cannot be distinguished under field conditions. These diseases can be identified by the isolation and identification of their causal agents or using molecular tools.

Both the diseases are primarily foliar diseases, but symptoms also appear on other aboveground plant parts. The most characteristic and diagnostic symptoms appear on trifoliate leaves. Small, water-soaked lesions may develop on cotyledons and/or primary leaves and stems of seedlings raised from infected seeds.

On leaves, the symptoms start as small, water-soaked spots on the lower surface. The spots develop on the leaf margin or between the veins and there can be numerous spots on a leaf or leaflet. As these spots enlarge, their centers become necrotic and dry. The spots are surrounded by narrow bands of lemon yellow tissue, which are diagnostic features of the disease. The yellow halo may be small or absent if the temperature goes above 27°C. As the disease progresses, the spots enlarge, and on susceptible varieties continue to enlarge until they merge with each other forming large dead areas of various shapes. Dark discolored lesions also develop along the main vein, indicating systemic infection. The dry centers of the spots may tear and fall out giving the leaves a tattered and shredded appearance. Badly infected leaves fall and severely infected plants defoliate (Figure 7.16a).

Water-soaked, occasionally sunken lesions appear on the stem, which enlarge longitudinally and turn brown. The lesions frequently break open at the surface exuding bacterial ooze. The occurrence of these lesions is more near the vicinity of the first or cotyledonary node. The lesions may girdle the stem at the cotyledonary node, and this phase is known as **joint-rot**. Generally, the plants break off at this point due to the weight of the plant or action of the wind. On petioles and pulvinus, initially the lesions are water-soaked, but may turn brown later on (Patel and Jindal 1972).

(a)

(b)

FIGURE 7.16 Common bacterial blight of bean: (a) Symptoms on aerial plant parts. (b) Symptoms on pods. (Courtesy of Robert Harveson and Howard Schwartz, www.plantmanagement.network.org.)

On pods, small, circular, water-soaked spots appear which enlarge to form sunken, brick-red lesions. Pod lesions ultimately turn brown and vary in size depending upon the age of the pod when infection occurs. In humid weather, a slimy bacterial exudate may be formed on pod lesions, which eventually dries giving the lesions a glazed appearance (Figure 7.16b). In severely infected pods, seeds may be shriveled, discolored, or reduced in size. The butter-yellow discoloration of seeds is more prominent on white-seeded varieties in comparison to seeds of colored varieties.

Though the most symptoms of common and fuscous blights are similar, the intensity of seed discoloration and brown discoloration at the stem nodes is more in the latter disease. In fuscous blight, the lesions on the stem are also more inclined to split the stem (Walker 1969).

Common blight is often found in association with fuscous blight. Since both the diseases occur simultaneously in many cases, it becomes difficult to assess the amount of yield loss caused by them individually.

7.23.3 Causal Organisms

Common blight: *Xanthomonas axonopodis* pv. *phaseoli* (*Xap*) (Smith 1897) Vauterin et al. 1995

Syn. *Xa. campestris* pv. *phaseoli*, *Xa. phaseoli*

Fuscous blight: *Xa. fuscans* subsp. *fuscans* (*Xff*) Schaad et al. 2007, subsp. nov.

Syn. *Xa. campestris* pv. *phaseoli* var. *fuscans*, *Xa. axonopodis* pv. *phaseoli* var. *fuscans*

According to Young et al. (2004), *Xa. axonopodis* pv. *phaseoli* var. *fuscans* is not a valid name because it does not conform either to the code (Rule 28 b) or to the standards (Standard 18.1).

On sucrose peptone agar and yeast extract glucose calcium carbonate agar (YGCA), the colonies of *Xap* are yellow, convex, slimy and mucoid. On media containing tyrosine or on sucrose peptone agar, *Xff* produces a brown diffusible pigment, melanin. *Xff* strains also produce a dark internal pigment and thus, can be distinguished from *Xap* on MXP medium. Both the bacteria are strict aerobes, non-sporing, and Gram-stain-negative rods. Xanthomonadins, having absorption peaks between 443 and 446 nm, are produced. The metabolism of glucose is strictly oxidative. Maximum temperature for growth is 38°C. The other properties include hydrolysis of starch, casein, and gelatin; lipolysis of Tween 80, proteolysis of milk, lack of pectinase activity in culture, and inability to reduce nitrates, to produce lysine dicarboxylase and to induce hypersensitive reaction in tobacco. *Xap* and *Xff* are sufficiently distinct genetically to be classified into different groups based on 16S–23S intergenic spacer regions, restriction fragment length polymorphism, analysis of genomic and plasmid DNA, pulsed-field gel electrophoresis, DNA-DNA hybridization, and amplified DNA polymorphism.

Lopes et al. (2008) developed a semi-selective medium to detect *Xa. axonopodis* pv. *phaseoli* from bean seeds. The recovery frequency of bacterial colonies in relation to nutrient agar varied between 30% and 112%, depending upon the isolate.

Pathogenic variation exists in both the pathogens, but evidence for existence of races, based on common bean differential genotypes, has yet to be found. *Xff* strains are generally more virulent than *Xap* strains, and strains of the same pathogen from different locations also vary in virulence. Mutlu et al. (2008), on the basis of evaluation of common bean genotypes against 69 strains of *Xap* and 15 strains of *Xff*, found that *Xff* strains were more virulent than *Xap* strains having the same origin, and pathogenic variation was more within *Xap* strains than within *Xff* strains. Opio and Male-Kigiwa (1995) reported eight races of *Xap* based on their reaction in tepary bean (*P. acutifolius*).

Despite causing identical symptoms, *Xap* and *Xff* are genetically and phenotypically distinct species. No race structure exists for common bean, although pathogenic variability has been observed, particularly for *Xap* (O'Leary et al. 2016).

To assess the relationship between *Xap* and *Xff*, O'Leary et al. (2016) sequenced whole genomes of five *Xap* and three *Xff* isolates representative of global diversity. Complete genomes of all eight isolates were ~5 Mb, with single chromosome and one to three plasmids. One plasmid was conserved among all the isolates, whereas a second was conserved in the three globally predominant *Xap* and all *Xff* isolates. A brown diffusible pigment, produced in culture by *Xff* strains as the result of secretion and oxidation of homogentisic acid, was associated with a frame shift in the gene encoding homogentisate dioxygenase (hmgA). Deletion of the hmgA from *Xap* isolates lacking this mutation resulted in production of brown pigment, while complementation of a *Xff* isolate with a functional hmgA reduced pigment production substantially. This provides biological evidence that brown pigment production in *Xff* is due to the lack of a functional hmgA gene.

7.23.4 Disease Cycle

The most important source of perpetuation of both the bacteria is infected or contaminated seed. Common blight bacterium can survive in the infected seed for 30 years. Internally infected and externally contaminated, but apparently healthy seeds, are a more important inoculum source than severely infected seeds. As low as 0.5% infected seeds are sufficient to cause epiphytotics. The minimum number of bacteria per seed required to colonize seedlings is 10^{3-4}. The infection or contamination of the seed occurs in different ways. Seed infection occurs via vascular system where bacteria move through the pod pedicel into the funiculus and also through the raphe into the seed coat. Local pod infection can also result in contamination and/or infection of seed. The bacteria move through the pod wall and contaminate the surface of developing seeds or penetrate beneath the seed coat.

He and Munkvold (2012) reported that vacuum extraction and centrifugation of seed extracts increased sensitivity for detection of *Xap* in common bean seed and the highest sensitivity was obtained with the 3 h vacuum extraction followed by centrifugation. The bacterium was detected in 48 of

70 samples using the 3 h vacuum extraction with centrifugation, whereas only 35 of 70 field samples tested positive using overnight soaking, a significant difference. These steps would be valuable modifications to the current methods approved by the International Seed Testing Association.

Bastas and Sahin (2017) examined 198 dry bean seed samples of six cultivars collected from 12 provinces of Central Anatolia Region of Turkey to assess these seed lots as primary inoculum sources of the major bacterial pathogens. It was found that 22.72%, 13.63%, 11.11%, 1.51%, and 0.5% of seed samples tested were contaminated with five seed-borne bacterial pathogens, namely, *Psm. savastanoi* pv. *phaseolicola* (*Psp*), *Psm. syringae pv. syringae* (*Pss*), *Xap*, *Xa. axonopodis* pv. *phaseoli* var. *fuscans* (*Xapf*), and *Cur. flaccumfaciens* pv. *flaccumfaciens* (*Cff*), respectively. The results showed that *Psp* and *Pss* were found together on cv. Cali, *Psp* and *Xap* on cv. Dermason and cv. Sira, and *Pss* and *Xap* on cv. Seker, cv. Dermason, and cv. Cali. It is evident that selection of pathogen-free seed is very important for preventing the spread and effective management of seed-borne bacterial pathogens of bean.

Infected plant residue is another source of survival, particularly when it is not buried in soil and remains undecomposed and also in areas where more than one crop of beans is taken from the same plot in a year. Resident bacteria on weed hosts and non-host plants also act as a source of survival. Gilbertson et al. (1989) developed a DNA probe from plasmid DNA of *Xap* for the rapid detection and identification of *Xap* and *Xff* from bean plants and host debris. A 3.4 kilo base EcoRI fragment of *Xap* plasmid DNA, having a sequence that is repeated throughout the genomes of both the pathogens, was more specific than total plasmid DNA for this purpose.

The penetration of bacteria into the host occurs through stomata, hydathodes, and wounds caused by insects, hailstorms, and cultural operations. The dissemination of the bacteria occurs through the infected seed, wind-splashed rain, sprinkler irrigation, and implements. Seed is an important mean for short and long distance spread of the bacteria. The occurrence of devastating blight epiphytotics after wind-driven hailstorms is the result of extensive spread of the bacteria by rain splashes and their entry through wounds caused by hailstorms. Overhead-irrigation is another effective agent for the secondary spread of the bacteria. *Xap* can also be disseminated on the bodies of insects, such as *Diaprepes abbreviatus* and *Cerotoma ruficornis*, which may also provide avenues for the pathogen entry by creating wounds while feeding (Kaiser and Vakili 1978). White flies (*Bemisia tabaci*), grasshoppers (*Melanoplus* sp.), Mexican bean beetles (*Epilachna varivestris*), and leaf miners have also been reported to transmit the bacterium.

When infected seeds germinate, the bacteria multiply and proceed as vascular pathogen throughout the large xylem vessels causing blocking of the vascular system and eventually wilting of the plants. In most cases, the vascular infection results from seed-borne inoculum, but it can also occur from penetration of bacteria through wounds. Stem cankers and leaf lesions may result from the systemic infection. Alternatively, the bacteria may be splashed from the infected plant debris onto the seedlings where they cause lesions on cotyledons and primary and trifoliate leaves under favorable conditions. Secondary spread of the bacteria from these lesions occurs through disseminating agents.

After entry into the host tissue through stomata and wounds, the bacteria invade the intercellular spaces of the parenchyma tissue, dissolving the middle lamella slightly in advance. The affected cells collapse and are penetrated by the bacteria leading to their disintegration and formation of lysigenous cavities. The bacteria then invade the xylem vessels in the stem and advance systemically in the larger xylem elements. The bacteria multiply in the xylem cells and when large cell masses are produced, the bacteria either rupture or dissolve the cell walls and reenter the parenchyma tissues. From the parenchyma tissue, the cells come out on the stem surface through stomata in the form of ooze and after dissemination, enter the same stem and other plant parts. Hence, the bacteria invade both parenchyma and vascular tissue.

The disease development is favored by warm humid weather. The optimum temperature for disease development is 24°C–32°C. Under favorable weather conditions, bacteria multiply on the host surface and reach a threshold level before the symptoms develop. This threshold population is around 5×10^6 bacteria per 20 cm^2 of leaf tissue. This requirement of a minimum bacterial population for symptom development is probably the reason why the symptoms are usually observed during the reproductive stage of plant development. When the weather conditions are unfavorable for disease development, the bacteria colonize the plants without causing symptoms and become widespread on the plants.

Akhavan et al. (2013) reported that the epiphytic population of *Xap* has a critical role in the development of the disease and subsequent epiphytotics. The epiphytic population of the bacterium in the field has two major parts; solitary cells (potentially planktonic) and biofilms which are sources for providing and refreshing the solitary cell components. Mode of irrigation significantly effects the epiphytic population of *Xap*. The mean epiphytic population size in the field having an overhead-irrigation system is significantly higher than that of having furrow irrigation. A significant positive correlation between the epiphytic population size of *Xap* and disease severity has been reported in both the overhead-irrigated ($r = 0.64$) and the furrow-irrigated ($r = 0.44$) fields.

Age of host plant, air temperature, and host nutrition influence common blight development. The plant is susceptible at all growth stages, but young leaves are more susceptible than old leaves. Polysaccharide produced by *Xap* acts as a pathogenicity factor and disease severity is directly correlated with its amount produced by the pathogen. It possibly interferes with movement of water and/or nutrients leading to production of lesions and wilting. da Silva et al. (2009) observed that bacterization of bean seeds with a fluorescent *Pseudomonas* isolate (DFs 842) for 5 h at 10°C and challenge inoculation with *Xap* caused metabolic alterations in bean plants due to the increase in total soluble protein content and polyphenol oxidase activity. The increase in polyphenol oxidase and decrease in peroxidase activity induced resistance in bean plants against *Xap*.

Extracellular ATP (eATP) can function as a signaling molecule to regulate a wide range of cellular processes. Jiao et al. (2016) investigated the role of eATP in mediating the change of photosystem II (PSII) photochemistry of the tissues of bean leaves infected with *Xa. campestris* pv. *phaseoli*. Infection of the leaves with *Xa. campestris* pv. *phaseoli* caused a significant decrease in the PSII maximal photochemical efficiency (F_v/F_m), the maximum quantum efficiency of PSII photochemistry at illumination (F_v'/F_m'), the PSII operating efficiency (Φ_{PSII}), the rate of non-cyclic electron transport through PSII (ETR), photochemical quenching (qP), and eATP level in the tissues of the infected leaves. However, the levels of non-photochemical quenching (qN) and the quantum yield of regulated energy dissipation of PSII (Y(NPQ)) were significantly increased. Application of exogenous ATP at 0·2 mm to uninfected leaves had no significant effect on any of the chlorophyll fluorescence parameters measured. However, in the tissues of infected leaves, the application of exogenous ATP alleviated the decreases of the F_v/F_m, F_v'/F_m', Φ_{PSII}, ETR, qP, and eATP level, and also abolished the increases of qN and Y(NPQ). The findings suggest that the change of PSII photochemistry by pathogen infection could be mediated by eATP.

Xie et al. (2012) compared the effectiveness of image analysis (IA) with an ordinal visual scale for quantitative measurement of disease severity, its application in quantitative genetic studies, and its effect on the estimates of genetic parameters. Studies were performed using eight backcross-derived families of common bean (n = 172) segregating for the molecular marker SU91, known to be associated with a QTL for resistance to common bacterial blight (CBB) and fuscous blight. Even though both IA and visual assessments were highly repeatable, IA was more sensitive in detecting quantitative differences between bean genotypes. The CBB phenotypic difference between the two SU91 genotypic groups was consistently more than 5-fold for IA assessments, but generally only 2- to 3-fold for visual assessments. The findings suggest that the visual assessment results in overestimation of the effect of QTL in genetic studies. This may be due to the lack of additivity and uneven intervals of the visual scale. Although visual assessment of disease severity is a useful tool for general selection in breeding programs, assessments using IA may be more suitable for phenotypic evaluations in quantitative genetic studies involving CBB resistance as well as other foliar diseases.

Donmez et al. (2013) screened 36 bean genotypes and two commercial cultivars, commonly grown in Erzurum and Erzincan provinces located in the Eastern Anatolia region of Turkey, for resistance to *Xap* under greenhouse and field conditions during 2001–2002. Among the 38 genotypes tested, only 36K was found to be resistant. In field, disease severity reduced the quality and quantity of bean seed.

Bhat et al. (2017) screened 41 bean genotypes by artificial inoculation during 2009 to 2010 in Srinagar, India, to identify genotypes resistant to *Xa. campestris* pv. *phaseoli*. Out of 41 genotypes screened, 2 were categorized as resistant, 2 as moderately resistant, 21 as moderately susceptible, 13 as susceptible, and 3 as highly susceptible.

The epidemiology of fuscous blight is similar to that of common blight. Darsonval et al. (2009b) have shown the involvement of type III secretion system of *Xff* in its multiplication in the phyllosphere and systemic colonization of bean plants, leading to its transmission to seeds. *hrp* regulatory genes of the bacterium play a major role in host colonization processes. Strains with mutations in *hrp* regulatory genes, *hrpG* and *hrpX*, were impaired in their phyllospheric growth. Unlike the wild-type strain, strains with mutations in *hrp* genes were also not transmitted to seeds by vascular pathway. However, transmission of the bacterium to seeds by floral structures occurred with both wild and mutant strains. Darsonval et al. (2009a) also evaluated the role of bacterial attachment and biofilm formation in host colonization processes for *Xff* and found that the mutations in five adhesion genes, namely, *pilA*, *fhab*, *xadA1*, *xadA2*, and *yapH*, identified in *Xff*, altered the ability of all the mutants to adhere to polypropylene or seed. *PilA* was involved in adhesin and transmission to seed, and mutation of *PilA* led to lower pathogenicity on bean. *YapH* was required for adhesion to seed, leaves, and abiotic surfaces, but not for in planta transmission to seed or aggressiveness on leaves. However, none of the known adhesins were required for transmission to seed through floral structures. All mutants tested, except in *yapH*, were altered in their vascular transmission to seed.

7.23.5 Management

The management of common and fuscous blights is difficult due to the lack of resistant cultivars and effective chemicals; and also seed-borne nature of the pathogens. However, the satisfactory control of both the diseases can be achieved by adopting the following integrated control measures.

1. Use of certified pathogen-free seed for sowing. Treat the seed with streptomycin/Streptocycline because the seed dust may contain the bacteria. Hot water treatment at 52°C for 20 min or hot air treatment [23–32 h in dry air (60°C) at 45%–55% RH] has been reported effective. The latter treatment does not appear to affect seed viability (Gondreau and Samson 1994)
2. Use of resistant varieties is the best method for the management of both the diseases. However, the breeding for disease resistance is complicated as different genes control resistance in pods and leaves. Great northern varieties, namely, Beryl, Harris, Ivory, and Marquis are resistant to common blight. In comparison to common bean, scarlet runner bean has a higher level of resistance, while the tepary bean has the highest level of resistance to common blight
3. Destruction of infected plant debris by deep ploughing and by burying it in the soil
4. Follow crop rotation for 3 years with non-host crops. In Romania, rotation with maize reduced 85% disease incidence

5. Escape the disease by choosing the suitable planting date
6. Destruction of volunteer bean plants early in the following year and of weed hosts
7. Avoid or limit the use of sprinkler irrigation
8. Do not use the run off contaminated water for irrigation
9. Avoid cultural operations when plants are wet
10. Preventive sprays of copper fungicides (Kocide, etc.) and antibiotics are helpful in reducing the disease severity as they reduce the epiphytic population of the bacteria. These sprays should be started early in the season and repeated at 7–10 days intervals. The use of antibiotics is not encouraged as it can result in the development of antibiotic resistant strains of the bacterium

    Application of Bion (0.05%) and BioZell-2000B (50% etheric oil of thyme, 20% oil of maize, 20% oil of anise, 10% oil of sesame) caused 68% and 50% reduction in common bean blight incidence along with 50% and 45% reduction in bacterial population, respectively (Abo-Elyouser 2006)
11. Giorgio et al. (2016) reported that 60 out of 162 bean rhizobacteria inhibited *in vitro* the growth of selected virulent strains of both *Xa. axonopodis*. pv. *phaseoli,* and its variety *fuscans*; and when applied to seeds before sowing, six of them reduced disease symptoms on bean in *in vivo* and greenhouse experiments

REFERENCES

Abo-Elyouser, K. A. M. 2006. Induction of systemic acquired resistance against common blight of bean (*Phaseolus vulgaris*) caused by *Xanthomonas campestris* pv. *phaseoli. Egypt. J. Phytopathol.* 34: 41–50.

Akhavan, A., Bahar, M., Askarian, H., Lak, M. R., Nazemi, A., et al. 2013. Bean common bacterial blight: Pathogen epiphytic life and effect of irrigation practices. Springerplus, 2: 41.

Bastas, K. K., and Sahin, F. 2017. Evaluation of seedborne bacterial pathogens on common bean cultivars grown in central Anatolia region, Turkey. *Eur. J. Plant Pathol.* 147: 239–253.

Bhat, H. A., Ahmad, K., Ahanger, R. A., Wani, S. H., Bhat, A. H., et al. 2017. Resistant resources and transmission of bacterial blight of common beans (*Phaseolus vulgaris* L.). *J. Pharmacogn. Phytochem.* 6: 380–383.

Darsonval, A., Darrasse, A., Durrand, K., Bureau, C., Cesbron, S., et al. 2009a. Adhesion and fitness in the bean phyllosphere and transmission to seed of *Xanthomonas fuscans* subsp. *fuscans. Mol. Plant-Microbe Interact.* 22: 747–757.

Darsonval, A., Darrasse, A., Meyer, D., Demarty, M., Durand, K., et al. 2009b. The type III secretion system of *Xanthomonas fuscans* subsp. *fuscans* is involved in the phyllosphere colonization process and in transmission to seeds of susceptible beans. *Appl. Environ. Microbiol.* 74: 2669–2678.

da Silva, E. G., Moura, A. B., Bacarin, M. A., and Deuner, C. C. 2009. Metabolic alterations on bean plants originated from microbiolization of seeds with *Pseudomonas* sp. and inoculated with *Xanthomonas axonopodis* pv. *phaseoli. Summa Phytopathol.* 35: 98–104.

Donmez, M. F., Sahin, F., and Elkoca, E. 2013. Identification of bean genotypes from turkey resistance to common bacterial blight and halo blight diseases. *Acta Sci. Pol.* 12: 139–151.

Gilbertson, R. L. and Maxwell D. P. 1992. Common blight of beans. pp. 18–39 in: *Plant Diseases of International Importance*. Vol. II, *Diseases of Vegetables and Oilseed Crops*. H. S. Chaube, U. S. Singh, A. N. Mukhopadhyay, and J. Kumar eds. Prentice Hall, Englewood Cliffs, NJ.

Gilbertson, R. L., Maxwell, D. P., Hagedorn, D. J., and Leong, S. A. 1989. Development and application of a plasmid DNA probe for detection of bacteria causing common bacterial blight of bean. *Phytopathology* 79: 518–525.

Giorgio, A., Cantore, P. L., Shanmugaiah, V., Lamorte, D., and Iacobellis, N. S. 2016. Rhizobacteria isolated from common bean in southern Italy as potential biocontrol agents against common bacterial blight. *Eur. J. Plant Pathol.* 144: 297–309.

Gondreau, C., and Samson, R. 1994. A review of thermotherapy to free plant material from pathogens, especially seeds from bacteria. *Crit. Rev. Plant Sci.* 13: 57–75.

He, Y., and Munkvold, G. P. 2012. Comparison of extraction procedures for detection of *Xanthomonas axonopodis* pv. *phaseoli* in common bean seed. *Plant Pathol.* 61: 837–843.

Jiao, Q.-S., Feng, H.-Q., Tian, W.-Y., Bai, J.-Y., Sun, K., et al. 2016. Extracellular ATP functions in alleviating the decrease of PSII photochemistry caused by the infection of *Xanthomonas campestris* pv. *phaseoli. Plant Pathol.* 65: 819–825.

Kaiser, W. J., and Vakili, N. G. 1978. Insect transmission of pathogenic xanthomonads to bean and cowpea in Puerto Rico. *Phytopathology* 68: 1057–1063.

Lopes, L. P., Alves, P. F. R., Zandona, C., Nunes, M. P., and Mehta, Y. R. 2008. A semi-selective medium to detect *Xanthomonas axonopodis* pv. *phaseoli* in bean seeds and its eradication through seed treatment with tolyfluanid. *Summa Phytopathol.* 34: 287–288.

Mutlu, N., Vidaver, A. K., Coyne, D. P., Steadman, J. R., Lambrecht, P. A., et al. 2008. Differential pathogenicity of *Xanthomonas campestris* pv. *phaseoli* and *X. fuscans* subsp. *fuscans* strains on bean genotypes with common blight resistance. *Plant Dis.* 92: 546–554.

O'Leary, M., Coaker, G., and Gilbertson, R. 2016. Complete genome sequences of *Xanthomonas axonopodis* pv. *phaseoli* and *X. fuscans* subsp. *fuscans* generated by long-read sequence technology (Abstr.). *Phytopathology* 106: S4.173.

Opio, F., and Male-Kigiwa, S. 1995. The status of bean breeding in Uganda. Breeding for disease resistance with emphasis on durability. *Proceedings Regional Workshop for Eastern, Central and Southern Africa*, October 2–6, 1994, Njoro, Kenya, pp. 110–113.

Osdaghi, E., and Zademohamad, A. A. 2016. *Phaseolus lunatus*, a new host of *Xanthomonas axonopodis* pv. *phaseoli* in Iran. *J. Phytopathol.* 164: 56–60.

Patel, P. N., and Jindal, J. K. 1972. Common and fuscous blights of bean. pp. 44–58 in: *Plant Bacteriology*. Vol. I, *Bacterial Diseases of Plants in India*. P. N. Patel ed. Indian Agricultural Research Institute, New Delhi, India.

Schaad, N. W., Postnikova, E., Lacy, G., Sechler, A., Agarkova, I., et al. 2007. List of new names and new combinations previously effectively, but not validly, published. *Int. J. Syst. Evol. Microbiol.* 57: 893–897. Effective Publication: Schaad, N. W., Postnikova, E., Lacy, G., Sechler, A., Agarkova, I., et al. 2006. Emended classification of xanthomonad pathogens on citrus. *Syst. Appl. Microbiol.* 29: 690–695.

Singh, S. P., and Miklas, P. N. 2015. Breeding common bean for resistance to common blight: A review. *Crop Sci.* 55: 971–984.

Smith, E. F. 1897. Description of *Bacillus phaseoli* n. sp. *Botanical Gazette* 24, 192.

Vauterin, L., Hoste, B. Kersters, K., and Swings, J. 1995. Reclassification of *Xanthomonas*. *Int. J. Syst. Bacteriol.* 45: 472–489.

Walker, J. C. 1969. *Plant Pathology*. 3rd ed. McGraw Hill Book Co., New York, 819 pp.

Xie, W., Yu, K., Pauls, K. P., and Navabi, A. 2012. Application of image analysis in studies of quantitative disease resistance, exemplified using common bacterial blight-common bean pathosystem. *Phytopathology* 102: 434–442.

Young, J. M., Bull, C. T., De Boer, S. H., Firrao, G., Saddler, G. E., et al. 2004. Names of plant pathogenic bacteria, 1864–2004 [http://www.isppweb.org/names_ bacterial_revised.asp]

7.24 HALO BLIGHT OF BEAN

Halo blight of bean (*Phaseolus vulgaris* L.) was first described by Burkholder (1926) from the United States. It has many similarities with common blight of bean and causes serious economic losses under favorable weather conditions. The disease is worldwide in distribution wherever beans are grown under cool and wet climatic conditions. The occurrence of the disease has not been reported from India. Under favorable weather conditions, epiphytotics resulting from a few infected seeds may cause substantial yield losses.

7.24.1 Host Range

In addition to common bean, the host range of the bacterium includes field and snap beans, especially cranberry, red kidney, and yellow-eyed beans, lima bean and scarlet runner bean. Artificially, the bacterium infects tepary bean, adzuki bean, *mung* bean, soybean, *Macroptilium atropurpureum*, *P. polyanthus*, *P. polystachyus*, and *Pueraria lobata*. Taylor et al. (1996), in addition to main host *P. vulgaris*, reported ten alternative natural hosts, which included ten legume species representing seven different genera, namely, *Cajanus cajan*, *Desmodium* sp., *Lablab purpureus*, *M. atropurpureum*, *Neonotonia wightii*, *P. acutifolius*, *P. coccineus*, *P. lunatus*, *Vigna angularis*, and *V. radiata*. Of these alternative natural hosts, *Desmodium* sp. constituted a new host record.

7.24.2 Symptoms

Most symptoms of halo blight are similar to those of common blight. The symptoms of halo blight appear on all the aboveground plant parts. Water-soaked, round to irregular lesions are produced on cotyledons of seedlings raised from infected and/or infested seeds. Yellowish-brown to reddish-brown, water-soaked lesions appear on primary leaves. Severely infected plants show stunting and die.

Small (about 1.5 mm in diameter), angular, greasy, water-soaked spots first appear on under surface of trifoliate leaves. As these spots enlarge and turn reddish-brown, characteristic light greenish-yellow haloes, 10 mm or more in diameter, often but not always, appear around the spots, giving the leaves a yellowish appearance. The haloes, more prominent on young expanding leaves, appear due to the production of a toxin, phaseolotoxin by the bacterium and are a diagnostic symptom of the disease (Figure 7.17a). At temperature above 27°C, haloes formed are very small or may be absent. In 2004, Oguiza and associates reported the bacterial strains that do not have the genetic information for toxin production. In severe infections, the infection may become systemic causing yellowing and death of new foliage. On stem, greasy, reddish-brown, elongated streaks appear. On pods, symptoms begin as small, water-soaked spots or streaks along the pod sutures. The spots, later on, turn reddish-brown and become sunken in the center (Figure 7.17b). Under humid conditions, a light cream- or silver-colored bacterial exudate may appear on the lesions. Seeds in infected pods are usually symptomless. However, in severely infected pods, seeds may be shriveled, discolored, or reduced in size.

Pod symptoms of halo and common blight are similar. Therefore, the identification of these diseases should be based on leaf symptoms and identification of the causal bacteria. Leaf symptoms of halo blight and bacterial brown spot may be confused due to the variation in margin/width of halo influenced by the temperature in the former disease. In case of brown spot, the spots are brown, necrotic having a narrow yellow halo.

(a)

(b)

FIGURE 7.17 Halo blight of bean: (a) Symptoms on aerial plant parts. (b) Symptoms on pod. (Courtesy of Robert Harveson and Howard Schwartz, www.plantmanagement.network.org.)

7.24.3 Causal Organism

Pseudomonas savastanoi pv. *phaseolicola* (Burkholder 1926) Gardan et al. 1992
Syn. *Psm. phaseolicola, Psm. syringae* pv. *phaseolicola*

The below given description of the bacterium is taken mainly from Burkholder (1926), who classified it as *Phytomonas medicaginis* var. *phaseolicola* nov. var.

Cells are rods with rounded ends, occur singly or in pairs, and measure 0.5–1.25 × 1.5–3.0 µm. Gram-stain-negative, non-sporing, not acid-fast, strict aerobes, and motile with a polar flagellum. Colonies on nutrient agar are white to colorless or at times creamy, concentrically ringed with edges undulate. The growth of the colonies is slow to moderate. The bacterium does not reduce nitrate, does not liquefy gelatin, does not produce H_2S and indole, and does not digest starch. Litmus milk is turned alkaline without curd formation and peptonization. Pectate gel pitting occurs at pH 4.0. Weak acid production occurs in dextrose and saccharose, but no acid from lactose and maltose and also no gas production from these sugars. Growth occurs on quinate and trigonelline, but not on mannitol, inositol, sorbitol, erythritol, L-tartrate, D-tartrate, L-lactate, anthranilate, betaine, and homoserine. Its ability to use rhamnose and raffinose as a sole source of carbon and sodium nitrate as a sole source of nitrogen varies with different strains. Maximum growth and production of phaseolotoxin occurs at 21°C–22°C. The thermal death point is around 49°C.

Taylor et al. (1996) carried out detailed race determination investigations on 175 selected isolates of *Psm. syringae* pv. *phaseolicola*, representative of the different geographical regions and hosts in which the pathogen was found, and identified nine races. Races 1, 2, 5, 6, and 7 were distributed worldwide, and race 6 was predominant. Other races were found mainly in Africa; races 3 and 4 in East/Central Africa and races 8 and 9 in Southern Africa. Most isolates were obtained from the main host, *P. vulgaris*, while the alternative natural hosts included ten legume species belonging to seven different genera.

7.24.4 Disease Cycle

Infected and/or infested seed is the most important source of perpetuation of the bacterium. The bacterium can survive for more than 4 years in infected seed. One contaminated seed in 16,000 is sufficient to cause severe epiphytotic under favorable weather conditions (Dillard and Legard 1991). Infection of bean seeds occurs through lesions formed on pods and bacteria penetrate the seed coats without causing any visible symptoms. Infestation of seeds occurs during threshing resulting from adherence of dust of diseased tissue to seed surface and to wounds. As a source of primary inoculum, both infected and infested seeds are equally important (Goto 1992).

Popović et al. (2014) described a diagnostic method based on nested-PCR, followed by ELISA and conventional bacteriology tests, for the rapid and reliable detection of 18 strains of *Psm. savastanoi* pv. *phaseolicola* isolated from infected bean leaves and seeds. The bacterium formed white, small, and flat colonies on nutrient agar, creamy white, flat, and circular on Milk-Tween agar, and light yellow, convex, and shiny on modified sucrose peptone agar. On the basis of results of DAS- and pre T-cell antigen (PTA)-ELISA with respect to reactivity to specific antibodies, all the analyzed strains belonged to *Psm. savastanoi* pv. *phaseolicola.* On inoculation, the bacterium formed greasy spots on bean pods and cotyledonary leaves and did not cause hypersensitive reaction on the leaves of tobacco and geranium.

Infected plant debris also acts as a source of survival of the bacterium. The survival in crop residue is maximum when it is on the soil surface. The dissemination of the bacterium occurs through wind-splashed rain, overhead irrigation, windstorms, and farm machinery. Severe outbreaks of disease often occur after heavy rainstorms. The disease is favored by cool to moderate temperature (16°C–23°C) and humid moist conditions. Temperatures above 27°C inhibit the development of yellow haloes and systemic chlorosis. When weather becomes warmer and drier, plants recover from the disease and new growth is healthy.

Hydrogen sulphide (H_2S) is an important signaling molecule in both animals and plants, despite its toxic nature. In plants, H_2S controls stomatal apertures, thereby altering the ability of bacteria to invade plant tissues. Bacteria are known to generate H_2S as well as being exposed to plant-generated H_2S. During their interaction with plants, pathogenic bacteria are known to undergo alterations in their genomic complement. *Psp* strain 1302A undergoes loss of a section of DNA known as a genomic island (PSPGI-1) when exposed to the resistance response of plants. Loss of PSPGI-1 from *Psp* 1302A enables the pathogen to overcome the plants resistance response and cause disease.

Neale et al. (2017), with the use of H_2S donor molecules, investigated changes induced in *Psp* 1302A genome, as demonstrated by excision of PSPGI-1. *Psp* 1302A cells were found to be resistant to low concentrations of H_2S. However, at sub lethal H_2S concentrations an increase in the expression of the PSPGI-1 encoded integrase gene (*xerC*), which is responsible for island excision, and a subsequent increase in the presence of the circular form of PSPGI-1 were detected. It shows that H_2S is able to initiate excision of PSPGI-1 from the *Psp* genome. Therefore, H_2S that may emanate from the plant has an effect on the genome structure of invading bacteria and their ability to cause disease in plants. Modulation of such plant signals may be a way to increase plant defense responses for crops in the future.

Tock et al. (2017) constructed high-resolution linkage maps for three recombinant inbred populations to map resistance to *Psm. syringae* pv. *phaseolicola* race 6 derived from the two common bean lines. This was complemented with a genome-wide association study of race 6 resistance in an Andean Diversity Panel of common bean. The genome-wide association study detected one QTL (HB5.1) on chromosome Pv05 for resistance to race 6 with significant influence on seed yield. Race 6 resistance from PI 150414 maps to a single major-effect QTL

(HB4.2) on chromosome Pv04 and confers broad-spectrum resistance to eight other races of the pathogen. Resistance segregating in a Rojo × CAL 143 population maps to five chromosome arms and includes HB4.2. The same HB5.1 QTL, found in both Canadian Wonder × PI 150414 and Rojo × CAL 143 populations, was effective against race 6, but lacks broad resistance. This study provides evidence for marker-assisted breeding for more durable halo blight control in common bean by combining alleles of race-non-specific resistance (HB4.2 from PI 150414) and race-specific resistance (HB5.1 from cv. Rojo).

Bozkurt and Soylu (2011) studied the reaction of different bean cultivars to nine races of *Psm. syringae* pv. *phaseolicola*. The virulent bacterial races produced water-soaked lesions in susceptible and moderately resistant cultivars, but in moderately resistant cultivars the symptoms were later associated with more tissue browning around the sites of inoculation. In contrast, the resistant response produced the characteristic HR, characterized as small discrete browning and tissue collapse at site of inoculation. No local cultivar was found completely resistant to all the races tested. Cultivars Sehirali-90 and Göynük-98 showed resistant or moderately resistant reaction to five races, while cultivar Karacaşehir-90 showed the same reaction to six races. Analysis of bacterial growth and of accumulation of isoflavonoid bean phytoalexin, phaseollin, carried out in plant tissues expressing compatible and incompatible interactions, revealed that HR was clearly associated with the restricted multiplication of bacteria during incompatible interactions. The pods treated with different races of the bacterium showed that a strong correlation was observed between the timing and extent of cell death including accumulation of phaseollin, being rapid and extensive in incompatible interactions compared to compatible interaction.

Donmez et al. (2013) screened 36 bean genotypes and two commercial cultivars, commonly grown in Erzurum and Erzincan provinces located in the eastern Anatolia region of Turkey, for resistance to *Psm. savastanoi* pv. *phaseolicola* under greenhouse and field conditions during 2001–2002. Among the 38 genotypes tested, only 36K was found to be resistant. In field, disease severity reduced the quality and quantity of bean seed.

7.24.5 Management

Management practices for the disease are the same as those for common and fuscous blights of bean.

REFERENCES

Bozkurt, İ. A., and Soylu, S. 2011. Determination of responses of different bean cultivars against races of *Pseudomonas syringae* pv. *phaseolicola*, causal agent of halo blight of bean. *Euphytica* 179: 417–425.

Burkholder, W. H. 1926. A new bacterial disease of the bean. *Phytopathology* 16: 915–928.

Dillard, H. R., and Legard, D. E. 1991. Bacterial diseases of beans. Fact sheet, Cornell University, Ithaca, New York [http://vegetablemdonline.ppath.cornell.edu/factsheets/Beans_Bacterial.htm.] Accessed December 31, 2018.

Donmez, M. F., Sahin, F., and Elkoca, E. 2013. Identification of bean genotypes from Turkey resistance to common bacterial blight and halo blight diseases. *Acta Sci. Pol.* 12: 139–151.

Gardan, L., Bollet, C., Abu Ghorrah, M., Grimont, F., Grimont, P. A. D. 1992. DNA relatedness among the pathovar strains of *Pseudomonas syringae* subsp. *savastanoi* Janse (1982) and proposal of *Pseudomonas savastanoi* sp. nov. *Int. J. Syst. Bacteriol.* 42: 606–612.

Goto, M. 1992. *Fundamentals of Bacterial Plant Pathology*. Academic Press, San Diego, CA, 342 pp.

Neale, H., Deshappriya, N., Arnold, D., Wood, M. E., Whiteman, M., et al. 2017. Hydrogen sulfide causes excision of a genomic island in *Pseudomonas syringae* pv. *phaseolicola*. *Eur. J. Plant Pathol.* 149: 911–921.

Popović, T., Balaž, J., and Stanković, S. 2014. A method for the rapid detection and identification of halo blight pathogen on common bean. *Arch. Biol. Sci.* 66: 1393–1400.

Taylor, J. D., Teverson, D. M., Allen, D. J., and Pastor-Corrales, M. A. 1996. Identification and origin of races of *Pseudomonas syringae* pv. *phaseolicola* from Africa and other bean growing areas. *Plant Pathol.* 45: 469–478.

Tock, A. J., Fourie, D., Walley, P. G., Holub, E. B., Soler, A., et al. 2017. Genome-wide linkage and association mapping of halo blight resistance in common bean to race 6 of the globally important bacterial pathogen. *Front. Plant Sci.* 8: 1–17.

7.25 BACTERIAL BROWN SPOT OF BEAN

Burkholder in 1930 described two bacterial diseases of bean and one of these was bacterial brown spot. The disease is favored by cool to moderate temperature and high moisture. Severe outbreaks of the disease occur when temperature is below 27°C and humidity above 95% prevails for 24 h or longer periods. The yield losses up to 55% have been reported in South Africa (Muedi et al. 2011).

7.25.1 Symptoms

The disease is more severe on young foliage and symptoms are rarely seen on older leaves. Initially, small, water-soaked spots appear on the lower surface of trifoliate leaves. A narrow, greenish-yellow diffuse border about 1.5 mm in diameter may appear around some of the lesions. The spots enlarge ranging from 3 to 8 mm in diameter and on maturity they give a typical **brown spot** appearance. Dead centers of the spots may fall out producing shot-hole symptoms (Figure 7.18a). On the stem, systemic, sunken, longitudinal lesions appear. Sunken, brown spots appear on the pods. If the infection occurs during early development of pods, infected pods may be twisted and kinked at the site of infection. Pale brown haloes, about 6 mm wide, occasionally appear around the spots on the pods (Figure 7.18b).

Leaf symptoms of bacterial brown spot may be confused with those of halo blight due to the variation in margin/width of halo influenced by temperature in the latter case.

7.25.2 Causal Organism

Pseudomonas syringae pv. *syringae* van Hall 1902
Syn. *Psm. syringae* pv. *japonica*

(a)

(b)

FIGURE 7.18 Bacterial brown spot of bean: (a) Symptoms on aerial plant parts. (b) Symptoms on pod. (Courtesy of Robert Harveson and Howard Schwartz, www.plantmanagement.network.org.)

Muedi et al. (2011) assessed the biochemical and genetic variability in *Psm. syringae* pv. *syringae* isolates collected from dry bean producing areas in South Africa using levan production, oxidase production, pectinolitic activity, arginine dihydrolase production and tobacco hypersensibility (LOPAT) tests and *SyrB* gene assessment. Biolog GN Microplates were used to assess carbon substrate utilization. The *SyrB* gene was present in 42% of isolates. The Biolog GN Microplates showed biochemical variation among isolates. Variable genomic patterns were observed in 48.5% of isolates by the BOX A1R primer and in 37.1% of isolates by the enterobacterial repetitive intergenic consensus (ERIC) 2 primer.

7.25.3 Disease Cycle

The bacterium survives in infected plant debris and infected seed, but the former is more important. Infected bean stems and pod pieces play an important role in the survival of bacterium. Disease epiphytotics are often traced to survival of the bacterium in infected crop residue. The bacterium survives on hairy vetch and other weeds, but these are not considered an important source of survival. Non-certified seed lots also carry the bacterium from one crop season to the other. These seed lots are also a potential means of dissemination of bacterium to short and long distance places including new areas. The dissemination of the bacterium also occurs through wind-splashed rain, overhead-irrigation, and farm equipment. Severe infections usually occur after major rainstorms.

Different strains of *Psm. syringae* pv. *syringae* causing diseases on various plant species grow as epiphytes on various plant species. *Psm. syringae* pv. *syringae*, the causal agent of bacterial brown spot also grows epiphytically on snap and dry beans without causing disease symptoms. The symptoms of bacterial brown spot on beans generally occur after the development of large epiphytic populations of the bacteria. Severe infections usually develop after major rainstorms, but the absence of symptoms before rainstorms does not mean that the bacteria were not present.

Muedi et al. (2015) assessed 27 common bean accessions over three seasons and locations for resistance to *Psm. syringae* pv. *syringae* using artificial inoculation technique. Plots were rated for disease reaction on a modified 1–9 CIAT (Centro Internacional de Agricultura Tropical) scale with 1 being resistant and 9 being susceptible. Disease ratings were used to construct the area under the disease progress curve (AUDPC). Highly significant genotype × environment interactions were recorded. VAX 4 and VAX 2 showed the lowest disease ratings of 1.8 and 2.0, and AUDPC values of 38.6 and 41.6, respectively, and hence, are potential resistance sources.

7.25.4 Management

Management practices given for common and fuscous blights are applicable to this disease also.

REFERENCES

Muedi, H. T. H., Fourie, D., and Mclaren, N. W. 2011. Characterisation of bacterial brown spot pathogen from dry bean production areas of South Africa. *Afr. Crop Sci. J.* 19: 357–367.

Muedi, H. T. H., McLaren, N. W., and Fourie, D. 2015. Reaction of selected common bean germplasm accessions to bacterial brown spot in South Africa. *Crop Prot.* 77: 87–93.

van Hall, C. J. J. 1902. *Bijdragen tot de kennis der Bakterieele Plantenzeikten.* Coöperatieve Drukkerij-vereeniging "Plantijn," Inaugural dissertation, Amsterdam, the Netherlands.

7.26 BACTERIAL WILT OF BEAN

Bacterial wilt of common bean (*Phaseolus vulgaris* L.), caused by *Curtobacterium flaccumfaciens* pv. *flaccumfaciens*, was first recognized and described as a new dry bean disease near Redfield, South Dakota from the farm of the Office of Forage Investigations after the 1921 growing season (Hedges 1922, 1926). The estimated yield loss was 25%. It then became one of the most problematic bacterial diseases in the USA, particularly throughout the irrigated high plains and Midwest. The disease was commonly found in dry bean production in western Nebraska during the 1960s and early 1970s. In 2003, the disease was found in two Nebraska great northern fields, and was widely observed throughout Colorado, Wyoming, and Nebraska from many fields during 2004–2006 (Harveson et al. 2005a, 2006). Affected fields were planted with beans, from many different dry bean markets and seed sources, including yellow, great northern, pinto, kidney, black, navy, pink, small red, and Anasazi (Harveson et al. 2006). It has also been reported from numerous countries representing widespread distribution across the world, including Alberta and Ontario provinces of Canada (Hsieh et al. 2002), Tunisia, Australia, Mexico, South Africa, and Brazil (Zaumeyer and Thomas 1957). The occurrence of the disease has also been reported from Turkey, Bulgaria, Greece, Hungary, Romania, Russia, Belgium, Poland, and the Ukraine (EPPO/CABI 1997, Zaumeyer and Thomas 1957), but it is now considered to be absent in these countries as the pathogen could not establish there. González et al. (2005) reported it from a few fields in southeastern Spain. Osdaghi et al. (2015) reported a yellow variant and Osdaghi and Lak (2015) reported an orange variant of *Cur. flaccumfaciens* pv. *flaccumfaciens* as the causal agents of the disease from Iran. Osdaghi et al. (2016) reported the occurrence of a red variant of *Cur. flaccumfaciens* infecting bean in Iran. In their feature article, Harveson et al. (2015) have found that the return of bean wilt throughout the central high plains of the US over the last decade is not due to a single factor, but due to a combination of new changes in cultural practices, environmental stresses, and unfamiliarity with the pathogen and its past history.

7.26.1 Host Range

Economically, the most important host is common bean (*P. vulgaris*). The bacterium also attacks scarlet runner bean (*P. coccineus*), lima bean (*P. lunatus*), and *urad* bean (*Vigna mungo*). The natural infection of the bacterium also occurs on Adzuki bean (*V. angularis*), *mung* bean (*V. radiata*), *cowpea* (*V. unguiculata*), hyacinth bean (*Lablab purpureus*), soybean (*Glycine max*), and pea (*Pisum sativum*) (EPPO/CABI 1997; Ishimaru et al. 2005). Gonçalves et al. (2017) determined the host range of *Cur. flaccumfaciens* pv. *flaccumfaciens* under field and greenhouse conditions and concluded that the cultivation of bean crop in succession with barley, black and white oats, canola, ryegrass, and wheat should not be recommended, mainly in areas with a history of bacterial wilt occurrence. In these cases, the better option for crop rotation during the winter is radish, a non-host for the bacterium.

7.26.2 Symptoms

The infection on young plants usually leads to their death. If plants survive an early attack, or are infected at a later growth stage, they may live throughout the season and produce mature seed. The characteristic symptom consists of wilting of leaves or parts of them, initially during the warmest hours of the day, followed by a recovery as the temperature drops in the evening. Wilting becomes permanent during the following days due to plugging of xylem vessels with bacterial masses, leading to reduction in water supply; the leaves turn brown and then drop. Infected plants in the central high plains of the USA additionally exhibit symptoms consisting of interveinal, necrotic lesions surrounded by bright-yellow borders, referred to as "**firing**" (Harveson and Schwartz 2007). These symptoms may be confused with those of common bacterial blight caused by *Xanthomonas axonopodis* pv. *phaseoli*. However, the lesions in bacterial wilt tend to be more wavy or irregular, and are also accompanied by wilting and often death of severely infected plants. The plants infected with common bacterial blight are rarely killed and wilting is not observed as a part of the disease process. Water-soaking of leaves is not always observed in wilt as it is commonly found in other bacterial diseases, namely, common bacterial blight, halo blight, and brown spot of beans. Occasionally, typical wilting symptoms may be absent and replaced by golden-yellow necrotic leaf lesions closely resembling those of common bacterial blight. The water-soaking does occur on the highly susceptible wilt-infected yellow bean leaves in Colorado. A preliminary field diagnosis of the disease should include both wilt and firing symptoms.

On pods, the disease is much more conspicuous than the common bacterial blight. Infection can occur on pod sutures, but it seldom results in formation of circular spots (Schwartz et al. 2005). The sutures may show dark discoloration, sometimes extending laterally. Seeds may be infected in apparently healthy pods due to movement of pathogen into developing seeds through the vascular system, following the sutures of the pods (Harveson et al. 2005a, 2005b). The water-soaked spots occasionally appear on young pods, the infected area turning either a yellowish-green or darker than the rest of the pod. On ripe pods, olive-green lesions in contrast to the yellow color of the normal pod; are more evident. It should be noted that seemingly vigorous plants may bear one or more shriveled shoots, or infected pods hidden by healthy foliage.

If the diseased plants survive to produce mature seed, they are often discolored as a result of bacterial infection and colonization. The discoloration is more in the white-seeded market classes such as navy and great northern, but less intense in cultivars with colored seed coats. There may be a small amount of yellow slime at the hilum, and seeds may be shriveled. Different color variants of the bacterium produce pink, orange, purple, and yellow discoloration in the seed coat of infected seeds.

7.26.3 Causal Organism

Curtobacterium flaccumfaciens pv. *flaccumfaciens* (Hedges 1922) Collins and Jones 1983
Syn. *Cur. flaccumfaciens*, *Corynebacterium flaccumfaciens* pv. *flaccumfaciens*, *Cor. flaccumfaciens* subsp. *flaccumfaciens*

The bacterium is an EPPO A2 (list No. 48) quarantine pest.

The bacterial cells are Gram-stain-positive, short, coryneform-shaped rods that characteristically bend or snap. The bacterium grows aerobically with optimal growth in culture occurring at 27°C–28°C. The colonies on yeast extract peptone glucose agar and nutrient broth yeast extract agar media after 2–3 days of incubation, are circular, smooth having entire margins and 2–4 mm in diameter; more often convex and translucent, but sometimes also flat and semi-opaque. Their pigmentation varies depending on temperature and pH and also varies with color variants. After 7–10 days of incubation on semi-selective medium (SSM) developed by Tegli et al. (1998), the colonies are 1–2 mm in diameter and are clearly detectable due to their yellow color in contrast to the violet background of the medium. A yellow halo surrounding the colony is sometimes present. Other bean seed-transmitted bacterial pathogens, namely, *Pseudomonas savastanoi pv. phaseolicola* and *Xa. axonopodis* pv. *phaseoli*, are unable to grow on SSM medium.

Hedges (1922) gave the bean wilt pathogen its species name after the description of its most distinguishing characteristic. The specific epithet, "*flaccumfaciens*" is derived from the Latin word "**flaccus**" meaning "**flabby or flaccid**," and "**faciens**" meaning "**making**," hence describing the pathogenic actions of the bacterium and resulting symptoms as "**wilt-making**."

Pathogen color variants have been reported that stain seeds, and are particularly conspicuous on white-seeded cultivars; including the original type strain yellow (Hedges 1926). Schuster designated the orange strain as *Cur. flaccumfaciens* pv. *auranticum* (Schuster and Christiansen 1957). The orange strains cannot be distinguished from the original yellow strain on the basis of morphological, physiological, and serological characteristics. It differs only in colony color. Another color variant, namely, purple strain was identified as *Cur. flaccumfaciens* pv. *violaceae* (Schuster and Sayre 1967). The purple variants tend to be less virulent, produce yellow colored colonies in culture, but produce a bright purple to blue water soluble pigment that diffuses into growth media. Over time, the pigments still remain in media, but the initial orange or yellow colonies are not as obscured, and can be visualized much easier than with younger cultures. A pink variant, highly virulent to dry beans was discovered and Harveson and Vidaver (2008) confirmed its identity on the basis of morphological and physiological characteristics.

7.26.4 Disease Cycle

Hedges (1926) provided the first evidence of seed-borne infection and transmission of the bacterium. The bacterium is externally and internally seed-borne, and infected seeds are a major source of perpetuation of the bacterium. The infected seeds are also a source of dispersal, both short and long distances. Due to a strong resistance to drying, the pathogen has been demonstrated to remain viable up to 24 years in seed stored under optimum conditions in the laboratory (EPPO/CABI 1997). The bacterium also survives in infected residue or weeds and the former is an important source of survival. It can also survive and remain pathogenic in soil for at least 2 years. Harveson et al. (2011) reported the presence of bacterial wilt isolates occurring with other crops, namely, soybean, wheat, corn, sunflower, and alfalfa grown in rotation with common bean in western Nebraska. These isolates were found in association with other bacterial diseases, suggesting survival in those crops' residues.

The bacterium cannot penetrate intact plant surfaces, but requires either natural openings in leaves or wounds for penetration. Storms with high velocity winds, driving rains, and hails provide the perfect opportunity for the bacterium to become established because they can cause wounds and physically move the pathogen and/or infected plant parts between and within fields. Any other physical damage to the bean plants from humans, animals, or farm equipment also enhances the disease severity.

Initial infection occurs when the pathogen enters the vascular system through either infected seed or through wounds occurring on leaves or stems. Wilt is most destructive after periods of plant stress accompanied with high temperatures exceeding 32°C (Ishimaru et al. 2005; Schwartz et al. 2005). In general, the disease develops more rapidly and causes greater damage during periods of severe plant stresses including high temperatures, moisture deprivation, or mechanical damage of any kind.

The systemic nature of the wilt was first demonstrated by Hedges (1922, 1926). In case of plants that survive and produce mature seeds, systemic infection enables the pathogen to move through the vascular system into developing pods and seed embryos, causing discoloration or staining of seeds. Staining of seeds can also occur in pods looking apparently healthy with no external symptoms.

The secondary spread of the pathogen occurs through wind-splashed rain and overhead-irrigation. Drip irrigation resulted in low disease incidence and higher yield compared to sprinkler irrigation, while the furrow irrigation generally gave the intermediate effects between the two methods of irrigation (Harveson et al. 2015). Mechanical wounds caused during cultural operation also facilitate the secondary spread by providing avenues for the entry of the bacterium.

Resistant cultivars: Due to lack of effective chemical control measures for bacterial wilt, the greater emphasis was placed on the breeding of resistant varieties. The resistant cultivar "Emerson" was developed in the early 1970s by the University of Nebraska specifically for controlling this disease. It is still available today, but it cannot be grown in all areas where the disease occurs. As a result of continuous breeding programs, many resistant lines/cultivars have been identified. Hsieh et al. (2005b) found great northern line L02E317 and cultivar Resolute and pinto lines L02B662 and 999S-2A highly resistant to yellow and orange variants of the pathogen. Conner et al. (2008) observed high levels of resistance to yellow, orange, and purple variants in the light red kidney bean cultivars AC Litekid, Chinook 2000, and Redkanner as well as dark red kidney bean cultivars Cabernet and Red Hawk. Urrea and Harveson (2014) identified PI 325691, a wild common bean, resistant to orange variant. These resistant sources can serve as valuable germplasm for breeding wilt resistant cultivars of common bean.

Hsieh et al. (2003) developed an indoor seed inoculation technique using hilum injury to screen common bean germplasm for resistance to bacterial wilt pathogen under controlled environments. The method was efficient and effective as it took only 14 days to complete the test.

7.26.5 Management

As no effective chemical measures are available for the management of the disease and the bacterium is seed-borne, the current management practices comprise of the following approaches.

1. Use healthy certified seed for planting. Two PCR tests for the detection of bacterium in bean seeds (Guimaraes et al. 2001; Tegli et al. 2002) are described. These tests are helpful for seed certification programs and quarantine tests
2. Treat seed with streptomycin or other effective bactericides
3. Grow resistant or tolerant varieties, if available
4. Rotate beans with non-host crops for 2–3 years
5. Destroy infected plant refuse after harvest by incorporating and burying in the soil
6. Eliminate volunteer bean plants at the early growth stage of plants
7. Stay out of bean fields when plants are wet
8. Avoid reusing irrigation water and overhead-irrigation
9. Preventive sprays of copper-based bactericides may be given during midvegetative or early flowering periods, depending upon weather and disease severity
10. Biological control: Huang et al. (2007) reported that treatment of naturally infected seeds of great northern bean cv. US1140 with *Rhizobium leguminosarum* pv. *viceae* R21 was effective in reducing the incidence and severity of wilt and increasing seedling emergence and height only for the seeds without visible symptoms (0% discoloration) or with light symptoms (1%–25% yellow discoloration), but not for seeds with moderate symptoms (26%–50% discoloration), severe symptoms (51%–75% discoloration), or very severe symptoms (76%–100% discoloration).

Treatment of hilum-injured bean seeds of great northern bean cv. US1140 or navy bean cv. Morden003 with a mixture of *Pantoea agglomerans* + *Cur. flaccumfaciens* pv. *flaccumfaciens* resulted in a high rate of colonization of seedlings by *Pa. agglomerans*, reduced frequency of infection by *Cur. flaccumfaciens* pv. *flaccumfaciens*, improved seedling emergence and height, and reduced disease severity, compared with seeds treated with the wilt pathogen alone (Hsieh et al. 2005a).

REFERENCES

Collins, M. D., and Jones, D. 1984. Validation of the publication of new names and new combinations previously effectively published outside the IJSB. *Int J. Syst. Bacteriol.* 34: 270–71. Effective publication: Collins, M. D., and Jones, D. 1983. Reclassification of *Corynebacterium flaccumfaciens*, *Corynebacterium betae*, *Corynebacterium oortii* and *Corynebacterium poinsettiae* in the genus *Curtobacterium*, as *Curtobacterium flaccumfaciens* comb. nov. *J. General Microbiol.* 129: 3545–3548.

Conner, R. L., Balasubramanian, P., Erickson, R. S., Huang, H.-C., and Mündel, H.-H. 2008. Bacterial wilt resistance in kidney beans. *Can. J. Plant Sci.* 88: 1109–1113.

EPPO/CABI. 1997. *Curtobacterium flaccumfaciens* pv. *flaccumfaciens*. pp. 991–94 in: *Quarantine Pests for Europe*. 2nd ed. I. M. Smith, D. G. McNamara, P. R. Scott, and H. Holderness eds. CAB International, Wallingford, UK.

Gonçalves, R. M., Schipanski, C. A., Koguishi, L., Soman, J. M., Sakate, R. K., et al. 2017. Alternative hosts of *Curtobacterium flaccumfaciens* pv. *flaccumfaciens*, causal agent of bean bacterial wilt. *Eur. J. Plant Pathol.* 148: 357–365.

González, A. J., Tello, J. C., and Rodicio, M. R. 2005. Bacterial wilt of beans (*Phaseolus vulgaris*) caused by *Curtobacterium flaccumfaciens* in southeastern Spain. *Plant Dis.* 89: 1361.

Guimaraes, P. M., Palmano, S., Smith, J. J., Grossi, de Sa' M. F., and Saddler, G. S. 2001. Development of a PCR test for the detection of *Curtobacterium flaccumfaciens* pv. *flaccumfaciens*. *Antonie van Leeuwenhoek* 80: 1–10.

Harveson, R. M., and Schwartz, H. F. 2007. Bacterial diseases of dry edible beans in the central high plains. *Plant Health Progress*. doi:10.1094/PHP-2007-0125-01-DG.

Harveson, R. M., Schwartz, H. F., Urrea, C. A., and Yonts, C. D. 2015. Bacterial wilt of dry-edible beans in the central high plains of the U.S.: Past, present, and future. *Plant Dis.* 99: 1665–1677.

Harveson, R. M., Schwartz, H. F., Vidaver, A. K., Lambrecht, P. A., and Otto, K. 2006. New outbreaks of bacterial wilt of dry beans in Nebraska observed from field infections. *Plant Dis.* 90: 681.

Harveson, R. M., Urrea, C., A., and Schwartz, H. F. 2011. Bacterial wilt of dry beans in western Nebraska. NebGuide Series No. G1562 (Revised). Coop. Ext. Serv., University of Nebraska-Lincoln.

Harveson, R. M., and Vidaver, A. K. 2008. A new color variant of the dry bean bacterial wilt pathogen (*Curtobacterium flaccumfaciens* pv. *flaccumfaciens*) found in western Nebraska. *Plant Health Prog*. doi:10.1094/PHP-2008-0815-01-BR.

Harveson, R. M., Vidaver, A. K., and Schwartz, H. F. 2005a. Bacterial wilt of dry beans in western Nebraska. NebGuide Series No. G05-1562-A. Coop. Ext. Serv., University of Nebraska-Lincoln.

Harveson, R. M., Watkins, J. E., Giesler, L. J., and Chaky, J. L. 2005b. Dry bean disease profiles I (foliar and bacterial diseases). Ext. Circular Series No. EC05-1893, University of Nebraska-Lincoln.

Hedges, F. 1922. A bacterial wilt of the bean caused by *Bacterium flaccumfaciens* nov. sp. *Science Washington* 55: 433–34.

Hedges, F. 1926. Bacterial wilt of beans (*Bacterium flaccumfaciens* Hedges), including comparisons with *Bacterium phaseoli*. *Phytopathology* 16: 1–22.

Hsieh, T. F., Huang, H. C., and Erickson, R. S. 2005a. Biological control of bacterial wilt of bean using a bacterial endophyte, *Pantoea agglomerans*. *J. Phytopathol.* 153: 608–614.

Hsieh, T. F., Huang, H. C., Erickson, R. S., Yanke, L. J., and Mundel, H.-H. 2002. First report of bacterial wilt of common bean caused by *Curtobacterium flaccumfaciens* in western Canada. *Plant Dis.* 86: 1275.

Hsieh, T. F., Huang, H. C., Mündel, H.-H., Conner, R. L., Erickson, R. S., et al. 2005b. Resistance of common bean (*Phaseolus vulgaris*) to bacterial wilt caused by *Curtobacterium flaccumfaciens* pv. *flaccumfaciens*. *J. Phytopathol.* 153: 245–249.

Hsieh, T. F., Huang, H. C., Mündel, H.-H., and Erickson, R. S. 2003. A rapid indoor technique for screening common bean (*Phaseolus vulgaris* L.) for resistance to bacterial wilt [*Curtobacterium flaccumfaciens* pv. *flaccumfaciens* (Hedges) Collins and Jones]. *Revista Mexicana de Fitopatología* 21: 370–374.

Huang, H. C., Erickson, R. S., and Hsieh, T. F. 2007. Control of bacterial wilt of bean (*Curtobacterium flaccumfaciens* pv. *flaccumfaciens*) by seed treatment with *Rhizobium leguminosarum*. *Crop Prot.* 26: 1055–1061.
Ishimaru, C., Mohan, S. K., and Franc, G. D. 2005. Bacterial wilt. pp. 50–52 in: *Compendium of Bean Diseases*. 2nd ed. H. F. Schwartz, J. R. Steadman, R. Hall, and R. L. Forster eds. American Phytopathological Society, St. Paul, MN.
Osdaghi, E., and Lak, M. R. 2015. Occurrence of a new orange variant of *Curtobacterium flaccumfaciens* pv. *flaccumfaciens*, causing common bean wilt in Iran. *J. Phytopathol.* 163: 867–871.
Osdaghi, E., Taghavi, S. M., Hamedi, J., and Mohammadipanah, F. 2015. Bacterial wilt of common bean (*Phaseolus vulgaris*) caused by *Curtobacterium flaccumfaciens* pv. *flaccumfaciens* in Iran. *Australasian Plant Dis. Notes* 10: 23.
Osdaghi, E., Taghavi, S. M., Hamzehzarghani, H., Fazliarab, A., Harveson, R. M., et al. 2016. Occurrence and characterization of a new red-pigmented variant of *Curtobacterium flaccumfaciens*, the causal agent of bacterial wilt of edible dry beans in Iran. *Eur. J. Plant Pathol.* 146: 129–145.
Schuster, M. L., and Christiansen, D. W. 1957. An orange-colored strain of *Corynebacterium flaccumfaciens* causing bean wilt. *Phytopathology* 47: 51–53.
Schuster, M. L., and Sayre, R. M. 1967. A coryneform bacterium induces purple-colored seed and leaf hypertrophy of *Phaseolus vulgaris* and other Leguminosae. *Phytopathology* 57: 1064–1066.
Schwartz, H. F., Franc, G. D., Hanson, L. E., and Harveson, R. M. 2005. Disease management. pp. 109–43 in: *Dry Bean Production and Pest Management*. H. F. Schwartz, M. A. Brick, R. M. Harveson, and G. D. Franc eds. Bull. No. 562A, Colorado State University, Cort Collins, CO.
Tegli, S., Sereni, A., and Surico, G. 2002. PCR-based assay for the detection of *Curtobacterium flaccumfaciens* pv. *flaccumfaciens* in bean seeds. *Lett. Appl. Microbiol.* 35: 331–337.
Tegli. S., Surico, G., and Esposito, A. 1998. Studi sulla diagnosi di *Curtobacterium flaccumfaciens* pv. *flaccumfaciens* nei semi di fagiolo. *Notiziario sulla Protezione delle Piante* 9: 63–71.
Urrea, C. A., and Harveson, R. M. 2014. Identification of sources of bacterial wilt resistance in common bean (*Phaseolus vulgaris*). *Plant Dis.* 98: 973–976.
Zaumeyer, W. J., and Thomas H. R. 1957. A monographic study of bean diseases and methods for their control. USDA Technical Bull. 868 (revised).

7.27 BACTERIAL BLIGHT OF COWPEA

Bacterial blight of cowpea [*Vigna unguiculata* (L.) Walp.] was originally described as canker of cowpea by Burkholder (1944) from USA. In India, it was first reported from Pune, Maharashtra by Patel and Diwan in 1950 and subsequently from many areas like, Tamil Nadu, Delhi, Punjab, Haryana, Uttar Pradesh, and Karnataka. The disease also occurs in China, Brazil, Nigeria, Tanzania, Thailand, Venezuela, Zimbabwe, Uganda, Senegal, Sudan, and South Africa.

The high mortality of the seedlings is responsible for major loss due to the disease. Plant mortality as high as 60% has been reported in variety California Black-Eye in Texas, USA. In India, 15%–20% seedling mortality was observed in certified seed production plots of Pusa Dofasli cultivar of the National Seeds Corporation of India in 1971 (Patel et al. 1972).

7.27.1 Host Range

Different host species have been reported to be infected artificially, but there is not much agreement among different workers on this aspect. It appears that cowpea is the main economic host under natural conditions. However, Sikirou and Wydra (2004) reported *Sphenostylis stenocarpa* (African yam bean) as potential host of the bacterium by artificial inoculation, for the first time. Gena et al. (2009) have reported successful inoculation and symptom production on French bean (*Phaseolus vulgaris*) and hyacinth bean (*Dolichos lablab*).

7.27.2 Symptoms

In seedlings developing from infected seeds, first symptoms appear on cotyledons or primary leaves. The infected cotyledons turn red and shrivel. On primary leaves, necrotic spots generally appear at the margin from where the bacterium moves to stem and then systemically proceeds to other plant parts. The severe infection damages cotyledons, primary leaves, and growing point resulting in the death of seedlings. In the surviving seedlings, the cankers usually develop at the points where cotyledons and primary leaves were attached. These cankers enlarge and turn brown, and ultimately the cankered stems break at this point due to the weight of plants and the action of fast winds.

The infection of trifoliate leaves results in the formation of light yellow, irregular to round spots (4–10 mm) having necrotic brown centers and deep yellow borders. These spots enlarge and may involve the whole leaf. Small veins within blighted areas turn reddish brown, become shriveled, and infection extends into the green leaf area. In severe infection, infection moves along the veins turning the area around the veins yellow. Later on, these leaves turn straw colored and defoliate prematurely.

On infected pods, dark green, water-soaked streaks appear along the sutures. Severely infected pods turn yellow, shrivel, and dry. The seeds produced in the severely infected pods are usually small, shriveled, and discolored.

7.27.3 Causal Organism

Xanthomonas axonopodis pv. *vignicola* (Burkholder 1944) Vauterin et al. 1995
Syn. *Xa. vignicola*, *Xa. campestris* pv. *vignicola*

Burkholder (1944) described the bacterium for the first time and named it as *Xanthomonas vignicola* sp. nov. With the introduction of pathovar system, it was changed to *Xa. campestris* pv. *vignicola* and later on, reclassified as *Xa. axonopodis* pv. *vignicola*. The below given description of the bacterium is taken primarily from Burkholder (1944).

Bacterial cells are Gram-stain-negative, non-sporing rods (0.5–0.9 × 1.0–2.8 μm) with rounded ends and motile with a single polar flagellum. Colonies on potato dextrose agar are large, mucoid, and pale yellow. The optimum temperature for growth is between 27°C and 30°C, maximum at 37°C, and minimum between 6°C and 9°C. The tests for starch hydrolysis, gelatin liquefaction, hydrogen sulphide production, and lipolytic activity are positive, while those for indole and ammonia production and nitrate reduction are negative. Also no growth occurs in sodium salts of lactic, formic, succinic, tartaric, and hippuric acids. Bacterial growth in citric and malic acid salts produces an alkaline reaction. Tyrosine and asparagine are not utilized as a combined source of carbon and nitrogen. The bacterium is aerobic and liquefies pectate medium. Khatri-Chhetri et al. (2003) reported that 55 strains of the bacterium obtained from 11 countries differed in metabolization of different carbon sources using the Biolog GN Microplate system; the strains belonging to different origins varied greatly, but the variation was not related to the virulence of the strains.

Wydra et al. (2004) developed the first semi-selective medium called cefazoline cellobiose-methionine medium for the isolation of the bacterium. Colonies of the bacterium on this medium were white, round, raised, and 0.2–1.8 mm in diameter after 4 days of incubation. The medium was helpful for isolation of the bacterium from cowpea plant and soil samples.

7.27.4 Disease Cycle

Infected seed is an important source of perpetuation of the bacterium. Volunteer plants also act as a source of inoculum. Seed may be externally or internally infected or may be both. Heavily infested seeds usually give rise to diseased seedlings even under dry conditions and primary infection through seed readily causes stem canker and results in high plant mortality. Khan (1996) observed 73.50% disease incidence in a crop raised from naturally infected seed. Soni (1982) reported that inoculation of cowpea plants at all growth stages, starting from seeds before sowing to flat green pod stage, led the pathogen to become seed-borne. Methods are available to detect the bacterium in the infected seeds (Soni and Thind 1987, 1991).

Claudius-Cole et al. (2014) used various modifications of the plating and soaking methods to detect *Xa. axonopodis* pv. *vignicola* from cowpea seeds, collected from five locations within two agro ecological regions in Nigeria. Percentage seed infection was significantly higher in Zaria, Minjibir (northern guinea savannah) and Iseyin (forest transition zone) compared to the other production locations, in both plating and soaking assay methods. The seed coat alone was the most ($P \leq 0.05$) colonized part of the seed by the pathogen followed by the cotyledons, while embryo was the least colonized. Pod inoculation was the most effective pathogenicity test followed by inoculation of young plants, while inoculation of the excised cotyledons, stems, and petioles was not reliable for expressing symptoms. Detached pods were equally efficient and took less time compared to inoculation of young plants to prove the pathogenicity of the bacterium. The detection of *Xa. axonopodis* pv. *vignicola* was better with methods that breach the seed coat or using the seed coat alone.

de Lima-Primo et al. (2015) reported that cowpea plants grown in greenhouse and field were inoculated with 10^5, 10^6, and 10^7 cfu mL^{-1} of *Xa. axonopodis* pv. *vignicola,* and seeds harvested from field plots were used to determine the seed-plant transmission of the bacterium. In the greenhouse, plants were subjected to leaf wetness duration periods of 0, 4, 8, 12, 16, 20, and 24 h after inoculation. In another experiment, soybean, common bean, and wild poinsettia were inoculated with *Xa. axonopodis* pv. *vignicola.* The plants inoculated with a concentration of 10^7 cfu mL^{-1} and subjected to 12 h of leaf wetness showed highest disease severity. Positive seed-plant transmission of the bacterium occurred in all the inoculum concentrations used. Hypersensitivity reaction was observed in wild poinsettia less than 24 h after inoculation.

The bacterium did not survive for more than 3 weeks in diseased plant debris under field conditions (Khan 1996), indicating that diseased plant debris does not play any role in its survival. Infected crop residue and some weeds may act as source of primary inoculum when two cowpea crops are sown consecutively (Sikirou and Wydra 2004).

Secondary spread of the bacterium occurs mostly through wind-splashed rain. The other means of secondary spread are overhead-irrigation and ground irrigation. When the planting is done in furrows running along the slope, irrigation/rain water spreads the bacterium while moving from the higher level to the lower level. Long distance spread or spread to the new areas occurs through seed. Zandjanakou-Tachin et al. (2007) reported the presence of the bacterium in the feces, intestines, and on the legs and mandibles of *Zonocerus variegatus* (grasshopper) and indicated the possibility of the transmission of the bacterium through this insect.

The disease is severe in humid and warm weather. Optimum temperature for disease development is 28°C–34°C. The disease is more severe in low-lying areas where water frequently stagnates even temporarily.

Rachael et al. (2015) screened 12 cowpea genotypes against three isolates collected from three different cowpea growing zones, namely, Makurdi, Guma, and Gboko local government areas in Benue State of Nigeria, designated as MKD388-1, GUM391, and GBK205-8, respectively, in screen house and field trials. Three genotypes, BOSADP, IT98K-1092–2, and IT98K-573-2-1 were consistently resistant in both pot and field experiments. Isolates did not differ significantly ($P < 0.05$) in their virulence means; rating 4.69 (MKD388-1), 4.50 (GUM391), and 4.33 (GBK205-8).

Singh and Patel (1977) reported that the resistance was governed by a single dominant gene in cowpea lines, namely, P 309, P 426, and P 910 and by a single recessive gene in Iron. In case of P 910 and Iron, the segregating populations showed some gradations in susceptible and resistant reactions indicating that in addition to major genes, some modifying factors(s) may be involved.

7.27.5 Management

1. As the infected seed is only the main source of inoculum of the bacterium, the production of pathogen-free seed or the elimination of the seed-borne inoculum will control the disease. Different methods of seed

treatment have been worked out by different workers. Jindal et al. (1989) recommended hot water (50°C for 30 min) or solar heat (soaking of seed in tap water from 8.00 A.M. to 12.00 noon and then spreading the seed in a thin layer exposed directly to sun rays for 5 h on a bright sunny day in summer) or Streptocycline (100 μg mL^{-1}) + Agallol (0.2%) or Streptocycline (100 μg mL^{-1}) + captan (0.2%) treatment. Jindal and Thind (1990a) found that *Erwinia herbicola* (recent name, *Pantoea agglomerans*) and *Penicillium oxalicum* were as effective as the above-mentioned four treatments in controlling seed-borne infection of the bacterium. These biocontrol agents also gave higher seed germination than hot water and chemical treatments

2. Secondary spread of the pathogen should be checked by applying chemicals or biocontrol agents. The sprays of copper and copper + maneb mixtures have been found effective in controlling the disease along with a significant increase in grain yield (Gitaitis et al. 1986). Jindal and Thind (1990b) have recomended the seed treatment with Streptocycline (100 μg mL^{-1}) + Agallol (2000 μg mL^{-1}) followed by three sprays of Streptocycline (100 μg mL^{-1}) + Bavistin (500 μg mL^{-1}) or two sprays of Streptocycline (100 μg mL^{-1}) + Bavistin (500 μg mL^{-1}) followed by third spray of Streptocycline (100 μg mL^{-1}) + Blitox 50-WP (2000 μg mL^{-1}). The combination of seed treatment and foliar sprays provided significantly better disease control and also gave a higher grain yield in comparison to seed treatment or spraying schedules alone. Later on, Jindal and Thind (1993) reported that foliar applications of *Bacillus* sp., *Er. herbicola*, and *P. oxalicum* gave significantly better control than Streptocycline (200 μg mL^{-1}), a potent and widely used bactericide
3. Grow disease resistant varieties. Two cultivars namely, Tuv 12349 and Tuv 15549 are resistant to the disease in the field and in greenhouse in Nigeria (Okechukwu and Ekpo 2004)

REFERENCES

Burkholder, W. H. 1944. *Xanthomonas vignicola* sp. nov. pathogenic on cowpea and beans. *Phytopathology* 34: 430–432.

Claudius-Cole, A. O., Ekpo, E. J. A. and Schilder, A. M. C. 2014. Evaluation of detection methods for cowpea bacterial blight caused by *Xanthomonas axonopodis* pv. *vignicola* in Nigeria. *Trop. Agric. Res. & Ext.* 17: 77–85.

de Lima-Primo, H. E., de L. Nechet, K., de A. Halfeld-Vieira, B., de Oliveira, J. R., Mizubuti, E. S. G., et al. 2015. Epidemiological aspects of cowpea bacterial blight. *Trop. Plant Pathol.* 40: 46–55.

Gena, M., Shah, R., Jadon, K. S., and Mali, B. L. 2009. Host range of *Xanthomonas axonopodis* pv. *vignicola*, the incitant of bacterial blight in cowpea. *Indian Phytopathol.* 62: 539–540.

Gitaitis, R. D., Bell, D. K. and Smittle, D. A. 1986. Epidemiology and control of bacterial blight and canker of cowpea. *Plant Dis.* 70: 187–190.

Jindal, K. K., and Thind, B. S. 1990a. Microflora of cowpea seeds and its significance in the biological control of seed borne infection of *Xanthomonas campestris* pv. *vignicola. Seed Sci. & Technol.* 18: 393–403.

Jindal, K. K., and Thind, B. S. 1990b. Control of bacterial blight of cowpea (*Vigna unguiculata* (L.) Walp.). *Trop. Pest Manag.* 36: 211–213.

Jindal, K. K., and Thind, B. S. 1993. Biological control of bacterial blight of cowpea (*Vigna unguiculata* (L.) Walp.). *Phytopathol. Medit.* 32: 193–200.

Jindal, K. K., Thind, B. S., and Soni, P.S. 1989. Physical and chemical agents for the control of *Xanthomonas campestris* pv. *vignicola* from cowpea seeds. *Seed Sci. Technol.* 17: 371–382.

Khan, A. N. A. 1996. Incidence and perpetuation of bacterial blight of cowpea caused by *Xanthomonas campestris* pv. *vignicola*. pp. 349–356 in: *Proceedings of the 9th International Conference on Plant Pathogenic Bacteria*. A. Mahadevan ed. Chennai, India.

Khatri-Chhetri, G. B., Wydra, K., and Rudolph, K. 2003. Metabolic diversity of *Xanthomonas axonopodis* pv. *vignicola*, causal agent of cowpea bacterial blight and pustule. *Eur. J. Plant Pathol.* 109: 851–860.

Okechukwu, R. U., and Ekpo, E. J. A. 2004. Sources of resistance to cowpea bacterial blight disease in Nigeria. *J. Phytopathol.* 152: 345–351.

Patel, P. N., Jindal, J. K., and Shekhawat, G. S. 1972. Bacterial blight (canker) of cowpea. pp. 79–83 in: *Plant Bacteriology*. Vol. I, *Bacterial Diseases of Plants in India*. P. N. Patel ed. Indian Agricultural Research Institute, New Delhi, India.

Rachael, D. T., Iheukwumere, C. C., and Lucky, O. 2015. Evaluation of selected cowpea genotypes for resistance to bacterial blight. *Int. J. Curr. Microbiol. App. Sci.* 4: 257–270.

Sikirou, R., and Wydra, K. 2004. Persistence of *Xanthomonas axonopodis* pv. *vignicola* in weeds and crop debris and identification of *Sphenostylis stenocarpa* as a potential new host. *Eur. J. Plant Pathol.* 110: 939–947.

Singh, D., and Patel, P. N. 1977. Studies on resistance in crops to bacterial diseases in India. VII. Inheritance of resistance to bacterial blight disease in cowpea. *Indian Phytopathol.* 30: 99–102.

Soni, P. S. 1982. Detection of Phytopathogenic Bacteria from Cowpea and Mung Seeds. PhD thesis, Punjab Agricultural University, Ludhiana, 180 pp.

Soni, P. S., and Thind, B. S. 1987. Detection *of Xanthomonas campestris* pv. *vignicola* (Burkh.) Dye from cowpea seeds and *X. campestris* pv. *vignaeradiatae* (Sabet et al.) Dye from mung bean seeds by growing on tests. *Phytopathol. Medit.* 26: 1–6.

Soni, P. S., and Thind, B. S. 1991. Detection of *Xanthomonas campestris* pv. *vignaeradiatae* (Sabet et al.) Dye from green gram seeds and *X. campestris* pv. *vignicola* (Burkh.) Dye from cowpea seeds with the help of bacteriophages. *Plant Dis. Res.* 6: 6–11.

Vauterin, L., Hoste, B., Kersters, K., and Swings, J. 1995. Reclassification of *Xanthomonas*. *Int. J. Syst. Bacteriol.* 45: 472–489.

Wydra, K., Khatri-Chhetri, G., Mavridis, A., Sikirou, R., and Rudolph, K. 2004. A diagnostic medium for the semi-selective isolation and enumeration of *Xanthomonas axonopodis* pv. *vignicola*. *Eur. J. Plant Pathol.* 110: 991–1001.

Zandjanakou-Tachin, M., Fanou, A., Le Gall, P., and Wydra, K. 2007. Detection, survival and transmission of *Xanthomonas axonopodis* pv. *manihotis* and *X. axonopodis* pv. *vignicola*, causal agents of cassava and cowpea bacterial blight, respectively, in/by insect vectors. *J. Phytopathol.* 155: 159–169.

7.28 BACTERIAL LEAF SPOT OF GREEN GRAM

Green gram [*Vigna radiata* (L.) Wilczek] suffers from two bacterial diseases, namely, bacterial leaf spot and halo blight. Out of these two diseases, the former is economically more important and widespread. Bacterial leaf spot of green gram or *mung* bean (local name of green gram in India) was first reported from China by Fang and associates in 1964. In India, its presence has been mentioned in the past, but detailed information on certain aspects of the disease has been published for the first time by Patel and his associates (Patel and Jindal 1972; Patel et al. 1972). The disease is widespread in green gram growing areas of India. In addition to China and India, it also occurs in Pakistan (Iqbal et al. 2003).

7.28.1 Host Range

Patel and Jindal (1972) reported that in addition to green gram the bacterium was moderately virulent on bean, lentil, and Indian bean (*Dolichos lablab*), but weakly virulent on lima bean (*Phaseolus lunatus*) and *P. bracteatus*. The progress of infection on bean was not as fast and destructive as with the common and fuscous blight pathogens, the actual pathogens of this host. On bean, the spots were water-soaked, but not raised as in green gram; they coalesced leading to the development of large necrotic patches.

7.28.2 Symptoms

The symptoms on cotyledons of the seedlings, produced by infected seeds, start as pinhead, circular, water-soaked spots, which are lighter green in color as compared to neighboring healthy tissue. The spots increase in size, turn brown, become necrotic and raised. In case of severe infection, the spots may coalesce to cover the whole leaf. Longitudinal lesions, slits, or cankers extending from soil level to the growing tip develop on the stems of seedlings grown from severely infected seeds.

On trifoliate leaves, brown, circular to irregular, and distinctly raised spots appear. Generally, the spots are dry and necrotic from the beginning. Initially, they appear on either surface as superficial eruptions, but gradually develop through the entire thickness of the leaflet and become corky and rough. After attaining some height, the eruptions begin to collapse due to disintegration of infected cells, resulting in the formation of rifts. Sometimes, the rifts are also caused due to the disintegration of leaf tissue by the action of wind. In older spots, the superficial eruptions collapse leaving behind saucer-shaped depressions (Figure 7.19a and b). In a few cases, the shot-hole symptoms are also observed. Generally, the spots are isolated, but in severe infections, they coalesce to form necrotic patches, but never develop into a typical leaf blight phase as in common blight of bean (Figure 7.16a). Lesions on petioles, petiolule, or pulvinus are brown, flat, or raised and may be elongated in the form of long streaks. Normally, the infected plants do not defoliate.

On stems, dark brown, elongated lesions are produced. In severely infected plants, dark brown, longitudinal slits or cankers develop on the stems, which often extend from soil level to the growing tip.

Lesions on pods are brown and raised, but are rarely seen in nature. In artificially inoculated pods, brown and raised lesions appear. Sometimes, a thin layer of shining gummy bacterial ooze can also be seen on the outer surface. On the inner side of green pods, brownish lesions are formed just beneath the bacterial ooze. The seeds touching these lesions also develop brown spots. The seeds produced in the infected pods are generally small, shriveled, and discolored. Most of the discolored seeds are brown and yellowish, but occasionally black also. The normal or effective disease development occurs only on the monsoon season crop. In the summer crop, the symptoms develop on the cotyledonary and first trifoliate leaves, but subsequent spread of the disease does not occur.

As the pathogen is largely parenchymatous, infected plants are not completely killed. However, severe infections result in

(a)

(b)

FIGURE 7.19 Bacterial leaf spot of green gram: (a) Symptoms on upper leaf surface. (b) Symptoms on lower leaf surface.

the necrosis of considerable leaf area leading to reduction in photosynthesis and general weakening of the plants.

7.28.3 Causal Organism

Xanthomonas axonopodis pv. *vignaeradiatae* (Sabet et al. 1969) Vauterin et al. 1995
Syn. *Xa. campestris* pv. *vignaeradiatae*

The description of *Xa. axonopodis* Starr and Garces 1950 emend. Vauterin et al. 1995 is given below. Cells are straight rods, Gram-stain-negative, non-sporing, and measure 0.4–0.6 × 0.8–2.0 µm; occur mostly singly or in pairs, occasionally short chains, and rarely in filaments. Motile by a single polar flagellum, obligately aerobic, and metabolism is strictly respiratory type. Colonies are usually yellow, smooth and butyrous, viscid or mucoid. Yellow pigments, xanthomonadins are produced. The viscous or mucoid growth of the culture is due to the production of an extracellular heteropolysaccharide called xanthan.

Catalase positive, oxidase is negative or weak. Urease and indole are not produced, and nitrate is not reduced to nitrite. Litmus milk is not acidified. Most of the strains utilize D-galactose, melibiose, acetic acid, DL-lactic acid, malonic acid, L-leucine, L-serine, and glycerol.

The members of this species are slow growing and metabolically less active than those of *Xa. campestris*. Both these species can be clearly differentiated on the basis of phenotypic characters. *Xa. axonopodis* is the most heterogeneous species currently recognized in genus *Xanthomonas* and includes several pathovars. In general, these pathovars are not distinguishable by their phenotypic characters, and their identification is reliable on the basis of their host range.

Xa. axonopodis pv. *vignaeradiatae* being a pathovar of *Xa. axonopodis* has the same morphological, cultural, and biochemical characters as those of the latter. In addition to abovementioned characteristics, *Xa. axonopodis* pv. *vignaeradiatae* produces H_2S, liquefies gelatin, hydrolyses starch, and ferments glucose, sucrose, mannose, xylose, lactose, and raffinose producing only acid (Patel and Jindal 1972). Thakur et al. (1977) confirmed the existence of six races in the bacterium. They further reported that: (1) resistance in a differential to a race is governed by a single gene, (2) gene governing resistance in a differential is dominant to its recessive allele governing susceptibility in another differential, and (3) that a gene or allele governing resistance or susceptibility in the differentials to different races does not overlap.

Patro and Jindal (2000) have shown the existence of plasmids in the bacterium and their correlation with virulence. The highly virulent isolate contained the maximum number of plasmids. The plasmid-borne genes were shown to govern virulence, exopolysaccharide production, antibiotic resistance, and pigmentation.

The bacterium produces exopolysaccharide, proteolytic, and cellulolytic enzymes which play role in pathogenesis. Higher proteolytic and cellulolytic activities were observed in susceptible green gram cv. Pusa Baisakhi after 120 h of inoculation with a virulent culture (Dutta and Jindal 2002).

7.28.4 Disease Cycle

Earlier, the bacterium was considered to be seed-borne, but Soni (1982) established the seed-borne nature of the bacterium and showed that seed is an important source of survival for the bacterium. He further reported that the bacterium survived for 1 year in the seeds and inoculation of the bacterium only at inflorescence and at flat green pod stage led the bacterium to become seed-borne. Deformity of seeds like shriveling, discoloration, and small size was only an indication, but not a sure test of seed-borne infection.

Soni and Thind (1987) compared different growing on tests for the detection of *Xa. axonopodis* pv. *vignaeradiatae* from green gram seeds and found that traditional growing on test was more suitable in monsoon season when conditions were highly favorable for disease development, while incubation growing on test proved more effective in summer crop season. They further reported that wounded seed growing on test was most effective for detecting the lowest level of seed-borne infection, and it detected 0.1% infected seeds in a seed lot. Bacteriophages have also been found effective for the detection of bacterium from shriveled and discolored green gram seeds (Soni and Thind 1991).

The bacterium enters the host through stomata and wounds, and is largely confined to the parenchyma tissue. Soni (1982), on the basis of recovery of the bacterium only from roots, primary, and trifoliate leaves, suggested the non-systemic nature of the bacterium. Secondary spread of the bacterium is mainly through wind-splashed rain. Dissemination to distant places and new areas is mainly through the infected seed.

Most of the commercially grown varieties in India including Pusa Baisakhi are susceptible to the bacterium, while Jalgaon 781 is resistant.

7.28.5 Management

1. As the bacterium perpetuates through infected seed, the use of bacteria-free seed and the eradication of seed-borne inoculum are the most effective and viable methods of disease control
2. Seed treatment with hot water at 52°C for 30 min, solar heat (for details see bacterial blight of cowpea), Streptocycline (100 µg mL^{-1}) + captan (0.2%), or Streptocycline (100 µg mL^{-1}) + Agallol (0.2%) have been found very effective for the eradication of seed-borne inoculum (Jindal and Thind 1987)
3. For controlling the secondary infection of the bacterium, four sprays, i.e., first two sprays of Streptocycline (100 µg mL^{-1}) + Bavistin (0.05%) followed by two sprays of Streptocycline (100 µg mL^{-1}) + Blitox (0.2%), at 10 days intervals, have been found effective by Jindal and Thind (1987). They also reported that the combination of these sprays with hot water treatment or solar heat treatment of seed proved most effective in controlling the disease and caused 69.9% and 72.5% reduction in disease intensity, respectively. Gour and Ahmed (1983) recommended three sprays

of Agrimycin-100 (0.025%) or Plantomycin (0.025%) + Bavistin (0.1%) for the control the disease. Borah et al. (2000) have advocated the use of a phylloplane bacteria, *Bacillus* sp. for the control of disease and recommended its preventive application. Rathore (2010), on the basis of 3 years pooled data, reported that streptocycline (0.01%) sprays gave the minimum disease intensity (6.1%) and highest grain and fodder yields (10.28 and 20.59 q/ha) compared to 24.9% disease intensity, 5.76 q/ha grain yield, and 12.18 q/ha fodder yield obtained in control.

REFERENCES

Borah, P. K., Jindal, J. K. and Verma, J. P. 2000. Biological management of bacterial leaf spot of mungbean caused by *Xanthomonas axonopodis* pv. *vignaeradiatae*. *Indian Phytopathol.* 53: 384–394.

Dutta, S., and Jindal, J. K. 2002. Hydrolytic enzymes and exopolysaccharide production of *Xanthomonas axonopodis* pv. *vignaeradiatae* inciting bacterial leaf spot of mungbean. *Annl. Plant Prot. Sci.* 10: 295–298.

Gour, R. B., and Ahmed, S. R. 1983. Studies on the chemical control of bacterial blight of moong incited by *Xanthomonas phaseoli* (Smith) Dowson. *Pesticides* 17: 22–23.

Iqbal, S. M., Zubair, M., Anwar, M., and Haqqani, A. M. 2003. Resistance in mungbean to bacterial leaf spot disease. *Mycopathologia* 1: 81–83.

Jindal K. K., and Thind, B. S. 1987. Control of Cercospora and bacterial leaf spot of green gram. *Indian J. Agric. Sci.* 57: 372–375.

Patel, P. N., and Jindal, J. K. 1972. Bacterial leaf spot and halo blight diseases of mung bean (*Phaseolus aureus*) and other legumes in India. *Indian Phytopathol.* 25: 517–525.

Patel, P. N., Jindal, J. K., and Singh, D. 1972. Studies on resistance in crops to bacterial diseases in India. IV. Resistance in mung bean (*Phaseolus aureus*) to *Xanthomonas phaseoli*. *Indian Phytopathol.* 25: 526–529.

Patro, T. S. S. K., and Jindal, J. K. 2000. Expression of plasmid-borne characters in *Xanthomonas axonopodis* pv. *vignaeradiatae*. *Trop. Agr. Res.* 12: 86–94.

Rathore, B. S. 2010. Efficacy of streptocycline and plant extracts against bacterial leaf spot disease caused by *Xanthomonas axonopodis* pv. *vignaradiatae* of greengram. *Indian Phytopathol.* 63: 384–386.

Sabet, K. A., Ishag, F., and Khalil, O. 1969. Studies on the bacterial diseases of Sudan crops VII. New records. *Ann. Appl. Biol.* 63: 357–369.

Soni, P. S. 1982. Detection of Phytopathogenic Bacteria from Cowpea and Mung Seeds. PhD thesis, Punjab Agricultural University, Ludhiana, India, 180 pp.

Soni, P. S., and Thind, B. S. 1987. Detection *of Xanthomonas campestris* pv. *vignicola* (Burkh.) Dye from cowpea seeds and *X. campestris* pv. *vignaeradiatae* (Sabet et al.) Dye from mung bean seeds by growing on tests. *Phytopathol. Medit.* 26: 1–6.

Soni, P. S., and Thind, B. S. 1991. Detection of *Xanthomonas campestris* pv. *vignaeradiatae* (Sabet et al.) Dye from green gram seeds and *X. campestris* pv. *vignicola* (Burkh.) Dye from cowpea seeds with the help of bacteriophages. *Plant Dis. Res.* 6: 6–11.

Starr, M. P., and Garcés, O. C. 1950. El agente causante de la gomosis bacterial del pasto imperial en Colombia. *Revista Facultad Nacional de Agronomia (Medellin)* 11: 73–83.

Thakur, R. P., Patel, P. N., and Verma, J. P. 1977. Studies on resistance in crops to bacterial diseases in India. XI. Genetic make-up of mungbean differentials of the races of bacterial leaf spot pathogen, *Xanthomonas phaseoli*. *Indian Phytopathol.* 30: 217–221.

Vauterin, L., Hoste, B., Kersters, K., and Swings, J. 1995. Reclassification of *Xanthomonas*. *Int. J. Syst. Bacteriol.* 45: 472–489.

7.29 HALO BLIGHT OF GREEN GRAM

Green gram [*Vigna radiata* (L.) Wilczek] or *mung* bean (vernacular name of green gram in India) suffers from four bacterial diseases, namely, bacterial leaf spot caused by *Xanthomonas axonopodis* pv. *vignaeradiatae*, halo blight caused by *Pseudomonas savastanoi* pv. *phaseolicola*, tan spot caused by *Curtobacterium flaccumfaciens* pv. *flaccumfaciens*, and bacterial blight caused by *Xa. axonopodis* pv. *phaseoli*. In India, halo blight is comparatively of less economic significance than bacterial leaf spot.

Patel and Jindal (1970) observed halo blight for the first time in 1969 at Indian Agricultural Research Institute, New Delhi on some *mung* bean lines resistant to bacterial leaf spot. Subsequently, Schmitthenner et al. (1971) reported a severe epiphytotic of the disease from Ohio in the USA. During 2009 to 2014, halo blight of *mung* bean caused by *Psm. syringae* pv. *phaseolicola* was found in three of the main Chinese geographic regions, which contain more than 90% of the *mung* bean growing areas in China. The identity of the bacterium was confirmed on the basis of phenotypic characteristics, the physiological and biochemical properties, pathogenicity tests, and fatty acid composition, in combination with specific polymerase chain reactions and 16S–23S ribosomal DNA sequence analyses. The results indicate that the bacterium is likely of epidemiological significance on *mung* bean in China (Sun et al. 2017). The disease also occurs in Australia, but the bacterium causing the disease has been identified as *Psm. savastanoi* pv. *phaseolicola*. In Australia, it was first reported from Queensland in the early 1980s (Ryley 2008).

7.29.1 Host Range

Schmitthenner et al. (1971) reported susceptible reaction on red kidney bean (*Phaseolus vulgaris*), lima bean (var. Fordhook 242), and soybean (var. Acme), in addition to *mung* bean. Schroth et al. (1971) also found that *mung* strain could infect few varieties of bean, lima bean, and soybean. Patel and Jindal (1972a) found that the infection on *P. atropurpureus*, *P. bracteatus*, *Dolichos lablab*, and *D. biflorus* was as severe as on *mung* bean. On *P. mungo*, *P. trilobus*, and *Lathyrus sativus*, only few spots with distinct haloes were produced. Among the non-legumes, *Ficus religiosa* was susceptible. Only *P. bracteatus* and *P. atropurpureus* are susceptible to both the pathogens (causal agents of common bean and *mung* bean halo blights). Bean plants showed hypersensitive reaction within 24 h when inoculated with the *mung* bean pathogen. Bean bacterium attacked 10 of the 15 *Phaseolus* spp. and two other legumes, whereas *mung* bean pathogen infected only 5 of 13 *Phaseolus* spp., four other legumes, and a non-legume

(Patel and Jindal 1972a). *Mung* bean strain of the bacterium has a distinct host range than those of bean strain, and they do not cross infect (Patel and Walker 1965).

7.29.2 Symptoms

On the seedlings developing from infected seeds, water-soaked spots appear first on cotyledonary leaves and then spread to the trifoliate leaves. Severe infection at this stage may result in the death of seedlings. On trifoliate leaves, the spots may be small or large, round or irregular, flat, and water-soaked. The spots are surrounded by halo-like zones of greenish-yellow tissue, which result from the toxin produced by the bacteria multiplying in the water-soaked lesions. Young developing leaves on severely infected plants show systemic chlorosis due to the translocation of the toxin. The formation of halo zones and development of systemic chlorosis are favored by low temperature (15°C–20°C) at nights as in halo blight of common bean. The spots on old leaves are sometimes angular and do not show distinct halo. Frequently, the infection spreads along the veins and veinlets turning them brown and necrotic and causing distortion of the developing leaflets.

On the petioles, long streaks appear which are first water-soaked, but later on turn red. Pulvinus and petiolules of the severely infected leaves appear brown and sunken. Generally, there is no defoliation of the plants. On the stem, first water-soaked spots appear which gradually enlarge and turn into reddish streaks extending longitudinally. Such stems often split. On pods, water-soaked lesions appear and turn brown with time.

7.29.3 Causal Organism

Pseudomonas savastanoi pv. *phaseolicola* (Burkholder 1926) Gardan et al. 1992 *mung* bean strain
Syn. *Psm. phaseolicola mung* bean strain, *Psm. syringae* pv. *phaseolicola mung* bean strain

The first isolation of the *mung* bean halo blight pathogen appears to have been made by D.C. Graham of Scotland in 1966 from *mung* bean seeds. He named the bacterium as *Pseudomonas phaseolicola,* but presumably did not work with it further. Schroth et al. (1971) compared this culture with *Psm. phaseolicola* from different hosts and found it to be distinct from bean isolates in utilization of several nutritional substrates and host range. Schmitthenner et al. (1971) found that the *mung* bean bacterium was similar to both *Psm. phaseolicola* and *Psm. glycinea* in cultural characteristics. As only the *mung* bean isolates were virulent to *mung* bean, they proposed that the *mung* bean bacterium be named as *mung* bean strain of *Psm. phaseolicola*. Keeping in view the host range and overall similarity in bacteriological properties of the *mung* halo blight pathogen to that of bean halo blight pathogen, Patel and Jindal (1972a) also identified it as *mung* strain of *Psm. phaseolicola*. Young et al. (1978) reclassified it as *Psm. syringae* pv. *phaseolicola mung* bean strain. Later on, Gardan et al. (1992) reclassified it as *Psm. savastanoi* pv. *phaseolicola mung* bean strain.

The below given description of the *mung* bean halo blight bacterium is based on the studies carried out by Patel and Jindal (1972b).

The colonies on nutrient agar are circular, white, glistening, flat, and entire, and produce green fluorescent pigment on King's medium B. Cells are Gram-stain-negative rods, motile with 1–4 polar flagella, and strictly aerobic. The bacterium utilizes glucose oxidatively, utilizes asparagine, and can grow in the presence of 4.0% sodium chloride. The hydrolysis of gelatin is slow and milk is peptonized without curd formation. The tests for production of ammonia, urease, and catalase, and starch hydrolysis are positive, while those for production of indole, acetoin, and H_2S, and reduction of nitrate are negative. The bacterium utilizes sodium propionate and succinate, but not sodium oxalate and tartrate.

Ryley (2008) reported that there are at least nine strains (races, pathotypes) of *Psm. savastanoi* pv. *phaseolicola* (casual agent of halo blight of *mung* bean in Australia), which has serious implications for resistance breeding; varieties that are immune to some strains may be susceptible to others.

7.29.4 Disease Cycle

The bacterium is seed-borne as reported by Schmitthenner et al. (1971), and its spread to different places occurs through seed. Not much information is available on this disease, and it is presumed that the disease cycle is similar to that of bean halo blight. The disease assumed serious proportions in Ohio, USA, and the crop raised from infected seed gave 60% less yield than the crop raised from healthy seed (Schmitthenner et al. 1971). The disease has not appeared in a severe form in India. Hence, there is a need to investigate if the environmental or other factors are responsible for its low incidence in India, particularly in *Kharif* crop.

7.29.5 Management

1. As the bacterium perpetuates only through infected seed, the use of pathogen-free seed is the best approach to manage the disease
2. There is no information available on seed treatment. However, it is possible that seed treatments recommended for the eradication of *Xa. axonopodis* pv. *vignaeradiatae* from *mung* bean seed may also work for this pathogen. Hence, these treatments should be tried to see their efficacy
3. Restricted use or avoidance of overhead-irrigation will help in reducing the incidence of the disease

REFERENCES

Burkholder, W. H. 1926. A new bacterial disease of the bean. *Phytopathology* 16: 915–928.

Gardan, L., Bollet, C., Abu Ghorrah, M., Grimont, F., and Grimont, P. A. D. 1992. DNA relatedness among the pathovar strains of *Pseudomonas syringae* subsp. *savastanoi* Janse (1982) and proposal of *Pseudomonas savastanoi* sp. nov. *Int. J. Syst. Bacteriol.* 42: 606–612.

Patel, P. N., and Jindal, J. K. 1970. Bacterial diseases in seed legumes in 1968 and 1969. Paper presented at the Fourth Annual Workshop Conference on Pulse Crops held at Ludhiana (Punjab) in April, 1970.

Patel, P. N., and Jindal, J. K. 1972a. Halo blight of mung (green gram). pp. 75–78 in: *Plant Bacteriology*. Vol. I, *Bacterial Diseases of Plants in India*. P. N. Patel ed. Indian Agricultural Research Institute, New Delhi, India.

Patel, P. N., and Jindal, J. K. 1972b. Bacterial leaf spot and halo blight diseases of mung bean (*Phaseolus aureus*) and other legumes in India. *Indian Phytopathol.* 25: 517–525.

Patel, P. N., and Walker, J. C. 1965. Resistance in *Phaseolus* to halo blight. *Phytopathology* 55: 889–894.

Ryley, M. 2008. Halo blight in mungbean - pathogen, hosts, infection, crop impact, management and future genetics and breeding options [https://grdc.com.au/resources-and-publications. Accessed January 12, 2019.]

Schmitthenner, A. F., Hoitink, H. A. J., and Kroetz, M. E. 1971. Halo blight of mungbean incited by a new strain of *Pseudomonas phaseolicola* (Abstr.). *Phytopathology* 61: 909.

Schroth, M. N., Vilma Vitanza, B., and Hildebrand, D. C. 1971. Pathogenic and nutritional variation in the halo blight group of fluorescent pseudomonads of bean. *Phytopathology* 61: 852–857.

Sun, S., Zhi, Y., Zhu, Z., Jin, J. Duan, C., et al. 2017. An emerging disease caused by *Pseudomonas syringae* pv. *phaseolicola* threatens mung bean production in China. *Plant Dis.* 101: 95–102.

Young, J. M., Dye, D. W., and Wilkie, J. P. 1978. Genus *Pseudomonas* Migula 1894. In: Young, J. M., Dye, D. W., Bradbury, J. F., Panagopoulos, C. G., and Robbs, C. F. 1978. A proposed nomenclature and classification for plant pathogenic bacteria. *N. Z. J. Agric. Res.* 21: 153–177.

7.30 BACTERIAL BLIGHT OF CLUSTERBEAN

Bacterial blight of clusterbean [*Cyamopsis tetragonoloba* (L.) Taub.] is a widespread disease, which is particularly destructive to the rainy season crop. The loss is mainly due to blightening and mortality of the seedlings in the early stage. Severe cankering of the stem and death of grown up plants also results in heavy losses. In *Kharif* 1971, the disease caused 50%–70% loss in yield amounting to rupees 3–5 lakhs in 600 acres of cv. Pusa Naubahar grown for certified seed production by National Seed Corporation of India.

Patel et al. (1953) reported a bacterial leaf spot of *guar* (local name for clusterbean in India) from Patna and Bombay and named the pathogen as *Xanthomonas cyamopsidis*. Later on, Patel and Patel (1958) described a severe leaf blight of *guar* from Anand, Gujarat and suggested the resemblance of its symptoms with those of black rot caused by *Xa. campestris* in crucifers. They presumed this disease different from that described by Patel et al. (1953) and hence, named the bacterium as *Xa. cyamophagus*. Srivastava and Rao (1963a) described a bacterial leaf and stem blight of *guar* and concluded that all the different types of symptoms are caused by the same pathogen. They also suggested the invalidation of *Xa. cyamophagus*.

Ren et al. (2014) reported bacterial blight of guar caused by *Xa. axonopodis* pv. *cyamopsidis* from China. The identity of the pathogen was confirmed on the basis of morphology, pathogenicity tests, 16S rDNA sequencing, and host plant specificity. The disease incidence was generally 40% to 50%, but as high as 80% was observed in several fields. The authors fear that bacterial blight may pose a threat to the economic sustainability of guar production in Xinjiang province of China.

Clusterbean suffers from another bacterial disease called bacterial leaf spot caused by *Pseudomnas syringae* pv. *syringae*.

7.30.1 Symptoms

Both leaf spot and blight phases of the disease appear simultaneously. Infection through stomata results in leaf spots, which are interveinal, round, water-soaked or oily in appearance, and well-defined on the dorsal surface of the leaf. They may enlarge or coalesce, ultimately leading to blight phase. Then the pathogen rapidly invades vascular tissue and causes flaccidity of the affected portions. These flaccid portions eventually become necrotic and turn brown. Several flaccid areas formed on a leaf, including those on the margins, coalesce resulting in blightening of the entire lamina.

Infections resulting from hydathodal penetrations appear as V-shaped areas at the leaf margins and progress toward the midrib. From the leaf, infection moves systemically through the petiole into the stem producing black longitudinal streaks on the stem. Infection of the stem results in wilting of the apical leaves and curvature of the stem. Finally the stem splits and breaks off.

On pods, black streaks appear, presumably resulting from local infection. In contrast to other systemic bacterial disease of legumes like common or fuscous blight of bean and cowpea blight, infection on suture is rarely seen. The seedlings, developing from infected seeds, show rapid infection of primary leaves from where the infection progresses to the tender stems, resulting in the death of seedlings.

7.30.2 Causal Organism

Xanthomonas axonopodis pv. *cyamopsidis* (Patel et al. 1953) Vauterin et al. 1995

Syn. *Xa. campestris* pv. *cyamopsidis*, *Xa. cyamopsidis*

The below given description of the bacterium is taken primarily from the publications of Patel et al. (1953) and Patel and Patel (1958).

Cells are short rods (0.4–0.8 × 1.2–1.5 µm), occur singly or rarely in pairs, encapsulated, facultatively anaerobic, nonsporing, not acid-fast, Gram-stain-negative, and motile by a polar flagellum. Colonies on potato dextrose agar are yellow, round having entire margins, shining, butyrous, pulvinate, and attain diameter of 1.1 cm after an incubation period of 4 days. Colonies on nutrient agar are mustard yellow, flat, smooth, glistening having entire margins, and grow up to 1.1 cm in diameter after 6 days of incubation.

The bacterium hydrolyses starch and casein, reduces litmus, produces hydrogen sulphide and ammonia, but not indole, and does not reduce nitrates. Acid is produced without gas from glucose, lactose, maltose, dextrose, xylose, and mannitol. The optimum temperature for growth ranges from 25°C to 31°C and thermal death point is 52°C. Kaur et al. (2010) reported that the bacterial growth in culture media is inhibited by aqueous leaf extracts of *Emblica officinalis* (aonla), *Eucalyptus citriodora*, and *Tagetes* sp. (marigold). *E. officinalis* leaf extracts exhibited maximum antibacterial activity followed by *E. citriodora*.

Kaur et al. (2005b) successfully employed different PCR-based techniques, such as, RAPD-PCR, rep-PCR, and IS*1112*-PCR to elucidate the pathogen population structure and intrapathovar relationship among a set of 25 *Xa. axonopodis* pv. *cyamopsidis* isolates. There was a significant variation in their pathogenicity and genetic structure. However, the pathogenicity data were found to be in line with the molecular data to some extent. RAPD markers revealed a high level of genetic diversity among the isolates used, in comparison to the other molecular markers employed. A weak correlation was observed among the various molecular markers employed, as each technique explores genetic variation differently. As there was no correlation between DNA amplification product patterns and geographical sites of isolation, it suggests that *Xa. axonopodis* pv. *cyamopsidis* has spread largely through exchange of infected plant germplasm.

Two pathogenic races, namely, race 0 and race 1 have been reported from USA. Race 1 infects cultivar Brooks, while race 0 does not. In India, this cultivar has been found highly susceptible to the bacterium under natural conditions, indicating that Indian strains are more virulent.

7.30.3 Disease Cycle

The bacterium is internally seed-borne. Seed is the most important source of perpetuation, particularly in areas where only one crop of *guar* is taken in a year.

Volunteer plants are another source of perpetuation where *guar* is planted in the same fields every year. This source becomes even more important when the variety grown is of shattering type. The pathogen does not survive in infected plant debris under north Indian conditions where the gap between two crops is of 9 months (Srivastava and Rao 1970).

The bacterium enters the host through stomata, hydathodes, or wounds, and its multiplication and movement in the host is largely vascular. The movement of the bacterium from the leaves, through the petiole, to the stem, and its further advancement to the seed clearly suggests its systemic nature. The disease is favored by high relative humidity and a warm temperature of 28°C–32°C, which usually prevail during monsoon in India. Therefore, the disease is usually more severe in July and August. Secondary spread of the bacterium occurs mostly through wind-splashed rain. Long distance spread or spread to the new areas occurs usually through infected seed. The disease is more severe in low-lying areas where water stagnates temporarily. The disease is also severe in furrow-sown fields where water contaminated with pathogen moves from one end to another end of the furrow in sloppy fields.

Shah et al. (1996) found that a temperature range of 20°C–28°C and relative humidity between 70% and 80% favored rapid progress of the disease. On the basis of epidemiological studies carried out for 5 years, from 1991 to 1995, they also developed a model to forecast the disease based on these parameters prevailing in the preceding week. Yadav and Nath (2006a) reported that a temperature range of 29°C–34°C and relative humidity above 68% favored the disease. They further reported that the maximum disease incidence of 58.3% was observed in the crop sown in the last week of July.

Kalaskar et al. (2014) investigated the activity of isoenzymees, peroxidase (PO), and PPO in healthy and bacterial blight infected samples of the 12 genotypes of clusterbean. The range of 0.42 to 1.07 and 0.08 to 0.36 in OD/min/g of fresh weight tissues in healthy plants and of 0.19 to 1.20 and 0.09 to 0.52 in OD/min/g of fresh weight tissues in bacterial blight infected plants was observed for PO and PPO activity, respectively. Decrease in the PO and PPO activity in bacterial blight infected susceptible plants compared to bacterial blight infected resistant plants was observed. The decrease in activity of PO (from 0.75 to 0.19) and of PPO (from 0.22 to 0.09) was maximum in the most susceptible genotype, PNB. The authors suggest that increase or no change in PO activity after infection can be considered as a marker for resistance to bacterial blight.

Most of the Indian improved varieties like Pusa Naubahar and Pusa Sadabahar are highly susceptible to the disease. The branching-type varieties are also susceptible, but they suffer less from the disease because the new shoots compensate to some extent for the dead shoots. In addition to *guar*, bean acts as a natural host of the bacterium (Patel and Jindal 1972). Kaur et al. (2004) evaluated ten genotypes under field conditions and found that cv. HG 75 was highly resistant and cv. PNB most susceptible to the disease.

7.30.4 Management

1. Use of pathogen-free seed and eradication of seed-borne inoculum are the most effective and practical means of disease management. The seeds from surface irrigated summer grown crops are free from infection (Srivastava and Rao 1963b, 1970)
2. Hot water treatment at 56°C for 10 min eradicates the seed-borne inoculum. Seed dip in Streptocycline solution (200 μg mL^{-1}) for 3 h proved highly effective for eradicating the seed-borne inoculum (Rathore 2000). Seed treatment with Streptocycline (500 μg mL^{-1}) followed by its foliar applications gave minimum disease incidence and highest grain yield (Yadav and Nath 2006b). Amin et al. (2017), on the basis of field trials conducted during *kharif* 2013 to 2015, reported that soaking of seed in Streptocycline (250 μg mL^{-1}) solution for 30 min plus first spray of Streptocycline at 250 μg mL^{-1} at just appearance of the disease and the second spray at 15th day after 1st spray gave the maximum yield and minimum disease intensity.

This treatment was closely followed by seed soaking in Streptocycline at 250 μg mL^{-1} for 30 min plus 1st spray of Streptocycline at 200 μg mL^{-1} at just disease appearance and the second at 15th day after 1st spray

3. Fresh leaf extracts of *Emblica officinalis* and *Eucalyptus* sp. have been demonstrated to be effective in reducing the disease under field conditions (Kaur et al. 2005a, 2006)
4. Use of resistant varieties

REFERENCES

Amin, A. M., Patel, N. R., Jaiman, R. K., Prajapati, D. B., and Prajapati, B. G. 2017. Management of bacterial blight in clusterbean. *Environ. Ecol.* 35(2C): 1223–1227.

Kalaskar, S. R., Shinde, A. S., Dhembre, V. M., Sheikh, W. A., Patil, V. S., et al. 2014. The role of peroxidase and polyphenoloxidase isoenzymes in resistance to bacterial leaf blight (*Xanthomonas campestris* pv. *cyamopsidis*) in clusterbean (*Cyamopsis tetragonoloba* (L.) Taub.). *J. Cell Tissue Res.* 14: 4577–4580.

Kaur, K., Jalali, I., and Chhabra, M. L. 2005a. Efficacy of botanicals for the management of bacterial blight of clusterbean under disease stress conditions. *Proceedings of the National Symposium on Plant Disease Diagnosis, Epidemiology and Management*, Hisar. 47pp.

Kaur, K., Jalali, I., and Pal, V. 2006. Factors affecting the bioefficacy of phytoextracts against *Xanthomonas axonopodis* pv. *cyamopsidis*. *Pl. Dis. Res.* 21: 164–167.

Kaur, K., Jalali, I., and Pal, V. 2010. Toxicity of botanicals against clusterbean bacterial blight pathogen. *J. Mycol. Pl. Pathol.* 40: 124–127.

Kaur, B., Purkayastha, S., Dilbaghi, N., and Chaudhury, A. 2004. Evaluation of clusterbean genotypes for resistance to bacterial blight. *Ann. Agri. Bio. Res.* 9: 213–215.

Kaur, B., Purkayastha, S., Dilbaghi, N., and Chaudhury, A. 2005b. Characterization of *Xanthomonas axonopodis* pv. *cyamopsidis*, the bacterial blight pathogen of cluster bean using PCR-based molecular markers. *J. Phytopathol.* 153: 470–479.

Patel, M. K., Dhande, G. W., and Kulkarni, Y. S. 1953. Bacterial leaf-spot of *Cyamopsis tetragonaloba* (L.) Taub. *Curr. Sci.* 22: 183.

Patel, P. N., and Jindal, J. K. 1972. Bacterial blight of guar. pp. 66–69 in: *Plant Bacteriology*. Vol. I, *Bacterial Diseases of Plants in India*. P. N. Patel ed. Indian Agricultural Research Institute, New Delhi, India.

Patel, A. J., and Patel, M. K. 1958. A new bacterial blight of *Cyamopsis tetragonaloba* (L.) Taub. *Curr. Sci.* 27: 258–259.

Rathore, B. S. 2000. Management of bacterial blight of clusterbean through seed dressers. *Plant Dis. Res.* 15: 89–92.

Ren, Y. Z., Yue, Y. L., Jin, G. X., and Du, Q. 2014. First report of bacterial blight of guar caused by *Xanthomonas axonopodis* pv. *cyamopsidis* in China. *Plant Dis.* 98: 840.

Shah, R., Dashora, P. K., Bhatnagar, M. K., and Mali, B. L. 1996. Forecasting model for bacterial blight of clusterbean. *Proceedings of the 9th International Conference on Plant Pathogens Bacteria*. A. Mahadevan ed. Chennai, India, pp. 525–527.

Srivastava, D. N., and Rao, Y. P. 1963a. Bacterial blight of guar. *Indian Phytopathol.* 16: 69–73.

Srivastava, D. N., and Rao, Y. P. 1963b. Seed transmission and control of bacterial blight of guar (*Cyamopsis tetragonaloba* (L.) Taub.). *Indian Phytopathol.* 16: 390–391.

Srivastava, D. N., and Rao, Y. P. 1970. Epidemiology and control of bacterial blight of guar (*Cyamopsis tetragonaloba* (L.) Taub.). Indian Phytopathol. Soc. Bull. No 6, 1–4.

Vauterin, L., Hoste, B., Kersters, K., and Swings, J. 1995. Reclassification of *Xanthomonas*. *Int. J. Syst. Bacteriol.* 45: 472–489.

Yadav, S., and Nath, R. 2006a. Effect of weather parameters and different sowing times on the occurrence of clusterbean bacterial blight. *J. Mycopathol. Res.* 44: 267–269.

Yadav, S., and Nath, R. 2006b. Efficacy of chemicals against bacterial blight of clusterbean. *J. Mycopathol. Res.* 44: 271–273.

7.31 BACTERIAL LEAF SPOT OF HYACINTH BEAN

Lablab bean, hyacinth bean, Indian bean, or Indian butter bean (*Lablab purpureus*, syn. *Dolichos lablab*, *L. niger*) is also called by another name, i.e., *sem* in India. The first natural occurrence of bacterial leaf spot on hyacinth bean was reported by Rangaswami and Gowda (1963) from Tamil Nadu, India. Subsequently, Patel and Jindal (1970) found it at Indian Agricultural Research Institute, New Delhi and Amravati in Maharashtra, India. The disease also occurs in Sudan (Sabet 1959).

7.31.1 Host Range

Rangaswami and Gowda (1963) reported that the bacterium infected *D. biflorus*, *Vigna catjang*, *Phaseolus vulgaris*, *Tophrosia purpurea*, and *Cyamopsis psoralioides*. However, Patel and Jindal (1973) did not obtain infection on the above-mentioned hosts except *P. vulgaris*. The susceptibility of *D. lablab* to *Xa. fuscans* subsp. *fuscans* and *Xa. axonopodis* pv. *glycines* has been reported. Whether *D. lablab* serves as a natural host of these pathogens is not known.

7.31.2 Symptoms

The disease is more severe during rainy season. The symptoms appear mainly on leaves, which are round or irregular, water-soaked spots. These spots eventually turn brown. Several such spots may coalesce. In a relatively dry weather, tissues surrounding closely situated spots may turn yellow and form large blotches, but it never leads to an extensive blightening. Due to severe spotting, leaves turn yellow and fall off prematurely. On stems, oblong, reddish-brown lesions, sometimes having greyish centers, appear. On pods, the spots are first water-soaked, but later on turn brown.

7.31.3 Causal Organism

Xanthomonas axonopodis pv. *phaseoli* (Smith 1897) Vauterin et al. 1995 hyacinth bean strain

7.31.4 Disease Cycle

The bacterium is seed-borne and also survives on vines growing perennially in the kitchen gardens. Warm and wet weather prevailing during monsoon season favors the disease development. In hyacinth or Indian bean, the bacterium is largely confined to parenchymatous tissue, while in common bean (*P. vulgaris*), it enters the vascular system. New growth on infected vines is generally free from infection under dry conditions. Considerable reduction in yield occurs when plants are badly infected or disease severity is high in monsoon season. Seedling mortality occurs very rarely (Patel and Jindal 1972).

Patel and Jindal (1973) evaluated 334 germplasm lines against the pathogen and found none of them to be immune or highly resistant. However, PLS 63, PLS 101, M.S. 9491, and D.L. 2434 showed moderate resistance as the lesions remained small, turned necrotic, and the infected leaves did not defoliate.

7.31.5 Management

1. As the bacterium is seed-borne, use of pathogen-free seed or treatment of infected seed can prove useful. Seed treatment recommended for bacterial blight of cowpea can be tried
2. As the bacterium is mainly confined to the parenchyma, foliar sprays of Streptocycline at 100 $\mu g\ mL^{-1}$ can also prove effective

REFERENCES

Patel, P. N., and Jindal, J. K. 1970. Bacterial diseases in seed legumes in 1968 and 1969. Paper presented at the Fourth Annual Workshop Conference on Pulse Crops held at Ludhiana (Punjab), India in April, 1970.

Patel, P. N., and Jindal, J. K. 1972. Bacterial leaf spot of *Dolichos lablab*. pp. 84–86 in: *Plant Bacteriology*. Vol. I, *Bacterial Diseases of Plants in India*. P. N. Patel ed. Indian Agricultural Research Institute, New Delhi, India.

Patel, P. N., and Jindal, J. K. 1973. Studies on resistance in crops to bacterial diseases in India. VI: Reaction of moth bean (*Phaseolus acontifolius*) and hyacinth bean (*Dolichos lablab*) lines to bacterial blight and leaf spot diseases incited by *Xanthomonas phaseoli* strains. *Indian Phytopathol.* 26: 607–608.

Rangaswami, G., and Gowda, S. S. 1963. On some bacterial diseases of ornamentals and vegetables in Madras state. *Indian Phytopathol.* 16: 74–85.

Sabet, K. A. 1959. Studies in the bacterial diseases of Sudan crops. III. On the occurrence, host range and taxonomy of the bacteria causing leaf blight diseases of certain leguminous plants. *Ann. Appl. Biol.* 47: 318–331.

Smith, E. F. 1897. Description of *Bacillus phaseoli* n. sp. *Botanical Gazette* 24: 192.

Vauterin, L., Hoste, B., Kersters, K., and Swings, J. 1995. Reclassification of *Xanthomonas*. *Int. J. Syst. Bacteriol.* 45: 472–489.

7.32 BACTERIAL LEAF SPOT AND STEM CANKER OF PIGEON PEA

Bacterial leaf spot and stem canker of pigeon pea (*Cajanus cajan*), also called *arhar* in India, was first reported from Bombay state by Kulkarni et al. (1950). Later on, it was reported from other states, namely, Tamil Nadu, Punjab, Uttar Pradesh, Madhya Pradesh, Delhi, and Haryana. The disease also occurs in China and Sudan.

7.32.1 Symptoms

The symptoms first appear as small, round, light brown, water-soaked spots on the upper surface of leaves, which later on turn dark brown, become angular in shape, and are surrounded by haloes. The upper surface of the spots becomes raised due to exudation and drying of the bacterial ooze. The spots may coalesce to form large lesions of 2 mm in diameter. Due to excessive spotting, the leaves turn yellow and defoliate prematurely. In severe infection, the spots appear all over the leaf, infection extending to main and lateral veins, and petioles; resulting in yellowing of the leaves and ultimately shedding of the leaves. On the main stem and lateral branches, dark brown cankers measuring 12.5 mm long and 2.5 mm wide appear. When these cankers are numerous and close to each other, they cause peeling of the bark. Generally, the cankers on the stem are not very deep and do not cause girdling of the stems. A gummy bacterial exudate comes out from these cankers. An early severe infection may cause death of the seedlings (Patel and Jindal 1972).

7.32.2 Causal Organism

Xanthomonas axonopodis pv. *cajani* (Kulkarni et al. 1950) Vauterin et al. 1995
Syn. *Xa. campestris* pv. *cajani*

Kulkarni et al. (1950) described the bacterium for the first time and named it as *Xanthomonas cajani* sp. nov. Dye (1978) changed it to *Xa.campestris* pv. *cajani,* and later on, Vauterin et al. (1995) reclassified it as *Xa. axonopodis* pv. *cajani.* The below given description of the bacterium is based primarily on Kulkarni et al. (1950).

Cells are short rods (0.9–1.4 × 1.3–2.2 µm). Gram-stain-negative, encapsulated, not acid-fast, non-sporing, and motile by a single polar flagellum. Colonies on potato dextrose agar are naphthaline yellow, shining, pulvinate, smooth having entire margins, and 1.5 cm in diameter after 7 days of incubation. Colonies on nutrient agar are round, slightly raised, yellow, shining, and attain a diameter of 0.7 cm after 4 days of incubation. The bacterium reduces litmus, peptonizes milk, liquefies gelatin, hydrolyses starch and casein, produces hydrogen sulphide and ammonia, but nitrate reduction, M.R. and V.P. tests are negative. The bacterium produces acid, but no gas from glucose, sucrose, dextrose, lactose, and maltose and does not utilize salicin. Citrates are utilized. Optimum temperature for growth is 26°C–32°C and thermal death point at 51°C. Growth is aerobic.

7.32.3 Disease Cycle

The bacterium is seed-borne. There is no information on its survival in the infected plant debris. The bacterium enters the host through stomata and wounds. The pathogen does not infect any other host (Kulkarni et al. 1952).

The disease is generally more severe during monsoon (July–September) when the relative humidity is 80%–90%, air temperature ranges from 24°C to 31°C, and the crop is succulent. New leaves formed after the rains are generally free from infection. When the pigeon pea is grown as a mixed crop with sorghum, the high humidity and low temperature resulting from shading by the sorghum plants also increase the disease severity.

7.32.4 Management

1. Use of resistant varieties is the most effective and economical method
2. Use of pathogen-free seed will ensure the disease-free crop because the infected seed is the only source of perpetuation of the bacterium. To eliminate the seed-borne infection, seed treatments used to manage bacterial blight of cowpea and bacterial blight of clusterbean can be tried.

References

Dye, D. W. 1978. Genus IX *Xanthomonas* Dowson 1939. In: Young, J. M., Dye, D. W., Bradbury, J. F., Panagopoulos, C. G. and Robbs, C. F. 1978. A proposed nomenclature and classification for plant pathogenic bacteria. *N. Z. J. Agric Res.* 21: 162–166.

Kulkarni, Y. S., Patel, M. K., and Abhyankar, S. G. 1950. A new bacterial leaf-spot and stem canker of pigeon pea. *Curr. Sci.* 19: 384.

Kulkarni, Y. S., Patel, M. K., and Abhyankar, S. G. 1952. A new bacterial leaf-spot and stem canker of pigeon pea. *Indian Phytopathol.* 5: 21–22.

Patel, P. N., and Jindal, J. K. 1972. Bacterial leaf spot and stem canker of pigeon pea. pp. 87–89 in: *Plant Bacteriology*. Vol. I, Bacterial Diseases of Plants in India. P. N. Patel ed. Indian Agricultural Research Institute, New Delhi, India.

Vauterin, L., Hoste, B., Kersters, K., and Swings, J. 1995. Reclassification of *Xanthomonas*. *Int. J. Syst. Bacteriol.* 45: 472–489.

7.33 Bacterial Wilt of Solanaceous Plants

Bacterial wilt caused by *Ralstonia solanacearum* is a highly destructive disease of many solanaceous and some non-solanaceous crops. It is considered the world's single most destructive bacterial plant disease. It is also considered the second most important disease of potato worldwide after late blight. It is also known by several other names like **brown rot of potato**, **southern bacterial wilt of tomato** (due to its prevalence in southern states of USA), **Moko disease of banana**, and **Granville wilt of tobacco**. The bacterium infects many solanaceous plants, namely, potato, tomato, eggplant, tobacco, and pepper, and some non-solanaceous hosts like banana, peanut, and Heliconia; geranium and other ornamental plants and weeds. More than 200 plant species belonging to 50 families are attacked by the bacterium. Tjou-Tam-Sin et al. (2017) reported for the first time bacterial wilt of cut roses (ornamental *Rosa* sp. cv. Armando) caused by *Ral. solanacearum*.

The disease was first described by E.F. Smith (1896) who also described the bacterium and named it as *Bacillus solanacearum*. The disease occurs in tropical, sub tropical, and temperate regions of the world. In the past, the disease has caused serious losses to different crops in different countries. The damage to the potato crop occurs on account of premature wilting of the standing crop and rotting of potato tubers both in field and stores. In Granville County of North Carolina, USA, tobacco crop suffered losses ranging from 25% to 100%, and in 1908 alone; the loss was above $100,000 for the entire state of North Carolina. During 1935–1945, the disease caused an annual loss of 2 million dollars to the tobacco crop (Kelman 1953). Around 1890, the disease almost eliminated the Moko plantain, a relative of banana in Trinidad; and later on, it destroyed 1 million plants on the Pacific coast of Costa Rica. Tomato and brinjal cultivation was abandoned due to the disease in some parts of the Madhya Pradesh, India. In India, losses ranging from 5% to 6% for tobacco, 50% for potato, and 62%–65% for brinjal have been reported (Rao 1972). According to Champoiseau and Allen (2009), the bacterium is responsible for an estimated loss of 1 billion US dollars worldwide every year, mostly to poor small-scale farmers in the tropical highlands.

In India, the disease was first reported by Cappel in 1892 from Pune district of the Maharashtra as bangle blight or bungdi disease of potato; although its bacterial nature was established in 1909 by Coleman. Hutchinson in 1913 reported it as Rangpur wilt of tobacco. It is the first bacterial disease recorded in India. The disease has not been noticed in the northwestern high hills (excluding Kumaon hills), where the mean monthly minimum temperatures during December–February remain below 0°C and in the northwestern and north central plains having mean monthly maximum summer temperatures above 40°C coupled with water deficit in soil continuously for a minimum period of 45 days. In India, race 1 is more predominant, and it has a wide genetic diversity. The strains of bacterium prevalent in India are also more virulent than those found in other countries particularly the US and South America.

Sakthivel et al. (2016) characterized 14 *Ral. solanacearum* strains obtained from tomato, brinjal, and chili pepper on the Andaman Islands, India, using a polyphasic approach. Strains were identified as *Ral. solanacearum* based on carbon substrate utilization profiling with Biolog similarity coefficients >0.82. The species identity was further confirmed by 16S ribosomal RNA and *recN* gene sequence analysis. Intraspecific identification of strains revealed the presence of race 1 biovar 3 and biovar 4. Both biovars caused wilting of plants with similar aggressiveness. All strains were identified as phylotype I, and multi-locus sequence typing revealed that the strains belong to a small number of clonal complexes that also comprise strains from mainland India, especially West Bengal and Kerala states.

Sagar et al. (2014) classified 75 bacterial strains, isolated from wilt infected potato plants from various potato growing regions of India, using traditional and molecular methods. The identity of all the strains was confirmed as *Ral. solanacearum* as expected single 280-bp fragment resulted in all the strains following PCR amplification using *Ral. solanacearum* specific universal primer pair 759/760. It was found that the Indian potato strains of *Ral. solanacearum* belong to three out of four phylotypes, namely: the Asian phylotype I, the American phylotype II, and the Indonesian phylotype IV. It is the first study regarding the diversity of *Ral. solanacearum* from potato in India using phylotype and sequevar scheme. It is also the first report of the occurrence of phylotype IV sequevar 8 (bv2T) strain of *Ral. solanacearum* causing potato bacterial wilt in midhills of Meghalaya in India.

7.33.1 Symptoms

As the bacterium infects many plant species, the symptoms differ on different hosts, although some of the symptoms are common on all the hosts.

Potato: Foliage symptoms include rapid wilting of leaves and stems, usually first visible at the warmest time of the day. Initially plants recover from wilt during the night time. The leaves may have bronze tint and petioles may develop epinasty. Eventually, plants fail to recover, become yellow and brown, necrotic, and die. As the disease progresses, a streaky brown discoloration appears on the stems above the soil line and also on the branches of the plants. Brown discoloration of the vascular system of the stems occurs, which can be seen by cutting the infected stems longitudinally or transversely. A white, slimy bacterial mass oozes from vascular bundles of the cut or broken stems. The bacterial ooze in the form of threads can be seen when cut potato stems are suspended in water. Under cool growing conditions, wilting and other foliage symptoms may not occur.

On tubers, external symptoms may or may not appear. However, in damp soils, the bacterial ooze often emerges from the eyes and stolen-end attachment of infected tubers. Soil may adhere to bacterial mass exuded from the tubers. Infected cut tubers reveal a browning and eventual necrosis of the vascular ring and immediately surrounding tissues. A few minutes after cutting, a creamy fluid exudate usually appears spontaneously on the vascular ring of the cut surface (Figure 7.20a). In case of ring rot of potato, tuber has to be squeezed in order to press out a mass of yellowish macerated vascular tissue and bacterial slime. Plants with foliar symptoms of wilt may produce healthy and diseased tubers, while those showing no symptoms of the disease may sometimes produce diseased tubers. Water-soaked lesions, 0.5–1.0 cm in diameter and 5 mm deep develop at the site of infected lenticels. The tuber skin remains intact, but becomes shriveled. A tuber cut through the lesion exposes a dirty white bacterial mass below the skin.

Tomato: The youngest leaves show the first sign of the disease as they become flaccid in appearance, usually at the warmest time of the day. Under less favorable conditions, the disease develops slowly. Stunting may occur (Figure 7.20b) and large number of adventitious roots develops on the stem. The vascular tissues of the stem show a brown discoloration (Figure 7.20c). Drops of white or yellowish bacterial ooze come out from the transverse sections of the stems.

Tobacco: On young seedlings, the wilt develops rapidly. The leaves of infected plants become light green and gradually turn yellow. The midrib and veins become flaccid and large leaves droop in an umbrella-like fashion. Necrotic areas may frequently occur between the veins. One of the main symptoms is the unilateral wilting. One-sided wilting of the entire plants and of half leaves is a common feature of the disease. In severe cases, the leaves wilt

(a)

(b)

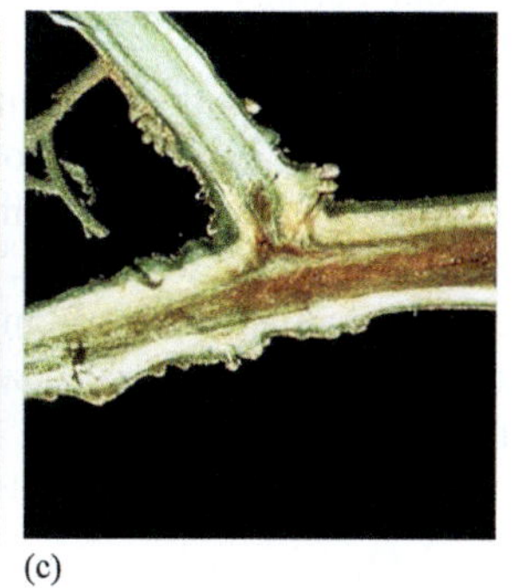
(c)

FIGURE 7.20 (a) Brown rot of potato: Cross section of an infected potato tuber shows greyish discoloration of vascular tissue and bacterial ooze. (Courtesy of Patrice Champoiseau_UF_Guatemala.) (b) Bacterial wilt of plants: Wilted tomato plants. (Courtesy of AVRDC—The World Vegetable Center.) (c) Bacterial wilt of plants: Longitudinally splitted tomato stem shows reddish brown discoloration. (Courtesy of AVRDC—The World Vegetable Center.)

without changing color and remain attached to the stem. When stems are cut open, the vascular tissue shows a brown discoloration.

Banana: Most characteristic symptom is the vascular discoloration in the inner leaves and fruit stalk (peduncle). On young and fast-growing plants, the youngest leaves turn pale-green or yellow and collapse. Lower leaves become yellow, droop, and dry out. Sometimes the first symptom is in the fruit, which appears distorted and pulp exhibits a dark brown discoloration. Premature yellowing and cracking occurs in infected banana fruits. Brown discoloration of vascular tissue is seen in cross sections of the stem. Extensive breakdown of pith occurs in the lower parts of the stem and taproot, and the stems become hollow. Young suckers may be blackened, stunted, or twisted. The pseudostems show brown vascular discoloration.

Geranium: First symptoms are wilting followed by yellowing of leaves, often sectorial. Plants develop a systemic wilt, but initially recover during cooler temperatures or at night. The lower leaves tend to wilt first and become chlorotic. Brown discoloration of the vascular tissue is often visible internally. In the later stages, leaves become brown and necrotic; plants never recover from wilt, desiccate, and die.

Eggplant: Most characteristic symptom is leaf drop, which occurs before there is much wilting.

Peanut: Foliage loses its green color, wilts suddenly, and turns dull black. Leaf yellowing is quite conspicuous.

Pepper: Plants rarely show true wilting. Leaves turn yellow and roll or curl before dropping. Vascular browning has been reported to be a characteristic symptom of the disease.

Shekhawat et al. (2000) found that vascular browning in the host depends on the ability of the infecting bacterial strains to produce brown pigment. The strains, which produce brown pigment in the cultures, produce vascular browning. The strains, which do not produce vascular browning, have been reported from Portugal, Kenya, and Australia, besides India.

Apart from the symptoms, the streaming test is a valuable diagnostic tool for quick identification of the disease in the field. It is performed by placing cut stem pieces of infected plants in clear water. After a few minutes, threads of a viscous white slime can be seen streaming from the cut end of the stem. These threads are of bacterial ooze exuding from the infected xylem vascular bundles.

7.33.2 Causal Organism

Ralstonia solanacearum (Smith 1896) Yabuuchi et al. 1996

Syn. *Burkholderia solanacearum*, *Pseudomonas solanacearum*

Safni et al. (2014) emended the description of *Ral. solanacearum* given by Yabuuchi et al. (1995) to include only strains belonging to phylotype II (Fegan and Prior 2005; Prior and Fegan 2005). The below given emended description of the bacterium is based on the data of Yabuuchi et al. (1995) and Safni et al. (2014). Cells are Gram-stain-negative and rod-shaped and may be motile or non-motile, but the type strain is motile. Catalase and oxidase tests are positive. The bacterium is able to grow at 28°C on casamino acid peptone glucose (CPG) medium. Nitrate is reduced to nitrite and some strains reduce nitrite to gas as well. Most strains can utilize citrate, hydrolyze Tween 80, and produce urease. Tests for arginine dihydrolase and lysine and ornithine decarboxylases are negative. Acid is produced oxidatively from glucose, mannose, fructose, glycerol, galactose, and sucrose. The major cellular fatty acids are summed feature 3 (C16: 1v7c/iso-C15: 0 2-OH), C16: 0, C18: 1v7c, summed feature 2 (iso-C16: 1 I/C14: 0 3-OH), C17: 0 cyclo, C18: 1 2-OH, C16: 1 2-OH, and C14: 0. The strains contain a signature nucleotide sequence, 59-AAGTTATGGACGGTGGAAGTC (Fegan and Prior 2005; Prior and Fegan 2005), in the 16S–23S rRNA ITS gene sequence. The DNA G + C content of the type strain is 66.6 mol% (HPLC method).

Using a polyphasic taxonomic approach on an extensive set of strains, Safni et al. (2014) provided evidence for a taxonomic and nomenclatural revision of members of *Ral. solanacearum* species complex. The data obtained from phylogenetic analysis of 16S–23S rRNA ITS gene sequences, 16S–23S rRNA ITS region sequences and partial endoglucanase (egl) gene sequences, and DNA-DNA hybridizations demonstrate that the *Ral. solanacearum* species complex comprises three genospecies. One of these includes the type strain of *Ral. solanacearum* and consists of strains of *Ral. solanacearum* phylotype II only. The second genospecies includes the type strain of *Ral. syzygii* and contains only phylotype IV strains. This genospecies is further divided into three distinct groups, comprising *Ral. syzygii*, the causal agent of Sumatra disease on clove trees in Indonesia, *Ral. solanacearum* phylotype IV strains isolated from different host plants mostly from Indonesia, and strains of the blood disease bacterium (BDB) that cause the banana blood disease, a bacterial wilt disease in Indonesia that affects bananas and plantains. The third genospecies is composed of *Ral. solanacearum* strains that belong to phylotypes I and III. As the differentiation of these genospecies is also supported by phenotypic data, the authors made the following taxonomic proposals: emendation of the descriptions of *Ral. solanacearum* and *Ral. syzygii* and descriptions of *Ral. syzygii* subsp. nov. for the current *Ral. syzygii* strains, *Ral. syzygii* subsp. *indonesiensis* subsp. nov. for the current *Ral. solanacearum* phylotype IV strains, *Ral. syzygii* subsp. *celebesensis* subsp. nov. for the BDB strains, and *Ral. pseudosolanacearum* sp. nov. for the strains of *Ral. solanacearum* phylotypes I and III.

The bacterium can be easily isolated from the symptomatic material on YPGA, a non-selective medium, or on Kelman's tetrazolium chloride medium. For isolating from symptomless material, SMSA medium modified by Elphinstone et al. (1996) is useful and has been used successfully in Europe. On solid agar media, individual bacterial colonies are usually visible after 36–48 h of growth at 28°C, and colonies of the normal or virulent type are white or cream-colored, irregularly shaped, highly fluidal, and opaque. The colonies of the mutant or non-virulent type, which appear occasionally, are uniformly round, smaller, butyrous, or dry. The colonies of virulent strains on tetrazolium chloride medium are irregularly shaped, fluidal, and white having pink centers, while those of non-virulent strains are round, smaller, and dark red. This medium is also used for the differentiation of virulent and weakly virulent/avirulent strains (Kelman 1954). The bacterium can be stored for many years at room temperature in sterilized tap, distilled, or deionized water. It can also survive for a long time at minus 80°C in liquid culture broth amended to 40% glycerol. Repeated transfers on agar media result in rapid loss of virulence and cultures stored at 4°C also lose viability.

The bacterium undergoes spontaneous conversion from fluidal wild-type to afluidal, avirulent, or weakly virulent state when stored in water or cultured in media. Such afluidal variants can be easily recognized by their characteristic colony morphology on tetrazolium agar medium. This conversion was considered unidirectional until the report of Shekhawat et al. (1992), which showed that the reversion back to fluidal virulent forms also occurred. The average of reversion rate was 3.45%. This reversion to virulent fluidal form may be playing a role in maintaining the virulence of the bacterium. The exposure of the bacterium to Terramycin and tetracycline increased the number of fluidal revertants by 5- to 16-fold. The incubation at higher temperatures of 35°C and 37°C, in comparison to optimum temperature of 30°C, also increased the reversion rate by 4.0- to 6.4-fold (Gadewar et al. 1999).

Genomes of three strains of *Ral. solanacearum* has been sequenced. The first genome sequenced was that of strain GMI 1000, which has a wide host range (Salanoubat et al. 2002). It is organized in two replicons, a chromosome of 3.7 Mb and a mega plasmid of 2.1 Mb, with a very similar G + C % (about 67%). The strains Mo IK2 (a strain of race 2, isolated from a banana tree) and 1609 (a race 3 strain, isolated from potato) have genomes, which are comparable to that of GMI 1000 (about 6 Mb, G + C % estimated at 67%).

The bacterium belongs to RNA homology group II (non-fluorescent) within the pseudomonads and within this group it is readily distinguished from other members by its failure to grow at 40°C. On the basis of RFLP and other genetic fingerprinting studies (Hayward 2000), the species is divided into Division I called the Asiaticum division (biovars 3, 4, and 5 originating in Asia) and Division II called the Americanum division (biovars 1, 2A, and 2T, originating in South America).

The species *Ral. solanacearum* is a complex taxonomic unit with a great degree of both phenotypic and genetic diversity. Thousands of genetically distinct strains of *Ral. solanacearum* have been characterized worldwide, revealing significant variation within the group. Therefore, *Ral. solanacearum* is considered a **species complex**. Unlike other phytopathogenic bacteria, race system of *Ral. solanacearum* is not based on gene-for-gene interactions, i.e., different cultivars carrying different *R* gene(s). Instead, the races are based on their pathogenicity on different host species. Although the biovar and race systems are widely accepted for the classification of *Ral. solanacearum*, there is no correlation between biovar and race; each race transects the biovars and each biovar contains various races. The only positive correlation between the biovar and race systems exists for biovar 2 and race 3. The differentiation of biovars is given in Tables 7.3 and 7.4.

Until today, five races of the bacterium have been identified on the basis of their ability to attack different host species. Race 1 occurs in tropical areas all over the world and is endemic in southern US. It has a wide host range and attacks tobacco, many other solanaceous crops, and many other hosts and floriculture crops. It infects hundreds of different plant species in 50 families.

Race 2 occurs mainly in tropical areas of South America, in the Caribbean, and Brazil, and infects bananas and *Heliconia* species, causing so-called Moko disease. In the Philippines, it causes bugtok disease on plantains.

Race 3 (biovar 2) is the most important race, occurs in sub tropical and temperate regions and at higher altitudes in the tropics. It is of special concern to potato industry in the US and Canada, because being a cold-temperature tolerant race it has caused serious disease problems in potato crop in other temperate countries of the world. It is also an extremely destructive pathogen of potato in the highland tropics of Africa,

TABLE 7.3
Tests for Differentiation of *Ral. solanacearum* Biovars

	Biovar				
Utilization of	**1**	**2**	**3**	**4**	**5**
Maltose	–	+	+	–	+
Lactose	–	+	+	–	+
Cellobiose	–	+	+	–	+
Mannitol	–	–	+	+	+
Sorbitol	–	–	+	+	–
Dulcitol	–	–	+	+	–

Source: OEPP/EPPO, *Bull. OEPP/EPPO Bull.*, 34, 173–178, 2004.
+ = Utilized, – = Not utilized.

TABLE 7.4
Tests for Differentiation of Sub phenotypes of Biovar 2 of *Ral. solanacearum*

Test	Biovar 2A RFLP group26	Biovar 2A RFLP group 27	Biovar 2T or 2N
Utilization of trehalose	–	+	+
Utilization of inositol	+	–	+
Utilization of D-ribose	–	–	+
Pectolytic activity	Low	Low	High

+ = Utilized, – = Not utilized.

Asia, and Latin America. This race was imported into the US from Guatemala in geranium cuttings on several occasions in 1999. In the United States, it was listed in 2002 as a Select Agent plant pathogen, and is now subjected to the strictest biosecurity regulations. Apart from its status as Select Agent, race 3 is considered a serious quarantine pest by the Animal and Plant Health Inspection Service of the US Department of Agriculture and by the European Plant Protection Organization (Champoiseau and Allen 2009). Government regulations in the US include a zero tolerance for this race. It attacks potato, tomato, occasionally *Pelargonium zonale*, aubergine, eggplant, and capsicum, some solanaceous weeds like black nightshade (*Solanum nigrum*), and bittersweet or climbing nightshade (*S. dulcamara*). Race 3 also infects a number of non-solanaceous weeds, which often do not show symptoms. This is the most predominant race and is mostly comprised of biovar 2A or RFLP group 26 with a worldwide distribution, RFLP group 27 found in Chile and Colombia, and biovar 2T, sometimes also called 2N, found in tropical areas in South America. Race 3 is not found in Canada. It is on the North American Plant Protection Organisation (NAPPO) list of A1 quarantine pests as well as Canada's list of quarantine pests. Milling et al. (2009) compared different traits of tropical, temperate (race 3 biovar 2), and warm-temperate strains of the bacterium in *in vitro* and *in vivo* and found that interaction with plants is required for expression of the temperate epidemiological trait of race 3 biovar 2. Race 4 affects ginger in most parts of Asia. Race 5 (biovar 5) infects mulberries in China.

Nowadays, the identification of *Ral. solanacearum* into races and biovars is considered unsatisfactory because it is not predictive and some groups (e.g., race 1) contain very large variation (Champoiseau and Allen 2009). A phylogenetically meaningful classification scheme developed, based on DNA sequence analysis (Fegan and Prior 2005; Prior and Fegan 2005), is considered better. In this classification, 23 sequevars and four phylotypes have been recognized within the species complex and these broadly reflect the ancestral relationships and geographical origins of the strains.

Molecular studies have shown *Ral. solanacearum* as a species complex consisting of highly diverse strains (Poussier et al. 2000; Fegan and Prior 2005). Molecular analyses using the nucleotide sequences of genes *egl* (encoding endoglucanase, a conserved virulence factor), *mut S* (a DNA mismatch repair enzyme), *hrp B* (a regulator of type III secretion), and the ITS region between the 16S and 23S ribosomal RNA genes have generated a phylogenetically meaningful sub division of the species complex into four phylotypes, which group strains by ancestral geographic origin. Phylotype I strains come from Asia, phylotype II strains from the Americas, phylotype III strains from Africa, and phylotype IV strains from Indonesia, which is the probable origin of the group (Poussier et al. 2000; Fegan and Prior 2005). These four groups are distinguished by a multiplex PCR reaction. Within phylotypes, strains are further clustered into sequevars based on nucleotide sequences of the *egl* and *mut S* genes (Poussier et al. 2000; Fegan and Prior 2005).

Schaad et al. (2001) developed real-time PCR for the detection and identification of *Ral. solanacearum* race 3 biovar 2. This technique was further refined by developing a real-time Bio-PCR by Ozakman and Schaad (2003). Real-time Bio-PCR was more sensitive (detected 20 cells/mL) than real-time PCR, and it detected *Ral. solanacearum* race 3 biovar 2 from latently infected potato tubers. Serological field kits, which incorporate highly specific monoclonal antibodies, are also commercially available (immunostrips-AgDia, CSL Pocket Diagnostics) for rapid on-site diagnosis. However, a confirmed diagnosis should always be obtained through laboratory tests on a pure culture. Priou et al. (2010) developed two sensitive, specific, and user-friendly serological methods, namely, double-antibody sandwich (DAS)-ELISA and indirect ELISA on nitrocellulose membrane after enrichment of the plant extract in a semi-specific broth, for detection of *Ral. solanacearum* (biovar 2A) in stems of symptomless potato plants 20 days before crop harvest. Detection probabilities were higher with double antibody sandwich enzyme-linked immunosorbent assay (DAS-ELISA). Being a reliable and user-friendly technique, DAS-ELISA can easily be used by national plant protection services and seed certification programs in developing countries.

Chandrasekhar and Umesha (2015) developed a species-specific PCR (sPCR) method to identify and detect isolates of *Ral. solanacearum*, causing bacterial wilt in chilli. PCR primers for *Ral. solanacearum* were identified by alignment of *hrpB* gene sequences and selection of sequences specific for *Ral. solanacearum* at their 3′ ends. The primers were found specific for *Ral. solanacearum*, as no PCR product was obtained when genomic DNA from other bacterial species including closely related *Ralstonia* species, were used as test species. Lone pair of primers (RshrpBF and RshrpBR) was designed using *hrpB* gene sequence, unique to *Ral. solanacearum* which amplified a predicted PCR product of 810 bp from 20 different isolates. To understand the evolutionary divergence of Indian *Ral. solanacearum* isolates, phylogenetic analysis was also attempted. Based on phylogenetic analysis, Indian isolates showed homology with the standard reference isolates from other countries, however, one new isolate showed complete evolutionary divergence by forming an out-group.

Stulberg et al. (2015) developed a multiplex PCR assay that identified *Ral. solanacearum* species complex strains, signaled whether the strain detected was a select agent (race 3 biovar 2 strains are considered select agents by the US government), and controlled false negatives associated with PCR inhibition or unsuccessful DNA extractions in one reaction. The authors identified unique sequences of non-phage-related DNA for the *Ral. solanacearum* species complex strains, and for Select Agent strains, using *in silico* genome subtraction. An internal plant DNA control assay was also designed and included. The multiplex PCR assay correctly identified 90 *Ral. solanacearum* species complex strains and 34 Select Agent strains, but did not recognize five out-group bacterial species. Moreover, the multiplex PCR assay facilitated the detection of plant DNA and *Ral. solanacearum* from infected tomato, potato, geranium, and tobacco plants. This rapid, accurate, and reliable detection assay can help US government officials to make timely and appropriate recommendations to exclude this bacterium from the United States.

Ral. solanacearum race 3 biovar 2 (R3bv2) sub group is a high-concern quarantine pathogen, while the related sequevar 7 group is endemic to the southeastern United States. Tran et al. (2016) compared the sensitivity, cost, and technical complexity of several assays to detect and distinguish R3bv2 and sequevar 7 strains of *Ral. solanacearum* in geranium, tomato, and potato tissue in the laboratory and in naturally infected tomato plants from the field. The sensitivity of PCR-based methods in infected geranium tissues was significantly improved by the use of Kapa3G Plant, a polymerase with enhanced performance in the presence of plant inhibitors. The cells of R3bv2 were killed within 60 min of application to Whatman FTA(R) nucleic acid-binding cards, suggesting that samples on FTA cards can be safely transported for diagnosis. On the whole, culture enrichment followed by dilution plating was the most sensitive detection method (10^1 cfu mL^{-1}), but it was also most laborious. Conducting PCR from FTA cards was faster, easier, and sensitive enough to detect approximately 10^4 cfu mL^{-1}, levels similar to those found in latently infected geranium plants.

Massart et al. (2014) developed a new multiplex real-time PCR assay to simultaneously detect *Clavibacter michiganensis* subsp. *sepedonicus* and *Ral. solanacearum*, listed as A2 pests in the EPPO region and as zero-tolerance quarantine organisms in the European Union, in a single assay. Additionally, the reliability of this molecular diagnostic test has been improved by the simultaneous amplification of an internal control, corresponding to a potato gene co-extracted from the sample. The polyvalence and the specificity of each set of bacterial primers and probes were evaluated on more than 90 bacterial strains. The detection limit of this triplex real-time protocol was similar to those observed with other molecular protocols previously developed for the individual detection of these bacteria. A concordance of 100% was obtained in a blind test mimicking the routine application of the technology. The new protocol represents a straight forward and convenient method potentially adapted to primary screening of potato tubers.

Albuquerque et al. (2015) selected 17 novel markers based on *Ral. solanacearum* specific protein domains and thoroughly validated for specificity and stability, both *in silico* and using "wet lab" assays. The validation assays based on PCR and hybridization revealed that the DNA regions selected as markers were unevenly distributed amongst the tested strains, with nine markers present throughout the species complex. The distribution of the remaining eight markers was highly variable between the different analyzed strains and allowed the attainment of strain-specific dot blot hybridization patterns, particularly informative for typing. The average probability value of each strain being positive for each of the 17 markers was calculated by an algorithm and used to obtain a dendrogram representing hierarchical clustering analysis of *Ral. solanacearum*, according to the similarity of their hybridization patterns. This method should prove to be a robust and straightforward procedure for genotyping members of the *Ral. solanacearum* species complex. Moreover, this quantitative hybridization approach will allow the construction of informative databases to determine new *Ral. solanacearum* genotypes and infer epidemiological patterns.

7.33.3 Disease Cycle

Ral. solanacearum is primarily soil-borne and also water-borne pathogen. Primary inoculum comes from soil, but there is no conclusive evidence that it is a persistent soil inhabitant. Top 30 cm soil contains the high population of the bacterium. The bacterium also survives in infected tubers including symptomless tubers, infected refuse, reservoir plants, and weeds. Although air-borne spread has never been reported, there is some evidence of epiphytic survival of the bacteria on the plant surfaces under conditions of high relative humidity. Depending upon biological and temperature conditions, the bacterium can survive from days to years in contaminated water, wet soils, or deep soil layers forming reservoir of inoculum from which it can disperse. In European countries, bittersweet or climbing nightshade (*S. dulcamara*), a solanaceous weed growing along waterways, is considered to be a source of primary inoculum. *Ral. solanacearum* in contaminated water in waterways infects the roots of this weed. The source of bacteria in waterways may come from waste generated by the processing of infected potatoes. Growers may pull *Ralstonia*-contaminated water from the waterways for irrigation, thereby inoculating their potato and tomato crops. Recent outbreaks of brown rot of potato in the United Kingdom were traced to *Ral. solanacearum*-infected *S. dulcamara* plants, which release the pathogen from roots into river waters that were subsequently used for potato irrigation. Hong et al. (2008) have also shown the survival of the bacterium in irrigation ponds in Gadsden County of North Florida (USA) and in common aquatic weeds, including *Hydrocotyle ranunculoides* (dollar weed) and *Polygonum pennsylvanicum* (Pennsylvania smart weed), growing in these ponds. Both weeds were latently infected and showed no symptoms of wilt. There was a positive correlation between ambient temperature and *Ral. solanacearum* population in irrigation water, as shown by previous research workers. In Australia, England, Kenya, and Sweden, race 3 biovar 2 was not detected in previously diseased fields after 2 years, suggesting that long-term survival in temperate regions is reduced.

The bacterium spreads through infected tubers, banana rhizomes and transplants, contaminated equipment, irrigation water, insects, and in banana through pruning knives. Its movement from plant to plant occurs through soil. Race 2 is transmitted by insects and has a high potential for natural spread. Race 3 spreads more easily with surface water when infected *S. dulcamara* grows with its roots floating in water. The long distance spread of the bacterium occurs through infected (latently) seed potatoes and other vegetative propagating plant materials. Among the hosts of the bacterium producing true seed, only groundnut seed has been shown to be naturally infected.

The bacterium enters the host through wounds caused by agricultural operations and pruning knives, and through the rifts caused by emergence of lateral roots. The feeding injuries caused by root and spiral nematodes also aid in the penetration of the pathogen. Besides, certain gall forming insects, cut worms, white grubs, and root invading parasitic fungi also provide avenues for the entry of bacterium. The secondary spread of the bacterium is not of much significance, as the bacterium does not become airborne. It occurs to a limited extent through irrigation water, insects, and contaminated equipment. The bacterial cells released in the soil after the death of diseased roots, invade the nearby roots.

Soil and air temperature, and soil moisture play an important role in disease development. Temperature and soil moisture have a synergistic effect on disease development. High temperature and high soil moisture alone cannot induce the disease. Light intensity, day length, and mineral nutrition also affect disease development. Race 3 has comparatively low optimum temperature of 27°C, while the range for maximum disease development is between 24°C and 35°C. The virulence of the bacterium decreases when temperature exceeds 35°C or falls below 12°C. The optimum temperature for disease development for races 1, 2, 4, and 5 is 35°C. Favorable temperature favors rapid multiplication of the bacterium in the host, its release, and infection of neighboring plants. High soil moisture (50%–100% water holding capacity) favors rapid disease development. It increases survival of the bacterium in soil, enhances multiplication and disease development and spread through soil. The disease is found in diverse soil types and in both acidic (pH 5.0–5.5) and alkaline (pH 7.0–8.5) soils.

The bacterium migrates first in the large xylem vessels, and in succulent stems, it multiplies in the intercellular spaces of the cortex and the pith causing lysigenous cavities. In potato tubers, the pathogen may produce an internal decay emanating from the vascular bundles and causing a breakdown in the bud tissue of the eyes.

The physiology of the host-parasite interaction has been excellently reviewed by Buddenhagen and Kelman (1964). Various biochemical and physiological changes occur in the host during pathogenesis, which are manifested in the form of internal and external symptoms on the host.

After entering the xylem vessels, the bacteria multiply and travel rapidly to the upper parts of the plants. This results in the production of large amounts of extracellular polysaccharide (EPS) slime. The presence of this slime along with large number of bacterial cells in the xylem vessels lead to reduction in sap flow resulting in wilting of the plants. The wilting can also occur due to rupturing of vessel elements on account of excessive hydrostatic pressure. The main virulence factor of *Ral. solanacearum* is EPS I, a long (>10^6 Da) polymer. Many diverse isolates of the bacterium produce EPS I (or a very similar polymer), and 85% of the EPS I exists as released, cell-free slime, while 15% remains in a cell surface-bound capsular form. EPS I accounts for >90% of its total exopolysaccharides produced in the plants. Mutants of *Ral. solanacearum* specifically blocked in synthesis of only EPS I rarely cause wilting or killing of plants, even if a large number of bacterial cells are directly injected into the stem. EPS I also promotes stem colonization, because EPS I-deficient mutants colonize stems much more slowly and to a much lesser extent in comparison to wild-types.

The bacterium produces ~10 major exoproteins that are readily detectable in addition to many others that are less abundant and/or conditionally produced. The bacterium uses two different systems to export polypeptides from its cytoplasm to the outside environment. These systems consist of a type II secretion system encoded by *eep* (enhanced expression of pheromone) genes, which exports most major exoproteins, including all the cell wall degrading exoenzymes, and a type III secretion system encoded by *hrp* genes, which appear to deliver toxic proteins directly into plant cells. Mutants lacking either system are severely reduced in colonization ability and/or multiplication in plants. These exoproteins are absolutely essential for the disease development.

Ral. solanacearum produces pectin methylesterase (PME) that removes methoxy groups from pectin, converting it into methanol and polygalacturonic acid units. An endopolygalacturonase, PehA (also called PglA), and two exopolygalacturonases, PehB (PglB) and PehC (PglC), attack the polygalacturonic acid units. In pectin-rich regions of plants and also in soil, these four enzymes breakdown pectin into oligomers and also provide monogalacturonic acid for the growth of the bacterium. PehB-deficient mutants show only slight reduction in virulence, while PehA-deficient mutants take at least two times more time than the wild-type strains to wilt and kill the plants. This reduction in virulence is even more pronounced when soil infestation is used for inoculation in comparison to stem or petiole inoculation. The role of PME in wilt development is not clear.

Ral. solanacearum produces two extracellular glucanases, Egl and CbhA *in vitro* and these enzymes attack β-1-4 glycosidic linkages of cellulose. Egl releases cellobiose, cellotriose, and cellotetraose from soluble cellulose, and CbhA releases cellobiose from the reducing ends of crystalline cellulose. Mutants deficient in Egl are slower to cause the disease than wild-type strains. It seems that Egl facilitates invasion of roots and/or penetration of xylem vessels by degrading cellulosic glucans in the cell walls.

Schell (2000) has reviewed the information on control of virulence and pathogenicity genes of the bacterium by an elaborate sensory network. Many of key extracytoplasmic virulence and pathogenicity factors are transcriptionally controlled by an extensive network of distinct, interacting signal transduction pathways. A five-gene *Phc* (phenotype conversion) system, which is the core of this sensory network, regulates exopolysaccharide, cell wall-degrading exoenzymes, and other factors in response to self-produced signal molecule that monitors the pathogen's growth status and environment. Four additional environmentally responsive two-component systems work independently, and with the *Phc* system to fine-tune virulence gene expression. Another critical system is *Prh* (proline rich homoeodomain), which transduces plant cell-derived signals through a six-gene cascade to activate deployment of the type III secretion pathway encoded by the *hrp* genes.

Aliye et al. (2015) evaluated 28 cultivars of potato under greenhouse and field conditions and found that under greenhouse conditions, all the cultivars were resistant to *Ral. solanacearum* with scores ranging from 0.77 to 1.17 of 5. However, under field conditions, only 2 of 28 cultivars found resistant under greenhouse conditions, showed adequate resistance to the pathogen, indicating the significant impact of environment on the activity of the pathogen and reaction of the crop. Percentage wilt incidence and latent infection showed significant ($P < 0.05$) positive correlation, with $r = 0.9438$. Therefore, evaluation of crop's performance based on the combination of the parameters like field wilt incidence and proportion of latent infection give us a better picture of the overall crop feat, than using wilt incidence as a sole parameter of evaluation as has been done in most studies. Moreover, the established correlation of latent infection with field wilt incidence will also be helpful in understanding the disease epidemiology and design effective management measures, accordingly.

Kumar et al. (2017) and Singh et al. (2018) performed virulence and pathogenicity assays on 6- to 7-day-old tomato seedlings using wild and mutant strains of *Ral. solanacearum* and found the technique as an easy and useful approach for investigating virulence functions of the pathogen at the seedling stage of host. The technique used was economical in terms of space, labor, and cost as well as the amount of bacterial inoculum required.

Du et al. (2017) evaluated *Ral. solanacearum* infection dynamics in pepper lines using bioluminescence imaging. The utility of the bioluminescence assay was validated by comparing *Ral. solanacearum* infection dynamics in real-time *in vivo* between resistant line BVRC 1 and susceptible line BVRC 25. Luminescence intensity was strongly correlated to bacterial population in planta ($R^2 = 0.88$). The distribution and multiplication of BL-Rs7 strain in resistant line BVRC 1 was conspicuously limited in plants inoculated in either roots or stem in comparison to susceptible line BVRC 25. These results suggest that pepper line BVRC 1 may resist colonization by interfering with *Ral. solanacearum* multiplication in the roots and stem.

Baichoo and Jaufeerally-Fakim et al. (2017) reported that salicylic acid and ethylene pathway marker genes were expressed at higher levels with a statistically significant difference ($P < 0.05$) in fold change expression compared to very little activation of the jasmonic acid pathway marker genes among the *Solanum* lines. The wild *Lycopersicon cerasiforme*, resistant lines CRA 66 and Hawaii 7996, and *S. commersonii* showed stronger activation of the salicylic acid and ethylene marker genes than the moderately resistant cultivar (MST 32/1) and the susceptible lines (Quatre carrées and Spunta). The marker genes reached their highest expression levels 4 h post-inoculation in the resistant and moderately resistant lines, while in the susceptible lines, it occurred 48 h after inoculation. The results indicate that salicylic acid and ethylene signaling pathways have a significant role in defense against *Ral. solanacearum*. The timing and magnitude of the upregulation of gene expression may determine the plant ability to put up a defense response against the pathogen.

Nakaho et al. (2017) found that on infiltration of leaves of resistant tomato cultivar LS-89 or susceptible cultivar Ponderosa with 10^6 cells mL^{-1} of *Ral. solanacearum*, cell death appeared in LS-89 at 18 and 24 h after infiltration. The proliferation of bacteria in the infiltrated leaf tissues of LS-89 was suppressed approximately 10%–30% of that of Ponderosa leaf tissues, and expression of the defense-related gene *PR-2* and HR marker gene *hsr203J* was induced in the infiltrated tissues. These results indicate that the response of LS-89 is a true HR, and induction of vascular HR in xylem parenchyma and pith cells surrounding xylem vessels seems to be associated with quantitative resistance of LS-89 to *Ral. solanacearum*.

Caldwell et al. (2017) used a combination of scanning electron microscopy and light microscopy to investigate *Ral. solanacearum* colonization in roots of soil-grown resistant and susceptible tomato cultivars at multiple time points after inoculation. It was found that colonization of the root vascular cylinder was delayed in resistant cultivar "Hawaii 7996" and that, once the bacteria entered the root vascular tissues, colonization in the vasculature was spatially restricted. The data suggest that resistance is due, in part, to the ability of the resistant cultivar to restrict bacterial root colonization in space and time.

Jiang et al. (2013) investigated the effect of calcium nutrition on tomato bacterial wilt and the regulation of resistance mechanisms. Plants grown in nutrient solution with calcium concentrations of 0.5, 5.0, and 25.0 mM were inoculated with *Ral. solanacearum* by the root dip method. The disease severities of 100%, 77.1%, and 56.8% were found in Ca concentrations of 0.5, 5.0, and 25.0 mM, respectively. Plant growth, i.e., height, stem diameter, and biomass was significantly better in high Ca concentration than in low Ca concentration. Tomato plants absorbed significantly more Ca in roots and shoots as the level of Ca in the nutrient solution increased. A negative correlation was found between Ca concentration, levels of hydrogen peroxide (H_2O_2), PO, and PPO in tomato plants, and disease severity, indicating that these chemicals played an important role in tomato resistance to bacterial wilt. The results suggested that Ca was involved in the regulation of H_2O_2 concentration, and activity of PO and PPO in tomato.

Youssef and Tartoura (2013) reported that compost provides effective protection against *Ral. solanacearum* via up-regulation of the capacity of the ascorbate-glutathione redox cycle and modulation of the cellular redox status, thereby eliminating reactive oxygen species damage and sustaining membrane stability.

7.33.4 Management

The management of the disease is very difficult due to the soil-borne nature and wide host range of the pathogen, high pathogenic variability and limited host resistance.

However, the following measures are helpful in reducing the disease incidence and severity.

1. Crop rotation of 5–7 years with non-host crops is recommended. Elhalag et al. (2015) reported that cowpea and maize are suitable for crop rotation with potato as the root exudates, containing asparagines, glutamine, and sulphur-containing methionine, from these crops enhance the sustainability of *Stenotrophomonas maltophilia*, the biocontrol agent of *Ral. solanacearum*. Deberdt et al. (2015) reported that the incidence of tomato wilt decreased by 86% and 60%, when tomato crop was sown after *Raphanus sativus* cv. Melody and *Crotalaria spectabilis*, respectively. Therefore, *C. spectabilis* and *R. sativus* cv. Melody can be used as previous crops to control bacterial wilt in ecological management strategies without drastic suppression of *Ral. solanacearum* population in stem tissues and in the rhizosphere
2. Altering soil pH based on the cropping period can be effective. For potatoes, lowering the pH to 4–5 in the summer and raising it to 6 in autumn helped to eradicate the pathogen (Smith et al. 1997). He et al. (2014) reported that addition of organic fertilizer and $CaCO_3$ increased soil pH and lowered the *Ral. solanacearum* population nearly 100 times and increased the Ca^{2+} content in tobacco plants significantly compared to treatments without $CaCO_3$. It suggests that $CaCO_3$ can serve as a potential soil amendment for the control of bacterial wilt caused by *Ral. solanacearum*
3. Use of resistant varieties: Development of resistant varieties saved the tobacco industry in the United States from near extinction around the middle of twentieth century. However, host plant resistance has not been very effective for the control of bacterial wilt on potato and tomato, since host resistance often varies among locations due to strain variation. In potato, resistance often fails at warmer temperatures. Grafting susceptible tomato cultivars onto resistant tomato or other solanaceous rootstock is effective on the commercial scale in Japan, Bangladesh, and the Philippines, but grafting has not been tested for use against race 3 biovar 2. Lin et al. (2008) obtained a significant control of tomato wilt by integrating the use of a resistant tomato rootstock (Hawaii 1996) with soil amendment consisting of urea and slaked lime
4. Destruction and removal of weed hosts like *S. dulcamara* with herbicides
5. Many attempts to manage the disease through soil fumigation and antibiotics have not given encouraging results. However, dipping of tomato seedlings in Streptocycline solution (1 g in 30 L of water) proved effective, and it gave better results than soil drench and soil drench + foliar sprays. Ji et al. (2007) obtained very effective control of tomato wilt and a significant increase in yield with the combined application of thymol (applied as preplant fumigation through drip irrigation at 73 kg/ha) and acibenzolar-*S*-methyl (foliar spray at 25 mg L^{-1}). Deberdt et al. (2018) observed no incidence of bacterial wilt in greenhouse tomato plants grown in soil treated with chemotype 3 (clove-scented) of *Pimenta racemosa* var. *racemosa* essential oil at a concentration of 0.14%. In the untreated control soil, 62% of plants showed symptoms of bacterial wilt. Treatment with chemotype 3 significantly increased the growth of tomato plants in comparison to untreated controls. These results suggest that chemotype 3 of *P. racemosa* var. *racemosa* essential oil is a good candidate for further development as a soil biofumigant for the control of tomato bacterial wilt
6. Imada et al. (2016) reported that magnesium oxide nanoparticles (MgO NP) induced systemic resistance in tomato plants against *Ral. solanacearum*. Salicylic acid-inducible *PR1*, jasmonic acid-inducible lipoxygenase (*LoxA*), ethylene-inducible *Osm*, and systemic resistance-related *GluA* were up-regulated in both the roots and hypocotyls of tomato plants after treatment of the plant roots with MgO NP. Histochemical analyses showed that β-1, 3-glucanase and tyloses accumulated in the xylem and apoplast of pith tissues of the hypocotyls after MgO NP treatment
7. *Psm. fluorescens* isolates and vascular arbuscular mycorrhizae, *Glomus mosseae* have given encouraging results for the control of the disease in the preliminary trials. Messiha et al. (2007a) obtained a significant control of brown rot of potato with *Stenotrophomonas maltophilia* in the Nile Delta area of Egypt. Sood and associates have suggested the integrated management of tomato wilt utilizing various cultural (soil amendments, soil solarization, adjusting time of transplanting, and biofumigation) and biological methods (host resistance, antagonistic rhizobacteria, and use of VAM) in Himachal Pardesh, India where race 1 biovar 3 of the bacterium is prevalent. Ambadar and Sood (2010) obtained 43% and 63% reduction in terminal wilt incidence of tomato through soil solarization for 10 and 8 weeks, respectively under Palampur (H.P., India) conditions.

Abo-Elyousr et al. (2012) investigated the induced defense responses caused by acibenzolar-*S*-methyl (ASM) and *Psm. fluorescens* alone or in combination against tomato bacterial wilt. Treatments of tomato seedlings with *Psm. fluorescens* or ASM caused 58% and 56% disease reduction, respectively. The highest disease reduction of 72% resulted from a combined application of both *Psm. fluorescens* and ASM. The application of ASM alone increased 64.3% seedlings biomass compared to infected control. Significant changes ($P \leq 0.05$) in the activities of polyphenol oxidase, β-glucosidase, and peroxidase were found. These results indicate that the management of tomato bacterial wilt can be enhanced by integrating foliar sprays and soil drench of ASM and *Ps. fluorescens* into the existing management schedule. This is the first report of the use of both ASM and *Psm. fluorescens* to control the tomato bacterial wilt disease under field conditions.

Kunwar et al. (2017) reported that grafting (susceptible tomato variety BHN 602 grafted onto a resistant rootstock BHN 998) alone or in combination with drip applications of ASM (178.6 μM) significantly reduced disease incidence and increased total marketable yield compared to non-grafted treatments. There were no significant differences between grafted plants with or without drip ASM applications in terms of bacterial wilt incidence or total marketable yield. However, it was demonstrated for the first time that foliar applications of ASM on grafted plants negatively affected the total marketable yield compared with drip applications of ASM on grafted plants or non-treated grafted control.

Erjavec et al. (2016) tested 150 protein extracts from 94 different basidiomycete and ascomycete wild mushroom species for antibacterial activity against *Ral. solanacearum*. In *in vitro* microtiter plate assays, 11 extracts completely inhibited bacterial growth and 4 extracts showed partial inhibition. On further testing of 5 of these 15 extracts, 3 extracts slowed disease progression and reduced disease severity in artificially inoculated tomato and potato plants. This is the first *in vitro* and *in vivo* study that demonstrates that mushroom protein extracts can be promising for treatment of bacterial wilt caused by *Ral. solanacearum*

8. Biological soil disinfestation (BSD-induction of anaerobic soil conditions by increasing microbial respiration through incorporation of fresh organic amendments and by reducing resupply of oxygen by covering with airtight plastic sheets) has also been tried to manage the disease. Messiha et al. (2007b) reported that BSD reduced the soil population of *Ral. solanacearum* race 3 biovar 2 by 92.5% to 99.9% in the field, and therefore it has the potential to become an important element in a sustainable and effective management strategy to control brown rot of potato especially in areas where the disease is endemic
9. Hot air treatment of ginger rhizomes for 30 min at 50°C is also effective.

Shekhawat et al. (2000), after reviewing the exhaustive work done on bacterial wilt of potato in India, have suggested the following measures for its management

1. Use of pathogen-free whole tubers for sowing
2. Application of stable bleaching powder at 12 kg/ha mixed with fertilizers in furrows at the time of sowing
3. Application of urea at 1000 kg/ha
4. Bacterization of seed tubers with *Bacillus subtilis*
5. Use of resistant varieties
6. Two to three year crop rotation with wheat, lupin, maize, paddy, and bean
7. Pre-treatment of seed tubers with 250 μg mL^{-1} of Streptocycline solution followed by two foliar sprays of 100 μg mL^{-1} concentration
8. Adjustment of date of planting and minimum tillage in the standing crop. Post-emergence earthing up in potatoes causes the maximum injury ranging from 10% to 50%
9. One or two deep ploughings of infested soil during May–June in plains or plateau and in October–November in hills

REFERENCES

Abo-Elyousr, K. A. M., Ibrahim, Y. E., and Balabel, N. M. 2012. Induction of disease defensive enzymes in response to treatment with acibenzolar-*S*-methyl (ASM) and *Pseudomonas fluorescens* Pf2 and inoculation with *Ralstonia solanacearum* race 3, biovar 2 (phylotype II). *J. Phytopathol.* 160: 382–389.

Albuquerque, P., Marcal, A. R. S., Caridade, C., Costa, R., Mendes, M. V., et al. 2015. A quantitative hybridization approach using 17 DNA markers for identification and clustering analysis of *Ralstonia solanacearum*. *Plant Pathol.* 64: 1270–1283.

Aliye, N., Dilbo, C., and Pillay, M. 2015. Understanding reaction of potato (*Solanum tuberosum*) to *Ralstonia solanacearum* and relationship of wilt incidence to latent infection. *J. Phytopathol.* 163: 444–455.

Ambadar, V. K., and Sood, A. K. 2010. Effect of solarization on tomato wilt incidence and population dynamics of *Pseudomonas fluorescens* and *Ralstonia solanacearum*. *J. Mycol. Plant Pathol.* 40: 120–123.

Baichoo, Z., and Jaufeerally-Fakim, Y. 2017. *Ralstonia solanacearum* upregulates marker genes of the salicylic acid and ethylene signaling pathways but not those of the jasmonic acid pathway in leaflets of *Solanum* lines during early stage of infection. *Eur. J. Plant Pathol.* 147: 615–625.

Buddenhagen, I. W., and Kelman, A. 1964. Biological and physiological aspects of bacterial wilt caused by *Pseudomonas solanacearum*. *Annu. Rev. Phytopathol.* 2: 203–230.

Caldwell, D., Kim, B.-S., and Iyer-Pascuzzi, A. S. 2017. *Ralstonia solanacearum* differentially colonizes roots of resistant and susceptible tomato plants. *Phytopathology* 107: 528–536.

Champoiseau, P. G., and Allen, C. 2009. *Ralstonia solanacearum* race 3 biovar 2: Tropical losses, temperate anxieties. APSnet Feature Story [http://www.apsnet.org/online/feature/ralstonia/]

Chandrasekhar, B., and Umesha, S. 2015. Detection, identification and phylogenetic analysis of Indian *Ralstonia solanacearum* isolates by comparison of partial *hrpb* gene sequences. *J. Phytopathol.* 163: 105–115.

Deberdt, P., Davezies, I., Coranson-Beaudu, R., and Jestin, A. 2018. Efficacy of leaf oil from *Pimenta racemosa* var. *racemosa* in controlling bacterial wilt of tomato. *Plant Dis.* 102: 124–131.

Deberdt, P., Gozé, E., Coranson-Beaudu, R., Perrin, B., Fernandes, P., et al. 2015. *Crotalaria spectabilis* and *Raphanus sativus* as previous crops show promise for the control of bacterial wilt of tomato without reducing bacterial populations. *J. Phytopathol.* 163: 377–385.

Du, H., Chen, B., Zhang, X., Zhang, F., Miller, S. A., et al. 2017. Evaluation of *Ralstonia solanacearum* infection dynamics in resistant and susceptible pepper lines using bioluminescence imaging. *Plant Dis.* 101: 272–278.

Elhalag, K. M., Emara, H. M., Messiha, N. A. S., Elhadad, S. A., and Abdallah, S. A. 2015. The relation of different crop roots exudates to the survival and suppressive effect of *Stenotrophomonas maltophilia* (PD4560), biocontrol agent of bacterial wilt of potato. *J. Phytopathol.* 163: 829–840.

Elphinstone, J. G., Hennessey, J., Wilson, J. K., and Stead, D. E. 1996. Sensitivity of different methods for the detection of *Pseudomonas solanacearum* in potato tuber extracts. *Bull. OEPP/EPPO Bull.* 26: 663–678.

Erjavec, J., Ravnikar, M., Brzin, J., Grebenc, T., Blejec, A., et al. 2016. Antibacterial activity of wild mushroom extracts on bacterial wilt pathogen *Ralstonia solanacearum. Plant Dis.* 100: 453–464.

Fegan, M., and Prior, P. 2005. How complex is the "*Ralstonia solanacearum* species complex?" pp. 449–461 in: *Bacterial Wilt: The Disease and the Ralstonia solanacearum Species Complex.* C. Allen, P. Prior, and A. C. Hayward eds. APS Press, St. Paul, MN.

Gadewar, A. V., Shekhawat, G. S., Chakrabarti, S. K., and Birhman, R. K. 1999. Temperature induced non-directed heritable changes in virulence of an unstable avirulent strain of *Ralstonia solanacearum. J. Indian Potato Assoc.* 26: 11–18.

Hayward, A. C. 2000. *Ralstonia solancearum. Encyclopedia of Microbiology.* 2nd ed. Vol. 4, Academic Press, London, UK, pp. 32–42.

He, K., Yang, S.-Y., Li, H., Wang, H., and Li, Z.-L. 2014. Effects of calcium carbonate on the survival of *Ralstonia solanacearum* in soil and control of tobacco bacterial wilt. *Eur. J. Plant Pathol.* 140: 665–675.

Hong, J. C., Momol, M. T., Jones, J. B., Ji, P., Olson, S. M., et al. 2008. Detection of *Ralstonia solanacearum* in irrigation ponds and aquatic weeds associated with ponds in North Florida. *Plant Dis.* 92: 1674–1682.

Imada, K., Sakai, S., Kajihara, H., Tanaka, S., and Ito, S. 2016. Magnesium oxide nanoparticles induce systemic resistance in tomato against bacterial wilt disease. *Plant Pathol.* 65: 551–560.

Ji, P., Momol, M. T., Rich, J. R., and Olson, S. M. 2007. Development of an integrated approach for managing bacterial wilt and root-knot on tomato under field conditions. *Plant Dis.* 91: 1321–1326.

Jiang, J.-F., Li, J.-G., and Dong, Y.-H. 2013. Effect of calcium nutrition on resistance of tomato against bacterial wilt induced by *Ralstonia solanacearum. Eur. J. Plant Pathol.* 136: 547–555

Kelman, A. 1953. The bacterial wilt caused by *Pseudomonas solanacearum. N.C. Agr. Exp. Stn. Tech. Bull.* 99: 194 pp.

Kelman, A. 1954. The relationship of pathogenicity of *Pseudomonas solanacearum* to colony appearance in a tetrazolium medium. *Phytopathology* 44: 693–695.

Kumar, R., Barman, A., Phukan, T., Kabyashree, K., Singh, N., et al. 2017. *Ralstonia solanacearum* virulence in tomato seedlings inoculated by leaf clipping. *Plant Pathol.* 66: 835–841.

Kunwar, S., Paret, M. L., Freeman, J. H., Ritchie, L., Olson, S. M., et al. 2017. Foliar applications of acibenzolar-s-methyl negatively affect the yield of grafted tomatoes in fields infested with *Ralstonia solanacearum. Plant Dis.* 101: 890–894.

Lin, C.-H., Hsu, S.-T., Tzeng, K.-C., and Wang, J.-F. 2008. Application of a preliminary screen to select locally adapted resistant rootstock and soil amendment for integrated management of tomato bacterial wilt in Taiwan. *Plant Dis.* 92: 909–916.

Massart, S., Nagy, C., and Jijakli, M. H. 2014. Development of the simultaneous detection of *Ralstonia solanacearum* race 3 and *Clavibacter michiganensis* subsp. *sepedonicus* in potato tubers by a multiplex real-time PCR assay. *Eur. J. Plant Pathol.* 138: 29–37.

Messiha, N., Diepeningen, A., Farag, N., Abdallah, S., Janse, J., et al. 2007a. *Stenotrophomonas maltophilia*: A new potential biocontrol agent of *Ralstonia solanacearum*, casual agent of potato brown rot. *Eur. J. Plant Pathol.* 118: 211–225.

Messiha, N., Diepeningen, A., Wenneker, M., Beuningen, A., Janse, J., et al. 2007b. Biological soil disinfestation (BSD), a new control method for potato brown rot, caused by *Ralstonia solanacearum* race 3 biovar 2. *Eur. J. Plant Pathol.* 117: 403–415.

Milling, A., Meng, F., Denny, T. P., and Allen, C. 2009. Interactions with hosts at cool temperatures, not cold tolerance, explain the unique epidemiology of *Ralstonia solanacearum* race 3 biovar 2. *Phytopathology* 99: 1127–1134.

Nakaho, K., Seo, S., Ookawa, K., Inoue, Y., Ando, S., et al. 2017. Involvement of a vascular hypersensitive response in quantitative resistance to *Ralstonia solanacearum* on tomato rootstock cultivar LS-89. *Plant Pathol.* 66: 150–158.

OEPP/EPPO. 2004. EPPO diagnostic protocols for regulated pests: *Ralstonia solanacearum. Bull. OEPP/EPPO Bull.* 34: 173–178.

Ozakman, M., and Schaad, N. W. 2003. A real-time Bio-PCR assay for detection of *Ralstonia solanacearum* race 3, biovar 2 in asymptomatic potato tubers. *Can. J. Plant Pathol.* 25: 232–239.

Poussier, S., Prior, P., Luisetti, J., Hayward, C., and Fegan, M. 2000. Partial sequencing of the *hrpB* and endoglucanase genes confirms and expands the known diversity within the *Ralstonia solanacearum* species complex. *Syst. Appl. Microbiol.* 23: 479–486.

Prior, P., and Fegan, M. 2005. Recent developments in the phylogeny and classification of *Ralstonia solanacearum. Acta Hortic.* (*ISHS*) 695: 127–136.

Priou, S., Gutarra, L., Aley, P., De Mendiburu, F., and Lique, R. 2010. Detection of *Ralstonia solanacearum* (biovar 2A) in stems of symptomless plants before harvest of the potato crop using post-enrichment DAS-ELISA. *Plant Pathol.* 59: 59–67.

Rao, M. V. B. 1972. Bacterial wilt due to *Pseudomonas solanacearum*. pp. 94–121 in: *Plant Bacteriology. Vol. I, Bacterial Diseases of Plants in India.* P. N. Patel ed. Indian Agricultural Research Institute, New Delhi, India.

Safni, I., Cleenwerck, I., De Vos, P., Fegan, M., Sly, L., et al. 2014. Polyphasic taxonomic revision of the *Ralstonia solanacearum* species complex: Proposal to emend the descriptions of *Ralstonia solanacearum* and *Ralstonia syzygii* and reclassify current *R. syzygii* strains as *Ralstonia syzygii* subsp. *syzygii* subsp. nov., *R. solanacearum* phylotype IV strains as *Ralstonia syzygii* subsp. *indonesiensis* subsp. nov., banana blood disease bacterium strains as *Ralstonia syzygii* subsp. *celebesensis* subsp. nov. and *R. solanacearum* phylotype I and III strains as *Ralstonia pseudosolanacearum* sp. nov. *Int. J. Syst. Evol. Microbiol.* 64: 3087–3103.

Sagar, V., Jeevalatha, A., Mian, S., Chakrabarti, S. K., Gurjar, M. S., et al. 2014. Potato bacterial wilt in India caused by strains of phylotype I, II and IV of *Ralstonia solanacearum. Eur. J. Plant Pathol.* 138: 51–65.

Sakthivel, K., Gautam, R. K., Kumar, K., Roy, S. D., Kumar, A., et al. 2016. Diversity of *Ralstonia solanacearum* strains on the Andaman Islands in India. *Plant Dis.* 100: 732–738.

Salanoubat, M., Genin, S., Artiguenave, F., Gouzy, J., Mangenot, S., et al. 2002. Genome sequence of the plant pathogen *Ralstonia solanacearum. Nature* 415: 497–502.

Schaad, N. W., Gaush, P., Postnikova, E., and Frederick, R. 2001. On-site one hour PCR diagnosis of bacterial diseases. *APS Congress*, 2001.
Schell, M. A. 2000. Control of virulence and pathogenicity genes of *Ralstonia solanacearum* by an elaborate sensory network. *Annu. Rev. Phytopathol.* 38: 263–292.
Shekhawat, G. S., Gadewar, A. V., and Chakrabarti S. K. 1992. Spontaneous phenotypic reversion from afluidal to fluidal state in strains of *Pseudomonas solanacearum. ACIAR Bact. Wilt Newslett.* 8: 5–6.
Shekhawat, G. S., Gadewar, A. V., and Chakravarti, S. K. 2000. Potato bacterial wilt in India. Technical Bulletin No. 38, Central Potato Research Institute, Shimla, India.
Singh, N., Phukan, T., Sharma, P. L., Kabyashree, K., Barman, A., et al. 2018. An innovative root inoculation method to study *Ralstonia solanacearum* pathogenicity in tomato seedlings. *Phytopathology* 108: 436–442.
Smith, E. F. 1896. A bacterial disease of the tomato, eggplant and Irish potato (*Bacillus solanacearum* nov. sp.). *U.S. Dept. Agr. Div. Veg. Phys. and Path. Bull.* 12: 1–28.
Smith, I. M., McNamara, D. G., Scott, P. R., and Holderness, M. 1997. *Ralstonia solanacearum.* in: *Quarantine Pests for Europe*. 2nd ed. EPPO/CABI.
Stulberg, M. J., Shao, J., and Huang, Q. 2015. A multiplex PCR assay to detect and differentiate select agent strains of *Ralstonia solanacearum. Plant Dis.* 99: 333–341.
Tjou-Tam-Sin, N. N. A., van de Bilt, J. L. J., Westenberg, M., Bergsma-Vlami, M., Korpershoek, H. J., et al. 2017. First report of bacterial wilt caused by *Ralstonia solanacearum* in ornamental *Rosa* sp. *Plant Dis.* 101: 378.
Tran, T. M., Jacobs, J. M., Huerta, A., Milling, A., Weibel, J., et al. 2016. Sensitive, secure detection of race 3 biovar 2 and native U.S. strains of *Ralstonia solanacearum. Plant Dis.* 100: 630–639.
Yabuuchi, E., Kosako, Y., Yano, I., Hotta, H., and Nishiuchi, Y. 1996. Validation of the publication of new names and new combinations previously effectively published outside the IJSB. *Int. J. Syst. Bacteriol.* 46: 625–626. Effective publication: Yabuuchi, E., Kosako, Y., Yano, I., Hotta, H., and Nishiuchi, Y. 1995. Transfer of two *Burkholderia* and an *Alcaligenes* species to *Ralstonia* gen. nov.: Proposal of *Ralstonia pickettii* (Ralston, Palleroni and Doudoroff 1973) comb. nov., *Ralstonia solanacearum* (Smith 1896) comb. nov. and *Ralstonia eutropha* (Davis 1969) comb. nov. *Microbiol. Immunol.* 39: 897–904.
Youssef, S. A., and Tartoura, K. A. H. 2013. Compost enhances plant resistance against the bacterial wilt pathogen *Ralstonia solanacearum* via up-regulation of ascorbate-glutathione redox cycle. *Eur. J. Plant Pathol.* 137: 821–834.

7.34 SOFT ROT OF FLESHY VEGETABLES AND FRUITS

Fleshy tissues of fruits and vegetables often undergo rotting in field, transit, and in storage. The bacteria are often associated with these rotting tissues. In majority of these cases, the bacteria involved are not plant pathogens. They are either saprophytes, i.e., they do not attack the living cells, or secondary parasites, which grow in tissues already damaged or weakened by pathogens or adverse environmental conditions. However, some bacteria are real plant pathogens that attack living plant tissues and cause soft rots. *Pectobacterium carotovorum* subsp. *carotovorum* (syn. *Erwinia carotovora* subsp. *carotovora*) and *Pe. atrosepticum* (syn. *Er. carotovora* pv. *atroseptica*, *Pe. carotovorum* subsp. *atrosepticum*) are the main soft rotting bacteria responsible for enormous losses due to soft rots. Soft rot of fleshy vegetables and fruits and blackleg of potato are described here. The other soft rotting bacteria of less economic importance include:

1. *Pseudomonas* species: *Psm. fluorescens*, *Psm. marginalis*, *Psm. cichorii*, and *Psm. viridiflava* are found associated with soft rot of fleshy fruits and vegetables, pink eye disease of potato, slippery skin disease of onion, and sour skin of onion. These bacteria cause soft rots at lower temperatures than *Pectobacterium* and *Bacillus* spp
2. *Ba. subtilis* and *Ba. megaterium* cause rotting of potato tubers and tobacco leaves. They cause rotting at higher temperatures compared to other soft rotting bacteria
3. *Clostridium* species cause rotting of potato tubers and tobacco leaves, and wet wood symptoms in poplar. These bacteria cause rotting under unaerobic conditions

Soft rot of fleshy fruits and vegetables is worldwide in occurrence, but is most prevalent in warm and humid climates. It occurs most commonly on fleshy vegetables like potato tubers, onion bulbs, roots of carrot, radish and turnip, and cabbage and cauliflower heads; and fleshy fruits such as tomato, eggplant, cucumber, watermelon, strawberry, peach, and papaya, celery petioles, and succulent leaves of lettuce and spinach. The losses may occur in the field, transit, and storage/market. The cumulative losses due to this disease are more than any other bacterial disease. L.R. Jones was the first person to report the soft rot of carrots from Vermont in 1901.

7.34.1 Symptoms

As the name indicates, it is a characteristic soft decay of fleshy organs of fruits and vegetables.

Soft rot symptoms on potato tubers start as small, water-soaked, tannish lesions on the surface, which enlarge rapidly in diameter and in depth. The affected inner tissue becomes soft, mushy, and watery, while its surface appears slightly depressed and wrinkled. Affected tissues become cream-colored and slimy, leading to disintegration and disorganization of plant cells. In certain cases, such as potato tubers and tomato fruits, the outer surface remains intact, while the inner entire contents change to a thin paste. The slimy mass consisting of bacterial cells and cell debris oozes out on the outer surface through cracks. The slimy mass turns tan, grey, or dark brown due to exposure to air. Within 3–4 days, a whole tuber or fruit may rot completely under favorable weather conditions. Onion bulbs, potato tubers, and cruciferous plants, when infected by soft rot bacteria, often emit an offensive sulphurous odor, while in most other plants, the infected tissues remain almost odorless until the secondary invaders grow on the rotten tissues and produce offensive smell (Agrios 2004).

Soft rot bacteria also invade swollen potato lenticels, which commonly swell when tubers are soaked in water or exposed to wet soil. The infected areas around the lenticels become slightly raised and darker in color than the potato skin. The lesions are initially ¼ inch in diameter and ¼–½ inch deep, but under favorable conditions, these lesions develop into soft rot. When low temperature prevails, the lenticel infection dries out, leaving a shallow spot with a chalky-white deposit under the normal skin color.

The aerial symptoms manifest as weak and chlorotic plants with margins of the leaflets curled upward. Irregular, brownish to black (but not as black as in blackleg of potato) lesions appear on the stems. These lesions enlarge causing a soft mushy rot of the stems. These symptoms are referred to as **aerial stem rot** or **bacterial stem rot**.

Succulent leaves are rarely infected in field, but their infection in storage or in plastic containers leads to their rapid rotting and disintegration.

7.34.2 Causal Organism

Pectobacterium carotovorum subsp. *carotovorum* (Jones 1901) Waldee 1945 emend. Hauben et al. 1999
Syn. *Erwinia carotovora* subsp. *carotovora*

Waldee (1945) created a new genus *Pectobacterium* to include soft rotting bacteria and suggested that all the soft rot causing species of the then genus *Erwinia* should be transferred to this genus. Graham (1964) also supported Waldee's proposal. Although Waldee's proposal did not get widespread acceptance, the name of *Pectobacterium* appeared in *Approved Lists of Bacterial Names* in 1980. Hauben et al. (1998) emended the description of genus *Pectobacterium* and its type species, *Pe. carotovorum* and also proposed five new sub species combinations including *Pe. carotovorum* subsp. *carotovorum*. The below given description of this sub species is taken from Hauben et al. (1998) and Gardan et al. (2003).

Cells are Gram-stain-negative rods with rounded ends and measure 0.5–1.0 × 1.0–3.0 µm. They occur mostly singly or in pairs, but chains are also formed, and are usually motile by means of peritrichous flagella. The bacterium produces H_2S from cysteine, liquefies gelatin, reduces nitrate, produces catalase, and degrades pectin. The tests for indole production, oxidase, and starch hydrolysis are negative. Acid is produced from α-methylglucoside, glucanate, inulin, and sorbitol. The bacterium assimilates 1-*O*-methyl α-galactopyranoside, 1-*O*-methyl β-galactopyranoside, and L-glutamate, but does not assimilate palatinose, *meso*-tartrate, or L-alanine. It can grow in the presence of 7% NaCl and can utilize citrate. The bacterium does not utilize hypoxanthin as nitrogen source. The bacterium is able to grow at 36°C. It does not produce reducing substances from sucrose. The G + C content of the DNA of the strains of the sub species ranges from 50.5 to 53.1 mol%. The selective medium of Beraha (1968) is helpful for the isolation of soft rot bacteria, while the differential medium of Friedman (1964) is useful in separating virulent and avirulent colonies.

In addition to *Pe. carotovorum* subsp. *carotovorum*, other phytopathogenic bacteria have also been shown to cause soft rot symptoms. Nabhan et al. (2013) identified a pectolytic soft rot pathogen with preference for monocotyledonous plants as *Pe. aroidearum* sp. nov. after several taxonomic analyses including DNA-DNA hybridization. Moretti et al. (2016) made the first official report of *Pe. aroidearum* as causal agent of soft rot on potato plants in Lebanon.

Pe. carotovorum subsp. *odoriferum* is generally considered to have a narrow host range and most often causes soft rot of chicory. Waleron et al. (2014) reported for the first time that *Pe. carotovorum* subsp. *odoriferum* occurs on a wide range of vegetables and has the ability to cause soft rot of potato during storage. All *Pe. carotovorum* subsp. *odoriferum* isolates caused soft rot symptoms on chicory and potato. Moreover, *Pe. carotovorum* subsp. *odoriferum* isolates caused more severe symptoms on potatoes at temperatures ranging from 15°C to 37°C compared to *Pe. atrosepticum* or *Pe. carotovorum* subsp. *carotovorum* isolates. Tissue maceration by *Pe. carotovorum* subsp. *odoriferum* isolates was highest at 28°C, and it was 2-fold higher compared to that caused by *Pe. carotovorum* subsp. *carotovorum* isolates.

Pe. atrosepticum, *Pe. wasabiae,* and *Pe. carotovorum* subsp. *brasiliense* have also been reported to cause soft rot symptoms (Pitman et al. 2010; Barzic and Com 2012; Onkendi and Moleleki 2014). Meng et al. (2017) reported cucumber stem soft rot caused by *Pe. carotovorum* subsp. *brasiliense* for the first time from China as well as in the world, which has a significant economic impact on cucumber production.

7.34.3 Disease Cycle

The bacterium perpetuates in infected plant refuse, fleshy organs in storage, in ponds and streams used for irrigation water, in infested soil, in contaminated stores/warehouses, and in pupae of several species of maggot fly. The contamination of potato tubers mostly occurs during harvest, handling, and washing. Plant tissues weakened, invaded, or killed by pathogens or by mechanical means are more susceptible to soft rot bacteria.

The penetration of bacteria occurs usually through wounds caused by cultural operations and insects. The bacteria can also penetrate through lenticels of freshly harvested tubers. The dissemination of bacteria occurs through wind-splashed rain, irrigation water, and insects. Several species of maggot fly are known to carry the bacteria.

Aerial stem rot infection usually results from bacteria present in the infested soil and brought by irrigation water, windblown rain, and insects. Stem infections occur through wounds and leaf scar avenues. In the field, the disease first appears on plants grown from previously infected seed pieces or healthy seed pieces contaminated from infested soil. The further spread and inoculation of bacteria into fleshy organs are facilitated greatly by insects (maggot flies). These bacteria can live in all growth stages of the insect. Adult insects lay eggs on decaying tissues. When the eggs hatch, larvae feed on rotting tissue, and their bodies become contaminated externally and internally with bacteria. These larvae inoculate the bacteria into healthy plant parts when they feed on them. The pupae and adult flies developing from internally contaminated larvae also become internally infested. These adult flies lay eggs smeared with soft rot bacteria. The larvae coming out from these eggs are contaminated and when they feed on healthy tissue, not only cause wounds, but also introduce the pathogen. By causing wounds, the maggots also break natural barrier of host resistance, thereby making even the resistant plants or plant parts susceptible.

Soft rot bacteria are typical wound parasites. After entering wounds, the bacteria feed and multiply on cell sap released by wounded cells. During multiplication, the bacteria produce a considerable amount of pectolytic enzymes and a lesser amount of cellulolytic enzymes. The pectolytic enzymes break down the pectic substances of middle lamella and cell walls. The pectic substances are present mainly in the middle lamella as calcium pectate. The pectolytic enzymes produced move ahead of bacteria and cause maceration of the host tissue by breaking the pectic substances of middle lamella and cell walls. Due to high osmotic pressure of the macerated tissue, water from the cells comes into the intercellular spaces leading to plasmolysis, collapse, and death of cells. Bacteria move into intercellular spaces of macerated tissue, multiply, produce pectolytic enzymes, which move ahead of bacteria and prepare that tissue for invasion. The invasion of host cells occurs at later stages when cell walls are dissolved partially. The infected plant parts become soft and are converted into a slimy mass containing numerous bacteria present in the cell sap. The slimy mass containing bacteria exudes through cracks or wounds formed on the outer surface of infected plant parts and spreads the bacteria to other fleshy organs present in the field or in stores.

A fairly high relative humidity is essential for the progress of the disease. When a decaying fleshy organ is placed in a dry atmosphere, the rotted organ dehydrates rapidly and further progress of the disease may be checked completely. The progress of the disease is best above 24°C. Oxygen depletion in potato tubers also favors soft rot. A film of water on potato tubers in wet soil or in stores depletes the oxygen from tubers.

duRaan et al. (2016) determined the cardinal growth temperatures for the species responsible for soft rot of potato *in vitro*. The optimal temperature of 31°C and temperature range of 20°C to 38°C for *Pe. carotovorum* subsp. *brasiliense* were similar to those recorded for *Pe. carotovorum* subsp. *carotovorum*. *Pe. wasabiae* grew at an optimal temperature of 29°C and range of 20°C–34°C. The authors suggested that the minimal variation in optimal temperatures between different species might be an indication that temperature ranges, rather than optimal temperatures, play an important role in disease development.

Kubheka et al. (2013) generated an mCherry-*Pe. carotovorum* subsp. *brasiliense*-tagged strain to study interactions between *Pe. carotovorum* subsp. *brasiliense* and potato plants. Confocal laser-scanning microscopy and *in vitro* viable cell counts showed that *Pe. carotovorum* subsp. *brasiliense* was able to penetrate roots of a susceptible potato cultivar as early as 12 h post-inoculation and migrated upward into aerial stem parts. A comparison of *Pe. carotovorum* subsp. *brasiliense* colonization patterns in tolerant and susceptible cultivars revealed that in the susceptible cultivar, *Pe. carotovorum* subsp. *brasiliense* cells colonized the xylem tissue, forming "biofilm-like" aggregates that led to occlusion of some of the vessels. In contrast, in the tolerant cultivar, *Pe. carotovorum* subsp. *brasiliense* appeared as free-swimming planktonic cells with no specific tissue localization. It suggests that resistance mechanisms in the tolerant cultivar limit aggregation of *Pe. carotovorum* subsp. *brasiliense in planta* leading to lack of symptom development.

Koh et al. (2013) investigated the effect of bromelain gene (*BL1*) of pineapple (*Ananas comosus* L. Merrill) on enhancing resistance to soft rot in transgenic *Brassica rapa* "Seoulbaechu." On inoculation of three homozygous T_2 lines with *Pe. carotovorum* subsp. *carotovorum*; BL8-2 line showed the lowest number of infected leaves in both wound and non-wound inoculated categories, while the non-infected line showed 100% infected leaves in both cases. The highest expression of *BL1* gene was also observed in BL8-2 homozygous line. Thus, the over-expressed *BL1* gene conferred enhanced resistance to soft rot in *B. rapa*.

Yang et al. (2011) characterized 2-aminobenzoic acid (2-AB) as a main induced systemic resistance (ISR) determinant from *Bacillus* sp. BS107 (BS107). The application of 2-AB at 2.3 mM significantly suppressed disease development, but exhibited no direct contact inhibition of pathogen. Root drench-treatment of tobacco seedlings with BS107 suppressed disease development caused by *Pe. carotovorum* subsp. *carotovotrum* SCC1. Reverse transcriptase-PCR analyses of tobacco leaves revealed up-regulation of the induced resistance marker genes such as *PR1a*, *PR1c*, *PR2*, and *PR4* by application of 2-AB on the root. It is evident that BS107 can play a role in promoting plant defenses by secretion of bacterial determinants including 2-AB for elicitation of ISR.

7.34.4 Management

The combination of sanitary measures and cultural practices can result in a very effective management of soft rot.

1. Remove all the debris from the warehouses and disinfect the floor and walls of the warehouses with copper sulphate (1 lb in 5 gal of water) or formaldehyde (1 pint in 10 gal of water)
2. All efforts should be made to avoid wounding of plants and their storage organs
3. Do not wash potatoes and other products before shipping or storage. If absolutely necessary, use chlorinated water for washing. The product must be dried before shipping or storage
4. Temperature (4°C) and humidity in stores should be kept low

5. Use pathogen-free potato seed and if possible, use whole tubers instead of seed pieces
6. Harvest potato crop only after the vines are completely dead to ensure tuber's skin maturity
7. Insecticides should be sprayed in the fields and in warehouses to kill the insects responsible for the spread of bacteria
8. Chemical sprays are generally not recommended except for the control of soft rot of tomatoes, for which Bordeaux mixture (0.8%) should be sprayed repeatedly. Currently, chlorine compounds are used to control post-harvest soft rot of fleshy fruits and vegetables. Mahovic et al. (2007) found that the application of chlorine dioxide gas (exposure to 99 mg ClO_2 for a 2- or 24-hour period) protected the harvested tomato fruits from soft rot
9. Rotate the susceptible crops with cereals and other non-host crops
10. da Silva et al. (2017) reported that $Ca(NO_3)_2$ applied as leaf spray and soil drench was effective in controlling soft rot of Chinese cabbage caused by *Pe. carotovorum* subsp. *carotovorum,* and it caused disease reduction up to 48.5% when sprayed onto the leaves at the rate of 0.15 g L^{-1}. A significant increase in the leaf calcium content was observed only in the plants that were sprayed with higher doses of $Ca(NO_3)_2$ and $CaCl_2$. Light microscopy examination revealed an increased number of chloroplasts and improved structuring of the palisade parenchyma in all the treatments, while transmission electron microscopy examination revealed an increased cell wall thickness that was especially evident for the 0.15 g L^{-1} $Ca(NO_3)_2$ treatment applied by leaf spraying and soil drenching
11. *Pantoea agglomerans* (*Er. herbicola* 252) has shown promising results in controlling both soft rot and blackleg diseases. Potato seed pieces/tubers were treated with antagonist before sowing. Lapidot et al. (2015) reported that *Paenibacillus dendritiformis* significantly reduced the maceration area of tuber slices infected by *Pe. carotovorum* subsp. *carotovorum*, significantly reduced disease indices in greenhouse experiments, and significantly increased tuber yield of infected plants in the field. This work demonstrates the potential of preliminary screening based on genome analysis to identify effective biocontrol agents

Gerayeli et al. (2018) reported that out of 235 *Bacillus* strains isolated from the rhizosphere of potato plants, 15 strains inhibited *in vitro* growth of *Pe. carotovorum* subsp. *carotovorum*. Antagonistic strains reduced the maceration capacity of soft rotting strains by 1.6–4.0 times. *Ba. pumilus* IrB8 and *Ba. amyloliquefaciens* IrB12 suppressed tuber maceration by 63.7% and 47.8%, respectively. The high levels of phenylalanine ammonia-lyase, polyphenol oxidase, peroxidase, and total phenols were obtained in a single application of IrB8 or IrB12 after 8 h. It is evident that IrB8 and IrB12 strains have the potential to be used for potato soft rot control.

7.35 BLACKLEG OF POTATO

Blackleg of potato has many similarities with soft rot disease. However, it occurs mainly on potatoes in contrast to the wide host range of soft rot bacteria. It is worldwide in occurrence. In the first decade of the twentieth century, the disease was independently studied/described by Appel in Germany, van Hall in the Netherlands, and Harrison in Canada. It used to cause considerable economic losses in all potato growing regions, but in the last one and a half decades, its incidence has greatly decreased in countries where micropropagated plantlets are used instead of seed tubers reducing the number of field generations.

Severe stand losses and blackleg symptoms were observed in Maine, USA during the 2014 potato season. Again in 2015, severe stand losses and blackleg symptoms were present in a number of states including Maine. PCR confirmed *Dickeya* spp. as being present in number of these instances. It is probable that *Dickeya* spp. have been in the Maine seed system, albeit at low levels, for over 5 years (Johnson 2016). Hao et al. (2016), on the basis of molecular analysis, identified *D. dianthicola* as the cause of the outbreak of blackleg of potato that occurred in Maine in 2015.

Recently, *D. solani* have been isolated most frequently from potato plants with the symptoms of blackleg and soft rot. *D. solani* strains were shown to cause more severe disease symptoms on potato plants than *D. dianthicola* especially at the higher temperature. They are also able to develop blackleg disease from lower inoculum levels. Polish *D. solani* strains showed higher activities of cell wall degrading enzymes than the Finnish and Israeli strains at all the tested temperatures, i.e., 18°C, 27°C, 37°C. This observation is correlated with the higher ability of Polish *D. solani* strains to cause soft rot. In addition, *D. solani* strains exhibited higher activity of the abovementioned enzymes and caused more severe potato tuber maceration in laboratory tests than the tested *D. dianthicola* strains (Golanowska et al. 2017). In Western Europe, *Pe. carotovorum* subsp. *brasiliense* is emerging as a causal agent of blackleg disease (van der Wolf et al. 2017).

Czajkowski et al. (2011) have briefly reviewed research on the causal agents of blackleg and soft rot of potato, namely, *Dickeya* and *Pectobacterium* species and the disease syndrome, including epidemiological and aetiological aspects. The review includes critical evaluation of control methods used in practice, i.e., the avoidance of the contamination of plants, in particular the use of seed testing programs and the application of hygienic procedures during crop production. The perspective of breeding and genetic modification to introduce resistance is dealt, and the review also includes the application of physical and chemical tuber treatments to reduce inoculum load and examines the possibility of biocontrol using antagonistic bacteria and bacteriophages.

7.35.1 Symptoms

Blackleg usually originates from a contaminated seed piece or tuber. Under favorable weather conditions, entire seed piece or tuber and sprouts emerging from it may rot in the soil, resulting in a poor stand. The non-emergence of plants, also called **blanking**, is a serious economic consequence of the disease. Under less favorable conditions, plants emerging from infected tubers show blackleg symptoms.

The symptoms usually start on the stems at the point of attachment to the tubers and extend upward. In the early growing season, a light brown to inky decay originating from the seed piece extends to the stem portions that are entirely below ground and the plant shows yellowing and stunting, and exhibits a stiffened and upright habit. However, the typical phase of the disease, i.e., blackleg usually develops after the plants have emerged from the soil. If there is sufficient moisture in the soil, the shoots arising from the infected tubers show watery decay of the cortical tissue and the shoots wilt. If the soil moisture is reduced, the base of the shoot develops a blackened, shriveled cortex, **blackleg phase** (Figure 7.21), and the growth of the shoot is stunted. The leaves turn yellow, margins of the leaflets roll upward and later, entire plant may wilt and often die. Stem decay and associated blackening is not always uniform throughout the entire length of the stem. Sometimes, the blackening of the tissue occurs near the soil level and then at a node or stem injury distant from the previous one, while the intervening tissue shows no external symptoms.

The tubers of the infected plants also become infected by the movement of pathogen through stolons. The dark-colored decay of the tubers starts at the stolon end, extends inward causing decay of pith tissue and turning the tuber into a hollow shell. These decaying tubers are often invaded by the secondary soil microorganisms, which modify the original disease symptoms. In improperly stored tubers, the pathogen causes decay at the sites of tuber lenticels and wounds. Under dry conditions, lenticel infections result in formation of restricted, small, sunken lesions around the lenticels. These restricted lesions are referred to as **hard rot** in the literature. However, under moist conditions and in association with weakly pathogenic or saprophytic microorganisms, lenticel infections can lead to decay of entire tubers. The flesh of the affected tubers first turns creamy, then greyish, and finally black.

FIGURE 7.21 Blackleg of potato caused by *Dickeya* sp. (Courtesy of Margaret T. McGrath, Cornell University, Ithaca, NY.)

Irregular cavities having black walls may extend through the center of the potato tubers. Near midseason, inky-black, slightly sunken lesions are formed at the stem end of the tubers.

7.35.2 Causal Organism

Pectobacterium atrosepticum (van Hall 1902) Gardan et al. 2003

Syn. *Erwinia carotovora* pv. *atroseptica*, *Pe. carotovorum* subsp. *atrosepticum*

van Hall in 1902, first named the bacterium as *Er. atroseptica*. Later, Dye in 1969, renamed it as *Er. carotovora* subsp. *atroseptica* and in 1978 as *Er. carotovora* pv. *atroseptica*. Hauben et al. (1999), on the basis of 16S rDNA sequences, reclassified it as *Pe. carotovorum* subsp. *atrosepticum*. Gardan et al. (2003) raised it to the species level and named it as *Pectobacterium atrosepticum*. The below given description of the bacterium is based on Hauben et al. (1998) and Gardan et al. (2003).

Cells are Gram-stain-negative rods having round ends and measure 0.5–1.0 × 1.0–3.0 μm. They occur mostly singly or in pairs, but chains occur as well and are usually motile by means of peritrichous flagella. Acid is produced from gluconate, α-methyl-glucoside, lactose, cellobiose, raffinose, and melibiose, but not from inulin, sorbitol, and D(+)-arabitol. The bacterium assimilates platinose, lactulose, L-alanine, 1–*O*-methyl β-galactopyranoside, and D-glucuronate, but does not assimilate *meso*-tartrate or D(-)-tartrate. The bacterium is unable to grow at 36°C and produces reducing compounds from sucrose. The G + C content of the DNA ranges from 51.3 to 53.1 mol%. Bell et al. (2004) sequenced the genome of the bacterium and demonstrated the presence of novel pathogenicity determinants in it.

Cigna et al. (2017) developed a PCR-sequencing tool to easily characterize the different *Pectobacterium* and *Dickeya* taxa. The *gapA* gene sequences from 53 published genomes were aligned and a phylogeny tree was constructed. A set of 35 signature nucleotides was discovered to distinguish genera, species, and sub species of *Pectobacterium* and *Dickeya*. Then, a PCR-primer couple was designed for amplifying the *gapA* gene in pectinolytic enterobacteria. The primers were tested on 22 isolates recovered from blackleg symptoms in several potato fields. Amplicons were sequenced and signature-nucleotides were analyzed. A phylogeny that includes *gapA* sequence specimens confirmed the taxonomical identification of these environmental isolates.

De Boer et al. (1987) recognized four different serological groups on the basis of antigenicity to the lipopolysaccharide. Serogroup I accounts for about 95 strains of this bacterium associated with blackleg in Canada, the United States, and several countries of Western Europe. Monoclonal antibodies specific to this group are available commercially and have been used to index seed tubers for the presence of the bacterium. Fragments of genomic DNA to be used as primers in the PCR have been identified for the detection of the bacterium.

Production of pectolytic enzymes, namely, pectate lyase, polygalacturonase, and pectin methyl esterase, is the primary, but not the sole pathogenicity factor. In addition, cellulase, xylanase, and protease activity are present, but their role in tissue maceration has not been studied in detail.

In addition to *Pe. atrosepticum*, other pathogenic bacteria have also been shown to be the cause of blackleg of potato. De Boer et al. (2012) detected *Pe. carotovorum* subsp. *brasiliensis*, *Pe. carotovorum* subsp. *carotovorum*, and *Pe. wasabiae*, including *Pe. atrosepticum*, in potato stems showing blackleg symptoms, collected from Canada, using species- and sub species-specific PCR. The tests included a new assay for *Pe. wasabiae* based on the phytase gene sequence. As mentioned above, Hao et al. (2016) identified *D. dianthicola* as the cause of blackleg of potato in Maine, USA. *D. solani* also causes blackleg and soft rot of potatoes.

7.35.3 Disease Cycle

The bacterium mainly survives in or on the contaminated potato tubers in the lenticels and in the damaged periderm tissue (Perombelon and Kelman 1980). It does not survive well in soil for more than 1 year; however, the survival in infected debris is possible. The dissemination of the bacterium is mainly through contaminated potato tubers. The other means of dispersal are wind-blown rain, seed planting equipment, cutting knives, and dipterous insects. The bacterium enters mainly through wounds and bruises caused on the potato tubers.

Temperature and moisture are the critical factors in the initiation and development of the disease. The extent of seed lot contamination and seed handling techniques also influence the severity of the disease. Most of the blackleg outbreaks have resulted from contaminated seed lots. Toth et al. (2003) showed that the disease development was directly related to seed tuber contamination and the threshold level of seed tuber contamination remains an important factor in predicting both disease development and progeny tuber contamination. High soil moisture and high temperature favor seed-piece decay and pre-emergence blackleg. Cool, wet soils at planting and high temperature after emergence of plants are favorable for blackleg phase. Even in the absence of disease, population of the bacterium may increase considerably in the root zone of the plants and may contaminate the progeny tubers. Therefore, visual evaluation of blackleg incidence is not a proper assessment of the level of contamination on progeny tubers.

Skelsey et al. (2016) applied spatial point pattern analysis to blackleg occurrence in seed potato crops in Scotland during 2010–2013 crop seasons to assess whether its distribution was random, regular, or aggregated, and the spatial scales at which these patterns occurred. These analyses provided the first quantitative evidence of localized and large-scale spatial clustering of potato blackleg.

7.35.4 Management

1. As the bacterium does not survive well in soil, the rotation with non-host crops is not very helpful in the management of the disease
2. Use of pathogen-free seed and seed handling techniques, which cause minimum injuries to the seed, are very effective
3. Disposal and burial of infected plant refuse and cull piles and disinfection of tools used in cutting and planting of potato tubers reduce the disease severity (De Boer 2001)
4. Proper wound healing and drying of potato tubers before storage and storage under proper conditions reduce the spread of the pathogen during storage

REFERENCES

Agrios, G. N. 2004. *Plant Pathology*. 5th ed. Elsevier Academic Press, San Diego, CA, 922 pp.

Barzic, M. R., and Com, E. 2012. Proteins involved in the interaction of potato tubers with *Pectobacterium atrosepticum*: A proteomic approach to understanding partial resistance. *J. Phytopathol.* 160: 561–575.

Bell, K. S., Sebaihia, M., Pritchard, L., Holden, M. T. G., Hyman, L. J., et al. 2004. Genome sequence of the enterobacterial phytopathogen *Erwinia carotovora* subsp. *atroseptica* and characterization of virulence factors. *Proc. Natl. Acad. Sci. USA* 101: 11105–11110.

Beraha, L. 1968. A rapid method for the preparation of a semi-soild agar medium for detection of pectolytic enzyme activity in *Erwinia carotovora*. *Plant. Dis. Reptr* 52: 167.

Cigna, J., Dewaegeneire, P., Beury, A., Gobert, V., and Faure, D. 2017. A *gapA* PCR-sequencing assay for identifying the *Dickeya* and *Pectobacterium* potato pathogens. *Plant Dis.* 101: 1278–1282.

Czajkowski, R., Pérombelon, M. C. M., van Veen, J. A., and van der Wolf, J. M. 2011. Control of blackleg and tuber soft rot of potato caused by *Pectobacterium* and *Dickeya* species: A review. *Plant Pathol.* 60: 999–1013.

da Silva Felix, K. C., da Silva, C. L., de Oliveira, W. J., Mariano, R. de L. R., and de Souza, E. B. 2017. Calcium-mediated reduction of soft rot disease in Chinese cabbage. *Eur. J. Plant Pathol.* 147: 73–84.

De Boer, S. H. 2001. Black leg/soft rot of potato. pp. 146–148 in: *Encyclopedia of Plant Pathology*. Vol. 1, O. C. Maloy, and T. D. Murray eds. John Wiley & Sons, New York.

De Boer, S. H., Li, X., and Ward, L. J. 2012. *Pectobacterium* spp. associated with bacterial stem rot syndrome of potato in Canada. *Phytopathology* 102: 937–947.

De Boer, S. H., Verdonck, L., Vruggink, H., Harju, P., Bång, H. O., et al. 1987. Serological and biochemical variation among potato strains of *Erwinia carotovora* subsp. *atroseptica* and their taxonomic relationship to other *E. carotovora* strains. *J. Appl. Bacteriol.* 63: 487–495.

duRaan, S., Coutinho, T. A., and van der Waals, J. E. 2016. Cardinal temperature differences, determined *in vitro*, between closely related species and subspecies of pectinolytic bacteria responsible for blackleg and soft rot on potatoes. *Eur. J. Plant Pathol.* 144: 361–369.

Friedman, B. A. 1964. Carbon source and tetrazolium agar to distinguish virulence in colonies of *Erwinia carotovora*. *Phytopathology* 54: 494–495.

Gardan, L., Gouy, C., Christen, R., and Samson, R. 2003. Elevation of three subspecies of *Pectobacterium carotovorum* to species level: *Pectobacterium atrosepticum* sp. nov., *Pectobacterium betavasculorum* sp. nov. and *Pectobacterium wasabiae* sp. nov. *Int. J. Syst. Evol. Microbiol.* 53: 381–391.

Gerayeli, N., Baghaee-Ravari, S., and Tarighi, S. 2018. Evaluation of the antagonistic potential of *Bacillus* strains against *Pectobacterium carotovorum* subsp. *carotovorum* and their role in the induction of resistance to potato soft rot infection. *Eur. J. Plant Pathol.* 150: 1049–1063.
Golanowska, M., Kielar, J., and Lojkowska, E. 2017. The effect of temperature on the phenotypic features and the maceration ability of *Dickeya solani* strains isolated in Finland, Israel and Poland. *Eur. J. Plant Pathol.* 147: 803–817.
Graham, D. C. 1964. Taxonomy of the soft rot coliform bacteria. *Annu. Rev. Phytopathol.* 2: 13–42.
Hao, J., Jiang, H., and Johnson, S. 2016. Detection and characterization of *Dickeya* species in the outbreak of blackleg disease of potato in Maine (Abstr.). *Phytopathology* 106: S2.2.
Hauben, L., Moore, E. R. B., Vauterin, L., Steenackers, M., Mergaert, J., et al. 1999. Validation of publication of new names and new combinations previously effectively published outside the IJSB. *Int. J. Syst. Bacteriol.* 49: 1. Effective Publication: Hauben, L., Moore, E. R. B., Vauterin, L., Steenackers, M., Mergaert, J., et al. 1998. Phylogenetic position of phytopathogens within the Enterobacteriaceae. *Syst. Appl. Microbiol.* 21: 384–397.
Johnson, S. 2016. *Dickeya*, a new potato pathogen in Maine and elsewhere (Abstr.). *Phytopathology* 106: S2.3.
Koh, Y.-J., Park, J.-I., Ahmed, N. U., Jung, H.-J., Kang, K.-K., et al. 2013. Enhancement of resistance to soft rot (*Pectobacterium carotovorum* subsp. *carotovorum*) in transgenic *Brassica rapa*. *Eur. J. Plant Pathol.* 136: 317–322.
Kubheka, G. C., Coutinho, T. A., Moleleki, N., and Moleleki, L. N. 2013. Colonization patterns of an mcherry-tagged *Pectobacterium carotovorum* subsp. *brasiliense* strain in potato plants. *Phytopathology* 103: 1268–1279.
Lapidot, D., Dror, R., Vered, E., Mishli, O., Levy, D., et al. 2015. Disease protection and growth promotion of potatoes (*Solanum tuberosum* L.) by *Paenibacillus dendritiformis*. *Plant Pathol.* 64: 545–551.
Mahovic, M. J., Tenney, J. D., and Bartz. J. A. 2007. Applications of chlorine dioxide gas for control of bacterial soft rot of tomatoes. *Plant Dis.* 91: 1316–1320.
Meng, X., Chai, A., Shi, Y., Xie, X., Ma, Z., et al. 2017. Emergence of bacterial soft rot in cucumber caused by *Pectobacterium carotovorum* subsp. *brasiliense* in China. *Plant Dis.* 101: 279–287.
Moretti, C., Fakhr, R., Cortese, C., De Vos, P., Cerri, M., et al. 2016. *Pectobacterium aroidearum* and *Pectobacterium carotovorum* subsp. *carotovorum* as causal agents of potato soft rot in Lebanon. *Eur. J. Plant Pathol.* 144: 205–211.
Nabhan, S., De Boer, S. H., Maiss, E., and Wydra, K. 2013. *Pectobacterium aroidearum* sp. nov., a soft rot pathogen with preference for monocotyledonous plants. *Int. J. Syst. Evol. Microbiol.* 63: 2520–2525.
Onkendi, E. M., and Moleleki, L. N. 2014. Characterization of *Pectobacterium carotovorum* subsp. *carotovorum* and *brasiliense* from diseased potatoes in Kenya. *Eur. J. Plant Pathol.* 139: 557–566.
Perombelon, M. C. M., and Kelman, A. 1980. Ecology of the soft rot erwinias. *Annu. Rev. Phytopathol.* 18: 361–387.
Pitman, A. R., Harrow, S. A., and Visnovsky, S. B. 2010. Genetic characterisation of *Pectobacterium wasabiae* causing soft rot disease of potato in New Zealand. *Eur. J. Plant Pathol.* 126: 423–435.
Skelsey, P., Elphinstone, J. G., Saddler, G. S., Wale, S. J., and Toth, I. K. 2016. Spatial analysis of blackleg-affected seed potato crops in Scotland. *Plant Pathol.* 65: 570–576.
Toth, I. K., Sullivan, L., Brierley, J. L., Avrova, A. O., Hyman, L. J., et al. 2003. Relationship between potato seed tuber contamination by *Erwinia carotovora* ssp. *atroseptica*, black leg disease development and progeny tuber contamination. *Plant Pathol.* 52: 119–126.
van der Wolf, J. M., de Haan, E. G., Kastelein, P., Krijger, M., de Haas, B. H., et al. 2017. Virulence of *Pectobacterium carotovorum* subsp. *brasiliense* on potato compared with that of other *Pectobacterium* and *Dickeya* species under climatic conditions prevailing in the Netherlands. *Plant Pathol.* 66: 571–583.
van Hall, C. J. J. 1902. Bijdragen tot de kennis der Bakterieele Plantenzeikten. Coöperatieve Drukkerij-vereeniging "Plantijn," Inaugural dissertation, Amsterdam, the Netherlands.
Waldee, E. L. 1945. Comparative studies of some peritrichous phytopathogenic bacteria. *Iowa St. Coll. J. Sci.* 19: 435–484.
Waleron, M., Waleron, K., and Lojkowska, E. 2014. Characterization of *Pectobacterium carotovorum* subsp. *odoriferum* causing soft rot of stored vegetables. *Eur. J. Plant Pathol.* 139: 457–469.
Yang, S. Y., Park, M. R., Kim. I. S., Kim, Y. C., Yang, J. W., et al. 2011. 2-Aminobenzoic acid of *Bacillus* sp. BS107 as an ISR determinant against *Pectobacterium carotovorum* subsp. *carotovotrum* SCC1 in tobacco. *Eur. J. Plant Pathol.* 129: 371–378.

7.36 COMMON SCAB OF POTATO

Common scab of potato was first discovered in 1825 in the United States, but its causal agent was identified as *Oospora scabiei* by Thaxter in 1891. In 1914, Güssow transferred the casual agent to Actinomycetes. Later on, Waksman and Henrici in 1943 transferred it to genus *Streptomyces* as *Streptomyces scabies*. In 1953, potato scab was found in Maine in soils having pH as low as 4.3, and it was named as **"acid scab"** Common scab of potato occurs in potato growing regions throughout the world. It is more prevalent in neutral or slightly alkaline soils and the severity increases during the dry years. Generally, the disease does not cause reduction in yield, but significant losses result from reduction in the market value of potato tubers. Severe infection may cause reduction in yield also and deep-pitted lesions increase losses resulting from wastage due to peeling. Also deep-pitted lesions, which are common in diseased tubers, affect processing quality of potatoes where the normal steam-peeling processes do not remove lesioned material. The disease is rated among the top five destructive diseases of potato by potato seed producers in the United States. It is also considered one of the most economically important diseases affecting potato around the world. According to Wilson (2004), the disease annually cost the Tasmanian (Australia) potato industry in excess of A$ 3.66 million or approximately 4% of the industry value.

7.36.1 Symptoms

The symptoms appear on the underground plant parts, mainly on the tubers. The symptoms are usually noticed late in the growing season or at harvest. Corky or scabby lesions of three types, namely, surface (shallow), raised (erumpent), and pitted (deep) appear on the surface of potato tubers. Sometimes surface lesions are also referred to as russeting **(russet scab)**, particularly on white-skinned potato varieties, as the general appearance resembles the skin of a russet-type tuber. The lesions start as small, superficial, roughened, circular spots, sometimes raised above and often slightly below the plane of the healthy skin. The spots enlarge, coalesce with

FIGURE 7.22 Common scab of potato: Scab lesions on potato tubers.

each other forming large scabby lesions. The lesions become corky due to abnormal proliferation of cells of the periderm of tubers. The lesions vary in shape and size and are slightly darker in color than the healthy skin (Figure 7.22). In **deep** or **pitted scab**, the lesions extend deep into the tuber and on an average may be 3–4 mm deep. These lesions are darker in color than the surface (shallow) lesions. The association of insects is suggested in the formation of pitted lesions (Walker 1969). The type of lesion formed is considered to be the outcome of the interaction of host response, aggressiveness of the pathogen strain, time of infection, and environmental conditions. Another scab, called **acid scab** is caused by *St. acidiscabies*. The symptoms of acid scab are similar, if not identical, to those of common scab. However, the acid scab occurs in soils having pH below 5.2, as well at higher pH levels, while the common scab is found in soils having pH range of 5.2–8.0.

Netted scab, caused by *St. reticulsiscabiei* is another disease of potato, which differs from common scab in the type of symptoms produced, causal agent (Bouchek-Mechiche et al. 2000a), host range, varietal reaction, and optimum soil temperature required for disease development (Bouchek-Mechiche et al. 2000b). Netted scab is reported only from European countries. The symptoms consist of superficial brown lesions on the tuber's skin and on the roots, resulting in yield reductions. Several potato cultivars are immune to the pathogen. However, the most popular cultivar for commercial cropping in continental Europe, such as Bintje, is susceptible. Sugarbeet is another host, but carrot and radish are not infected. The differences in climatic requirements, i.e., soil temperature and moisture for the development of common and netted scabs suggest that netted scab may become a problem in areas or in seasons unfavorable for common scab, if susceptible cultivars are grown.

Gouws and McLeod (2012) observed a distinct atypical symptom on potato in isolated production regions (Western Free State and Mpumalanga) in South Africa since early 2010. Deep longitudinal fissures (3–12 mm) containing scab-like lesions were observed on the surface area of tubers of several potato cultivars (Mondial, BP1, and Buffelspoort). The lesions on cv. Mondial were interesting since this cultivar is tolerant to typical common scab in South Africa. *St. scabiei* reference isolate caused typical circular, raised, brown, corky common scab lesions. A previously unreported *Streptomyces* sp. is the causal agent of a new lesion type, **fissure scab**, on potato in South Africa.

7.36.2 Causal Organisms

The following *Streptomyces* species have been reported as causal agents of potato scabs:

Streptomyces acidiscabies Lambert & Loria 1989a
St. europaeiscabiei Bouchek-Mechiche et al. 2000a
St. luridiscabiei Park et al. 2003
St. niveiscabiei Park et al. 2003
St. puniciscabiei Park et al. 2003
St. reticuliscabiei Bouchek-Mechiche et al. 2000a
St. scabiei (syn. *St. scabies*) (*ex* Thaxter 1892) Lambert & Loria 1989b
St. stelliscabiei Bouchek-Mechiche et al. 2000a
St. turgidiscabies Miyajima et al. 1998

Leiminger et al. (2013) reported *St. bottropensis* causing potato common scab for the first time from Germany, and Guo et al. (2014) reported *St. galilaeus* from China. Yang et al. (2017) reported *St. diastatochromogenes* causing potato common scab in China.

St. acidiscabies and *St. reticulisiscabiei* cause acid scab and netted scab, respectively, while the remaining species cause common scab. Among these species, *St. scabiei* is the most frequently isolated species from common scab lesions and is worldwide in distribution. Unlike most other plant pathogenic procaryotes, *Streptomyces* species consist of slender, branched, prostrate mycelium containing few or no septa. The diameter of the hyphae is about 1 micrometer.

St. scabiei: The formation of grey-colored spores in spiral chains is a characteristic feature of the species. The spores are smooth, cylindrical, and measure 0.5 × 0.9–1.0 μm. About 20 or more spores are formed in a chain. Brown melanoid pigment is formed in spores on tyrosine nutrient agar and peptone iron agar. The septa formation on sporogenous hyphae starts from the tip and proceeds toward the base. As the spores mature, the septa constrict, eventually breaking the cells at the tip. On germination, the spores produce one or two germ tubes, which grow into mycelium.

The organism is tyrosinase positive and gives Gram-stain-positive reaction. It utilizes L-arabinose, D-fructose, D-glucose, D-mannitol, rhamnose, sucrose, D-xylose, and raffinose, but not xanthine as a sole source of carbon. It is sensitive to 20 μg mL^{-1} streptomycin and 0.5 μg mL^{-1} crystal violet. The lowest pH for its growth is 5.0, and it's G + C content of the DNA is 70% mol (Goto 1992).

St. acidiscabies: The spores are formed as flexuous chains consisting of about 20 or more spores in a chain. The spores are smooth, white, but reddish on certain media having high pH and measure 0.4 × 0.5–1.1 μm. The organism utilizes L-arabinose, D-fructose, D-glucose, D-mannose, D-rhamnose, D-xylose, and sucrose, but not raffinose as a sole source of carbon. The lowest pH for its growth is 4.0. It is tolerant to 20 μg mL^{-1} streptomycin and 0.5 μg mL^{-1} crystal violet. The G + C content of the DNA is 71% mol. The species is distinguished from other pathogenic *Streptomyces* species by the production of red pigment, white to reddish spore mass, and resistance to high acidic reactions.

Despite the morphological differences and genetic diversity in scab-causing species, it seems that these species share pathogenicity genes and mechanism(s) of pathogenicity. These pathogenicity genes are clustered in a large block called pathogenicity island. The pathogenicity island described in *St. turgidiscabies* is a single 425 kb restriction endonuclease fragment which contains genes thaxtomin (*txtA*, *txtB*, *txtC*, and *nos*) for synthesis of the pathogenicity determinant thaxtomin near one end and genes for pathogenicity factors tomatinase (*tomA*) and a necrosis inducing protein (*nec1*) near the other end. These three loci are also found in a similar large contiguous chromosomal region of *St. acidiscabies* and *St. scabiei* (Wanner 2006).

The pathogenicity island genes have been found to be transferred horizontally from pathogenic species to saprophytic species resulting in the emergence of new pathogenic forms. The independent emergence of *St. acidiscabies* and *St. turgidiscabies* in the USA and Japan is the result of this phenomenon (Bouchek-Mechiche et al. 2006).

Flores-Gonzalez et al. (2008) developed a PCR-based method to directly detect pathogenic *Streptomyces* causing common scab of potato using primers designed to amplify a fragment of the *txtA* and *txtB* genes, which are pathogenicity determinants in the main pathogenic *Streptomyces* species. They detected pathogenic *Streptomyces* from tuber lesions in 70 samples out of 84 naturally infected samples comprising 19 potato cultivars collected during 2000–2004 from the Netherlands, the UK, France, Germany, and Spain by PCR. The results of pathogen detection by PCR were confirmed by isolation. All the pathogenic isolates showed the basic general phenotypic traits of the *St. scabiei* phenetic cluster. RFLP analysis of amplified rRNA sequences, together with carbon sources utilization and repetitive BOX profiles, allowed most isolates to be assigned to *St. europaeiscabiei*, which emerged as the main cause of potato common scab in Western Europe. Another PCR assay (SYBR green quantitative real-time) developed by Qu et al. (2008) using primers designed from the *txtAB* (genes encoding thaxtomin synthetase) operon allows rapid, accurate, and cost effective quantification of pathogenic *Streptomyces* strains in potato tubers and in soil. The assay was specific for pathogenic *Streptomyces* strains and the detection limit was 10 fg of the target DNA, or one genome equivalent. The amount of pathogenic *Streptomyces* DNA in total DNA extracts from 1 g asymptomatic and symptomatic tubers quantified ranged from 10^1 to 10^6 pg.

Corrêa et al. (2015) showed that the PCR-RFLP of *atpD* gene technique can be used for diagnosis and identification of phytopathogenic *Streptomyces* strains, representing a rapid and inexpensive tool. Moreover, this gene can be useful for phylogenetic analysis of *Streptomyces* spp. associated with potato scab.

7.36.3 Disease Cycle

St. scabiei can survive indefinitely, as a saprophyte, in most soils except the most acidic ones. Its population in soil in the absence of hosts can be reduced, but cannot be eliminated completely as it reproduces to some extent on organic matter present in the soil. In contrast, *St. acidiscabies* is a poor soil inhabitant and does not survive as well as *St. scabiei*. *St. scabiei* can also survive on infected potato tubers and in infected plant debris.

The pathogenic species are spread through infected potato seed tubers, soil water, and soil carried by wind and implements. The penetration into the host occurs through lenticels, wounds, stomata, and directly through the cuticle in freshly harvested and young tubers. Young tubers are more susceptible to infection than older ones. Khatri et al. (2010) reported that direct inoculation of tubers with a spore suspension of *St. scabiei* resulted in disease development, demonstrating that infection could be initiated in a soilless media. Tubers were most susceptible to infection between 3 and 20 days after tuber initiation, confirming that this early period of tuber formation is critical to disease development.

The infection by *St. scabiei* usually starts with the onset of tuberization. The optimum temperature for tuber infection is 20°C–22°C, while the infection range varies from 10°C to 31°C. After penetration, the pathogen grows between or through a few layers of peridermal cells, these cells die due to the pathogenic action, and the bacterium then feeds saprophytically on them producing a metabolite. The metabolite promotes rapid division of living cells surrounding the lesion. This results in the formation of several layers of corky (suberized) cells that isolate the pathogen and surrounding tuber cells. After the first layer of suberized cells is penetrated by the bacterium, the second layer is formed, and it goes on until several such layers are formed in the growing season. As these layers are pushed outward and sloughed off, the bacterium grows and multiplies in additional dead cells, thereby leading to formation of large scab lesions. The depth of lesion is influenced by the host variety, time of infection, soil pH, soil moisture, and invasion of the scab lesion by secondary invaders including insects. Insect attack usually results in breakdown of the corky layers and allows the pathogen to invade the tuber deeply. The following factors influence the disease severity.

7.36.3.1 Soil pH

Soil pH from 5.2 to 8.0 favors the development of common scab, whereas the acid scab can also occur up to pH 4.0.

7.36.3.2 Soil Moisture

Maintaining soil moisture near field capacity (above −0.4 bars) during 2–6 weeks period following tuber initiation will reduce the incidence of the disease. Bacteria that grow at high soil moisture outgrow *St. scabiei* on the tuber surface.

7.36.3.3 Soil Type and Nutrition

Light-textured soils and those containing high organic matter are conducive to disease development.

None of the potato cultivars is immune, although a very few show highly resistant reaction. The other hosts include radish, carrot, turnip, beet, and sweet potato.

The disease cycle of *St. acidiscabies* is similar to that of *St. scabiei* except that the former can grow in the more acidic soils, i.e., up to pH 4.0.

Nahar et al. (2018) reported that the genetic background of potato cultivars influenced the abundance of pathogenic *Streptomyces* spp., with five to six times more abundant *Streptomyces* spp. in rhizosphere soil of susceptible cultivars compared with tolerant cultivars, which would result in substantially more inoculum left in the field after harvest.

The nitrated dipeptide phytotoxin, thaxtomin, inhibits cellulose biosynthesis in expanding plant tissues, stimulates Ca^{2+} spiking, and causes cell death. The thaxtomin production by *Streptomyces* species, which cause common, acid, and pitted scabs, the quantitative relationship between toxin production and virulence, and the lack of its production by saprophytic species support the hypothesis that this toxin is a pathogenicity determinant on potato tubers (Loria et al. 2006). Thaxtomin is not associated with the development of netted scab lesions, suggesting that *St. reticulsiscabiei* might possess some other pathogenicity determinants. Thaxtomin production is also not found associated with species, which cause russet scab.

Lerat et al. (2010) reported that addition of both cellobiose (0.5%) and suberin (0.1%) in a starch-containing minimal medium doubled bacterial growth and triggered thaxtomin A production, which correlated with the upregulation (up to 342-fold) of genes involved in thaxtomins synthesis. The addition of either suberin or cellobiose alone did not affect these parameters. Suberin appeared to stimulate the onset of secondary metabolism, which is a prerequisite to the production of molecules such as thaxtomin A, while cellobiose induced the biosynthesis of this secondary metabolite.

Tegg and Wilson (2010) observed no association between resistance to common scab disease and tolerance to thaxtomin A toxicity. Disease resistant cultivars "Russet Burbank" and "Atlantic" were sensitive and tolerant to thaxtomin A toxicity, respectively. Similarly, disease susceptible cultivars "Bismark" and "Tasman" showed susceptibility and tolerance to thaxtomin A. It is evident that while thaxtomin A is critical to disease expression, reaction to this toxin is only one component influencing resistance to common scab disease, and many other anatomical, physiological, or biochemical factors are critical to defense against this disease. Fyans et al. (2016) also provided evidence that phytotoxins other than thaxtomin A may also contribute to the development of common scab by *Streptomyces* spp.

Lapaz et al. (2017) analyzed the presence of *Streptomyces* pathogenicity island (PAI) genes, including genes encoding for *txtA*, *txtB* synthetase, *tomA*, and a necrosis protein (*nec1*). Among the pathogenic isolates, 50% isolates contained the four pathogenicity genes, 33% had an atypical composition of PAI marker genes, and 17% did not contain any gene. The absence of the genes reported to be involved in thaxtomin biosynthesis (*txtA*, *txtB*) was confirmed by whole-genome sequencing of two representative strains of this group. The finding suggests the participation of other virulence factors in plant pathogenicity.

Hiltunen et al. (2011) reported that 18 genotypes selected based on high sensitivity or tolerance, using shoot growth as the criterion, were multiplied *in vitro*, and evaluation of their progeny of ca. 6,500 tubers for resistance to common scab caused by *St. turgidiscabies* and *St. scabies* in a glasshouse and in three different fields showed that these genotypes differed in scab indices and disease severity ($P < 0.0001$). The relative shoot height *in vitro* (thaxtomin A used at 0.5 μg mL^{-1}) and the scab index in the field showed significant correlation ($r_s = -0.463$, $P = 0.0528$, $n = 18$), also consistent with the results obtained under controlled conditions in the glasshouse. Therefore, the *in vitro* bioassay may be used to discard scab-susceptible genotypes and elevate the overall levels of common scab resistance in the potato breeding populations.

7.36.4 Management

1. Use of resistant varieties is the most effective strategy for the management of potato scab. There is a good correlation between resistance to common and acid scab among potato cultivars. Wilson et al. (2009) have suggested a somatic cell selection approach using phytotoxin, thaxtomin A as a positive selection agent. The best disease-resistant variant, among the 13 disease-resistant variants obtained from cv. Iwa, showed on an average 85%–86% lower disease score (or 91%–92% less estimated tuber surface cover) than the parent cultivar. Wilson et al. (2010) further advocated the use of somatic cell selection with thaxtomin A to obtain stable and extremely resistant variants of potato cv. Russet Burbank to common scab
2. Maintaining soil moisture levels near field capacity by light and frequent irrigation during 2–6 weeks period following tuberization will inhibit/reduce infection
3. Severity of common scab is significantly reduced in soils by maintaining pH at 5.2 or below (by using fertilizers like ammonium sulphate which produce acidic reaction), but this measure is associated with other drawbacks. Firstly, the number of crops, which can be grown in soils having such a low pH, is greatly reduced. Secondly, plant nutrients are best available in soils having pH levels near 6.5
4. Rotation with small grain crops, corn, and alfalfa reduces the soil borne inoculum. In Punjab, India, potato rice rotation almost eliminated the disease, which at one time was very severe
5. Use of disease-free seed and treatment of infected seed will prevent the introduction of the pathogen into disease-free fields. It also reduces the disease severity in infested fields

6. Biological control with *Streptomyces* phages has shown promising results
7. Green manuring reduces soil-borne inoculum by encouraging antagonistic microorganisms
8. Tegg et al. (2008) have suggested a novel approach for controlling the disease by foliar application of 2, 4-D at 0.90 mM (0.20 g L^{-1}). 2, 4-D is not inhibitory to the growth of *St. scabiei,* but it reduces the toxicity of phytotoxin, thaxtomin A, essential for induction of common scab. The application of 2, 4-D can be incorporated into the integrated management strategy of the disease
9. *Pseudomonas fluorescens* LBUM223 controls common scab of potato, caused by *St. scabies*, through phenazine-1-carboxylic acid (PCA) production. Arseneault et al. (2013) reported that under soil conditions, PCA production by *Pseudomonas* sp. LBUM223 does not control common scab by inhibiting growth of *St. scabies* in the geocaulosphere, but instead, reduces *St. scabies* thaxtomin A production, required for pathogenesis, leading to reduced virulence of *St. scabies*. Arseneault et al. (2016) reported that *Psm. fluorescens* LBUM223 must colonize the potato geocaulosphere at high levels (10^7 bacteria g^{-1} of soil) in order to achieve biocontrol of common scab through increased PCA production
10. Tomihama et al. (2016) showed that rice bran (RB) amendment reduced potato scab by repressing the pathogenic *Streptomyces* population in young tubers. Amplicon sequencing analyses of 16S ribosomal RNA genes from the rhizosphere microbiome revealed that RB amendment dramatically changed bacterial composition and led to an increase in the relative abundance of Gram-stain-positive bacteria such as *Streptomyces* spp., and this was negatively correlated with the disease severity. Most actinomycete isolates derived from the RB-amended soil showed antagonistic activity against pathogenic *St. scabiei* and *St. turgidiscabies* on R2A medium. Some of the *Streptomyces* isolates suppressed the disease on inoculation onto potato plants in a field experiment. These results suggest that RB amendment increases the levels of antagonistic bacteria against the scab pathogens in the potato rhizosphere. Hiltunen et al. (2017) investigated the effect of repeated applications of an antagonistic *Streptomyces* strain 272 on common scab suppressiveness in a long-term field trial over 5 years. After a single application of strain 272, efficient disease suppression did not persist in the soil to the following growing season. However, when strain 272 was applied in 3 or more consecutive years, the soil remained suppressive to scab for at least 2 years beyond the last application, suggesting that, with repeated applications, it may be possible to enhance development of scab suppression in soil

REFERENCES

Arseneault, T., Goyer, C., and Filion, M. 2013. Phenazine production by *Pseudomonas* sp. LBUM223 contributes to the biological control of potato common scab. *Phytopathology* 103: 995–1000.

Arseneault, T., Goyer, C., and Filion, M. 2016. Biocontrol of potato common scab is associated with high *Pseudomonas fluorescens* LBUM223 populations and phenazine-1-carboxylic acid biosynthetic transcript accumulation in the potato geocaulosphere. *Phytopathology* 106: 963–970.

Bouchek-Mechiche, K., Gardan, L., Andrivon, D., and Normand, P. 2006. *Streptomyces turgidiscabies* and *Streptomyces reticulsiscabiei*: One genomic species, two pathogenic groups. *Int. J. Syst. Evol. Microbiol.* 56: 2771–2776.

Bouchek-Mechiche, K., Gardan, L., Normand, P., and Jouan, B. 2000a. DNA relatedness among strains of *Streptomyces* pathogenic to potato in France: Description of three new species, *S. europaeiscabiei* sp. nov. and *S. steliiscabiei* sp. nov. associated with common scab, and *S. reticulsiscabiei* sp. nov. associated with netted scab. *Int. J. Syst. Evol. Microbiol.* 50: 91–99.

Bouchek-Mechiche, K., Pasco, C., Andrivon, D., and Jouan, B. 2000b. Differences in host range, pathogenicity to potato cultivars and response to soil temperature among *Streptomyces* species causing common and netted scab in France. *Plant Pathol.* 49: 3–10.

Corrêa, D. B. A., Salomão, D., Rodrigues-Neto, J., Harakava, R., and Destéfano, S. A. L. 2015. Application of PCR-RFLP technique to species identification and phylogenetic analysis of *Streptomyces* associated with potato scab in Brazil based on partial *atpD* gene sequences. *Eur. J. Plant Pathol.* 142: 1–12.

Flores-Gonzalez, R., Velasco, I., and Montes, F. 2008. Detection and characterization of *Streptomyces* causing potato common scab in Western Europe. *Plant Pathol.* 57: 162–169.

Fyans, J. K., Bown, L., and Bignell, D. R. D. 2016. Isolation and characterization of plant-pathogenic *Streptomyces* species associated with common scab-infected potato tubers in Newfoundland. *Phytopathology* 106: 123–131.

Goto, M. 1992. *Fundamentals of Bacterial Plant Pathology*. Academic Press, San Diego, CA, 342 pp.

Gouws, R., and McLeod, A. 2012. Fissure scab, a new symptom associated with potato common scab caused by a *Streptomyces* sp. in South Africa. *Plant Dis.* 96: 1223.

Guo, F. L., Zhang, H. Y., Yu, X. M., Zhao, W. Q., Liu, D. Q., et al. 2014. First report of *Streptomyces galilaeus* associated with common scab in China. *Plant Dis.* 98: 683.

Hiltunen, L. H., Alanen, M., Laakso, I., Kangas, A., Virtanen, E., et al. 2011. Elimination of common scab sensitive progeny from a potato breeding population using thaxtomin A as a selective agent. *Plant Pathol.* 60: 426–435.

Hiltunen, L. H., Kelloniemi, J., and Valkonen, J. P. T. 2017. Repeated applications of a nonpathogenic *Streptomyces* strain enhance development of suppressiveness to potato common scab. *Plant Dis.* 101: 224–232.

Khatri, B. B., Tegg, R. S., Brown, P. H., and Wilson, C. R. 2010. Infection of potato tubers with the common scab pathogen *Streptomyces scabiei* in a soilless system. *J. Phytopathol.* 158: 453–455.

Lambert, D. H. and Loria, R. 1989a. *Streptomyces acidiscabies* sp. nov. *Int. J. Syst. Bacteriol.* 39: 393–396.

Lambert, D. H. and Loria, R. 1989b. *Streptomyces scabies* sp. nov. *Int. J. Syst. Bacteriol.* 39: 387–392.

Lapaz, M. I., Huguet-Tapia, J. C., Siri, M. I., Verdier, E., Loria, R., et al. 2017. Genotypic and phenotypic characterization of *Streptomyces* species causing potato common scab in Uruguay. *Plant Dis.* 101: 1362–1372.

Leiminger, J., Frank, M., Wenk, C., Poschenrieder, G., Kellermann, A., et al. 2013. Distribution and characterization of *Streptomyces* species causing potato common scab in Germany. *Plant Pathol.* 62: 611–623.
Lerat, S., Simao-Beaunoir, A.-M., Wu, R., Beaudoin, N., and Beaulieu, C. 2010. Involvement of the plant polymer suberin and the disaccharide cellobiose in triggering thaxtomin A biosynthesis, a phytotoxin produced by the pathogenic agent *Streptomyces scabies*. *Phytopathology* 100: 91–96.
Loria, R., Kers, J., and Joshi, M. 2006. Evolution of plant pathogenicity in *Streptomyces*. *Annu. Rev. Phytopathol.* 44: 469–487.
Miyajima, K., Tanaka, F., Takeuchi, T., and Kuninaga, S. 1998. *Streptomyces turgidiscabies* sp. nov. *Int. J. Syst. Bacteriol.* 48: 495–502.
Nahar, K., Goyer, C., Zebarth, B. J., Burton, D. L. and Whitney, S. 2018. Pathogenic *Streptomyces* spp. abundance affected by potato cultivars. *Phytopathology* 108: 1046–1055.
Park, D. H., Kim, J. S., Kwon, S. W., Wilson, C., Yu, Y. M., et al. 2003. *Streptomyces luridiscabiei*, sp. nov., *Streptomyces puniciscabiei* sp. nov. and *Streptomyces niveiscabiei* sp. nov., which cause potato common scab disease in Korea. *Int. J. Syst. Bacteriol.* 53: 2049–2054.
Qu, X., Wanner, L. A., and Christ, B. J. 2008. Using the *txtAB* operon to quantify pathogenic *Streptomyces* in potato tubers and soil. *Phytopathology* 98: 405–412.
Tegg, R. S., Gill, W. M., Thompson, H. K., Davies, N. W., Ross, J. J. et al. 2008. Auxin-induced resistance to common scab disease of potato linked to inhibition of thaxtomin A toxicity. *Plant Dis.* 92: 1321–1328.
Tegg, R. S., and Wilson, C. R. 2010. Relationship of resistance to common scab disease and tolerance to thaxtomin A toxicity within potato cultivars. *Eur. J. Plant Pathol.* 128: 143–148.
Thaxter, R. 1892. Potato scab. Annual Report of the Connecticut Agricultural Experiment Station for 1891, pp. 153–160.
Tomihama, T., Nishi, Y., Mori, K., Shirao, T., Iida, T., et al. 2016. Rice bran amendment suppresses potato common scab by increasing antagonistic bacterial community levels in the rhizosphere. *Phytopathology* 106: 719–728.
Walker, J. C. 1969. *Plant Pathology*. 3rd ed. McGraw-Hill Book Co., New York, 819 pp.
Wanner, L. A. 2006. A survey of genetic variation in *Streptomyces* isolates causing potato common scab in the United States. *Phytopathology* 96: 1363–1371.
Wilson, C. R. 2004. A summary of common scab disease of potato research from Australia. pp. 198–214 in: *Proceedings of the International Potato Scab Symposium*. S. Naito, N. Kondo, S. Akino et al. eds. Hokkaido University Sapporo, Japan.
Wilson, C. R., Luckman, G. A., Tegg, R. S., Yuan, Z. Q., Wilson. A. J., et al. 2009. Enhanced resistance to common scab of potato through somatic cell selection in cv. Iwa with the phytotoxin thaxtomin A. *Plant Pathol.* 58: 137–144.
Wilson, C. R., Tegg, R. S., Wilson, A. J., Luckman, G. A., Eyles, A., et al. 2010. Stable and extreme resistance to common scab of potato obtained through somatic cell selection. *Phytopathology* 100: 460–467.
Yang, F. Y., Yang, D. J., Zhao, W. Q., Liu, D. Q., and Yu, X. M. 2017. First report of *Streptomyces diastatochromogenes* causing potato common scab in China. *Plant Dis.* 101: 243.

7.37 RING ROT OF POTATO

The disease was first observed in Germany in 1906, where Spieckermann described it in 1913. In early 1930s, the disease spread to Canada and the United States, and by the 1940s, it was found in most major potato growing regions of North America. The rapid spread occurred mainly through the infected tubers. The disease caused heavy losses in Canada and the United States and in some instances the losses were even up to 100%. After 1940s, yield losses due to the disease have been considerably reduced by the adoption of zero tolerance in seed certification standards in most seed producing areas, including the United States, Canada, and the European Union. Many countries, including the United States have attempted unsuccessfully, the eradication of the disease. The main reason for the persistence of the disease is the unnoticed presence of the pathogen in apparently healthy tubers due to lack of any symptoms. The disease still poses a continuous threat to potato seed industry. The occurrence of the disease has been reported from Turkey also (Altundağ et al. 2009).

7.37.1 Symptoms

In continuous cool weather, the symptoms are not observed before harvest. In warm weather, the symptoms usually appear after midseason. Severely infected tubers may not sprout or shoots may be killed before emergence from soil. The shoots, which emerge, appear stunted, bear less number of leaves, and finally die.

When the disease develops late in the season, often only one or two stems in a hill show symptoms. Affected stems show stunting, and wilt during the day and recover at night. The wilting progresses slowly in the humid weather. The wilting and yellowing usually start from the lower leaves and progress upward. The interveinal areas of the leaflets turn yellowish; leaf margins become necrotic and roll upward (Figure 7.23a). Brown stripes appear on the stems, which may extend to leaf petioles. In advanced stages of disease development, longitudinally splitted stems show brown discoloration of vascular tissue. When an infected stem is cut across at the base, a milky white to creamy bacterial exudate can be seen at the cut end by squeezing the stem. The deterioration of roots, particularly young roots, also occurs. The extent and type of symptoms can vary with the environment and host cultivar. However, in some cases, there are no above ground symptoms.

The infection of tubers occurs through stolons. In tubers, sometimes the symptoms may not appear until harvest and appear later in storage. In most cases, there are no symptoms on the surface of tubers except in severely infected tubers. When an infected tuber is cut across the stem end, a creamy-yellow to light brown discoloration of the vascular ring, about ¼ inch below the skin, is seen along with bacterial ooze that may be increased by squeezing the tuber. As the disease progresses, a creamy-yellow or light brown crumbly or cheesy rot develops in the vascular ring and on squeezing the tuber, a soft pulpy exudate comes out from the diseased areas (Agrios 2004). This is the most **characteristic symptom** of the disease and the name **"ring rot"** (Figure 7.23b) has been derived from this diagnostic symptom. As the disease advances further, bacterial exudate becomes more copious and breakdown of the tissue of vascular ring occurs forming macroscopic cavities. Due to dehydration, the affected areas may become crumbly

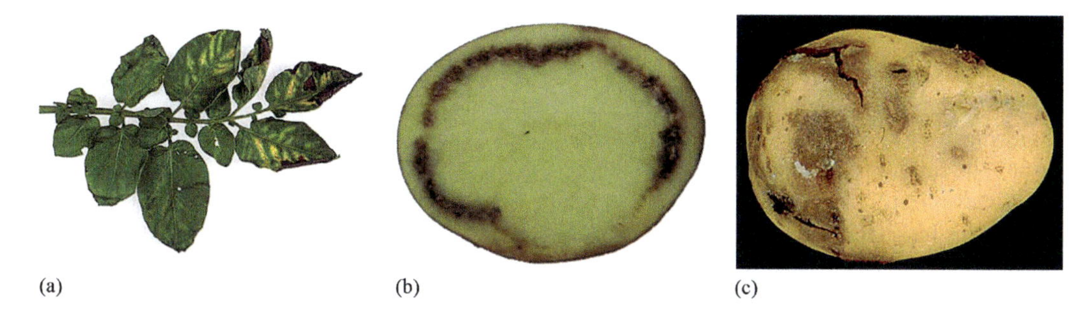

FIGURE 7.23 Ring rot of potato: (a) Symptoms on aerial plant parts. (b) Cross section of potato tuber showing vascular discoloration. (c) Discoloration and cracking of potato tuber skin. (Courtesy of Solke H. De Boer, Agriculture & Agri-Food, Ottawa, Canada.)

and powdery; or soft, mushy, and foul-smelling, if invaded by soft rot bacteria, which invariably invade. In the advanced stage, the rot involves pith of the tuber, leaving a hollow, but a firm shell of the tuber. The rot also extends to the cortex of tubers leading to formation of canker lesions on the surface of tubers and cracking of skin (Figure 7.23c). In stores, the disease continues to progresses in tubers, occasionally forming cracks on the surface and finally causing an extensive rot. Symptoms of ring rot in the vascular tissue of infected tubers are not always as evident/prominent as described above, appearing as only a broken, sporadically appearing dark line, or as a continuous yellowish discoloration. Ring rot has some overlapping symptoms with brown rot.

Diagnosis: Infected tubers may grow for several generations in the field without producing any visible symptoms, hence making the diagnosis of the disease difficult. Tomato and eggplant have been used as indicator plants for the diagnosis of the disease, as foliar symptoms on these hosts appear within 2–3 weeks in comparison to 4–6 weeks on potato plants. Earlier, starting from late 1980s, serological techniques like latex agglutination test, ELISA, and immunofluorescent antibody staining (IFAS) were routinely used for the diagnosis of ring rot. Nowadays, new techniques like PCR developed for detecting specific DNA sequences found in the chromosome or plasmid of *Clavibacter sepedonicus* are being used for this purpose. Smith et al. (2008) integrated an internal reaction control into a TaqMan PCR assay for the detection of this bacterium. The reaction control, cloned into plasmid pCmsC4, consisted of a sequence unrelated to *Cla. michiganensis* subsp. *sepedonicus* flanked by primer sequences used in the TaqMan PCR, thus eliminating the need for multiplexing. The use of the reaction control plasmid permitted the validation of negative results, and thus facilitated the use of TaqMan real-time PCR in the routine testing of diagnostic samples for *Cla. michiganensis* subsp. *sepedonicus*. This method takes care of the drawbacks like occurrence of false negative results due to reagent or thermocyler failure and amplicon carry-over into samples or reagents leading to false positive results, associated with some of the earlier described methods. Another real-time PCR assay developed by Gudmestad et al. (2009) uses the cellulase A gene sequence as the basis for primer design. Cellulase A primers were specific to *Cla. michiganensis* subsp. *sepedonicus* and did not react with any other coryneform bacteria or bacteria pathogenic to potato. The assay was more sensitive than Cms50/72A PCR primers, immunofluorescence antibody, and enzyme-linked immunosorbent assays for the detection of the bacterium from infected and symptomless potato tubers. The assay proved to be a very robust detection tool with the added advantage of detecting only virulent strain of the bacterium.

7.37.2 Causal Organism

Clavibacter sepedonicus (Spieckermann and Kotthoff 1914) Li et al. 2018

Syn. *Cla. michiganensis* subsp. *sepedonicus*, *Corynebacterium michiganense* subsp. *sepedonicum*, *Cor. michiganense* pv. *sepedonicum*, *Cor. sepedonicum*

For many years, the bacterium was classified in the genus *Corynebacterium*. However, due to considerable differences in cell wall and DNA base composition, it was placed in a newly created genus, *Clavibacter*. The below given description of the bacterium has been taken mainly from Davis et al. (1984).

As the bacterium grows slowly in culture media, its isolation is difficult from the infected plant material. It is rarely isolated when the samples contain less than 10^5 cells/cm^2 or a gram of host tissue. The colonies on nutrient agar appear within 3–6 days and are white to slightly yellow. Cells are Gram-stain-positive, club-shaped rods that vary from small coccoids through rod shapes and may appear as snapping, V-shaped cells. Cells are non-motile, non-sporing, obligate aerobes and measure 0.4–0.6 × 0.8–1.2 µm. Optimal temperature for growth is 18°C–21°C, while the maximum is 30°C. Starch hydrolysis is positive, but neither galatin liquefaction nor production of pectinolytic enzymes occurs. Acid is produced from mannitol, but not from lactose, trehalose, inositol, and mannitol. Pathogenic strains vary in virulence and production of extracellular slime. Both mucoid and non-mucoid colony types have been isolated from potato.

In nature, the bacterium infects only potato, although it can colonize roots of sugar beet (Rowe et al. 1995).

7.37.3 Disease Cycle

Infected tubers are the main source of perpetuation of the bacterium. The bacterium also survives in the infected plant refuse and volunteer plants growing from infected tubers left in the soil from the previous season. The survival also occurs for 10–24 months in dried bacterial slime spread from infected tubers and stems on crates, baskets, storage bins, sacks, and warehouses, and also on the surface of cultivating, harvesting, and grading equipment. The survival is longest under cool and dry conditions.

The dissemination of the bacterium occurs through cutting knives, planters, particularly picker-type planters, chewing insects like Colorado potato beetles and flea beetles, and irrigation water. A knife that cuts an infected potato tuber can spread the bacterium to the subsequent 20–100 healthy seed pieces. Secondary spread occurs by the bacteria released from the wounds caused by chewing insects and cultural practices. Secondary spread is of much less significance in comparison to primary spread from infected tubers. The penetration of the bacterium into the host occurs only through wounds. The bacterium can remain in the host for a long time without producing visible symptoms. A symptomless tuber carrying the bacterium internally may produce symptomless plants and symptomless tubers, but actually all the three stages are infected. This is particularly true when the plants grow under continuous cool conditions. Favorable soil temperature for disease development is 19°C–28°C. In plants growing from infected tubers, the bacteria move slowly through the vascular tissue and reach all the lower portions of the growing plant. Bacteria also move out of the vessels into the surrounding parenchyma tissues, where they multiply and cause their breakdown forming cavities. From the parenchymatous tissue, the bacteria again move to the newly formed xylem vessels. Late in the season, the bacteria migrate from the stem to the newly formed tubers via stolons, and these infected tubers serve as a source of inoculum for the next crop.

Steinmöller et al. (2013) examined the risk of dissemination of *Cla. michiganensis* subsp. *sepedonicus* through potato residues from processing industries. The residues were categorized into different risk categories ranging from category 0 (no risk of dissemination) to category 4 (high risk of dissemination). The residues not heated during processing and used in agriculture, e.g., as fertilizer, were pooled in the highest risk category 4. The residues that were sanitized before use in agriculture by composting or pasteurization were still classified as probably high risk (risk category 3) because no information was available concerning the inactivation of the pathogen by these treatments. The samples contaminated with *Cla. michiganensis* subsp. *sepedonicus* were composted for 6 days at 70°C and for 13 days at 55°C, and pasteurized for 90 min at 70°C. The viable bacteria detected by bioassay on eggplants, and cultivation on semi-selective media from plant sap forming characteristic colonies, were found in all the treatments. It is evident that these sanitation treatments are not effective to inactivate the bacterium, and the above given classification of treated residues in category 3 (probably high risk) is thus confirmed.

7.37.4 Management

The management of the disease is primarily through the exclusion and eradication of the pathogen. The following measures are effective in controlling or minimizing the disease.

1. Use of certified pathogen-free seed. In the US and Canada, certified seed has a zero tolerance for ring rot
2. Destruction of infected plant debris and volunteer plants
3. Follow crop rotation with non-host plants for 3 years
4. If ring rot is present on a farm, a thorough clean up must be done. Dispose of all infected potato tubers away from potato production areas. Clean all surfaces of warehouses and farming equipment by removing all dirt, debris, trash, and plant refuse and then wash with a strong detergent in hot water applied by a high pressure washer. After cleaning, disinfect all warehouses and equipment with a disinfectant (1,000–2,000 μg mL^{-1} chlorine solution, 5% Lysol solution, or 2 pounds of copper sulphate dissolved in 10 gallons of water). For proper disinfection, the disinfectant must be present for at least 10 min, preferably 20–30 min, on any surface being treated. Disinfect also crates, bins, bags, etc.
5. While cutting seed tubers, frequently disinfect the cutting knives with a disinfectant or use rotary knives, which continuously run through a disinfectant solution, which should be frequently renewed
6. Use assist- or cup-feed planters instead of picker-type planters
7. Control insects such as Colorado potato beetles, plant bugs, aphids, and leafhoppers, which may transmit the bacterium
8. As the highly resistant cultivars also support sizable populations of the bacterium and act as potentially silent carriers of the disease, the use of resistant cultivars is not recommended (Ishimaru 2001)

REFERENCES

Agrios, G. N. 2004. *Plant Pathology*. 5th ed. Elsevier Academic Press, San Diego, CA, 922 pp.

Altundağ, Ş., Karahan, A., Kilinc, A. O., and Özakman, M. 2009. First report of *Clavibacter michiganensis* subsp. *sepedonicus* causing bacterial ring rot of potato in Turkey. *Plant Pathol.* 58: 794.

Davis, M. J., Gillaspie, A. G. Jr., Vidaver, A. K., and Harris, R. W. 1984. *Clavibacter*: A new genus containing some phytopathogenic coryneform bacteria, including *Clavibacter xyli* subsp. *xyli* sp. nov., subsp. nov. and *Clavibacter xyli* subsp. *cynodontis* subsp. nov., pathogens that cause ratoon stunting disease of sugarcane and Bermudagrass stunting disease. *Int. J. Syst. Bacteriol.* 34: 107–117.

Gudmestad, N. C., Mallik, I., Pasche, J. S., Anderson, N. R., and Kinzer, K. 2009. A real-time PCR assay for the detection of *Clavibacter michiganensis* subsp. *sepedonicus* based on the cellulase A gene sequence. *Plant Dis.* 93: 649–659.

Ishimaru, C. 2001. Ring rot of potato. pp. 863–64 in: *Encyclopedia of Plant Pathology*. Vol. 2, O. C. Maloy, and T. D. Murray eds. John Wiley & Sons, New York.
Li, X., Tambong, J., Yuan, K. (X.), Chen, W., Xu, H., et al. 2018. Reclassification of *Clavibacter michiganensis* subspecies on the basis of whole-genome and multi-locus sequence analyses. *Int. J. Syst. Evol. Microbiol.* 68: 234–240.
Rowe, R. C., Miller, S. A., and Riedel, R. M. 1995. Bacterial ring rot of potatoes. Extension Fact Sheet, Plant Pathology. Ohio State University USA.
Smith, D. S., De Boer, S. H., and Gourley, J. 2008. An internal reaction control for routine detection of *Clavibacter michiganensis* subsp. *sepedonicus* using a real-time TaqMan PCR-based assay. *Plant Dis.* 92: 684–693.
Spieckermann, A., and Kotthoff, P. 1914. Untersuchungen über die Kartoffelpflanze und ihre Krankheiten. I. Die Bakterienringfäule der Kartoffelpflanze. *Landwirtschaftliche Jahrbücher, Berlin* 46, 659–732.
Steinmöller, S., Möller, P., Bandte, M., and Büttner, C. 2013. Risk of dissemination of *Clavibacter michiganensis* ssp. *sepedonicus* with potato waste. *Eur. J. Plant Pathol.* 137: 573–584.

7.38 BLACK ROT OF CRUCIFERS

Black rot (bacterial wilt) is one of the most destructive diseases of cruciferous crops. The disease was first reported by Garman in 1891 from Lexington Kentucky, USA on cabbage, and Pammel first described the pathogen from rutabaga in Iowa in 1895. The pathogen was one of the first bacterium shown to be seed-borne by Harding et al. (1904). According to Williams (1980), it is the most important worldwide disease of crucifers, attacking all cultivated brassicas, radishes, and numerous cruciferous weeds. It affects many cultivated and wild crucifers. Cabbage, cauliflower, and kale are the most susceptible hosts while broccoli, mustards, rape, rutabaga, and turnip are susceptible hosts. Radish, *gobhi sarson* (canola), and Chinese *sarson* are also infected. In warm and wet climate, the losses due to black rot may exceed 50% on some crops. It is particularly more severe in developing countries, in the tropical and sub tropical regions where small-scale and uneducated farmers often use their self-produced seed to raise the transplants (Schaad 2001).

The disease is worldwide in occurrence where cabbage is grown. In India, it was first reported by Patwardhan in 1928. According to Patel and Kulkarni (1953), the pathogen was not only introduced in India, but it was repeatedly brought with the imported seed. Several epiphytotics of the disease on cabbage and cauliflower have been reported from West Bengal, Rajasthan, and Himachal Pradesh. In 1969, National Seeds Corporation of India suffered an estimated loss of Rs. 5,00,000.00 from a 10 hectare seed production plot of cauliflower laid in Himachal Pradesh, due to combined infection of black rot and stump rot (*Erwinia* sp.) (Trivedi and Patel 1972).

7.38.1 Symptoms

The most characteristic symptom of the disease is the formation of V-shaped chlorotic lesions, with the base of V toward midrib, on the margins of leaves developing from hydathodal infections. Lesions enlarge and spread toward midrib of the leaf and the affected areas later turn brown and dry. In the chlorotic areas, veins and veinlets turn black, and this network of black venation can be seen more clearly when the infected leaf is held against light (Figure 7.24a). The bacteria spread through veins to midrib of the leaf from where it moves to stem through petiole. From the stem, the bacteria move upward and downward to other leaves and roots. The chlorotic lesions may appear anywhere on the leaf lamina from the systemic infection moving from stem to midrib of the leaves. Infected leaves show stunted growth and may fall off prematurely. The cross sections of infected stems and petioles show black rings due to the infection of water-conducting vessels. The chlorotic lesions also appear from invasions occurring through wounds caused by insects or other means. Under ideal conditions, the symptoms appear 10–14 days after entry of the bacterium.

Cabbage and cauliflower heads are also invaded and become discolored. Fleshy tissue of radish, turnip, knolkhol, and rutabaga infected by black rot show vascular discoloration and later internal break down. Such fleshy tissues are often invaded by soft rot bacteria and undergo rotting emitting a foul smell.

On seedlings, developing from infected seeds or those developing from healthy seeds, but infected by bacteria present in plant refuse or soil, the symptoms appear as black lesions on the margins of the cotyledons. The lesions enlarge and

(a)

(b)

FIGURE 7.24 Black rot of cabbage: (a) V-shaped lesions on field plants. (Courtesy of Margaret T. McGrath, Cornell University, Ithaca, NY.) (b) V-shaped lesions on transplants. (Courtesy of Sally Miller, www.ohioonline.osu.edu.)

coalesce and the cotyledons turn yellow to black. The infected cotyledons shrivel, hang down, and drop prematurely, but before this the bacteria move into stem and young leaves. Affected seedlings wilt, collapse, and die (Figure 7.24b).

In addition to *Xanthomonas campestris* pv. *campestris*, four other pathovars of *Xa. campestris*, namely, *armoraciae*, *aberrans*, *barbareae*, and *incanae* have been reported to cause non-vascular leaf spots (some also cause sunken lesions on stem and petioles) on many cruciferous and some non-cruciferous plants. McCulloch (1929) was the first to report *Xa. campestris* pv. *armoraciae* (syn. *Bacterium campestris* var. *armoraciae*) as the causal agent of non-vascular leaf spots on horseradish (*Armoracia rusticana*), but weakly pathogenic on cabbage and cauliflower (causing only leaf spots, but no lesions on stems and petioles), and non-pathogenic on kale, mustarad, and radish. White (1930) reported for the first time *Xa. campestris* pv. *raphani* (syn. *Bac. vesicatorium* var. *raphani*) as the cause of bacterial leaf spot of radish, turnip, cabbage, cauliflower, kale, Brussels sprout, and mustard, but non-pathogenic to horseradish. Although these pathovars generally cause non-vascular infections, vascular infection by *Xa. campestris* pv. *aberrans* (Vicente et al. 2006) and by *Xa. campestris* pv. *armoraciae* and *Xa. campestris* pv. *incanae* (Vicente et al. 2001) have also been reported.

7.38.2 Causal Organism

Xanthomonas campestris pv. *campestris* (Pammel 1895) Dowson 1939
Syn. *Xa. campestris*

Cells are small, Gram-stain-negative, non-sporing rods having a single polar flagellum. It is catalase and hydrogen sulphide positive, but negative for oxidase, nitrate reduction, and indole production. The colonies on nutrient agar and yeast dextrose carbonate agar are yellow, convex, mucoid, and shining. It produces a yellow extracellular polysaccharide called xanthan on media containing glucose. The bacterium hydrolyzes starch, and this character is used for its easy recognition on Schaad's selective medium. Acid is produced within 21 days in Dye's medium C from arabinose, mannose, galactose, trehalose, cellobiose, and fructose.

It prefers neutral or alkaline medium, and its growth is inhibited or suppressed by acids. The optimum growth occurs at 30°C, while lethal temperature is around 50°C (Onsando 1992). Complete genome sequencing has been done for two strains, namely, ATCC 33913 (CFBP 5241) and 8004 (CFBP 6650) of the bacterium.

Lugo et al. (2013) evaluated 56 native isolates of *Xa. campestris* pv. *campestris* collected from 12 farming districts of Trinidad along with seven reference strains for resistance to copper in buffered (pH 7.0) and unbuffered (pH 5.6) nutrient agar media. Thirty-four and 33 native isolates were highly resistant to copper (growth on $\geq$200 μg mL^{-1} copper) in buffered and unbuffered media, respectively; whereas all the reference strains were highly susceptible to copper. The mean minimum inhibition concentration for the 56 native isolates was 224.6 μg mL^{-1} copper indicating that high levels of copper resistance are present in *Xa. campestris* pv. *campestris* in Trinidad. There was a strong association between length of time of continuous applications of copper formulations and proportion of the native *Xa. campestris* pv. *campestris* with resistance to copper (Pearson's r = 0.96; Spearman's r = 0.93). However, there was no association between resistance to copper and aggressiveness of the pathogen at 10 days after inoculation.

The bacterium is composed of genetical and serological heterogeneous groups of strains. The variability in *Xa. campestris* pv. *campestris* continuously endangers cultivars with a narrow genetic base for resistance.

Vicente et al. (2001), on the basis of reaction of the pathogen to cultivars of *Brassica oleracea*, *B. rapa*, *B. napus*, *B. carinata*, and *B. juncea*, identified six races of the pathogen. Races 1 and 4 are predominant on *B. oleracea* crops. Based on four avirulence genes of *Xa. campestris* pv. *campestris* and four matching *Brassica* resistance genes, they also proposed a gene for gene hypothesis to explain the interaction of races and differential cultivars. Fargier and Manceau (2007) described three new races, hence making the total number of races to nine. However, they added no new avirulence gene in bacterial genotypes and no new resistance gene in plant genotypes to those proposed in the gene-for-gene model of Vicente et al. (2001). Mulema et al. (2012) race-typed 60 of the 250 isolates of *Xa. campestris* pv. *campestris* obtained from black rot samples collected in 2009 mainly from *B. oleracea* crops (broccoli, cabbage, cauliflower, and kales), and using a differential set of *Brassica* spp., found only two races, race 1 occurring in Kenya and Tanzania and race 4 occurring in Kenya, Tanzania, and Uganda. Another race, race 5 was observed from one isolate obtained from a *B. rapa* sample collected from Tanzania in 2003. Genomic fingerprinting with repetitive-PCR revealed clusters that did not show significant correlations between isolates and geographical location, isolates and host adaptation, or isolates and race. However, it did demonstrate existence of genetic differences within the Kenya, Tanzania, and Uganda *Xa. campestris* pv. *campestris* population indicating that it is not a similar clonal population of the same genetic background.

Chidamba and Bezuidenhout (2012) determined genetic diversity in *Xa. campestris* pv. *campestris* using 28 isolates obtained from cabbage samples collected from Gauteng, Mpumalanga, and North West Provinces of South Africa. Four differential *Brassica* cultivars with known avirulence genes were used for race typing using spray inoculation method. Four races, namely, 1, 3, 4, and 6 were identified. Of the 28 isolates, four were identified as race 1, two as race 3, 19 as race 4, and three as race 6. Repetitive DNA polymerase chain reaction-based fingerprinting using ERIC- and BOX-primers was used to assess the genetic diversity. Generated fingerprints of the bacterium were relatively similar. As the cluster analysis did not strictly group isolates based on their geographical origin, it suggests a limited diversity in *Xa. campestris* pv. *campestris* strains occurring in cabbage producing regions of South Africa.

Lema et al. (2012) identified seven races of *Xa. campestris* pv. *campestris* based on race-typing tests of 75 isolates of the bacterium obtained from infected samples collected from fields of northwestern Spain. The identified races were 1, 4, 5, 6, 7, 8, and 9; race 4 was the most frequent in *B. oleracea* and race 6 in *B. rapa* crops. Cluster analysis derived from the combined fingerprints showed four groups, but no clear relationship to race, crop, or geographical origin was found. Rep-PCR analysis was not found a reliable method for the differentiation of these races. Hence, the race typing of this pathogen should be done by using the differentials of *Brassica* spp. or another alternative approach.

Singh et al. (2016) obtained 217 isolates of *Xa. campestris* pv. *campestris* from 12 different black rot-infected crucifer crops from 19 states of India and using a set of seven differential crucifer hosts, identified races 1, 4, and 6. Among these races, race 1 followed by race 4 dominated most of the states of India.

Cruz et al. (2017) reported the occurrence of races 4, 6, and 7 and also described two new races, i.e., race 10 and 11 from Portugal.

7.38.3 Disease Cycle

The bacterium survives in infected seed (internally and externally), plant refuse, and numerous weeds. Infected cruciferous weeds growing nearby the crop are an important source of carryover and infection of the subsequent crops. Most of these weeds do not show typical disease symptoms, and many show no symptom at all under most favorable conditions for disease development. If the weed population around the seed production fields becomes infected, it becomes extremely difficult to produce the pathogen-free seed.

Silva et al. (2017) studied the survival of *Xa. campestris* pv. *campestris* in phyllosphere and rhizosphere of 26 weed species belonging to 14 families between August 2014 and October 2015 in Brazil. *Lepidium virginicum* and *Raphanus raphanistrum* (*Brassicaceae*) demonstrated great potential for the survival of *Xa. campestris* pv. *campestris* in the phyllosphere, and the bacterium was isolated after 56 and 70 days, respectively. Low variation between maximum and minimum temperatures, high rainfall, and high relative humidity at specific times of the year contributed to longer survival periods in the phyllosphere of some weed species. The bacterium survived in the rhizosphere of *R. raphanistrum* only, from where it was isolated up to 28 days. No relation was found between climatic factors and survival in the rhizosphere.

Krauthausen et al. (2018) reported that *Xanthomonas* isolates from shepherd's purse (*Capsella bursa-pastoris*) may not constitute a serious risk to serve as primary inoculum for black rot epidemics in Brassicas. However, *Xanthomonas* isolates from wallflower (*Erysimum cheiri*) have this potential. Most of the isolates from cruciferous weeds and the wallflower isolates do not fit into the actual classification of *Xa. campestris* pathovars.

The local spread of the pathogen is mainly through wind-splashed rain, while the distant spread is through infected seed and diseased nursery plants. As low as 0.03% infected seed can result in the development of an epiphytotic. The other means of dissemination include wind-blown detached leaves, dust particles, insects, agricultural equipment, and workers. The bacterium penetrates mainly through hydathodes and wounds caused by insects and cultural operations. The stomatal penetration occurs in cotyledons or very young leaves. In developed leaves, the penetration through stomata occurs in highly susceptible crops like cauliflower, in water-soaked tissue, or when inoculum is applied with a suitable detergent.

The optimum temperature for disease development is 25°C–28°C. The bacteria can multiply at 16°C–18°C, but the progress of disease is very slow and that too with no visible symptoms. Free moisture in the form of rain or dew on plant surface is required for infection and disease development. Under favorable conditions, the symptoms may appear 8–10 days after infection. Generally, the epiphytotics occur due to favorable weather conditions, but occasionally these can be traced to an infected seed lot or infected nursery bed.

After penetration through stomata and hydathodes, the bacteria multiply and progress intercellularly until they reach and invade the spiral vessels. The bacteria multiply in the vascular tissue and produce a mucilaginous extracellular polysaccharide, xanthan. The multiplication of bacterial cells and accumulation of xanthan cause plugging of xylem vessels impeding the normal flow of water and nutrients. The bacteria then spread in the vessels throughout the plant reaching the seed via xylem of funiculus. In the infected seeds, the bacteria remain protected below the seed coat. External seed infestation occurs either through local lesions on siliques or during threshing. Infected xylem vessels disintegrate at places and the bacteria move to the surrounding parenchyma cells, killing and disintegrating these cells and forming cavities. In wet weather, the bacteria often ooze to the surface of leaves through hydathodes and wounds from where they are disseminated by rain splashes, wind, and insects. Unfavorable weather conditions and the absence of disseminating agents may result into disappearance of the visible symptoms. In these plants, the bacteria progress slowly within the vascular tissue, and the visible symptoms appear on the return of favorable weather conditions.

Nuñez et al. (2017) developed a diagrammatic scale, with eight levels of severity, ranging from 0.19% to 48.8%, to evaluate black rot severity on kale (*B. oleraceae* var. *acephala*) leaves. More than 95% of the leaves assessed showed severity levels ranging from 0.1% to 21%, and 5% of the leaves showed severities higher than 22%. The validation of the scale was performed by ten inexperienced evaluators, and the data were analyzed with two methods: linear regression and Lin's statistics. In the absence of scale, most evaluators overestimated disease severity, but the use of the scale resulted in increased precision, accuracy, repeatability, and reproducibility of the estimates according to both validation methods. The proposed diagrammatic scale proved useful for assessments of black rot severity in kale leaves, and it may be of interest to researchers performing studies on epidemiology or breeding for resistance.

Schaad and Kendirck (1975) developed a laboratory technique capable of detecting low number of *Xa. campestris* pv. *campestris* cells in crucifer seeds. The method was capable of detecting a single diseased seed per 1,000 inoculated seeds and was superior to several other methods tested. Schaad (1978) also demonstrated the use of immunofluorescence for rapid identification of the bacterium in crucifer seeds. The indirect immunofluorescence test (using antiserum to ribosomes) was better than the direct test for quick identification of the bacterium. Zaccardelli et al. (2007) also developed a method for detection and identification of the bacterium from crucifer seeds and from naturally infected and inoculated plants. The method is based on PCR amplification of fragments from the pathogenicity-associated gene *hrcC* (a PCR product of 519 bp). The method was very specific for *Xa. campestris* pv. *campestris* because no amplicons were obtained from 39 pure cultures of phytopathogenic bacteria including four pathovars of *Xa. campestris*, two pathovars of *Xa. axonopodis,* and species of *Pseudomonas* and *Pectobacterium*. Moreover, PCR reactions were also negative for 50 unidentified environmental isolates purified from the surface of crucifers.

Van Der Wolf and Van Der Zouwen (2010) reported that inoculation of cauliflower blossom with *Xa. campestris* pv. *campestris*, by pollination with blue bottle flies (*Calliphora vomitoria*) as a vector, can result in seed infestation. In 2 years of poly-tunnel experiments, fly inoculation of cauliflower resulted in high densities of *Xa. campestris* pv. *campestris* (>10,000 cfu g^{-1}) in approximately 30% of seed batches of a breeding line, both before and after seed treatment with warm water. The presence of the bacterium in seed derived from fly-inoculated plants was confirmed in a grow-out test. In a 1-year trial, no seed infection was found in warm water-treated seed derived from cv. Opaal, an open pollinated cultivar. No indication of internal seed infection was found after blossom inoculation with *Xa. campestris* pv. *armoraciae.*

Van der Wolf et al. (2013) spray-inoculated cauliflower plants with *Xa. campestris* pv. *campestris* at the 8-leaf stage or during flowering. The examination of seed samples by dilution plating for deep-seated infection following hot water treatment detected *Xa. campestris* pv. *campestris* in 61% of the 23 seed samples harvested from plants inoculated during flowering. However, symptom development in seedlings raised from the seeds could not be confirmed in a grow-out test under favorable conditions for infection. The bacterium was not detected in 59 seed samples harvested from leaf-inoculated plants with the exception of one sample from plants inoculated at peduncle formation. In another experiment, the spray-inoculation of flowers resulted in infection of 52% flowers leading to 0.18% infected seeds, and 1%–10% of the infected siliques contained infected seeds.

Laala et al. (2015) devised a technique named seed-qPCR for the detection of living bacterial cells of *Xa. campestris* pv. *campestris* in seed lots. The technique is based on an enrichment of bacterial population associated with infected seeds by seed germination coupled with real-time Taq-man PCR after extraction of the target DNA. The technique is inexpensive and can detect even one contaminated seed among 10,000 healthy seeds. The seed-qPCR method combines an efficient extraction based on bacterial multiplication on seedlings with a sensitive technique qPCR for the detection of bacteria in seed lots.

7.38.4 Management

Management of the disease is based mainly on the eradicative and preventive measures given below:

1. Use certified pathogen-free seed
2. Hot water treatment (50°C for 30 min) of all crucifer seeds is essential. However, Carisse et al. (1999) have recommended treatment of cabbage, broccoli, and brussels sprouts seed at 50°C for 25 min and those of cauliflower, turnip, rutabaga, and kale for 15 min. Immediately after the treatment, wash the seed in cold water and allow it to dry. Proper hot water treatment also helps to eliminate seed-borne pathogens of blackleg, anthracnose, downy mildew, Alternaria leaf spot, and Fusarium yellows. Another recommendation is soaking of seed in 0.5% cupric acetate dissolved in 0.005 N acetic acid at 40°C for 20 min (Schaad et al. 1980) or in 10% household bleach at 50°C for 20 min
3. Nursery beds should be at least ¼ mile away from the transplanted fields
4. Three year rotation should be practiced both for nursery beds and transplanted fields
5. Avoid sprinkler irrigation or use it in moderation
6. Agricultural equipment and tools should be disinfected with bleach after use
7. Destroy the plant refuse by burning or by deeply burying in the soil immediately after harvest. High soil moisture hastens the decomposition of plant refuse
8. Control the insects, which spread the pathogen and destroy the cruciferous weeds, which serve as source of inoculum for the bacterium
9. Maintain balanced soil fertility in both seedbed and main field
10. Spray the crop with copper fungicides at 10-day intervals
11. Use of biocontrol agents: Wulff et al. (2002) have shown the possibility of managing the disease with an endophytic strain of *Bacillus subtilis* (strain BB). The control was better in dry and short rainy seasons (in comparison to main rainy season) and in clay loam soil (in contrast to sandy loam soil). Liu et al. (2015) tested individual plant growth-promoting rhizobacterial (PGPR) strains in greenhouse assays, and mixtures of PGPR strains in greenhouse and field tests for the control of black rot in Chinese cabbage. In the field test, all the treatments significantly reduced the disease incidence at 3 weeks after transplanting, the head disease severity at the harvest time, and increased the marketable yield compared to the non-bacterized control
12. Grow resistant varieties. Two cabbage cultivars, namely, Early Fuji and Badger are resistant, while cultivar, White rock of cauliflower is tolerant to the pathogen

REFERENCES

Carisse, O., Wellman-Desbiens, E., Toussaint, V., and Otis, T. 1999. *Preventing Black Rot*. Horticultural Research and Development Centre, Quebec, Canada.

Chidamba, L., and Bezuidenhout, C. C. 2012. Characterisation of *Xanthomonas campestris* pv. *campestris* isolates from South Africa using genomic DNA fingerprinting and pathogenicity tests. *Eur. J. Plant Pathol.* 133: 811–818.

Cruz, J., Tenreiro, R., and Cruz, L. 2017. Assessment of diversity of *Xanthomonas campestris* pathovars affecting cruciferous plants in Portugal and disclosure of two novel *X. campestris* pv. *campestris* races. *J. Plant Pathol.* 99: 403–414.

Dowson, W. J. 1939. On the systematic position and generic names of the Gram negative bacterial plant pathogens. *Zentralblatt für Bakteriologie, Parasitekunde und Infektionskrankheiten 2* 100, 177–193.

Fargier, E., and Manceau, C. 2007. Pathogenicity assays restrict the species *Xanthomonas campestris* into three pathovars and reveal nine races within *X. campestris* pv. *campestris*. *Plant Pathol.* 56: 805–818.

Harding, H. A., Stewart, F. C., and Prucha, M. J. 1904. Vitality of the cabbage black rot germ on cabbage seed. *N. Y. Agr. Exp. Sta. Bull.* 251: 177–194.

Krauthausen, H.-J., Hörner, G., Zimmermann, S., Voegele, R. T., and Brändle, F. 2018. Competence of *Xanthomonas campestris* from cruciferous weeds and wallflower (*Erysimum cheiri*) to induce black rot in cabbage. *Eur. J. Plant Pathol.* 151: 275–289.

Laala, S., Bouznad, Z., and Manceau, C. 2015. Development of a new technique to detect living cells of *Xanthomonas campestris* pv. *campestris* in crucifers seeds: The seed-qPCR. *Eur. J. Plant Pathol.* 141: 637–646.

Lema, M., Cartea, M. E., Sotelo, T., Velasco, P., and Soengas, P. 2012. Discrimination of *Xanthomonas campestris* pv. *campestris* races among strains from northwestern Spain by *Brassica* spp. genotypes and rep-PCR. *Eur. J. Pl. Pathol.* 133: 159–169.

Liu, K., Garrett, C., Kloepper, J. W., and McInroy, J. A. 2015. Induction of systemic resistance in Chinese cabbage against black rot by plant growth-promoting rhizobacteria. *Phytopathology* 105: S2.7.

Lugo, A. J., Elibox, W., Jones, J. B., and Ramsubhag, A. 2013. Copper resistance in *Xanthomonas campestris* pv. *campestris* affecting crucifers in Trinidad. *Eur. J. Plant Pathol.* 136: 61–70.

McCulloch, L. 1929. A bacterial leaf spot of horseradish caused by *Bacterium campestris* var. *armoraciae*, n. var. *J. Agric. Res.* 38: 269–287.

Mulema, J. M. K., Vicente, J. G., Pink, D. A. C., Jackson. A., Chacha, D. O., et al. 2012. Characterization of isolates that cause black rot of crucifers in East Africa. *Eur. J. Plant Pathol.* 133: 427–438.

Nuñez, A. M. P., Monteiro, F. P., Pacheco, L. P., Rodríguez, G. A. A., Carvalho, C., et al. 2017. Development and validation of a diagrammatic scale to assess the severity of black rot of crucifers in kale. *J. Phytopathol.* 165: 195–203.

Onsando, J. M. 1992. Black rot of crucifers. pp. 243–252 in: *Plant Diseases of International Importance*. Vol. II, *Diseases of Vegetables and Oilseed Crops*. H. S. Chaube, U. S. Singh, A. N. Mukhopadhyay, and J. Kumar eds. Prentice Hall, Englewood Cliffs, NJ.

Pammel, L. H. 1895. Bacteriosis of rutabaga. (*Bacillus campestris* n. sp.). *Bull. Iowa State College Agri. Expt. Sta.* 27: 130–134.

Patel, M. K., and Kulkarni, Y. S. 1953. A review of bacterial plant disease investigation of India. *Indian Phytopathol.* 6: 131–40.

Schaad, N. W. 1978. Use of direct and indirect immunofluorescence tests for identification of *Xanthomonas campestris*. *Phytopathology* 68: 249–252.

Schaad, N. W. 2001. Black rot of crucifers. pp. 145–46 in: *Encyclopedia of Plant Pathology*. Vol. 1, O. C. Maloy, and T. D. Murray eds. John Wiley & Sons, New York.

Schaad, N. W., Gabrielson, R. L., and Mulanax, M. W. 1980. Hot acidified cupric acetate soaks for eradication of *Xanthomonas campestris* from crucifer seeds. *Appl. Erviron. Microbiol.* 39: 803–807.

Schaad, N. W., and Kendrick, R. 1975. A qualitative method for detecting *Xanthomonas campestris* in crucifer seed. *Phytopathology* 65: 1034.

Silva, J. C., Silva Júnior, T. A. F., Soman, J. M., Tomasini, T. D., Sartori, M. M. P., et al. 2017. Survival of *Xanthomonas campestris* pv. *campestris* in the phyllosphere and rhizosphere of weeds. *Plant Pathol.* 66: 1517–1526.

Singh, D., Rathaur, P. S., and Vicente, J. G. 2016. Characterization, genetic diversity and distribution of *Xanthomonas campestris* pv. *campestris* races causing black rot disease in cruciferous crops of India. *Plant Pathol.* 65: 1411–1418.

Trivedi, B. M., and Patel, P. N. 1972. Black rot of crucifers. pp. 122–33 in: *Plant Bacteriology*. Vol. I, *Bacterial Diseases of Plants in India*. P. N. Patel ed. Indian Agricultural Research Institute, New Delhi, India.

Van Der Wolf, J. M., and Van Der Zouwen, P. S. 2010. Colonization of cauliflower blossom (*Brassica oleracea*) by *Xanthomonas campestris* pv. *campestris*, via flies (*Calliphora vomitoria*) can result in seed infestation. *J. Phytopathol.* 158: 726–732.

van der Wolf , J. M., van der Zouwen, P. S., and van der Heijden, L. 2013. Flower infection of *Brassica oleracea* with *Xanthomonas campestris* pv.*campestris* results in high levels of seed infection. *Eur. J. Plant Pathol.* 136: 103–111.

Vicente, J. G., Conway, J., Roberts, S. J., and Taylor, J. D. 2001. Identification and origin of *Xanthomonas campestris* pv. *campestris* races and related pathovars. *Phytopathology* 91: 492–499.

Vicente, J. G., Everett, B., and Roberts, S. J. 2006. Identification of isolates that cause a leaf spot disease of brassicas as *Xanthomonas campestris* pv. *raphani* and pathogenic and genetic comparison with related pathovars. *Phytopathology* 96: 735–745.

White, H. E. 1930. Bacterial spot of radish and turnip. *Phytopathology* 20: 653–662.

Williams, P. H. 1980. Black rot: A continuing threat to world crucifers. *Plant Dis.* 64: 736–742.

Wulff, E. G., Mguni, C. M., Mortensen, C. N., Keswani, C. L., and Hockenhull, J. 2002. Biological control of black rot (*Xanthomonas campestris* pv. *campestris*) of brassicas with an antagonistic strain of *Bacillus subtilis* in Zimbabwe. *Eur. J. Plant Pathol.* 108: 317–325.

Zaccardelli, M., Campanile, F., Spasiano, A., and Merighi, M. 2007. Detection and identification of the crucifer pathogen, *Xanthomonas campestris* pv. *campestris*, by PCR amplification of the conserved Hrp/type III secretion system gene *hrcC*. *Eur. J. Plant Pathol.* 118: 299–306.

BACTERIAL DISEASES OF TOMATO

Many bacterial diseases occur on tomato (*Solanum lycopersicum* L). Among them, the important ones are bacterial spot, which also occurs on pepper (*Capsicum annuum* L.), bacterial speck, bacterial canker and bacterial wilt, and pith necrosis. All the four diseases are described below.

7.39 BACTERIAL SPOT OF TOMATO AND PEPPER

Bacterial spot of tomato and pepper is a serious disease of both the crops. Localized epiphytotics occur in areas where warm and moist conditions prevail during the crop season. It is a major problem in tropical and sub tropical regions. Especially under overhead-irrigation, the disease can cause enormous losses both on tomato and pepper. The disease is much less damaging in greenhouses. Crop losses occur due to defoliation resulting in yield reductions and severe spotting of fruits leading to reduction in their market value or making the fruits unsalable (Jones 2001).

Bacterial leaf spot of chillies was first observed in the USA in 1912 and of tomatoes in 1917. The first detailed account of bacterial spot of tomato was published by Doidge in 1920, and she named the causal bacterium as *Bacterium vesicatorium*. The specific name was derived in part from the development of vesicles in host cells during pathogenesis. Gardner and Kendric in 1923 showed that the same bacterium also caused a leaf spot of pepper. Approximately at the same time, Higgins described bacterial spot of pepper in detail. Since then, the disease has been reported from many countries including Canada, Central and South America, China, Australia, Austria, Bulgaria, France, Germany, Niger, Nigeria, Philippines, Poland, Romania, Russia, Senegal, Spain, Sudan, Taiwan, Tanzania, Thailand, Turkey, Zambia, Zimbabwe, Caribbean, Israel, Hungary, New Zealand, Italy, Yugoslavia, and Southeast Asia. *Xanthomonas gardneri* (group D strains) is predominant in Ontario, Canada. Due to its efficient overwintering under southern Ontario conditions, it is considered to be an aggressive form. The bacterium was responsible for the epiphytotic of the disease in Ontario processing tomato fields in 2000. The occurrence of this species has also been reported from Brazil. In India, bacterial spot was first observed on pepper from Pune in 1948 by Patel et al. (1950). It is now present in most of the tomato and pepper growing areas of the country.

7.39.1 Host Range

In addition to tomato and pepper, the bacterial spot pathogens infect chilli pepper (*C. frutescens*), currant tomato (*S. pimpinellifolium*), Aztec tobacco (*Nicotiana rustica*), *Datura tatula*, nightshade (*S. douglassii*), black nightshade (*S. nigrum*), buffalo bur (*S. rostratum*), and bittersweet or climbing nightshade (*S. dulcamara*). The survival of bacterial spot pathogens on these hosts may be an important source of primary inoculum in specific localities.

Araújo et al. (2015) reported that plants of *Nicandra physaloides*, *S. americanum,* and *Euphorbia heterophylla* showing leaf lesions were found naturally growing among tomato plants infected with bacterial spot in commercial tomato fields in Brazil. Isolates from *N. physaloides* and *S. americanum* were identified as *Xa. perforans* and those from *E. heterophylla* as *Xa. gardneri*. Each of them was able to artificially infect and cause symptoms on the three weed species and tomato plants.

7.39.2 Symptoms

The symptoms appear on all the aboveground plant parts of tomato and pepper. On tomato leaves, small (up to 3 mm in diameter), irregular and dark brown to black, greasy lesions appear. After a few days, the lesions are often surrounded by yellow haloes. The centers of the lesions dry out and frequently tear. Lesions coalesce to form big irregular dead areas, which give a scorched appearance. On mature plants, the lesions are more on fully expanded and older leaves. The leaves having numerous lesions turn yellow and may defoliate (Figure 7.25a). Defoliation begins from lower leaves, and it is generally more on pepper plants. The shot-hole symptoms

(a)

(b)

FIGURE 7.25 Bacterial spot of tomato: (a) Symptoms on foliage. (b) Symptoms on fruits. (Courtesy of M. Babadoost, Illinois Fruit and Vegetable News, Vol. 13, No. 8.)

caused on tomato leaves are usually associated with *Xa. perforans*. On stem and petioles, large black lesions are formed. Infection of flower parts, particularly on pedicel, generally results in severe blossom drop/blossom abortion resulting in yield loss and split sets of fruit.

On green tomato fruits, small, water-soaked, slightly raised spots appear. The spots increase in size up to 4–6 mm in diameter and become dark brown, scabby, and slightly sunken. Initially, the spots may be surrounded by greenish-white haloes, which eventually disappear (Figure 7.25b). The symptoms do not appear on ripe fruits.

On pepper, the spots on leaves, petioles, stem, and fruits are distinctly raised. Under conditions of high rainfall and high temperature, the lesions on pepper may take the form of blight.

The symptoms of bacterial spot and bacterial speck on tomato are similar, except on fruits, and are often confused. Symptoms of bacterial speck on green fruits appear as black, slightly raised stipplings, which cause lesions less than 1.5 mm in diameter.

7.39.3 Causal Organisms

Xanthomonas euvesicatoria Jones et al. 2004
Xanthomonas gardneri (*ex* Sutic 1957) Jones et al. 2004
Xanthomonas perforans Jones et al. 2004
Xanthomonas vesicatoria (*ex* Doidge 1920) Vauterin et al. 1995
Syn. *Xa. campestris* pv. *vesicatoria*

The taxonomy of the bacteria causing bacterial spot of tomato and pepper has undergone many changes. In 1921, Ethel Doidge identified it as *Bacterium vesicatorium*. It was reclassified as *Pseudomonas vesicatoria* in 1925, as *Phytomonas vesicatoria* in 1930, and as *Xanthomonas vesicatoria* in 1939. With the introduction of pathovar system in 1978, it was changed to *Xa. campestris* pv. *vesicatoria*. Vauterin et al. (1995) sub divided the bacterium into two species, placing the strongly amylolytic strains in *Xa. vesicatoria* and non-amylolytic strains in *Xa. axonopodis* pv. *vesicatoria*. Jones et al. (2000) found that xanthomonads pathogenic to tomato, pepper, or both consist of at least four phenotypic groups, namely, A, B, C, and D. Later on, Jones et al. (2004) renamed group A strains as *Xa. euvesicatoria*, group B strains as *Xa. vesicatoria*, group C strains as *Xa. perforans* sp. nov., and group D strains as *Xa. gardneri* on the basis of DNA:DNA hybridization data. Group A strains reclassified here as *Xa. euvesicatoria* represent the strains originally identified by Doidge in 1921 as *Bac. vesicatorium* referring to them as feebly amylolytic. The specific epithet "*gardneri*" was proposed by Sutic in 1957, who identified a bacterial disease of tomato from Yugoslavia. However, this bacterium has not been a major plant pathogen except in Ontario, Canada since its discovery. Strains of *Xa. euvesicatoria* and *Xa. vesicatoria* are widely distributed compared to those of other two species (Jones et al. 1998).

Xa. euvesicatoria and *Xa. vesicatoria* differ from each other in utilization of cis-aconitic acid, reaction to monoclonal antibodies, amylolytic and pectolytic activity, whole cell protein profiles, rep-PCR genomic fingerprinting, and reaction on differential hosts (races).

The bacteria are Gram-stain-negative rods, measure 0.7–1.0 × 2.0–2.4 μm, motile by a single polar flagellum, and strictly aerobic. They grow slowly on nutrient agar and produce colonies, which are circular, shinning, wet, entire, and Naples yellow. Xanthomonadin, a water-insoluble yellow pigment is produced.

A 420-bp fragment of the *hrpB7* gene was amplified by PCR from 75 strains representing *Xa. euvesicatoria*, *Xa. vesicatoria*, *Xa. perforans*, and *Xa. gardneri*; the PCR products were sequenced and phylogenetic analysis revealed that *hrpB7* is highly conserved within each species, with a single-nucleotide polymorphism among the *Xa. vesicatoria* strains. Four probes and two primer sets were designed to target the four bacterial spot pathogens based on their *hrpB7* gene sequences. In order to simultaneously detect the four bacterial spot pathogens, the four probes and two primer sets were optimized for a multiplex real-time TaqMan PCR assay and the optimized multiplex assay proved to be highly specific to the four bacterial spot pathogens. As the optimized multiplex assay facilitated the identification of each bacterial spot pathogen from pure cultures and infected plant tissue, it holds a great potential as a diagnostic tool (Strayer et al. 2016a).

Races: Stall et al. (2009) have given an excellent account of races of bacterial spot pathogens that infect pepper and tomato, resistance genes and their corresponding avirulence genes involved in imparting resistance to both the hosts, and occurrence of hypersensitive and quantitative forms of resistance. They also emphasized the need to intensify the research on quantitative or multi-genic resistance and to transfer this resistance to commercial cultivars of tomato and pepper.

Pathogenic variation in bacterial spot pathogens is clearly established. Some strains are pathogenic only to tomato or pepper, while others infect both the hosts. Both hypersensitive and quantitative forms of resistance to bacterial spot pathogens occur in tomato and pepper. Five resistance genes involved in HR in pepper and four in tomato have been identified. The evolution of resistance in tomato and pepper to bacterial spot pathogens has been slow in development. The genes in tomato and pepper that impart resistance against these pathogens are listed in Table 7.5. The origin of these genes and their corresponding avirulence genes present in the bacteria are also given.

The occurrence of pathogenic races in bacterial spot pathogens is known. Some of these races attack both the hosts. There are 11 races (0–10), which attack pepper. Differential reaction of these races to four resistance genes is given in Table 7.6. The resistance of plants of accession PI 163192 of pepper was the result of an HR to the bacterial spot pathogen. This was the first report of an HR occurring in resistant host plants to a bacterial plant pathogen. Initially, two races attacking pepper were indentified. The bacterium pathogenic on plants (accession PI 163192) having resistance gene *Bs1* was designated as race 1, while the prevalent bacteria in the field that caused bacterial leaf spot were designated as race 2. Later on, an avirulence gene cloned from race 2 caused race 1 to be hypersensitive in pepper leaves containing the *Bs1* gene. This work strongly supported the gene-for-gene hypothesis of

TABLE 7.5
Resistances in Tomato and Pepper and Bacterial Effectors That Interact with Them

	Plant		Bacterium	
Resistance Gene	**Source**	**Species**	**Effector**	**Location**
Pepper				
Bs1	PI 163192	*Capsicum annuum*	*avrBs1*	Plasmid
Bs2	PI 260435	*C. chacoense*	*avrBs2*	Chromosome
Bs3	PI 271322	*C. annuum*	*avrBs3*	Plasmid
Bs4	PI 235047	*C. pubescens*	*avrBs4*	Plasmid
bs5	PI 163192 or	*C. annuum*	ND	ND
	PI 271322	*C. annuum*	ND	ND
bs6	PI 163192 or	*C. annuum*	ND	ND
	PI 271322	*C. annuum*	ND	ND
BsT	Commercial pepper	*C. annuum*	*avrBsT*	Plasmid
Tomato				
rx1, rx2, rx3	Hawaii 7988	*Solanum lycopersicum*	*avrRxv*	Chromosome
Xv3	PI 128216 and Hawaii 7981	*S. pimpinellifolium* and *S. lycopersicum*	*avrXv3*	Chromosome
Xv4	716	*S. pennellii*	*avrXv4*	Chromosome
Bs4	Commercial tomato	*S. lycopersicum*	*avrBs4*	Plasmid

Source: Stall, R.E. et al., *Annu. Rev. Phytopathol.*, 47, 265–284, 2009.
ND = Not determined.

TABLE 7.6
Differential Reaction of Races of Xanthomonads to Pepper Resistance Genes

	Resistance Gene[a]			
Pathogen Race	***Bs1***	***Bs2***	***Bs3***	***Bs4***
0	HR[b]	HR	HR	HR
1	Sus	HR	HR	HR
2	HR	HR	Sus	Sus
3	Sus	HR	Sus	HR
4	Sus	Sus	HR	HR
5	HR	Sus	Sus	Sus
6	Sus	Sus	Sus	HR
7	Sus	HR	HR	Sus
8	Sus	HR	Sus	Sus
9	Sus	Sus	HR	Sus
10	Sus	Sus	Sus	Sus

Source: Stall, R.E. et al., *Annu. Rev. Phytopathol.*, 47, 265–284, 2009.
[a] Sources of resistance genes are listed in Table 7.5.
[b] HR refers to hypersensitive reaction which is involved in resistance, Sus refers to susceptible reaction.

disease resistance established by Flor in 1955 as a resistance gene *Bs1* in the host (cv. Early Calwonder 10R), and an avirulence gene (*avrBs1*) in the bacterium was necessary for resistance to bacterial leaf spot. Subsequently *Bs2*, *Bs3*, and *Bs4* resistance genes were identified. Each of these single dominant genes, i.e., *Bs1* to *Bs4* can be overcome individually by specific races of the pathogens in the field-grown plants. To improve the level of resistance in the host, three genes, namely, *Bs1*, *Bs2*, and *Bs3* from *C. annuum* were transferred to a single plant of Early Calwonder (ECW) and designated as Early Calwonder 123. Each gene was inherited independently of the others. The combination of all the three resistance genes in a plant provided a level of disease less than on the universally susceptible control to strains of race 6, which is pathogenic on plants with all the three genes. The combination of *Bs1* and *Bs2* provided some control against race 6, while the combination of *Bs1* and *Bs3* was not effective. None of the combinations provided a satisfactory disease control.

A Hawaiian tomato accession (H 7998) showed a highly resistant reaction to the bacterial spot pathogen. The resistant plants exhibited an HR, but the resistance was not inherited as a single dominant gene as in most other hypersensitive resistances. Three genes were involved in this case and these genes were designated as *rx1*, *rx2*, and *rx3*. The inoculation of pathogen strains from Brazil on H 7998 plants produced two types of reactions. Some strains produced the susceptible reaction and these strains were designated as race T2, and the strains giving HR on H 7998 were designated as race T1 (Table 7.7). T2 strains were abundant in Brazil, Asia, Argentina, and midwestern states of United States, but were never found in Florida. As T2 strains had many characteristics different from those of T1 strains, they were identified as *Xa. vesicatoria*. Race T1 remained the sole pathogen of bacterial spot of tomato in Florida for at least 30 years. In 1991, H 7998 showed susceptible reaction to bacterial spot pathogen. This bacterium showed pectolytic and amylolytic activity and was similar to previously discovered T2 strains. However, unlike the T2 strains, it caused HR in many PI accessions of *S. pimpinellifolium*. Therefore,

TABLE 7.7
Differential Reactions of Races of Xanthomonads on Tomato Resistance Genes

Pathogen Race[a]	Resistance Gene[b]		
	rx1, rx2, rx3	*Xv3*	*Xv4*
T1	HR[c]	Sus	Sus
T2	Sus	Sus	Sus
T3	Sus	HR	HR
T4	Sus	Sus	HR

Source: Stall, R.E. et al., *Annu. Rev. Phytopathol.*, 47, 265–284, 2009.
[a] Letter T before each race indicates the reaction of races from tomato plants.
[b] Sources of resistance genes are listed in Table 7.5.
[c] HR refers to hypersensitive reaction which is involved in resistance, Sus refers to susceptible reaction.

this strain was designated as race T3. A near-isogenic line containing *Xv3* was developed by backcrossing plants with the resistance gene from PI 128216 into FL 7060 seven times. This line was designated as FL 216 and is useful in determining races of the tomato bacterial spot pathogens along with H 7998. In 1991, T3 strains comprised only a small percentage of bacterial spot pathogen's population in Florida. It gradually eliminated the T1 race in Florida and in 2006 no culture of T1 was isolated. The elimination of T1 strains by T3 strains was thought to be the result of an antagonism of T3 strains to T1 strains caused by bacteriocin production by the T3 strains (Tudor-Nelson et al. 2003). At least three bacteriocins are produced by T3 strains and at least one of them is antagonistic to all the T1 strains tested. The *avr* gene associated with T3 hypersensitivity was cloned and designated as *avrXv3*. Tomato strains with this *avr* gene also initiate hypersensitivity in pepper plants. Due to genetic differences of race T3 strains from the other strains of xanthomonads that cause bacterial spot of tomato, T3 strains were identified as *Xa. perforans*.

The strains of *Xa. perforans* were found in Florida which proved pathogenic on plants having gene *Xv3*. These strains were designated as race T4. The cultivation of grape tomatoes in Florida apparently provided the selection pressure to cause the development of T4 strains or their increase over T3 strains. In a survey of races on tomato in 2006 in Florida, approximately 77% of the strains collected were of the T4 race and the rest were of T3 race. The *avr* gene in race T4 that incites HR in plants having *Xv4* gene was cloned and designated as *avrXv4*. Both the T3 and T4 races contain *avrXv4*. Probably, race T4 originated by mutation of *avrXv3* in race T3 because race T3 contains both *avrXv3* and *avrXv4* genes, and T3 and T4 are genetically similar.

The durability of a resistance gene based on HR in a plant depends on the stability of the *avr* gene in the pathogen. Many mechanisms operate in bacterial spot pathogens that result in instability of *avr* genes. Many *avr* genes of bacterial spot pathogens are contained in self-transmissible plasmids. The loss of such a plasmid can result in the loss of an *avr* gene. Another mechanism of inactivation of *avr* genes is addition of an insertion element into the *avr* gene coding sequence. Total genomic sequencing of one strain of the bacterial spot pathogen revealed 58 insertion elements in the chromosome and another eight on four plasmids of the bacterium. One of these elements was *IS476*, which is associated with inactivation of *avrBs1* in race 2 of pepper strains. The frequency of these changes affects race composition among pathogen populations and determines the durability of the corresponding plant resistance.

7.39.4 Disease Cycle

The bacteria survive in diseased plant debris, volunteer plants, and as epiphytes on weed species. Epiphytic survival on weeds seems to be of minimal importance as a primary source of inoculum. These bacteria have also been found to persist as latent infections on host leaves and furthermore to survive there under unfavorable conditions. This is important in distribution of these pathogens on transplants that may not have visible symptoms. The role of plant debris in the survival of bacteria is more important in temperate regions where the crop is raised in summer months and the debris does not decompose readily due to low temperature. The bacteria also survive on seed as contaminants, and the seed contamination occurs during the seed extraction process.

The local or short distance spread of the bacteria occurs through wind-splashed rain, rain splashes, and overhead-irrigation. The dissemination to distant places and new areas occurs through infested seed and diseased transplants.

The bacteria penetrate into the host through stomata and wounds caused by wind-blown sand, insects, mechanical means, and breaking of hairs. The entry of the bacteria in the fruits occurs through cracks in the epidermis and through fruit hairs, which are present only on immature fruits.

The disease development is favored by high relative humidity and overhead-irrigation. The optimum temperature for disease development is 24°C–30°C. The bacteria may occur in low populations as epiphytes on the surface of host plants. At the onset of favorable conditions, these low populations increase rapidly and cause infection. Marcuzzo et al. (2009) made the following observations on the epidemiological aspects of bacterial spot of tomato in Cacador, Brazil. The epiphytotic onset of the disease coincided with physiological maturation of the first fruit clusters. Bacterial leaf spot continued to develop even in dry period, possibly leaf wetness from dew drops favored disease development. The bacterial population on leaf surface varied during the first 10 weeks after planting depending on weather, but it remained stable after the appearance of the first symptoms of the disease.

Dutta et al. (2014) investigated colonization of pepper blossoms by *Xa. euvesicatoria.* The bacterial populations reached up to 10^5 cfu/blossom on stigmas and ranging from 10^5 to 10^6 cfu/blossom on stylar and ovary tissues 96 h after inoculation. Inoculation of blossoms resulted in asymptomatic pepper fruit, but a positive correlation was found between inoculum concentration and percentage of infested seed lots. Real-time PCR detected *Xa. euvesicatoria* in 39% of the seed lots assayed and viable colonies were recovered from 35% of them. Successful transmission occurred in 16% of the seed lots tested.

Giovanardi et al. (2018) investigated the impact of disease outbreaks on the phytosanitary quality of seeds for two pathosystems, i.e., tomato-*Xa. vesicatoria* and pepper-*Xa. euvesicatoria.* Seed contamination levels in pepper ranged from 34 to 100 cfu g^{-1}, and the contamination rate ranged from 1.50% to 3.17% for *Xa. euvesicatoria*, while the tomato seeds produced were not infected by *Xa. vesicatoria.* During seedling grow-out assays and the second cropping year, no symptoms were observed in either tomato or pepper plants. It is evident that the calculated pepper seed contamination rate for *Xa. euvesicatoria* appeared to be less than the threshold necessary to initiate a disease outbreak. Finally, all seeds obtained during the second cropping year were uninfected.

Sun et al. (2014) investigated the feasibility of using image analysis to estimate foliar disease severity of bacterial spot in tomato. Five diseased leaves from each plant, spray-inoculated with bacterial spot race T3 (*Xa. perforans*), were randomly collected and scanned to obtain digital images 21 days after inoculation. The disease severity (% leaf area) was measured using image analysis. Marker-trait association analysis identified four quantitative trait loci conferring resistance to race T3 in resistant line, PI 114490 using data obtained from image analysis, the Horsfall-Barratt (HB) category scale data, and HB midpoint converted values. However, the disease severity was slightly underestimated using the HB category scale, and the phenotypic variation explained by each marker was overestimated using the HB category data compared to the image analysis-measured disease severity data. It is evident that image analysis can provide a consistent, accurate, and reliable method compared to the HB scale to estimate disease severity for genetic studies of foliar bacterial spot in tomato.

Duan et al. (2015) prepared and evaluated a standard area diagram (SAD) set to assess bacterial spot severity on tomato leaves caused by *Xanthomonas* spp. The proposed SAD set contains illustrations of leaves with 12 distinct disease severity values from 0.5% to 90% and was validated by 12 raters without any experience of plant disease evaluation. Regression analysis and Lin's concordance correlation (ρ_c) analysis of estimated (without the SADs) versus true disease severity (based on image analysis) showed that the precision and accuracy improved for all raters using the SADs. The SAD set improved accuracy (correction factor, $Cb = 0.91$ and 1.00, without and with the SADs, respectively) and agreement ($\rho_c = 0.83$ and 0.96, without and with the SADs, respectively) of the estimates of severity. Severity estimates using the SADs were also more reliable (coefficient of determination, $R^2 = 0.74–0.94$ without the SADs, and $R^2 = 0.82–0.95$ with the SADs, and intra-class correlation $\rho = 0.85$ without the SADs and $\rho = 0.91$ using the SADs). The SAD set proposed in this study will improve the accuracy and reliability of estimates of bacterial spot severity on tomato leaves for inexperienced raters and can be used for assessing severity of other bacterial diseases on tomato leaves.

Araújo et al. (2017) reported that out of 204 strains obtained from 33 counties (22 with processing tomatoes and 11 with fresh-market tomatoes) of Brazil, *Xa. perforans* predominated among the strains (92%) and was present in most counties. Moreover, this species was also prevalent in most areas of both fresh-market tomatoes (63.6% of counties surveyed) and processing tomatoes (95.4% of counties surveyed). High genetic diversity was observed within *Xa. perforans*, with 137 BOX-PCR haplotypes. Fifteen strains (7.5%) were identified as *Xa. gardneri*, which was found mostly in fresh-market fields located at regions with altitude higher than 900 m, and only one strain of *Xa. euvesicatoria* (0.5%) was found in a processing tomato field.

Safaie Farahani and Taghavi (2018) reported that rutin, a polyphenolic substance, at concentration of 2 mM induced resistance in tomato against *Xa. perforans* and reduced the disease severity of bacterial spot, but showed no antibacterial activity *in vitro* against the bacterium. Pre-treatment of tomato plants with rutin led to induction of *PR-5*, *PAL*, and *LOX* genes compared to control and rutin-induced resistance against *Xa. perforans* in tomato might be mediated through stimulation of some defense genes such as *PR-5*, *PAL,* and *LOX*.

7.39.5 Management

For an effective management of the disease it is desirable to integrate the following control measures.

1. Use of pathogen-free seed. If the pathogen-free seed is not available, it must be treated by any of the following treatments
   a. Hot water treatment at 50°C for 25 min after pre-warming the seed at 38°C in water for 10 min. Immediately after the treatment, put the seed in cold water for 5 min (Leboeuf et al. 2005)
   b. Acetic acid (0.8%) steep for 24 h or 5% HCl for 5–10 h at 21°C
   c. Shahrtash and Lewis Ivey (2015) treated tomato and pepper seed artificially infested with a mixture of *Xa. vesicatoria*, *Xa. euvesicatoria*, and *Xa. perforans* (~10^4 cfu mL^{-1}) with hot water (pepper seeds for 30 min at 51°C; tomato seeds for 25 min at 50°C) or chlorine (20% for one minute), which represent industry standards for disinfesting seed, or sodium selenate (765 mg L^{-1} for 30 min). All the three treatments significantly reduced the incidence of infested pepper and tomato seed compared to the non-treated infested control. The incidence of infested pepper seeds following treatment with sodium selenate was significantly higher than that of chlorine and hot water treated seed. There was no difference in the incidence of infested seed between chlorine treated and sodium selenate treated tomato seed. Hot water treatment of pepper and tomato seed was the most effective in reducing *Xanthomonas* spp. on seed. The presence of *Xanthomonas* spp.

was confirmed using colony-pick PCR with *Xanthomonas*-specific PCR primers. None of the treatments significantly reduced or delayed germination.

2. Use of disease-free transplants, and this is achieved by growing the nursery where the crop production does not occur
3. Rotation with non-host plants for 2–3 years
4. Destroy all the volunteer plants
5. Avoid cull piles near greenhouse and field locations
6. Use of chemicals: Use of copper compounds and streptomycin has created a serious problem of development of pathogen strains resistant to these chemicals. In order to control copper resistant strains, a combination of copper and mancozeb is used, and these chemicals should be applied as preventive sprays after transplanting. Strayer et al. (2016b) reported that in greenhouse studies, the application of Ag-dsDNA-GO, a silver-based nanocomposite, at either 75 or 100 μg mL^{-1} prior to artificial inoculation of tomato plants with *Xa. perforans* significantly reduced disease severity when compared with copper-mancozeb and negative controls ($P = 0.05$). The study highlights the potential of Ag-dsDNA-GO as an alternative to copper in tomato transplant production. Ag-dsDNA-GO also exhibited antibacterial activity against copper-tolerant *Xa. vesicatoria*, *Xa. euvesicatoria*, and *Xa. gardneri* strains.

   Paret et al. (2013) reported that application of TiO_2/Zn; AgriTitan (nanoscale titanium dioxide doped with zinc) at ≈500 to 800 μg mL^{-1} significantly reduced bacterial spot (*Xa. perforans*) severity compared with untreated and copper control in greenhouse studies on naturally and artificially infected transplants. Protection was similar to the grower standard, copper + mancozeb. The use of TiO_2/Zn at ≈500 to 800 μg mL^{-1} also significantly reduced disease incidence in three of the four field trials compared with untreated and copper control and was comparable to or better than the grower standard. The treatments did not cause any adverse effects on tomato yield in any of the field trials.

   Integration of foliar sprays of ammonium lignosulfonate (2% or 4%) and potassium phosphate (25 mM) in the standard bactericide management programs has also been suggested for improving the disease management efficiency (Abbasi et al. 2002)
7. ASM (trade names Actigard 50WG and CGA 245704) is a plant activator that induces systemic acquired resistance in many crops against a number of pathogens. Louws et al. (2001) showed that it (35 g a.i. ha^{-1}) can be used as a viable alternative to copper-based bactericides for field management of bacterial spot, and also for speck of tomato, where copper resistant populations of these pathogens predominate. Huang and Vallad (2018) investigated the effects of soil application rates of ASM on bacterial spot of tomato and the expression levels of the two pathogenesis-related genes, *PR1a* and *PR1b*, in leaf tissues. ASM at 18.8 mg L^{-1}, corresponding to the labeled rate, was sprayed on tomato seedlings and soil application was made at 0.84 and 10 mg L^{-1}. The soil application of ASM at 10 mg L^{-1} consistently reduced the final disease severity and disease progress compared to the untreated control. The expression levels of *PR1a* and *PR1b* in the leaf tissues were significantly induced by both soil and foliar applications of ASM. The soil applications of ASM at 10 mg L^{-1} markedly reduced disease progress compared to the control and copper standard. Although the control efficiency of soil applications of ASM depends on rates used, the study suggests that ASM can be used as soil applications to induce tomato resistance against bacterial spot.

Obradovic et al. (2004) obtained better control of the disease with the combination of phage and ASM than phage, ASM, or copper-mancozeb alone. Obradovic et al. (2005) also suggested the integrated use of acibezolar-*S*-methyl and phages as an alternative management strategy for the disease.

Phage treatment, integrated with other practices, is widely used in greenhouse and production fields in Florida as a part of standard integrated management program for this disease. Based on the results of greenhouse and field trials, Omniytics, Inc., in Salt Lake city, UT was the first to obtain registration from Environmental Protection Agency (EPA) of USA for using host specific phages on tomatoes in greenhouses and production fields in Florida as a part of standard integrated management program for the control of the disease.

Ji et al. (2006) also obtained significant control of bacterial spot and speck of tomato with combined application of *Pseudomonas syringae* str. Cit 7 and *Psm. fluorescens* str. 89B-61.

7.40 BACTERIAL CANKER AND WILT OF TOMATO

Bacterial canker of tomato (*Solanum lycopersicum* L.) was first observed by Erwin F. Smith in 1909 from Grand Rapids, Michigan, USA, and he named it as Grand Rapids disease. However, Smith soon renamed the disease as bacterial canker of tomato. By 1920s, the occurrence of the disease was reported from Central and South America, Africa, Asia, Australia, and Europe. Now the disease is worldwide in occurrence and occurs wherever tomatoes are grown. Although usually sporadic in its occurrence, it is so destructive in nature that both field and greenhouse crops, especially the latter, suffer heavily from the disease. It is a quarantined pathogen and Canada, USA, European Union (EPPO A2 quarantine pest), and some other countries enforce zero tolerance for import and export of plant material infected with the pathogen. Janse (2005) has listed 60 countries where the disease occurs. Yield losses ranging from 20% to 70% have been reported from France and USA.

7.40.1 Host Range

In addition to tomato, the bacterium attacks *S. douglassii* (nightshade), *S. nigrum* (black nightshade), and *S. triflorum*.

7.40.2 Symptoms

Tomato plants are susceptible at all the growth stages. A wide variety of symptoms, resulting from superficial or localized and systemic infections, are produced. **Wilting** is the most common symptom in spite of different names of the disease.

Generally, there are no symptoms on the seedlings, but under conditions favorable to the disease, infected seedlings show weak and stunted growth or may be killed. Wilting is first seen on the older leaves, and the leaves may curl downward before wilting. Wilting of leaflets often occurs on one side of the petiole, but soon the leaflets on the other side also wilt and wilting progresses involving the whole plant. The progress of the wilt is more rapid in younger plants. A light brown discoloration of vascular tissues is visible when wilted stems, branches, and petioles are split opened vertically. In stems, discoloration is more prominent at nodes and just above the soil line. In the advanced stages of disease development, cankers develop on stems, petioles, and midribs in the form of brownish streaks. These streaks split open exposing necrotic tissue in the pith and cortex. The pith turns granular to mealy and cavities are formed. In severely infected plants, yellow ooze may exude from the cut end of a stem when it is squeezed. The seed-borne inoculum or invasion of vascular system by the pathogen entering through wounds result in systemic infection leading to wilting and canker formation.

Superficial/localized infections resulting from the entry of epiphytic bacteria through stomata, hydathodes, and broken trichomes (hairs) produce the following symptoms. Bacteria entering through hydathodes cause marginal necrosis of leaves, and these brown necrotic areas (**leaf firing**) may extend until the entire leaflet is shriveled. Round to irregular spots also appear on the leaves. Similar spots may also be noted on stem and petioles (Figure 7.26a).

On green fruits, first small, white, superficial spots appear. These spots enlarge up to 1–3 mm in diameter and develop a raised dark brown center surrounded by a distinct white halo (Figure 7.26b). Due to their resemblance to bird's eye, they are called **bird's-eye spots.** However, their occurrence is erratic, and they do not always develop in disease outbreaks. Their occurrence reduces the market value of the fruits and sometimes making them unsalable (Braun 2001). Systemic infection results in darkening of the vascular tissues within the fruit. The bacteria can multiply in the vascular bundles within the fruits up to the seed. This leads to the formation of yellowish strands from the stem to the seeds and also the internal infection of the seed. The fruit may show a black peppering at the vascular bundles under the calyx scar.

7.40.3 Causal Organism

Clavibacter michiganensis (Smith 1910) Li et al. 2018
Syn. *Cla. michiganensis* subsp. *michiganensis*, *Corynebacterium michiganense* subsp. *michiganense*, *Cor. michiganense* pv. *michiganense*

E.F. Smith was the first to isolate and characterize the bacterium, and he named it as *Aplanobacter michiganense*. This name was changed several times before it was named as *Corynebacterium michiganense*, a name that remained in use for several decades. Davis et al. (1984) proposed the reclassification of *Cor. michiganense* and its five sub species, namely, *michiganense*, *insidiosum*, *sepedonicum*, *nebraskense*, and *tessellarium*, *Cor. iranicum*, *Cor. tritici*, and *Cor. rathayi* in the new genus *Clavibacter*. Li et al. (2018) reported that the whole genome analysis based on average nucleotide identity and digital DNA-DNA hybridization as well as multi-locus sequence analysis of seven housekeeping genes support raising each of the *Cla. michiganensis* sub species to species level. Therefore, the recent name of the causal organism of the disease is *Cla. michiganensis* and not *Cla. michiganensis* subsp. *michiganensis* as proposed by Davis et al. (1984).

The bacterium grows well on nutrient agar containing 5% glucose and potato dextrose agar producing yellow mucoid colonies. Cells are Gram-stain-positive, non-motile,

(a)

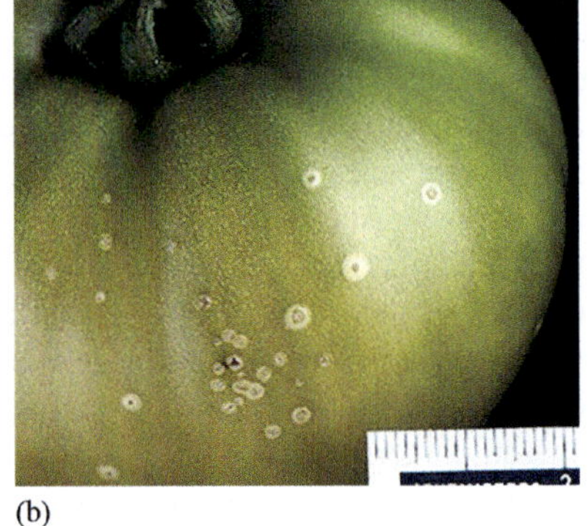
(b)

FIGURE 7.26 Bacterial canker of tomato: (a) Wilted plants showing necrosis and death of leaves. (b) Bird's eye-like spots on fruit. (Courtesy of Diane Cuppels, Agriculture & Agri-Food, Ottawa, Canada.)

non-sporing, irregularly shaped rods. Gelatin is weakly liquefied and H_2S is produced. Acid is produced from mannose, fructose, and sucrose, but not from dulcitol, sorbitol, inulin, and α-methyl-D-glucoside. Acetate, citrate, fumarate, malate, and succinate are used as a sole source of carbon. The genes involved in pathogenicity have been identified and characterized. The following characteristics of this bacterium differentiate it from other species of *Clavibacter*: tolerance to sodium chloride up to 6%, H_2S production, weak liquefaction of gelatin and utilization of mannose, cellobiose, and succinate, but not of mannitol and propionate as the sole source of carbon.

Ftayeh et al. (2011) developed a new selective and highly sensitive medium for isolation of *Cla. michiganensis* subsp. *michiganensis* (hereafter referred to as *Cmm*) from seed and latently infected plants. The new medium (BCT) proved superior to all published semi-selective media (D2, KBT, D2ANX, SCM, mSCM, CMM1, mCNS, and EPPO) for *Cmm* detection. BCT is designated as a selective medium because: (i) its mean plating efficiency amounts to ≤89% within 7 days for all 30 *Cmm* strains from different sources tested; (ii) the high selectivity, because of inhibition to an extent of 98%–100% of accompanying bacterial species occurring on tomato plants and seed, or bacteria obtained from culture collections; and (iii) the remarkable detection sensitivity. Because of high selectivity, 8 cfu of *Cmm* in field plant homogenates containing 12,750 cfu of accompanying saprophytes were detected on BCT. Moreover, BCT also supported growth of the closely related *Cla. michiganensis* subsp. *insidiosus*, *nebraskensis*, and *tessellarius*. The authors have recommended the new medium for *Cmm* detection in tomato seed and in symptomless tomato plantlets to improve the control of bacterial canker of tomato.

Yasuhara-Bell et al. (2013) used LAMP to specifically identify *Cmm*, using LAMP primers developed to detect *micA*, a chromosomally stable gene that encodes a type II lantibiotic, michiganin A. A total of 409 bacterial strains (351 *Cmm* and 58 non-*Cmm*) from a worldwide collection were tested with LAMP to determine its specificity. LAMP results were compared with genetic profiles established using PCR amplification of seven genes (*dnaA*, *ppaJ*, *pat-1*, *chpC*, *tomA*, *ppaA*, and *ppaC*). *Cmm* strains produced eight distinct profiles. The LAMP reaction identified all *Cmm* strains and distinguished them from other *Cla. michiganensis* sub species and non-*Clavibacter* bacteria. LAMP has advantages over immunodiagnostic and other molecular detection methods due to its specificity and isothermal nature, which allows for easy field application. The LAMP reaction is also not affected by as many inhibitors as PCR. Therefore, LAMP has a potential to provide an easy, one-step test for rapid identification of *Cmm*.

Yasuhara-Bell et al. (2014) reported that three genes, clavicidin (*clvA*) and two associated genes (*clvF* and *clvG*, encoding a putative ABC-type multi-drug transport system ATPase and permease component, respectively) are unique and highly conserved in *Cmm*. Loop-mediated amplification of *clvA* and PCR of *clvA*, *clvF*, and *clvG* showed that these genes were present only in *Cmm* and not in other *Clavibacter* species/sub species, as well as other genera of plant-associated bacteria. Sequences of *clvA*, *clvF*, and *clvG* from 48 geographically diverse strains of *Cmm* were analyzed phylogenetically to determine sequence variation. Maximum parsimony, neighbor-joining, and maximum likelihood analyses placed strains into sub groups irrespective of their geographical origins. Multi-locus sequence analysis of these three genes produced the same results. The data support the use of the *clvA* gene for identification of *Cmm*, using either a previously designed LAMP assay or PCR with additional primers developed in this study. The genes investigated here are novel targets for specific detection of *Cmm* and have numerous molecular diagnostic applications.

Sen et al. (2015) reported that *Cmm* strains show great variability in virulence and are usually described as being hypervirulent, hypovirulent, or non-virulent. The bacterium lacks a type III secretion system, and only a few virulence factors have been experimentally determined from the many putative virulence factors. As its molecular mode of infection is unknown, researchers have avoided intensive work on this pathogen. Genetic plant mechanisms conferring resistance to the bacterium are apparently complex, and breeders have yet to develop disease-resistant cultivars.

7.40.4 Disease Cycle

Infected seed is an important source of perpetuation of the bacterium, and it survives in the inner most layers of the seed coat and as an external contaminant. The infection of the seed results from systemic infection of the plant, while external contamination occurs during extraction of seed. One infected seed in 10,000 may be enough for the outbreak of the disease under favorable conditions. The bacterium also survives on volunteer tomato plants and weed hosts.

Seed-borne inoculum plays an important role in the outbreak of the disease. The contamination of seeds with only a few bacterial cells that are too below the detection levels; give rise to high number of infected seedlings. The seed lots having a very few infested seeds pose a serious problem in greenhouse crops. No doubt the certified seed reduces the chances of infection, but is not a guarantee for a disease-free seed. When contaminated seeds germinate, the bacteria enter the cotyledons through wounds, probably through broken trichomes. The bacteria move systemically through the xylem from which it enters the phloem, pith, and cortex. The optimum conditions for disease development include high moisture/relative humidity and temperature of 24°C–32°C. There is usually a long latent period of several weeks in mature plants.

The detection of pathogen in seed lots is required for seed certification. PCR is considered the most sensitive method for the detection of plant pathogenic bacteria. Classical PR primers derived from the *pat-1* gene in plasmid pCM2 of *Cmm* were developed for its detection, but the primers reacted with 75% of the strains tested.

Infected seedlings are one of the most important primary sources of inoculum of the bacterium in commercial production plots. It was found that the pathogen dispersed spatially from root-inoculated source seedlings and colonized the leaf surfaces of surrounding seedlings to distances of 65–75 cm. A sub irrigation system reduced, but did not prevent dispersal of the bacterium. Infections by epiphytic *Cmm* populations can occur under a wide range of temperature conditions, seedling ages, and foliar wetness periods. There is a need to minimize the primary sources of infection as well as developing new strategies to reduce the dispersal during the nursery stage (Frenkel et al. 2016).

The pathogen can survive in the infected plant debris in soil up to 2 or 3 years. However, the survival in plant debris is related to the extent of decomposition of the debris in the soil. Woody stems buried deep in soil usually remain intact and help in survival of the bacterium. Vega and Romero (2016) reported that *Cmm* survived in host debris left on the soil surface in greenhouse (Buenos Aires—La Plata, Argentina) for 120–260 days for crop production cycles that ended in winter and for 45–75 days for those that ended in summer. In stems or roots buried in winter, this period was 45–75 days. The authors concluded that host debris, including roots, might be an important primary inoculum source of the pathogen in greenhouses.

The penetration of the host occurs through wounds caused by cultural operations, broken trichomes, and natural openings like stomata and hydathodes. On tomato fruits, the wounds caused by wind-blown sand, breaking of hairs, and insect punctures also act as avenues for the entry of the pathogen. Secondary spread of the bacterium occurs mainly through wind-splashed rain and in greenhouses through sprinkler irrigation. Secondary spread also occurs through contaminated equipment and cultural operations like pruning, staking, and training of vines. The bacteria growing epiphytically on healthy and diseased plants serve as a major source of inoculum for the secondary spread.

Bach et al. (2003) developed a real-time TaqMan PCR assay based on the primers specific to the internal transcribed spacer sequence between 16S and 23S rDNA, a common region in all the five sub species of *Cla. michiganensis* and a sub species specific TaqMan probe for the detection of *Cmm*. However, both the abovementioned PCR assays do not distinguish between the living and dead cells of the bacterium and false positive results obtained with dead cells limit the use of PCR in risk analysis of seed lots. A quantitative real-time PCR assay devised by binding the bacterial DNA with ethidium monoazide has taken care of this problem as the method allows the selective quantification of only viable cells (Luo et al. 2008). The primers and TaqMan probe, based on 16S–23S rDNA spacer sequences, are highly specific for *Cmm* at the sub species level. The detection limit of the direct real-time PCR is 10^3 cfu mL^{-1} in samples and with an apparent sensitivity of 2 cfu of target cells in PCR reaction solution.

Lelis et al. (2014) investigated the colonization and disease development of axenically grown tomato plants by *Cmm*. A spontaneous rifampicin resistant strain of *Cmm* tagged with a marker that expressed a green fluorescent protein (GFP) in a stable way and which possessed a similar virulence to the parental strain was used. Symptoms such as canker, wilting, and growth reduction, typical for *Cmm*, developed in *in vitro* plants. The presence of *Cmm* in vascular and parenchymatic tissue of *in vitro* tomato plants was confirmed by epifluorescence stereo- and confocal laser-scanning microscopy. The study shows that *in vitro* tomato plants can be effectively used for detailed studies on interactions between *Cmm* and its host, in particular if a GFP-tagged strain of the pathogen is used.

Xu et al. (2012) reported that a virulent, stable, constitutively bioluminescent *Cmm* strain, BL-*Cmm* 17 coupled with an *in vivo* imaging system allowed visualization of the *Cmm* colonization process in tomato seedlings in real-time. The dynamics of bacterial infection in seedlings through wounds were compared under two humidity regimes, i.e., 45% and 83%. Bacteria multiplied rapidly in cotyledon petioles after clip inoculation, and moved in the stem toward both root and shoot. Luminescent signals were also observed in roots of tomato seedling over time, and root development was reduced in inoculated plants maintained under both humidity regimes. Wilting was more severe in seedlings at 83% relative humidity. A strong positive correlation between light intensity and bacterial population *in planta* suggests that bioluminescent *Cmm* strains will be useful in evaluating the efficacy of bactericides and host resistance.

Pathogenicity of *Cmm* is dependent on plasmid-borne virulence factors and serine proteases located on the chromosomal *chp*/*tomA* PAI. Chalupowicz et al. (2012) examined the colonization patterns and movement of *Cmm* during tomato infection using a GFP-labeled strain. A plasmid expressing GFP in *Cmm* was constructed and found to be stable *in planta* for at least 1 month. Confocal laser-scanning microscopy (CLSM) of inoculated stems showed that the pathogen extensively colonized the lumen of xylem vessels and preferentially attached to spiral secondary wall thickening of the protoxylem. Acropetal movement of the wild-type strain *Cmm* NCPPB382 (*Cmm*382) in tomato resulted in an extensive systemic colonization of the whole plant reaching the apical region after 15 days, whereas *Cmm*100 (lacking the plasmids pCM1 and pCM2) or *Cmm*27 (lacking the *chp*/*tomA* PAI) remained confined to the area surrounding the inoculation site. *Cmm*382 formed biofilm-like structures composed of large bacterial aggregates on the interior of xylem walls as observed by confocal laser-scanning microscopy and scanning electron microscopy. These findings suggest that virulence factors located on the *chp*/*tomA* PAI or the plasmids are required for effective movement of the pathogen and for the formation of cellular aggregates in tomato.

7.40.5 Management

The management of bacterial canker of tomato poses a significant challenge due to the lack of effective chemicals and resistant varieties and due to the systemic nature of the pathogen. The following multi-faceted approach, which mainly prevents or reduces the primary source of inoculum, will reduce the incidence and severity of the disease.

1. Use only certified pathogen-free seed, obtained from canker-free plants, for planting. Seed should be extracted only by the standard acetic acid extraction

method or by the fermentation process. The seed extracted by these methods is free from seed coat contamination, but the embryonic infection of the seed is not completely controlled. Hence, the seed must be treated with hot water or disinfected with acid or chlorine. In hot water treatment, the seed is pre-warmed in water at 38°C for 10 min and then kept in hot water at 50°C for 25 min. Immediately after the treatment, put the seed in cold water for 5 min. The disinfection of seed with acid is done by dipping the seed in 0.8% solution of acetic acid for 24 h at 21°C and with chlorine by dipping the seed in 1% sodium hypochorite (20% bleach) for 20–40 min. For proper seed disinfection, pre-soaking of seed in water, ranging from 1–18 h is essential
2. Plant only certified disease-free transplants grown under a vigorous inspection program
3. Ensure the destruction of infected plant debris by its thorough decomposition. Soil solarization by covering the soil with transparent polythene sheets for 6 weeks proved effective in reducing the disease in plastic tunnels (Antoniou et al. 1995)
4. Follow the crop rotation for at least 3 years
5. Destroy the volunteer tomato plants and weeds belonging to the family *Solanaceae*
6. Sterilize seedbeds and soil in the greenhouse to destroy bacteria. Steam sterilization is preferred over methyl bromide
7. Pruning knives and other tools used in the greenhouse should be disinfected regularly. Quaternary ammonium compounds should be preferred over sodium hypochlorite or laundry bleach as a greenhouse disinfectant (Zitter 1985)
8. Fixed copper sprays can help in protecting the crop, particularly in superficial or localized infections. The combination of copper and mancozeb is better than copper alone. Preventive sprays which delay the disease development/outbreak until after the main fruit set will minimize the yield loss because fruit infection, responsible for major yield loss, occurs only on green immature fruits (less than 3 cm in diameter).

   The application of acibenzolar-*S*-methyl (benzo-[1, 2, 3]-thiadiazole-7-carbothioic acid-*S*-methyl ester; Bion 50WG) activates defense response of the host against the bacterium. Baysal et al. (2003) reported reduction in disease severity up to 76.3% along with reduction in bacterial growth up to 68.2% in ASM treated plants. The enhanced resistance was due to the significant increase in peroxidase and chitinase activity
9. Barda et al. (2015) reported that the extracellular metabolites secreted by an epiphytic fungus *Pseudozyma aphidis* inhibited *Xanthomonas campestris* pv. *vesicatoria*, *Xa. campestris* pv. *campestris*, *Pseudomonas syringae* pv. *tomato*, *Erwinia amylovora*, *Cla. michiganensis*, and *Agrobacterium tumefaciens in vitro*. The application of *P. aphidis* spores on tomato plants in the greenhouse reduced 60% incidence and 35% severity of bacterial canker. Moreover, infected plants treated with *P. aphidis* were 25% taller than control plants. It was found that *P. aphidis* activated *PR1a*—and other pathogenesis—related genes in tomato plants and could trigger an induced-resistance response against *Cla. michiganensis* that proceeded in a salicylic acid-independent manner, as shown using NahG-transgenic tomato plants
10. Wittmann et al. (2016) developed transgenic tomato plants by transferring the endolysin gene (*lys*) of bacteriophage CMP1. Endolysin gene encodes a peptidase that effectively reduces *Cla. michiganensis* population by specifically hydrolyzing its murein. The gene was transferred into tomato plants by *Agrobacterium*-mediated transformation. The presence of the gene was verified by PCR and the gene product was confirmed in immunoblots, and it stably expressed over three generations. Transgenic tomato plants did not show disease symptoms after infection with *Cmm*, but the significantly reduced populations of bacterial cells were identified in xylem sap and leaf extracts

7.41 BACTERIAL SPECK OF TOMATO

Bacterial speck of tomato (*Solanum lycopersicum* L.) caused by *Pseudomonas syringae* pv. *tomato* was first described by Okabe in 1933 from Taiwan. The economic importance of the disease was realized in late 1970s when it caused serious losses to winter tomato crop of 1977–1978 in southern Florida and in 1978 in southern transplant fields and tomato production areas of northern United States. Cool and moist conditions favor the development of the disease.

7.41.1 Host Range

Tomato is the only economically important host of the pathogen. The bacterium has also been isolated from a wide range of crop and weed species such as *Arabidopsis thaliana*, *Stellaria media*, *Lamium amplexicaule,* and *Gnaphalium* spp.

7.41.2 Symptoms

The symptoms appear on all the aboveground plant parts. Small, dark brown specks, 1–2 mm in diameter, often surrounded by discrete yellow haloes, appear on the leaves. Speck lesions coalesce to form blighted areas leading to considerable leaf distortion. On field plants, the lesions are more concentrated at the leaf margins (Figure 7.27a). Leaf lesions of bacterial speck are difficult to distinguish from those of bacterial spot and young lesions of early blight caused by *Alternaria solani*. Unlike bacterial leaf spot, the leaf lesions of bacterial speck are not greasy, but in bacterial speck, the leaf curling is more severe than bacterial spot.

(a)

(b)

FIGURE 7.27 Bacterial speck of tomato: (a) Symptoms on leaf. (From www.aggie-horticulture.tamu.edu.) (b) Symptoms on fruits. (Courtesy of M. Babadoost, Illinois Fruit and Vegetable News, Vol. 13, No. 8.)

In severe infection, irregular, dark brown lesions also appear on stem and petioles, and blossoms also get blighted. Lesions on stem and petioles also cannot be distinguished from those caused by bacterial spot. Infection of 3- to 5-leaf stage seedlings results in stunting of seedlings and considerable yield losses.

On immature green fruits (less than 3 cm in diameter), dark brown to black, slightly raised lesions, but smaller (1.5 mm in diameter) than those of bacterial spot appear, which may be surrounded by dark green haloes. The lesions are very superficial and do not crack or become scabby like those of bacterial spot. Ripe fruits are not susceptible to infection, but the black lesions formed on unripe fruits exist even after ripening (Figure 7.27b).

7.41.3 Causal Organism

Pseudomonas syringae pv. *tomato* (Okabe 1933) Young et al. 1978
Syn. *Psm. tomato*

The bacterial cells are straight, Gram-stain-negative rods and measure 0.7–1.0 × 1.8–2.8 µm. It produces cream-colored colonies with a diffusible green fluorescent pigment on King's medium B. It is strictly aerobic, motile having 1–2 polar flagella, produces levan on nutrient agar containing 5% sucrose, but oxidase and arginine dihydrolyase tests are negative. It differs from a closely related tomato pathogen, *Psm. syringae* pv. *syringae* by its fatty acid profile, ability to cause pits in pectate gel at pH 4.6, and inability to utilize erythritol or DL-lactate as a sole source of carbon (Cuppels 2001). The G + C content of DNA is 58–60 mol%. The bacterium elicits a hypersensitive reaction on tobacco, but does not cause rotting of potato tubers. Two races of the bacterium, namely, race 0 and race 1 are known. Yan et al. (2008) have proposed the division of *Psm. syringae* pv. *tomato* into two pathovars.

The bacterium has been the focus of intensive research and hence extensively used to study the molecular mechanisms of host responses to infection because *Psm. syringae* pv. *tomato* strain DC 3000 is the pathogen of a model plant, *Arabidopsis thaliana*. A major part of information about the plant immune system has been obtained by studying the interaction of *Psm. syringae* pv. *tomato* DC 3000 with its two hosts, *A. thaliana* and tomato. The molecular basis of pathogenesis in *Psm syringae* pv. *tomato* is based on the coordinated expression of multiple pathogenicity and virulence factors such as toxins, extracellular proteins, and polysaccharides; and translocation of proteins into plant cells by the type III (hrp) secretion system. The contribution of individual virulence factors to parasitism and disease development varies significantly between strains (Preston 2000). The complete genome sequence of *Psm. syringae* pv. *tomato* strain DC 3000 determined by Feil et al. (2005) revealed that the genome (6.5 mbp) is composed of a circular chromosome and two plasmids.

7.41.4 Disease Cycle

The bacterium survives in diseased plant debris, on roots and foliage of non-host plants including weeds. The bacterium can survive in soil for more than 1 year. The bacterium can also survive in the crevices and cavities of seed coat. The survival in/on seed can be for a long time (over 20 years). The dispersal of the bacterium occurs through wind-driven rain, overhead-irrigation, and farm equipment. It may move considerable distances in the water droplets or aerosols generated by rainstorms. The pathogen also spreads through contaminated seed and diseased transplants.

The penetration of the bacterium into the host is mainly through wounds, stomata, and hydathodes. The studies have shown that stomata can play an active role in limiting bacterial invasion as part of plant innate immune system (Melotto et al. 2008). As a counter defense, *Psm. syringae* pv. *tomato* strain DC 3000 uses the virulence factor coronatine to actively open stomata. Hail and windstorms are often associated with increased disease severity.

The disease is favored by cool (21°C) and moist (high R.H. and prolonged period of free moisture) weather. Resident population of the bacterium present on tomato plants can cause symptoms in 3–5 days under favorable weather conditions and under field conditions the symptoms can develop within 6–10 days. Fruits that develop during cool and moist weather can be severely infected. Fruits are most susceptible to infection from anthesis until they reach a diameter of 3 cm. Grandular and non-grandular leaf hairs present on ovaries are gradually lost after anthesis, leaving openings in young fruit's epidermis. The bacterial cells associated with leaf hairs enter through these openings and cause fruit infection.

Psm. syringae pv. *tomato* produces a toxin called coronatine, which acts as a virulence factor and is responsible for the yellow haloes surrounding leaf lesions and the stunting of young seedlings. Bacterial populations that do not produce this toxin grow poorly on host plants and cause little damage. A fragment of gene cluster controlling coronatine production has been used as a DNA probe for disease diagnosis and epidemiological studies of the pathogen. Zaccardelli et al. (2005) developed a rapid method for the detection and identification of the bacterium from tomato plants and seeds based on the PCR amplification of *Psm. syringae* pv. *tomato* strain DC 3000 *hrpZpst* gene. The results confirmed the high specificity observed using pure cultures. Cuppels et al. (2006) developed PCR-based lesion assays using crude DNA extracts and primer sets (COR 1/2 for bacterial speck and BSX 1/2 for bacterial spot) for the detection of bacterial speck and bacterial spot pathogens.

Xin and He (2013) have reviewed the work to characterize the molecular mechanisms by which *Psm. syringae* pv. *tomato* DC 3000 causes disease in plants. Genomic analysis shows that the bacterium carries a large repertoire of potential virulence factors, including proteinaceous effectors that are secreted through the type III secretion system and a polyketide phytotoxin called coronatine, which structurally mimics the plant hormone jasmonate. The findings concerning the pathogenesis of *Psm. syringae* pv. *tomato* DC3000 has not only provided several conceptual advances in understanding how a bacterial pathogen employs type III effectors to suppress plant immune responses and promote disease susceptibility, but has also facilitated the discovery of the immune function of stomata and key components of jasmonate signaling in plants. The concepts derived from the study of pathogenesis of this bacterium may prove useful in understanding pathogenesis mechanisms of other plant pathogens.

7.41.5 Management

1. Use of resistant varieties is the most effective control of the disease. A bacterial speck resistance gene, *Pto* (single dominant) discovered in an Ontario breeding line ONT 7710 of *S. pimpinellifolium* confers resistance to race 0, but not to race 1 of the pathogen. This was the first plant resistance gene to be isolated and characterized (Martin et al. 1993)
2. Use pathogen-free seed or treat the seed with hot water (pre-warm at 38°C for 10 min followed by 50°C for 25 min) or 0.8% acetic acid or 1% sodium hypochlorite (20% bleach)
3. Use of disease-free transplants
4. Destruction of diseased plant debris, weeds, and volunteer plants
5. Use of drip irrigation to reduce leaf wetness
6. Crop rotation for 3 years
7. Protective sprays of fixed copper compounds are effective. As copper resistance is widespread among the *Psm. syringae* pv. *tomato* populations in main tomato growing areas, the copper compounds should be combined with mancozeb or chlorothalonil. The sprays should be given at regular intervals starting from anthesis until the first-formed fruits attain the size of 3 cm in diameter
8. Treatment of tomato seeds (artificially inoculated with *Psm. syringae* pv. *tomato*) with a combination of mild chemo-thermal treatment, inoculation of seed with *Azospirillum brasiliense* (a plant growth-promoting bacterium), and later, a single foliar application of a copper bactericide, nearly eliminated the disease even when the plants were grown under mist for 6 weeks. However, these treatments were ineffective when tried individually (Bashan and de-Bashan 2002)
9. For the use of ASM in disease management, refer to bacterial spot of tomato and pepper

7.42 PITH NECROSIS OF TOMATO

Pith necrosis of tomato (*Solanum lycopersicum* L.) is a vascular disease that can affect tomato plants at any growth stage and in different crop systems, such as greenhouses and production fields. The causal bacterium, *Pseudomonas corrugata* is considered a weak pathogen on tomato plants that are growing too fast. The economic losses due to disease are rare, but may occur under favorable conditions for the disease (Scarlett et al. 1978; Sesto et al. 1996). Sometimes, the disease can be confused with bacterial canker.

The disease was first described by Scarlett et al. (1978) from UK. Since then it has been reported from Albania, Argentina, Belarus, Brazil, Canada, Denmark, France, Germany, Greece, India, Israel, Italy, Japan, Latvia, Macedonia, the Netherlands, New Zealand, Norway, Poland, Portugal, Russia, South Africa, Sweden, Switzerland, Syria, Tanzania, Turkey, and USA.

In southern Italy, some serological variation in strains associated with the disease has been observed. A related bacterium has also been isolated from tomato in Italy, which was named *Psm. mediterranea* (Catara et al. 2002). Xu et al. (2013) have also reported *Psm. mediterranea* causing tomato pith necrosis in the United States. Both *Psm. corrugata* and *Psm. mediterranea* have been reported causing tomato pith necrosis in Greece (Trantas et al. 2015).

Yildiz et al. (2004) reported *Psm. viridiflava* as the cause of tomato stem and pith necrosis from Brazil and according to them the disease also occurs in New Zealand, Italy, USA, Greece, Morocco, Japan, Turkey, and Bulgaria. *Pseudomonas* species causing pith necrosis symptoms on tomato and pepper collected from different areas of Argentina were identified as *Psm. corrugata*, *Psm. viridiflava*, and *Pseudomonas* spp. (Alippi et al. 2003). Aiello et al. (2013) have reported pith necrosis of tomato due to *Xanthomonas perforans* from south-eastern Sicily (Italy). Seventeen fluorescent *Pseudomonas* strains causing pith necrosis were designated as three unnamed *Pseudomonas* genomospecies- FP1, FP2, and FP3 (FP, fluorescent pseudomonads) (Sutra et al. 1997).

Trantas et al. (2013) reported *Psm. cichorii* as the causal agent of pith necrosis of tomato and presented its biochemical and molecular characterization. A detailed characterization of the genetic variability among strains of *Psm. cichorii* was obtained using BOX-PCR and MLSA utilizing three housekeeping genes (*gyrB*, *rpoD*, and *rpoB*). These findings show the emergence of a new genomovar of *Psm. cichorii*, another indication for the genetic heterogeneity of the species.

7.42.1 Host Range

Tomato is the main host. Sweet pepper (*Capsicum annuum*) and *Chrysanthemum* spp. are also host of the bacterium (Fiori 1992).

7.42.2 Symptoms

The disease occurs randomly within fields, and initial symptoms are usually observed at the time the first fruit clusters reach the mature green stage. Initial symptoms include chlorosis (yellowing) of young leaves and shoots followed by wilting of the infected shoots in the upper part of the plant canopy. This wilting is usually associated with internal browning (or necrosis) of the basal portion of the stem. Thereafter, plants wilt and rapidly collapse. On the stem, large, slightly sunken, and glassy to black necrotic patches develop that also show cracks. When symptomatic stem is cut open longitudinally, the pith (center of the stem) may be hollow, and/or degraded and often has distinct chambers (ladder like appearance) and often exhibits a dark discoloration (Figure 7.28). Profuse development of adventitious roots can be associated with the affected pith areas, and the stem may appear swollen. The pith necrosis may spread into the entire stem of the plant and to the petioles. Bacterial slime may ooze from leaf scars. Infected stems may shrink, crack, or collapse.

Plants may die if the lower stem is affected; however, the disease usually does not progress, and plants outgrow the condition. Plants affected with pith necrosis neither exhibit the marginal necrosis of leaflets nor the bird's-eye spotting of the fruit, characteristic of bacterial canker.

FIGURE 7.28 Pith necrosis of tomato. (Courtesy of Gerald Holmes, California Polytechnic State University at San Luis Obispo, Bugwood.org.)

7.42.3 Causal Organism

Pseudomonas corrugata (*ex* Scarlett et al. 1978) Roberts and Scarlett 1981 sp. nov.

The below given description of the bacterium is as emended by Sutra et al. (1997). The bacterium is a Gram-stain-negative, non-sporing rod and strictly aerobic. The cells are motile having multiple polar flagella. Colonies on YBGA are wrinkled or smooth and frequently produce yellow to brown pigments. This organism is non-fluorescent on King's medium B, and oxidase positive, does not produce levan, is not pectolytic, reduces nitrates into nitrites, and accumulates PHB. About 80% of the strains are arginine dihydrolase positive after 15 days of incubation, and about 20% produce hypersensitive reaction on tobacco plants. The reactions of hydrolysis of Tween 80 and of gelatin are variable. The levels of DNA relatedness of *Psm. corrugata* strains with type strain CFBP 2431 range from 66% to 99%. The G + C content of the DNA is 58.4–60.8 mol%. Regarding the utilization and non-utilization of various substrates/chemical compounds by *Psm. corrugata* strains, the readers may refer to Sutra et al. (1997).

Psm. corrugata strain 2140R has been used successfully as a biocontrol agent for Take-all disease of wheat caused by *Gaeumannomyces graminis* var. *tritici*. The bacterium produces a lipopetide siderophore, named corrugatin (Risse et al. 1998).

7.42.4 Disease Cycle

The bacterium has been isolated from seed (Kritzman 1991). The bacterium is also soil-borne and can survive in the soil (Scortichini 1989). Furthermore, it has been isolated from irrigation and recirculation water (Scarlett et al. 1978; Sadowska-Rybak et al. 1997). The disease may spread by manipulations during the growing season. Pith necrosis usually occurs, especially on steam-sterilized soils under high nitrogen fertilization and high humidity and low night temperatures. Disease incidence and severity is favored by high nitrogen fertilization, cool temperatures at night, high humidity, and plastic mulch.

The bacterium has been observed in the rhizosphere of lucerne and wheat and as an endophyte in strawberry and grapevine (Lukezic 1979; Roberts and Brewster 1991; Bell et al. 1995). In Argentina, all *Psm. corrugata* strains tested were copper-resistant, while *Psm. viridiflava* strains were more variable (Alippi et al. 2003).

7.42.5 Management

1. The best practice is prevention of disease by avoiding the use of excessive amounts of nitrogen in tomato cultivation, especially early in the season when nights are still cool
2. Disease can be controlled, and diseased plants even cured, by reducing nitrogen fertilization and proper

addition of potassium and calcium to strengthen the plants (Scarlett et al. 1978)

3. The use of plant activators such as acibenzolar-*S*-methyl (Actigard) have resulted in 55% disease reductions, but the application must be started before the appearance of symptoms
4. Seed treatment with 1% sodium hypochlorite for 1 min eliminates seed-borne infection (Yildiz et al. 2004)

REFERENCES

Abbasi, P. A., Soltani, N., Cuppels, D. A., and Lazarovits, G. 2002. Reduction of bacterial spot disease severity on tomato and pepper plants with foliar applications of ammonium lignosulfonate and potassium phosphate. *Plant Dis.* 86: 1232–1236.

Aiello, D., Scuderi, G., Vitale, A., Firrao, G., Polizzi, G., et al. 2013. A pith necrosis caused by *Xanthomonas perforans* on tomato plants. *Eur. J. Plant Pathol.* 137: 29–41.

Alippi, A. M., Dal Bo, E., Ronco, L. B., López, M. V., López, A. C., et al. 2003. *Pseudomonas* populations causing pith necrosis of tomato and pepper in Argentina are highly diverse. *Plant Pathol.* 52: 287–302.

Antoniou, P. P., Tjamos, E. C., Andreou, M. T., and Panagopoulos, C. G. 1995. Effectiveness, modes of action and commercial application of soil solarization for control of *Clavibacter michiganensis* subsp. *michiganensis* of tomatoes. *Acta Hortic.* (*ISHS*) 382: 119–128.

Araújo, E. R., Costa, J. R., Ferreira, M. A. S. V., and Quezado-Duval, A. M. 2017. Widespread distribution of *Xanthomonas perforans* and limited presence of *X. gardneri* in Brazil. *Plant Pathol.* 66: 159–168.

Araújo, E. R., Costa, J. R., Pontes, N. C., and Quezado-Duval, A. M. 2015. *Xanthomonas perforans* and *X. gardneri* associated with bacterial leaf spot on weeds in Brazilian tomato fields. *Eur. J. Plant Pathol.* 143: 543–548.

Bach, H. J., Jessen, I., Schloter, M., and Munch, J. C. 2003. A TaqMan-PCR protocol for quantification and differentiation of the phytopathogenic *Clavibacter michiganensis* subspecies. *J. Microbiol. Methods* 52: 85–91.

Barda, O., Shalev, O., Alster, S., Buxdorf, K., Gafni, A., et al. 2015. *Pseudozyma aphidis* induces salicylic-acid-independent resistance to *Clavibacter michiganensis* in tomato plants. *Plant Dis.* 99: 621–626.

Bashan, Y., and de-Bashan, L. E. 2002. Reduction of bacterial speck (*Pseudomonas syringae* pv. *tomato*) of tomato by combined treatments of plant growth-promoting bacterium, *Azospirillum brasilense*, streptomycin sulphate, and chemo-thermal seed treatment. *Eur. J. Plant Pathol.* 108: 821–829.

Baysal, O., Soylu, E. M., and Soylu, S. 2003. Induction of defence-related enzymes and resistance by the plant activator acibenzolar-*S*-methyl in tomato seedlings against bacterial canker caused by *Clavibacter michiganensis* subsp. *michiganensis*. *Plant Pathol.* 52: 747–753.

Bell, C. R., Dickie, G. A., and Chan, J. W. Y. F. 1995. Variable response of bacteria isolated from grapevine xylem to control grape crown gall disease in planta. *Am. J. Enol. Viticult.* 46: 499–508.

Braun, E. 2001. Bacterial canker of tomato. pp. 76–77 in: *Encyclopedia of Plant Pathology*. O. C. Maloy, and T. D. Murray eds. John Wiley & Sons, New York.

Catara, V., Sutra, L., Morineau, A., Achouak, W., Christen, R., et al. 2002. Phenotypic and genomic evidence for the revision of *Pseudomonas corrugata* and proposal of *Pseudomonas mediterranea* sp. nov. *Int J. Syst. Evol. Microbiol.* 52: 1749–1758.

Chalupowicz, L., Zellermann, E.-M., Fluegel, M., Dror, O., and Eichenlaub, R. 2012. Colonization and movement of GFP-labeled *Clavibacter michiganensis* subsp. *michiganensis* during tomato infection. *Phytopatholgy* 102: 23–31.

Cuppels, D. A. 2001. Bacterial speck of tomato. pp. 82–83 in: *Encyclopedia of Plant Pathology*. Vol. 1, O. C. Maloy, and T. D. Murray eds. John Wiley & Sons, New York.

Cuppels, D. A., Louws, F. J., and Ainsworth, T. 2006. Development and evaluation of PCR-based diagnostic assays for the bacterial speck and bacterial spot pathogens of tomato. *Plant Dis.* 90: 451–458.

Davis, M. J., Gillaspie, A. G. Jr., Vidaver, A. K., and Harris, R. W. 1984. *Clavibacter*: A new genus containing some phytopathogenic coryneform bacteria, including *Clavibacter xyli* subsp. *xyli* sp. nov., subsp. nov. and *Clavibacter xyli* subsp. *cynodontis* subsp. nov., pathogens that cause ratoon stunting disease of sugarcane and Bermudagrass stunting disease. *Int. J. Syst.Bacteriol.* 34: 107–117.

Doidge, E. M. 1920. A tomato canker. *J. Dept. Agric. Union S. Afr.* 1: 718–721.

Duan, J., Zhao, B., Wang, Y., and Yang, W. 2015. Development and validation of a standard area diagram set to aid estimation of bacterial spot severity on tomato leaves. *Eur. J. Plant Pathol.* 142: 665–675.

Dutta, B., Gitaitis, R., Sanders, H., Booth, C., Smith, S., et al. 2014. Role of blossom colonization in pepper seed infestation by *Xanthomonas euvesicatoria*. *Phytopathology* 104: 232–239.

Feil, H., Feil, W. S., Chain, P., Larimer, F., DiBartolo, G., et al. 2005. Comparison of the complete genome sequences of *Pseudomonas syringae* pv. *syringae* B728a and pv. *tomato* DC3000. *Proc. Natl. Acad. Sci. USA* 102: 11064–11069.

Fiori, M. 1992. A new bacterial disease of chrysanthemum: A stem rot by *Pseudomonas corrugata* Roberts et Scarlett. *Phytopathol. Mediterr.* 31: 110–114.

Frenkel, O., Bornestein, M., Shulhani, R., Sharabani, G., Sofer, M., et al. 2016. Secondary spread of *Clavibacter michiganensis* subsp. *michiganensis* in nurseries and the conditions leading to infection of tomato seedlings. *Eur. J. Plant Pathol.* 144: 569–579.

Ftayeh, R. M., von Tiedemann, A., and Rudolph, K. W. E. 2011. A new selective medium for isolation of *Clavibacter michiganensis* subsp. *michiganensis* from tomato plants and seed. *Phytopathology* 101: 1355–1364.

Giovanardi, D., Biondi, E., Ignjatov, M., Jevtić, R., and Stefani, E. 2018. Impact of bacterial spot outbreaks on the phytosanitary quality of tomato and pepper seeds. *Plant Pathol.* 67: 1168–1176.

Huang, C.-H., and Vallad, G. E. 2018. Soil applications of acibenzolar-*S*-methyl induce defense gene expression in tomato plants against bacterial spot. *Eur. J. Plant Pathol.* 150: 971–981.

Janse, J. D. 2005. *Phytobacteriology: Principles and Practice*. CABI Publishing, Wallingford, UK, 360 pp.

Ji, P., Campbell, H. L., Kloepper, J. W., Jones, J. B., Suslow, T. V., et al. 2006. Integrated biological control of bacterial speck and spot of tomato under field conditions using foliar biological control agents and plant growth-promoting rhizobacteria. *Biol. Control* 36: 358–367.

Jones, J. B. 2001. Bacterial spot of tomato. pp. 83–84 in: *Encyclopedia of Plant Pathology*. Vol. 1, O. C. Maloy, and T. D. Murray eds. John Wiley & Sons, New York.

Jones, J. B., Bouzar, H., Stall, R. E., Almira, E. C., Roberts, P. D., et al. 2000. Systematic analysis of xanthomonads (*Xanthomonas* spp.) associated with pepper and tomato lesions. *Int. J. Syst. Evol. Microbiol.* 50: 1211–1219.

Jones, J. B., Lacy, G. H., Bouzar, H., Stall, R. E., and Schaad, N. W. 2004. Reclassification of the xanthomonads associated with bacterial spot disease of tomato and pepper. *Syst. Appl. Microbiol.* 27: 755–762.

Jones, J. B., Stall, R. E., and Bouzar, H. 1998. Diversity among xanthomonads pathogenic on pepper and tomato. *Annu. Rev. Phytopathol.* 36: 41–58.

Kritzman, G. 1991. A method for detection of seedborne bacterial diseases in tomato seeds. *Phytoparasitica* 19: 133–141.

Leboeuf, J., Cuppels, D., Dick, J., Pitblado, R., Loewen, S., et al. 2005. Bacterial diseases of tomato: Bacterial spot, bacterial speck, bacterial canker. Ministry of Agriculture, Food and Rural Affairs, Ontario, Canada [http://www.omafra.gov.on.ca/english/crops/facts/05-069.htm]

Lelis, F. M. V., Czajkowski, R., de Souza, R. M., Ribeiro, D. H., and van der Wolf, J. M. 2014. Studies on the colonization of axenically grown tomato plants by a GFP-tagged strain of *Clavibacter michiganensis* subsp. *michiganensis*. *Eur. J. Plant Pathol.* 139: 53–66.

Li, X., Tambong, J., Yuan, K. (X.), Chen, W., Xu, H., et al. 2018. Reclassification of *Clavibacter michiganensis* subspecies on the basis of whole-genome and multi-locus sequence analyses. *Int. J. Syst. Evol. Microbiol.* 68: 234–240.

Louws, F. J., Wilson, M., Campbell, H. L., Cuppels, D. A., Jones, J. B., et al. 2001. Field control of bacterial spot and bacterial speck of tomato using a plant activator. *Plant Dis.* 85: 481–488.

Lukezic, F. L. 1979. *Pseudomonas corrugata*, a pathogen of tomato, isolated from symptomless alfalfa roots. *Phytopathology* 69: 27–31.

Luo, L. X., Walters, C., Bolkan, H., Liu, X. L., and Li, J. Q. 2008. Quantification of viable cells of *Clavibacter michiganensis* subsp. *michiganensis* using a DNA binding dye and a real-time PCR assay. *Plant Pathol.* 57: 332–337.

Marcuzzo, L. L., Becker, W. F., and Fernandes, J. M. C. 2009. Some epidemiological aspects of bacterial spot (*Xanthomonas* spp.) of tomato in Cacador/SC, Brazil. *Summa Phytopathol.* 35: 132–135.

Martin, G. B., Brommonschenkel, S. H., Chunwongse, J., Frary, A., Ganal, M. W., et al. 1993. Map-based cloning of a protein kinase gene conferring disease resistance in tomato. *Science* 262: 1432–1436.

Melotto, M., Underwood, W., and He, S. Y. 2008. Role of stomata in plant innate immunity and foliar bacterial diseases. *Annu. Rev. Phytopathol.* 46: 101–122.

Obradovic, A., Jones, J. B., Momol, M. T., Balogh, B., and Olson, S. M. 2004. Management of tomato bacterial spot in the field by foliar application of bacteriophages and SAR inducers. *Plant Dis.* 88: 736–740.

Obradovic, A., Jones, J. B., Momol, M. T., Olson, S. M., Jackson, L. E., et al. 2005. Integration of biological control agents and systemic acquired resistance inducers against bacterial spot on tomato. *Plant Dis.* 89: 712–716.

Okabe, N. 1933. Bacterial diseases of plants occurring in Formosa. II. Bacterial leaf spot of tomato. *J. Soc. Trop. Agric. Taiwan* 5, 26–36.

Paret, M. L., Vallad, G. E., Averett, D. R., Jones, J. B., and Olson, S. M. 2013. Photocatalysis: Effect of light-activated nanoscale formulations of TiO_2 on *Xanthomonas perforans* and control of bacterial spot of tomato. *Phytopathology* 103: 228–236.

Patel, M. K., Kulkarni, Y. S., and Dhande, G. W. 1950. Bacterial spot of chillies. *Indian Phytopathol.* 3: 95–97.

Preston, G. M. 2000. *Pseudomonas syringae* pv. *tomato*: The right pathogen, of the right plant, at the right time. *Mol. Plant Pathol.* 1: 263–275.

Risse, D., Beiderbeck, H., Taraz, K., Budzikiewicz, H., and Gustine, D. 1998. Corrugatin, a lipopetide siderophore from *Pseudomonas corrugata*. *Z. Naturforsch.* 53c: 295–304.

Roberts, J. T., and Scarlett, C. M. 1981. Validation of the publication of new names and new combinations previously effectively published outside the *IJSB*. *Int. J. Syst. Bacteriol.* 31: 215–218. Effective publication: Scarlett, C. M., Fletcher, J. T., Roberts, P., and Leliott, R. A. 1978. Tomato pith necrosis caused by *Pseudomonas corrugata* n. sp. *Ann. Appl. Biol.* 88: 105–114.

Roberts, W. P., and Brewster, C. M. 1991. Identification of wheat rhizosphere bacteria inhibitory to root growth. *Australasian Plant Pathol.* 20: 47–51.

Sadowska-Rybak, M., Pein, B., and Buettner, C. 1997. Transmission of *Pseudomonas corrugata* by watering and by nutrient solutions of low tide-high tide-irrigation-systems. *Gesunde. Pflanzen* 49: 226–229.

Safaie Farahani, A., and Taghavi, S. M. 2018. Rutin promoted resistance of tomato against *Xanthomonas perforans*. *Eur. J. Plant Pathol.* 151: 527–531.

Scarlett, C. M., Fletcher, J. T., Roberts, P., and Lelliott, R. A. 1978. Tomato pith necrosis caused by *Pseudomonas corrugata* n. sp. *Ann. Appl. Biol.* 88: 105–114.

Scortichini, M. 1989. Occurrence in soil and primary infection of *Pseudomonas corrugate* Roberts and Scarlett. *J. Phytopathol.* 125: 33–40.

Sen, Y., van der Wolf, J., Visser, R. G. F., and van Heusden, S. 2015. Bacterial canker of tomato: Current knowledge of detection, management, resistance, and interactions. *Plant Dis.* 99: 4–13.

Sesto, F., Areddia, R., and Catara, V. 1996. Infections of *Pseudomonas corrugata* in tomato plantlets in the nursery. *Informatore Fitopatologico* 46: 62–64.

Shahrtash, M., and Lewis Ivey, M. L. 2015. Evaluation of sodium selenate as a seed disinfestant to eradicate *Xanthomonas vesicatoria*, *X. euvesicatoria* and *X. perforans* on pepper and tomato seeds (Abstr.). *Phytopathology* 105: S2.9–2.10.

Smith, E. F. 1910. A new tomato disease of economic importance. *Science, Washington* 31: 794–796.

Stall, R. E., Jones, J. B., and Minsavage, G. V. 2009. Durability of resistance in tomato and pepper to xanthomonads causing bacterial spot. *Annu. Rev. Phytopathol.* 47: 265–284.

Strayer, A. L., Jeyaprakash, A., Minsavage, G. V., Timilsina, S., Vallad, G. E., et al. 2016a. A multiplex real-time PCR assay differentiates four *Xanthomonas* species associated with bacterial spot of tomato. *Plant Dis.* 100: 1660–1668.

Strayer, A., Ocsoy, I., Tan, W., Jones, J. B., and Paret, M. L. 2016b. Low concentrations of a silver-based nanocomposite to manage bacterial spot of tomato in the greenhouse. *Plant Dis.* 100: 1460–1465.

Sun, H., Wei, J., Zhang, J., and Yang, W. 2014. A comparison of disease severity measurements using image analysis and visual estimates using a category scale for genetic analysis of resistance to bacterial spot in tomato. *Eur. J. Plant Pathol.* 139: 125–136.

Sutra, L., Siverio, F., Lopez, M. M., Hunault, G., Bollet, C., et al. 1997. Taxonomy of *Pseudomonas* strains isolated from tomato pith necrosis: Emended description of *Pseudomonas corrugata* and proposal of three unnamed fluorescent *Pseudomonas* genomospecies. *Int. J. Syst. Bacteriol.* 47: 1020–1033.

Trantas, E. A., Sarris, P. F., Mpalantinaki, E. E., Pentari, M. G., Ververidis, F. N., et al. 2013. A new genomovar of *Pseudomonas cichorii*, a causal agent of tomato pith necrosis. *Eur. J. Plant Pathol.* 137: 477–493.

Trantas, E. A., Sarris, P. F., Pentari, M. G., Mpalantinaki, E. E., Ververidis, F. N., et al. 2015. Diversity among *Pseudomonas corrugata* and *Pseudomonas mediterranea* isolated from tomato and pepper showing symptoms of pith necrosis in Greece. *Plant Pathol.* 64: 307–318.

Tudor-Nelson, S. M., Minsavage, G. V., Stall, R. E., and Jones, J. B. 2003. Bacteriocin-like substances from tomato race 3 strains of *Xanthomonas campestris* pv. *vesicatoria*. *Phytopathology* 93: 1415–1421.

Vauterin, L., Hoste, B., Kersters, K., and Swings, J. 1995. Reclassification of *Xanthomonas*. *Int. J. Syst. Bacteriol.* 45: 472–489.

Vega, D., and Romero, A. M. 2016. Survival of *Clavibacter michiganensis* subsp. *michiganensis* in tomato debris under greenhouse conditions. *Plant Pathol.* 65: 545–550.

Wittmann, J., Brancato, C., Berendzen, K. W., and Dreiseikelmann, B. 2016. Development of a tomato plant resistant to *Clavibacter michiganensis* using the endolysin gene of bacteriophage CMP1 as a transgene. *Plant Pathol.* 65: 496–502.

Xin, X.-F., and He, S. Y. 2013. *Pseudomonas syringae* pv. *tomato* DC 3000: A model pathogen for probing disease susceptibility and hormone signaling in plants. *Annu. Rev. Phytopathol.* 51: 473–498.

Xu, X., Baysal-Gurel, F., and Miller, S. A. 2013. First report of tomato pith necrosis caused by *Pseudomonas mediterranea* in the United States and *P. corrugata* in Ohio. *Plant Dis.* 97: 988.

Xu, X., Rajashekara, G., Paul, P. A., and Miller, S. A. 2012. Colonization of tomato seedlings by bioluminescent *Clavibacter michiganensis* subsp. *michiganensis* under different humidity regimes. *Phytopathology* 102: 177–184.

Yan, S., Liu, H., Mohr, T. J., Jenrette, J. Chiodini, R., et al. 2008. Role of recombination in the evolution of Model plant pathogen *Pseudomonas syringae* pv. *tomato* DC3000, a very atypical tomato strain. *Appl. Environ. Microbiol.* 74: 3171–3181.

Yasuhara-Bell, J., Kubota, R., Jenkins, D. M., and Alvarez, A. M. 2013. Loop-mediated amplification of the *Clavibacter michiganensis* subsp. *michiganensis micA* gene is highly specific. *Phytopathology* 103: 1220–1226.

Yasuhara-Bell, J., Marrero, G., and Alvarez, A. M. 2014. Genes *clvA*, *clvF* and *clvG* are unique to *Clavibacter michiganensis* subsp. *michiganensis* and highly conserved. *Eur. J. Plant Pathol.* 140: 655–664.

Yildiz, H. N., Aysan, Y., Sahin, F., and Cinar, O. 2004. Potential inoculum sources of tomato stem and pith necrosis caused by *Pseudomonas viridiflava* in the Eastern Mediterranean Region of Turkey. *Z. Pflanzenkr. Pflanzenschutz* 111: 380–387.

Young, J. M., Dye, D. W., and Wilkie, J. P. 1978. Genus *Pseudomonas* Migula 1894. In: Young, J. M., Dye, D. W., Bradbury, J. F., Panagopoulos, C. G., and Robbs, C. F. 1978. A proposed nomenclature and classification for plant pathogenic bacteria. *N. Z. J. Agric. Res.* 21: 153–177.

Zaccardelli, M., Spasiano, A., Bazzi, C., and Merighi, M. 2005. Identification and *in planta* detection of *Pseudomonas syringae* pv. *tomato* using PCR amplification of $hrpZ_{pst}$. *Eur. J. Plant Pathol.* 111: 85–90.

Zitter, T. A. 1985. *Bacterial Diseases of Tomato*. Department of Plant Pathology, Cornell University, Ithaca, NY.

7.43 BACTERIAL BLIGHT OF PEA

Pseudomonas syringae pv. *pisi* causes bacterial blight of pea (*Pisum sativum* L.), and it was first reported by Sackett (1916) from Colorado, USA. It is a serious disease of pea in cool and wet weather. In 2004, under favorable weather conditions, the disease caused close to 90% plant mortality in some fields in Castilla y León region, the main pea growing area of Spain. Although the epiphytotics of the disease in Australia are infrequent and may be 1 in 10 years, the field pea crop suffered a loss of 10% due to an epiphytotic in 1992. In Wimmera region of Australia, the disease incidence exceeding 30% has also been reported. The yield losses as high as 70% have been reported on an early sown pea cv. Sugar Snap from Horsham, northwest Victoria, Australia. The disease also occurs in Armenia and India.

Psm. syringae pv. *syringae* has also been reported as causal agent of bacterial blight of pea (Martín-Sanz et al. 2011, 2012a). Martín-Sanz et al. (2011) have also reported that *Psm. syringae* pv. *syringae* is as frequent as *Psm. syringae* pv. *pisi* as the cause of bacterial diseases in pea.

7.43.1 Host Range

In addition to *P. sativum*, *Psm. syringae* pv. *pisi* infects *P. sativum* var. *arvense*, *Lathyrus odoratus*, *L. latifolius*, *Vicia bengalensis*, *Lablab niger*, and *Vigna* sp. (Bradbury 1998).

7.43.2 Symptoms

The symptoms appear on all the aboveground plant parts. The initial symptoms are discrete, dark green, water-soaked spots on leaves, pods, and stems. In cool and wet weather, the spots enlarge and coalesce to cover large areas. In warm and dry weather, lesion development is checked resulting in smaller lesions and water-soaking also becomes less obvious. With age, lesions become darker and finally necrotic (Figure 7.29a and b). Older lesions on pods are brown and papery. In severely infected pods, the seeds may be covered with bacterial slime and pathogen may penetrate the seeds through the funicle and micropyle. When the infection takes place at the seedling stage, the entire crop may be destroyed.

7.43.3 Causal Organisms

Pseudomonas syringae pv. *pisi* (Sackett 1916) Young et al. 1978

Syn. *Psm. pisi*

Psm. syringae pv. *syringae* van Hall 1902

The below given description of *Psm. syringae* pv. *pisi* is taken mainly from Martín Sanz et al. (2005).

Bacterial cells are Gram-stain-negative, non-sporing rods, measuring 0.7 × 2.0–3.0 µm, and motile by one to five polar flagella. Most strains produce a yellowish-green diffusible pigment on King's medium B, which fluoresces blue under ultraviolet light. Moderate growth occurs in 24 h on agar slants, which is filiform, glistening, and greyish-white. The tests for nitrate reduction, starch hydrolysis, indole production, and H_2S production are negative. An alkaline reaction is produced in litmus milk and soft curd formed is cleared. Gelatin is liquefied, but there is no lipolytic activity. Acid, but no gas is

(a) (b)

FIGURE 7.29 Bacterial blight of pea: (a) Vegetable pea. (Courtesy of Julien Lanson, www.inra.frhyp3.) (b) Field pea. (Courtesy of Philip Wright, Department of Primary Industries, Orange, Australia.)

produced from glucose, sucrose, and galactose. The bacterium is chemoorganotroph and obligate aerobe. Optimum growth occurs at 26°C–28°C and at pH 6.5–7.5. The bacterium has the typical characteristics of a LOPAT group 1[a] green fluorescent pseudomonad. The bacterium can be detected from the infected seedlings with the help of ELISA. An avirulence gene designated as *avrPpiB* from race 3 has been cloned and sequenced.

Martín-Sanz et al. (2013) characterized a collection of 88 putative *Psm. syringae* pv. *syringae* strains, including 39 strains isolated from pea, by rep-PCR, MLST, and *syrB* amplification and evaluated these strains for pathogenicity and virulence. rep-PCR data grouped the strains from pea into two groups (1B and 1C) together with strains from other hosts; a third group (1A) was formed exclusively with strains isolated from non-legume species. The inoculations of these strains on two pea cultivars showed that the strains from groups 1A and 1C were less virulent than strains from group 1B, suggesting a possible pathogenic specialization in this group. These findings show the existence of genetically and pathogenically distinct *Psm. syringae* pv. *syringae* strain groups from pea, which can be useful for the diagnostic and epidemiology of the pathogen and also for resistance breeding.

Out of 298 bacterial isolates collected from pea cultivars, landraces, and breeding lines in North-Central Spain over several years, 225 of the isolates were identified as *Psm. syringae*, either pv. *pisi* (110 isolates) or pv. *syringae* (112 isolates) on the basis of biochemical-physiological characteristics and molecular markers. It shows that *Psm. syringae* pv. *syringae* is as frequent as *Psm. syringae* pv. *pisi* as the cause of bacterial diseases in pea. Out of 225 isolates, 222 were pathogenic on pea. Simultaneous infections of *Psm. syringae* pv. *pisi* and pv. *syringae* in the same fields were observed, suggesting the importance of resistance to both the pathovars in future commercial cultivars. The search for resistance among pea genotypes suitable for production in this region of Spain or as breeding material, identified the presence of resistance genes for all the *Psm. syringae* pv. *pisi* races except race 6. The pea cultivars Kelvendon Wonder, Cherokee, Isard, Iceberg, Messire, and Attika were found suitable sources of resistance to *Psm. syringae* pv. *syringae* (Martín-Sanz et al. 2011).

Martín-Sanz et al. (2012b) examined a collection of 91 strains of *Psm. syringae* pv. *pisi* that included 53 strains from recent outbreaks in Spain and from 14 other countries and characterized them on the basis of 55 nutritional tests, genetic analysis (rep-PCR, amplification of AN3 and AN7 specific markers, and MLST), and pathogenicity on the differential pea cultivars to identify races. Principal component analysis and distance dendrograms confirmed the existence of two genetic lineages within this pathovar, which are clearly discriminated by the AN3/AN7 markers, rep-PCR, and MLST. The strains from races 1 and 7 amplified the AN3 marker, those from races 2, 6, and 8 amplified AN7, while strains of races 3, 4, and 5 amplified either AN3 or AN7. The Spanish collection diversity reflects the variability found in the worldwide collection, suggesting multiple introductions of the bacteria into Spain by contaminated seed lots.

There are eight races of the pathogen, which have been identified on the basis of their reactions to eight differentials. Race 2 is the most frequently distributed worldwide followed by races 6 and 4. Gene-for-gene relationship between pea cultivars and pathogenic races is given in Table 7.8.

The identification of seven races of *Psm. syringae* pv. *pisi* based on a set of eight different pea cultivars allowed a gene-for-gene relationship to be proposed based on six matching pairs of resistance genes in the host and avirulence genes in the pathogen (Taylor et al. 1989; Bevan et al. 1995). On the basis of segregation among F_2 populations of crosses between differential cultivars sequentially inoculated with races of *Psm. syringae* pv. *pisi*, Bevan et al. (1995) got evidence for four and possibly six putative resistance (*R*)/avirulence (*A*) gene pairs. *R1*, *R2*, and *R3* are dominant resistance alleles at single loci; *R4* is a dominant allele at a single locus, which exhibits variable expression possibly dependent on genetic background. There is evidence that *R3* and *R4* are at linked loci. Homology tests provided evidence that the alleles *R2*, *R3*, and *R4* were present in more than one cultivar. Two other alleles, *R5* and *R6*, were postulated to explain the observed segregation in certain crosses. They further inferred that out of races 2, 3,

TABLE 7.8
Gene-for-Gene Relationship between Pea Cultivars and Races of *Pseudomonas syringae* pv. *pisi*

							Avirulence Gene						
Resistance Gene (*R*)							**1**	**2**	**3**	**4**	**5**	**6**	**7**
							1	.	.	.	.	.	.
							.	2	.	.	2	.	2
							3	.	3	.	.	.	3
							4	.	.	4	4	.	4
							.	.	.	.	(5)	.	.
							(6)	.	.	.	(6)	.	.
Kelvendon Wonder	.	.	.	.	.	.	+	+	+	+	+	+	+
Early Onward	.	2	.	.	.	.	+	−	+	+	−	+	−
Belinda	.	.	3	.	.	.	−	+	−	+	+	+	−
Hurst's Greenshaft	.	.	.	4	.	(6)	−	+	+	−	−	+	−
Partridge	.	.	3	4	.	.	−	+	−	−	−	+	−
Sleaford Triumph	.	2	.	4	(5)	.	−	−	+	−	−	+	−
Vinco	1	2	3	.	(5)	.	−	−	−	+	−	+	−
Fortune	.	2	3	4	.	.	−	−	−	−	−	+	−

Source: Bevan, J.R. et al., *Plant Pathol.* 44: 98–108, 1995.
+ = susceptible response, − = hypersensitive reaction, the genes in brackets are partially confirmed, . = gene absent.

and 4, each carries a different single avirulence gene, race 6 carries no apparent avirulence gene, and race 7 carries at least *A2*, *A3*, and *A4*. Race 1 carries *A1*, *A3*, *A4*, and possibly *A6*, and race 5 carries *A2*, *A4*, and possibly *A5* and *A6*.

7.43.4 Disease Cycle

Seed-borne inoculum is the main source of perpetuation of the bacterium. The bacterium is internally and externally seed-borne and can survive for at least 3 years. Infected plant debris, particularly on soil surface and volunteer plants are other sources of inoculum. The survival time of bacterium is reduced by burying the debris. In southern Australia, it was reduced from 2 years to 11 months when the stubbles were buried 10 cm deep in the soil.

The penetration of the bacterium occurs through stomata, but mainly through wounds caused by hail, frost, storms, sand blasting, and cultural operations. Ice-nucleating activity of the epiphytic cells of the bacterium increases frost damage. The local spread of the bacterium occurs through wind-splashed rain and overhead-irrigation. The long distance spread is through the infected seed. Cool and wet or humid weather favors the disease, while warm and dry weather reduces the severity of the disease.

Epiphytic phase is the main characteristic of the life cycle of the bacterium. The bacterium survives, multiplies, and spreads epiphytically in the crop. The inoculation of race 2 on susceptible cultivars resulted in the reduction of its population during the first 3 days, but then it increased 1,000 times more than the initial count without producing any symptoms. When it was inoculated on resistant cultivars, the bacteria did not multiply and the population levels decreased slightly. It indicates that the development of the resident phase is race specific (Grondeau et al. 1996).

Soil water content and level of seed-borne inoculum greatly influence the transmission of the bacterium from seed to the seedlings. High inoculum level and high soil moisture content increase the number of infected seedlings and the number of lesions per plant (Roberts et al. 1996).

Stem inoculation technique is generally used to screen pea germplasm assuming that the responses would be same in other plant parts. However, Elvira-Recuenco (2000) showed that a resistant response in the stem was not always associated with the resistance in the leaf and/or pod. This was found for both race-specific and non-race-specific resistance. Elvira-Recuenco et al. (2003) found that race-specific resistance genes conferred resistance in the stem in a consistent way, but the leaf and pod response varied depending on the matching resistance and avirulence gene combination. *R2* generally conferred resistance in all the plant parts. *R3* and *R4* singly did not confer complete resistance in leaf and pod, but *R3* in combination with *R2* or *R4* enhanced leaf and pod resistance. Race-non-specific resistance conferred stem resistance to all the races, leaf and pod resistance to races 2, 5, and 7, and variable reactions in leaves and pods to races 1, 3, 4, and 6.

Martín-Sanz et al. (2012a) studied the effects of frost on disease severity caused by *Psm. syringae* pv. *pisi* and *Psm. syringae* pv. *syringae* and found that frost tolerance was effective in lowering the disease effects caused under frost-stress conditions, even in the absence of disease resistance genes. However, the highest degree of this protection is reached when frost tolerance and disease-resistance genes are combined in the same genetic background.

7.43.5 Management

1. The cheapest and most effective method is the use of resistant cultivars. Many cultivars contain race-specific resistance genes for all the known races except the race 6
2. The next best approach is the use of the pathogen-free seed. The following measures help in producing the pathogen-free seed
    a. Production of seed in arid areas
    b. Avoidance of sprinkler irrigation
    c. Field inspection from midflowering to late pod-fill stage. Removal of statutory certification controls in UK in 1993 greatly increased the incidence of seed-borne infection (Reeves et al. 1996). Seed test is available to detect seed-borne infection in the seed. It can detect the seed-borne infection level of less than 0.05%.
3. The following measures reduce the disease incidence by checking the spread of the pathogen
    a. Use clean farm equipment
    b. Avoid sprinkler irrigation
    c. Destroy infected plant debris and volunteer plants

4. Crop rotation for 2 years
5. In Australia, the disease is successfully managed by adopting an integrated disease management approach consisting of crop rotation, use of pathogen-free seed, avoidance of planting in areas prone to frequent frosts or extreme wet weather, crop hygiene, and avoiding early sowing. Seed treatment and application of foliar bactericides have limited use in controlling the disease (Hollaway et al. 2007)

REFERENCES

Bevan, J. R., Taylor, J. D., Crute, I. R., Hunter, P. J., and Vivian, A. 1995. Genetics of specific resistance in pea (*Pisum sativum*) cultivars to seven races of *Pseudomonas syringae* pv. *pisi*. *Plant Pathol.* 44: 98–108.

Bradbury, J. F. 1998. *Pseudomonas pisi. IMI Descriptions of Fungi and Bacteria*. Sheet 126, CAB International, Wallingford, UK.

Elvira-Recuenco, M. 2000. Sustainable Control of Pea Bacterial Blight: Approaches for Durable Genetic Resistance and Biocontrol by Endophytic Bacteria. PhD thesis, Wageningin University, the Netherlands (cited by Martín Sanz et al. 2005).

Elvira-Recuenco, M., Bevan, J. R., and Taylor, J. D. 2003. Differential responses to pea bacterial blight in stems, leaves and pods under glasshouse and field conditions. *Eur. J. Plant Pathol.* 109: 555–564.

Grondeau, C., Mabiala, A., Ait-Oumeziane, R., and Samson, R. 1996. Epiphytic life is the main characteristic of the life cycle of *Pseudomonas syringae* pv. *pisi*, pea bacterial blight agent. *Eur. J. Plant Pathol.* 102: 353–363.

Hollaway, G. J., Bretag, T. W., and Price, T. V. 2007. The epidemiology and management of bacterial blight (*Pseudomonas syringae* pv. *pisi*) of field pea (*Pisum sativum*) in Australia: A review. *Aus. J. Agric. Res.* 58: 1086–1099.

Martín Sanz, A., Garcia Vaquero, C. A., and Caminero Saidana, C. 2005. Pea bacterial blight: A problem in Castilla y León. Grain Legumes No. 41: 8–9.

Martín-Sanz, A., Palomo, J. L., Pérez de la Vega, M., and Caminero, C. 2011. Identification of pathovars and races of *Pseudomonas syringae*, the main causal agent of bacterial disease in pea in North-Central Spain, and the search for disease resistance. *Eur. J. Plant Pathol.* 129: 57–69.

Martín-Sanz, A., Pérez de la Vega, M., and Caminero, C. 2012a. Resistance to *Pseudomonas syringae* in a collection of pea germplasm under field and controlled conditions. *Plant Pathol.* 61: 375–387.

Martín-Sanz, A., Pérez de la Vega, M., Murillo, J., and Caminero, C. 2012b. Genetic, biochemical and pathogenic diversity of *Pseudomonas syringae* pv. *pisi* strains. *Plant Pathol.* 61: 1063–1072.

Martín-Sanz, A., Pérez de la Vega, M., Murillo, J., and Caminero, C. 2013. Strains of *Pseudomonas syringae* pv. *syringae* from pea are phylogenetically and pathogenically diverse. *Phytopathology* 103: 673–681.

Reeves, J. C., Hutchins, J. D., and Simpkins, S. A. 1996. The incidence of races of *Pseudomonas syringae* pv. *pisi* in UK pea (*Pisum sativum*) seed stocks, 1987–1994. *Plant Variet. Seeds* 9: 1–8.

Roberts, S. J., Ridout, M. S., Peach, L., and Brough, J. 1996. Transmission of pea bacterial blight (*Pseudomonas syringae* pv. *pisi*) from seed to seedling: Effects of inoculum dose, inoculation method, temperature and soil moisture. *J. Appl. Bacteriol.* 81: 65–72.

Sackett, W. G. 1916. A bacterial stem blight of field and garden peas. *Bull. Colo. Agr. Expt. Sta.* 218: 1–43.

Taylor, J. D., Bevan, J. R., Crute, I. R., and Reader, S. L. 1989. Genetic relationship between races of *Pseudomonas syringae* pv. *pisi* and cultivars of *Pisum sativum*. *Plant Pathol.* 38: 364–375.

Young, J. M., Dye, D. W., and Wilkie, J. P. 1978. Genus *Pseudomonas* Migula 1894. In: Young, J. M., Dye, D. W., Bradbury, J. F., Panagopoulos, C. G., and Robbs, C. F. 1978. A proposed nomenclature and classification for plant pathogenic bacteria. *N. Z. J. Agric. Res.* 21: 153–717.

7.44 BACTERIAL WILT OF CUCURBITS

Bacterial wilt of cucurbits affects many cucurbitaceous species. Cucumbers and muskmelons are more susceptible to the disease. Squash, pumpkins, gourds, and wild cucurbits are also infected, while the watermelon is resistant or rarely attacked. The disease is present throughout the United States and in Canada east of the Rocky Mountains. It also occurs in Europe, Japan, and South Africa. The disease has also been reported from New Mexico on pumpkin and watermelon by Sanogo et al. (2011). Rojas et al. (2015) have reviewed the research work to understand the cucurbit bacterial wilt pathosystem by recounting the early findings and updating our understanding of the disease cycle, including pathogen and vector biology. The authors have also highlighted the research areas that could lead to more efficient and ecologically based management of bacterial wilt.

7.44.1 SYMPTOMS

The earliest symptom of the disease is drooping of one or more leaves followed by drooping and wilting of all the leaves of the infected vine. Wilt progresses through lateral shoots involving the entire plant. Wilted leaves shrivel and dry up and infected stems turn pale and become soft. Before shriveling and drying, cucumber and muskmelon leaves may turn dull green, while those of squash and pumpkin become dull greyish-green and later turn bronze. The entire plant may wilt within 2 weeks after infection. Later on, wilted plants dry up and die (Figure 7.30a). Squash and pumpkin plants do not show rapid wilting, but may be dwarfed producing excessive blossoms and branches.

A diagnostic feature of the disease is the white bacterial ooze, which comes out from the cut ends when infected stems are cut and pressed between the fingers. This sticky bacterial mass can be drawn into fine threads by touching the cut ends of the stem with a finger and gently pulling the finger away from the cut end (Figure 7.30b). A positive test is the confirmation of the disease, but a negative test does not indicate that the disease is absent. Another diagnostic test is to put a freshly cut infected stem in a glass of water and watch it. If bacterial wilt is present, the water will start turning milky at the cut end of the stem and ultimately making the whole water milky.

The infection of the squash fruits usually occurs through infected vines. However, occasionally it may be caused by beetles feeding on the blossoms and the rind of fruits. In the

(a)

(b)

FIGURE 7.30 Bacterial wilt of cucumber: (a) Wilted plants. (Courtesy of Gerald Holmes, California Polytechnic State University at San Luis Obispo, Bugwood.org.) (b) Strands of viscous bacterial ooze drawn from cut stem ends. (Courtesy of M.P. Hoffmann.)

stored squash fruits, slimy rot progresses internally without any external symptoms. With the further progress of the internal rot, dark spots or blotches appear on the outer surface, which enlarge and coalesce. Later on, soft rotting microorganisms invade these fruits and completely destroy these fruits (Agrios 2004).

7.44.2 Causal Organism

Erwinia tracheiphila (Smith 1895) Bergey et al. 1923.
Syn. *Er. uredovora*

The following description of the bacterium is taken primarily from Hauben et al. (1998). Cells are facultatively anaerobic rods, Gram-stain-negative, and motile by peritrichous flagella. Growth factors are required for growth and metabolism is fermentative. The bacterium grows very poorly on nutrient agar, but moderately well on yeast extract glucose chalk agar and glucose nutrient agar. H_2S is produced, but gelatin is not liquefied. The tests for oxidase, urease, starch hydrolase, arginine dihydrolase, casein hydrolysis, phenylalanine deaminase, and nitrate reduction are also negative.

Strains grow on arginine, glucosamine, leucine, phenylalanine, citrulline, and tyrosine, but not on betaine, choline, hydroxypyroline, thymine, tryptamine, and xanthin as nitrogen sources. The strains are sensitive to ampicillin, amoxycillin, carbenicillin, cephalexine, and kanamycin but not to fusidinic acid, methicillin, and sulfafurazol. The G + C content of the DNA of members of the species ranges from 50.0 to 52.0 mol%.

Rojas et al. (2013) investigated genetic variability in 12 *Er. tracheiphila* strains, obtained from muskmelon, cucumber, or summer squash, by rep-PCR and cross inoculation studies on different hosts. The authors provided the first evidence of genetic diversity within the pathogen and suggested that strain specificity is associated with plant host.

7.44.3 Disease Cycle

As the bacterium dies within 1–2 months in the dead plants, its survival in the diseased plant debris is overruled. The bacterium overwinters (survives) in the digestive tract of hibernating adults of two cucumber beetles, namely, striped cucumber beetle (*Acalymma vittatum*) and spotted cucumber beetle (*Diabrotica undecimpunctata howardi*). The former is more common; and in North Dakota, it is the main spreading agent of the bacterium.

In the spring, infected beetles feed on the leaves causing deep wounds and deposit the bacteria in the wounds along with their feces. The bacteria enter the xylem vessels exposed by wounds, multiply in the xylem vessels, and spread to other parts of the plant. All the beetles do not carry the bacterium and not all the wounds act as avenue of infection. Only deep feeding wounds in which xylem vessels are exposed serve as avenues of infection. After entering the xylem vessels, bacteria multiply rapidly producing polysaccharides and gums. The accumulation of bacterial cells, polysaccharides, and gums, along with formation of tyloses results in the blockade of vascular system hindering the normal flow of water and nutrients. The plugging of xylem vessels is so extensive that it allows less than one-fifth of normal water flow, hence making it the primary cause of wilting. The rate of multiplication and the period for which the bacterium multiplies differ in susceptible and resistant cucumber cultivars. The generation time in susceptible and resistant cucumbers is 5.4 and 11.2 h, respectively. The bacterial population starts decreasing on fourth day in the resistant cucumber, whereas in the susceptible cucumber the peak population reaches on the sixth day (Watterson et al. 1972).

Vrisman et al. (2016) studied host colonization dynamics of *Er. tracheiphila* using a bioluminescent strain (TedCu10-BL#9). Inoculated *Cucumis melo* plants showed bioluminescent bacteria in the inoculated leaf and petiole beginning 1 day post-inoculation (DPI) and the bacteria spread to roots via the stem by 2 DPI, reached the plant extremities 4 DPI, and the plants wilted 6 DPI. However, *Cucurbita* plants inoculated with TedCu10-BL#9 did not wilt, even at 35 DPI. Bioluminescent bacteria were detected 6 DPI in the main stem of squash and pumpkin plants, which harbored approximately 10^4 and 10^1 cfu g^{-1}, respectively, without symptoms.

Liu et al. (2018) provided experimental evidence that host age impacts the rate of symptom development in cucurbit bacterial wilt and that movement of *Er. tracheiphila* within muskmelon plants occurs primarily in the downward direction.

Rojas and Gleason (2012) monitored the survival of *Er. tracheiphila* on muskmelon leaves under various temperature and moisture conditions using a rifampicin-resistant strain of the bacterium. Survival of *Er. tracheiphila* on wet muskmelon leaves depended on temperature ($P < 0.01$), with the greatest survival at 10°C and 15°C and least at 30°C and 35°C. Leaf wetness also impacted survival; an initial 12-hour dry period resulted in a 1,000- to 10,000-fold reduction in population size, followed by stabilization of the surviving population. It is evident that *Er. tracheiphila* can survive on muskmelon leaves under a wide range of environmental conditions, suggesting that epiphytic populations might serve as a reservoir of inoculum for infections.

The beetles spread the bacteria through their contaminated mouth parts when they feed on healthy plants after feeding on infected plants. Infected wild cucurbits can also serve as source of primary inoculum and from these plants the spread occurs through beetles. Cucumber beetles prefer wilted plants for feeding resulting in an increase in number of contaminated beetles. The spread of the pathogen is fast under normal moisture conditions because the activity of beetles decreases in prolonged cool and rainy weather. In extremely dry weather, the bacteria cannot penetrate exposed vascular system due to the lack of film of water. Each contaminated beetle can inoculate several healthy plants after one feeding on an infected plant.

7.44.4 Management

1. There is no cure for the cucurbit wilt. Therefore, control is aimed at prevention of the disease by controlling cucumber beetles
2. Control of cucumber beetles with insecticide should be started as soon as the first beetle is seen in the crop and sprays continued until the end of season at weekly intervals. Use any of the fast killing insecticides like methoxychlor, diazinon, and carbryl (Sevin). The carbryl should be sprayed late in the day to prevent honeybee mortality. Organic control options for cucumber beetles include foliar application of kaolin clay (Surround WP), insecticidal soap containing neem oil, and rotenone 1% dust
3. Cover the plants with row covers, cones, or other barriers to prevent feeding of cucumber beetles on vines. Row covers should be placed over the plants early in the season and their edges sealed to prevent the entry of the beetles. Row covers should be removed to allow pollination by insects. Rojas et al. (2011) studied the feasibility of delaying removal of row covers to suppress bacterial wilt of muskmelon and found that in years in which bacterial wilt occurred, delayed removal of row covers for 10 days resulted in the highest returns. However, when bacterial wilt was absent, the delayed-removal row-cover treatments had the lowest returns. Results of the sensitivity analysis indicated that delayed removal of row covers for 10 days could be a cost-effective component of an integrated bacterial wilt suppression strategy for muskmelon where bacterial wilt occurs in ≥50% of production seasons
4. Place yellow sticky cups after every 20 feet along the rows to trap the beetles. Replace these traps when they become dusty and non-sticky
5. Remove and destroy the diseased plants to prevent further spread of the disease
6. In fall, remove and destroy the diseased plant debris to reduce sites for beetle hibernation and to minimize the survival of other cucurbit pathogens
7. The cultivars resistant to bacterial wilt have little or no bitterness in their fruits. Cucurbitacin B and C are the compounds responsible for bitterness in the fruits, and these compounds also attract cucumber beetles. Therefore, the cultivars having no bitterness in fruits are most likely to be least susceptible to bacterial wilt
8. To prevent squash rot in storage, harvest the fruits only from healthy vines and store them in clean and fumigated stores
9. Do not plant cucurbits near weedy woods or brush where the cucumber beetles are likely to overwinter

REFERENCES

Agrios, G. N. 2004. *Plant Pathology*. 5th ed. Elsevier Academic Press, San Diego, CA, 922 pp.

Bergey, D. H., Harrison, F. C., Breed, R. S., Hammer, B. W., and Huntoon, F. M. 1923. *Bergey's Manual of Determinative Bacteriology*. 1st ed. Williams & Wilkins Co., Baltimore, MD.

Hauben, L., Moore, E. R. B., Vauterin, L., Steenackers, M., Mergaert, J., et al. 1999. Validation of publication of new names and new combinations previously effectively published outside the IJSB. *Int. J. Syst. Bacteriol.* 49: 1. Effective publication: Hauben, L., Moore, E. R. B., Vauterin, L., Steenackers, M., Mergaert, J., et al. 1998. Phylogenetic position of phytopathogens within the *Enterobacteriaceae*. *Syst. Appl. Microbiol.* 21: 384–397.

Liu, Q., Beattie, G. A., Saalau Rojas, E., and Gleason, M. L. 2018. Bacterial wilt symptoms are impacted by host age and involve net downward movement of *Erwinia tracheiphila* in muskmelon. *Eur. J. Plant Pathol.* 151: 803–810.

Rojas, E. S., Batzer, J. C., Beattie, G. A., Fleischer, S. J., Shapiro, L. R., et al. 2015. Bacterial wilt of cucurbits: Resurrecting a classic pathosystem. *Plant Dis.* 99: 564–574.

Rojas, E. S., Dixon, P. M., Batzer, J. C., and Gleason, M. L. 2013. Genetic and virulence variability among *Erwinia tracheiphila* strains recovered from different cucurbit hosts. *Phytopathology* 103: 900–905.

Rojas, E. S., and Gleason, M. L. 2012. Epiphytic survival of *Erwinia tracheiphila* on muskmelon (*Cucumis melo* L.). *Plant Dis.* 96: 62–66.

Rojas, E. S., Gleason, M. L., Batzer, J. C., and Duffy, M. 2011. Feasibility of delaying removal of row covers to suppress bacterial wilt of muskmelon (*Cucumis melo*). *Plant Dis.* 95: 729–734.

Smith, E. F. 1895. *Bacillus tracheiphilus* sp. nov., die Ursache des Verwelkens verschiedener Cucurbitaceen. *Centralblatt für Bakteriologie und Parasitenkunde* 2 1, 364–373.

Vrisman, C. M., Deblais, L., Rajashekara, G., and Miller, S. A. 2016. Differential colonization dynamics of cucurbit hosts by *Erwinia tracheiphila*. *Phytopathology* 106: 684–692.

Sanogo, S., Etarock, B. F., and Clary, M. 2011. First report of bacterial wilt caused by *Erwinia tracheiphila* on pumpkin and watermelon in New Mexico. *Plant Dis.* 95: 1583.

Watterson, J. C., Williams, P. H., and Durbin, R. D. 1972. Multiplication and movement of *Erwinia tracheiphila* in resistant and susceptible cucurbits. *Plant Dis. Reptr* 56: 949–953.

7.45 ANGULAR LEAF SPOT OF CUCURBITS

Angular leaf spot of cucurbits (also called angular leaf spot of cucumber) caused by *Pseudomonas syringae* pv. *lachrymans* is a widespread disease. The disease affects many cucurbit species, but it is more severe on cucumber, honeydew melon, and zucchini squash. The disease also occurs on muskmelon, watermelon, cantaloupe, vegetable marrow, squashes, and pumpkins. Yield loss occurs mainly due to less number of fruits and less weight per fruit due to reduced assimilation of infected leaves. Losses in processing cucumbers may go as high as 50% or more in wet seasons when control measures are not undertaken. Angular leaf spot of cucumber was first reported from the United States in 1913, and the causal agent was identified in 1915 by Smith and Bryan. Since then, the disease has spread to many countries. Janse (2005) has given a comprehensive list of countries where the disease occurs, and the list includes Algeria, Argentina, Australia, Brazil, Canada, China, Columbia, Egypt, widespread in Europe, Gabon, India, Iran, Israel, Japan, Jordan, Kazakhstan, Kenya, Korea, Laos, Mexico, New Zealand, Philippines, Singapore, South Africa, Zimbabwe, Tajikistan, Thailand, Turkey, USA, Uzbekistan, and Venezuela. In temperate regions of the world, the disease can cause up to 37% and 40% reduction in fruit number and weight, respectively, besides rendering some fruits as unmarketable culls. During wet conditions, the disease renders over 80% of leaves infected by the 10th day of the first infection. In India, the disease has been found in Kashmir Valley where its incidence and intensity ranges between 23.35% to 74.45% and 10.50% to 26.02%, respectively (Bhat et al. 2007).

Dutta et al. (2013a) reported *Xanthomonas cucurbitae* causing angular, necrotic spots with chlorotic halos on watermelon leaves, with no observable symptoms on fruits, from Georgia, USA. Dutta et al. (2013b) reported an outbreak of *Xa. cucurbitae* on pumpkin from Georgia causing severe symptoms on leaves and fruits. Leaves displayed small (2–3 mm), angular, water-soaked, yellow lesions, while fruits had small (2–3 mm), sunken, circular, dry lesions.

Newberry et al. (2016) collected 27 *Ps. syringae* strains causing foliar lesions and blighting on watermelon, cantaloupe, and squash in Florida, Georgia, and California over several years. On the basis of genetic profiles obtained through both MLSA (*gyrB*, *rpoD*, *gapA*, and *gltA*) and BOX-PCR (BOXA1R), they identified 26 of the *Ps. syringae* strains to be distributed among three clades within genomospecies 1, and phylogenetically distinct from genomospecies 2 member *Ps. syringae* pv. *lachrymans*. Considerable genetic diversity among *Ps. syringae* strains infecting cucurbits is associated with the same disease, and reflects the larger ecological diversity of *Ps. syringae* populations from genomospecies 1.

7.45.1 Symptoms

In cucumber, the symptoms appear on leaves, petioles, stems, and fruits. On cotyledons, soft, transparent, round to irregular lesions are formed. Initially small, round to irregular, water-soaked spots appear on cucumber leaves. The spots enlarge until their spread is limited by large veins; hence the spots become angular in shape. The spots on the upper leaf surface turn whitish-grey to brown, while on the lower surface they appear gummy and shiny. In wet weather, bacterial ooze comes out from spots in the form of droplets on the lower leaf surface. In dry weather, bacterial ooze dries to form a thin, papery, white layer. As the infected areas die and dry, they shrink and tear away from the healthy leaf tissue forming large irregular holes in the leaves (Agrios 2004). Young fully expanded leaves are more susceptible than older leaves (Figure 7.31). On stem and petioles, light yellow to light brown lesions appear, which exude abundant bacterial ooze. The severe infection of stem may lead to blightening and killing of the whole plant.

On squash leaves, brown spots of varying sizes appear. The spots are surrounded by yellow haloes, and the leaf area next to yellow halo on the lower leaf surface appears water-soaked following wet weather.

On watermelon leaves, the symptoms start as small, circular, dark lesions surrounded by yellow haloes. As the spots enlarge, they become angular having white centers and involve considerable area of the leaf.

Water-soaked, circular, usually superficial spots, much smaller than those on leaves, appear on the cucumber fruits. After the death of infected tissue, the centers of the lesions become chalky white and may show star-like cracks. Soft rot bacteria and fungi enter through these cracks and cause rotting

FIGURE 7.31 Angular leaf spot of cucurbits: Angular spots on leaves. (Courtesy of Gerald Holmes, California Polytechnic State University at San Luis Obispo, Bugwood.org.)

of the fruit accompanied by foul small. Fruit infection can also occur in the transit or storage. The infection of young fruits can cause deformity and bending of fruits. The infection of young fruits can also result into their dropping from the vines. During a wet weather, droplets of milky bacterial ooze frequently appear on the infection sites. The running down of the slime along the fruits causes secondary infection of the fruit (Anonymous 2002).

7.45.2 Causal Organism

Pseudomonas syringae pv. *lachrymans* (Smith and Bryan 1915) Young et al. 1978
Syn. *Psm. lachrymans*

Cells are short rods, non-sporing, Gram-stain-negative, motile by more than one polar flagellum. Poly-β-hydroxybutyrate granules are not accumulated, arginine dihydroxylase and oxidase are not produced, nitrates are reduced, protocatechuate is cleaved, and levan is formed on nutrient agar containing 5% sucrose. Gelatin is liquefied, arbutin and esculin are hydrolyzed, β-glucosidase test is positive, and pectate gel pitting occurs at pH 4.8. Growth occurs on inositol, mannitol, sorbitol, erythritol, L-tartrate, trigonelline, and quinate, but not on D-tartrate, L-lactate, anthranilate, homoserine, and betaine (Goto 1992).

Significant variability in virulence, aggressiveness, and genome structure has been reported among the numerous strains of the bacterium that have been identified (Olczak-Woltman et al. 2007). Earlier Krivchenko and Medvedeva (1985) reported 16 races of the bacterium and occurrence of resistance against these races in cucumber cultivars.

Słomnicka et al. (2015) determined the diversity of pathogenic *Psm. syringae* pv. *lachrymans* and pv. *syringae* strains isolated from cucurbit plants and found that pathogenicity tests of 22 strains on cucumber revealed remarkable differences in their virulence level. Genetic characterization using MLST loci, as well as ITS1, ERIC, and REP fingerprinting, allowed the grouping of strains into four phylogenetic groups. After amplicon sequencing and BLAST alignment, the collected strains complied with *Psm. syringae* pathovars: *lachrymans* and *syringae* and with saprophytic *Psm. fluorescens*. The strains of *Psm. syringae* pv. *lachrymans* may be categorized into at least two types differing in both disease symptoms and molecular characteristics. The findings confirm that MLST-loci-based grouping of strains corresponds well to grouping based on pathogenicity tests.

Słomnicka et al. (2018) reported the genome sequence of *Psm. syringae* pv. *lachrymans* strain 814/98, highly virulent to cucumber. The genome size was estimated to be 6.58 Mb, with 57.97% GC content. In all, 6,024 genes encoding proteins and 92 genes encoding RNAs were identified in this genome. The comparisons with the available sequenced genomes of pathovar *lachrymans* as well as with other *Psm. syringae* pathovars revealed the presence of three unique plasmids and 24 type III effector proteins in strain 814/98. The phylogenetic analyses of MLST loci and type III effector proteins clearly showed the existence of two distinct clusters of strains within pathovar *lachrymans*, which were grouped into either phylogroup 1 or 3, supporting non-monophyly within this pathovar.

7.45.3 Disease Cycle

The main sources of perpetuation of the bacterium are infected and/or infested seeds. In infected seeds, the bacterium resides in the seed coat and can survive for more than 2 years. Bhat (2009) reported that the bacterium survived up to 21 months in the infected seed, but the inoculum load of the seeds and the number of diseased seedlings decreased with the storage period. The inoculum load and the number of diseased seedlings immediately after the crop harvest were 5.3×10^3 cfu-g^{-1} of seed and 19.44%, respectively, while the respective figures after 21 month's storage were 2.03×10^3 cfu g^{-1} of seed and 7.48%. The bacterium also survives in the infected plant debris. Bhat (2009) also reported that in Kashmir Valley, the bacterium survived for 8 months in plant debris lying on the soil surface and buried in the soil, but the survival in plant debris on the soil surface was better in comparison to that buried in soil. The dispersal of the pathogen occurs through wind-splashed rain, sprinkler irrigation, insects (spotted cucumber beetle, southern corn rootworm), hands of workers, and farm machinery. Wind-blown soil, infected plant residue, and irrigation water contaminated with the bacterium also act as dispersing agents. The penetration of the bacterium occurs through stomata and wounds.

The optimum conditions for disease development are high relative humidity and a temperature range of 23°C–28°C. High relative humidity is more important than temperature for the outbreak of disease and a severe damage occurs when relative humidity is above 90%. Lesions formed when relative humidity is below 85%, remain small, and do not develop into typical angular spots. Excessive nitrogen fertilizers favor the disease development.

Olczak-Woltman et al. (2009) studied the inheritance of resistance to the disease based on disease severity (number and size of necrotic and chlorotic lesions on the infected leaves) and the presence or absence of a chlorotic halo around the necrotic spots and found that the disease severity appears to be controlled by multiple genes, and the heritability of the resistance was estimated to be 53%. They further reported that the presence or absence of the chlorotic halo was to be governed by a single gene, with the presence of the halo (the susceptible phenotype) being a dominant character, and also identified a RAPD marker linked to the gene conferring the chlorotic halo. Genetic distance between the marker, OP-AO07, a polymorphic 420 bp amplicon in the DNA of the susceptible plants, and the locus encoding the chlorotic halo was estimated to be 13 cm.

7.45.4 Management

1. Use pathogen-free and certified seed. Never use the seed from an infected crop. In case of a non-certified seed, treat the seed in hot water containing calcium propionate at 4.4 oz gal^{-1} of water or acidic cupric acetate at 6.7 oz gal^{-1} of water at 50°C for 20 min. Treatment with hot water alone is not effective, but this treatment also does not completely eliminate the pathogen
2. Follow crop rotation with non-host crops for 3 years
3. Destroy the infected plant residue by burning or by burying
4. Limit the use of sprinkler irrigation
5. Avoid excessive use of nitrogenous fertilizers
6. Avoid cultural operations and handling of plants when they are wet
7. Minimize wounds to the plants, particularly to the cucumber fruits. Use smooth and flexible containers for storing fruits after harvest
8. Control cucumber beetles and other insects by following a regular spray schedule
9. Give protected sprays of cooper hydroxide (0.2%) or mancozeb (0.2%) at weekly intervals
10. Root application of *Trichoderma asperellum* (T-203) caused accumulation of phytoalexins, induced systemic resistance against the bacterium in cucumber plants, and caused 80% reduction in disease severity (Yedidia et al. 2003)

REFERENCES

Agrios, G. N. 2004. *Plant Pathology*. 5th ed. Elsevier Academic Press, San Diego, CA, 922 pp.

Anonymous. 2002. Angular leaf spot of cucurbits. Report on Plant Diseases No. 919. Department of Crop Sciences, University of Illinois, Urbana, IL, 3 pp.

Bhat, N. A. 2009. Survival of *Pseudomonas syringae* pv. *lachrymans*- incitant of angular leaf spot of cucumber under temperate conditions of Kashmir valley. *Indian Phytopathol.* 62: 429–434.

Bhat, N. A., Masoodi, S. D., Ahmad, M., and Zargar, M. Y. 2007. Occurrence and severity of angular leaf spot of cucumber in Kashmir valley. *Ann. Plant Prot. Sci.* 15: 410–413.

Dutta, B., Gitaitis, R. D., Lewis, K. J., and Langston, D. B. 2013a. A new report of *Xanthomonas cucurbitae* causing bacterial leaf spot of watermelon in Georgia, USA. *Plant Dis.* 97: 556.

Dutta, B., Gitaitis, R. D., Sanders, F. H., Booth, C., Smith, S., et al. 2013b. First report of bacterial leaf spot of pumpkin caused by *Xanthomonas cucurbitae* in Georgia, United States. *Plant Dis.* 97: 1375.

Goto, M. 1992. *Fundamentals of Bacterial Plant Pathology*. Academic Press, San Diego, CA, 342 pp.

Janse, J. D. 2005. *Phytobacteriology: Principles and Practice*. CABI Publishing, Wallingford, UK, 360 pp.

Krivchenko, V. I., and Medvedeva, N. I. 1985. Intraspecific differentiation of the bacterial pathogen of cucumber *Pseudomonas lachrymans*. *Sbornik Nauchnykh Trudov po Prikladnoi Botanike, Genetike i Selektsii* 92: 92–97.

Newberry, E. A., Jardini, T. M., Rubio, I., Roberts, P. D., Babu, B., et al. 2016. Angular leaf spot of cucurbits is associated with genetically diverse *Pseudomonas syringae* strains. *Plant Dis.* 100: 1397–1404.

Olczak-Woltman, H., Bartoszewski, G., Mądry, W., and Niemirowicz-Szczytt, K. 2009. Inheritance of resistance to angular leaf spot (*Pseudomonas syringae* pv. *lachrymans*) in cucumber and identification of molecular markers linked to resistance. *Plant Pathol.* 58: 145–151.

Olczak-Woltman, H., Masny, A., Bartoszewski, G., Plucienniczak, A., and Niemirowicz-Szczytt, K. 2007. Genetic diversity of *Pseudomonas syringae* pv. *lachrymans* strains isolated from cucumber leaves collected in Poland. *Plant Pathol.* 56: 373–382.

Słomnicka, R., Olczak-Woltman, H., Bartoszewski, G., and Niemirowicz-Szczytt, K. 2015. Genetic and pathogenic diversity of *Pseudomonas syringae* strains isolated from cucurbits. *Eur. J. Plant Pathol.* 141: 1–14.

Słomnicka, R., Olczak-Woltman, H., Oskiera, M., Schollenberger, M., Niemirowicz-Szczytt, K., et al. 2018. Genome analysis of *Pseudomonas syringae* pv. *lachrymans* strain 814/98 indicates diversity within the pathovar. *Eur. J. Plant Pathol.* 151: 663–676.

Smith, E. F., and Bryan, M. K. 1915. Angular leaf-spot of cucumbers. *J. Agric. Res.* 5: 465–476.

Yedidia, I., Shoresh, M., Kerem, Z., Benhamou, N., Kapulnik, Y., et al. 2003. Concomitant induction of systemic resistance to *Pseudomonas syringae* pv. *lachrymans* in cucumber by *Trichoderma asperellum* (T-203) and accumulation of phytoalexins. *Appl. Environ. Microbiol.* 69: 7343–7353.

Young, J. M., Dye, D. W., and Wilkie, J. P. 1978. Genus *Pseudomonas* Migula 1894. In: Young, J. M., Dye, D. W., Bradbury, J. F., Panagopoulos, C. G., and Robbs, C. F. 1978. A proposed nomenclature and classification of plant pathogenic bacteria. *N. Z. J. Agric. Res.* 21: 153–177.

7.46 YELLOW VINE DISEASE OF CUCURBITS

Yellow vine disease affects squash, pumpkins, watermelon, muskmelon (cantaloupe), and other members of the cucurbit family. The disease first appeared on squash and pumpkins in the Cross Timbers Vegetation Area of Texas and Oklahoma in 1988. By 1991, it caused large-scale losses to watermelon and cantaloupe in this region. The presence of the disease has since been confirmed in the states of Arkansas, Colorado, Kansas, Nebraska, Massachusetts, Michigan, Missouri, Connecticut, Ohio, and Tennessee in the USA. In states that have a history of the disease, the disease incidence was often spotty within a field and adjacent fields may or may not be affected. It may be absent in some years or may cause widespread crop failure in other years. Besler and Little (2015) reported the disease on pumpkin and summer squash from Georgia and Sikora et al. (2012) reported it on watermelon and yellow squash from Alabama.

7.46.1 Host Range

Susceptible hosts include squash, watermelon, pumpkin, and cantaloupe. Pumpkin and squash are more susceptible to infection than cantaloupe and watermelon, although this could largely be due to feeding and ovipositional preference exhibited by squash bugs (Bonjour et al. 1990; Bruton et al. 1998).

7.46.2 Symptoms

The symptoms of the disease normally appear 10–15 days prior to fruit maturity and are similar in cantaloupe, pumpkin, watermelon, and squash. The color of the leaves changes from green to lime-yellow, and then to bright yellow. Affected plants gradually decline and exhibit a blighted appearance within 7–10 days. In some cases, immature plants may not turn yellow, but wilt and collapse in a day or so (Figure 7.32). Fruits and flowers on affected plants are not distorted, but watermelon fruits lose their chlorophyll very quickly. The distinguishing characteristic of the affected plants is that the phloem, which normally has a clear or translucent appearance, becomes honey-colored, particularly in the crown area. Although foliar symptoms are similar to vine decline symptoms caused by soil-borne pathogens, root degeneration occurs only in the later stages of disease development.

The symptoms can vary in different plant species and with age of plant at the time of infection. The disease may be confused with other diseases, such as bacterial wilt or Fusarium wilt. Secondary invasion of roots and lower stems by other microbes can further complicate its field diagnosis.

7.46.3 Causal Organism

Serratia marcescens Bizio 1823
Syn. *Bacillus marcescens*

The bacterium is a ubiquitous, rod-shaped, motile, facultative anaerobic, Gram-stain-negative enterobacterium with strains occupying diverse ecological niches. On nutrient agar, the colonies are round, smooth, entire, convex, creamy white, and range from 1.5 to 2.0 mm in diameter after 4 days of growth at 28°C. The temperature range for growth is from 5°C to 40°C and pH range is from 5 to 9. The tests for indole and H_2S production, urea hydrolysis, phenylalanine deaminase, and oxidase are negative; and positive for catalase production, lipase, deoxyribonuclease, lysine decaroxylase, Voges–Proskauer, methyl red, gelatin hydrolysis, and nitrate reduction. The acid is produced from glucose, sucrose, maltose, and mannitol, but not from lactose.

FIGURE 7.32 Yellow vine disease of cucurbits (squash). (Courtesy of Jim Jasinski, Department of Extension, Ohio State University, Columbus, OH.)

The bacterium can degrade tryptophan and citrate. One of the end products of tryptophan degradation is pyruvic acid, which is then incorporated into different metabolic processes of the bacterium. A final product of citrate degradation is carbon, which shows that the bacterium can use citrate as a carbon source. Its ability to hydrolyze casein differentiates it from other Gram-stain-negative bacteria. Another determinative test of *Se. marcescens* is its ability to produce lactic acid by oxidative and fermentative metabolism.

Primarily a human pathogen, *Se. marcescens* is involved in hospital-acquired infections, particularly catheter-associated bacteremia, urinary tract infections, and wound infections.

The phenotyping experiments and rep-PCR and DNA-DNA hybridization investigations carried out by Zhang et al. (2003) demonstrated that cucurbit yellow vine disease-associated strains of *Se. marcescens* cluster together in a group significantly different from other strains of the species. Zhu et al. (2007) provided the first molecular characterization of 16S rDNA and 16S–23S rDNA intergenic spacer from *Se. marcescens* strains. The method was specific, sensitive, and accurate, providing a new technique for differentiating different strains from the same species.

Avila et al. (1998) designed three primers, YV1, YV2, and YV3, using the nucleotide sequence of the procaryotic DNA. Fragments of 0.64 and 1.43 kb were amplified with primers YV1-YV2 and primers YV1-YV3, respectively, from symptomatic plants. None of the primer sets yielded fragments from asymptomatic plants, unrelated bacteria, or selected soil-borne fungal pathogens of cucurbits. Phylogenetic analysis indicated that the procaryote is a gamma-3 proteobacterium. The consistent association of the 0.64 and 1.43 kb fragments with symptomatic plants suggests that the gamma-3 proteobacterium may be the causal agent of yellow vine disease of cantaloupe, squash, and watermelon.

Zhang et al. (2005) made use of a suppressive subtractive hybridization (SSH) strategy, using *Se. marcescens* strain Z01-A, isolated from yellow vine-affected zucchini as the tester and rice endophytic *Se. marcescens* strain R02-A (IRBG 502) as the driver. SSH revealed 48 sequences, ranging from 200 to 700 bp, that were present in Z01-A, but absent in R02-A. Arrayed dot hybridization showed that the conservation of subtracted sequences among cucurbit pathogenic and non-pathogenic *Se. marcescens* strains varied. Thirty-four sequences were present only in pathogenic strains. Primers were designed based on one Z01-A-specific sequence, A79, and used in a multiplex PCR to discriminate between *Se. marcescens* strains pathogenic on cucurbits and those from other ecological niches.

Rascoe et al. (2003) reported that *Se. marcescens* strains isolated from watermelon and zucchini differed substantially from strains of the same species isolated from other environmental niches. Cucurbit strains also formed a distinct cluster, separate from other strains, when their fatty acid methyl ester profiles were analyzed. In substrate utilization assays (BIOLOG, Vitek, and API 20E), the cucurbit strains lacked a number of metabolic functions characteristic for *Se. marcescens*, failing to catabolize 25–30 compounds that were utilized by *Se. marcescens* reference strains. These biological differences may reflect gene loss or repression that occurred as the cucurbit strains adapted to life as an intracellular parasite and plant pathogen. In greenhouse inoculation studies, strains of *Se. marcescens* isolated from diseased cucurbits grew endophytically within pumpkin plants without inducing visible symptoms, suggesting that the plant pathogenic strains of *Se. marcescens* may have descended from a plant endophyte (Besler 2014).

7.46.4 Disease Cycle

The bacterium survives the winter in overwintering adult squash bugs (*Anasa tristis* De Geer). Besler (2014) reported that squash bugs harbored *Se. marcescens* throughout the year, with the highest detection rate (50%) found in overwintering squash bugs. The pathogen is spread to young cucurbits in the spring when the adults begin to feed. As the squash bug sucks sap from the plant's phloem, the bacterium is introduced. Once the bacterium invades the phloem, it multiplies, colonizes, and eventually clogs the vascular tissue. Young seedlings in the first true leaf stage of development are more susceptible to pathogen transmission than older seedlings. As the symptoms progression is usually very slow, the spread of the pathogen between plants within the field does not contribute much to disease severity. Usually symptoms are not detected until just prior to harvest.

Bruton et al. (2003) experimentally confirmed that the squash bug can transmit *Se. marcescens*. The *Se. marcescens*-*A. tristis* relationship described by them was the first instance in which the squash bug has been identified as a vector of a plant pathogen. Squash bugs overwinter as adults under plant debris, around buildings, or under rocks. When adults emerge in the spring, they fly to growing cucurbit plants to feed and mate. All the cucurbit crops are susceptible, but squash bugs show a preference for squashes and pumpkins.

Wayadande et al. (2005) reported that fifth instar nymphs of *A. tristis* successfully transmitted the bacterium after molting to the adult stage. The ability of the vector to transmit *Se. marcescens* after molting to the adult stage suggests that the hemocoel acts as the site of retention of transmissible bacteria. Pair et al. (2004) demonstrated that the squash bug harbors *Se. marcescens* in its overwintering state. The squash bug–*Se. marcescens* overwintering relationship reported herein greatly elevates the pest status of squash bug and places more importance on development of integrated strategies for reducing potential overwintering and emerging squash bug populations.

7.46.5 Management

Effective control of the disease is completely dependent on early management of squash bugs, beginning at emergence or transplanting of the seedlings. Timely operations are the key to successful squash bug control and eliminating squash bugs is the key to disease management.

1. Handpicking the larvae and nymphs is difficult, but can be accomplished by holding a bag under the affected leaves, cutting the leaf from the plant and dropping it into the bag. Disposing of the infected plant material away from the garden slows the spread of the disease. Removal of the eggs with a cotton swab dipped in alcohol and their disposal is effective in small gardens
2. Spray insecticides to control squash bug as soon as the transplants establish or seedlings emerge in the field. Early insecticide sprays should target overwintering adults on young plants. Multiple foliar sprays are needed for extended periods of control. Systemic insecticides used for cucumber beetle control suppress squash bug for weeks. Foliar sprays targeting newly hatched nymphs are more effective. Both carbaryl, sold under the brand name Sevin, and permethrin, the active ingredient in Ambush, are reported to be effective in controlling squash bugs
3. Use of trap crop is also very effective in controlling the disease. Squash bugs are most attracted to Hubbard squash, summer squash, pumpkins, watermelons, muskmelons, cucumbers, and butternut squash in decreasing order. In Texas, many growers have successfully used early-planted straight neck summer squash ("Lemon Drop" or "Hyrific") as a trap crop in the border rows of their watermelon fields to attract and control squash bugs to manage the disease. Trap crop plants should be 2–3 weeks older than the main crop to attract the bugs. Time the trap crop spray just prior to main crop emergence and give the second application if needed, at the first true leaf stage of the main crop. It will also control cucumber beetles and bacterial wilt

REFERENCES

Avila, F. J., Bruton, B. D., Fletcher, J., Sherwood, J. L., Pair, S. D., et al. 1998. Polymerase chain reaction detection and phylogenetic characterization of an agent associated with yellow vine disease of cucurbits. *Phytopathology* 88: 428–436.

Besler, K. R. 2014. Epidemiology and Management of Cucurbit Yellow Vine Disease, and Characterization of the Causal Agent *Serratia marcescens*. MSc thesis, University of Georgia.

Besler, K. R., and Little, E. L. 2015. First report of cucurbit yellow vine disease caused by *Serratia marcescens* in Georgia. *Plant Dis*. 99: 1175.

Bizio, B. 1823. Lettera di Bartolomeo Bizio al chiarissimo canonico Angelo Bellani sopra il fenomeno della polenta porporina. *Biblioteca Italiana o sia Giornale di Letteratura, Scienze e Arti* (*Anno VIII*) 30: 275–295.

Bonjour, E. L., Fargo, W. S., and Rensneri, P. E. 1990. Ovipositional preference of squash bugs (Heteroptera: Coreidae) among cucurbits in Oklahoma. *J. Econ. Entomol.* 83: 943–947.
Bruton, B., Mitchell, F., Fletcher, J., Pair, S., Wayadande, A., et al. 2003. *Serratia marcescens*, a phloem-colonizing, squash bug-transmitted bacterium: Causal agent of cucurbit yellow vine disease. *Plant Dis.* 87: 937–944.
Bruton, B. D., Fletcher, J., Pair, S. D., Shaw, M., and Sittertz-Bhatkar, H. 1998. Association of a phloem-limited bacterium with yellow vine disease in cucurbits. *Plant Dis.* 82: 512–520.
Pair, S. D., Bruton, B. D., Mitchell, F., Fletcher, J., Wayadande, A., et al. 2004. Overwintering squash bugs harbour and transmit the causal agent of cucurbit yellow vine disease. *J. Econ. Entomol.* 97: 74–78.
Rascoe, J., Berg, M., Melcher, U., Mitchell, F. L., Bruton, B. D., et al. 2003. Identification, phylogenetic analysis, and biological characterization of *Serratia marcescens* strains causing cucurbit yellow vine disease. *Phytopathology* 93: 1233–1239.
Sikora, E. J., Bruton, B. D., Wayadande A. C., and Fletcher, J. 2012. First report of the cucurbit yellow vine disease caused by *Serratia marcescens* in watermelon and yellow squash in Alabama. *Plant Dis.* 96: 761.
Wayadande, A., Bruton, B., Fletcher, J., Pair, S., and Mitchell, F. 2005. Retention of cucurbit yellow vine disease bacterium *Serratia marcescens* through transstadial molt of vector *Anasa tristis* (Hemiptera: Coreidae). *Ann. Entomol. Soc. Am.* 98: 770–774.
Zhang, Q., Melcher, U., Zhou, L., Najar, F. Z., Roe, B. A., et al. 2005. Genomic comparison of plant pathogenic and nonpathogenic *Serratia marcescens* strains by suppressive subtractive hybridization. *Appl. Environ. Microbiol.* 71: 7716–7723.
Zhang, Q., Weyant, R., Steigerwalt, A. G., White, L. A., Melchur, U., et al. 2003. Genotyping of *Serratia marcescens* strains associated with cucurbit yellow vine disease by repetitive elements-based polymerase chain reaction and DNA-DNA hybridization. *Phytopathology* 93: 1240–1246.
Zhu, H., Zhou, W. Y., Xu, M., Shen, Y. L., and Wei, D. Z. 2007. Molecular characterization of *Serratia marcescens* strains by RFLP and sequencing of PCR-amplified 16S rDNA and 16S-23S rDNA intergenic spacer. *Lett. Appl. Microbial.* 45: 174–248.

7.47 CROWN GALL OF PLANTS

Most of the bacterial plant pathogens cause necrotic or hypoplastic effect on the host tissue. However, crown gall bacterium (*Agrobacterium tumefaciens*) has an opposite effect, i.e., it causes hyperplasia and hypertrophy of the affected tissue. The disease is worldwide in occurrence and the causal bacterium has the widest host range of any plant pathogen. It infects a great variety of dicotyledonous and some gymnospermous plants. However, it does not infect monocots (grasses). More than 643 plant species belonging to 331 genera and 93 families are infected by the bacterium. The disease is more severe on all the stone fruits. Apple, pear, blueberry, grapes, and some ornamental plants such as rose are also severely infected. Many hosts are only known through artificial inoculation. Tomato acts as a very good indicator host, but has never been found as a natural host. The detailed host range can be found in De Cleene and De Ley (1976) and Bradbury (1986) and also on http://www.cabicompendium.org/cpc/home.asp.

The disease is known to occur in the USA since 1870. In 1897, Cavara from Italy was the first to isolate the bacterium from stem galls of grapes and prove its pathogenicity. However, between 1892 and 1894, numerous investigators from Europe and America worked on the disease without isolating the bacterium. Smith and Townsend (1907) described the bacterium and named it as *Bacterium tumefaciens*, which later on, was changed to *Agrobacterium tumefaciens* by Conn in 1942.

In India, the occurrence of the disease was reported by Singh in 1943, but its authentic report was made by Durgapal (1971) who isolated the bacterium and proved its pathogenicity. The disease occurs on cherry, peach, plum, and pecan nut (*Carya illinoensis*).

Two other diseases, having some similarities with crown gall are caused by two other *Agrobacterium* species. *Ag. rhizogenes* causes hairy root of apple and some other plants. *Ag. rubi* causes galls on canes of *Rubus* spp. (brambles). Hairy root of apple is described separately.

7.47.1 Symptoms

The galls or overgrowths occur commonly on the sub terranean roots and crowns of trees and shrubs (Figure 7.33). They may also be formed on stem, branches, leaves, and leaf-veins.

FIGURE 7.33 Crown gall of plants. (Courtesy of missouribotanicalgarden.org.)

Aerial galls appear generally on grapevines, raspberry, and blackberry. Aerial galls are not common on fruit trees, although they can occur. Galls are usually formed in late spring or early summer.

Young galls are more or less spherical, white or flesh-colored, and soft. As they grow, their surface becomes convoluted. In advanced stages, outer tissue of the galls turns dark brown to black due to death and decay of peripheral cells. Breakdown of peripheral tumor tissue releases crown gall bacteria into the soil. As the galls age, they become hard, rough, woody, and more cleft. Relatively soft galls are disintegrated, while those consisting of a corky exterior and a hard woody interior persist on the host. The size of galls may vary from 0.5–10 cm in diameter, but in extreme cases they may be 30 cm or more in diameter.

Occasionally, complex tumors called **teratomas** are formed. These tumors may give rise to roots and buds, and the buds give rise to shoots carrying leaves. In some hosts, secondary tumors develop at points distant from the primary tumors. These tumors are sterile, i.e., they do not contain bacteria and their tissue can be grown on chemically defined media. When tissues of such tumors are transplanted into a healthy plant, sterile tumors are produced. Crown galls or tumors caused by *Ag. tumefaciens* differ from other hyperplastic diseases that the former result from uncontrolled growth or proliferation of host cells due to the transformation of infected cells, caused by the pathogen.

Aerial plant parts of mildly infected plants may appear normal. However, in case of severe infection, the plants are stunted, unthrifty, and bear small and chlorotic leaves. In general, infected plants are more susceptible to adverse environmental conditions, particularly to winter injury. The yield of severely infected plants is reduced and most severely infected plants and vines may die.

Crown gall of grapes is an important disease of wine grapes in cold climates. In grapes, the galls are formed mostly on the aerial parts of the vines. The galls on grape roots are not typical; however, the bacteria can induce a localized necrosis of roots. Most grape cultivars are susceptible to the disease. The incidence of grape crown gall from year to year is generally related to the severity of the preceding winter and the age of the vines. For more details on the disease, the readers can refer to the review article by Burr and Otten (1999).

Cane gall: Small protuberances or elongate ridges appear on the fruiting canes of *Rubus* spp. somewhat before the middle of growing season. They increase in size and may completely cover sections of the surface, especially in the lower portion of the cane. The gall tissue is whitish at first, but turns brown with age. As the season advances, severely infected canes split open, the leaves wither, and the berries dry up.

The simultaneous occurrence of crown gall, hairy root, and wound overgrowths is commonly found in nature. Wound overgrowths, non-parasitic in nature, are generally confused with crown gall since they may occur along with the latter. Such overgrowths often occur on apple trees at the margin of root and scion, particularly when an imperfect union has been made (Walker 1969).

Besides crown gall, some plant pathogenic bacteria produce galls on twigs, limbs, and trunks of plants. In the latter case, bacterial pockets are found in light green to light yellow soft parenchymatous tissues surrounded by white woody tissues. Hence, the section of a tumor or gall should be examined for identification/differentiation of the disease.

7.47.2 Causal Organism

Agrobacterium tumefaciens (Smith and Townsend 1907) Conn 1942
Syn. *Bacterium tumefaciens*, *Rhizobium radiobacter*

For detailed taxonomic discussion and reclassification of *Agrobacterium* species into *Rhizobium*, see Young et al. (2001). Garrity et al. (2005) have recognized both *Agrobacterium* and *Rhizobium* as separate genera in the family Rhizobiaceae. *Ag. tumefaciens* is used here as a valid name.

Bacterial cells are Gram-stain-negative, non-sporing rods measuring 0.6–1.0 × 1.5–3.0 μm, and motile by 1–6 peritrichous flagella. Aerobic, chemooganotrophs, soil inhabitants, and possess a respiratory type of metabolism with oxygen as the terminal electron acceptor. Colonies are generally circular, smooth, convex, and colorless to light beige. Growth on media containing carbohydrates is usually copious and slimy. *Ag. tumefaciens* strongly absorbs Congo red and aniline blue in contrast to little or no absorption by *Ag. rhizogenes*. *Ag. tumefaciens* produces an abundant growth on sodium selenate agar and calcium glycerophosphate medium with mannitol in contrast to no or very little growth by *Ag. rhizogenes*. The mol% of G + C of the DNA is 57–63 (T_m).

The genome of *Ag. tumefaciens* str. C 58 consists of a circular chromosome (2.8 Mb), a linear chromosome (2.0 Mb), and two plasmids, namely, tumor-inducing plasmid called Ti-plasmid (206.479 kbp) and At-plasmid called the cryptic plasmid. The pathogenic ability, i.e., the ability to cause galls/tumors on the host plants is governed by the Ti-plasmid. The strains, which lack Ti-plasmid are non-pathogenic. The plants infected with *Ag. tumefaciens* str. C 58 produce nopaline type of opine. If the bacterium is grown near its maximum temperature of 30°C, then the bacterium loses its Ti-plasmid as well as its pathogenicity. Plasmid At of str. C 58 has been shown to be involved in the metabolism of opines and conjugation. The genome of str. C 58 was simultaneously sequenced by Goodner et al. (2001) and Wood et al. (2001).

Al-Karablieh and Khlaif (2002) found that the detection of *Ag. tumefaciens* from infected and symptomless plants by detecting *tmr* gene was more efficient than the routine isolation of the bacterium because the former method was more sensitive.

Ag. tumefaciens is an excellent genetic engineering tool. The DNA transformation capabilities of the bacterium have been extensively exploited in biotechnology as a means of inserting foreign genes into plants. Nester (2008) has rightly called it a **natural genetic engineer** and also a **household name** in plant genetic engineering and plant molecular genetics.

Ag. vitis Ophel and Kerr 1990 causes crown gall of grapes. Earlier, the strains of agrobacteria, causing crown gall of grapes, were classified as biovar 3 of *Ag. tumefaciens*. Taxonomic evaluation of biovar 3 strains by Ophel and Kerr (1990) resulted in the naming of a new species, *Ag. vitis*. The bacterium is insensitive to agrocin 84 produced by *Ag. radiobacter* strain K 84. In contrast to *Ag. tumefaciens*, it has a limited host range. According to Kawaguchi (2009), *Rhizobium vitis* (syn. *Ag. vitis*) strains are genetically heterogeneous and can be separated into several genetic groups. Although *Ag. vitis* is mainly responsible for causing crown gall of grapes, *Ag. rhizogenes* and *Ag. tumefaciens* also infect grapes. Out of 56 tumorigenic Spanish grapevine strains of *Agrobacterium* spp., three strains were identified as *Ag. rhizogenes* (biovar 2), three as *Ag. tumefaciens* (biovar 1), and the remaining 50 as *Ag. vitis* (biovar 3) (Palacio-Bielsa et al. 2009). The strains of these three species were clearly distinguished on the basis 16S rRNA gene sequencing. However, no universal primer pair was found for the detection of these three species.

Johnson et al. (2013) developed real-time PCR assays employing DNA extraction methods that included magnetic capture hybridization, immunomagnetic separation, and extraction with the Mo Bio Powerfood kit for the detection of *Ag. vitis*. The assays incorporating magnetic capture hybridization or immunomagnetic separation followed by real-time PCR were 10,000-fold more sensitive than direct real-time PCR when tested using boiled bacterial cell suspensions, having detection thresholds of 10^1 cfu mL^{-1} compared with 10^5 cfu mL^{-1}. DNA extraction with the Powerfood DNA extraction kit was 10-fold more sensitive than direct real-time PCR, with a detection threshold of 10^4 cfu mL^{-1}. All the three assays were able to detect *Ag. vitis* in healthy-appearing grapevine cuttings taken from infected vines.

Süle et al. (2012) examined 40 strains of tumorigenic and non-tumorigenic strains of *Ag. tumefaciens*, *Ag. rhizogenes*, *Ag. rubi*, and *Ag. vitis* using capillary isoelectric focusing. The isoelectric points of the investigated strains clearly differentiated these strains according to their respective species. The isoelectric points values obtained for *Ag. tumefaciens*, *Ag. rhizogenes*, *Ag. rubi*, and *Ag. vitis* were 2.2, 4.0, 2.15, and 2.6, respectively. The differentiation corresponded to the phenotypic, PCR, and fatty acid characterizations. Strains of *Rhizobium* species differed from *Agrobacterium* strains in their isoelectric points values. The merits of capillary isoelectric focusing over the phenotypic determinative tests, PCR, and fatty acid analysis are its fast speed (15 min), relative simplicity, and very small amount of chemicals used. This rapid and simple method is a major improvement over the classical methods of separation of *Agrobacterium* species and should prove useful for rapid characterization of *Agrobacterium*-like colonies isolated from plant tumors for epidemiological and generic diversity studies.

7.47.3 Disease Cycle

The bacterium is soil inhabitant and remains viable in soil for several years. It can also survive in infected plant debris, on and around root surfaces utilizing nutrients leaked, and in the outer portion of primary galls. The bacterium is widely distributed on the living planting stock. The bacterium penetrates the host through wounds, fresh and deep wounds are more conducive for infection. The wounding of the plants results from pruning, cultural operations, growth cracks, grafting, and chewing insects. The dissemination of the bacterium occurs through wind-splashed rain and chewing insects. Abundant bacterial cells present on the surface of galls are washed down by rainwater and carried away by the flowing water. Long distance spread of the pathogen occurs through the infected nursery and planting material, bacterial cells present on the surface or inside the host tissue. The tumor formation occurs between 10°C and 33°C. In some hosts like chrysanthemum, the bacterium also causes systemic infection.

Infection process and the host pathogen interaction in this disease have been elucidated thoroughly. Nester (2008) has given a comprehensive account of host-parasite interaction leading to the development of tumors. A single bacterial cell can cause infection under favorable conditions. The wounding of host cells results in the release of cell sap containing phenolic and sugar compounds. The bacterial cells are attracted by these chemicals such as acetosyringone, swim through soil water with the help of flagella, and attach to the wounded plant cells. Attachment of bacteria to plant cells is accomplished through two steps. After an initial weak and reversible attachment, the bacteria synthesize cellulose fibrils that anchor them to the wounded plant cells. Four main genes involved in this process are *chvA*, *chvB*, *pscA*, and *att*. The products of the first three genes appear to be involved in the actual synthesis of the cellulose fibrils. These fibers also anchor the bacteria to each other, resulting in the formation of a microcolony. After production of cellulose fibrils, a Ca^{2+} dependent outer membrane protein called rhicadhesin is produced that also helps in sticking the bacteria to the cell wall.

A large fragment of Ti-plasmid, called transfer DNA or T-DNA is excised from the plasmid, which after insertion into the host cell gets incorporated at a semi-random location into the plant genome. For transferring T-DNA, *Ag. tumefaciens* uses a type IV secretion system, involving the production of T-pilus. Two-component sensor system of Vir A/Vir G is able to detect phenolic signals, particularly acetosyringone and hydroxyacetosyringone released by wounded plant cells. This results in a signal transduction event, activating the expression of 11 genes within the Vir B operon, which are responsible for the formation of T-pilus. First of all, the Vir B propilin is formed, which is a polypeptide of 121 amino acids and is processed by the removal of 47 residues to form a T-pilus sub unit. The sub unit becomes circular by the formation of a peptide bond between two ends of the polypeptide. Products of the other Vir B genes transfer the sub units across the plasma membrane. Evidence provided by various studies shows that all the genes from Vir B6 to Vir B10 might encode components of the transporter. An ATP-ase is also required for the active transport of the sub units (Wikipedia 2008).

Before transferring T-DNA, it is removed from the Ti-plasmid. A Vir D1/D2 complex cuts the DNA at the left and right border sequences. The Vir D2 protein is covalently attached to the 5′ end. Vir D2 contains a motif that leads to the nucleoprotein complex being targeted to the type IV secretion system. In the cytoplasm of the host, T-DNA complex gets coated with Vir E2 proteins, which are exported through the type IV secretion system independently from the T-DNA complex. Nuclear localization signals located on the Vir E2 and Vir D2 are recognized by the importin alpha protein, which then associates with importin beta and the nuclear pore complex to transfer the T-DNA into the nucleus. Another protein, VIP 1 also plays an important role in the process, possibly by acting as an adapter to bring the Vir E2 to the importin. After entering the nucleus, VIP 2 may target the T-DNA to the areas of chromatin that are being actively transcribed so that the T-DNA can integrate into the host genome (Wikipedia 2008).

Loyter et al. (2005) put forth the hypothesis that *Arabidopsis thaliana* Vir E2 interacting protein 1, which associates with the T-complexes by binding to their Vir E2 component, acts as a molecular link between the T-complex and the histone constituents of the host chromatin. The hypothesis was based on their observations of interaction of *Arabidopsis thaliana* Vir E2 interacting protein 1 with the histones in *in vitro* and *in planta*. Gelvin (2010), in his review article, has highlighted the role of plant proteins involved in *Agrobacterium*-mediated genetic transformation. Integration of T-DNA into a plant genome establishes a permanent transformation event, permitting stable expression of T-DNA-encoded transgenes. The transformation process is complex and requires participation of numerous plant proteins. The review deals with current information on plant proteins that contribute to *Agrobacterium*-mediated transformation, the roles of these proteins in the transformation process, and the modern technologies that have been used to elucidate the cell biology of transformation.

T-DNA region contains genes, which cause gall formation. Among them, genes *iaam* and *laah* encode for the synthesis of indole acetic acid via the IAM pathway. As this biosynthetic pathway is not used in many plants for the production of IAA, it indicates that the plants have no molecular means of regulating it, and the IAA will be produced constitutively. The *ipt* and other genes encode the sythensis of cytokinins. The production of IAA and cytokinins stimulates cell division and gall formation and explains the hormonal independence of crown gall tissue. T-DNA region also contains genes, which encode the synthesis of opines. Opines, namely, nopaline, octopine, agropine, and agrocinopine are present in the crown gall tissue and not in the healthy plant tissue. The infecting bacterium utilizes these opines for its nourishment as a source of energy, carbon, and usually nitrogen. Strains of *Ag. tumefaciens* are frequently classified according to the opines synthesized. Although the presence of the bacterium is necessary for the induction of crown gall, once the gall is initiated, the bacterium can be withdrawn without affecting the growth of galls.

Kawaguchi (2015) reported that a non-pathogenic strain of *Ag. vitis*, ARK-1, limits the development of crown gall of grapevine. Co-inoculation of grapevine shoots with ARK-1 and the tumorigenic (Ti) strain VAT03-9 with cell suspension having 1:1 cell ratio resulted in significantly lower expression of the virulence genes *virD2* and *virE2* of VAT03-9, 1 day after inoculation compared with expression levels when shoots were inoculated only with VAT03-9. ARK-1 began to suppress the VAT03-9 population 7 days after inoculation. The biological control activity of ARK-1 seems to be based on the suppression of essential virulence genes.

Li et al. (2015) designed the primer pair ipt 3F/ipt 3R on the basis of the *ipt* gene, and this primer pair effectively amplified a 247 bp DNA fragment from the transfer DNA sequences of tumorigenic *Agrobacterium*. Using this primer pair, a sensitive real-time PCR assay was developed for quantifying tumorigenic *Agrobacterium* in soil as low as 10^2 cfu g^{-1}. The detection of tumorigenic *Agrobacterium* in 19 soil samples with this assay revealed $\leq 10^2$ cfu g^{-1} in maize or uncultivated soil and 10^3 to 10^6 cfu g^{-1} in soil planted with peach or cherry. The relationship between the disease severity and the tumorigenic *Agrobacterium* density in soil was verified by an artificial inoculation test. The disease incidence of 30%–50% and the disease index of 20–30 were found in soil inoculated with *Ag. tumefaciens* at 10^3 cfu g^{-1}. The disease incidence increased to >80% when the inoculation density of *Ag. tumefaciens* was increased to $>10^5$ cfu g^{-1} of soil. These results indicated that tumorigenic *Agrobacterium* density in soil was positively correlated with the incidence of crown gall on peach seedlings, and the presence of 10^3 bacterial cells per gram of soil were unsuitable for peach planting. Determination of tumorigenic *Agrobacterium* density can improve the prediction of crown gall disease outbreaks as well as aid in the management of the disease.

Fuller et al. (2017) designed oligonucleotide primers and probes to target *virD2* for use in a molecular diagnostic tool, which relies on isothermal amplification and lateral-flow-based detection. The oligonucleotide tools were tested in the assay and evaluated for detection limit and specificity in detecting alleles of *virD2*. One set of primers that successfully amplified *virD2* when used with an isothermal recombinase was selected. Both the tested probes had detection limits in picogram amounts of DNA. Probe 1 could detect all tested pathogenic isolates that represented the maximum diversity of *virD2*. Finally, the coupling of lateral-flow detection to the use of these oligonucleotide primers in isothermal amplification made the process simple and easy, and alleviated reliance on specialized tools necessary for molecular diagnostics. The assay is advancement for the rapid molecular detection of pathogenic *Agrobacterium* spp.

Yakabe et al. (2014) examined the possibility of Paradox (*Juglans hindsii* × *J. regia*) walnut seed to acquire *Ag. tumefaciens* based on harvest method used, prior to planting. Paradox seeds were collected directly from the mother tree without contacting the soil or gathered after lying on the orchard floor for up to 28 days, for a 2-year period from two commercial nurseries. *Ag. tumefaciens* was never detected in or on the 2,650 seeds collected directly from the mother tree. Both virulent and avirulent *Ag. tumefaciens* strains were detected in and on the husk of nuts incubated on the orchard floor at a frequency directly proportional to the time spent on the orchard floor. Irrespective of *Ag. tumefaciens* contamination in or on the husk, *Ag. tumefaciens* was never detected in the seed interior. Avoiding or reducing soil-borne populations of *Ag. tumefaciens* at the time of seed collection can play an important role in managing crown gall.

Johnson et al. (2016), using a highly sensitive magnetic capture hybridization procedure along with real-time PCR, measured the distribution of tumorigenic *Ag. vitis* in dormant canes and green shoots of grapevines. Tumorigenic *Ag. vitis* was distributed from the basal to apical nodal and internodal tissues of canes as well as in non-lignified green shoots. In 2013, *Ag. vitis* was detected in up to 17% of shoot tips and 52% of meristems of greenhouse-grown plants initiated from known *Ag. vitis* contaminated cuttings. A lower frequency of detection was observed from surface disinfected shoot tips (7%) in comparison to non-disinfected tips (37%), suggesting an epiphytic survival on green tissues. In 2014, vines propagated from cuttings collected from crown gall infected vines from a different vineyard yielded lower incidences of *Ag. vitis* from shoot tips, and the bacterium was not detected in meristems. Tumorigenic *Ag. vitis* was also detected in cuttings of wild grapevines that were collected both adjacent to and far removed from commercial vineyards.

Hao et al. (2017) investigated the effect of tumorigenic and non-tumorigenic *Ag. vitis* strains on grape graft take and found that tumorigenic *Ag. vitis* negatively affects graft strength and subsequent plant growth.

7.47.4 Management

As there is no cure of infected plants, prevention of infection is essential. Therefore, the main emphasis should be laid on the production of clean nursery stock.

1. Raise the disease-free nursery by adopting the following measures
   a. Nursery should be raised in the soil, free from the pathogen and chewing insects
   b. Destroy or dispose of the infected stock
   c. Well-fitted grafts wrapped with antiseptic tape should be used. Budding should be preferred over grafting
   d. Inspection and certification of the stock should be mandatory
   e. Avoid or minimize the wounding of nursery stock
   f. Budding and grafting tools should be properly disinfected
2. Application of antibiotics and other bactericides has generally not been successful, partially due to their short-lived activity in soil and partially due to presence of the bacterium in the host tissue
3. Soil fumigation tried in nurseries has also met with poor success mainly due to the occurrence of bacterium in the plant tissue. However, in some cases, the incidence of crown gall has increased after fumigation resulting from the elimination of microorganisms which compete with the crown gall bacterium (Burr 2001)
4. *Ag. radiobacter* strain K 84 is very effective in controlling crown gall. The seeds, cuttings, and roots of young seedlings are dipped in cell suspension of strain K84 before sowing. Since its first release in 1973 in Australia, it has been adopted in Europe, Africa, North America, and South America and is used on commercial scale for the control of crown gall on stone fruits and roses. It was the first case of a microorganism to be used commercially for the control of a plant disease. However, in 1973, some reports from Greece indicated its non-effectiveness for controlling crown gall disease (Kerr 1980)

Ag. radiobacter str. K84 produces a bacteriocin called agrocin 84, which is inhibitory to many pathogenic strains of *Agrobacterium.* The biosynthesis of agrocin 84 is encoded by a 48 kb plasmid, called pAgK84. It is a conjugative plasmid and can be transferred to the agrobacteria including plant pathogens. The transfer of this plasmid makes the recipient bacterium resistant to agrocin 84, thereby resulting in non-control of the recipient bacterium with str. K84. This might has been one of the reasons for the failure of control of crown gall in Greece. Another failure of biological control in one peach nursery in Italy, due to transfer of pAgK84 from *Ag. rhizogenes* K84 to natural pathogenic agrobacteria has also occurred (Raio et al. 2009)

The transfer of pAgK84 to plant pathogenic agrobacteria posed a potential threat to the continued success of the biological control of crown gall. Prof. Allen Kerr of Waite Agricultural Research Institute, Adelaide, Australia in collaboration with Dr. Farrand of University of Illinois identified the region of DNA that mediates the transfer of the plasmid pAgK84 to other bacteria. Prof. Kerr's team deleted this region and constructed a deletion mutant (transfer deficient, Tra) and named it as strain K1026. In June 1987, strain K1026 was released for experimental field trials in Waite Agricultural Research Institute, Adelaide, Australia. It was only the third recombinant DNA organism to be field-tested anywhere in the world. In 1988, Department of Agriculture, New South Wales, Australia released strain K1026 for commercial use as a pesticide, under the trade name "**No Gall**". Strain K1026 was the first recombinant DNA organism to be released for commercial use anywhere in the world. Strain K1026 has not been approved for release in any other country, except Australia, where strain K84 is still used for control. Strain K1026 is not intended to be used on fruit bearing crops. *Ag. tumefaciens* biovar 2 is more sensitive to biocontrol by strain K84 or K1026 of *Ag. radiobacter.*

Ag. radiobacter strain K84 is widely used for the control of crown gall disease. Sharma and Gupta (2017) have shown the possibility of using *Ag. radiobacter* strain UHFBA-218, a native strain isolated from rhizosphere soil of cherry root stock Colt (Solan, Himachal Pradesh, India) against *Ag. tumefaciens* isolate C 58. Frequency of conjugal transfer of agrocin genes and acquisition of virulence genes (*VirD2* and *ipt*) was found to be minimum in strain UHFBA-218 compared to standard strain K84. Transconjugants obtained from co-inoculation of *Ag. radiobacter* strain UHFBA-218 + *Ag. tumefaciens* isolates (C 58, I_1, and I_2) showed partial acquisition of virulence genes.

Sharma et al. (2016) prepared a white stone powder-based formulation (Sol Gall) by mixing *Brevibacillus brevis* and *B. choshinensis* in 1:1 ratio. Its soil drench application gave complete control of crown gall disease in peach nursery plants.

The following precautions should be taken to ensure effective control with biocontrol agents.

1. Do not mix the biocontrol agent with chlorinated water
2. Do not expose the control agent to direct sunlight
3. Do not expose the control agent to either very high or freezing temperatures

Strains K84 and K1026 are effective in controlling crown gall caused only by nopaline type strains of *Ag. tumefaciens* and *Ag. rhizogenes* and not effective for crown gall of grapes caused by *Ag. vitis*. Dipping of basal ends of grape cuttings in cell suspension of *Rahnella aquatilis* HX2 reduced the incidence of grape crown gall to 30.8% compared to 93.5% in control in a 3-year field trial (Chen et al. 2007). F 2/5, a non-tumorigenic strain of *Ag. vitis* was also found effective for the control of crown gall of grapes. F 2/5 produces an antibiotic that inhibits growth of many tumorigenic *Ag. vitis* strains *in vitro*, but antibiotic minus mutants of F 2/5 were found to be as effective as the wild-type strain for controlling crown gall (Burr et al. 1997). F 2/5 gives better disease control when applied 24 h prior to pathogen inoculation. Kawaguchi (2009) obtained an effective control of grapevine crown gall with a non-pathogenic *Rhbi. vitis* (syn. *Ag. vitis*) strain VAR03-1. The soaking of roots of grapevine seedlings in a cell suspension of strain VAR03-1 before planting in infested soil caused significant reduction in tumor formation in a greenhouse test and also gave an effective control in the field. The established population of VAR03-1 strain was 10^6 cfu g^{-1} in the grapevine rhizosphere soil, and it persisted for 2 years.

Ag. vitis strains ARK-1, ARK-2, and ARK-3 isolated from graft unions of nursery stock of grapevine were compared with *Ag. vitis* strain VAR03-1, a biocontrol agent effective against crown gall mentioned above. Among these three strains, strain ARK-1 inhibited maximum tumor formation, established its populations on roots of grapevine tree rootstock, and persisted on roots for a year. None of the three strains produced a halo of inhibition against *Ag. vitis* (Ti) strain on yeast mannitol agar (YMA) medium. Strain ARK-1 did not reduce tumor incidence on the stems of grapevine when the strain was dead or only culture filtrate was used. These findings show the possibility that these three strains inhibit grapevine crown gall *in planta* by a different mechanism than that of VAR03-1. As strain ARK-1 was most effective in inhibiting tumor formation on grapevine, it appears to be a promising new agent to control grapevine crown gall (Kawaguchi and Inoue 2012).

Kawaguchi et al. (2015) reported that soaking the roots of apple, Japanese pear, peach, rose, and tomato in a cell suspension of *Ag.* (=*Rhizobium*) *vitis* ARK-1 before planting into soil infected with tumorigenic *Agrobacterium* spp. reduced the number of plants developing crown gall tumors. Meta-analyses of the results from six, four, and four field trials of apple, Japanese pear, and peach, respectively, conducted during 2010–2013, showed integrated risk ratio (IRR) after treatment with ARK-1 to be 0.38 for apple crown gall, 0.16 for Japanese pear crown gall, and 0.20 for peach crown gall, indicating that the disease incidence was significantly reduced by ARK-1 treatment. Meta-analyses of the results from three greenhouse trials of both rose and tomato showed IRR after treatment with ARK-1 to be 0.29 for rose crown gall and 0.16 for tomato crown gall, indicating that the disease incidence was significantly reduced by ARK-1 treatment.

The submersion of dormant grape cuttings in water bath at 55°C for 30 min eradicates the inoculum from the cuttings except the survival of low levels of inoculum in tissues near the galls. The most effective way of producing *Ag. vitis*-free grapevines is by propagating vines from shoot tips *in vitro* because the bacterium does not systemically invade grape shoots. The establishment of such a planting requires the selection of an area where soil is free from *Ag. vitis*.

REFERENCES

Al-Karablieh, N., and Khlaif, H. 2002. Occurrence and distribution of crown gall disease in Jordan. *Phytopathol. Mediterr.* 41: 226–234.

Bradbury, J. F. 1986. *Guide to the Plant Pathogenic Bacteria*. CAB International, Wallingford, UK.

Burr, T. 2001. Crown gall. pp. 263–265 in: *Encyclopedia of Plant Pathology*. Vol. 1, O. C. Maloy, and T. D. Murray eds. John Wiley & Sons, New York.

Burr, T. J., and Otten, L. 1999. Crown gall of grape: Biology and disease management. *Annu. Rev. Phytopathol.* 37: 53–80.

Burr, T. J., Reid, C. L., Tagliati, E., Bazzi, C., and Süle, S. 1997. Biological control of grape crown gall by strain F2/5 is not associated with agrocin production or competition for attachment sites on grape cells. *Phytopathology* 87: 706–711.

Chen, F., Guo, Y. B., Wang. J. H., Li, J. Y., and Wang, H. M. 2007. Biological control of grape crown gall by *Rahnella aquatilis* HX2. *Plant Dis.* 91: 957–963.

Conn, H. J. 1942. Validity of the genus *Acaligenes*. *J. Bacteriol.* 44: 353–360.

De Cleene, M., and De Ley, J. 1976. The host range of crown gall. *Bot. Rev.* 42: 389–466.

Durgapal, J. C. 1971. A preliminary note on bacterial diseases of temperate plants in India. *Indian Phytopathol.* 24: 379–382.

Fuller, S. L., Savory, E. A., Weisberg, A. J., Buser, J. Z., Gordon, M. I., et al. 2017. Isothermal amplification and lateral-flow assay for detecting crown-gall-causing *Agrobacterium* spp. *Phytopathology* 107: 1062–1068.

Garrity, G. M., Bell, J. A., and Lilburn, T. 2005. The revised road map to the manual. pp. 159–200 in: *Bergey's Manual of Systematic Bacteriology*. 2nd ed. Vol. 2, *The Proteobacteria, Part A, Introductory Essays*. D. J. Brenner, N. R. Krieg, J. T. Staley, and G. M. Garrity eds. Springer-Verlag, New York.

Gelvin, S. B. 2010. Plant proteins involved in *Agrobacterium*-mediated genetic transformation. *Annu. Rev. Phytopathol.* 48: 45–68.

Goodner, B., Hinkle, G., Gattung, S., Miller, N., Blanchard, M., et al. 2001. Genome sequence of the plant pathogen and biotechnology agent *Agrobacterium tumefaciens* C58. *Science* 294: 2323–2328.
Hao, L., Kemmenoe, D., Canik Orel, D., and Burr, T. 2017. Investigation of tumorigenic and non-tumorigenic *Agrobacterium vitis* strains on grape graft take (Abstr.). *Phytopathology* 107: S2.8.
Johnson, K. L., Cronin, H., Reid, C. L., and Burr, T. J. 2016. Distribution of *Agrobacterium vitis* in grapevines and its relevance to pathogen elimination. *Plant Dis*. 100: 791–796.
Johnson, K. L., Zheng, D., Kaewnum, S., Reid, C. L., and Burr, T. 2013. Development of a magnetic capture hybridization real-time PCR assay for detection of tumorigenic *Agrobacterium vitis* in grapevines. *Phytopathology* 103: 633–640.
Kawaguchi, A. 2009. Studies on the diagnosis and biological control of grapevine crown gall and phylogenetic analysis of tumorigenic *Rhizobium vitis*. *J. Gen. Plant Pathol*. 75: 462–463.
Kawaguchi, A. 2015. Biological control agent *Agrobacterium vitis* strain ARK-1 suppresses expression of the *virD2* and *virE2* genes in tumorigenic *A. vitis*. *Eur. J. Plant Pathol*. 143: 789–799.
Kawaguchi, A., and Inoue, K. 2012. New antagonistic strains of nonpathogenic *Agrobacterium vitis* to control grapevine crown gall. *J. Phytopathol*. 160: 509–518.
Kawaguchi, A., Inoue, K., and Tanina, K. 2015. Evaluation of the nonpathogenic *Agrobacterium vitis* strain ARK-1 for crown gall control in diverse plant species. *Plant Dis*. 99: 409–414.
Kerr, A. 1980. Biological control of crown gall through production of agrocin 84. *Plant Dis*. 64: 25–30.
Li, Q., Guo, R.-J., Li, S.-D., Li, S.-F., and Wang, H.-Q. 2015. Determination of tumorigenic *Agrobacterium* density in soil by real-time PCR assay and its effect on crown gall disease severity. *Eur. J. Plant Pathol*. 142: 25–36.
Loyter, A., Rosenbluh, J., Zakai, N., Li., J., Kozlovsky, S. V., et al. 2005. The plant VirE2 interacting protein 1. A molecular link between the *Agrobacterium* T-complex and the host cell chromatin. *Plant Physiology* 138: 1318–1321.
Nester, E. 2008. *Agrobacterium*: The natural genetic engineer: 100 years later. APSnet Feature Story [https://www.apsnet.org/edcenter/apsnetfeatures/Pages/Agrobacterium.aspx]. Accessed December 25, 2018.
Ophel, K., and Kerr, A. 1990. *Agrobacterium vitis* sp. nov. for strains of *Agrobacterium* biovar 3 from grapevines. *Int. J. Syst. Bacteriol*. 40: 236–241.
Palacio-Bielsa, A., Gonzalez-Abolafio, R., Alvarez, B., Lastra, B., Cambra, M. A., et al. 2009. Chromosomal and Ti plasmid characterization of tumorigenic strains of three *Agrobacterium* species isolated from grapevine tumours. *Plant Pathol*. 58: 584–593.
Raio, A., Peluso, R., Puopolo, G., and Zoina, A. 2009. Evidence of pAgK84 transfer from *Agrobacterium rhizogenes* K84 to natural pathogenic *Agrobacterium* spp. in an Italian peach nursery. *Plant Pathol*. 58: 745–753.
Sharma, A., and Gupta, A. K. 2017. New insights in the biological control of crown gall through native *Agrobacterium radiobacter* strain UHFBA-218. *Plant Dis. Res*. 32: 137–152.
Sharma, A., Gupta, A. K., Singh, D., Khosla, K., and Mahajan, R. 2016. Biological control of crown gall disease in peach and cherry nursery plants by white stone powder based formulation (Sol Gall) of *Brevibacillus* spp. *Indian Phytopathol*. 69: 231–236.
Smith, E. F., and Townsend, C. O. 1907. A plant tumour of bacterial origin. *Science* N.S. 26: 671–273.
Süle, S., Horká, M., Matoušková, H., Kubesová, A., Salplachta, J., et al. 2012. Characterization of *Agrobacterium* species by capillary isoelectric focusing. *Eur. J. Plant Pathol*. 132: 81–89.
Walker, J. C. 1969. *Plant Pathology*. 3rd ed. McGraw-Hill Book Co., New York, 819 pp.
Wikipedia. 2008. *Agrobacterium tumefaciens*- Wikipedia, the free encyclopedia [https://en.wikipedia.org/wiki/Agrobacterium_tumefaciens]. Accessed December 28, 2018.
Wood, D. W., Setubal, J. C., Kaul, R., Monks, D. E., Kitajima, J. P., et al. 2001. The genome of the natural genetic engineer *Agrobacterium tumefaciens* C58. *Science* 294: 2317–2323.
Yakabe, L. E., Parker, S. R., and Kluepfel, D. A. 2014. Incidence of *Agrobacterium tumefaciens* biovar 1 in and on "Paradox" (*Juglans hindsii* × *Juglans regia*) walnut seed collected from commercial nurseries. *Plant Dis*. 98: 766–770.
Young, J. M., Kuykendall, L. D., Martinez-Romero, E., Kerr, A., and Sawada, H. 2001. A revision of *Rhizobium* Frank 1889, with an emended description of the genus, and the inclusion of all species of *Agrobacterium* Conn 1942 and *Allorhizobium undicola* de Lajudie et al. 1998 as new combinations: *Rhizobium radiobacter, R. rhizogenes, R. rubi, R. undicola* and *R. vitis*. *Int. J. Syst. Evol. Microbiol*. 51: 89–103.

7.48 HAIRY ROOT OF APPLE

Hairy root is another important disease of apple though it is less serious than crown gall. It is also known by other names such as "**woolly root**" and "**woolly knot**" and was long considered merely a form of crown gall. The pathogen infects *Malus pumila*, *Rosa* spp., *Rubus* spp.; rose, pear, raspberry, cotoneaster, hollyhock, honey locust, honey suckle, mulberry, pea vine, peach, quince, Russian olive, and spirea. Artificially inoculated host range extends to more than 37 plant species belonging to 30 genera and 15 families (De Cleene and De Ley 1981). The disease is widely distributed in apple nurseries and orchards in the USA. It is also found on rose in California and Texas states. The disease also occurs in Japan, Bulgaria, and France. Pulawska and Kałużna (2012) reported the occurrence of the disease on different host plants from Poland. Kuzmanović et al. (2013) reported its occurrence on apricot trees from central Serbia.

7.48.1 Symptoms

The disease commonly occurs on grafted apple nursery trees, which are usually 1, 2, or 3 years old. The production of large number of small roots either directly from stems or from roots or from localized hard swellings, that often occur at graft unions of root and scion, is the characteristic feature of the disease (Walker 1969). The production of roots may be as profuse as that of witches' broom. It can appear simultaneously with crown gall on the same plant and the symptoms can be confused in the early stage. On rose, the hairy roots originate at the ends of cuttings or at disbud scars. These malformations may be fleshy at first, but eventually become fibrous. Infected apple trees become stunted.

7.48.2 Causal Organism

Agrobacterium rhizogenes (Riker et al. 1930) Conn 1942 emend. Sawada et al. 1993
Syn. *Ag. radiobacter* biovar *rhizogenes*, *Ag. radiobacter* pv. *rhizogenes*

The below given description of the bacterium is taken from Sawada et al. (1993) and Hayward and Waterston (1998). Cells are Gram-stain-negative, non-sporing rods measuring 0.6–1.0 × 1.5–3.0 μm, motile by means 1–4 peritrichous flagella. Aerobic, possessing a strictly respiratory type of metabolism. Colonies are circular, convex, smooth, and non-pigmented to light beige. Growth on carbohydrate containing media is usually accompanied by copious, extracellular polysaccharide slime. Optimum temperature for growth is between 25°C and 28°C and no growth occurs at 35°C. Strains require growth factors like biotin and do not grow in simple mineral salts medium. Citrate and L-tyrosine are utilized. Acid is produced from dulcitol and α-methyl-D-glucoside in the Ayers-Rupp-Johnson's synthetic medium containing 0.02% yeast extract. 3-Ketolactose is not produced from lactose, and the arginine dihydrolase test is also negative. A large Ti-plasmid or a hairy root-inducing plasmid is detected in many strains. The mol% G + C of the DNA is 59.0–63.0 (T_m). Like *Ag. tumefaciens*, it is also used as a tool in genetic engineering, though less frequently.

7.48.3 Disease Cycle

The epidemiology and other aspects of the disease are comparatively much less worked out than that of crown gall. Like *Ag. tumefaciens*, *Ag. rhizogenes* is also an exclusively wound pathogen. It is readily disseminated on nursery stock, which may become infected from soil or by tools contaminated with the pathogen when cuttings and graftings are made. On rose, the bacteria may penetrate roots through injuries caused by the root-lesion nematode, *Pratylenchus vulnus*. The bacterium survives in steamed or untreated soil for more than 1 year. The virulence of the bacterium has been maintained in the culture for over 19 years. The other fundamental aspects of the disease cycle are similar to those of crown gall.

Bosmans et al. (2016) developed a SYBR Green-based quantitative real-time PCR assay based on the *rol* locus of the root inducing-plasmid for detection and quantification of *Agrobacterium* biovar 1 strains in hydroponic systems. The assay designed based on all *rolB* sequences currently available in GenBank, and developed using a collection of both target and non-target strains, was specific for rhizogenic *Agrobacterium* biovar 1 strains. Based on a calibration with artificially contaminated water samples mimicking hydroponic conditions, unknown bacterial concentrations could be accurately quantified in water samples from surveys carried out in different greenhouses. The detection limit obtained was of less than one cell mL^{-1} of water, following filtration of a volume of 500 mL. The qPCR assay is suitable for routine detection of rhizogenic *Agrobacterium* biovar 1 strains in aqueous samples. Moreover, the assay can be used for pathogen assessment at the pre-symptomatic stage, allowing sufficient time to undertake the necessary preventive measures to control the disease in hydroponics before the infection of plants.

7.48.4 Management

Management approaches are the same as those for crown gall.

REFERENCES

Bosmans, L., Paeleman, A., Moerkens, R., Wittemans, L., Van Calenberge, B., et al. 2016. Development of a qPCR assay for detection and quantification of rhizogenic *Agrobacterium* biovar 1 strains. *Eur. J. Plant Pathol.* 145: 719–730.

Conn, H. J. 1942. Validity of the genus *Acaligenes*. *J. Bacteriol.* 44: 353–360.

De Cleene, M., and De Ley, J. 1981. The host range of infectious hairy-root. *Bot. Rev.* 47: 147–194.

Hayward, A. C., and Waterston, J. M. 1998. *Agrobacterium rhizogenes*. *IMI Descriptions of Fungi and Bacteria*. Sheet 41, CAB International, Wallingford, UK.

Kuzmanović, N., Ivanović, M., Prokić, A., Gasić, K., Blagojević, N., et al. 2013. Identification and characterization of *Agrobacterium* spp. isolated from apricot in Serbia. *Eur. J. Plant Pathol.* 137: 11–16.

Pulawska, J., and Kałużna, M. 2012. Phylogenetic relationship and genetic diversity of *Agrobacterium* spp. isolated in Poland based on *gyrB* gene sequence analysis and RAPD. *Eur. J. Plant Pathol.* 133: 379–390.

Riker, A. J., Banfield, W. M., Wright, W. H., Keitt, G. W., and Sagen, H. E. 1930. Studies on infectious hairy root of nursey apple trees. *J. Agric. Res.* 41: 507–540.

Sawada, H., Ieki, H., Oyaizu, H., and Matsumoto, S. 1993. Proposal for rejection of *Agrobacterium tumefaciens* and revised descriptions for the genus *Agrobacterium* and for *Agrobacterium radiobacter* and *Agrobacterium rhizogenes*. *Int. J. Syst. Bacteriol.* 43: 694–702.

Walker, J. C. 1969. *Plant Pathology*. 3rd ed. McGraw-Hill Book Co., New York, 819 pp.

7.49 FIRE BLIGHT OF POME FRUIT TREES

Fire blight is probably the most devastating bacterial disease of some pome fruits and also the most feared by growers (Thomson 2001). It is able to kill complete pome fruit orchards within a single growing season (Vanneste 2000). It is most destructive on pears, but certain apple and quince varieties are also highly susceptible. Crabapple (*Malus* spp.), firethorn (*Pyracantha* spp.), and hawthorn (*Crataegus* spp.) are also frequently attacked. The bacterium also occasionally attacks rock spray (*Cotoneaster* spp.). The bacterium infects more than 180 plant species including fruit trees, wild and ornamental plants, most of them belonging to family *Rosaceae*. Kim et al. (2001) reported a similar disease on Asian pear, *Pyrus pyrifolia* from Korea and Japan caused by *Erwinia pyrifoliae*. Lopez et al. (2011) reported *Er. piriflorinigrans* sp. nov. as the cause of pear blossom necrosis in València, Spain. The bacterium was identified using a polyphasic approach. Myung et al. (2016) reported fire blight of apple caused by *Er. amylovora* from Korea.

Fire blight is of historical importance because this was the first plant disease shown to be caused by a bacterium. Professor T.J. Burrill from the University of Illinois, USA, through a series of experiments starting from 1878 to 1884 proved that the disease is caused by a bacterium. It is considered that fire blight originated from America and was known to be there for many years before it was discovered in the Hudson Valley of New York in 1794. The disease spread to most areas of the US by the 1900s. Since then, it has spread to the whole of Europe, except Latvia and Ukraine, Iran, Israel, Jordan, Lebanon, North America, New Zealand, and Turkey. The disease is known to occur in more than 40 countries. So far there is no confirmed report of its occurrence in India. Shtienberg et al. (2015) have given an account of the existence of fire blight of pear in Israel for the last 30 years; highlighting unique characteristics that play a crucial role in the epidemiology of the disease under Israeli conditions, summarized the conclusions derived from the local experience, and presented future perspectives regarding fire blight management. The take-home message of the Israeli fire blight story is that the battle against this hazardous disease is never ending.

A survey in California in 1976 estimated a loss of $4.7 million due to fire blight, while in southwest Michigan alone the loss in 1991 was $3.8 million. Again in 2000, a severe outbreak of fire blight of apples occurred in Michigan, which resulted into a loss of $42 million due to the removal of 350,000–450,000 apple trees covering approximately 1,550–2,300 acres (Longstroth 2001). In 1998, apple and pear growers in Washington and northern Oregon suffered a loss of $68 million due to fire blight damage.

7.49.1 Symptoms

The symptoms are found on all the aboveground plant parts as well as on roots. The infected blossoms, leaves, twigs, and branches give a burnt or scorched appearance, hence the name fire blight. The symptoms appear in the spring when the trees start the new growth. The first sign is the exudation of watery, light tan bacterial ooze from the twig, branch, or trunk cankers (holdover cankers) formed in the previous season(s). The ooze turns dark after exposure to air, spreads to form a thin glistening layer, or dries down to form hard beads or films. The following different types or phases of symptoms appear. However, not all the types or phases appear in all the orchards or years or areas.

7.49.1.1 Blossom Blight

It is the most devastating phase of the disease. It is usually the first symptom to appear within 1 to 2 weeks after bloom. Infected blossoms first appear water-soaked and then they begin to shrivel, wilt, and eventually turn brown to black (Figure 7.34a). The infected blossoms may fall, but often remain hanging on the trees. Infection of a single flower in the cluster usually results in the death of whole spur. During the warm and humid weather, bacteria come out on the surface of infected blossoms in the form of ooze and are disseminated by pollinating insects . This phase is critical for the outbreak of the fire blight epiphytotics because the bacteria can be readily disseminated throughout the orchard by pollinating insects.

7.49.1.2 Fruit Blight

Usually immature and unripened fruits are infected. Infection occurs through infected spur or directly through wounds or lenticels of the fruits. The infected fruits become water-soaked, turn brown, shrivel, and finally turn black. The infected fruits usually remain clinging to the trees for several months. Bacterial ooze comes out on the surface of infected fruits in warm and humid weather (Figure 7.34b).

7.49.1.3 Shoot or Twig Blight

Also called **"blight strikes"**, these symptoms develop on actively growing vegetative shoots. Twigs or shoots may be infected directly or infection from blossoms may move into them. Infected shoots wilt from the tip downward and form a diagnostic **"shepherd's crook"** (hooked) at the end. The bark of infected twigs turns brownish-black and is soft in the beginning, but shrinks and hardens later on. The leaves on the infected twigs turn black and remain clinging to the twig (Figure 7.34c). This phase is more destructive to young and vigorously growing trees.

7.49.1.4 Leaf Blight

In addition to blightening of leaves on infected twigs, leaf blight occurs from independent infection of leaves through wounds or through stomata and hydathodes. The necrotic, brown to black lesions appear along the midrib or main veins, between the veins, and along the margins. As the

(a)

(b)

(c)

FIGURE 7.34 Fire blight of apple: (a) Blossom blight. (Courtesy of Illinois Fruit and Vegetable News, Vol. 14, No. 7.) (b) Symptoms on Bartlett pear fruit. (Courtesy of Sugar David, www.plant-disease.ippc.orst.edu.) (c) Shoot blight. (Courtesy of Florida Division of Plant Industry, Florida Department of Agriculture and Consumer Services, Bugwood.org.)

blackening spreads, the leaves curl and shrivel, hang downwards, and generally remain attached to the hooked and blighted twigs.

7.49.1.5 Trauma Blight

The symptoms are similar to shoot blight, but are usually more random and widespread throughout the orchard. The symptoms result on account of damage caused to the host by traumatic events like hail, frost, or storm. The wounds caused by these traumatic events allow the bacteria to penetrate the host, bypassing the normal defenses even in the resistant hosts.

7.49.1.6 Canker Blight

It is also called by other names like "**body blight**", "**limb**", "**trunk blight**", depending on which part the infection occurs. Cankers are formed as a result of bacteria travelling systemically into the wood tissue of the tree. From infected twigs and fruit spurs, infection moves down to the branches, scaffold limbs, and trunks, forming cankers. In the beginning, the bark where canker is formed appears water-soaked and later on, becoming discolored, sunken, and dry. The cankers formed on large limbs and trunk appear slightly to deeply depressed and occasionally develop cracks at the margins. When a canker enlarges and girdles the branch, the portion of the branch above the cankers dies. Many small cankers, less than 2.5 cm in diameter, usually not sunken and rarely cracked, often appear on small twigs. These usually go unnoticed due to their small size, but are a major source of inoculum for infection in the next spring. One active, overwintering canker contains sufficient bacteria to severely contaminate flower blossoms of trees present in ¼ to ½ acre. Cankers without cracks are termed indefinite and are more viable source of inoculum.

7.49.1.7 Collar and Root Blight

When the cankers develop at the base of the trunk, it is called the **collar blight**. The formation of cankers at collar or ground level results in an early death of plants.

The infection of the root by the bacterium results in **root blight** or **rootstock blight**. The primary route of entry of the bacterium into the rootstock is internally, often through apparently healthy limbs and trunks of trees that had only a few infected blossoms or shoots of the scion. After reaching the susceptible rootstock, the bacteria initiate the formation of new cankers that completely girdle and kill the tree in one to a few months. Some of the rootstocks, namely, M-26 and M-9, generally preferred in tree size controlling orchards, are highly susceptible to fire blight pathogen.

7.49.2 Causal Organism

Erwinia amylovora (Burrill 1882) Winslow et al. 1920
Syn. *Bacterium amylovorum*

The bacterium is rod-shaped, Gram-stain-negative, and motile with peritrichous flagella, and requires nicotinic acid as a growth factor. The colonies on nutrient agar or King's agar medium B are smooth, creamy-white, and round with a glistening shine. Miller and Schroth selective (MSS) medium, which utilizes sorbitol as the carbon source, is suitable for the isolation of the bacterium due to elimination of common bacterial species. The modification of MSS medium by Ishimaru and Klos (1984) is helpful for isolation of the bacterium in the presence of high population of *Pantoea agglomerans*. The optimum temperature for growth is around 27°C. Gelatin is liquefied, levan is produced from sucrose, but nitrate reduction and H_2S production are negative. Lactate, citrate, formate, and trehalose are utilized as a sole source of carbon. The G + C content of the DNA is 53.6–54.1 mol%.

The bacterium contains a 40 kb segment of DNA that contains two large regions termed *hrp* and *dsp*. Genes present in these regions code for harpin and probably for other proteins that are responsible for pathogenicity, hypersensitivity, and perhaps host range. An extracellular polysaccharide called amylovoran and type III secretion system are two major, but separate virulence factors of the bacterium. The other virulence factors include lipopolysaccharides like levan, glucan, and levansucrase. However, Billing (2011) has stated that neither the extracellular polysaccharide produced by *Er. amylovora* is a toxin, nor responsible for the wilting symptom seen in the early stages of the disease.

Tian et al. (2017) reported that *Er. amylovora* type VI secretion system (TVISS) plays a role in bacterial competition and virulence possibly by influencing exopolysaccharide production. Of the 33 deletion mutants generated within three T6SS clusters of *Er. amylovora*, 19 exhibited reduced virulence on immature pear fruit as compared with that of the wild-type strain. Among them, 6, 1, and 12 genes belonged to T6SS-1, T6SS-2, and T6SS-3 clusters, respectively. Interestingly, these 19 mutants also produced less amylovoran or levan or both.

Most strains of the bacterium contain a non-conjugative plasmid called the ubiquitous plasmid pEA29. The loss of this plasmid results in the reduction of virulence of the bacterium and the strains cured of this plasmid show thiamine auxotrophy. The presence of pEA29 influences the expression of chromosomal amylovoran biosynthetic operon during infection. The amylovoran is a pathogenicity determinant of the fire blight pathogen, and thiamine biosynthesis is required for the high level expression of the amylovoran biosynthetic operon during infection. Among the other plasmids found in *Er. amylovora*, plasmid pEL60 is of special interest because it is an IncL/M plasmid with a high degree of similarity to the backbone of pCTX-M3, a multiple antibiotic resistance plasmid identified in the opportunistic human pathogen *Citrobacter freundii*. pEL60 is a cryptic plasmid, encoding only genes involved in replication, maintenance, and conjugative transfer, while pCTX-M3 contains two additional insertions totaling 29.6 kb that encode a number of antibiotic resistance genes. Koczan et al. (2009) also stressed the importance of amylovoran and levan in pathogenesis of *Er. amylovora*. They showed that biofilm formation played a critical role in the pathogenesis; amylovoran was required for biofilm formation and levan contributed to biofilm formation. Scanning electron microscopic analysis revealed an extensive biofilm formation in xylem vessels after inoculation with *Er. amylovora* wild-type cells. On the other hand, Δams mutant cells were non-pathogenic and died rapidly following inoculation. Lee et al. (2010) reported that virulence factor expression levels influenced the disease severity caused by wild isolates of *Er. amylovora* on apple trees. On apple seedlings, three virulence characteristics, namely, amylovoran production, biofilm formation, and growth in immature apple fruit accounted for >70% of the variation in disease severity, while in greenhouse grown trees, amylovoran production, biofilm formation, growth in immature apple fruit, hypersensitive cell death elicitation, and sorbitol utilization accounted for >75% of the variation in disease severity.

The first genome of *Er. amylovora* to be published was the genome of the French isolate CFBP 1430 (Smits et al. 2010). Currently, 15 genome sequences of *Er. amylovora* strains are publicly available, of which three isolates are reported to specifically be pathogenic on *Rubus* spp. (Mann et al. 2013; Smits et al. 2014).

Smits et al. (2017), in their review article, reported that the publication of the first *Er. amylovora* genome has greatly accelerated and advanced our understanding of the fire blight pathogen. Further, the publication of multiple genomes made it clear that chromosomal diversity is relatively small and that most of the pan-genome variance is attributable to plasmids. In addition to gaining a more detailed view of the known virulence factors, genomics has enabled new breakthrough studies of virulence regulation mechanisms. Additionally, several niche adaptation and ecological fitness factors, though not directly influencing virulence, have been studied in greater detail, providing novel insights into the physiology and ecology of the bacterium. Moreover, application of genome data has yielded improved diagnostics and enabled population studies at different geographic scales.

7.49.3 Disease Cycle

The bacteria present in the dormant or holdover cankers, formed on the twigs, branches, and stems in the previous year, are the main source of perpetuation. The bacteria present at the margins of the cankers are more active. The bacteria present in buds and apparently healthy wood also contribute to the survival, but to a lesser extent. Weißhaupt et al. (2016) showed a potential role of fruit mummies (19.9% samples infected) and buds in overwintering as well as a source of primary inoculum for dissemination of the pathogen early in the growing season. Blossoms of non-host plants growing close to infected trees were also shown to be colonized by *Er. amylovora* and to enable epiphytic survival and propagation of bacteria. Sobiczewski et al. (2017) reported that *Er. amylovora* has an ability to survive as a semi-necrotroph or necrotroph, which allows its overwintering in dead apple leaves.

In the next spring, when weather warms up, the bacteria in the cankers begin to multiply and invade the healthy adjoining bark. During humid or wet weather, bacterial cells ooze out through the cracks and lenticels, and this coincides with the opening of pear blossoms. Due to sweet and sticky nature of bacterial ooze, various insects like bees, flies, and ants are attracted and get smeared with the ooze when they feed on it, and spread it to all the flowers they visit subsequently. Splashing rain also spreads the bacteria from oozing cankers to flowers and other plant parts. Under arid conditions, the bacterial ooze dries to form long, narrow aerial strands, which are spread by wind currents.

Slack et al. (2017) examined microbiologically *Er. amylovora* exopolysaccharide ooze and found on an average 10^8 cfu μL^{-1} in ooze droplets on an average. Examination of apple host tissue at the site of emergence of ooze droplets using scanning electron microscopy revealed that ooze was not exuding through natural openings, but it was found on erumpent mounds and small (10 μm) tears in tissue. It is evident that *Er. amylovora* induced wounds in tissue provided the exit holes for ooze extrusion from the host. Analyses of *Er. amylovora* populations in ooze droplets and within the stems from which ooze droplets emerged indicated that approximately 9% of the total bacterial population from infected stems is diverted to ooze. Gene expression analyses revealed that *Er. amylovora* cells in stem sections located above ooze droplets and in ooze droplets were actively expressing critical pathogenicity genes such as *hrpL*, *dspE*, and *amsK*. The ooze that emerges from the host tissue acts as a source of large, concentrated populations of *Er. amylovora*, and as the cells in ooze droplets are expressing genes required for pathogenesis, they are already primed for infection as and when they are dispersed from ooze to new infection courts.

The bacteria present on the stigmas of flowers do not infect immediately, but grow epiphytically. Stigmas have a moist, nutrient-rich surface, which provides a favorable habitat for the multiplication of bacteria, increasing the number of cells up to 10^{6-7} per stigma. These bacteria are again spread to other flowers by rain or honeybees or other insects before the onset of environmental conditions favorable for infection. A mean temperature of 15.5°C is required for the epiphytic growth of the bacteria on the stigmas. With the help of rain or dew, the bacteria move from stigmas to the hypanthium or floral cup of the flower and enter the flowers through secretory cells (nectarthodes). After penetration into flowers, the bacteria multiply rapidly, producing necrotic symptoms.

Flowers can become infected within minutes after a rain or heavy dew when the average daily temperature is 16°C or higher. From the infected flower, infection spreads down the pedicel and then into the fruit spur. Infection of the spur causes the death of all the flowers, fruits, and leaves borne on it. Shoot blight usually develops after blossom blight. The bacteria from infected flowers are disseminated by splashing rain and they enter leaves, twigs, and branches. The penetration of leaves occurs mostly through wounds caused by insects, hailstorms, strong winds, but less frequently through stomata and hydathodes. From the leaf, infection spreads to the petiole and then to the twig or branch. Twigs and branches are penetrated directly through the lenticels and wounds and indirectly via flower or leaf infections. Once inside the host tissue, the bacteria move intercellularly through the parenchyma cells and in some cases through the xylem and phloem leading to systemic infection. In systemic infection, surprisingly, a single flower infection can cause the death of a tree in a few weeks. Unlike other wilt bacteria, *Er. amylovora* moves rapidly from the vessels to the other tissues and causes the death of cells of these tissues. The blight and canker symptoms result from the death of these tissues.

Ideal conditions for infection and disease development include wet or humid weather coupled with daytime temperature ranging from 24°C to 29°C and night temperature above 13°C. High soil fertility including excessive nitrogen, which promotes succulent growth, also favors disease development. Vigorously growing shoots are most severely infected. The infection progresses very rapidly in the succulent tissue and may advance 10–30 cm in a day (Thomson 1992).

Tancos et al. (2017) detected *Er. amylovora* in asymptomatic budwood in trees more than 20 m from trees with fire blight symptoms. In some seasons, there were significant ($P \leq 0.05$) differences in the incidence of *Er. amylovora* in asymptomatic budwood collected from symptomatic trees and those up to 20 m from them. The examination of individual bud dissections revealed that *Er. amylovora* was within the tissue beneath the bud scales containing the meristem. There is an urgent need to better understand the factors that lead to the presence of *Er. amylovora* in bud tissues to ensure the production of pathogen-free apple trees.

Billing (2011) has explained why views on host invasion by *Er. amylovora* differ. There are limited observations on orchard pear and apple trees; in most experimental studies, young apple shoots on potted plants were used. Tissue maturity at the site of shoot inoculation is of prime importance. If xylem vessels are damaged, inoculum may be sucked into the vessels and the bacteria will multiply there. In younger tissue, there is less suction pressure. No one has suggested that migration in bark tissue cannot be a major route, and means of escape from mature xylem vessels have not been demonstrated and remain a matter for speculation.

Wöhner et al. (2018) confirmed the earlier finding that deletion of the TIIISS abolishes virulence of *Er. amylovora*. The authors also established a new gene-for-gene relationship between the effector protein Eop1 and the fire blight resistant ornamental apple cultivar Evereste and the wild species *Malus floribunda* 821.

Pirc et al. (2009) developed specific, sensitive, and reliable TaqMan real-time PCR assays for the detection of *Er. amylovora* targeting its chromosomal DNA (*amsC* gene and ITS region). The real-time PCR assays in combination with an automated DNA-extraction method based on magnetic beads (QuickPick™) reliably detected at least 10^3 cells mL^{-1} (*c*. four cells per reaction) of the pathogen from blighted woody plant material. The assays were also able to detect the bacterium from symptomless plants including cases where subsequent isolation of the bacterium failed, giving an absolute quantification of *Er. amylovora* before and after enrichment in liquid media, which provided proof for viability and the ability of the bacterium to multiply.

Gosch et al. (2012) described a LAMP of DNA assay for genomic DNA of *Er. amylovora*, which relies on a highly specific primer design that does not allow amplification from typical DNA sources of the orchard biological system and from sample handling. The assay enabled the fast detection of down to approximately 20 cfu of pure *Er. amylovora* or 100 fg genomic DNA (corresponding to approximately 25 *Er. amyolovora* equivalents) per reaction in 45 min. The assay is suitable for field testing as fast and reliable detection of *Er. amylovora* in orchard samples is achieved with naked eye through a visual color change with hydroxynaphthol blue.

Moradi et al. (2012) developed a reliable and rapid pathogen detection protocol that utilizes LAMP for detection of *Er. amylovora*, using primers derived from the highly conserved fragment of the chromosomally *amsH* gene. Testing of 208 naturally infected samples revealed that the specificity value of LAMP was 84%, while conventional and nested PCR could detect only 59% and 73% of the whole collection. The specificity assay also showed that the LAMP protocol is species-specific for detection of *Er. amylovora* even in interspecies analysis.

Hinze et al. (2016) described an improved protocol to be used directly on whole bacteria in the field based on a real-time PCR approach previously published. The method allows for early detection and quantification of the pathogen prior to the occurrence of first symptoms. There was a clear correlation between bacterial abundance and subsequent disease development, however, in some cases, no disease symptoms could be observed despite a pathogen load of up to 3.4×10^6 cells per blossom. Integration of the amount of pathogen detected into refined prediction algorithms may allow for the improvement of applied forecasting models, finally permitting a better abatement of fire blight.

Certain orchard management practices, adopted in the US in the past, have increased the frequency of more devastating outbreaks of fire blight. These are briefly summarized as below:

1. **High tree density:** Instead of 100–200 apple trees per acre, now 500–1,000 trees per acre are planted
2. **Tree size control:** In order to accomplish high tree densities, tree size-controlling rootstocks, namely, M-9 and M-26 are widely used. Both these rootstocks are highly susceptible to the disease (Steiner 1998)
3. **Susceptible varieties:** Apple varieties, namely, Gala, Fuji, Gingergold, Jonagold, and Braeburn grown due to their demand in the fresh market, are highly susceptible to the disease. The combination of these varieties on highly susceptible rootstocks like M-9 and M-26 further aggravates the problem
4. **Training system:** New tree training systems, adopted to make high-density plantings more productive, push the trees into early bearing. In this process, some natural physiological mechanisms that resist the progress of infection are reduced.

7.49.4 Management

No single control measure is effective. Therefore, an integrated and aggressive fire blight management program should be followed. The following measures will help in minimizing the disease severity.

1. **Sanitation**
   a. During the winter, all blighted twigs and branches, cankers, and even entire trees, if needed, should be cut and burnt. The infected twigs and branches should be cut at least 20–25 cm below the visible injury or canker.
   b. The orchard should be inspected in the early season to locate the cankers, which were not removed during the previous pruning because these become more prominent once the bacteria start oozing from them. These cankers should be removed by cutting the twigs and branches at least 30 cm below the point of visible infection. The cutting tools should be disinfected after every cut with 10% commercial sodium hypochlorite solution.
2. **Horticultural practices**
   a. To reduce excessive succulency, trees should be grown in soil having balanced nutrition and normal dose of nitrogen. The trees should also receive limited pruning.
   b. A proper insect control should be followed in the post-bloom period to control the insects responsible for the spread of bacteria.
   c. Moderately resistant and resistant varieties, if available, should be grown in areas where disease is destructive.
   d. Growth-regulating acylcyclohexanedione compounds, namely, prohexadione-calcium (trade names Apogee® and Regalis®) and trinexapac-ethyl (trade names Moddus® and Primo®) offer a new, promising, and safe method for induction of resistance in pome fruit trees against fire blight and other pathogens and also for maintaining vegetative and reproductive equilibrium. The application of these compounds activates plant defense reactions leading to formation and storage of a biocidal compound, luteoforol in the treated tissue. The release of this compound as a result of pathogen infection causes the death of the pathogen and the adjacent plant cells. Spinelli et al. (2004) reported that the combination of *Pa. agglomerans* and acylcyclohexanediones significantly reduced both blossom blight incidence and the percentage of shoot infection. The enhanced protection achieved with the combination resulted from the reduction in blossom blight incidence caused by the former and prevention of shoot infection due to the activation of plant defenses by the latter. Spinelli et al. (2007) found that both prohexadione-calcium and trinexapac-ethyl caused significant reduction in blossom blight incidence leading to significant reduction in shoot blight incidence. Both the compounds also caused 60%–66% reduction in the migration of the bacterium in the plant tissue.

      Johnson and Temple (2017) evaluated ASM for the control of fire blight on 3- to 14-year-old pear and apple trees. Suspensions of ASM (7.5–22.5 g of active ingredient per liter plus silicone surfactant) were painted onto a 30- to 45-cm length of branch tissue immediately below primary pruning cuts or sprayed onto an 80- to 100-cm length of central trunk. In some experiments, a second ASM treatment was made in late June to early July in conjunction with secondary pruning of redeveloped cankers. Over 5 years of field experiments, trees treated with ASM yielded 62% less diseased wood at the time of secondary and tertiary canker removal compared with non-ASM-treated trees. Tree mortality and proportion of pruning cuts where fire blight redeveloped were also reduced by ASM treatment. Systemic acquired resistance induced by ASM application could prove an aid to pruning therapy in young fire blight susceptible pear and apple trees where, after canker removal, disease symptoms frequently redevelop owing to residual cells of the pathogen distributed within symptomless portions of the tree.
3. **Chemicals**: Various chemicals, namely, copper oxychloride, Bordeaux mixture (0.5%), and streptomycin (100 μg-mL^{-1}) have been used for the control of the disease. Bordeaux mixture and copper oxychloride, as dormant sprays, have offered some, but not much, protection from the disease. Moreover, copper fungicides may cause a russeting or scarring of the fruit surface (Teviotdale 2003). Elkins et al. (2015) reported that epiphytic populations of *Er. amylovora* on flowers at petal fall were significantly suppressed in three of four seasons in fixed copper + horticultural oil treated plots compared with only horticultural oil treated plots. The copper sanitation may add value to a fire blight management program by delaying the increase of epiphytic populations of the bacterium in flowers.

   Streptomycin is highly effective against blossom blight and also against trauma blight, but the timings of its sprays are critical for effective control because this antibiotic is active only for 3 days after its application. For maximum benefit, streptomycin must be applied on the day when an infection event occurs or a day after it. A delay in spraying for 24 h will result in 10%–20% reduction in control. For the prediction of infection event accurately, fire blight forecasting models, namely, **MARYBLYT™** and **CougarBlight** can be effectively used.

The continuous use of streptomycin in many areas has resulted into the development of bacterial strains resistant to this antibiotic, hence making its use ineffective in these areas. Pathogen strains resistant to streptomycin were first observed in California pear orchards in 1971 and in Washington and Oregon in 1972 and then in the western United States as well as in Michigan and Missouri. Russo et al. (2008) reported the occurrence of these resistant strains from New York. Streptomycin resistance has also been reported from British Columbia (Canada), Israel, Lebanon, Mexico, and New Zealand. Tancos and Cox (2016) monitored the development of streptomycin resistance in *Er. amylovora* in New York State from 2011 to 2014 using clustered regularly interspaced short palindromic repeat (CRISPR) spacer sequencing. They hypothesized that eradication efforts in 2002 failed, and the resistant bacterial populations continued to spread throughout the state, resistance may also be developing at the same time and may be due to the introduction of *Er. amylovora* resistant to streptomycin from other regions of the United States. The increased understanding, as to how streptomycin resistant *Er. amylovora* isolates are introduced, evolve, or have become established, afforded by CRISPR profiling has been useful for disease management and restricting the movement of streptomycin resistance in New York. In such areas, another broad-spectrum antibiotic, oxytetracycline can be effectively used. For best results, it should be applied within 24 h before infection and also within 12–24 h after the infection. McGhee and Sundin (2011) suggested kasugamycin as an alternate antibiotic for fire blight management, but also cautioned about the possible role of non-target bacteria in the future transfer of a kasugamycin resistance to *Er. amylovora*.

Messenger, a biochemical pesticide has shown some promise in controlling blossom blight phase when used in combination with streptomycin sprays. It is not bactericidal, but acts by activating the systemic acquired resistance genes of the host, for which about 5–7 days are needed. Therefore, its application should be made sufficiently in advance of blossom blight infection.

4. **Biological control**: Johnson and Stockwell (1998) have highlighted the use of biocontrol agents for the management of fire blight. Blight-Ban A506 (a product of *Pseudomonas fluorescens* strain A 506) is now available commercially for its control. Two applications of Blight-Ban caused 40%–60% reduction in blossom blight. In another trial, its single application caused 50% reduction in fire blight, but when this application was followed by weekly applications of streptomycin, the reduction in disease was 70%. The reduction in frost injury was also more in the combination treatment. Stockwell et al. (2010) reported that *Pa. vagans* C9-1 suppressed fire blight in all five trials in which the pathogen was applied as lyophilized cells, but in none of the trials when applied as freshly harvested cells. A mixture of *Ps. fluorescens* A506 and *Pa. vagans* C9-1 also suppressed fire blight, but the magnitude of disease suppression in all field trials (averaging 32%) was less than that attained with *Pa. vagans* C9-1 alone (42%).
Stockwell et al. (2011) hypothesized that strain *Psm. fluorescens* A506 diminishes the biocontrol activity of *Pa. vagans* strain C9-1 and *Pa. agglomerans* strain Eh252 for fire blight, because A506 produces an extracellular protease in culture that degrades the peptide antibiotics produced by C9-1 and Eh252. This hypothesis was tested in five replicated field trials using strain A506 and A506 *aprX::Tn5*, an extracellular protease-deficient mutant, as individuals and combined with C9-1 or Eh252 for biological control of fire blight. On average, mixtures containing A506 *aprX::Tn5* were superior to those containing the wild-type strain, confirming that the extracellular protease of A506 diminished the biological control activity of C9-1 and Eh252 in situ. The mixtures of A506 *aprX::Tn5* and C9-1 or Eh252 were also superior to oxytetracycline or single biocontrol strains in suppressing pear fire blight. It is evident that certain biological control agents are mechanistically incompatible, in which one strain interferes with the mechanism by which a second strain suppresses plant disease. The mixtures composed of mechanistically compatible strains of biological control agents can suppress disease more effectively than individual biological control agents.
Mikiciński et al. (2016) evaluated *Psm. graminis* strain 49M isolated from the apple phyllosphere under laboratory and greenhouse conditions and found that it was equally or even more effective to protect apple blossoms and apple terminal shoots than the reference strains A506 of *Psm. fluorescens* and C9-1 of *Pa. vagans*, as well as the bioproducts Blight-Ban A506, Blossom Protect, and Hortocyna 18 SP (streptomycin), used for comparison. However, its effectiveness on pear fruitlets was slightly less, but significantly better (over 10%) than that of Blight-Ban A506. On its introduction onto apple blossoms in orchards, it effectively colonized blossoms during the entire bloom period in different weather conditions, including rain. The strain showed no pathogenicity toward plant tissue (pear fruitlets, apple blossoms, tobacco leaves) and is therefore a potential candidate for the development of a new biopesticide against fire blight.
Svircev et al. (2006) obtained an effective control of the disease, comparable to streptomycin, using *Pa. agglomerans* (a biocontrol agent of the pathogen)

and phages specific to *Er. amylovora*. In this case, *Pa. agglomerans* was also used to deliver and sustain phages on the blossom surface. Nagy et al. (2015) reported that two phage strains, i.e., ΦEa104 and H5K were able to translocate in apple seedlings in parts above ground level following their application to the roots and also in roots after spraying them onto the stem and leaves. A water suspension of phages effectively decreased severity of *Er. amylovora* infection in apple seedlings following treatment of roots or aerial plant parts and application to the cotyledon, as judged by symptom bonitation. The demonstration of *in planta* translocation of *Er. amylovora*-specific bacteriophages, and their role in reducing the fire blight severity may significantly contribute to a better control of *Er. amylovora* and promote further investigations on penetration and translocation of phages into plants.

5. **Transgenic plants**: Transgenic apple lines, overexpressing the apple gene *MpNPR 1*, have shown increased resistance to fire blight pathogen (Malony et al. 2006).

REFERENCES

Billing, E. 2011. Fire blight. Why do views on host invasion by *Erwinia amylovora* differ? *Plant Pathol.* 60: 178–189.

Burrill, T. J. 1882. The bacteria: An account of their nature and effects, together with a systematic description of the species. *Report of the Board of Trustees of the Illinois Industrial University* 11, 93–157.

Elkins, R. B., Temple, T. N., Shaffer, C. A., Ingels, C. A., Lindow, S. B., et al. 2015. Evaluation of dormant-stage inoculum sanitation as a component of a fire blight management program for fresh-market Bartlett pear. *Plant Dis.* 99: 1147–1152.

Gosch, C., Gottsberger, R. A., Stich, K., and Fischer, T. C. 2012. Blue [Ea]LAMP—A specific and sensitive method for visual detection of genomic *Erwinia amylovora* DNA. *Eur. J. Plant Pathol.* 134: 835–845.

Hinze, M., Köhl, L., Kunz, S., Weißhaupt, S., Ernst, M., et al. 2016. Real-time PCR detection of *Erwinia amylovora* on blossoms correlates with subsequent fire blight incidence. *Plant Pathol.* 65: 462–469.

Ishimaru, C., and Klos, E. J. 1984. New medium for detecting *Erwinia amylovora* and its use in epidemiological studies. *Phytopathology* 74: 1342–1345.

Johnson, K. B., and Stockwell, V. O. 1998. Management of fire blight: A case study in microbial ecology. *Annu. Rev. Phytopathol.* 36: 227–248.

Johnson, K. B., and Temple, T. N. 2017. Induction of systemic acquired resistance aids restoration of tree health in field-grown pear and apple diseased with fire blight. *Plant Dis.* 101: 1263–1268.

Kim, W. S., Hildebrand, M., Jock, S., and Geider, K. 2001. Molecular comparison of pathogenic bacteria from pear trees in Japan and the fire blight pathogen Erwinia amylovora. *Microbiology* 147: 2951–2959.

Koczan, J. M., McGrath, M. J., Zhao, Y., and Sundin, G. W. 2009. Contribution of *Erwinia amylovora* exopolysaccharides amylovoran and levan to biofilm formation: Implications in pathogenicity. *Phytopathology* 99: 237–244.

Lee, S. A., Ngugi, H. K., Halbrendt, N. O., O'Keefe, G., Lehman, B., et al. 2010. Virulence characteristics accounting for fire blight disease severity in apple trees and seedlings. *Phytopathology* 100: 539–550.

Longstroth, M. 2001. The 2000 fire blight epidemic in Southwest Michigan apple orchards. *Compact Fruit Tree* 34: 16–19.

Lopez, M. M., Rosello, M., Llop, P., Ferrer, S, Christen, R., et al. 2011. *Erwinia piriflorinigrans* sp. nov., a novel pathogen that causes necrosis of pear blossoms. *Int. J. Syst. Evol. Microbiol.* 61: 561–567.

Malony, M., Borejsza-Wysocka, M. M., Aldwinckle, H. S., Jin, Q. L., and He, S. Y. 2006. Transgenic apple lines over-expressing the apple gene MpNPR1 have increased resistance to fire blight. *Acta Hortic.* (*ISHS*) 704: 521–526.

Mann, R. A., Smits T. H. M., Buhlmann A., Blom J., Goesmann A., et al. 2013. Comparative genomics of 12 strains of *Erwinia amylovora* identifies a pan-genome with a large conserved core. *PLoS One* 8(2): e55644.

McGhee, G. C., and Sundin, G. W. 2011. Evaluation of kasugamycin for fire blight management, effect on nontarget bacteria, and assessment of kasugamycin resistance potential in *Erwinia amylovora*. *Phytopathology* 101: 192–204.

Mikiciński, A., Sobiczewski, P., Puławska, J., and Maciorowski, R. 2016. Control of fire blight (*Erwinia amylovora*) by a novel strain 49M of *Pseudomonas graminis* from the phyllosphere of apple (*Malus* spp.). *Eur. J. Plant Pathol.* 145: 265–276.

Moradi, A., Nasiri, J., Abdollahi, H., and Almasi, M. 2012. Development and evaluation of a loop-mediated isothermal amplification assay for detection of *Erwinia amylovora* based on chromosomal DNA. *Eur. J. Plant Pathol.* 133: 609–620.

Myung, I.-S., Lee, J.-Y., Yun, M.-J., Lee, Y.-H., Lee, Y.-K., et al. 2016. Fire blight of apple, caused by *Erwinia amylovora*, a new disease in Korea. *Plant Dis.* 100: 1774.

Nagy, J. K., Schwarzinger, I., Künstler, A., Pogány, M., and Kiraly, L. 2015. Penetration and translocation of *Erwinia amylovora*-specific bacteriophages in apple—A possibility of enhanced control of fire blight. *Eur. J. Plant Pathol.* 142: 815–827.

Pirc, M., Ravnikar, M. Tomlinson, J., and Dreo, T. 2009. Improved fire blight diagnostics using quantitative real-time PCR detection of *Erwinia amylovora* chromosomal DNA. *Plant Pathol.* 58: 872–881.

Russo, N. L., Burr, T. J., Breth, D. I., and Aldwinckle, H. S. 2008. Isolation of streptomycin-resistant isolates of *Erwinia amylovora* in New York. *Plant Dis.* 92: 714–718.

Shtienberg, D., Manulis-Sasson, S., Zilberstaine, M., Oppenheim, D., and Shwartz, H. 2015. The incessant battle against fire blight in pears: 30 years of challenges and successes in managing the disease in Israel. *Plant Dis.* 99: 1048–1058.

Slack, S. M., Zeng, Q., Outwater, C. A., and Sundin, G. W. 2017. Microbiological examination of *Erwinia amylovora* exopolysaccharide ooze. *Phytopathology* 107: 403–411.

Smits, T. H. M., Duffy, B., Sundin, G. W., Zhao, Y. F., and Rezzonico, F. 2017. *Erwinia amylovora* in the genomics era: From genomes to pathogen virulence, regulation, and disease control strategies. *J. Plant Pathol.* 99 (Special issue): 7–23.

Smits, T. H. M., Guerrero-Prieto V. M., Hernandez-Escarcega G., Rezzonico F., Blom J., et al. 2014. Comparative genomics of *Erwinia amylovora* isolates from Mexico. *Acta Hortic.* (*ISHS*) 1056: 173–177.

Smits, T. H. M., Rezzonico F., Kamber T., Blom J., Goesmann A., et al. 2010. Complete genome sequence of the fire blight pathogen *Erwinia amylovora* CFBP 1430 and comparison to other *Erwinia* spp. *Mol. Plant-Microbe Inter.* 23: 384–393.

Sobiczewski, P., Iakimova, E. T., Mikiciński, A., Węgrzynowicz-Lesiak, E., and Dyki, B. 2017. Necrotrophic behaviour of *Erwinia amylovora* in apple and tobacco leaf tissue. *Plant Pathol.* 66: 842–855.

Spinelli, F., Vanneste, J. L., Ciampolini, F., Cresti, M., Rademacher, W., et al. 2007. Potential and limits of acylcyclohexanediones for the control of blossom blight in apple and pear caused by *Erwinia amylovora. Plant Pathol.* 56: 702–710.

Spinelli, F., Vanneste, J. L., Cornish, D., Yu, J., and Costa, G. 2004. Growth-regulating acylcyclohexanediones, trinexapac-ethyl and prohexadione-calcium, decrease blossom blight incidence in pome fruits. *Acta Hortic.* (*ISHS*) 704: 245–248.

Steiner, P. W. 1998. Problems in managing fire blight in high density orchards on M-9 and M-26 rootstocks. *Presented in the Annual Meeting of the Va./W.Va. State Horticultural Societies*, Roanoke, VA.

Stockwell, V. O., Johnson, K. B., Sugar, D., and Loper, J. E. 2010. Control of fire blight by *Pseudomonas fluorescens* A506 and *Pantoea vagans* C9-1 applied as single strains and mixed inocula. *Phytopathology* 100: 1330–1339.

Stockwell, V. O., Johnson, K. B., Sugar, D., and Loper, J. E. 2011. Mechanistically compatible mixtures of bacterial antagonists improve biological control of fire blight of pear. *Phytopathology* 101: 113–123.

Svircev, A. M., Lehman, S. M., Kim, W., Barszcz, E., Schneider, K. E., et al. 2006. Control of the fire blight pathogen with bacteriophages. *Proceedings of 1st International Symposium on Biological Control of Bacterial Plant Disease*, Seeheim/Darmstadt, Germany, 259: Dtsch. Bibl., CIP-Einh.aufn.

Tancos, K. A., Borejsza-Wysocka, E., Kuehne, S., Breth, D., and Cox, K. D. 2017. Fire blight symptomatic shoots and the presence of *Erwinia amylovora* in asymptomatic apple budwood. *Plant Dis.* 101: 186–191.

Tancos, K. A., and Cox, K. D. 2016. Exploring diversity and origins of streptomycin-resistant *Erwinia amylovora* isolates in New York through CRISPR spacer arrays. *Plant Dis.*100: 1307–1313.

Teviotdale, B. L. 2003. Fire blight. IPM Education and Publications, Univ. Calif. Agric. Nat. Res. Publ. 7414.

Thomson, S. V. 1992. Fire blight of apple and pear. pp. 32–65 in: *Plant Diseases of International Importance*. Vol. III, *Diseases of Fruit Crops*. J. Kumar, H. S. Chaube, U. S. Singh, and A. N. Mukhopadhyay eds. Prentice Hall, Englewood Cliffs, NJ.

Thomson, S. V. 2001. Fire blight. pp. 459–61 in: *Encyclopedia of Plant Pathology*. O. C. Maloy, and T. D. Murray eds. John Wiley & Sons, New York.

Tian, Y., Zhao, Y., Shi, L., Cui, Z., Hu, B., et al. 2017. Type VI secretion systems of *Erwinia amylovora* contribute to bacterial competition, virulence, and exopolysaccharide production. *Phytopathology* 107: 654–661.

Vanneste, J. L. 2000. *Fire Blight: The Disease and Its Causative Agent, Erwinia Amylovora*. CABI, Wallingford, UK.

Weißhaupt, S., Köhl, L., Kunz, S., Hinze, M., Ernst, M., et al. 2016. Alternative inoculum sources for fire blight: The potential role of fruit mummies and non-host plants. *Plant Pathol.* 65: 470–483.

Winslow, C. E. A., Broadhurst, J., Buchanan, R. E., Krumwiede, C. J., Rogers, L. A., et al. 1920. The families and genera of the bacteria. Final report of the Committee of the Society of American Bacteriologists on characterization and classification of bacterial types. *J. Bacteriol.* 5: 191–229.

Wöhner, T. W., Richter, K., Sundin, G. W., Zhao, Y., Stockwell, V. O., et al. 2018. Inoculation of *Malus* genotypes with a set of *Erwinia amylovora* strains indicates a gene-for-gene relationship between the effector gene *eop1* and both *Malus floribunda* 821 and *Malus* "Evereste." *Plant Pathol.* 67: 938–947.

7.50 CITRUS CANKER

Citrus canker is one of the most devastating diseases of citrus (*Citrus* spp.) trees. It is also one of the most feared diseases in the United States and some other countries. Losses occur due to the reduction in yield, including premature fruit drop and also reduction in market value of the infected fruits. Sometimes badly infected fruits may become unsalable. Fruits produced in canker-endemic regions suffer from a restricted market due to the enforcement of international plant quarantines.

The disease was first observed in 1913 in Florida. Clara Hasse identified the causal bacterium as *Pseudomonas citri* in 1915. The presence of the disease on citrus specimens collected from India (on *C. medica* from Dehra Dun, Uttarakhand, India in 1827–1831) and Indonesia (on *C. aurantifolia* in 1842–1844) and deposited in Kew Herbarium suggests that the disease probably originated in India or Indonesia. The reference to its Chinese origin is also found in the literature (Goto 1992). The disease is endemic in Southeast Asia and Japan from where it has spread to many citrus-growing countries. The disease occurs in severe form in seasons and/or regions characterized by warm and humid weather conditions. According to Goto (1992), the disease is particularly epiphytotic and damaging on seedlings and young trees, especially after storms (hurricanes) accompanied with rain under warm conditions. Full-grown trees show much less disease severity and damage.

Citrus canker has a long history in Florida, where it was repeatedly introduced and eradicated. It was introduced in Florida in 1912 along with infected nursery plants from Japan and from there it spread to all the Gulf States of the US and beyond. It took about 20 years to eradicate the disease from Florida by destroying and burning of more than three million nursery and a quarter million fruit bearing trees (Agrios 2004). It took 17 more years (until 1949) to eliminate the disease from the entire United States. The appearance of a bacterial leaf spot (*Xanthomonas* leaf spot or citrus bacterial spot), resembling to citrus canker in symptoms, in 1984 and the appearance of actual citrus canker in 1986 in Manatee County of Florida forced the government to start a new series of eradicative measures, which continued until 1992 and resulted in the destruction of more than 20 million nursery and orchard trees. Florida was declared free of citrus canker in early 1994, but unfortunately the disease appeared again in urban Miami in October 1995, in commercial citrus in Manatee County in June 1997 and subsequently in many other counties of the state. In spite of removal and destruction of 1.56 million commercial trees and 600,000 dooryard citrus trees statewide, the infected area increased to 1701 km^2 until March 2002.

Citrus canker is also present in Oceania, Argentina, Brazil, Congo, Democratic Republic, Madagascar, Mauritius, Gabon, the Netherlands, Paraguay, and Uruguay. The eradication programs have succeeded in eradicating the disease from Australia, South Africa, New Zealand, and Mozambique. The latest outbreak and eradication of citrus canker in Australia occurred in 1991. In Brazil, the disease was noticed in 1957 from where it subsequently spread to Uruguay, Paraguay, and Argentina. The disease has become established in these countries in spite of the attempts to eradicate it. However, the eradication efforts in Brazil have kept the disease under reasonable control (incidence below 0.22%) in São Paulo state, the major citrus producing state of the country. Several protocols for eradication were adopted in São Paulo state since the first detection of the disease in 1957. Among these, the one carried out from 1999 to 2009 was the most effective for maintaining the disease at a very low level throughout the state. That protocol mandated periodic inspections and removal of symptomatic and asymptomatic trees based on disease incidence in the block. After 2009, the eradication protocols were less stringent and the disease incidence continued to increase (Behlau et al. 2014).

In addition to the abovementioned countries, the disease is also present in Bolivia, Senegal, Mali, Burkina Faso, Tanzania, Iran, Saudi Arabia, Yemen, Bangladesh, Comoros, Malaysia, and Vietnam, but absent from Europe (Ference et al. 2018). It is one of the most important diseases of the international quarantine and all citrus-producing countries free of citrus canker strictly prohibit the import of citrus plants and fruits from disease affected countries.

7.50.1 Host Range

Xa. citri subsp. *citri* infects a large number of citrus species. Among the most susceptible hosts are grapefruit, Mexican (key) lime and lemons, trifoliate oranges, some sweet oranges, and limes. In India, *kagzi* lime is the most susceptible host. These hosts suffer badly from the disease in moist tropical and sub tropical climates. Mandarin oranges like satsuma, Cleopatra, and ponkan are resistant. The other natural hosts include *Swinglea glutinosa*, *Fortunella margarita*, *F. japonica*, and *Severinia buxifola*.

7.50.2 Symptoms

The symptoms appear on all the aboveground plant parts. Young citrus tissues during the last half of the expansion phase of the growth are at maximum susceptibility. Water-soaked, translucent, tiny spots, usually dark green with raised convex surface first appear on lower surface and then on upper surface of the leaves. Initially, the spots are corky having oily appearance, but later on turn brown and are surrounded by glossy, dark brown margins. As the disease advances, the spots turn greyish or whitish, center of the spot ruptures exposing a light brown spongy mass and forming a crater-like depression. Yellow haloes are usually formed around such mature lesions. The centers of the spots may fall down giving a shot-hole effect. The lesions are round to irregular in shape and measure 5–10 mm in diameter. As the disease severity increases, the spots coalesce giving a scabby appearance (Figure 7.35a). The formation of lesions on midrib or petioles of leaves results in more defoliation than the scattered lesions only on the leaf blade. The feeding activities of Asian citrus leaf miner, *Phyllocnistis citrella*, increase the disease severity on leaves by exposing the leaf mesophyll tissues to splashed inoculum and thereby increasing the leaf infection.

As the disease progresses, the symptoms also appear on twigs, branches, fruits, and thorns. The lesions on twigs are observed generally on more susceptible plant species like grapefruit, lime, sweet orange, and trifoliate orange, etc. The lesions on twigs and branches are raised, corky, and surrounded by an oily or water-soaked margin, but no chlorotic haloes. The lesions may vary in size from 1 to several centimeters in length (Figure 7.35b). Severely infected twigs start drying from tip to backward and severe defoliation occurs.

The symptoms on fruits are similar to those found on leaves except that yellow haloes are not always present. The lesions are rough and corky, having more prominent crater-like depressions and remain confined to the rind. Severely infected fruits become highly blemished and fetch very low price or may become unmarketable (Figure 7.35c).

The vigorously growing citrus trees are highly susceptible to infection as each growth flush of the tree during a growing season is accompanied by a period of increased susceptibility. When leaves, twigs, and fruits reach maturity, they become more resistant to infection due to the thickening of the cuticle and hardening of the tissue.

Until now, five types of citrus cankers or related forms have been reported affecting *Citrus* species. **Asiatic** or **true citrus canker** or **canker/cancrosis A**, originally found in Asia, is by far the most widespread and severe form of the disease

FIGURE 7.35 Citrus canker: (a) Leaf spots. (b) Twig cankers. (c) Lesions on fruit.

(Gottwald et al. 2002). It also affects large number of *Citrus* species. The symptoms described above are of a typical true citrus canker.

False canker or **cancrosis B**, first found in South America, occurs mainly on lemons and to a lesser extent on lime, sour orange, sweet orange, and citron, but not on grapefruit. It is found in Argentina, Paraguay, and Uruguay.

Cancrosis/canker C or **Mexican lime canker**, originally found in São Paulo state of Brazil, primarily affects Mexican lime and to a lesser extent lemon. In Mexican lime, it is most severe on Galego lemon, but causes only slight infection on Tahiti lemon.

Rodoriguez et al. (1985) reported a disease causing pustules on leaves and young shoots of Mexican lime, but not infecting fruits, and on other citrus species along the Pacific coast of Mexico in 1981. It is called **canker D**.

In 1984, another bacterial disease of citrus called nursery-form canker was found in a nursery in Florida and subsequently in many other nurseries of the state (Schoulties et al. 1987). The symptoms appeared on leaves and shoots of nursery plants, but not on trees. The lesions produced were flat, but occasionally sunken, necrotic having an extremely water-soaked appearance and chlorosis, but never looked like erupted canker lesions. The host range of the bacterium included grapefruit, sweet and trifoliate oranges, tangor, tangerine, tangalo, mandarin, citrumelo, and citrange. The disease was called as *Xanthomonas* leaf spot or citrus bacterial spot, and the pathogen was named as *Xa. campestris* pv. *citrumelo*. Now the disease is called as **canker E**. The strains of pathogen causing cankers B, C, and D were designated as *Xa. campestris* pv. *aurantifolia*.

Citrus canker can be confused with other diseases like scab, anthracnose, leprosies, and melanose, but the following characteristic symptoms of citrus canker will distinguish it from other diseases (Patel and Desai 1972).

1. The lesions are raised on both the surfaces of the leaves
2. The lesions have a distinguished margin
3. The lesions have crater-like depressions in the center, which can be seen more clearly with the help of a hand lens

7.50.3 Causal Organisms

Xanthomonas citri subsp. *citri* (*ex* Hasse 1915) Gabriel et al. 1989 subsp. nov. (syn. *Xa. citri*, *Xa. campestris* pv. *citri*, *Xa. axonopodis* pv. *citri*) causes Asiatic or true citrus canker or canker A.

Xa. fuscans subsp. *aurantifolii* Schaad et al. 2007 subsp. nov. (syn. *Xa. campestris* pv. *aurantifolii*) causes cankers B, C, and D.

Xa. alfalfae subsp. *citrumelonis* Schaad et al. 2007 subsp. nov. (syn. *Xa. campestris* pv. *citrumelo*) causes canker E.

The bacterium causing Asiatic or true citrus canker was first identified as *Pseudomonas citri* by Hasse in 1915. After the establishment of genus *Xanthomonas*, it was transferred to this genus as *Xanthomonas citri*. With the introduction of pathovar system, it was changed to *Xa. campestris* pv. *citri*, and in 1995, Vauterin et al. reclassified it as *Xa. axonopodis* pv. *citri*. The important morphological, physiological, and biochemical characters of this bacterium described by Goto (1992) are given below.

Cells are Gram-stain-negative, non-sporing, straight rods, and measure 0.5–0.75 × 1.5–2.0 μm. They are predominantly single, aerobic, non-encapsulated, having a polar flagellum and chemoorganotrophic. The cells do not accumulate poly-β-hydroxybutyrate granules and the metabolism of glucose is oxidative. The colonies on solid media are circular, convex, having an entire margin. The colonies are creamy-yellow and straw yellow on potato glucose agar and on nutrient agar, respectively, due to the production of xanthomonadins. Catalase test is positive, while oxidase is negative or weak. Methionine or cysteine is essential for growth, but serine inhibits the growth. Asparagine is not used as a sole source of carbon and nitrogen. The bacterium hydrolyzes starch, esculin, casein, and Tween 80; liquefies gelatin and pectate gel and produces tyrosinase. The tests for nitrate reduction, arginine dihydrolase, urease, and amino acid dicarboxylases are negative. Glucose, sucrose, fructose, maltose lactose, galactose, xylose, mannose, glycerol, dextrin, starch, and citrate are utilized as the sole sources of carbon. Optimum temperature for growth is 28°C.

Until now, five different types of citrus cankers or related forms have been reported, and these have been named as cankers A, B, C, D, and E. Schaad et al. (2005) studied 44 strains of bacteria representing causal agents of all these five citrus canker types, and on the basis of DNA-DNA relatedness assays, sequence of 16S–23S intergenic spacer regions, and AFLP analysis grouped them into three distinct genotypes: taxon I included all "A" strains, taxon II contained all "B", "C", and "D" strains and taxon III contained all "E" strains. They named taxon I strains as *Xanthomonas smithii* subsp. *citri* (ex Hasse 1915) sp. nov. nom. rev. comb. nov., taxon II strains as *Xanthomonas fuscans* subsp. *aurantifolii* (*ex* Gabriel et al. 1989) sp. nov. nom. rev. comb. nov., and taxon III strains as *Xanthomonas alfalfae* subsp. *citrumelo* (*ex* Riker and Jones) Gabriel et al. 1989 sp. nov. nom. rev. comb. nov.

Schaad et al. (2006) emended the classification of xanthomonad pathogens on citrus as their earlier publication (Schaad et al. 2005) contained nomenclature errors. *Xa. smithii* subsp. *citri* and *Xa. alfalfae* subsp. *citrumelo* were reclassified as *Xa. citri* subsp. *citri* and *Xa. alfalfae* subsp. *citrumelonis*, respectively. *Xa. fuscans* subsp. *aurantifolii* and *Xa. alfalfae* subsp. *citrumelonis* share many common characteristics. The below given description of these three pathogens is taken from Schaad et al. (2006).

7.50.3.1 *Xa. citri* subsp. *citri*

The colonies of *Xa. citri* subsp. *citri* appear on YDC and FS agars after 40–44 and 56–60 h of incubation, respectively at 28°C–30°C. The bacterium utilizes arabinose, maltose, lactose, cellobiose, mannitol, and aspartic acid; hydrolyzes pectate, liquefies gelatin, and causes an alkaline hydrolysis of litmus milk. The bacterium produces large amounts of EPS both in culture media and in host tissue. The cells of the bacterium are rapidly killed when diluted in water, i.e., $<10^6$ cells mL^{-1}. Therefore, the EPS protects the bacterial cells in water and from desiccation in air (in the bacterial ooze). Strains of *Xa. citri* subsp. *citri* are susceptible to bacteriophages CP1 and CP2, whereas those of *Xa. fuscans* subsp. *aurantifolii* are not. The bacterium may be distinguished from *Xa. campestris* and most other *Xanthomonas* species, sub species, and pathovars by DNA-DNA reassociation assays, ITS sequencing, rep-PCR profiles, and phenotypic traits.

7.50.3.2 *Xa. fuscans* subsp. *aurantifolii*

The colonies of *Xa. fuscans* subsp. *aurantifolii* appear on YDC and FS agars after 50–60 and 70–76 h of incubation, respectively, at 28°C–30°C. The strains of the bacterium produce a water soluble brown pigment on several common agar media including YDC. The bacterium utilizes lactose, cellobiose, and mannitol, and precipitates litmus milk. *Xa. fuscans* subsp. *aurantifolii* is distinguished from *Xa. citri* subsp. *citri* and *Xa. alfalfae* subsp. *citrumelonis* as it precipitates litmus milk and hydrolyzes gelatin. *Xa. fuscans* subsp. *aurantifolii* does not utilize maltose or hydrolyze pectate, whereas *Xa. citri* subsp. *citri* does. *Xa. fuscans* subsp. *aurantifolii* is differentiated from most other *Xanthomonas* species, sub species, and pathovars by DNA-DNA reassociation assays, rep-PCR profiles, ITS sequences, and phenotypic traits.

7.50.3.3 *Xa. alfalfae* subsp. *citrumelonis*

The colonies of the bacterium on YDC and FS agars appear after 30–34 and 40–44 h of incubation, respectively, at 28°C–30°C. *Xa. alfalfae* subsp. *citrumelonis* utilizes raffinose, whereas *Xa. citri* subsp. *citri* does not. *Xa. alfalfae* subsp. *citrumelonis* can be differentiated from *Xa. fuscans* subsp. *aurantifolii* due to its more rapid growth on agar media, liquefaction of gelatin, and utilization of maltose. It is distinguished from *Xa. citri* subsp. *citri* by its ability to utilize raffinose, to produce acid from cellobiose and mannitol, and faster growth on YDC and FS agars. *Xa. alfalfae* subsp. *citrumelonis* is distinguished from *Xa. campestris* pv. *campestris* and other *Xanthomonas* species, sub species, and pathovars by DNA-DNA reassociation assays, rep-PCR profiles, ITS sequences, and phenotypic traits. These three sub species can be distinguished from each other by serological assays.

Kositcharoenkul et al. (2011) developed a single-tube nested PCR for the detection of *Xa. citri* subsp. *citri*. The assay targets the *pthA* gene of *Xa. citri* subsp. *citri* and utilizes different annealing temperatures for the two primer pairs. The assay reliably detected as few as 10^2 cells, was unaffected by the presence of PCR inhibitors and was 10- and 8,500-fold more sensitive than standard PCR and ELISA, respectively. The use of a washing method for DNA extraction increased the sensitivity compared to direct extraction from leaf tissue. The single-tube nested PCR detected *Xa. citri* subsp. *citri* from 67 samples out of 90 samples collected from affected pomelo orchards compared to 54 samples detected by standard PCR. It was also able to detect *Xa. citri* subsp. *citri* from samples with and without symptoms. This assay can be used as a rapid and sensitive technique for routine *Xa. citri* subsp. *citri* detection in field samples for surveillance of citrus canker.

Yu et al. (2017) reported that the primer sets XcAsF/XcAswR, XaBF/XaBR, and XaCF1/XaCR1 specifically differentiated *Xa. citri* subsp. *citri* A* and AW, *Xa. fuscans* subsp. *aurantifolii* B, and *Xa. fuscans* subsp. *aurantifolii* C strains in PCR, respectively. With quantitative real-time PCR analysis, it was possible to detect a range of 1 to 10^7 bacterial cell(s), which will be helpful to detect the species at pre-symptom developmental stage of quarantine period of the plant. PCR/primer efficiency was also demonstrated by using raw bacterial cells that will reduce considerable time and cost needed for genomic DNA isolation. The designed molecular markers and qPCR method will be useful for routine diagnosis of *Xanthomonas* species, which will improve the citrus canker disease management.

7.50.4 Disease Cycle

As the citrus trees are perennial, the bacterium survives on the infected trees, mainly in the lesions formed on branches and stems. The bacteria in stem lesions can remain viable for a long period of 5–7 years. The bacterium is frequently found in soil, on citrus roots, or weeds growing in orchards. Its population on weed roots does not decline in winter. The bacterium can survive for 2–3 months in leaf lesions, if the infected leaves are kept under dry conditions. The soil heavily infected with *Xa. citri* subsp. *citri* supports the inoculum level sufficient for infection only for a few weeks during the citrus growing season. However, until now, there is no direct evidence that the bacteria surviving in low numbers on weed hosts or in soil can lead to the development of epiphytotics.

The bacteria multiply in lesions formed on leaves, twigs, and fruits. During wet weather, the bacteria ooze out from these lesions and are dispersed to new growth and other plants. Wind-driven rain is the main natural dispersal agent, and the wind speed >18 mph helps in the stomatal penetration of bacteria. The rainstorms also increase the water congestion of leaf tissue and citrus foliage can hold 7 microliters of water per centimeter square of leaf area. Even one cell forced through stomatal openings in water-congested tissues can lead to successful infection. Pria et al. (2006) found that temperature had a greater effect than leaf wetness duration on the incidence of Asiatic citrus canker in Brazil. As 100% disease incidence occurred during 4 h leaf wetness period, they suggested that minimum leaf wetness period is less than 4 h. Wind-driven rain is the main agent for short distance or local spread of the bacteria, but in some cases it may also cover a considerable distance. In Argentina, the wind-blown inoculum was detected up to 32 m from the infected trees. Greater wind speeds consistently disperse more bacteria. In Florida, severe rainstorms and tropical storms carried the inoculum up to a distance of 7 miles. Long distance spread occurs generally through infected nursery plants, rainwater, or floods. Strong winds and rainstorms increase the disease severity leading to disease outbreaks by causing injuries to the host tissue, increasing humidity, and by aiding in the dispersal of the inoculum. However, the size of injuries does not affect the disease severity because a single bacterium can cause the infection under most favorable conditions (Goto 1992). Prolonged humidity for about 12 h after artificial inoculation significantly increases the infection rate, whereas for stomatal infection, it is essential only for several minutes. The healing of wounds affects the chances of infection, and the healing is much faster at higher temperatures around 25°C.

The penetration of the bacterium into the host tissue occurs through stomata and more frequently through wounds caused by pruning, thorns, and insects. The penetration through stomata generally occurs in the young tissue. The number and size of stomata does not affect the disease severity because the number of lesions does not differ significantly on leaves having more and wider stomata than on those having less and narrower stomata. Instead, the size of the lesion increases in susceptible varieties compared to resistant ones. The wounds caused by citrus leaf miners also serve as avenues for the entry of the bacterium and even the penetration of the resistant hosts also occurs due to the overcoming of the natural barriers of the host. The contaminated larvae of leaf miner also transport the bacteria through feeding galleries. The leaf miner wounds differ from natural wounds in that they take more time for healing, i.e., 10–14 days in comparison to 1 day for natural wounds, and also require lower inoculum dose (1/1,000 of the dose required for infection through natural openings) for infection. On some trees, citrus canker infections are confined entirely to leaf miner wounds. A study conducted in South Broward county of Florida in 2002 showed that approximately 60% of the diseased citrus canker leaves were injured by leaf miners, and the number of canker lesions increased significantly on the leaf miner infested leaves in comparison to leaves infected through stomata. Christiano et al. (2007) showed that wounds caused by Asian citrus leaf miner (*P. citrella*) on leaves of Tahiti lime led to more disease severity than mechanical wounds and intact leaves. The maximum disease severity of 87% on leaves having pupal stage injury and of 63% on those having third instar stage injury, which was five times higher than that on intact leaves, was due to greater amount of damage that exposes mesophyll cells to infection.

The optimum temperature for infection of host lies between 20°C and 30°C. Excessive nitrogen fertilization leads to more canker development due to flushing of more shoots and succulency of host tissue. The disease is most severe on young trees that flush many angular shoots than on mature or old trees. The angular shoots that develop from summer to autumn usually have a prolonged disease cycle due to the presence of young and susceptible shoots for a long time. The fruits also show infection for several months because their continuous growth provides suitable stomata for infection.

Francis et al. (2010) developed a detached leaf protocol for rapid screening of germplasm for resistance to citrus canker (*Xa. citri* subsp. *citri*) and citrus bacterial spot (*Xa. alfalfae* subsp. *citrumelonis*) to evaluate limited quantities of leaf material. The detached leaf assay was found useful for the characterization and differentiation of lesion phenotype for both the pathogens permitting rapid screening of germplasm resistance based on the quantification of number of lesions and bacterial concentration.

Braido et al. (2014) developed two SAD sets i.e., five-diagram set depicting five disease severities (0.5%, 2.0%, 8.0%, 27.0%, and 40.0%) and six-diagram set depicting six disease severities (0.5%, 1.0%, 3.0%, 9.0%, 20.0%, and 40.0%) to improve the accuracy and reliability of visual estimates of Asiatic citrus canker on unripe (green) fruits of sweet orange. Both the sets significantly improved the accuracy and reliability of estimates, but the five-diagram set was significantly more accurate and reliable than the six-diagram set. Agreement measured by Lin's concordance correlation coefficient, was 0.220 to 0.913 when SADs were not used, 0.814 to 0.955 when five-diagram set was used, and 0.863 to 0.925 when six-diagram set was used. Although the SAD set was developed for sweet orange, doubtlessly it has the applicability to other citrus trees, including grapefruit.

Braido et al. (2015) also devised two SAD sets, i.e., five-diagram set depicting five disease severities (0.7%, 2.0%, 7.0%, 21.0%, and 39.0%) and six-diagram set depicting six disease severities (0.7%, 2.0%, 4.0%, 10.0%, 21.0%, and 39.0%) as aids for raters to improve the accuracy and reliability of visual estimates of canker severity on ripe fruits of sweet orange. Both five- and six-diagram sets improved the accuracy and reliability of raters' estimates, but five-diagram set resulted in significantly more accurate and reliable estimates than the six-diagram set, and is thus recommended for use. Agreement, measured by Lin's concordance correlation coefficient (ρ_c) was 0.735 to 0.906 when SAD sets were not used, 0.931 to 0.985 when five-diagram set was used, and 0.879 to 0.979 when six-diagram set was used. Mean inter-rater reliability (R^2) was 0.638 when SAD sets were not used, 0.889 when five-diagram set was used, and 0.834 when six-diagram set was used.

Hu et al. (2016) compared the transcriptional profiles between citrus leaf tissue inoculated with *Xa. citri* subsp. *citri* 306 and a *pthA4*-deletion mutant strain (*Xa. citri* subsp. *citri* 306Δ*pthA4*) at 6, 48, and 120 h post-inoculation (hpi). PthA4-mediated infection significantly upregulated expression of genes involved in cell wall degradation and modification, DNA packaging, G-protein, protein synthesis, sucrose metabolism, and cell division functions at both 48 and 120 hpi, while the downregulated genes were mainly enriched in photosynthesis, transport, secondary metabolism, cytochrome P450, and various plant defense-associated mechanisms. To validate microarray results, gene expression of 26 genes representing genes associated with cell wall, immunity system, and carbohydrate metabolism was confirmed using quantitative reverse-transcription polymerase chain reaction. Expression patterns of these genes at 48 and 120 hpi were consistent with the microarray results. Putative effector binding element for PthA4 (EBEPthA4) in the promoter regions of multiple genes upregulated by PthA4 was also identified, to which PthA4 might bind directly to control their gene expression. The study provides a dynamic picture of citrus genes regulated by PthA4 during the *Xa. citri* subsp. *citri* infection of citrus leaves at different stages and will be useful in further understanding the virulence mechanism of *Xa. citri* subsp. *citri* and identifying potential targets of PthA4.

Zhang et al. (2010) generated transgenic "Duncan" grapefruit (DG) and "Hamlin" sweet orange (Ham) expressing the Arabidopsis *NPR1* gene (*AtNPR1*), which is a key positive regulator of the long-lasting broad-spectrum resistance known as systemic acquired resistance (SAR). The overexpression of *AtNPR1* in citrus increased resistance to citrus canker, and the resistance was related with the expression levels of *AtNPR1* in the transgenic plants. The line DG 42-2 with the highest expression level of *AtNPR1* was also the most resistant, and it developed significantly fewer lesions accompanied by a 10-fold reduction in *Xa. citri* subsp. *citri* population. The lesions developed on DG 42-2 were also smaller and darker than those on the control and lacked callus formation. These lesion phenotypes resembled those on canker resistant kumquats and canker susceptible citrus trees treated with SAR-inducing compounds. It is evident that over-expression of *AtNPR1* in citrus is a promising approach for development of more resistant cultivars to citrus canker.

Kobayashi et al. (2017) obtained transgenic "Pera" sweet orange [*C. sinensis* (L.) Osbeck] plants, constitutively expressing the sarcotoxin IA peptide fused to the PR1a signal peptide from *Nicotiana tabacum* for secretion in the intercellular space, by *Agrobacterium*-mediated transformation using thin sections of mature explants. The evaluation of citrus canker resistance in leaves of transgenic and non-transgenic plants was performed by inoculations of *Xa. citri* subsp. *citri* by infiltration and spraying. The *Xa. citri* subsp. *citri* population was up to 2 log unit lower in leaves of the transgenic plants compared to those of non-transgenic controls. Incidence of canker lesions was significantly higher in non-transformed controls (>10 lesions/cm^2) than in the transgenic plants (<5 lesions/cm^2) after injection infiltration or spraying with *Xa. citri* subsp. *citri* inoculum. Accumulation of sarcotoxin IA peptide in sweet orange tissue did not cause any deleterious effects on the growth and development of the transgenic plants, indicating that this approach is suitable to provide resistance to citrus canker.

7.50.5 Management

1. The enforcement of strict quarantine is the best method of control in disease-free areas. *Xa. citri* subsp. *citri* is an EPPO A 1 quarantine pest. Every effort should be made to eradicate the bacterium from such areas when it is introduced or found in these areas
2. In areas where the disease is endemic, the following integrated management measures should be adopted to contain it
   a. Use of disease-free nursery and production of pathogen-free budwood are the foundations of an integrated management program
   b. Use of resistant or field resistant cultivars is the most important feature of the integrated program. Eighty-five percent of citrus production in Japan is from the cultivation of a moderately resistant cultivar, Unshu (satsuma) orange
   c. Cultural practices like use of windbreaks and pruning of diseased summer and autumn shoots are very effective. Pruning should be done during the dry season to prevent the spread of the pathogen. Windbreaks are most effective for the control of disease on susceptible varieties
   d. On susceptible or highly susceptible cultivars and in areas prone to disease epiphytotics, prevention of primary infection on spring shoots is very essential for the management of the disease
   e. Periodical sprays of insecticides should be given to control the citrus leaf miner
   f. Goto (1992) has given the criteria on the basis of which an epiphytotic can be predicted 1–2 months in advance. Based on that, the citrus growers can be advised to take timely steps for an effective chemical control
   g. Preventive sprays of copper-based bactericides are used throughout the world to manage citrus canker. Copper sprays alone may control the disease on resistant or moderately resistant cultivars, while the control on susceptible or highly susceptible cultivars will require the integration of several control measures. The timing and number of copper sprays will vary with weather conditions, age of trees, and reaction of host cultivars, but in general, 3–5 copper sprays are sufficient on cultivars having intermediate level of resistance. A nucleic acid sequence-based amplification protocol using primers for *gum* D mRNA has been developed to assess the viability of this pathogen to a bactericidal treatment. The method is rapid, specific, and sensitive, and is able to detect viable bacterial cells using a hybridization device, which allows the visualization of the results in only 30 min (Scuderi et al. 2010)

In India, Patel and Desai (1970) found that the disease was effectively controlled within 3 years by removing the infected twigs every year and spraying the trees with 1% Bordeaux mixture four times during the year. The first spray should be given in November–December, second in February–March, third before monsoon or first fortnight of June, and fourth in July–August or when there is break in rains.

The continuous use of copper bactericides has resulted in the appearance of copper tolerant strains of the bacteria in Argentina. The copper resistant strains of *Xa. citri* pv. *citri* have also been reported from Réunion, France (Richard et al. 2017). In such areas, copper-mancozeb, a mixture of copper and mancozeb, should be used instead of copper alone. Behlau et al. (2012) reported that there is a low, but significant likelihood of horizontal gene transfer of copper resistance genes from other xanthomonads or epiphytic bacteria to *Xa. citri* subsp. *citri* in nature.

Graham et al. (2016) evaluated antimicrobial activity of two structurally different nano-zinc oxide materials, plate-like Zinkicide SG4 and particulate Zinkicide SG6, against *Xa. citri* subsp. *citri*. In field trials conducted in southeast Florida in 2014 and 2015, Zinkicide SG4 and SG6 reduction of grapefruit canker incidence exceeded that of cuprous oxide and cuprous oxide/zinc oxide bactericides. No sign of phytotoxicity to the fruit rind was observed during either season. Antimicrobial activity of Zinkicide for protection of leaves and fruit against *Xa. citri* subsp. *citri* was comparable or exceeded that for commercial copper and zinc oxide formulations, which may be attributed to translaminar movement of Zinkicide.

Yu et al. (2016) demonstrated that bismerthiazol can effectively control citrus canker by both inhibiting the growth of *Xa. citri* subsp. *citri* and triggering the plant's host defense response through the expression of several pathogenesis-related genes (*PR1*, *PR2*, *CHI*, and *RpRd1*) and the non-expression of PR genes (*NPR1*, *NPR2*, and *NPR3*) in "Duncan" grapefruit, especially at early treatment times. Additionally, bismerthiazol induced the expression of the marker genes *CitCHS* and *CitCHI* in the flavonoid pathway and of the *PAL1* gene in the salicylic acid (SA) biosynthesis pathway at different times. Bismerthiazol also induced the expression of the priming defense-associated gene *AZI1*. The results show that the induction of the defense response in "Duncan" grapefruit by bismerthiazol may involve the SA signaling pathway and the priming defense and that bismerthiazol may serve as an alternative to copper bactericides for the control of citrus canker.

Balogh et al. (2008) obtained an effective and consistent control of the disease with a bacteriophage. The phage application caused 59% reduction in disease severity. However, phage treatment was ineffective, when applied with skim milk. Ibrahim et al. (2017) reported that applications of phages formulated with skim milk and sucrose, in combination with ASM as soil drench under field conditions significantly decreased incidence of Asiatic citrus canker by 14.8% (trial 1) and 16.8% (trial 2) compared with untreated control plants. Overall, the inoculated plants treated with bacteriophages + ASM combination showed significant disease reduction under greenhouse and field conditions. The bacteriophages + ASM combination tested in these trials can be an effective tool in the integrated management programs of Asiatic citrus canker.

Caicedo et al. (2016) reported that bacteria belonging to *Pseudomonas* and *Bacillus* genera have a strong ability to reduce the severity of citrus canker in sweet orange. The reduction in disease severity was associated with alteration in bacterial attachment and biofilm formation, the factors known to contribute to *Xa. citri* subsp. *citri* virulence. These quorum-quenching (interference with cell to cell signaling) bacteria may represent a highly valuable tool in the biological control and offer an alternative to the traditional copper treatment currently used to treat citrus canker.

REFERENCES

Agrios, G. N. 2004. *Plant Pathology*. 5th ed. Elsevier Academic Press, San Diego, CA, 922 pp.

Balogh, B., Canteros, B. I., Stall, R. E., and Jones, J. B. 2008. Control of citrus canker and citrus bacterial spot with bacteriophages. *Plant Dis*. 92: 1048–1052.

Behlau, F., Barelli, N. L., and Belasque, J., Jr. 2014. Lessons from a case of successful eradication of citrus canker in a citrus-producing farm in São Paulo State. *Brazil. J. Plant Pathol*. 96: 561–568.

Behlau, F., Canteros, B. I., Jones, J. B., and Graham, J. H. 2012. Copper resistance genes from different xanthomonads and citrus epiphytic bacteria confer resistance to *Xanthomonas citri* subsp. *citri*. *Eur. J. Plant Pathol*. 133: 949–963.

Braido, R., Gonçalves-Zuliani, A. M. O., Janeiro, V., Carvalho, S. A., Belasque, J. Jr., et al. 2014. Development and validation of standard area diagrams as assessment aids for estimating the severity of citrus canker on unripe oranges. *Plant Dis*. 98: 1543–1550.

Braido, R., Gonçalves-Zuliani, A. M. O., Nocchi, P. T. R., Belasque, J. Jr., Janeiro, V., et al. 2015. A standard area diagram set to aid estimation of the severity of Asiatic citrus canker on ripe sweet orange fruit. *Eur. J. Plant Pathol*. 141: 327–337.

Caicedo, J. C., Villamizar, S., Ferro, M. I. T., Kupper, K. C., and Ferro, J. A. 2016. Bacteria from the citrus phylloplane can disrupt cell-cell signalling in *Xanthomonas citri* and reduce citrus canker disease severity. *Plant Pathol*. 65: 782–791.

Christiano, R. S. C., Pria, M. D., Jesus, W. C. Jr., Parra, J. R. P., Amorim, L., et al. 2007. Effect of citrus leaf-miner damage, mechanical damage and inoculum concentration on severity of symptoms of Asiatic citrus canker in Tahiti lime. *Crop Prot*. 26: 59–65.

Ference, C. M., Gochez, A. M., Behalu, F., Wang, N., Graham, J. H., et al. 2018. Recent advances in the understanding of *Xanthomonas citri* ssp. *citri* pathogenesis and citrus canker disease management. *Mol. Plant Pathol*. 19: 1302–1318.

Francis, M. I., Peña, A., and Graham, J. H. 2010. Detached leaf inoculation of germplasm for rapid screening of resistance to citrus canker and citrus bacterial spot. *Eur. J. Plant Pathol*. 127: 571–578.

Gabriel, D. W., Kingsley, M. T., Hunter, J. E., and Gottwald, T. 1989. Reinstatement of *Xanthomonas citri* (ex Hasse) and *X. phaseoli* (ex Smith) to species and reclassification of all *X. campestris* pv. *citri* strains. *Int. J. Syst. Bacteriol*. 39: 14–22.

Goto, M. 1992. Citrus canker. pp. 170–208 in: *Plant Diseases of International Importance*. Vol. III, *Diseases of Fruit Crops*. J. Kumar, H. S. Chaube, U. S. Singh, and A. N. Mukhopadhyay eds. Prentice Hall, Englewood Cliffs, NJ.

Gottwald, T. R., Graham, J. H., and Schubert, T. S. 2002. Citrus canker: The pathogen and its impact. APSnet Feature Story [http://www.plantmanagementnetwork.org/php] review/citrus canker.

Graham, J. H., Johnson, E. G., Myers, M. E., Young, M., Rajasekaran, P., et al. 2016. Potential of nano-formulated zinc oxide for control of citrus canker on grapefruit trees. *Plant Dis*. 100: 2442–2447.

Hasse, C. H. 1915. *Pseudomonas citri*, the cause of citrus canker. *J. Agric. Res*. 4: 97–100.

Hu, Y., Duan, S., Zhang, Y., Shantharaj, D., Jones, J. B., et al. 2016. Temporal transcription profiling of sweet orange in response to PthA4-mediated *Xanthomonas citri* subsp. *citri* infection. *Phytopathology* 106: 442–451.

Ibrahim, Y. E., Saleh, A. A., and Al-Saleh, M. A. 2017. Management of Asiatic citrus canker under field conditions in Saudi Arabia using bacteriophages and acibenzolar-*S*-methyl. *Plant Dis*. 101: 761–765.

Kobayashi, A. K., Vieira, L. G. E., Bespalhok Filho, J. C., Leite, R. P. Jr., Pereira, L. F. P., et al. 2017. Enhanced resistance to citrus canker in transgenic sweet orange expressing the sarcotoxin IA gene. *Eur. J. Plant Pathol.* 149: 865–873.

Kositcharoenkul, N., Chatchawankanphanich, O, Bhunchoth, A., and Kositratana, W. 2011. Detection of *Xanthomonas citri* subsp. *citri* from field samples using single-tube nested PCR. *Plant Pathol.* 60: 436–442.

Patel, R. S., and Desai, M. V. 1970. Control of citrus canker. *Indian J. Hort.* 27: 93–98.

Patel, R. S., and Desai, M. V. 1972. Citrus canker. pp. 184–91 in: *Plant Bacteriology*. Vol. I, *Bacterial Diseases of Plants in India*. P. N. Patel ed. Indian Agricultural Research Institute, New Delhi, India.

Pria, M. D., Christiano, R. C. S., Furtado, E. L., Amorim, L., and Filho, A. B. 2006. Effect of temperature and leaf wetness duration on infection of sweet oranges by Asiatic citrus canker. *Plant Pathol.* 55: 657–663.

Richard, D., Tribot, N., Boyer, C., Terville, M., Boyer, K., et al. 2017. First report of copper-resistant *Xanthomonas citri* pv. *citri* pathotype A causing Asiatic citrus canker in Réunion, France. *Plant Dis.* 101: 503.

Rodoriguez, S. G., Stapleton, J. J., and Civerolo, E. L. 1985. *Xanthomonas campestris* involved in Mexican lime bacteriosis in Colima, Mexico. *Proceeding of 6th International Conference on Plant Pathogenic Bacteria*, College Park, MD, pp. 658–662.

Schaad, N. W., Postnikova, E., Lacy, G., Sechler, A., Agarkova, I., et al. 2005. Reclassification of *Xanthomonas campestris* pv. *citri* (ex Hasse 1915) Dye 1978 forms A, B, C, D and E as *X. smithii* subsp. *citri* (ex Hasse) sp. nov. nom. rev. comb. nov.; *X. fuscans* subsp. *aurantifolii* (ex Gabriel 1989) sp. nov. nom. rev. comb. nov., and *X. alfalfae* subsp. *citrumelo* (ex Riker and Jones) Gabriel et al., 1989 sp. nov. nom. rev. comb. nov.; *X. campestris* pv. *malvacearum* (ex Smith 1901) Dye 1978 as *X. smithii* subsp. *smithii* sp. nov. comb. nov nom. nov.; *X. campestris* pv. *alfalfae* (ex Riker and Jones, 1935) Dye 1978 as *X. alfalfae* subsp. *alfalfae* (ex Riker *et al.*, 1935) sp. nov. nom. rev.; and "var. *fuscans*" of *X. campestris* pv. *phaseoli* (ex Smith, 1897) Dye 1978 as *X. fuscans* subsp. *fuscans* sp. nov. *Syst. Appl. Microbiol.* 28: 494–518.

Schaad, N. W., Postnikova, E., Lacy, G., Sechler, A., Agarkova, I., et al. 2007. List of new names and new combinations previously effectively, but not validly, published. *Int. J. Syst. Evol. Microbiol.* 57: 893–97. Effective Publication: Schaad, N. W., Postnikova, E., Lacy, G., Sechler, A., and Agarkova, I., et al. 2006. Emended classification of xanthomonad pathogens on citrus. *Syst. Appl. Microbiol.* 29: 690–695.

Schoulties, C. L., Civerolo, E. L., Miller, J. W., Stall, R. E., Krass, C. J., et al. 1987. Citrus canker in Florida. *Plant Dis.* 71: 388–395.

Scuderi, G., Golmohammadi, M., Cubero, J., Loòpez, M. M., Civilleri, G., et al. 2010. Development of a simplified NASBA protocol for detecting viable cells of the citrus pathogen *Xanthomonas citri* subsp. *citri* under different treatments. *Plant Pathol.* 59: 764–772.

Yu, S. M., Ramkumar, G., and Lee, Y. H. 2017. Detection of *Xanthomonas citri* subsp. *citri* A*, AW and *X. fuscans* subsp. *aurantifolii* B, C using PCR and real-time PCR. *J. Plant Pathol.* 99: 461–467.

Yu, X., Armstrong, C. M., Zhou, M., and Duan, Y. 2016. Bismerthiazol inhibits *Xanthomonas citri* subsp. *citri* growth and induces differential expression of citrus defense-related genes. *Phytopathology* 106: 693–701.

Zhang, X., Francis, M. I., Dawson, W. O., Graham, J. H., Orbović, V., et al. 2010. Over-expression of the Arabidopsis *NPR1* gene in citrus increases resistance to citrus canker. *Eur. J. Plant Pathol.* 128: 91–100.

7.51 CITRUS HUANGLONGBING

It is the most serious and devastating disease of citrus. The disease is also called by various other names such as **citrus greening disease, yellow dragon disease**, or **yellow shoot disease**. Yellow dragon disease and yellow shoot disease are the English translations of Chinese word "**huanglongbing**". International Organization of Citrus Virologists, in its congress held in China in 1995, decided to adopt the original Chinese name "huanglongbing" (HLB) as the official name for the disease.

Southern China has long been considered as the area of origin for the Asian form of the disease. However, Gottwald et al. (2007) are of the view that earlier descriptions of citrus dieback in India suggest that the disease was present there earlier. Capoor (1963) attributed citrus die-back to citrus tristeza virus after biological indexing studies, but Raychaudhuri et al. (1969, 1974) later showed that huanglongbing was a major component of the problem. Beattie et al. (2006) have hypothesized that the disease might have originated in Africa, possibly in a symptomless host such as *Verpris lanceolata*.

The disease is widespread in Asia (eastern, central, and southern), Africa, Arabian Peninsula, and Brazil. The disease was detected in Florida state on 24 August 2005 in a pummelo tree in a commercial tropical fruit nursery in Florida City. Since then, it has spread to most of the southern Florida. In Brazil, the disease was first reported in 2004 from São Paulo state and both Asian and American forms of the disease occur there. According to Pietersen et al. (2010), until to date only '*Candidatus* Liberibacter africanus' is associated with the disease in commercial citrus in South Africa in addition to '*Ca.* Li. africanus subsp. capensis', which has been found only in South Africa infecting an indigenous rutaceous species, *Calodendrum capense* (cape chestnut). The disease has not been reported from Australia, Mexico, Central America, and the Mediterranean citrus-growing areas. The citrus-growing countries, where the disease is absent, are actively guarding against the entry of the pathogen.

Bove (2006) has rightly stated that HLB is the most important, serious, severe, destructive, and devastating disease of citrus in the world. There are no known sources of resistance and management of the disease is difficult and expensive. The reduction in yield can reach to the extent of 30% to 100% depending on proportion of affected canopy. The affected orchards become economically unviable in 7–10 years after planting. About 100 million trees have been affected and destroyed in many countries of South and Southeast Asia, Indonesia, the Philippines, India, Arabian Peninsula, and South Africa. Since 2004, more than 500 thousand trees have been officially destroyed in Brazil due to HLB and around 300–400 thousand trees unofficially destroyed by commercial citrus growers (Gottwald et al. 2007). Catastrophic losses due to the disease from India during 1960s were reported by Fraser et al. (1966). During the past 27 years, eight reviews have been published on the disease (da Graca 1991; Garnier and Bove 1993; da Graca and Korsten 2004; Halbert and Manjunath 2004; Bove 2006; Gottwald et al. 2007; Gottwald 2010; Wang and Trivedi 2013; Wang et al. 2017).

The review published by Gottwald et al. (2007) deals with epidemiology of HLB and the implications of temporal increase and spatial spread of the disease on prospects for disease control and mitigation. The review also contains the concise information on the main features and symptoms of the disease, the causal organisms, and their vectors. The second review published by Gottwald (2010) deals with the epidemiological information of the disease. Wang and Trivedi (2013), in their review article, have presented information on the research on citrus, '*Ca.* Li. asiaticus', and psyllid interactions with specific emphasis on evolutionary relationships among '*Ca.* Liberibacter spp.', genetic diversity, host range, genome analysis, transmission, virulence mechanisms, and the ecological importance of HLB. The latest review by Wang et al. (2017) deals with the progress made in understanding the HLB disease pyramid, and how the microbiome affects the HLB disease pyramid including the interaction between HLB and the citrus microbiome, the interaction between liberibacters and psyllids, the interaction between liberibacters and gut microbiota in psyllids, and the effect of HLB on selected above and below ground citrus pathogens. Their implications for disease management are also discussed.

Saha et al. (2017) have designed a web portal with information for consumers and growers as well as genomics resources for citrus, Asian citrus psyllid, and '*Ca.* Lib. asiaticus'. This is an open source software project with the goal of developing a system for storing, integrating, and analyzing all-omics data related to huanglongbing. The portal can be accessed at https://citrusgreening.org/.

7.51.1 Host Range

The bacteria infect most citrus species, hybrids, and many citrus relatives. Most severely infected hosts are sweet oranges, mandarins, mandarin hybrids, and tangelo (Halbert and Manjunath 2004). HLB in India and Africa is more severe on mandarins and sweet oranges, while limes, lemons, pummelos, and grapefruit show tolerant reaction. However, Asian form of HLB found in Taiwan and Florida and the American form reported from Brazil affect all commercial citrus species severely. It is possible that strains of HLB pathogen adapt to citrus species over a period of time, which is evident from the fact that grapefruit and pummelo, once considered resistance to the pathogen in Taiwan are now quite susceptible.

Some citrus relatives, namely, *Atalantia missionis*, *Microcitrus australasica*, *Swinglea glutinosa*, *Clausena indica*, *Balsamocitrus dawei*, and *Severinia buxifolia* are also severely infected. The other citrus relatives infected are *Limonia acidissima*, *Verpris lanceolata*, and *Murraya paniculata*. There is disagreement in the literature about the status of *M. paniculata* as a host of HLB pathogens. It is a host in Brazil, but not in Taiwan. *Catharanthus roseus* (pink periwinkle) and *Nicotiana xanthii* (tobacco) are the non-rutaceous hosts reported. Tomato has also been shown the host of '*Ca.* Li. asiaticus' and the bacterium was transmitted through dodder (Duan et al. 2008).

7.51.2 Symptoms

The symptoms vary with host species/cultivar, age of host, time of infection, and biotic and abiotic factors that affect the growth of trees. The most characteristic foliage symptoms are the blotchy mottling and yellowing of leaves (Figure 7.36a). In blotchy mottling, the leaves develop a pattern of yellow and dark green areas lacking clear limits between the colors. The patterns of blotchy mottling are asymmetrical on two halves of the leaf. The small yellow leaves appear on a single shoot/branch or on a section of the tree canopy and yellowing gradually spreads to the entire tree (Figure 7.36b). The appearance of a yellow shoot or shoots has given the name "**yellow shoot disease**". The newest leaves may show symptoms, similar to zinc deficiency with yellow venation, while the older leaves have the characteristic greening mottle. The other symptoms are reduced foliage, severe die-back of twigs or limbs, poor flowering, and stunting of trees.

The fruits on diseased trees are small, poorly colored, and often misshaped. Seeds in symptomatic fruits are often small, dark, and aborted; and the fruits remain green on stylar end even after maturity (Figure 7.36c), hence the name **greening**. The fruit development can be asymmetrical and lopsided (Figure 7.36d). The diseased fruits have a sour bitter taste and drop prematurely. The following symptoms of HLB help in differentiating it from chronic citrus decline caused by citrus tristeza virus, blight, and citrus variegated chlorosis.

1. Yellowing of leaf midrib and veins, and slight swelling of veins in early stage of the disease
2. Seeds are aborted in small and misshaped fruits. The fruits, especially small, have a bitter taste compared to sweet taste found in other diseases

The above given symptoms are sufficient to diagnose the disease under field conditions, but final confirmation by other methods is required for regulatory measures. Before the development of molecular techniques, the diagnosis of the disease was done by biological indexing, thin layer chromatography for determining the presence of gentisoyl-β-glucoside, although not all HLB-infected citrus species produce it, use of monoclonal antibodies, and use of indicator plants like sweet orange, Duncan grapefruit, and sweet tangor seedlings. A staining technique employing gentisic acid has been found very useful in the survey work. Recently developed techniques such as DNA hybridization and PCR are rapid, sensitive, and specific for the diagnosis of HLB in the laboratory. PCR is now the main confirmatory test and is routinely used in many laboratories including those in Florida and Brazil. Das et al. (2007) confirmed the presence of the disease in northeastern

FIGURE 7.36 Citrus huanglongbing: (a) Leaf symptoms. (b) Yellow shoot. (Courtesy of Tim Gottwald). (c) Greening on stylar end of fruits. (Courtesy of Tim Gottwald). (d) Asymmetrical lopsided fruit development.

region of India by PCR technique. Manjunath et al. (2008) have developed a TaqMan-based real-time quantitative PCR method to detect '*Ca.* Li. asiaticus' from nymphs and adults of *Diaphorina citri* collected from apparently healthy and diseased citrus trees. This finding is highly significant in a way that the presence of the bacterium in an area can be detected one to several years before the development of disease symptoms and hence the strategies for disease management can be initiated in an advance. Li et al. (2008) using TaqMan real-time PCR established a universal standard curve (based on DNA extracted from naturally infected plants) for quantification of the pathogen in various tissues of different citrus species planted in different geographical regions for studying the host pathogen interactions and epidemiology, and for devising disease management strategies.

There are three types of HLB, namely, Asian, African, and American. The symptoms of Asian and African forms are generally the same, but the Asian form is considered more severe because it causes more extensive and severe die-back eventually killing the trees. The bacterium causing Asian form is also more heat tolerant. The symptoms of African form are suppressed by long exposure to temperature above 30°C. Hence, it is usually found in citrus at elevations above 700 m, while Asian form occurs in lower lying, hotter areas. American form appears to resemble closely the Asian form in symptom expression and severity, but its causal agent is also intolerant to heat like that of African type.

Chen et al. (2009) have shown the presence of a phytoplasma related to '*Ca.* Phytoplasma asteris' in citrus infected with '*Ca.* Li. asiaticus'. Of the 141 samples collected from Guangdong province, P. R. China, 69 (48.9%) samples were positive for both '*Ca.* Phy. asteris' and '*Ca.* Li. asiaticus'. After the transmission from symptomatic citrus to periwinkle (*Catharanthus roseus*) via dodder (*Cuscuta campestris*), both phytoplasma and '*Ca.* Li. asiaticus' were detected from the infected periwinkle. In addition to yellowing/mottling, the infected periwinkle plants showed symptoms of virescence and phyllody which are commonly associated with phytoplasma diseases.

7.51.3 Causal Organisms

'*Candidatus* Liberibacter asiaticus' Jagoueix et al. 1994 causes Asian form of HLB.

'*Ca.* Liberibacter africanus' Jagoueix et al. 1994 causes African form of HLB.

'*Ca.* Liberibacter africanus subsp. capensis' Garnier et al. 2000 causes HLB of cape chestnut (*Calodendrum capense*).

'*Ca.* Liberibacter americanus' Teixeira et al. 2005 causes American form of HLB.

The above mentioned four pathogens are fastidious, phloem-limited, and Gram-stain-negative bacteria. '*Ca.* Li. asiaticus' requires higher temperature than '*Ca.* Li. africanus' for optimum expression of the disease.

Roberts et al. (2015) extracted total DNA from *Clausena anisata*, *Vepris lanceolata*, and *Zanthoxylum capense* trees and tested for the presence of liberibacters by a generic Liberibacter TaqMan real-time PCR assay. Liberibacters present in positive samples were characterized by amplifying and sequencing *rplj*, *omp*, and 16S rRNA gene regions. Phylogenetic analysis of the *rplj* and *omp* gene regions revealed unique clusters for liberibacters associated with each tree species. The following names were proposed for these novel liberibacters: '*Candidatus* Liberibacter africanus subsp. clausenae', '*Candidatus* Liberibacter africanus subsp. vepridis', and '*Candidatus* Liberibacter africanus subsp. zanthoxyli' infecting *C. anisata*, *V. lanceolata*, and *Z. capense*, respectively.

For many years, HLB was considered to be caused by mineral deficiencies/toxicities or water-logging. Later on, the transmission of the casual agent through grafting and insects led to the conclusion that it was a virus. The electronic microscopic studies revealed the presence of mycoplasma-like organisms (MLOs) in the phloem sieve elements of infected plants, but not in healthy plants. However, Garnier and Bove in 1978, after close examination of these organisms, found that they have thicker envelops than MLOs and hence they are true bacteria.

Sechler et al. (2009) designed a medium called Liber A to culture all the three '*Ca.* Liberibacter spp.' that cause HLB. The medium containing citrus vein extract and a growth factor supported the growth of '*Ca.* Liberibacter spp.' for four or five single colony transfers before the decline of viability. The colonies of '*Ca.* Li. asiaticus' were irregularly shaped, convex, and 0.1–0.3 mm after 3–4 days of incubation. Two strains of '*Ca.* Li. asiaticus' and one of '*Ca.* Li. americanus' grown on this medium proved pathogenic to citrus and were isolated from non-inoculated tissues of inoculated trees and seedlings 9 and 2 months later, respectively. The identity of the bacteria was confirmed by real-time PCR and 16S rDNA sequencing. This is the first report of the culturing and pathogenicity of '*Ca.* Li. asiaticus' and '*Ca.* Li. americanus' associated with HLB.

Duan et al. (2009) have reported the complete genome sequence of '*Ca.* Li. asiaticus', one of the causal agents of HLB. A complete circular genome of this bacterium was obtained by metagenomics, using the DNA extracted from a single '*Ca.* Li. asiaticus'-infected psyllid. The 1.23 mb genome has an average G + C content of 36.5%. A high percentage of genes involved in both cell motility (4.5%) and active transport in general (8.0%) may be contributing to its virulence. The bacterium appears to have a limited ability for aerobic respiration and is likely auxotrophic for at least five amino acids. As the bacterium is intracellular in nature, it lacks type III and type IV secretion systems as well as typical free-living or plant-colonizing extracellular degradative enzymes. However, it seems that it has all type I secretion system genes needed for both multi-drug efflux and toxin effector secretion. This is the first genome sequence of a fastidious alphaproteobacteria that is both an intracellular plant pathogen and insect symbiont. Wulff et al. (2009) have reported rRNA operons and genome size of another liberibacter, '*Ca.* Li. americanus'. The sequences of the entire 23S rRNA and 5S rRNA genes from '*Ca.* Li. americanus' have been obtained, using a consensus primer designed on known $tRNA^{Met}$ sequences of rhizobia. The size of the '*Ca.* Li. americanus' genome was determined by pulsed field gel electrophoresis (PFGE), using '*Ca.* Li. americanus'-infected periwinkle plants for bacterial enrichment, and was found to be close to 1.31 mbp. In order to determine the number of ribosomal operons on the '*Ca.* Li. americanus' genome, probes designed to detect the 16S rRNA gene and the 3′ end of the 23S rRNA gene were developed and used for southern hybridization with I-*Ceu*I-treated genomic DNA. These results suggest that there are three ribosomal operons in a circular genome. '*Ca.* Li. americanus' is the first liberibacter species for which such data are available.

Tomimura et al. (2009) investigated the genetic diversity and relationship among '*Ca* Li. asiaticus' isolates from different hosts and distinct geographical areas in southeast Asia. Genetic diversity was estimated by sequencing four well-characterized DNA fragments, namely, the 16S rDNA and 16S/23S intergenic spacer regions, the outer membrane protein (*omp*) gene region, the *trm*U-*tuf*B-*sec*E-*nus*G-*rp*/KAJL-*rpo*B region (gene cluster region), and the bacteriophage-type DNA polymerase region. The sequences of the 16S rDNA and 16S/23S intergenic spacer regions were identical in all the isolates, but nucleotide substitutions were observed in both the *omp* gene and the gene cluster regions. The extended bacteriophage-type DNA polymerase sequences acquired by thermal asymmetric interlaced polymerase chain reaction provided the most sequence diversity among the isolates. Phylogenetic analysis of the bacteriophage-type DNA polymerase sequences revealed three clusters in the southeast Asian population of the pathogen and all the Indonesian isolates were clustered in one group. Remaining two clusters were not correlated with geographic distribution and the differences in genetic sequences also did not reflect differences in the hosts, i.e., mandarin or pummelo from which they were isolated. The bacteriophage-type DNA polymerase region would be useful for molecular differentiation among southeast Asian isolates of '*Ca.* Li. asiaticus'.

Ghosh et al. (2015) characterized the variability in the Indian '*Ca.* Li. asiaticus' based on the tandem repeats at the genomic locus CLIBASIA_01645 and categorized it into four classes based on the tandem repeat number (TRN); Class I (TRN $\leq$ 5), Class II (TRN $>$ 5 $\leq$ 10), Class III (TRN $>$ 10 $\leq$ 15), and Class IV (TRN $>$ 15). It was found that the Indian population of '*Ca.* Li. asiaticus' is more diverse than compared to Florida and Guangdong populations, which showed less diversity. While TRN5 and TRN7 genotypes dominated Florida and Guangdong populations, respectively, the Indian '*Ca.* Li. asiaticus' populations with TRN copy numbers 9, 10, 11, 12, and 13 were widely distributed throughout the country. Additionally, TRN2 and TRN17 genotypes were also observed among the Indian '*Ca.* Li. asiaticus' populations. The predominant '*Ca.* Li. asiaticus' genotypes from the northeastern region of India were TRN6 and TRN7 (53.12%) and surprisingly similar to neighboring South China populations. Citrus cultivars showed no preference for any specific '*Ca.* Li. asiaticus' genotype.

7.51.4 Disease Cycle

As the host trees infected by the bacteria are perennial in nature, the bacteria survive in the infected hosts. The pathogen is spread by infected nursery plants, by grafting diseased bud wood, infected dodder, and two species of psyllid insects, namely, Asian citrus psylla, *Diaphornia citri* and African citrus psylla, *Trioza erytreae*. Natural transmission of Asian and American HLB is by *D. citri,* while that of African HLB is by *T. erytreae*. The temperature tolerance and sensitivities of these two vector species match those of HLB type they are naturally associated with; i.e., *D. citri* is heat tolerant, whereas *T. erytreae* is sensitive to high temperature. However, under experimental conditions, both the insect species can transmit both the pathogens causing Asian and African HLB. Adults of both the insect species can acquire and transmit the bacteria, but the acquisition periods reported by different workers vary greatly. An acquisition period of 15–30 min and 5 h minimum for *D. citri* and 24 h for *T. erytreae* has been reported. Nymphs of both the species acquire the bacteria, which are transmitted by the adults in a persistent manner. There is also evidence for transovarial transmission by *T. erytreae*. Other species of psyllid have been found infecting citrus, but none of them have been found to act as a vector. Coletta-Filho et al. (2014) reported that pathogen acquisition by *D. citri* was positively associated with plant infection level and time since inoculation, with acquisition occurring as early 60 days after inoculation, while the trees showed first disease symptoms around 200 days after inoculation. These results suggest that there is ample potential for psyllids to acquire the pathogen from trees during the asymptomatic phase of infection. If so, this could limit the effectiveness of tree rouging as a disease management tool and would likely explain the rapid spread observed for the disease in the field.

'*Ca*. Li. asiaticus' (*C*Las) is transmitted in a persistent, circulative, and propagative manner by *D. citri*. Killiny et al. (2017) showed that the levels of ATP and many other nucleotides were significantly higher in *C*Las-infected than healthy psyllids. These results indicated that *C*Las stimulated *D. citri* to produce more ATP and many other energetic nucleotides, while it may inhibit their consumption by *D. citri*. As a result of ATP accumulation, the adenylated energy charge (AEC) increased and the AMP/ATP and ADP/ATP ratios decreased in *C*Las-infected *D. citri* psyllids. The survival analysis confirmed a shorter life span for *C*Las-infected *D. citri* psyllids. In conclusion, *C*Las alters the energy metabolism of its psyllid vector, *D. citri*, in order to secure its need for energetic nucleotides.

Beloti et al. (2018) did not obtain the successful transmission of '*Ca*. Li. asiaticus' by *D. citri* to *Murraya koenigii* (curry leaf) seedlings and rated it as immune. As *M. koenigii* is attractive for *D. citri* and is immune to infection by '*Ca*. Li. asiaticus'. They postulated that *M. koenigii* is an excellent candidate for use as a trap crop to improve the management of the insect vector and consequently of huanglongbing.

Ammar et al. (2011) detected '*Ca*. Li. asiaticus' by fluorescence *in situ* hybridization (FISH) and confocal laser scanning microscopy in the filter chamber, midgut, Malpighian tubules, hemolymph, salivary glands, ovaries, and in muscle and fat tissues of *D. citri* that acquired '*Ca*. Li. asiaticus' from infected plants, as well as in the phloem of infected citrus leaves. '*Ca*. Li. asiaticus' appeared as pleomorphic bodies or short thin rods that were much more dispersed and individually distinct in citrus leaf phloem and in *D. citri* hemolymph, but more densely aggregated in cells of the salivary glands and other *D. citri* organs and tissues. It is the first *in situ* demonstration of '*Ca*. Li. asiaticus' infection in various psyllid organs and tissues and shows the near-systemic infection of *D. citri* by '*Ca*. Li. asiaticus'. Canale et al. (2017) reported that long latent period and persistence of transmission are indirect evidences of circulative propagation of '*Ca*. Li. asiaticus' in *D. citri*.

Johnson et al. (2014) emphasized the significance of early root infection to tree health and suggested a model for phloem-limited bacterial movement from the initial insect feeding site to the roots where it replicates, damages the host root system, and then spreads to the rest of the canopy during subsequent leaf flushes. Hilf and Lewis (2016) reported that grafting with intact asymptomatic and HLB-symptomatic leaves resulted in infection of 61 of 78 (78%) and 35 of 41 (85%) plants with '*Ca*. Li. asiaticus', respectively. Also inoculum consisting of the leaf petiole only or only an inoculum tissue remnant under the bark of the receptor tree resulted in 6 of 12 (50%) and 7 of 31 (23%) infected trees, respectively. The majority of grafted plants developed the foliar blotchy mottle symptom considered diagnostic for HLB, while some plants also displayed the stunted, chlorotic shoots for which the disease is named. The qPCR data together with the symptoms displayed demonstrated that individual leaves from infected trees can serve as effective inoculum sources for transmission and propagation of '*Ca*. Li. asiaticus' via grafting. Although, the HLB pathogens are bacteria, they do not spread by casual contamination of personnel and tools or by wind and rain.

Davis et al. (2016) determined the distribution of '*Ca*. Li. asiaticus' within bark tissue of mature infected citrus trees using the bark samples collected at the graft union and at standard intervals in the canopy above and root system below. The highest populations of '*Ca*. Li. asiaticus' resided in the small diameter branches and canopy tissue of the sampled trees; while populations decreased by 2–3 logs in the larger diameter trunk, graft union, and upper root zone. There are several possible explanations for the distinct concentration pattern found in the infected trees that could assist in identifying and improving methods of control for HLB. Lopes et al. (2013) reported that '*Ca*. Li. asiaticus' titers and the percentages of '*Ca*. Li. asiaticus'-positive psyllids were lower in plants growing in the warmer temperature regime (24°C–38°C) than in plants maintained in the cooler regimes. It shows that the lower incidence and slower spread of '*Ca*. Li. asiaticus' in warmer regions of São Paulo state of Brazil are related to the influence of ambient temperatures on titers of '*Ca*. Li. asiaticus' in leaves.

Flores-Sánchez et al. (2017) evaluated the fit of diffusion and classic disease gradient models to large-scale HLB spatial data originated from initial foci to improve sampling, monitoring, and control strategies for *D. citri*, vector of '*Ca*. Li. asiaticus'. The diffusion model provided the best fit among the family of time-gradient curves (r^2 = 0.90–0.99) due to the flexibility of a three-parameter model. The gradients were well conformed to the model in a 25–82.6 km range, having the east-west direction the longest effect. Out of four transect routes, namely, Yuc-1, Yuc-2, QRoo-1, and QRoo-2, selected based on the directionality of the prevailing winds and foci location of HLB infected plants, Yuc-2 and QRoo-2 transects showed 82.6 and 43.9 km gradients with a diffusion coefficient of 0.15 and 0.09, respectively. The study constitutes the first quantitative evidence of the regional spread of '*Ca.* Li. asiaticus' from a single focus and the application of a flexible model that improved the fit and allowed to better compare different gradients. These results are useful to determine the size of regional areas of *D. citri* control, a management program currently implemented in Mexico to combat HLB.

Hartung et al. (2010) could not prove the seed-plant transmission of '*Ca.* Li. asiaticus' and were also unable to confirm an earlier report of transmission of the bacterium through seed. Hilf et al. (2013) observed bacterial cells with the morphology and physical dimensions appropriate for '*Ca.* Li. asiaticus' in phloem sieve elements in the vascular bundle of grapefruit seed coats using transmission electron microscopy (TEM). Fluorescence in situ hybridization (FISH) analyses using probes complementary to the '*Ca.* Li. asiaticus' 16S rRNA gene revealed bacterial cells in the vascular tissue of intact seed coats of grapefruit and pummelo and in fragmented vascular bundles excised from grapefruit seed coats. The physical measurements and the morphology of individual bacterial cells were consistent with those ascribed in the literature to '*Ca.* Li. asiaticus'. However, no bacterial cells were observed in seed preparations from fruits of non-infected trees. A small library of clones amplified from seed coats from a non-infected tree using degenerate primers targeting procaryote 16S rRNA gene sequences contained no '*Ca.* Li. asiaticus' sequences, while 95% of the sequences in a similar library from DNA from seed coats from an infected tree were identified as '*Ca.* Li. asiaticus', providing molecular genetic corroboration that the bacterial cells observed by TEM and FISH in seed coats from infected trees were of '*Ca.* Li. asiaticus'.

Kim et al. (2009) have reported microscopic and microarray analyses of response of sweet orange (*C. sinensis*) to '*Ca.* Li. asiaticus' infection. The microarray analysis indicated that the pathogen infection affected expression of 624 genes, and the proteins encoded by some of these genes were associated with sugar metabolism, plant defense, phytohormone, and cell wall metabolism. The anatomical analyses indicated that HLB bacterium infection caused phloem disruption, sucrose accumulation, and plugged sieve pores. The phloem blockade occurred due to the plugging of sieve pores rather than the formation of bacterial cell aggregates because '*Ca.* Li. asiaticus' does not form aggregates in citrus. The up-regulation of three key starch genes including ADP-glucose pyrophosphorylase, starch synthase, glucose-bound starch synthase, and starch debranching enzyme likely contributed to accumulation of starch in HLB-affected leaves. The up-regulation of *pp2* gene is related to callose deposition to plug sieve pores in the infected plants.

Ramadugu et al. (2016) screened 65 *Citrus* accessions and 33 accessions belonging to 20 other genera closely related to *Citrus* for resistance to '*Ca.* Li. asiaticus' and found two immune, six resistant, and 14 tolerant accessions. Most citrus cultivars were considered susceptible: 15 citrons, lemons, and limes retained leaves in spite of the disease status. Resistance and high levels of field tolerance were observed in many genera other than *Citrus*. Disease resistance/tolerance was observed in Australian citrus relative genera *Eremocitrus* and *Microcitrus*, which are sexually compatible with citrus and may be useful in future breeding trials to impart resistance against '*Ca.* Li. asiaticus' to cultivated citrus.

7.51.5 Management

1. Enforcement of quarantines to prevent the entry of the pathogens into pathogen-free areas
2. Use of pathogen-free propagating plant material
3. Removal and destruction of infected trees as soon as they are noticed. In South Africa, a management strategy consisting of production of pathogen-free bud wood and nursery trees, psyllid control, and removal of infected trees was implemented. Le Roux et al. (2006) found that as a result of this strategy, the disease incidence of 38% found in 1960s dropped to less than 1% in 2006. When HLB was reported in Brazil, citrus industry in São Paulo state of Brazil had already moved to bud wood isolation, a pathogen-free nursery certification program, nursery industry isolation, and requiring all nursery production to be in secure insect-proof screen houses. The Florida Department of Agriculture and Consumer Services (FDACS) strengthened its existing bud wood nursery certification program by expanding testing for HLB and movement of certified bud wood sources to an isolated greenhouse location in northern Florida outside the commercial production area. The FDACS has desired all the field-grown nursery production to cease and that no field-grown citrus nursery trees to be sold after January 2008. Further, all subsequent nursery production must be in insect-proof greenhouses or screen houses with a requisite 1 mile buffer where no HLB-infected commercial or residential citrus has been detected (Gottwald et al. 2007)
4. Tetracycline injections or sprays reduced foliar and fruit incidence ranging from 10% to 97% (da Graca 1991). However, Buitendag and von Broembsen (2003) have discouraged the use of antibiotics for the control of the disease because only partial success was achieved. Zhang et al. (2011) reported that the combined application of penicillin and streptomycin via trunk injection or root soaking eliminated or suppressed the '*Ca.* Li. asiaticus' bacterium in the HLB-affected citrus plants. Zhang et al. (2010) devised an optimized system using '*Ca.* Li. asiaticus'-infected periwinkle cuttings to screen chemicals effective for the control of this bacterium. The vermiculite was used as a growth medium and a fertilization routine consisted of half-strength

Murashige and Tucker medium supplemented with naphthalene acetic acid (4 μg mL^{-1}) and indole-3-butyric acid (4 μg mL^{-1}). The presence of '*Ca.* Li. asiaticus' in the infected cuttings was detected using both nested PCR and quantitative real-time PCR. This system allowed a plant regeneration rate of 60.6% for '*Ca.* Li. asiaticus'-infected cuttings in contrast to <1% regeneration rate with water alone. When treated with penicillin G sodium at 50 μg mL^{-1}, all plants regenerated from '*Ca.* Li. asiaticus'-infected cuttings were free from infection of '*Ca.* Li. asiaticus'. Treatment with 2, 2-dibromo-3-nitrilopropionamide at 200 μL L^{-1} also significantly reduced the percentage of infected plants and the titer of the bacterium. Li et al. (2016) reported that three or four applications of ascorbic acid (AA) (60–600 μM), β-aminobutyric acid (BABA) (0.2–1.0 mM), 2,1,3-benzothiadiazole (BTH) (1.0 mM), 2,6-dichloroisonicotinic acid (0.1 mM), non-metabolizable glucose analog 2-deoxy-D-glucose (2-DDG) (100 μM), BABA (1.0 mM) + BTH (1.0 mM), BTH (1.0 mM) + AA (600 μM), and BTH (1.0 mM) + 2-DDG (100 μM) during each season, slowed down the growth of '*Ca.* Li. asiaticus' in the infected plants and reduced HLB disease severity by approximately 15% to 30% compared to the non-treated control, depending on the age and initial HLB severity of infected trees. These treatments also conferred positive effect on fruit yield and quality. These findings indicate that plant defense inducers may be part of a useful strategy for the management of citrus HLB.

Hu and Wang (2016) evaluated the spatiotemporal dynamics of oxytetracycline hydrochloride (OTC) and its effectiveness for controlling HLB via trunk injection. Uniform distribution of OTC throughout tree canopies and root system was achieved 2 days after trunk injection. High concentrations of OTC (>850 μg kg^{-1}) remained in leaf and root for at least 1 month and moderate concentrations (>500 μg kg^{-1}) persisted for more than 9 months. Reduction of '*Ca.* Li. asiaticus' populations in root system and leaves of OTC treated trees were over 95% and 99% (i.e., 1.76 and 2.19 log reduction) between 2 and 28 days post-injection. The health of trees receiving OTC treatment improved, fruit yield increased, and juice acidity reduced compared to water-injected control even though the differences were not statistically significant during the test period. It is evident that trunk injection of OTC can be used as an effective measure for integrated management of citrus HLB. Hu et al. (2018) evaluated eight plant defense activators and three antibiotics in three field trials to control HLB by trunk injection of young and mature sweet orange trees. Four trunk injections of salicylic acid, oxalic acid, acibenzolar-*S*-methyl, and potassium phosphate, provided significant control of the disease by suppressing '*Ca.* Li. asiaticus' titer and disease progress. Trunk injection of penicillin, streptomycin, and oxytetracycline hydrochloride resulted in excellent control of the disease. In general, antibiotics were more effective in reduction of '*Ca.* Li. asiaticus' titer and symptom development than plant defense activators. These treatments also increased yield and improved fruit quality

5. Thermotherapy of bud wood at 47°C for 4 h also gave 100% control (da Graca 1991). However, this method is considered impracticable for a large-scale use. Hoffman et al. (2013) reported that continuous thermal exposure to 40°C to 42°C for a minimum of 48 h was sufficient to significantly reduce titer or eliminate '*Ca.* Li. asiaticus' bacteria entirely from HLB-affected citrus seedlings. Results of qPCR after treatment illustrated significant decreases in the '*Ca.* Li. asiaticus' bacterial titer, combined with healthy vigorous growth by all surviving trees. Repeated qPCR testing confirmed that previously infected, heat-treated plants showed no detectable levels of '*Ca.* Li. asiaticus', while untreated control plants remained highly infected. The treatment may prove useful for the control of '*Ca.* Liberibacter'-infected nursery and greenhouse plants
6. The control of psyllid vectors through chemicals is probably the easiest strategy to control HLB. Biological control of psyllid vectors has also been tried and found successful in some cases. Two hymenopterous ectoparasites, namely, *Tamarixia dryl* and *T. radiatus* parasitize nymphs of both the psyllid vectors. On Reunion Island, these parasites significantly reduced the psyllid populations and lessened the damage of HLB. However, the biological control with these parasites in Taiwan was not effective due to the presence of indigenous populations of hyperparasites, which attack these hymenopterous biocontrol agents. As these hyperparasites are not present on Reunion Island, the biocontrol is more effective. In Florida, the introduction of *T. radiatus* caused 4%–70% reduction in psyllid populations. Another internal parasitoid, *Diaphorencyrtus aligarhensis*, has also been found to attack *D. citri*. Unfortunately, the use of insecticides to control psyllid vectors of HLB also reduces the population of bicontrol insects (Gottwald et al. 2007)
7. In an international workshop on HLB held in Ishigaki, Japan, a novel strategy to control HLB was approved. The strategy is to interplant citrus with guava, which apparently inhibits the psyllid vector. The inhibition can be by attracting the vector to guava and killing it when it feeds, presumably by some toxin or by repelling or confusing the vector through the release of volatile compounds. The exact effect is not known, but preliminary data obtained are quite encouraging (Beattie et al. 2006)

REFERENCES

Ammar, E.-D., Shatters, R. G. Jr., and Hall, D. G. 2011. Localization of '*Candidatus* Liberibacter asiaticus', associated with citrus huanglongbing disease, in its psyllid vector using fluorescence *in situ* hybridization. *J. Phytopathol.* 159: 726–734.

Beattie, G. A. C., Holford, P., Mabberley, D. J., Haigh, A. M., Bayer, R., et al. 2006. Aspects and insights of Australia-Asia collaborative research on Huanglongbing. *Proceedings of International Workshop for the Prevention of Citrus Greening Disease in Severely Infected Areas.* Int. Res. Div., Agriculture, Forestry and Fisheries Research Council Secretariat, Ministry of Agriculture, Forestry and Fisheries, Tokyo, Japan, pp. 47–64.

Beloti, V. H., Alves, G. R., Coletta-Filho, H. D., and Yamamoto, P. T. 2018. The Asian citrus psyllid host *Murraya koenigii* is immune to citrus huanglongbing pathogen '*Candidatus* Liberibacter asiaticus'. *Phytopathology* 108: 1089–1094.

Bove, J. M. 2006. Huanglongbing: A destructive, newly-emerging, century-old disease of citrus. *J. Plant Pathol.* 88: 7–37.

Buitendag, C. H., and von Broembsen, L. A. 2003. Living with citrus greening disease. pp. 269–271 in: *Proceeding of 12th Conference of the International Organization of Citrus Virologists.* P. Moreno, J. V. da Graca, and L. W. Timmer eds. IOCV, Riverside, CA.

Canale, M. C., Tomaseto, A. F., Haddad, M. d. L., Coletta-Filho, H. D., and Lopes, J. R. S. 2017. Latency and persistence of '*Candidatus* Liberibacter asiaticus' in its psyllid vector, *Diaphorina citri* (Hemiptera: Liviidae). *Phytopathology* 107: 264–272.

Capoor, S. P. 1963. Decline of citrus trees in India. *Bull. Nat. Inst. Sci. India* 24: 48–64.

Chen, J., Pu, X., Deng, X., Liu, S., Li, H., et al. 2009. A phytoplasma related to '*Candidatus* Phytoplasma asteris' detected in citrus showing huanglongbing (yellow shoot disease) symptoms in Guangdong, P. R. China. *Phytopathology* 99: 236–242.

Coletta-Filho, H. D., Daugherty, M. P., Ferreira, C., and Lopes, J. R. S. 2014. Temporal progression of '*Candidatus* Liberibacter asiaticus' infection in citrus and acquisition efficiency by *Diaphorina citri. Phytopathology* 104: 416–421.

da Graca, J. V. 1991. Citrus greening disease. *Annu. Rev. Phytopathol.* 29: 109–136.

da Graca, J. V., and Korsten, L. 2004. Citrus huanglongbing: Review, present status and future strategies. pp. 229–245 in: *Diseases of Fruits and Vegetables.* Vol. I, S. A. M. H. Naqvi ed. Kluwer Academic Press, Dordrecht, the Netherlands.

Das, A. K., Rao, C. N., and Singh, S. 2007. Presence of citrus greening (huanglongbing) disease and its psyllid vector in North-eastern region of India confirmed by PCR technique. *Curr. Sci.* 92: 1759–1763.

Davis, H., Orrock, J., and Johnson, E. 2016. Distribution of '*Candidatus* Liberibacter asiaticus' within bark tissue of mature infected citrus trees (Abstr.). *Phytopathology* 106: S2.8.

Duan, Y., Zhou, L., Hall, D. G., Li, W., Doddapaneni, H., et al. 2009. Complete genome sequence of citrus huanglongbing bacterium, '*Candidatus* Liberibacter asiaticus' obtained through metagenomics. *Mol. Plant-Microbe Interact.* 22: 1011–1020.

Duan, Y. P., Gottwald, T., Zhou, L. J., and Gabriel D. W. 2008. First report of dodder transmission of '*Candidatus* Liberibacter asiaticus' to tomato (*Lycopersicon esculentum*). *Plant Dis.* 92: 831.

Flores-Sánchez, J. L., Mora-Aguilera, G., Loeza-Kuk, E., López-Arroyo, J. I., Gutiérrez-Espinosa, M. A., et al. 2017. Diffusion model for describing the regional spread of huanglongbing from first-reported outbreaks and basing an area wide disease management strategy. *Plant Dis.* 101: 1119–1127.

Fraser, L. R., Singh, D., Capoor, S. P., and Nariani, T. K. 1966. Greening virus, the likely cause of citrus dieback in India. *FAO Plant Prot. Bull.* 14: 127–130.

Garnier, M., and Bove, J. M. 1993. Citrus greening disease and the greening bacterium. pp. 212–219 in: *Proceedings of 12th Conference of the International Organization of Citrus Virologists.* P. Moreno, J. V. da Graca, and L. W. Timmer eds. IOCV, Riverside, CA.

Garnier, M., Jagoueix-Eveillard, S., Cronje, P. R., Le Roux, H. F., and Bové, J. M. 2000. Genomic characterization of a Liberibacter present in an ornamental rutaceous tree, *Calodendrum capense*, in the Western Cape province of South Africa. Proposal of '*Candidatus* Liberibacter africanus subsp. capensis.' *Int. J. Syst. Evol. Micrbiol.* 50: 2119–2125.

Ghosh, D. K., Bhose, S., Motghare, M., Warghane, A., Mukherjee, K., et al. 2015. Genetic diversity of the Indian populations of '*Candidatus* Liberibacter asiaticus' based on the tandem repeat variability in a genomic locus. *Phytopathology* 105: 1043–1049.

Gottwald, T. R. 2010. Current epidemiological understanding of citrus huanglongbing. *Annu. Rev. Phytopathol.* 48: 119–139.

Gottwald, T. R., da Graca, J. V., and Bassanezi, R. B. 2007. Citrus huanglongbing: The pathogen and its impact. APSnet Feature Story [http://www.apsnet.org/ huanglongbing]

Halbert, S. E., and Manjunath, K. L. 2004. Asian citrus psyllids (Sternorrhyncha: Psyllidae) and greening disease in citrus: A literature review and assessment of risk in Florida. *Florida Entomologist* 87: 330–354.

Hartung, J. S., Halbert, S. E., Pelz-Stelinski, K., Brlansky, R. H., Chen, C., et al. 2010. Lack of evidence for transmission of '*Candidatus* Liberibacter asiaticus' through citrus seed taken from affected fruit. *Plant Dis.* 94: 1200–1205.

Hilf, M. E., and Lewis, R. S. 2016. Transmission and propagation of '*Candidatus* Liberibacter asiaticus' by grafting with individual citrus leaves. *Phytopathology* 106: 452–458.

Hilf, M. E., Sims, K. R., Folimonova, S. Y., and Achor, D. S. 2013. Visualization of '*Candidatus* Liberibacter asiaticus' cells in the vascular bundle of citrus seed coats with fluorescence in situ hybridization and transmission electron microscopy. *Phytopathology* 103: 545–554.

Hoffman, M. T., Doud, M. S., Williams, L., Zhang, M.-Q., Ding, F., et al. 2013. Heat treatment eliminates '*Candidatus* Liberibacter asiaticus' from infected citrus trees under controlled conditions. *Phytopathology* 103: 15–22.

Hu, J., Jiang, J., and Wang, N. 2018. Control of citrus huanglongbing via trunk injection of plant defense activators and antibiotics. *Phytopathology* 108: 186–195.

Hu, J., and Wang, N. 2016. Evaluation of the spatiotemporal dynamics of oxytetracycline and its control effect against citrus huanglongbing via trunk injection. *Phytopathology* 106: 1495–1503.

Jagoueix, S., Bové, J. M., and Garnier, M. 1994. The phloem-limited bacterium of greening disease of citrus is a member of the α subdivision of the Proteobacteria. *Int. J. Syst. Bacteriol.* 44: 379–386.

Johnson, E. G., Wu, J., Bright, D. B., and Graham, J. H. 2014. Association of '*Candidatus* Liberibacter asiaticus' root infection, but not phloem plugging with root loss on huanglongbing-affected trees prior to appearance of foliar symptoms. *Plant Pathol.* 63: 290–298.

Killiny, N., Hijaz, F., Ebert, T. A., and Rogers, M. E. 2017. A plant bacterial pathogen manipulates its insect vector's energy metabolism. *Appl. Environ. Microbiol.* 83: e03005–e03016.

Kim, J. S., Sagaram, U. S., Burns, J. K., Li, J.-L., and Wang, N. 2009. Response of sweet orange (*Citrus sinensis*) to '*Candidatus* Liberibacter asiaticus' infection: Microscopy and microarray analyses. *Phytopathology* 99: 50–57.
Le Roux, H. F., van Vuuren, S. P., and Manicom, B. Q. 2006. Huanglongbing in South Africa. *Proceedings of Huanglongbing-Greening International Workshop*, Ribeirao Preto, Brazil, pp. 5–9.
Li, J., Trivedi, P., and Wang, N. 2016. Disease control and pest management field evaluation of plant defense inducers for the control of citrus huanglongbing. *Phytopathology* 106: 37–46.
Li, W., Li, D., Twieg, E., Hartung, J. S., and Levy, L. 2008. Optimized quantification of unculturable '*Candidatus* Liberibacter spp.' in host plants using real-time PCR. *Plant Dis*. 92: 854–861.
Lopes, S. A., Luiz, F. Q. B. F., Martins, E. C., Fassini, C. G., Sousa, M. C., et al. 2013. '*Candidatus* Liberibacter asiaticus' titers in citrus and acquisition rates by *Diaphorina citri* are decreased by higher temperature. *Plant Dis*. 97: 1563–1570.
Manjunath, K. L., Halbert, S. E., Ramadugu, C., Webb, S., and Lee, R. F. 2008. Detection of '*Candidatus* Liberibacter asaticus' in *Diaphorina citri* and its importance in the management of citrus huanglongbing in Florida. *Phytopatholohy* 98: 387–396.
Pietersen, G., Arrebola, E., Breytenbach, J. H. J., Korsten, L., le Roux, H. F., et al. 2010. A surrey for '*Candidatus* Liberibacter' species in South Africa confirms the presence of only "*Ca*. L. africanus" in commercial citrus. *Plant Dis*. 94: 244–249.
Ramadugu, C., Keremane, M. L., Halbert, S. E., Duan, Y. P., Roose, M. L., et al. 2016. Long-term field evaluation reveals huanglongbing resistance in *Citrus* relatives. *Plant Dis*. 100: 1858–1869.
Raychaudhuri, S. P., Nariani, T. K., Gosh, S. K., Viswanath, S. M., and Kumar, D. 1974. Recent studies on citrus greening in India. pp. 53–57 in: *Proceedings of 6th Conference of the International Organization of Citrus Virologists*. L. G. Weathers, and M. Cohen eds. University of California, Division of Agricultural Sciences, Riverside, CA.
Raychaudhuri, S. P., Nariani, T. K., and Lele, V. C. 1969. Citrus dieback problem in India. *Proc. 1st Int. Citrus Symp*. 3: 1433–1437.
Roberts, R., Steenkamp, E. T., and Pietersen, G. 2015. Three novel lineages of '*Candidatus* Liberibacter africanus' associated with native rutaceous hosts of *Trioza erytreae* in South Africa. *Int. J. Syst. Evol. Microbiol*. 65: 723–731.
Saha, S., Flores, M., Hosmani, P., Fernandez-Pozo, N., Brown, S., et al. 2017. Citrusgreening.org—A systems biology resource for vector biologists (Abstr.). *Phytopathology* 107: S2.11–12.
Sechler, A., Schuenzel, E. L., Cooke, P., Donnua, S., Thaveechai, N., et al. 2009. Cultivation of '*Candidatus* Liberibacter asiaticus', '*Ca*. L. africanus' and '*Ca*. L. americanus' associated with huanglongbing. *Phytopathology* 99: 480–486.
Teixeira, D. C., Saillard, C., Eveillard, S., Danet, J. L., da Costa, P. I., et al. 2005. '*Candidatus* Liberibacter americanus', associated with citrus huanglongbing (greening disease) in São Paulo State, Brazil. *Int. J. Syst. Evol. Microbiol*. 55: 1857–1862.
Tomimura, K., Miyata, S., Furuya, N., Kubota, K., Okuda, M., et al. 2009. Evaluation of genetic diversity among '*Candidatus* Liberibacter asiaticus' isolates collected in Southeast Asia. *Phytopathology* 99: 1062–1069.
Wang, N., Stelinski, L. L., Pelz-Stelinski, K. S., Graham, J. H., and Zhang, Y. 2017. Tale of the huanglongbing disease pyramid in the context of the citrus microbiome. *Phytopathology* 107: 380–387.
Wang, N., and Trivedi, P. 2013. Citrus huanglongbing: A newly relevant disease presents unprecedented challenges. *Phytopathology* 103: 652–665.
Wulff, N. A., Eveillard, S., Foissac, X., Ayres, A. J., and Bové, J.-M. 2009. rRNA operons and genome size of '*Candidatus* Liberibacter americanus', a bacterium associated with citrus huanglongbing in Brazil. *Int. J. Syst. Evol. Microbiol*. 59: 1984–1991.
Zhang, M., Duan, Y., Zhou, L.,Turechek, W. W., Stover, E., et al. 2010. Screening molecules for control of citrus huanglongbing using an optimized regeneration system for '*Candidatus* Liberibacter asiaticus'-infected periwinkle (*Catharanthus roseus*) cuttings. *Phytopathology* 100: 239–245.
Zhang, M., Powell, C. A., Zhou, L., He, Z., Stover, E., et al. 2011. Chemical compounds effective against the citrus huanglongbing bacterium '*Candidatus* Liberibacter asiaticus' in planta. *Phytopathology* 101: 1097–1103.

7.52 CITRUS STUBBORN DISEASE

The symptoms of the disease reflect the name of the disease as the diseased trees remain stunted and fail to grow despite the best efforts of the orchardists. It is a serious disease of certain citrus species and occurs in hot dry areas. It is considered as a greatest threat to the production of grapefruit and sweet oranges in California and some Mediterranean countries. Due to slow progress of the symptoms, the spread of the disease often goes unnoticed. The loss occurs due to reduction in yield and low market value of the produce.

The disease was first reported from California in 1915 on navel oranges. It affected most commercial citrus varieties in California, but grapefruit, tangelos, and navel and Valencia oranges were most severely infected. During the mid 1950s and early 1960s, the disease appeared on more than 2 million trees, incidence ranging from 50% to 100% in some orchards in California. This large-scale infestation of trees caused an alarm for the research scientists and the investigations for the management of the disease were intensified. Since its first appearance, the disease has spread to many countries including western United States, eastern Mediterranean Basin, Middle East, Brazil, and northern Africa. The disease is not present in Florida and strict quarantines are enforced to regulate the entry of citrus material into the state. The presence of the disease in Australia is doubtful.

Etiological studies during 1930s showed that the causal agent could be transmitted by grafting. For many years, the disease was thought to be caused by a virus due to its graft transmissibility and non-isolation of a pathogen. The suppression of the disease by tetracycline and not by penicillin demonstrated by Igwegbe and Calavan (1970) and later on electron microscopic observations indicating the presence of mycoplasma-like organisms in the phloem element of diseased, but not healthy citrus trees indicated that the causal agent is a member of the class Mollicutes. Fudl-Allah et al. (1972) cultured an organism from infected citrus trees, followed by its characterization and naming (named as *Spiroplasma citri*) by Saglio et al. (1973).

7.52.1 Host Range

Citrus is the main economic host of *Sp. citri*. Most citrus species and cultivars are infected, some of them symptomlessly.

Sweet oranges, tangelos, grapefruit, mandarins, and their hybrids show maximum disease severity. Trifoliate oranges and their hybrids, acid limes, and lemons show mild symptoms on artificial inoculation. Many dicots and some monocots including weeds (several species of family *Brassicaceae*) are also the hosts of *Sp. citri,* but many of them do not show any symptom. Non-citrus host species include pink periwinkle (*Catharanthus roseus*), London rocket (*Sisymbrium irio*), horse radish (*Armoracia rusticana*), wild radish (*Raphanus raphanistrum*), turnip, Bermuda grass, Johnson grass, foxglove, rice, peach, cherry, pear, sesame, zinnia, aster, and marigold. The bacterium also causes brittle root of cultivated horse radish.

7.52.2 Symptoms

The symptoms of the disease are most pronounced under warm conditions. In cool weather, even highly susceptible cultivars may remain symptomless. The symptoms vary greatly and often some of the symptoms are expressed at one time. The leaves on infected trees are small, cupped, abnormally upright, often mottled, and chlorotic (Figure 7.37a).

Excessive defoliation of affected trees occurs in the winter. Mottled appearance and veinal chlorosis of leaves can be confused with symptoms caused by citrus tristeza virus and/or the cachexia viroid (Brown 1992). Twigs and branches on the affected trees generally show upright growth. Shortening of internodes and production of more shoots may result into witches' broom appearance. Some of the affected twigs dieback and the bark may be abnormally thickened. Wilting of affected branches is a characteristic symptom in the advanced stage of the disease. The disease is rarely lethal, but the trees may show severe stunting. The affected trees may bloom prolifically more than once during the season, many fruits drop prematurely leading to reduction in yield.

Affected trees usually produce smaller and fewer fruits, which are generally lopsided and resemble acorns (Figure 7.37b). The rind of such fruits is abnormally thick and green at the base and thin and colored at the tip. Fruits from affected trees are usually sour, bitter, or insipid. The seeds of such fruits are poorly developed and frequently aborted (Agrios 2004). The symptoms on mature trees are generally less conspicuous. Symptoms of the disease are persistent, remaining evident even when the tree is top-worked with healthy bud wood, hence the name "**stubborn**." Some hosts do not show any symptoms, while others show only mild stunting and deformation of the leaves. Non-citrus hosts may show symptoms of stunting and other deformities. The titer of *Sp. citri,* but not the bacterial genotype is, at least in part, responsible for the severity of the disease in affected citrus trees. Quantitative PCR showed that the spiroplasma titer was more than 6,000 times higher in fruits from severely symptomatic than from mildly symptomatic trees (Mello et al. 2010).

As the symptoms of citrus stubborn are readily confused with other diseases/maladies, the proper identification of disease and its confirmation by greenhouse tests is essential for proper diagnosis of the disease. The diagnosis of the disease in field should be carried out in the warm weather, when the day temperature continuously exceeds 27°C, because the symptoms are most prominent at this temperature. Its confirmation should be done by grafting on indicator plants and by culturing the causal organism. Duncan or Marsh grapefruit and Madam Vinous sweet orange are very good indicator hosts, and the grafting should be done on the seedlings of these hosts in a warm greenhouse (day temperature 27°C–35°C and night temperature 21°C–24°C). Three growth flushes should be observed for leaf symptoms (misshapen and mottling) and stunting of seedlings. The confirmation of the disease can also be done by identification of the pathogen on the basis of helical morphology and motility by dark field microscopy and/or serological techniques like ELISA. Nowadays, DNA probes and PCR techniques are routinely used in many laboratories.

The detection of *Sp. citri* by PCR has been done using primers based on gene sequences for spiralin, the most abundant membrane protein of the bacterium. However, these primers lack the sensitivity needed for consistent pathogen detection from field citrus trees. Because the spiralin gene is present as a single copy in the chromosome of all *Sp. citri* strains examined, PCR detection based on multiple-copy membrane protein genes like P89 or P58 will be certainly more efficient. Yokomi et al. (2008) developed a PCR technique for the detection of *Sp. citri* using primers based on sequences of the P89 putative adhesion gene and P58 putative adhesion multi-gene of the bacterium. PCR results matched with

(a)

(b)

FIGURE 7.37 Citrus stubborn: (a) Symptoms on aerial plant parts. (b) Lopsided development of sweet orange fruit. (Courtesy of J.M. Bové, INRA Centre de Recherches de Bordeaux, Bugwood.org.)

those of pathogen culturing 85%–100% of the time depending on the primers used. Moreover, the method also detected the bacterium from 5% to 10% of trees giving negative results by culturing of the pathogen, suggesting that PCR performed as well or better than pathogen culturing. However, real-time PCR developed with detection limits estimated to be between 10^{-4} and 10^{-6} ng by serial dilution of a recombinant *Sp. citri* plasmid into DNA extracts from healthy sweet orange proved to be the best method for detection.

Shi et al. (2014) developed a novel diagnostic method for the detection of *Sp. citri*, using a *Sp. citri*-secreted protein as the detection marker. Using mass spectrometry analysis, the authors identified a unique secreted protein from *Sp. citri* that is highly expressed in the presence of citrus phloem extract. ScCCPP1, an antibody generated against this protein, was able to distinguish *Sp. citri*-infected citrus and periwinkle from healthy plants. Besides, the antiserum can be used to detect the pathogen using a simple direct tissue print assay without the need for sample processing or specialized laboratory equipment and may be suitable for field surveys. The finding provides a novel concept of using pathogen-secreted protein as a marker for diagnosis of a citrus bacterial disease and can probably be applied to other plant diseases. Wang et al. (2015) developed two primer sets, Php-orf1 and Php-orf3, from conserved prophage sequences in *Sp. citri* genome to improve sensitivity for qPCR detection of *Sp. citri* from infected citrus trees. These sequences improved at least 10-fold sensitivity for the detection of *Sp. citri*-infected trees and no false-positive samples were detected with any of the primer sets. The enhanced sensitivity resulted from the higher copy number of prophage genes in *Sp. citri* genome and, thus, improved citrus stubborn disease diagnosis from field samples.

7.52.3 Causal Organism

Spiroplasma citri Saglio et al. 1973

The bacterium is motile, helical in shape, filamentous, and lacks a true cell wall. Helical filaments are usually 100–200 nm in diameter and 2–4 μm in length. Colonies on solid media containing 20% horse serum and 0.8% Noble agar (Difco) are umbonate and 60–150 μm in diameter. Moderate turbidity is produced in liquid cultures. Acid is produced from glucose and fructose, but delayed from mannose. Arginine hydrolysis and phosphatase activity are positive and tetrazolium chloride is anaerobically reduced. Optimum temperature for growth is 32°C. The mol% G + C of the DNA is 25–27 (T_m, BD). DNA hybridization tests between *Sp. citri* and other established species of the genus show no significant genetic relatedness (<1%). Its viability is increased by passage through insect vectors (leafhoppers).

7.52.4 Disease Cycle

The pathogen survives in infected hosts. It is found in the phloem of infected hosts. The pathogen is spread naturally in citrus orchards by several leafhoppers, like *Scaphytopius nitridus*, *Circulifer tenellus* (beet leafhopper), and *Neoaliturus haemoceps*. The transmission (spread) of the pathogen also occurs through budding and grafting, but neither through seed nor mechanically. The natural spread of the pathogen is most rapid in young citrus orchards.

Among the vectors, *C. tenellus* is the most efficient vector in the nature. It has a wide host range including numerous weeds of family *Brassicaceae*. *Sp. citri* multiplies in *C. tenellus* before its transmission to the host. After ingestion by the leafhopper, the pathogen traverses the gut wall, enters the hemocoel, and multiplies during the latent period ranging from days to weeks. Then the cells move to salivary glands where they continue to multiply in the cell cytoplasm. Finally, the pathogen cells are deposited with salvia into the phloem cells of host plants. In some insects, symptoms such as shortened life span, reduction in egg numbers, and weakening of flight muscles and other organs are produced. It indicates that the insects also act as hosts for the pathogen. However, there is no evidence of transovarial transmission (Fletcher 2001).

Colonization of phloem sieve tube cells by the pathogen results in the occlusion and necrosis of these cells. However, the sieve cells are not uniformly invaded; some sieve tubes have no or a few cells, while others are completely occluded. Occlusion of tubes interferes with the movement of nutrients leading to manifestation of some of the disease symptoms. Loss of chlorophyll may result due to the breakdown of anthocyanin pigments. Stunting can occur either due to the reduction in the availability of nutrients or variation in the normal balance of plant growth regulators. *Sp. citri* has been reported to cause changes in the levels of both auxins and cytokinins.

7.52.5 Management

The following measures should be taken to manage the disease.

1. Enforcement of strict quarantines to prevent the entry of the pathogen in disease-free areas like Australia and Florida state of USA
2. Use of pathogen-free bud wood and rootstocks (Anonymous 2003)
3. Early detection, removal, and destruction of diseased trees
4. Destroy all the Brassica weed hosts in field borders and within groves to limit the spread of the pathogen to healthy trees. Potential insect vectors on the weeds should be killed before their removal
5. As tetracycline treatment of young citrus trees has shown encouraging results, it should be extended to commercial scale to test its viability

REFERENCES

Agrios, G. N. 2004. *Plant Pathology*. 5th ed. Elsevier Academic Press, San Diego, CA, 922 pp.

Anonymous 2003. Citrus stubborn disease. University of California Agriculture and Natural Resources, IPM [http://ipm.ucanr.edu/PMG/r107101211.html]. December 31, 2018.]

Brown, L. G. 1992. *Stubborn Disease of Citrus*. Plant Pathology Circular No. 351. Florida Department of Agriculture and Consumer Services, Division of Plant Industry, Gainesville, FL.

Fletcher, J. 2001. Citrus stubborn disease. pp. 222–224. in: *Encyclopedia of Plant Pathology*. Vol. 1, O. C. Maloy, and T. D. Murray eds. John Wiley & Sons, New York.

Fudl-Allah, A. E.-S. A., Calavan, E. C., and Igwegbe, E. C. K. 1972. Culture of a mycoplasmalike organism associated with stubborn disease of citrus. *Phytopathology* 62: 729–731.

Igwegbe, E. C. K., and Calavan, E. C. 1970. Occurrence of mycoplasmalike bodies in phloem of stubborn-infected citrus seedlings. *Phytopathology* 60: 1525–1526.

Mello, A. F. S., Yokomi, R. K., Melcher, U., Chen, J. C., and Fletcher, J. 2010. Citrus stubborn severity is associated with *Spiroplasma citri* titre but not with bacterial genotype. *Plant Dis.* 94: 75–82.

Saglio, P., Lhospital, M., Laflèche, D., Dupont, G., Bové, J. M., et al. 1973. *Spiroplasma citri* gen. and sp. n.: A mycoplasma-like organism associated with "stubborn" disease of citrus. *Int. J. Syst. Bacteriol.* 23: 191–204.

Shi, J., Pagliaccia, D., Morgan, R., Qiao, Y., Pan, S., et al. 2014. Novel diagnosis for citrus stubborn disease by detection of a *Spiroplasma citri*-secreted protein. *Phytopathology* 104: 188–195.

Wang, X., Doddapaneni, H., Chen, J., and Yokomi, R. K. 2015. Improved real-time PCR diagnosis of citrus stubborn disease by targeting prophage genes of *Spiroplasma citri*. *Plant Dis.* 99: 149–154.

Yokomi, R. K., Mello, A. F. S., Saponari, M., and Fletcher, J. 2008. Polymerase chain reaction-based detection of *Spiroplasma citri* associated with citrus stubborn disease. *Plant Dis.* 92: 253–260.

7.53 CITRUS VARIEGATED CHLOROSIS

Citrus variegated chlorosis (CVC) also called citrus X disease was first reported from southwestern part of the state of Minas Gerais and São Paulo state of Brazil in 1987 (Rossetti et al. 1990; Lee et al. 1991). Its rapid spread was a cause of concern to the Brazilian citrus industry. The disease is caused by the xylem-limited bacterium *Xylella fastidiosa* subsp. *pauca*. In 2000, it was estimated that in the state of São Paulo, Brazil, 34% of the 200 million citrus trees had symptoms of the disease and by 2001, the number of symptomatic trees increased to 36%, with 24% having severe symptoms of leaf chlorosis and small fruits (Anonymous 2002). CVC is currently present in approximately 40% of citrus plants in Brazil and causes an annual loss of around 120 million US dollars to the Brazilian citrus industry (Gonçalves et al. 2012). It has also been reported from Argentina (Brlansky et al. 1991), Paraguay, and Costa Rica. In central and South America, *Xyl. fastidiosa* subsp. *pauca* has become very noxious due to the rapid spread (most likely via distribution of infected planting material) of CVC in citrus, leading to more than a third of all trees in the area having symptoms of CVC (Janse and Obradovic 2010). Janse and Obradovic (2010) have further cautioned that the bacterium is a real and emerging threat for Europe, not only for grapes and citrus, but also for stone fruits (almond, peach, and plum) and oleander (e.g., glassy-winged sharpshooter likes to feed on oleander) because it is difficult to prevent it from entering and also difficult to control once established, deserving more attention than up until now.

Rossetti et al. (1990) were the first to show by electron microscopy that a xylem-limited bacterium, probably a strain of *Xyl. fastidiosa*, was present in all symptomatic leaves and fruits tested, but not in similar tissues from symptomless trees. Chagas et al. (1992) confirmed these results. While both the reports strongly suggest that *Xyl. fastidiosa* may be the causal agent of CVC, fulfillment of Koch's postulates is necessary to ascertain that a particular pathogen is the cause of the disease. Chang et al. (1993) fulfilled the Koch's postulates by culturing the CVC bacterium in cell-free media and produced the disease symptoms with the isolated bacterium. Sweet orange seedlings inoculated with a pure culture of the CVC bacterium supported multiplication of the bacterium, which became systemic within 6 months after inoculation and was reisolated from the inoculated seedlings. Symptoms characteristic of the disease developed 9 months after inoculation. They also devised an ELISA system for the detection of the CVC bacterium in citrus tissues.

7.53.1 Host Range

Sweet orange [*Citrus sinensis* (L.) Osbeck] especially Pera, Hamlin, Natal, and Valencia cultivars, and all commonly used rootstocks in Brazil, i.e., *C. limonia*, *C. reshni*, and *C. volkameriana* are susceptible.

7.53.2 Symptoms

The symptoms may occur on twigs, leaves, and fruits. The infected plants do not usually die. Symptoms can manifest on trees from nursery stage to over 10 years old, but are more obvious on 3- to 6-year-old trees. Symptoms in older trees are generally restricted to a few branches. Leaf chlorosis, similar to zinc deficiency or greening disease, begins in the middle and upper part of the tree canopy. Older leaves display conspicuous chlorotic "**variegation**" on the upper surface. The characteristic symptoms are small interveinal chlorotic spots, resembling zinc deficiency, on the upper side of mature leaves, which correspond to slightly raised, gummy, and light to dark brown lesions on the underside of leaves. Silva-Stenico et al. (2009) observed the deficiency of P and K in symptomatic trees and high concentrations of Fe, Mn, and Zn in chlorotic areas, although the previous studies revealed the deficiency of zinc in leaves. This is the first report showing a correlation between chlorotic citrus leaves and higher concentrations of Fe, Mn, and Zn in infected plants.

On a newly affected tree, symptoms generally appear in sectors, whereas old infections are more uniform throughout the canopy. Severely affected trees frequently have protruding branches with small leaves and fruit, and a degree of defoliation. Small fruits also occur on branches without foliar symptoms. Normal fruit thinning does not occur and fruit size is significantly reduced, often no more than one-third the diameter of healthy fruit, having increased sugar content and very hard rind. Very hard rind damages juicing machines, therefore, it results in rejection of fruit lots having a significant number of symptomatic fruits by the processing plants. Fruits on infected trees also ripen earlier. The growth rate of affected trees is greatly reduced and twigs and branches may wilt. In chronic stage, the disease can cause stunting and die-back of twigs. Hopkins et al. (1991) reported the occurrence of the bacterium in roots also. Pera, Hamlin, Natal, and Valencia sweet oranges are the most susceptible, but tangerines and Tahiti lime are resistant and the latter remain symptomless even when growing close to severely affected sweet orange trees.

7.53.3 Causal Organism

Xylella fastidiosa subsp. *pauca* Schaad et al. 2004

Bacterial colonies become visible on periwinkle wilt medium after 5 days and on CS20 and PD2 agar media after 7–10 days of incubation. The bacterial cells are Gram-stain-negative, rod-shaped, having rippled walls, non-flagellated, strictly aerobic, and measure 0.2–0.4 × 1.4–3.0 μm (Chang et al. 1993). The optimum temperature for growth is 26°C–28°C. As the cells are small in size, they become visible only under dark field or phase contrast microscopy.

Due to its potential risks and absence in Europe, *Xyl. fastidiosa* subsp. *pauca* is on the A1 Quarantine list of EPPO.

Simpson et al. (2000) reported the complete genome sequence of *Xyl. fastidiosa* clone 9a5c, the causal agent of CVC. The genome comprises a 52.7% GC-rich 2,679,305-base-pair (bp) circular chromosome and two plasmids of 51,158 bp and 1,285 bp. They assigned putative functions to 47% of the 2,904 predicted coding regions and predicted efficient metabolic functions, with sugars as the principal energy and carbon source, supporting existence in the nutrient-poor xylem sap. The mechanisms associated with pathogenicity and virulence involve toxins, antibiotics, and ion sequestration systems, as well as bacterium–bacterium and bacterium–host interactions mediated by a range of proteins. Orthologues of some of these proteins have only been identified in animal and human pathogens; their presence in *Xyl. fastidiosa* indicates that the molecular basis for bacterial pathogenicity is both conserved and independent of host. At least 83 genes are bacteriophage-derived and include virulence-associated genes from other bacteria, providing direct evidence of phage-mediated horizontal gene transfer. It was the first whole genome sequence of a plant pathogenic bacterium. The sequences of three *Xylella* strains (from almond, oleander, and citrus) are now available at Integrated Genomics.

Van Sluys et al. (2003) reported the genome sequence of *Xyl. fastidiosa* (Temecula strain), isolated from a naturally infected grapevine with Pierce's disease (PD) in a wine-grape-growing region of California. Comparative analyses with a previously sequenced *Xyl. fastidiosa* strain 9a5c causing CVC (Simpson et al. 2000) revealed that 98% of the PD *Xyl. fastidiosa* Temecula genes are shared with those of CVC *Xyl. fastidiosa*. Genomic differences are limited to phage-associated chromosomal rearrangements and deletions that also account for the strain-specific genes present in each genome. Genomic islands, one in each genome, were identified, and their presence in other *Xyl. fastidiosa* strains was analyzed. They concluded that these two organisms have identical metabolic functions and are likely to use a common set of genes in plant colonization and pathogenesis, permitting convergence of functional genomic strategies. Comparative analyses of the complete sequences and annotations of PD *Xyl. fastidiosa* and a *Xyl. fastidiosa* representative of the CVC group isolated from Brazil revealed that these strains exhibit remarkably limited genomic variability and share 95.7% amino acid identity in equivalent regions. There are only three genomic rearrangements, two identified genomic islands, 41 PD *Xyl. fastidiosa,* and 152 CVC *Xyl. fastidiosa* strain-specific genes, and some genes harboring frame shifts.

By cloning and sequencing specific RAPD products, Pooler and Hartung (1995) developed pairs of PCR primers that can be used to detect *Xyl. fastidiosa* in general and *Xyl. fastidiosa* that causes CVC specifically. They also identified a CVC-specific region of the *Xyl. fastidiosa* genome that contains a 28-nucleotide insertion and single base changes that distinguish CVC and grape *Xyl. fastidiosa* strains. The screening for size difference among RAPD products was most efficient rather than presence/absence of a specific rapid band to use RAPD products to develop specific primers.

7.53.4 Disease Cycle

The bacterium multiplies and spreads in the xylem of leaves, stems, and roots. Therefore, the survival of the bacterium occurs within the infected trees. The spread of the bacterium from tree to tree occurs through insect vectors. Medium and long distance dissemination also occurs through infected planting material. There is a 9–12 months incubation period before symptoms appear. Although the pathogen is considered not to be seed-borne, transmission from seeds to seedlings of sweet orange has been reported (Li et al. 2003). Coletta-Filho et al. (2014) conducted experiments for 7 years to prove seed-to-seedling transmission of *Xyl. fastidiosa*, but found no positive PCR detection of the bacterium in seedlings propagated from seeds infected with *Xyl. fastidiosa*, demonstrating the lack of seed-to-seedling transmission. The bacterium was found colonizing the fruit (exocarp, central axis, and mesocarp) and the seed parts (seed coat and endosperm plus embryo).

Sharpshooter leafhoppers (*Cicadellidae*) have been found to transmit the bacterium in Brazil. Of 16 sharpshooter species tested in Brazil, 12 have been identified as vectors of CVC bacterium (Roberto et al. 1996). Most species transmitted at an efficiency rate of lower than 5%; however, two species, *Macugonalia leucomelas* and a *Bucephalogonia* sp., transmitted at rates of 11% and 17%, respectively (Lopes 2000). Almeida et al. (2005) reported that out of 11 sharpshooter vectors found in Brazil, *Acrogonia citrina*, *B. xanthophis*, *Dilobopterus costalimai*, *M. leucomelas*, and *Oncometopia fascialis* are the most important. The transmission trials indicate that vector transmission is much less efficient in citrus compared to that of grape.

Marucci et al. (2005) examined the effects of *C. sinensis* infection by *Xyl. fastidiosa*, on the feeding preferences of two sharpshooter vectors, *D. costalimai* and *O. facialis*. Both the species preferred healthy plants for feeding compared to symptomatic plants. However, *O. facialis* did not discriminate between healthy citrus and symptomless infected plants. Feeding by *D. costalimai* was markedly reduced when confined on CVC symptomatic plants, but not on asymptomatic infected ones. The ingestion rate by *O. facialis* was not affected by the presence of CVC symptoms. These results suggest that citrus trees with early infections (asymptomatic) by *Xyl. fastidiosa* may be more effective as inoculum sources for CVC spread by insect vectors than those with advanced symptoms.

Brlansky et al. (2002) reported the ability of sharpshooter leafhopper *O. nigricans* to transmit the CVC strain of *Xyl. fastidiosa* in Florida, USA. Transmission was verified by specific polymerase chain reaction-based assays as the symptom development in the greenhouse was not a reliable indicator of transmission. Individual insects were able to transmit the bacterium. This information on sharpshooter transmission of CVC bacterium is needed to assess the threat posed by the CVC disease to the citrus industries in the United States as this insect is found frequently feeding on citrus in Florida.

Niza et al. (2015) studied the colonization processes of a pathogenic GFP-labeled *Xyl. fastidiosa* citrus strain in sweet orange (susceptible) and tangor (resistant) parents and two resulting hybrids and found that the bacterium showed increased populations and movement in the susceptible genotypes, but slower compared to other hosts such as grapevine. The predominant pitted stem morphology in citrus makes the bacterial movement difficult. In susceptible genotypes, *Xyl. fastidiosa* can move from the primary to the secondary xylem vessels, while in resistant plants it remains confined to the primary xylem vessels. Induction of lignification occurs earlier in the resistant genotypes in the presence of the pathogen and represents a genetic mechanism that leads to formation of a physical barrier, impairing bacterial colonization.

Gonçalves et al. (2012), on the basis of a three season's trial, gave a negative exponential model for the relationships between yield versus CVC severity and yield versus AUDPC. In addition, the relationship between yield versus CVC severity and canopy volume was fitted by a multivariate exponential model. The use of the AUDPC variable showed practical limitations when compared with the variable CVC severity. The parameter values in the relationship of yield-CVC severity were similar for all treatments unlike in the multivariate model. Consequently, the yield-CVC intensity relationship (with 432 data points) could be described by one single model: $y = 114{\cdot}07 \exp(-0.017\,x)$, where y is yield (symptomless fruit weight in kg), and x is disease severity ($R^2 = 0.45$; $P < 0.01$). After comparing two methods of CVC quantification in the field, Gonçalves et al. (2011) concluded that the incidence quantification of CVC through the symptomatic branches showed high repeatability among the raters compared to use of a descriptive rating scale with four ratings (commonly used for the disease quantification) and suggested the use of incidence of symptomatic branches as variable for CVC quantification.

Oliveira et al. (2002) developed a real-time qPCR assay to quantify *Xyl. fastidiosa* in naturally and artificially infected citrus. *Xyl. fastidiosa* cell number detected in the leaves increased according to the age of the leaf, and bacteria were not detected in the upper midrib section in young leaves, indicating temporal and spatial distribution patterns of bacteria, respectively. In addition, *Xyl. fastidiosa* cell number quantified in leaves of "Pera" orange and "Murcott" tangor reflected the susceptible and resistant status of these citrus cultivars, respectively. None of the 12 endophytic citrus bacteria or the four strains of *Xyl. fastidiosa* non-pathogenic to citrus that were tested showed an increase in the fluorescence signal during qPCR. In contrast, all ten CVC causing strains exhibited an increase in fluorescence signal, thus indicating the specificity of this qPCR assay. qPCR provides a powerful tool for studies of different aspects of the *Xylella*-citrus interactions and can be incorporated into breeding programs in order to select CVC resistant plants more quickly.

Li et al. (2013) developed two TaqMan-based assays, one for targeting the 16S rDNA signature region for the identification of *Xyl. fastidiosa* at the species level and another for the specific identification of the CVC strains and systematically validated both the assays in comparison with the primer/probe sets from four previously published assays on one platform and under similar PCR conditions. The species specific assay detected all *Xyl. fastidiosa* strains and did not amplify any other citrus pathogen or endophyte tested. The CVC-specific assay detected all CVC strains, but did not amplify any non-CVC *Xyl. fastidiosa* nor any other citrus pathogen or endophyte evaluated. Both sets were multiplexed with a reliable internal control assay targeting host plant DNA, and their diagnostic specificity and sensitivity remained unchanged. The internal control provided quality assurance for DNA extraction, performance of PCR reagents, platforms, and operators. The limit of detection for both assays was equivalent to 2–10 cells of *Xyl. fastidiosa* per reaction for field citrus samples. Petioles and midribs of symptomatic leaves of sweet orange provided the best materials for detection of the pathogen as these harbored the highest populations of *Xyl. fastidiosa*. The new species specific assay will be invaluable for molecular identification of *Xyl. fastidiosa* at the species level, and the CVC-specific assay will be very powerful for the specific identification of *Xyl. fastidiosa* strains that cause citrus variegated chlorosis.

7.53.5 Management

As chemical curative control of the disease is not available, the emphasis is laid mainly on prevention, use and development of resistant varieties, cultural and hygienic measures including protection of nursery stock from infection and nursery certification, and chemical and biological vector control. These methods, however, are often partly successful because the bacterium has many symptomless hosts, including weeds, ornamentals, and other crops and possibly also still unknown vectors. Pruning and removal of symptomatic parts of trees to remove inoculum sources is labor intensive and time consuming, and also partly successful because of introduction of the pathogen from neighboring areas.

Cultural practices: Stress is often a determining factor in the development of symptoms once a plant gets infected with the bacterium. Therefore, cultural practices including, adequate nutrition should be directed toward the healthy growth of plants. Iron deprivation possibly provides a way to reduce disease severity by preventing biofilm formation in the xylem vessels (Toney and Koh 2006). The cultural practices, like cultivar selection and weed control in and around orchards have proved to be effective (source: http://edis.ifas.ufl.edu/in174).

Vector control: Vector control by biological agents and insecticides is important and should be tried, if practicable.

REFERENCES

Almeida, R. P. P., Blua, M. J., Lopes, J. R. S., and Purcell, A. H. 2005. Vector transmission of *Xylella fastidiosa*: Applying fundamental knowledge to generate disease management strategies. *Ann. Entomol. Soc. USA* 98: 775–786.

Anonymous. 2002. CVC challenges Fundecitrus. pp. 16–19 in: *Fundecitrus: Relatorio de Gestao* 1995–2002.

Brlansky, R. H., Damsteegt, V. D., and Hartung, J. S. 2002. Transmission of the citrus variegated chlorosis bacterium *Xylella fastidiosa* with the sharpshooter *Oncometopia nigricans*. *Plant Dis.* 86: 1237–1239.

Brlansky, R. H., Davis, C. L., Timmer, L. W., Howd, D. S., and Contreras, J. 1991. Xylem-limited bacteria in citrus from Argentina with symptoms of citrus variegated chlorosis. (Abstr.) *Phytopathology* 81: 1210.

Chagas, C. M., Rossetti, V., and Beretta, M. J. G. 1992. Electron microscopy studies of a xylem-limited bacterium in sweet orange affected with citrus variegated chlorosis disease in Brazil. *J. Phytopathol.* 134: 300–312.

Chang, C. J., Garnier, M., Zreik, L., Rossetti, V., and Bove, J. M. 1993. Culture and serological detection of the xylem-limited bacterium causing citrus variegated chlorosis and its identification as a strain of *Xylella fastidiosa*. *Curr. Microbiol.* 27: 137–142.

Coletta-Filho, H. D., Carvalho, S. A., Silva, L. F. C., and Machado, M. A. 2014. Seven years of negative detection results confirm that *Xylella fastidiosa*, the causal agent of CVC, is not transmitted from seeds to seedlings. *Eur. J. Plant Pathol.* 139: 593–596.

Gonçalves, F. P., Lourenço, S. A., Stuchi, E. S., Hau, B., and Amorim, L. 2011. Comparative analysis for quantification of Citrus Variegated Chlorosis in the field. *Sci. Agric.* (*Piracicaba, Braz.*) 68: 562–565.

Gonçalves, F. P., Stuchi, E. S., Lourenço, S. A., Hau, B., and Amorim, L. 2012. Relationship between sweet orange yield and intensity of citrus variegated chlorosis. *Plant Pathol.* 61: 641–647.

Hopkins, D. L., Bistline, F. W., Russo, L. W., and Thompson, C. M. 1991. Seasonal fluctuation in the occurrence of *Xylella fastidiosa* in root and stem extracts from citrus with blight. *Plant Dis.* 75: 145–147.

Janse, J. D., and Obradovic, A. 2010. *Xylella fastidiosa*: Its biology, diagnosis, control and risks. *J. Plant Pathol.* 92: S1.35–S1.48.

Lee, R. F., Derrick, K. S., Beretta, M. J. G., Chagas, C. K., and Rossetti, V. 1991. Citrus variegated chlorosis: A new destructive disease of citrus in Brazil. Citrus Industry 72: 10–13, 15.

Li, W.-B., Pria, W. D. Jr., Lacava, P. M., Qin, X., and Hartung, J. S. 2003. Presence of *Xylella fastidiosa* in sweet orange fruit and seeds and its transmission to seedlings. *Phytopathology* 93: 953–958.

Li, W., Teixeira, D. C., Hartung, J. S., Huang, Q., Duan, Y., et al. 2013. Development and systematic validation of qPCR assays for rapid and reliable differentiation of *Xylella fastidiosa* strains causing citrus variegated chlorosis. *J. Microbiol. Methods.* 92: 79–89.

Lopes, J. R. S. 2000. Transmission of *Xylella fastidiosa* in citrus and ecology of its transmission vectors. Intern. Soc. Citric. Congr. 2000: 64.

Marucci, R. C., Lopes, J. R. S., Vendramim, J. D., and Corrente, J. E. 2005. Influence of *Xylella fastidiosa* infection of citrus on host selection by leafhopper vectors. *Entomologia Experimentalis et Applicata* 117: 95–103.

Niza, B., Coletta-Filho, H. D., Merfa, M. V., Takita, M. A., and de Souza, A. A. 2015. Differential colonization patterns of *Xylella fastidiosa* infecting citrus genotypes. *Plant Pathol.* 64: 1259–1269.

Oliveira, A. C., Vallim, M. A., Semighini, C. P., Araújo, W. L., Goldman, G. H., et al. 2002. Quantification of *Xylella fastidiosa* from citrus trees by real-time polymerase chain reaction assay. *Phytopathology* 92: 1048–1054.

Pooler, M. R., and Hartung, J. S. 1995. Specific PCR detection and identification of *Xylella fastidiosa* strains causing citrus variegated chlorosis. *Curr. Microbiol.* 31: 377–381.

Roberto, S. R., Coutinho, A., Lima, J. E. O., Miranda, V. S., and Carlos, E. F. 1996. Transmissao de *Xylella fastidiosa* pleas cigarrinhas *Dilobopterus costalimai*, *Acogonia terminalis* e *Oncometopia facialis* (Hemiptera: Cicadellidae) em citros. *Fitopatol. Bras.* 21: 517–518.

Rossetti, V., Garnier, M., Bove, J. M., Beretta, M. J. G., Teixeira, A. R. R., et al. 1990. Presence de bacteries dans le xyleme d'orangers atteints de chlorose variegee, une nouvelle maladie des agrumes au Brasil. C. R. Acad. Sci. (Paris) 310: 345–49. Series III.

Schaad, N. W., Postnikova, E., Lacy, G., Fatmi, M., and Chang, C.-J. 2004. *Xylella fastidiosa* subspecies: *X. fastidiosa* subsp. *piercei* subsp. nov., *X. fastidiosa* subsp. *multiplex* subsp. nov., and *X. fastidiosa* subsp. *pauca* subsp. nov. *Syst. Appl. Microbiol.* 27: 290–300.

Silva-Stenico, M. E., Pacheco, F. T., Pereira-Filho, E. R., Rodrigues, J. L., Souza, A. N., et al. 2009. Nutritional deficiency in citrus with symptoms of citrus variegated chlorosis disease. *Bra. J. Biol.* 69: 859–864.

Simpson, A. J., Reinach, F. C., Arruda, P., Abreu, F. A., Acencio, M., et al. 2000. The genome sequence of the plant pathogen *Xylella fastidiosa*. The Xylella fastidiosa Consortium of the Organization for Nucleotide Sequencing and Analysis. *Nature* 406: 151–157.

Toney, J., and Koh, M. 2006. Inhibition of *Xylella fastidiosa* biofilm formation via metal chelators. Journal of the Association for Laboratory Automation 11: 30–32.

Van Sluys, M. A., de Oliveira, M. C., Monteiro-Vitorello, C. B., Miyaki, C. Y., Furlan, L. R., et al. 2003. Comparative analyses of the complete genome sequences of pierce's disease and citrus variegated chlorosis strains of *Xylella fastidiosa*. *J. Bacteriol.* 185: 1018–1026.

7.54 BACTERIAL CANKER AND GUMMOSIS OF STONE FRUIT TREES

Bacterial canker and gummosis affects mainly stone fruit trees. The disease is most severe on sweet cherry (*Prunus avium*); particularly on young trees and the trees under stress are much more susceptible than healthy trees. The disease is also a major limitation for timber production from wild cherry. Pome fruits like apple and pear, rootstocks, and some ornamental trees are also infected, but to a lesser extent. Out of *Pseudomonas syringae* pv. *syringae* and *Psm. syringae* pv. *morsprunorum*, the causal agents of the disease, the latter has a limited host range and is confined mainly to tart cherry and plum. The disease also occurs on sour cherry (*P. cerasus*), apricot (*P. armeniaca*), almond (*P. amygdalus*), and peach (*P. persica*). *Psm. syringae* pv. *syringae* has a wide host range and is more virulent on sweet cherry and other stone fruits. The disease is also known by other names like **bud blast**, **blossom blast**, **spur and twig blight**, **sour sap**, and **die-back**.

Disease outbreaks are usually associated with prolonged cold wet weather accompanied with late spring frost. *Psm. syringae* pv. *morsprunorum* is widespread in Europe and is present in eastern North America and South Africa. In South Africa, bacterial canker caused by *Psm. syringae* pv. *syringae* is one of the most important diseases of stone fruit trees, and annual losses probably exceed 10 million US dollars (Hattingh et al. 1989).

Third pathogen, namely, *Psm. syringae* pv. *persicae* causing leaf spots, cankers, and gummosis of fruits on peach trees has been reported from France. It killed more than one million trees in France. One or more of these pathogens have also been reported from Australia, Canada, India, Japan, Lebanon, UK, and USA.

7.54.1 Symptoms

The symptoms appear on all the aboveground plant parts. The most characteristic symptom is the formation of cankers accompanied by the exudation of gum. The exudation of gum occurs in most cankers early in the growing season. The gum comes out through injuries or cracks in the bark and runs down on the surface of limbs. New cankers are generally formed in late winter and early spring, and these generally develop on trunk, limbs, and twigs. Cankers usually develop at the bud union, in pruning wounds, or when the pathogen invades woody tissue following infection of blossoms or dormant buds. The spread of the cankers is mostly upward and slightly downwards and to the sides. The cankered areas are slightly sunken and often darker than the surrounding healthy bark. On cherry, the cankers are typically elongated in shape. Cankers expand rapidly in spring and may girdle trunks, branches, and twigs. Leaves on the terminal portions of cankered limbs and branches become curled, droop, first turn light green, and then yellow, and finally die. Within a few weeks, the branch or the plant parts above the cankers also die. The removal of cankered bark exposes the underlying bright orange to brown tissue, often emitting a strong sour smell. The cankers on the shoots also develop from infections occurring through leaf scars. Infected young shoots wilt and die.

On leaves, water-soaked spots appear, which are small, circular to angular in shape, and often surrounded by chlorotic haloes in the early stage. Later on, spots become brown, dry, and brittle, and eventually fall out forming shot-holes or giving a tattered appearance to the infected leaves. The spots may also coalesce, especially at the margins of the leaves, to form large patches of dead tissue. In sweet cherry, leaf spots typically occur on the margins of the leaves. However, symptoms on leaves are apparently limited to some fruit cultivars growing in regions where summers are wet and humid (Hattingh and Roos 1992).

Infected dormant buds turn brown and fail to open or break in spring, and this phase is referred to as **"dormant bud blast"** or **"dead bud condition."** Small cankers often develop at the base of these buds. Cherry, apricot, and pear generally suffer heavily from dormant bud blast in certain areas. Some infected dormant buds open in spring, but they collapse in early summer exhibiting wilted leaves and dried-up fruits. Both leaf and flower buds are equally damaged. In severe cases, 90% or more buds on a tree may be killed.

The infection of the blossoms occurs in spring, and the infected blossoms turn brown to black and die. The blossom infection usually spreads into the fruit-bearing spur causing the collapse of the whole blossom cluster. During full bloom, the poor bloom of the infected trees is quite evident.

On green cherry fruits, the lesions are chocolate-brown, surrounded by water-soaked tissue. As the lesions and fruit age, the affected tissues collapse forming deep black depressions in the flesh with margins turning yellow to red. Elliptical and brown lesions having water-soaked margins appear on fruit stalks. Fruit infections occur sporadically, but these can be severe when prolonged wet and cold weather prevails during or shortly after bloom (Jones 2001).

The symptoms of bacterial canker can be confused with those of Cytospora canker, but these can be distinguished on the basis of differences in the canker margin. There is a distinct margin at the top and bottom of Cytospora cankers, while the margins of bacterial cankers are irregular and streaks of discolored strands frequently extend into healthy tissue.

7.54.2 Causal Organisms

Pseudomonas syringae pv. *syringae* van Hall 1902
Syn. *Psm. syringae* pv. *japonica*
Psm. syringae pv. *morsprunorum* (Wormald 1931) Young et al. 1978
Psm. syringae pv. *persicae* (Prunier et al. 1970) Young et al. 1978
Syn. *Psm. morsprunorum* f. sp. *persicae*

The cells are straight rods, measure 0.7–1.2 × 1.5 µm, Gram-stain-negative, aerobic, motile having polar flagella, and do not accumulate poly-β-hydroxybutyrate as a carbon reserve. The bacteria give negative reaction for oxidase and arginine dihydrolase. Colonies on King's medium B are circular, smooth, and glistening with entire or lobate margins, and produce light green fluorescent pigment. Pathovars *syringae* and *morsprunorum* can be differentiated from each other on the basis of GATTa (gelatin liquefaction, esculin hydrolysis, tyrosinase activity, and tartrate utilization) scheme. *Psm. syringae* pv. *syringae* is positive for the first two tests and negative for the last two tests, while reverse is true for *Psm. syringae* pv. *morsprunorum*. LOPAT scheme can also be used for the differentiation of these pathovars. Both the pathovars elicit hypersensitive reaction on tobacco leaves.

Most strains of *Psm. syringae* pv. *syringae* produce the phytotoxin called syringomycin, which plays a role in the virulence of the pathogen. Two genes, namely, *syrA* and *syrB* involved in the production of this toxin have been identified and analyzed in strain B 301D of the bacterium (Xu and Gross 1988). The study further revealed that *syrA* is a regulatory gene and is required for both syringomycin production and pathogenicity. Both the genes are required for the formation of proteins Sr4 and Sr5, which are believed to be components of the syringomycin synthetase complex. Probes and PCR primers for genes required for syringomycin toxin are highly reliable for the detection of pathogenic strains of this pathovar.

Many strains of *Psm. syringae* pv. *syringae* act as ice-nuclei. *Psm. syringae* pv. *syringae* contains plasmid-mediated genes for copper resistance. Many strains of *Psm. syringae* pv. *morsprunorum* contain genes for coronatine production. The complete genome sequencing of *Psm. syringae* pv. *syringae* strain B 728a by Feil et al. (2005) revealed that the genome (6.1 mb) contains a circular chromosome and no plasmid. Among the 976 protein-encoding genes, there are genes that encode for syringopeptin, syringomycin, indole acetic acid biosynthesis, arginine degradation, and production of ice-nuclei. Certain unique genes such as, for ectoine synthase, DNA repair, and antibiotic production may contribute to the epiphytic fitness and stress tolerance of the bacterium.

Abbasi et al. (2013) studied genetic diversity among 78 Iranian strains of *Psm. syringae* pv. *syringae* by extracting genomic DNA from strains and using it in rep-PCR and IS50-PCR analysis. Cluster analysis was performed using UPGMA. The strains of the bacterium were separated into nine distinguishable genotypic groups by the combination data set of both rep-PCR and IS50-PCR at 73% similarity level. There was no significant correlation between genetic diversity and geographical origin of the isolates. These results indicate that a combination of rep-PCR and IS50-PCR fingerprinting can be used as a high resolution genomic fingerprinting method for elucidating diversity among the strains of this bacterium. The existence of a considerable genetic diversity among the strains of bacterium causing canker of stone fruit trees in Iran, reported here, will be of immense use in the development of resistant genotypes against this bacterial pathogen.

Third pathogen, *Psm. syringae* pv. *persicae* is more closely related to *Psm. syringae* pv. *morsprunorum*. It does not produce fluorescent pigment on King's medium B, but it does on casamino acid sucrose agar.

7.54.3 Disease Cycle

The bacteria perpetuate in the margins of active cankers, infected buds, and systemically in the vascular systems of certain hosts. The bacteria also survive epiphytically on blossoms, buds, and leaves of healthy and infected plants, and also on non-host plants including weeds. The bacterial population on blossoms can build up to 10^{4-6} cfu/blossom (Kennelly et al. 2007). In fact, the concept of a plant pathogen surviving as an epiphyte on healthy plant tissue was first proposed in studies of *Psm. syringae* pv. *morsprunorum*. Epiphytic populations of these bacteria also establish on the surface of leaves. The bacteria present in the buds colonize the new leaves as they emerge in spring. Environmental conditions influence the multiplication of bacteria causing fluctuations in population levels. During adverse atmospheric conditions like warm, dry spells, epiphytic bacteria enter through stomata and colonize the intercellular spaces of spongy parenchyma. On reaching the uninvaded stomatal chambers, they multiply profusely, and the cell masses extrude through stomata. It appears that epiphytic populations are constantly replenished in this way (Hattingh and Roos 1992). In plum, *Psm. syringae* pv. *syringae* spreads from colonized blossoms, which are not killed, to the developing seeds and survives in the seed.

The dissemination of the bacteria occurs through wind-splashed rain, water splashes, pruning tools, and infected budwoods and nursery plants. The penetration of the bacteria occurs through wounds, stomata, and injuries caused by frost and severe storms on blossoms, buds, and leaves. Spotts et al. (2010) showed that all the seven types of injuries and wounds, whether natural or operational, facilitated the infection of sweet cherry by *Psm. syringae* pv. *syringae*. Infection of inoculated wounds made in spring and summer (heading cuts when trees were planted, scoring cuts, and summer pruning) resulted in maximum canker incidence and severity. Inoculation of heading cuts resulted in the highest tree mortality of 86%. Bacterial cankers did not develop in summer or winter when active cankers were cut, and the same pruning tool was used immediately to make heading cuts on healthy trees. Heading cuts became resistant to infection after 1 week in summer and 3 weeks in winter.

The infection and progress of disease is greatly influenced by environmental factors. Free water on leaf surface and high relative humidity are essential for at least 24 h for significant leaf infection to occur. Symptoms appear after about 5 days at 21°C–27°C. The disease is usually severe after cold winters and prolonged spring rains. A warning system based on temperature and free moisture conditions for the prediction of blossom blight infection in the field has been developed (Latorre et al. 2002). The development of the cankers is governed largely by the weather and the growth of the host. The cankers expand in the fall and winter, but the maximum development occurs in the period between the end of cold weather and the beginning of the rapid tree growth in the spring. The active growth of the host and high temperature in the spring result in the formation of callus tissue around the cankers, which checks the further advance of the cankers. Some cankers are inactivated permanently, while others restart the growth when favorable conditions for the growth of bacteria return. Fresh infections during the active growing season hardly cause any significant loss as their development is quickly arrested by the callus tissue.

Frost injury predisposes the host tissue to infection, but the infection is governed by the extent of wet weather during the thawing period. Many strains of *Psm. syringae* pv. *syringae* act as ice-nuclei and hence cause frost injury to plants by forming ice at relatively higher freezing temperatures. These bacteria also produce bacteriocins, which inhibit the growth of non-ice nucleation-active bacteria, thereby making their own growth less competitive. Frost injury predisposes the buds, flowers, and young leaves to infection. Infection of buds starts at the base of outer bud scales and spreads killing the bud tissue leading to their death. The infection may spread downwards killing the stem tissues around the base of bud. The bacteria penetrate flowers through natural openings and wounds and spread quickly through floral parts under humid conditions. The infection may advance into the spur and further to the twig. Young and succulent leaves are frequently infected during cool and wet springs and the bacteria penetrate through stomata. Bacteria spread intercellularly and cause collapse and death of cells leading to formation of necrotic and angular spots. The infections on mature leaves are rare. Under humid conditions, the bacteria ooze out from infected plant parts through stomata, lenticels, and wounds formed in the infected tissue and are spread to other plant parts by various agencies.

The limbs generally get infected during the fall and early winter. The entry of bacteria occurs through natural and pruning wounds, leaf scars, and through the bases of infected buds and spurs. After entry into the host, bacteria multiply and move intercellularly into the bark and then further into the ray parenchyma of the phloem and xylem. In advanced stages of infection, bacteria breakdown parenchyma cells forming cavities full of bacterial cells. Occasionally, the bacteria invade the xylem vessels, but they do not move for long distances in these vessels (Agrios 2004).

Donmez et al. (2010), on the basis of artificial inoculation screening, reported that apricot cultivars Hasanbey, Çöloğlu, Soğanci, and Şekerpare were resistant, while Şam, Tokaloğlu (Erzincan), and Erken Ağerik were susceptible to *Psm. syringae* pv. *syringae*.

7.54.4 Management

Currently, the effective management of the disease is not possible due to the lack of effective chemicals and biocontrol agents, lack of host resistance, and the endophytic nature of the pathogen during some phases of the diseases cycle. However, the following cultural practices and protective measures will certainly reduce the incidence and severity of the disease.

1. Most if not all, factors that reduce the vigor of trees probably favor disease development. Hence, avoiding or reducing the effect of these stress factors will certainly minimize the severity of the disease
2. Plant only the healthy nursery trees in the orchards
3. Use only the healthy budwood for propagation
4. Susceptible varieties should be propagated on rootstocks resistant to the pathogens and grafting should be as high as possible. Mazzard seedlings, particularly the F12-1, are the most resistant rootstocks
5. Prune the cankered limbs at least 15 cm below the visible cankers. Pruning should not be done in early spring and fall, as the bacteria are active during this period. Pruning tools should be disinfected frequently with 10% clorox (household bleach) or shellack thinner (70% ethyl alcohol). Cankers on trunks and large branches can be removed by cauterization with a hand held propane burner. The flame is directed particularly at the margins of the canker until the under lying tissue crackles and chars. It has been used with some success in New Zealand
6. Destroy badly infected young plantings
7. Protective sprays of fixed copper compounds (copper oxychloride at 8–12 lb A^{-1}) or Bordeaux mixture (12:12:100) has resulted in partial control of the disease in nursery and orchards. However, pathogen strains resistant to copper are common in orchards in Pacific Northwest and other areas with a history of copper use. Moreover, copper is phytotoxic to most stone fruits, particularly to sweet cherry and its repeated applications aggravate it further. Cherry nursery plants should be sprayed before digging for planting, and in the orchard; sprays should be given in the fall and in the early spring before blossoming. Air blast sprayer should be preferred over other sprayers as it would blow off all except terminal leaves from the trees and thereby reduce the chances of leaf infection through leaf scars
8. ASM (sold under the trade name, Actigard in the United States and Bion in Europe) is a compound that enhances the natural defenses of plants by inducing systemic acquired resistance. Monthly applications of ASM moderately reduced bacterial decline of hazelnut caused by *Psm. syringae* pv. *syringae* in Italy
9. Avoid ring nematodes infested soils for planting orchards, as the high population of these nematodes is generally associated with high disease incidence.

REFERENCES

Abbasi, V., Rahimian, H., and Tajick-Ghanbari, M. A. 2013. Genetic variability of Iranian strains of *Pseudomonas syringae* pv. *syringae* causing bacterial canker disease of stone fruits. *Eur. J. Plant Pathol.* 135: 225–235.

Agrios, G. N. 2004. *Plant Pathology*. 5th ed. Elsevier Academic Press, San Diego, CA, 922 pp.

Donmez, M. F., Karlidag, H., and Esitken, A. 2010. Identification of resistance to bacterial canker (*Pseudomonas syringae* pv. *syringae*) disease on apricot genotypes grown in Turkey. *Eur. J. Plant Pathol.* 126: 241–247.

Feil, H., Feil, W. S., Chain, P., Larimer, F., DiBartolo, G., et al. 2005. Comparison of the complete genome sequences of *Pseudomonas syringae* pv. *syringae* B728a and pv. *tomato* DC 3000. *Proc. Natl. Acad. Sci. USA* 102: 11064–11069.

Hattingh, M. J., and, Roos, I. M. M. 1992. Bacterial canker of stone fruits. pp. 394–404 in: *Plant Diseases of International Importance*. Vol. III, Diseases of Fruit Crops. J. Kumar, H. S. Chaube, U. S. Singh, and A. N. Mukhopadhyay eds. Prentice Hall, Englewood Cliffs, NJ.

Hattingh, M. J., Roos, I. M. M., and Mansvelt, E. L. 1989. Infection and systemic invasion of deciduous fruit trees by *Pseudomonas syringae* in South Africa. *Plant Dis.* 73: 784–789.

Jones, A. L. 2001. Bacterial canker of stone fruit trees. pp. 74–76 in: *Encyclopedia of Plant Pathology*. Vol. 1, O. C. Maloy, and T. D. Murray eds. John Wiley & Sons, New York.

Kennelly, M. M., Cazorla, F. M., de Vicente, A., Ramos, C., and Sundin, G. W. 2007. *Pseudomonas syringae* diseases of fruit trees: Progress toward understanding and control. *Plant Dis.* 91: 4–17.

Latorre, B. A., Rioja, M. E., and Lillo, C. 2002. The effect of temperature on infection and a warning system for pear blossom blast caused by *Pseudomonas syringae* pv. *syringae*. *Crop. Prot.* 21: 33–39.

Prunier, J., Luisetti, J., and Gardan, L. 1970. Études sur les bactérioses des arbres fruitiers. II. Caractérisation d'un *Pseudomonas* non-fluorescent agent d'une bactérioses nouvelle du pêcher. *Ann. Phytopathol.* 2: 181–197.

Spotts, R. A., Wallis, K. M., Serdani, M., and Azarenko, A. N. 2010. Bacterial canker of sweet cherry in Oregon- infection of horticultural and natural wounds, and resistance of cultivar and rootstock combinations. *Plant Dis.* 94: 345–350.

van Hall, C. J. J. 1902. *Bijdragen tot de kennis der Bakterieele Plantenzeikten*. Coöperatieve Drukkerij-vereeniging "Plantijn," Inaugural dissertation, Amsterdam, the Netherlands.

Wormald, H. 1931. Bacterial diseases of stone fruit trees in Britain. III. The symptoms of bacterial canker in plum trees. *J. Pomol. Hort. Sci.* 9: 239–256.

Xu, G. W., and Gross, D. C. 1988. Physical and functional analyses of the *syrA* and *syrB* genes involved in syringomycin production by *Pseudomonas syringae* pv. *syringae*. *J. Bacteriol.* 170: 5680–5688.

Young, J. M., Dye, D. W., and Wilkie, J. P. 1978. Genus *Pseudomonas* Migula 1894. In: Young, J. M., Dye, D. W., Bradbury, J. F., Panagopoulos, C. G., and Robbs, C. F. 1978. A proposed nomenclature and classification for plant pathogenic bacteria. *N. Z. J. Agric. Res.* 21: 153–177.

7.55 BACTERIAL SPOT OF STONE FRUITS

Bacterial spot, also referred to as **bacterial leaf spot**, **bacteriosis**, **bacterial shot hole**, or **shot hole**, is caused by *Xanthomonas arboricola* pv. *pruni*. The disease was first described by E.F. Smith in 1903 on plums in Michigan, USA. Although primarily a problem of peach (*Prunus persica*) and nectarine, the disease also occurs on apricots (*P. armeniaca*), plums (*P. domestica*) and, to a lesser extent on cherries (*P. avium*, *P. cerasus*, *P. laurocerasus*, *P. japonica*) and other stone fruits. Besides cultivated species, all the ornamental *Prunus* species and hybrids can be affected by this bacterium (Stefani 2010). In a review article, Lamichhane (2014) has provided an overview of *Xa. arboricola* diseases of stone fruit, almond, and walnut trees and has discussed the current and future management strategies.

Losses due to bacterial spot occur on account of reduction in fruit yield and quality, resulting from fruit infection. Up to half of the fruit on susceptible varieties may become unsalable. Several consecutive years of near or complete defoliation may reduce the vigor and yield of the trees. Additionally, extensive defoliation and twig die-back results in stunting and gradual loss of leaders from season to season. Such weakened trees are more subjected to winter injury. Losses are more in light and low-fertility soils as less vigorous trees are more susceptible to the disease than vigorous trees. Bacterial spot is of most concern in regions with annual rainfall greater than 51 cm per year.

There is not much information regarding damages or economic losses caused by the disease outbreaks (Roselló et al. 2012) barring some reports. A severe epiphytotic of bacterial spot of peach occurred in the southeastern US (Werner et al. 1986). The most severe epiphytotics have been reported on the Sino-Japanese plum group (*P. salicina* and *P. japonica*) and their hybrids, peach (*P. persica* and hybrids) and nectarine (*P. persica* var. *nectarina*) in Spain (Ritchie 1995; Stefani 2010). In the United States, Dunegan (1932) observed that 25%–75% of the fruits could show lesions in neglected peach orchards. According to Stefani (2010), an epiphytotic in a commercial plum orchard of northern Italy affecting 30% fruits of cv. Golden Plum, could easily result in crop losses estimated over 11,200 € per ha, while it was 9,500 € for fruits of cv. Angeleno. Estimated crop losses in Australia during years favorable to the disease were 3.1 million Australian dollars annually on *Prunus* species (Stefani 2010).

The disease has been reported in most of the stone fruit-producing countries from Africa (South Africa, Morocco, and Zimbabwe); America (Bermuda, Canada, Mexico, USA, Argentina, Brazil, and Uruguay); Asia (China, India, Japan, North Korea, South Korea, Lebanon, Pakistan, Saudi Arabia, Taiwan, and Tajikistan); Europe (Bulgaria, France, Italy, Austria, Moldova, Montenegro, the Netherlands, Romania, Russia, Slovenia, Spain, Switzerland, and Ukraine); and Oceania (Australia and New Zealand) (Young 1977; Jindal et al. 1989; Panić et al. 1998; Anonymous 2006a, 2006b, 2009; Roselló 2007; Roselló et al. 2007; Pothier et al. 2010). The disease has also been reported on Japanese plum (*P. salicina*) from Taiwan (Shen et al. 2013).

7.55.1 Symptoms

The most obvious leaf symptoms are yellow, chlorotic leaves having numerous, small spots (1–5 mm) on the leaves. The leaf spots are always angular as a result of being delimited by the veins of the leaf. Initially, the spots appear as water-soaked, angular spots, which are generally visible when viewed with a light source behind the leaf. In about 3 days, the lesion becomes visible with reflected light. As the infection progresses, the spots enlarge and turn deep purple to rusty brown or black. Within 1–2 weeks, the center of the lesion is "walled off" by the leaf and drops out, resulting in a "**shot-hole**" or tattered appearance (Figure 7.38a). Leaf lesions are much more common at the distal ends of

(a)

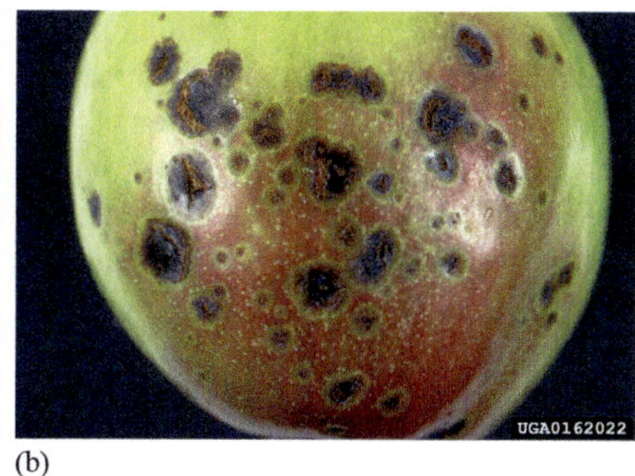

(b)

FIGURE 7.38 Bacterial spot of stone fruits: (a) Symptoms on plum leaf. (b) Symptoms on plum fruit. (Courtesy of U. Mazzucchi, Università di Bologna, Bugwood.org.)

the leaves and around major leaf veins. This occurs because the water film is thicker in such areas, and these regions of the leaf dry more slowly. Severely infected leaves soon turn yellow and drop prematurely. On susceptible varieties, a few lesions on leaves can result in severe defoliation, while tolerant varieties may require many more lesions to cause defoliation. Under heavy disease pressure, only the youngest leaves remain on the tree. Leaves are most susceptible before becoming fully expanded. Leaf symptoms usually are first visible 5–14 days after infection. Rapid symptom expression is dependent on warm temperatures. Severe leaf loss early in the summer reduces fruit size and weakens the trees. The foliar lesions caused by pesticide sprays or other injuries are usually circular in shape and do not have a water-soaked appearance in contrast to water-soaked angular lesions caused by the pathogen.

On peaches and nectarines, twig symptoms usually consist of two types of cankers, i.e., spring and summer cankers. **Spring cankers** develop on young twigs produced the previous summer. Initially, these cankers appear as water-soaked, slightly darkened blisters, extending 1–2 cm either side of leaf and flower buds; these affected buds usually fail to open. These cankers appear about the time the first leaves appear. These cankers seldom extend more than half way around the twigs, but an encircled twig will die-back. When a canker extends downward from the terminal bud, which fails to open, it is termed a "**black tip**."

Summer cankers develop on green twigs of the current season's growth. They usually occur later in the summer after leaf spots are evident. The lesions begin as water-soaked, dark purplish spots between the nodes. With time, they enlarge, turn brown to purple-black, become slightly sunken, and are round to elliptical in shape. Cankers caused by the peach scab fungi (*Venturia carpophila*, *Cladosporium carpophilum*, and *Cladosporium* spp.) are similar, except that these are slightly raised.

On fruits, the earliest symptoms are generally observed about 3 weeks after petal fall. Small, round, olive-brown to black spots are formed. They are usually sunken and frequently surrounded by a water-soaked margin. Spots may slowly enlarge and merge to cover large irregular areas on the fruit (Figure 7.38b). Infections that occur approximately 4 weeks after petal fall develop into large, open lesions that extend deep into the fruit flesh, sometimes almost to the pit by harvest. In contrast, infections that occur after initiation of pit-hardening usually remain near the fruit surface. These shallow lesions may coalesce, resulting in cracking of the skin. Brown rot fungi (*Monilinia fructicola* and *M. laxa*) can easily enter these cracks and become established. The lesions on some cultivars frequently exude a yellowish gum after rainy periods. Fruits infected at an early stage of development are usually the most malformed. On peaches, spots usually form on the side exposed to the sun.

On plum and apricot, leaf symptoms are very similar to those on peach, with pronounced shot-hole effect. Cankers, when formed, are perennial, in contrast to peach. They continue to develop in 2- to 3-year-old twigs and become much bigger and more devastating than in peach. The symptoms on plum fruits are quite different from those on peach. Large, circular, sunken, greasy spots which darken to black lesions may form on some varieties; on others, small pit-like lesions are common. The large lesions may later crack open. On cherry, leaf symptoms are similar to those on peach, but not of importance. Fruit infection resulting in malformed fruit is more prevalent. Bacteria may be found throughout the fruit (Janse 2005).

On almond, leaf symptoms are similar to those observed on peach. In early spring, infected fruits show sunken corky lesions with oozing gum that streams or clumps. The sunken lesions become raised following dehydration of the mesocarp in summer. Sometimes, circular dark spots appear on the endocarp and may affect the nut. Infected fruits either fall prematurely or remain attached to the trees and mummify over time. Mummified fruits harbor viable bacteria and serve as potential inoculum sources (Roselló et al. 2012).

7.55.2 Causal Organism

Xanthomonas arboricola pv. *pruni* (Smith 1903) Vauterin et al. 1995
Syn. *Xa. pruni*, *Xa. campestris* pv. *pruni*

The species name "*arboricola*," from Latin "tree dwellers," was first attributed by Vauterin et al. (1995) and justified by including bacteria that live and cause diseases of tree species.

It is a quarantine pathogen for the European Union and in the European and Mediterranean Plant Protection Organization list (EPPO A2 List of pest recommended for regulation).

There seems to be no strain differentiation between hosts (Scortichini et al. 1996). The phylogenetic relationships, based on the *rpoD* housekeeping gene, among a worldwide collection of *Xa. arboricola* revealed that strains isolated from *Prunus* spp. (pv. *pruni*) constitute a homogeneous group different from all other pathovars, which in turn are quite heterogeneous (Rademaker et al. 2005). The diversity among *Xa. arboricola* pv. *pruni* strains isolated from *Prunus* spp. is very low. Based on multiple-locus variable-number analysis, strains isolated from diseased cherry laurel in the Netherlands were clustered in two groups (Bergsma-Vlami et al. 2012). Based on fluorescent amplified fragment length polymorphism and housekeeping genes, Boudon et al. (2005) analyzed strains isolated from infected stone fruits in western Europe and the United States and reported a low diversity among these strains. They also found the same bacterial genotype in five countries of three continents. Another study, based on the housekeeping genes from a worldwide collection of strains, further validated a lack of relationship among strains due to geographic origin (Hajri et al. 2012). This almost clonal global population can be associated with the limited genetic diversity of its host (Zaccardelli et al. 1999; Boudon et al. 2005).

Pothier et al. (2011) sequenced the whole genome of *Xa. arboricola* pv. *pruni* strain from Europe (Italy, CFBP 5530), being the first for this species. The chromosome is 4.85 Mb with 65.6% GC ratio and 3912 predicted CDS. The bacterium has a unique 41.2 kb plasmid with a 62.3% GC ratio and 45 CDS. Automatic annotation using several sequenced *Xanthomonas* genomes as templates, and partial manual annotation, identified a suite of 21 type III secretion system effectors, iron acquisition, and other putative virulence/ecological fitness determinants, many apparently unique to *Xa. arboricola*. Efficient database framework for comparative genome analyses using BLAST score ratios comparisons against *Xa. axonopodis* pv. *citri* str. 306 and *Xa. campestris* pv. *campestris* LMG8004 indicated a pangenome of 2,786 CDS and 848 singletons in the bacterium. Applied genomics has identified over 90 variable number tandem repeats, several of which are currently being used for its biodiversity analysis and related sub species and design of improved diagnostics.

Garita-Cambronero et al. (2014) reported the draft genome sequence of *Xa. arboricola* pv. *pruni* strain Xap33 (CITA33), isolated from symptomatic almond leaves (*P. amygdalus* cv. Guara) cultivated in Aragón, Spain. 16S rRNA comparison analysis performed with BLASTn detected 99%–100% identity to other *Xanthomonas* strains, including the *Xa. arboricola* pv. *juglandis* strain LMG 747. In addition, whole-genome nucleotide comparison using BLASTn against 214 complete and draft genomes from several *Xanthomonas* strains determined that the Xap33 genome sequence has 99% identity to the draft genome sequences of *Xa. arboricola* pv. *pruni* strains MAFF301420, MAFF301427, and MAFF31562. This information will allow genetic analysis and manipulation of *Xa. arboricola* pv. *pruni* Xap33, as well as future comparative studies with other *Xa. arboricola* pathovars. It will also help in understanding the molecular mechanisms in the plant–pathogen interactions related to bacterial spot disease on almond.

Ballard et al. (2011) developed a Bio-PCR protocol for the detection of *Xa. arboricola* pv. *pruni* and showed that primer pair 29F/R, designed from cloned sequence, was to be the only one capable of differentiating *Xa. arboricola* pv. *pruni* from all other *Xa. arboricola* pathovars. Bühlmann et al. (2013) developed a rapid, sensitive detection assay for *Xa. arboricola* pv. *pruni*, which showed unrivaled specificity with the 79 tested strains and distinguished it from a phylogenetically close pathovar such as *Xa. arboricola* pv. *corylina*. The sensitivity of this test was comparable to that of a previously reported TaqMan™ assay at 10^3 cfu mL^{-1}, while the unrivaled speed of LAMP technology enabled a positive result to be obtained in <15 min. The developed assay can be used with real-time fluorescent detectors for quantitative results as well as with DNA-staining dyes to function as a simplified strategy for on-site pathogen detection.

7.55.3 Disease Cycle

The bacterium overwinters in peach twigs infected late in the summer. These infections are not visible until next spring, when they appear as spring cankers. The bacterium also overwinters in association with buds and in leaf scars that become infected during leaf drop the previous season. Leaf scar infections usually result in spring cankers. The bacteria are spread from the cankers by dripping dew and in splashing and/or wind-blown rain to the newly emerging leaves. Hard, driving rains are more important than gentle rains in initiating new infections. This explains why bacterial spot is often more severe on one side of the tree than on the other. The bacteria infect expanding leaves, green fruit, and twigs. Three week period following petal fall is critical for early-season fruit infection and establishment of inoculum on new foliage. Rainfall during this period is favorable for infection. The bacterium requires water congestion for infection to occur. Moisture plays a critical role in the infection process particularly when it causes plant tissues to become water congested (Zehr et al. 1996; Battilani et al. 1999). Zehr et al. (1996) reported that both leaf wetness and water congestion can markedly influence the establishment of infection. In a greenhouse study they found that only slight necrosis developed when water-congested leaves were allowed to dry immediately after inoculation or to remain wet up to 18 h. However, necrosis levels increased once leaf surface remained wet for 24–48 h. This demonstrates that the duration of leaf wetness influences the severity of disease. Newly emerged leaves are immune to infection as they do not contain air spaces between the cells.

Penetration occurs through stomata, lenticels, and wounds when surface moisture is present. Repeated infections occur throughout the growing season on all susceptible parts as long as the environment is favorable for disease development. Warm temperatures (21°C–29°C) with light rains, heavy dews, and windy weather are most conducive for disease development and spread. The disease makes little progress during hot and dry summer weather. In the late summer and fall, bacteria are blown to young and succulent stems, where penetration occurs through the leaf scars left by late-maturing leaves or through lenticels. Some of the infections initiated in the fall (which may not show signs of canker formation) serve as inoculum for early spring disease development, thus completing the disease cycle. Wind and wind-blown sand can increase the severity of disease by creating wounds for the bacterial entry. A dramatic increase in secondary infections can be seen after violent summer storms.

Disease incidence is highly correlated with cultural practices and stress caused by unfavorable soil and climatic conditions. The disease is more common and most severe in areas where stone fruits are grown on light, sandy soils and the environment is humid and warm during the growing season. Moreover, nutrient deficiency and excessive mineral fertilization predispose plants to the pathogen. The occurrence of spring frost can create micro-lesions on plant tissues through which the bacterium may enter and establish infection (Matthee and Daines 1968; Zehr et al. 1996; Scortichini 2010).

Morales et al. (2018) developed a model for predicting disease development and successfully evaluated it, based on the relationship between disease severity and the accumulated heat expressed in cumulative degree day (CDD). Optimal temperature for growth of epiphytic populations ranged from 20°C to 30°C under leaf wetness. The incubation at 25°C and 30°C gave the shortest incubation periods of 7.7 and 5.9 days, respectively. Incubation periods of 150, 175, and 280 CDD were required for 5%, 10%, and 50% of disease severity, respectively.

Ritchie (2011) found that there is a period when fruits are very susceptible or bacterial inoculum is at an optimal level, and the environment at these times is critical for infection of fruit. He further observed that analysis of weather records showed an association of bacterial spot incidence and severity on peach fruit with occurrence of frequent periods of precipitation from mid-March to mid-May and particularly when precipitation occurred in April. In North Carolina where this work was carried out, this period coincides with petal-fall through pit-hardening. Fruit infections that occurred during this period usually caused lesions that extended deep into the fruit flesh. Lesions on fruit that developed from infections near or after pit-hardening were shallow or mostly limited to the fruit skin.

Socquet-Juglard et al. (2012) evaluated following four inoculation techniques in the greenhouse: (1) infiltration in the leaves using a needleless syringe, (2) injuries caused by a needle, (3) injuries caused by scissors, and (4) dipping shoots of the plant into the inoculum. All inoculation techniques induced symptoms, but infiltration technique was most efficient as it quickly caused severe necrosis and leaf drop. The needle injuries technique was also efficient, but it produced mild symptoms, while the scissors injury technique was less efficient as it produced very mild symptoms. Though, shoot dipping technique was the least efficient in symptom induction, it yielded more quantitative data, e.g., number of spots per leaf, or damaged surface, calculated disease incidence, and resistance index, required for evaluating the level of resistance of a cultivar. Therefore, they used this technique to screen six apricot cultivars for resistance to bacterial spot.

There are no established treatment thresholds for this pathogen. However, since infections occur only when the leaves are wet and the amount of disease increases exponentially, the decision to begin a protective spray program should be based on the following three observations: (1) if the disease on the present site in the past affected the quality of more than 5% of the fruit, (2) if 20% or more leaves show lesions in the current season, and (3) if new lesions have developed in the last week. Protective treatments need to be continued until 2 to 3 weeks before harvest, but can be safely suspended during periods of extended dry weather.

7.55.4 Management

It is very difficult to control the disease on highly susceptible and also on moderately susceptible varieties under optimal environmental conditions for disease development. The following measures should be taken to minimize the losses from the disease.

1. Management of the disease should start with host resistance when available for cultivars adapted to the growing area. Some of the resistant peach varieties include Bell of Georgia, Biscoe, Harken, Loring, Madison, Ranger, Redhaven, Redskin, Sunhaven, Clayton, Cador, and Senitel, and of plum are Robusto, Segundo, Rubysweet, and Bruce
2. Maintain proper vigorous growth (but not excessive shoot growth) of the trees by providing optimum soil fertility and following proper cultural practices
3. Employ appropriate windbreaks and ground covers to minimize blowing of sand within and surrounding the orchards
4. Allow for good air circulation by avoiding low-lying orchard sites and by adhering to correct pruning practices. Prune during dry weather in the latter part of the dormant season
5. Production and use of healthy planting materials is of utmost importance. Indeed, early detection of the pathogen in nursery materials is the only way to avoid its further movement to the orchards. There are several recently developed early detection methods for *Xa. arboricola* (Lamichhane and Varvaro 2014a), but further efforts are needed to improve the sensitivity of these methods
6. Periodic monitoring and quantification of epiphytic populations using effective techniques (Lamichhane and Varvaro 2014b) is needed. Once the population density exceeds the economic threshold, preventive treatments using copper-based compounds should be

applied to reduce inoculum levels. The application of these compounds should also follow events causing wounds on plants, including autumn leaf fall and lesions caused by spring frost and hail

7. Chemicals for bacterial spot control available in the market [copper-containing compounds and oxytetracycline (FireLine 17WP, Mycoshield 17WP)] are very limited (Horton et al. 2011). Other materials evaluated have provided at best minimal disease control (Ritchie 2005, 2006, 2007). Even the copper-containing compounds and oxytetracycline do not provide complete control of the disease and, in years favorable for disease development, may not provide control that is commercially acceptable. However, the sprays of these chemicals may help to reduce the levels of fruit and leaf infection, but for better results, they must be applied preventatively. Copper-containing compounds are used as dormant sprays and relatively higher doses of copper are used initially, which are significantly reduced as the new growth emerges. The strategy for using copper (as dormant spray) is to place a bactericidal barrier on the surface of the tree before the bacteria emerge from overwintering sites. Peaches and other stone fruits are very sensitive to copper, thus extreme care must be taken when copper is used during the growing season. Also, copper can accumulate on the trees if adequate rainfall fails to occur between applications. Some injury to existing leaves, as reddish spots and shot-holes with some very mild defoliation, is expected if an effective rate of copper is being used. Because of risk for foliar injury, most copper materials are not registered for use as post-bloom or more than two applications post-bloom and hence, to be used only at low rates. In orchards having bacterial spot, avoid spraying when foliage is wet as this can readily disseminate the bacteria.

   It is difficult to recommend a spraying schedule for bacterial spot that fits every orchard. Ritchie (2011) has given the following spraying schedule (based on using 100 gal water per acre, with a range of 75–125 gal) for a moderately to highly susceptible cultivar in an orchard where bacterial spot occurs. As rates of copper vary in different formulations, the rates based on metallic copper equivalent (MCE) are given below.

   - Dormant bud to bud-swell: copper 1.0–2.5 lb MCE per acre
   - Bud-burst (top of buds have opened) to 1/4″ emerged green leaf tissue: 0.75–1.25 lb MCE per acre
   - Pink to 25% bloom: 0.5–1.0 lb MCE per acre
   - First petal-fall to 50% petal fall: 0.25–0.75 lb MCE per acre

   As the injury from copper is at high risk at the following growth stages, low doses of copper given below are recommended.

   - First shuck-split to 50% shuck-split: FireLine 17WP or Mycoshield 0.75–1.0 lb may be tank-mixed with 0.10–0.20 lb MCE per acre
   - Shucks off: FireLine 17WP or Mycoshield 0.75 lb may be tank-mixed with 0.10–0.20 lb MCE per acre

   Additional sprays, if needed based on weather and disease conditions, may be given. Discontinue the use of copper, if unacceptable injuries occur. Antibacterial activity of oxytetracycline has a short period of a few days and is readily washed off with rainfall (Christiano et al. 2010).

   Although copper-based formulations satisfactorily control epiphytic bacterial populations, they are ineffective once the pathogen enters the plant. Another constraint to the use of these products on *Prunus* spp. is damage to leaves and fruit. Because of these negative effects, alternatives to copper compounds are being sought. Alternatives have included the application of chemicals such as acibenzolar-*S*-methyl, an inducer of systemic acquired resistance and prohexadione calcium, a growth regulator to control bacterial diseases

8. Kawaguchi et al. (2014) obtained significant reduction in leaf spot and fruit spot incidence on peach trees with non-pathogenic *Xa. campestris* strains AZ98101 and AZ98106. Biondi et al. (2009) also advocated the use of a bacterial antagonist for the biological control of bacterial leaf/fruit spot of stone fruits
9. Zaccardelli et al. (1994) have shown the potential of bacteriophages for the control of the disease

REFERENCES

Anonymous. 2006a. EPPO standards PM 7/64. (1) Diagnostics *Xanthomonas arboricola* pv. *pruni*. *Bulletin OEPP/EPPO Bulletin* 36: 129–133.

Anonymous. 2006b. Distribution maps of quarantine pests for Europe [http://pqr.eppo.org/datas/XANTPR/XANTPR.pdf]

Anonymous. 2009. First report of *Xanthomonas arboricola* pv. *pruni* in the Netherlands on ornamental *Prunus laurocerasus*. *EPPO Reporting Service* 9: 2009/178 [http://archives.eppo.org/EPPOReporting/2009/Rse-0909.pdf]

Ballard, E. L., Dietzgen, R. G., Sly, L. I., Gouk, C., Horlock, C., et al. 2011. Development of a Bio-PCR protocol for the detection of *Xanthomonas arboricola* pv. *pruni*. *Plant Dis.* 95: 1109–1115.

Battilani, P., Rossi, V., and Saccardi, A. 1999. Development of *Xanthomonas arboricola* pv. *pruni* epidemics on peaches. *J. Plant Pathol.* 81: 161–171.

Bergsma-Vlami, M., Martin, W., Koenraadt, H., Teunissen, H., Pothier, J. F., et al. 2012. Molecular typing of Dutch isolates of *Xanthomonas arboricola* pv. *pruni* isolated from ornamental cherry laurel. *J. Plant Pathol.* 94: S129–S135.

Biondi, E., Dallai, D., Brunelli, A., Bazzi, C., and Stefani, E. 2009. Use of a bacterial antagonist for the biological control of bacterial leaf/fruit spot of stone fruits. *IOBC Bull.* 43: 277–281.

Boudon, S., Manceau, C., and Nottéghem, J.-L. 2005. Structure and origin of *Xanthomonas arboricola* pv. *pruni* populations causing bacterial spot of stone fruit trees in western Europe. *Phytopathology* 95:1081–1088.

Bühlmann, A., Pothier, J. F., Tomlinson, J. A., Frey, J. E., Boonham, N., et al. 2013. Genomics-informed design of loop-mediated isothermal amplification for detection of phytopathogenic *Xanthomonas arboricola* pv. *pruni* at the intraspecific level. *Plant Pathol.* 62: 475–484.

Christiano, R. S. C., Reilly, C. C., Miller, W. P., and Scherm, H. 2010. Oxytetracycline dynamics on peach leaves in relation to temperature, sunlight, and simulated rainfall. *Plant Dis.* 94: 1213–1218.

Dunegan, J. C. 1932. The bacterial spot disease of the peach and other stone fruits. Technical Bulletin US Department of Agriculture 273: 1–53.

Garita-Cambronero, J., Sena-Vélez, M., Palacio-Bielsa, A., and Cubero, J. 2014. Draft genome sequence of *Xanthomonas arboricola* pv. *pruni* strain Xap33, causal agent of bacterial spot disease on almond. *Genome Announc.* 2(3): e00440–e00514.

Hajri, A., Pothier, J. F., Fischer-Le, Saux, M., Bonneau, S., Poussier, S., et al. 2012. Type three effector gene distribution and sequence analysis provide new insights into the pathogenicity of plant pathogenic *Xanthomonas arboricola*. *Appl. Environ. Microbiol.* 78: 371–384.

Horton, D., Brannen, P., Bellinger, B., Lockwood, D., and Ritchie, D. Sr. eds. 2011. Southeastern Peach, Nectarine and Plum Pest Management and Culture Guide [http://www.ent.uga.edu/peach/PeachGuide.pdf]

Janse, J. D. 2005. *Phytobacteriology: Principles and Practice*. Cabi Publishing, Wallingford, UK, 360 pp.

Jindal, K. K., Sharma R. C., and Gupta V. K. 1989. Chemical control of bacterial leaf spot and fruit gummosis caused by *Xanthomonas campestris* pv. *pruni* in almond (*Prunus dulcis*). *Indian J. Agri. Sci.* 59: 754–755.

Kawaguchi, A., Inoue, K., and Inoue, Y. 2014. Biological control of bacterial spot on peach by nonpathogenic *Xanthomonas campestris* strains AZ98101 and AZ98106. *J. Gen. Pl. Pathol.* 80: 158–163.

Lamichhane, J. R. 2014. *Xanthomonas arboricola* diseases of stone fruit, almond, and walnut trees: Progress towards understanding and management. *Plant Dis.* 98: 1600–1610.

Lamichhane, J. R., and Varvaro, L. 2014a. *Xanthomonas arboricola* disease of hazelnut: Current status and future perspectives for its management. *Plant Pathol.* 63: 243–254.

Lamichhane, J. R., and Varvaro, L. 2014b. Comparison of methods used in the recovery of phylloplane bacteria: A case study of *Pseudomonas savastanoi* pv. *savastanoi* applied to the phylloplane of *Olea europaea* sub-species. *J. Plant Prot. Res.* 54: 22–27.

Matthee, F. N., and Daines, R. H. 1968. Effect of soil types and substrate aeration on stomatal activity, water diffusion pressure deficit, water congestion, and bacterial infection of peach and pepper foliage. *Phytopathology* 58: 1298–1301.

Morales, G., Moragrega, C., Montesinos, E., and Llorente, I. 2018. Environmental and inoculum effects on epidemiology of bacterial spot disease of stone fruits and development of a disease forecasting system. *Eur. J. Plant Pathol.* 152: 635–651.

Panić, M., Jovanović, O., Antonijević, D., and Miladinović, Z. 1998. The first appearance of bacterial plant pathogen *Xanthomonas arboricola* pv. *pruni* in Yugoslavia. *Zaštija bilja* 49: 285–294.

Pothier, J. F., Pelludat, C., Bünter, M., Genini, M., Vogelsanger, J., et al. 2010. First report of the quarantine pathogen *Xanthomonas arboricola* pv. *pruni* on apricot and plum in Switzerland. *Plant Pathol.* 59: 404.

Pothier, J. F., Smits, T. H., Blom, J., Vorhoelter, F., Goesmann, A., et al. 2011. Complete genome sequence of the stone fruit pathogen *Xanthomonas arboricola* pv. *pruni*. (Abstr.). *Phytopathology* 101: S144.

Rademaker, J. L. W., Louws, F. J., Schultz, M. H., Rossbach, U., Vauterin, L., et al. 2005. A comprehensive species to strain taxonomic framework for *Xanthomonas*. *Phytopathology* 95: 1098–111.

Ritchie, D. F. 1995. Bacterial spot. pp. 50–52 in: *Compendium of Stone Fruit Diseases*. J. M. Ogawa, E. I. Zehr, G. W. Bird, D. F. Ritchie, K. Uriu, and J. K. Uyemoto eds. APS Press, St. Paul, MN.

Ritchie, D. F. 2005. Applications of copper materials and Mycoshield for bacterial spot control on peaches, 2004. *Fungic. & Nematic. Tests* 60: STF011. Am. Phytopathol. Soc. [http://www.apsnet.org/Pages/default.aspx]

Ritchie, D. F. 2006. Copper materials, FlameOut, ProPhyt, and Serenade ASO for bacterial spot management on peaches, 2005. *Fungic. & Nematic. Tests* 61: STF007. Am. Phytopathol. Soc. [http://www.apsnet.org/Pages/default.aspx]

Ritchie, D. F. 2007. Potential bacterial spot control with copper, oxytetracycline, biologicals, and plant-defense activators, 2006. Plant Dis. Management Rep. 1: STF012. Am. Phytopathol. Soc. [http://www.apsnet.org/Pages/default.aspx]

Ritchie, D. F. 2011. Learning from peach bacterial spot epidemics: Potential strategies for reducing fruit losses. *Presented at the 2011 Great Lakes Fruit, Vegetable and Farm Market EXPO*, Grand Rapids, MI.

Roselló, M. 2007. Detección de bacterias fitopatógenas de cuarentena y caracterización de aislados relacionados, en frutales de la Comunitat Valenciana. PhD thesis, University of Valencia, Spain.

Roselló, M., Santiago, R., Cambra, M. A., García-Vidal, S., Morente, C., et al. 2007. Outbreaks of *Xanthomonas arboricola* pv. *pruni* in Spain. Diagnostic and monitoring of bacterial diseases of stone fruits and nuts. *COST873 WG1/WG2 Joint Meeting, Angers, France*: 4 [http://www.cost873.ch/5_activities/meeting_detail.php?ID=36]

Roselló, M., Santiago, R., Palacio-Bielsa, A., García-Figueres, F., Montón, C., et al. 2012. Current status of bacterial spot of stone fruits and almond caused by *Xanthomonas arboricola* pv. *pruni* in Spain. *J. Plant Pathol.* 94: S1.15–21.

Scortichini, M. 2010. Epidemiology and predisposing factors of some major bacterial diseases of stone and nut fruit trees species. *J. Plant Pathol.* 92: S173–S178.

Scortichini, M., Janse, J. D., Rossi, M. P., and Derks, J. H. J. 1996. Characterization of *Xanthomonas campestris* pv. *pruni* strains from different hosts by pathogenicity tests and analysis of whole-cell fatty acids and whole-cell proteins. *J. Phytopathol.* 144: 69–74.

Shen, Y. M., Huang, T. C., Chao, C. H., and Liu, H. L. 2013. First report of bacterial spot caused by *Xanthomonas arboricola* pv. *pruni* on Japanese plum in Taiwan. *Plant Dis.* 97: 835.

Smith, E. F. 1903. Observations on a hitherto unreported bacterial disease, the cause of which enters the plant through ordinary stomata. *Science, Washington* 17: 456–457.

Socquet-Juglard, D., Patocchi, A., Pothier, J. F., Christen, D., and Duffy, B. 2012. Evaluation of *Xanthomonas arboricola* pv. *pruni* inoculation techniques to screen for bacterial spot resistance in peach and apricot. *J. Plant Pathol.* 94: S1.91–96.

Stefani, E. 2010. Economic significance and control of bacterial spot/canker of stone fruits caused by *Xanthomonas arboricola* pv. *pruni*. *J. Plant Pathol.* 92: S1.99–103.

Vauterin, L., Hoste, B., Kersters, K., and Swings, J. 1995. Reclassification of *Xanthomonas*. *Int. J. Syst. Evol. Microbiol.* 45: 472–489.

Werner, D. J., Ritchie, D. F., Cain, D. W., and Zehr, E. I. 1986. Susceptibility of peaches and nectarines, plant introductions, and other *Prunus* species to bacterial spot. *Hort. Sci.* 21: 127–130.

Young, J. M. 1977. *Xanthomonas pruni* in almond in New Zealand. *J. Agric. Res.* 20: 105–107.

Zaccardelli, M., Ceroni, P., and Mazzucchi, U. 1999. Amplified fragment length polymorphism fingerprinting of *Xanthomonas arboricola* pv. *pruni*. *J. Plant Pathol.* 81: 173–179.

Zaccardelli, M., Saccardi, A., Gambin, E., Minardi, P., and Mazzucchi, U. 1994. *Xanthomonas campestris* pv. *pruni* bacteriophages on peach trees and their potential use for biological control. *Proceedings of VIII International Conference on Plant Pathogenic Bacteria*, Versailles, France, June 9–12, 1992, pp. 875–878.

Zehr, E. I., Shepard, D. P., and Bridges, W. C. Jr. 1996. Bacterial spot of peach as influenced by water congestion, leaf wetness duration, and temperature. *Plant Dis.* 80: 339–341.

7.56 BACTERIAL CANKER OF MANGO

The disease is also called **bacterial black spot of mango** (*Mangifera indica* L.). Doidge in 1909 reported the disease for the first time from South Africa. Since then, it has been reported from Australia, the Comoro Islands, Kenya, Japan, Malaysia, Mauritius, New California, Pakistan, the Philippines, Reunion Island, Taiwan, Thailand, and the United Arab Emirates. The disease probably occurs throughout southern and eastern Africa and in most parts of Asia. The disease also occurs in Brazil (Gagnevin and Pruvost 2001). In India, it was first reported by Patel et al. (1948) as bacterial leaf spot and is prevalent in Uttar Pradesh, Bihar, Maharashtra, Tamil Nadu, Delhi, Haryana, and Punjab. Yasuhara-Bell et al. (2013) reported the occurrence of bacterial black spot of mango for the first time from Hawai'i or anywhere in the United States. Sanahuja et al. (2016) confirmed the occurrence of bacterial canker of mango for the first time in the Americas. Zombré et al. (2017) reported bacterial canker of mango from Togo. Uribe-Lorío and Wang (2017) reported *Erwinia billingiae* as a causal agent of bacterial canker on mango trees in Costa Rica. The phylogenetic analysis of 16S rRNA and concatenated MLST genes showed the Costa Rican strain closely related to *Er. billingiae*.

The disease is a limiting factor in mango production because it is very difficult to control it, particularly on susceptible varieties. It can be very destructive in areas where high temperature and rainfall occur concomitantly. Most of the commercial varieties are highly susceptible to the pathogen, and the infection results in drastic yield losses. These losses result from premature fruit drop, reduction in fruit quality, severe defoliation of plants, and fruit rot in storage. Insect and fungal invasions through fruit cankers further aggravate the loss. The fruit infections ranging from 50% to 80% are common on highly susceptible cultivars. In 1966 and 1967, severe epiphytotics of the disease caused almost 100% fruit loss on most susceptible cultivars in most of the mango growing areas of South Africa. The fruit loss was mainly due to the fruit drop and non-marketability of the fruit. The estimated loss alone in 1966 was approximately US $1 million (Gagnevin and Pruvost 2001).

In India, the pathogen has been observed to cause 20%–50% fruit infection in cv. Langra and 10%–85% in seedling mangoes under natural conditions. In hybrids and cv. Bangalora, 37%–40% fruits showed infection, of which 20%–25% dropped. In many hybrids, even 100% fruit infection was observed. The infected fruits, including those invaded by insects and fungi rot in the orchard during ripening (Shekhawat and Patel 1972). Kishun (1982) observed 20%–80%, 10%–70%, 10%–40%, and 10%–55% fruit drop in Alphanso, Pairi, Totapuri, and local cultivars, respectively, and the maximum fruit drop in all the cultivars occurred 30–60 days after fruit set. He further reported that the fruit rot in storage was 5%–80%, 11%–67%, 10%–100%, and 5%–80% in Alphanso, Pairi, Totapuri, and local varieties, respectively.

7.56.1 Host Range

Xanthomonas citri pv. *mangiferaeindicae* have been isolated from cashew (*Anacardium occidentale* L.) in India, but these strains were lost. Non-pigmented strains were also isolated from ambarella in the French West Indies and from Brazilian pepper (*Schinus terebinthifolius* Raddi) in Reunion. The ambarella strains are not only taxonomically distinct from mango isolates, but are also much less aggressive on mango. In contrast, isolates from Brazilian pepper cannot be distinguished from mango isolates by most phenotypic and genetic analyses, but they are also less aggressive on mango compared to mango isolates. Hence, both the ambarella and Brazilian pepper strains do not pose any serious threat to mango.

Zombre et al. (2016) reported the first outbreak of a cashew bacterial disease in Burkina Faso (Western Africa) where *Xa. citri* pv. *mangiferaeindicae* recently emerged on mango. A comprehensive molecular characterization, based on multi-locus sequence analysis, supplemented with pathogenicity assays of isolates obtained during the outbreak, indicated that the causal agent on cashew in Burkina Faso was *Xa. citri* pv. *mangiferaeindicae* and not *Xa. citri* pv. *anacardii*. Pathogenicity data supported by population biology in Burkina Faso suggested a lack of host specialization. Therefore, the inoculum from each crop can be potentially harmful to both the host species. Symptoms induced on cashew leaves and fruit by *Xa. citri* pv. *mangiferaeindicae* and non-pigmented strains of *Xa. citri* pv. *anacardii* are similar, although the causative bacteria are genetically different.

7.56.2 Symptoms

The symptoms appear on the aerial parts of the mango trees. Leaf and fruit symptoms are most common, but symptoms on twigs and branches occur on highly susceptible cultivars or when the infection is very severe. On leaves, the symptoms appear as minute, water-soaked, irregular spots, which enlarge, turn brown or dark brown, become angular due to delineation by veins, and finally cankerous and raised. Individual spots are usually smaller than 0.5 cm^2. Most of the lesions are surrounded by chlorotic haloes, which are more pronounced on young leaves and may extend 5 mm beyond the necrotic areas. On old leaves, the haloes are narrow and can be best seen when the leaf is held against light. Several lesions may coalesce to form larger necrotic cankerous patches. In old lesions, the upper epidermis becomes loose, greyish white, and sometimes peels-off. Severe leaf infection may result in yellowing of the leaves and defoliation (Figure 7.39).

On fruits, the symptoms appear as small, water-soaked spots on lenticels. These spots later become star-shaped, develop into cankers, which are much more raised and darker than those on leaves, and often burst open releasing a highly contagious gummy substance containing bacterial cells. Insect larvae and fungi enter fruits through the openings of these lesions and cause rotting and fruit drop. Often, a "**tear stain**" infection pattern is observed on the fruit. Many times fruits in bunches look healthy, but their close examination reveals the presence of lesions at the points where they touch each other.

On twigs and branches, the lesions are raised having longitudinal rifts, predominantly formed on the under surface, but without any gummy ooze. Several lesions may run together forming longitudinal scars. The disease is not systemic. Bud necrosis and inflorescence cankers are rarely observed and only on highly susceptible cultivars.

FIGURE 7.39 Bacterial canker of mango: Leaf symptoms.

Susceptibility of the plant organs varies with time for a given cultivar. Young leaves are resistant, likely because they do not have functional stomata. However, they become very susceptible when they enlarge, with lesions appearing just after leaf hardening. Leaf susceptibility then decreases with time. Fruit susceptibility increases over time and is highest during the month preceding harvest. This correlates with maximum receptivity of the lenticels.

7.56.3 Causal Organism

Xanthomonas axonopodis pv. *mangiferaeindicae* (Patel et al. 1948) Ah-You et al. 2007 comb. nov.
Syn. *Xa. campestris* pv. *mangiferaeindicae*, *Xa. citri* pv. *mangiferaeindicae*

Patel et al. (1948) first described the bacterium and named it *Pseudomonas mangiferae-indicae* sp. nov. Robbs et al. (1974) reclassified the bacterium as *Xa. campestris* subsp. *mangiferae-indicae* sp. nov. With the introduction of pathovar system for the classification of plant pathogenic bacteria, it was changed to *Xa. campestris* pv. *mangiferaeindicae*. Later on, Ah-You et al. (2007) changed it to *Xa. axonopodis* pv. *mangiferaeindicae*. The name *Xa. citri* pv. *mangiferaeindicae* given by Ah-You et al. (2009) does not confirm to Standards 17 and 21.

The bacterium has all the characteristics of genus *Xanthomonas* except the production of yellow pigment. It produces non-pigmented (creamy-white) colonies on agar media. However, a few yellow-pigmented strains have been isolated from mango in Brazil, Florida, South Africa, and Reunion Island. Atypical strains, including yellow-pigmented strains collected from various countries and the non-pigmented strains isolated from mango in Brazil and from ambarella (*Spondias cytherea* Sonn.) in the French West Indies are phenotypically and genetically distinct from *Xa. citri* pv. *mangiferaeindicae*. The below given description of the bacterium is taken primarily from Manicom and Wallis (1984).

Bacterial cells are Gram-stain-negative, aerobic, non-sporulating rods (0.4–0.5 × 1.0–1.5 μm), and motile by means of a single polar flagellum. Colonies on nutrient agar or King's medium B are round, shallow, convex having entire margins. The colony color is initially smoke grey, but soon changes to creamy-white. On potato dextrose agar, the colonies are circular, smooth, glistening, pulvinate having entire margins and measure 1.0–1.5 cm in diameter after 7 days of incubation. The bacterium is strictly oxidative. The optimum temperature for growth is 27°C, but the growth can occur up to 37°C. The thermal death point is around 55°C.

The tests are positive for esculin hydrolysis, caseinase, catalase, cellulase, gelatinase, hydrogen sulphide, levan and lipase production, and starch hydrolysis, but negative for acetoin production, arginine dihydrolase activity, indole and ammonia production, 2-ketogluconate production, nitrate reduction, cytochrome oxidase, poly-β-hdroxybutyrate synthesis, urease production, growth on asparagine as sole source of C and N, soft rot of potatoes, and hypersensitive reaction on tobacco. The bacterium produces acid from glucose, sucrose, xylose, galactose, mannose, D-ribose, trehalose, cellobiose, melibiose, melezitose, esculin, dextrin, glycerol, and sorbitol when growth occurs utilizing these compounds. SXTPA, a semi-selective medium was developed by Chand and Kishun (1995) for the isolation of the bacterium. Two semi-selective media, namely, KC and NCMT3 having plating efficiencies ranging from 76% to 104% and from 78% to 132%, respectively, have also been developed for the isolation of the bacterium (Pruvost et al. 2005). In the latter case, the repeated isolations of the pathogen were possible even from asymptomatic leaves.

Most of the strains produce heat stable toxin (Kishun and Shukla 2000). Gagnevin et al. (1997) assessed the genetic diversity among 138 strains of the bacterium isolated from mango, ambarella, and pepper tree in 14 countries and found that 11 strains did not belong to *Xa. campestris* pv. *mangiferaeindicae*. They further revealed that the remaining 127 strains formed four groups, of which the group with the greatest diversity consisted of strains from southeast Asia where mango originated. Another group contained only strains isolated from pepper trees in Reunion, indicating that pepper tree may not be a collateral host for bacterial leaf spot pathogen.

Ah-You et al. (2007), on the basis of pathogenic variation and host range, divided and reclassified *Xa. campestris* pv. *mangiferaeindicae* into three pathovars, namely, *Xa. axonopodis* pv. *mangiferaeindicae* pathogenic on mango and cashew, *Xa. axonopodis* pv. *anacardii* pathogenic on cashew, and *Xa. axonopodis* pv. *spondiae* pathogenic on ambarella and mombin. These three pathovars also correspond to three genetically divergent groups of *Xa. axonopodis* identified by Rademaker on the basis of AFLP data. Ah-You et al. (2009), using amplified fragment length polymorphism, multi-locus sequence analysis, and DNA-DNA hybridization for genotypic classification of *Xanthomonas* pathovars associated with the plant family *Ancardiaceae*, identified causal agents of mango bacterial canker and cashew bacterial spot as *Xa. citri* pv. *mangiferaeindicae* and *Xa. citri* pv. *anacardii*, respectively.

7.56.4 Disease Cycle

The bacterium survives in leaf lesions and in cankers formed on the twigs and branches, but the survival is more efficient in leaf lesions. Three-month-old leaf lesions on a susceptible cultivar may contain 10^7 cfu per lesion, while the population in 18-month-old lesions was 10^5 cfu per lesion. The population of the bacterium in leaf lesions decreases relatively slowly under controlled conditions. The role of twig cankers is also important in the survival of the bacterium as these persist on the trees while the infected leaves prematurely defoliate.

The population of the bacterium on mango tree surfaces can average as high as 10^6 cfu per leaf or fruit. Surface population varies with climatic conditions, but is correlated with the number of lesions per leaf or fruit. Humid conditions favor the epiphytic growth. The presence of free water allows the release and redistribution of bacteria from the ruptured epidermis. The population quickly decreases to undectable levels during dry conditions. Populations of 10^{4-6} cfu per fruit can be found on young fruits under natural conditions and probably these constitute inoculum associated with latent infection, which will be responsible for symptoms when the fruit receptivity increases. Kishun and Chand (1993) have reported the epiphytic survival of the bacterium on 18 weeds found in and around the mango orchards.

Long distance dissemination or dissemination to the pathogen-free areas occurs through contaminated or infected propagating plant material. Seed-plant transmission of the bacterium has not been proved. Wind-driven rains and overhead-irrigation are important for local or short distance dissemination of the bacterium. In India, Kishun (1986) reported that *Myllocerus discolor* var. *variegata* (ashy weevil), *Orthaga evadrusalis* (leaf webber), and *Cantheconidia furcellata* (bug) transmitted the pathogen from the diseased to healthy plants. Later on, Kishun and Chand (1989) extended the list of such insects to 22 including the first two insects mentioned above.

The bacterium enters the host through stomata, lenticels, and wounds. Under North Indian conditions, the infection on fruits starts by the end of April, and the disease progresses considerably even during summer months. On leaves and twigs, fresh infections develop only after the rains, and the severity of the disease is maximum in July–August. During winter, the disease remains practically dormant.

7.56.5 Management

1. Use of disease-free nursery is the most effective approach for the management of the disease. Both the rootstocks and scion wood used should be free from the pathogen. However, it is very difficult to determine whether the planting material is free from the pathogen or not due to latent infection of the pathogen or presence of its epiphytic population
2. Good nursery management practices including disinfection of bud wood, using bactericides, protection of plants from stormy winds with windbreaks, destruction of sources of inoculum, and reduction of epiphytic bacterial populations by copper sprays should be followed
3. Chemical control alone is generally ineffective, but better control of the disease has been achieved by following an integrated approach that includes destroying inoculum sources by pruning, use of windbreaks and drip irrigation, and regularly spraying copper fungicides. In South Africa, more than 20 annual sprays of copper are given on susceptible cultivars. There is every likelihood that it will result in the development of pathogen population resistant to the copper. Three sprays of Streptocycline (100 μg mL^{-1}) at 15 days intervals, starting when

fruits are of pea seed size, have been tried in India and found effective in some cases. Fruit stones used for seedlings should be dipped in Streptocycline solution (100 μg mL^{-1}) for 4 h before sowing (personal communication from Ram Kishun)
4. Attempts have been made to manage the disease by the use of antagonistic bacteria (Kishun et al. 2006), where the single application of these bacteria reduced the disease
5. Use of resistant varieties is another very effective method of disease management, but unfortunately there is hardly any cultivar with complete resistance to the pathogen. Heidi and Sensation cultivars are partially resistant to the pathogen. In India, Bombay Green has been found resistant to the pathogen under both natural and artificial inoculation conditions

REFERENCES

Ah-You, N., Gagnevin, L., Chiroleu, F., Jouen, E., Neto, J. R., et al. 2007. Pathological variations within *Xanthomonas campestris* pv. *mangiferaeindicae* support its separation into three distinct pathovars that can be distinguished by amplified fragment length polymorphism. *Phytopathology* 97: 1568–1577.

Ah-You, N., Gagnevin, L., Grimont, P. A. D., Brisse, S., Nesme, X., et al. 2009. Polyphasic characterization of xanthomonads pathogenic to members of the Anacardiaceae and their relatedness to species of *Xanthomonas*. *Int. J. Syst. Evol. Microbiol.* 59: 306–318.

Chand, R., and Kishun, R. 1995. A semi selective medium for the isolation of bacterial pathogen *Xanthomonas campestris* pv. *mangiferaeindicae*. *African Plant Prot.* 1: 45–48.

Gagnevin, L., Leach, J. E., and Pruvost, O. 1997. Genomic variability of the *Xanthomonas* pathovar *mangiferaeindicae*, agent of mango bacterial black spot. *Appl. Environ. Microbiol.* 63: 246–253.

Gagnevin, L., and Pruvost, O. 2001. Epidemiology and control of mango bacterial black spot. *Plant Dis.* 85: 928–935.

Kishun, R. 1982. Loss in mango fruit due to bacterial canker *Xanthomonas mangiferaeindicae*. *Proceedings of 5th International Conference on Plant Pathogenic Bacteria*, Cali, Colombia, pp. 181–184.

Kishun, R. 1986. Role of insects in transmission and survival of *Xanthomonas campestris* pv. *mangiferaeindicae*. *Indian Phytopathol.* 39: 509–511.

Kishun, R., and Chand, R. 1989. Mechanical transmission of *Xanthomonas campestris* pv. *mangiferaeindicae* through insects. *Indian J. Plant Pathol.* 7: 112–114.

Kishun, R., and Chand, R. 1993. Epiphytic survival of *Xanthomonas campestris* pv. *mangiferaeindicae* on weeds and its role in MBCD. *Plant Dis. Res.* 9: 35–40.

Kishun, R., Mishra, D., Ram, R. A., and Verma, A. K. 2006. Management of mango bacterial canker through antagonists. *J. Eco-friendly Agric.* 1: 54–56.

Kishun, R., and Shukla, R. S. 2000. Evaluation of *Xanthomonas campestris* pv. *mangiferaeindicae* for toxins and their phytotoxicity. *Proceedings of International Conference on Integrated Plant Disease. Management for Sustainable Agriculture*, New Delhi, India, pp. 743–744.

Manicom, B. Q., and Wallis, F. M. 1984. Further characterization of *Xanthomonas campestris* pv. *mangiferaeindicae*. *Int. J. Syst. Bacteriol.* 34: 77–79.

Patel, M. K., Moniz, L., and Kulkarni, Y. S. 1948. A new bacterial disease of *Mangifera indica* L. *Curr. Sci.* 17: 189–190.

Pruvost, O., Roumagnac, P., Gaube, C., Chiroleu, F., and Gagnevin, L. 2005. New media for the semiselective isolation and enumeration of *Xanthomonas campestris* pv. *mangiferaeindicae*, the causal agent of mango bacterial black spot. *J. Appl. Microbiol.* 99: 803–815.

Robbs, C. F., Ribeiro, R. de L. D., and Kimura, O. 1974. Sobre el posicao taxonomica de *Pseudomonas mangiferaeindicae* Patel et al. 1948, agente causal da "Mancha bacteriana" das folhas da Mangira (*Mangifera indica* L.). *Arq. Univ. Fed. Rural Rio de Janeiro* 4: 11–14.

Sanahuja, G., Ploetz, R. C., Lopez, P., Konkol, J. L., Palmateer, A. J., et al. 2016. Bacterial canker of mango, *Mangifera indica*, caused by *Xanthomonas citri* pv. *mangiferaeindicae*, confirmed for the first time in the Americas. *Plant Dis.* 100: 2520.

Shekhawat, G. S., and Patel, P. N. 1972. Bacterial canker of mango. pp. 192–195 in: *Plant Bacteriology*. Vol. I, *Bacterial Diseases of Plants in India*. P. N. Patel ed. Indian Agricultural Research Institute, New Delhi, India.

Uribe-Lorío, L., and Wang, A. W. 2017. Identification of *Erwinia billingiae* as a causal agent of bacterial canker on mango (*Mangifera indica*) trees in Costa Rica (Abstr.). *Phytopathology* 107: S4.12–13.

Yasuhara-Bell, J., de Silva, A. S., Alvarez, A. M., Shimabuku, R., and Ko, M. 2013. First report in Hawai'i of *Xanthomonas citri* pv. *mangiferaeindicae* causing bacterial black spot on *Mangifera indica*. *Plant Dis.* 97: 1244.

Zombre, C., Sankara, P., Ouédraogo, S. L., Wonni, I., Boyer, K., et al. 2016. Natural infection of cashew (*Anacardium occidentale*) by *Xanthomonas citri* pv. *mangiferaeindicae* in Burkina Faso. *Plant Dis.* 100: 718–723.

Zombré, C., Wonni, I., Ouédraogo, S. L., Kpemoua, K. E., Assignon, K., et al. 2017. First report of *Xanthomonas citri* pv. *mangiferaeindicae* causing mango bacterial canker on *Mangifera indica* in Togo. *Plant Dis.* 101: 503.

7.57 BACTERIAL BLIGHT OF POMEGRANATE

Bacterial blight is a serious disease of pomegranate (*Punica granatum* L.). The disease was first reported in India from Delhi in 1952 by Hingorani and Mehta (1952), and they named it as bacterial leaf spot of pomegranate. Later on, Hingorani and Singh (1959) reported it from Bangalore (Karnataka) in 1959. The disease was of minor importance until 1991, when it appeared in severe form in an experimental plot of Indian Institute of Horticultural Research, Bangalore, resulting in 60%–80% yield losses (Chand and Kishun 1991). Presently, the disease occurs widely, and its outbreaks have been recorded in all major pomegranate-growing states of India, namely, Karnataka, Maharashtra, and Andhra Pradesh.

The bacterium was first noticed in select farms in Bellary district of Karnataka in the 1980s; it started spreading rapidly in the early 2000s and assumed epiphytotic proportions in the last couple of years. It has caused severe damage and destroyed 90% of the cultivated area in the districts of Bagalkot, Belgaum, Bellary, Bijapur, Chitradurga, Gulbarga, Koppal, Raichur, and Tumkur in Karnataka. The occurrence of the disease was also reported from western Maharashtra particularly from Sholapur district, and the disease was observed throughout the year on pomegranate trees (Dhandar et al. 2004). A field survey undertaken in the major pomegranate growing region of western Maharashtra revealed the presence of the disease in 20.0%–88.0% orchards (Raghuwanshi et al. 2013). The disease occurred in epiphytotic proportions during 2001 in Anantapur and Mahabubnagar districts of Andhra Pradesh and is still on the increase resulting in huge losses (Subramanyam 2011). Losses caused by bacterial blight were also recorded in Hanumangarh district of Rajasthan in 2009. The disease has also been reported from Jorhat region of Assam (Bora and Kataki 2014). Today, the disease has become a threat to pomegranate production in all the three major pomegranate growing states, namely, Maharashtra, Karnataka, and Andhra Pradesh. All the commercial grown cultivars in Karnataka, namely, Ganesh, Bhagwa, Mridula, and Arakata are susceptible to the disease (Sharma et al. 2011).

Besides India, the disease has been reported from Pakistan (Akhtar and Bhatti 1992) and South Africa, where it was noticed in 2007 (Petersen et al. 2010).

7.57.1 Host Range

Pomegranate is considered to be the only host of the bacterium.

7.57.2 Symptoms

The pathogen attacks all the aboveground plant parts including flowers, leaves, twigs, stem, buds, and fruits, but it is more destructive when fruits are infected. The initial symptoms appear as water-soaked translucent, irregular to circular, minute black spots on leaves. Gradually, the centers of the spots become necrotic and turn dark brown. In severe cases, spots coalesce and produce a large patch that may result in shedding of infected leaves. Dark spots develop on stem nodes, causing cracking and breaking of the branches (Figure 7.40a). Bacterial ooze is found on infected plant parts in humid weather.

Initially, spots on fruits are black, round, and surrounded by bacterial ooze. Under favorable conditions, the spots enlarge, become raised, turn to dark brown lesions having indefinite margins, and result in splitting of the fruits (Figure 7.40b). The fruits are more vulnerable to the infection than leaf, as evidenced by more disease incidence and severity on fruits, irrespective of season, location, and variety (Raju et al. 2011). The infection on fruits may cause up to 90% yield reduction. Disease build up is rapid from July to September. Severity increases during June and July and reaching the maximum in September and October and then declines.

7.57.3 Causal Organism

Xanthomonas axonopodis pv. *punicae* (Hingorani and Singh 1959) Vauterin et al. 1995

Colonies on nutrient agar, after 3 days of incubation at 27°C ± 1°C, are yellow, mucoid, circular, convex, glistening, and raised. Cells are Gram-stain-negative rods, measure 0.45 × 0.75–3.0 μm; occur singly or in chains, having a single polar flagellum. No endospores or capsule is formed. The optimum temperature for the growth of bacterium is 30°C; thermal death point is around 52°C. Optimum pH is 7.2 and pH range is 5.5–8.0. Kale et al. (2012) reported that restriction digestion of plasmids confirmed that this bacterium has a mega plasmid of size around 200 kbp typical of genus *Xanthomonas*.

Mondal and Mani (2012) reported that nanocopper at 0.2 μg mL^{-1} concentration, which is 10,000 times lower than that usually recommended for copper oxychloride, suppressed the growth of bacterium. They further reported that scanning electron microscopy revealed cell wall degradation in nanocopper-treated bacterial cells that failed to colonize plant tissues as well as to produce the characteristic intense water-soaking. *In vitro* evaluation of different botanicals and bactericides revealed the superior efficacy of bactrinashak at 0.05% (20 mm inhibition zone) followed by Streptocycline at 0.05%, and garlic extract at 5% was next best (Benagi and Ravi Kumar 2011).

Type III secretion system effectors, referred to as *Xanthomonas* outer proteins (Xops), are known to be key factors required for bacterial colonization in distinct eukaryotic cells. Kumar and Mondal (2013) generated a *xopN* null mutant of *Xa. axonopodis* pv. *punicae* (Xap *ΔxopN*). Infiltration of Xap *ΔxopN* in pomegranate leaves revealed

(a)

(b)

FIGURE 7.40 Bacterial blight of pomegranate: (a) Symptoms on aerial plant parts. (b) Symptoms on fruits. (Courtesy of V.I. Benagi.)

3-fold reduction in water-soaked areas on lesions (based on detailed image analysis on blight lesions) compared to that of with wild Xap. The *in planta* population count of Xap *ΔxopN* was reduced approximately 32-fold compared to wild strain. In addition, the Xap *ΔxopN* induced more callose deposition in infected pomegranate leaves. Mondal and Mani (2009) generated genomic fingerprints of *Xa. campestris* pv. *punicae* (recent name, *Xa. axonopodis* pv. *punicae*) using ERIC sequence-based primers and established a correlation between genotypic groups based on ERIC fingerprints and pathogenicity of the isolates. A unique 900 bp amplicon present in all the highly virulent isolates was identified that can be used as genetic marker to screen isolates for virulence.

7.57.4 Disease Cycle

The bacterium overwinters in infected stems, fruits, and fallen leaves. The penetration of the bacterium occurs through wounds and natural openings. Wind-splashed rain, irrigation water, insects, humans, and contaminated pruning tools spread the bacterium locally. Long distance dissemination of bacterium occurs through infected nursery plants and fruits. The disease build up is influenced by increased day temperature, low humidity, and rain. The increase in day temperature (38.6°C) and afternoon relative humidity of 30.4% along with cloudy weather and intermittent rainfall favored the disease initiation and further spread of the bacterium (Kumar et al. 2009). Sharma et al. (2011) recorded more disease severity due to higher infection rate (0.2/unit/day) in Mrigabahar crop regulated in June–July (summer rainy season) as compared to Hasthabahar crop regulated in September–October (autumn season), which had an infection rate of 0.08/unit/day, thereby, indicating the brisk spread of the bacterium in the rainy season. The pathogen also overwinters in infected leaves of neem and bail planted along the bunds of pomegranate fields (Kumar et al. 2009).

In recent years, wireless sensor networks (WSNs) have attracted much attention due to their immense potential applications. The technological development in WSN has made it possible to use in monitoring and control of weather and crop parameter in precision agriculture. Bacterial blight of pomegranate is an important weather related disease that can potentially cause heavy losses in climates with high rainfall and high temperature imparting long period of leaf wetness. Patil et al. (2012) developed a WSN system for use in precision agriculture applications, where real-time data of weather and crop parameters like environmental temperature, leaf wetness, etc. were sensed from the WSN deployed in pomegranate orchard and sent to the base station and stored in the database. The infection index computations were carried out based on sample data, but for any package of recommended practices as guidelines for entrepreneurs; data for 3–4 seasons are required. The farmers and scientist can assess and forecast the disease based on the value of infection index and can take necessary measures to manage the disease.

Priya et al. (2016) compared two genotypes, 318734 (resistant) and Ruby (highly susceptible) for biochemical and histological parameters to understand the defense mechanism. Significantly higher accumulation of defense related metabolites, *viz.* total phenol, flavonoid, and antioxidant contents, was observed in resistant genotype 318734. Fewer numbers of stomatal pores that served as portals of entry for plant pathogens were also recorded in this genotype. Resistance observed in genotype 318734 might be due to an incompatible interaction between host and pathogen compared to other genotypes. This is the first report of putative resistance sources in pomegranate against *Xa. axonopodis* pv. *punicae*.

7.57.5 Management

Use of disease-free planting material, removal and burning of infected plant parts, and phytosanitary cultivation techniques are the important measures that can reduce the disease severity to a great extent. A large number of chemicals have been tried by various workers with varying success. Alternate use of these chemicals will greatly help in reducing the chances of development of pathogen strains resistant to the chemicals.

Four actinomycetes, namely, *Streptomyces fradiae*, *St. avermitilis*, *St. cinnamonensis*, and *St. canus*, isolated from *Glomus mosseae* spores, improved shoot and root growth of pomegranate and increased biomass by 68%–277% in 3 months period which may be attributed to production of varying levels of growth hormones (Poovarasan et al. 2013). The authors have also hinted at the possibility of using mycorrhizae associated actinomycetes as bioinoculant for growth promotion and for controlling bacterial blight in pomegranate cultivation.

Five sprays of Bromopal (500 μg mL^{-1}) + copper oxychloride (2,000 μg mL^{-1}) followed by application of zinc sulphate + magnesium sulphate + boron + lime (1.0 g each per liter) 1 day after the spray of chemicals, or Streptocycline (500 μg mL^{-1}) + copper oxychloride (2,000 μg mL^{-1}) at 10 days intervals reduced the disease incidence to 19.08% and 19.65%, respectively, compared to 78.65% in control. Bromopal (500 μg mL^{-1}) + copper oxychloride (2,000 μg mL^{-1}) treatment gave maximum yield of 8.03 t ha^{-1} followed by 7.74 t ha^{-1} given by Streptocycline (500 μg mL^{-1}) + copper oxychloride (2,000 μg mL^{-1}) in comparison to 1.34 t ha^{-1} recorded in untreated control (Kumar et al. 2009). Benagi and Ravi Kumar (2011) reported that field applications, over the seasons, of bactrinashak, Streptocycline, or bacterimycin (each at 0.05% concentration) in combination with copper oxychloride at 0.25% were equally effective in reducing the disease severity.

In another study, eight sprays of Streptocycline (500 μg mL^{-1}) + copper oxychloride (2,000 μg mL^{-1}) and Bromopal (500 μg mL^{-1}) + copper oxychloride (2,000 μg mL^{-1}) reduced the disease incidence to 25.5% and 33.3%, respectively, compared to 78.5% in control. The maximum mean yield of 9.3 t ha^{-1} was recorded in Streptocycline (500 μg mL^{-1}) + copper oxychloride (2,000 μg mL^{-1}) followed by 8.50 t ha^{-1} in Bromopal (500 μg mL^{-1}) + copper oxychloride (2,000 μg mL^{-1}) in contrast to 2.95 t ha^{-1} in untreated check (Ravi Kumar et al. 2011). Lokesh et al. (2014) reported that five sprays of Streptocycline (300 μg mL^{-1}) + Ampiclox (500 μg mL^{-1}), at fortnightly intervals in a field trial, caused maximum reduction in disease incidence and disease severity followed by Streptocycline (500 μg mL^{-1}) + copper oxychloride (0.3%).

Maity et al. (2017) made an attempt to supplement the chemical control through improving the systemic acquired resistance of plants using nutrients such as nitrogen and salicylic acid as elicitor. The foliar application of salicylic acid at the rate of 300 μg mL^{-1} after soil application of nitrogen at 100% of the recommended dose gave the least bacterial blight infection with average disease severity of 11.58%, followed by without nitrogen application, under challenge inoculation of the pathogen. The integration of this strategy for activation of defense mechanism of plant with Streptocycline application resulted in further reduction of bacterial blight disease incidence and severity under field condition to very low levels, i.e., 3.84% and 18.57%, respectively.

Lalithya et al. (2017) found that foliar application of ethylene at 200 μg mL^{-1} resulted in significant reduction in disease incidence in leaves (from 33.01% to 4.87%), fruits (from 25.77% to 2.05%), and twigs (from 1.76% and 0.97% to 0%) compared to untreated plants. Treatment with salicylic acid and paclobutrazol also resulted in reduction of disease incidence. Ethylene treatment also increased total phenols and anthocyanin content in rind as well as in arils accompanied with enhanced fruit yield. The enhanced disease tolerance along with an increased accumulation of defense compounds like phenols and anthocyanins by ethylene, salicylic acid, and paclobutrazol treatments resulted in improving fruit yield and quality of pomegranate fruits. As the application of these compounds contributed in inducing disease resistance along with improvement in developmental and reproductive parameters of pomegranate, their use could be a viable option for cost-effective and viable pomegranate farming.

Yenjerappa et al. (2011) recorded 15.8% and 13.8% bacterial blight index in September first and second fortnightly pruned crops, respectively, compared to 60.7% blight index in November second fortnightly pruned crop. Minimum disease index of 6.59% was recorded in the crop given 6 months rest, and the highest disease index of 80.36% was recorded in the crop given 1 month rest. An integrated approach, involving various components like selection of hastabahar treatment, proper training, sanitation, use of micro nutrients, use of organics and antibiotics, along with copper compounds followed in farmers' fields for 3 years (2008–2009 to 2010–2011) over six locations brought down the disease severity from 69% to 3.0%. These control measures also increased the fruit yield from 5.2 t ha^{-1} to 10 t ha^{-1}, giving an increase of 4.8 t ha^{-1} worth Rs. 4,80,000.00 (Benagi et al. 2012).

REFERENCES

Akhtar, M. A., and Bhatti, M. H. R. 1992. Occurrence of bacterial leaf spot of pomegranate in Pakistan. *Pakistan J. Agric. Res.* 13: 95–97.

Benagi, V. I., and Ravi Kumar, M. R. 2011. Present status of pomegranate bacterial blight and its management. *Acta Hortic.* (*ISHS*) 890: 475–480.

Benagi, V. I., Ravi Kumar, M. R., and Nargund, V. B. 2012. Threat of bacterial blight on pomegranate in India- Mitigation by an integrated approach. pp. 113–116 in: *II International Symposium on the Pomegranate*. P. Melgarejo, and D. Valero eds. CIHEAM/ Universidad Miguel Hernández, Zaragoza, Spain (Options Méditerranéennes: Série A. Séminaires Méditerranéens; n. 103).

Bora, L. C., and Kataki, L. 2014. *Xanthomonas axonopodis* pv. *punicae*—A new threat to pomegranate plants in Assam. *Indian J. Hill Farm.* 27: 100–101.

Chand, R., and Kishun, R. 1991. Studies on bacterial blight (*Xanthomonas campestris* pv. *punicae*) of pomegranate. *Indian Phytopathol.* 44: 370–372.

Dhandar, D. G., Nallathambi, P., Rawal, R. D., and Sawant, D. M. 2004. Bacterial leaf and fruit spot: A new threat to pomegranate orchards in Maharashtra state. *Presented in 26th Annual Conference and Symposium ISMPP*, Goa University, Goa, India, pp. 39–40.

Hingorani, M. K., and Mehta, P. P. 1952. Bacterial leaf spot of pomegranate. *Indian Phytopathol.* 5: 55–56.

Hingorani, M. K., and Singh, N. J. 1959. *Xanthomonas punicae* sp. nov. on *Punica granatum* L. *Indian J. Agril. Sci.* 29: 45–48.

Kale, P. B., Chimote V. P., Raghuwanshi K. S., Kale A. A., Jadhav A. S., et al. 2012. Microbial, biochemical, pathogenicity and molecular characterization of *Xanthomonas axonopodis* pv. *punicae* from pomegranate. *J. Pure Appl. Microbiol.* 6: 1699–1706.

Kumar, R., and Mondal, K. K. 2013. XopN-T3SS effector modulates *in planta* growth of *Xanthomonas axonopodis* pv. *punicae* and cell-wall-associated immune response to induce bacterial blight in pomegranate. *Physiol. Mol. Plant Pathol.* 84: 36–43.

Kumar, R., Shamarao Jahagirdar, M. R., Yenjerappa, S. T., and Patil, H. B. 2009. Epidemiology and management of bacterial blight of pomegranate caused by *Xanthomonas axonopodis* pv. *punicae*. *Acta Hortic.* (*ISHS*) 818: 291–296.

Lalithya, K. A., Manjunatha, G., Raju, B., Kulkarni, M. S., and Lokesh, V. 2017. Plant growth regulators and signal molecules enhance resistance against bacterial blight disease of pomegranate. *J. Phytopathol.* 165: 727–736.

Lokesh, R., Kumaranag, K. M., Chandrashekar, N., and Khan, A. N. A. 2014. *In vivo* efficacy of some antibiotics against bacterial blight of pomegranate caused by *Xanthomonas axonopodis* pv. *punicae*. *Int. Res. J. Biol. Sci.* 3: 31–35.

Maity, A., Sharma, J., Sarkar, A., More, A. K., Pal, R. K., et al. 2017. Salicylic acid mediated multi-pronged strategy to combat bacterial blight disease (*Xanthomonas axonopodis* pv. *punicae*) in pomegranate. *Eur. J. Plant Pathol.* 1–15. doi:10.1007/s10658-017-1333-3

Mondal, K. K., and Mani, C. 2009. ERIC-PCR-generated genomic fingerprints and their relationship with pathogenic variability of *Xanthomonas campestris* pv. *punicae*, the incitant of bacterial blight of pomegranate. *Curr. Mirobiol.* 59: 616–620.

Mondal, K. K., and Mani, C. 2012. Investigation of the antibacterial properties of nanocopper against *Xanthomonas axonopodis* pv. *punicae*, the incitant of pomegranate bacterial blight. *Annal. Microbiol.* 62: 889–893.

Patil, P., Kulkarni, U., Desai, B. L., Benagi, V. I., and Naragund, V. B. 2012. Wireless sensor network for the management of bacterial blight of pomegranate. *Proceedings of Agro-Informatics and Precision Agriculture*, India, pp. 270–276.

Petersen, Y., Mansvelt, E. L., Venter, E., and Langenhoven, W. E. 2010. Detection of *Xanthomonas axonopodis* pv. *punicae* causing bacterial blight on pomegranate in South Africa. *Australasian Plant Pathology* 39: 544–546.

Poovarasan, S., Mohandas, S., Paneerselvam, P., Saritha, B., and Ajay, K. M. 2013. Mycorrhizae colonizing actinomycetes promote plant growth and control bacterial blight disease of pomegranate (*Punica granatum* L. cv. Bhagwa). *Crop Prot.* 53: 175–181.

Priya, B. T., Murthy, B. N. S., Gopalakrishnan, C., Artal, R. B., and Jagannath, S. 2016. Identification of new resistant sources for bacterial blight in pomegranate. *Eur. J. Plant Pathol.* 146: 609–624.

Raghuwanshi, K. S., Hujare, B. A., Chimote, V. P., and Borkar, S. G. 2013. Characterization of *Xanthomonas axonopodis* pv. *punicae* isolates from western Maharashtra and their sensitivity to chemical treatments. *Bioscan* 8: 845–850.

Raju, J., Benagi, V. I., Naragund, V. B., and Ashtaputre, S. A. 2011. Survey for the incidence and severity of bacterial blight in pomegranate caused by *Xanthomonas axonopodis* pv. *punicae*. *Karnataka J. Agric. Sci.* 24: 570–572.

Ravi Kumar, M. R., Wali, S. Y., Benagi, V. I., Patil, H. B., and Patil, S. S. 2011. Management of bacterial blight of pomegranate through chemicals/antibiotics. *Acta Hortic.* (*ISHS*) 890: 481–484.

Sharma, K. K., Jadhav, V. T., and Sharma, J. 2011. Present status of pomegranate bacterial blight caused by *Xanthomonas axonopodis* pv. *punicae* and its management. *Acta Hortic.* (*ISHS*) 890: 513–522.

Subramanyam, K. 2011. Pomegranate bacterial blight disease situation in Andhra Pradesh. *Acta Hortic.* (*ISHS*) 890: 523–527.

Vauterin, L., Hoste, B., Kersters, K., and Swings, J. 1995. Reclassification of *Xanthomonas*. *Int. J. Syst. Bacteriol.* 45: 472–489.

Yenjerappa, S. T., Nargund, V. B., and Jawada, R. S. 2011. Management of bacterial blight of pomegranate through cultural methods. *Acta Hortic.* (*ISHS*) 890: 485–490.

7.58 PIERCE'S DISEASE OF GRAPEVINE

Pierce's disease (hereafter referred to as PD), named after N.B. Pierce, is a lethal disease of grapevines (*Vitis vinifera* L.). It was first observed in 1884 around Anaheim when a mysterious epiphytotic affected vines in southern California. Earlier, the disease was believed to be caused by a virus. In 1973, a xylem-limited rickettsia-like bacterium was implicated with the disease on the basis of an electron-microscopic examination of the infected plant tissue. Later on in 1978, Davis and associates isolated the bacterium and proved its pathogenicity.

The disease occurs in the United States, Mexico, Argentina, Brazil, Paraguay, Costa Rica, Slovenia, and Venezuela. In North America, PD is mainly confined to the areas having mild winters. The disease is less prevalent where winter temperatures are cold. It is present in southern United States of Florida, Texas, and California, and in the eastern it extends up to Virginia. In Florida and other southeastern states, the disease has prohibited the commercial cultivation of European grape varieties. With the introduction and establishment of glassy-winged sharpshooter in California after 1990, status of the disease changed from minor to severe with enormous increase in losses (Anonymous 2000). Su et al. (2013) reported Pierce's disease of grapevines caused by *Xylella fastidiosa* for the first time from Taiwan, also confirming it as a first report from the Asian continent.

Chatterjee et al. (2008), in their review article, have highlighted the progress made in understanding the process by which *Xyl. fastidiosa* spreads within the xylem vessels of susceptible plants as well as the traits that contribute to its acquisition and transmission by sharpshooter vectors. Its complex lifestyle as both a plant and insect colonist involves traits that are in conflict with these stages, thus apparently necessitating the use of a gene regulatory scheme that allows cells expressing different traits to co-occur in the plant.

Janse and Obradovic (2010), in their mini review, have given an elaborate account of host species attacked by *Xyl. fastidiosa* and its sub species. The authors also concluded that *Xyl. fastidiosa* is a real and emerging threat for Europe, not only for *Vitis* and *Citrus,* but also for stone fruits (almond, peach, and plum) and oleander. It is difficult to prevent it from entering and difficult to control once established, hence, deserving more attention than up until now. Resistance in European grapes is scarce or even absent. Vector control in the USA has not proved to be very effective. Cultural practices to keep plants in optimum condition are of importance, but not sufficient and the use of avirulent strains for cross-protection is still in its infancy.

7.58.1 Host Range

Strains of *Xyl. fastidiosa* subsp. *fastidiosa* causing PD have a very wide host range, and it has been further expanded by including the hosts showing positive insect transmission of the bacterium. Hopkins (1989) stated that many grasses, legumes, and perennial trees belonging to at least 28 families are infected by the bacterium, but most of them do not show any symptom. Hill and Purcell (1997) found that 94 out of 155 tested species were the host of the bacterium. Wistrom and Purcell (2005), on the basis of greenhouse studies, showed that out of 29 plant species tested, 27 were the host of the bacterium. Common sunflower, cocklebur, annual bur-sage, morning glory, horseweed, sacred datura, poison hemlock, and faba bean were the most frequently infected hosts. Purcell (2006) has compiled a comprehensive list of host species of PD strains.

There are other strains of the bacterium than PD strain. These strains cause different diseases in different hosts in different regions. These diseases include phony peach disease, alfalfa dwarf disease, citrus variegated chlorosis, plum leaf scald, and leaf scorch diseases of coffee, oleander, oak, elm, maple, mulberry, plum, and sycamore. Some of these strains also cross infect hosts.

7.58.2 Symptoms

The symptoms of PD observed in summer and fall are more reliable for identification than spring symptoms. The vines infected in spring show first symptom of water stress in mid-summer. The margins of leaves become slightly yellow or red, in white and red varieties, respectively. As the infection progresses, leaf margins dry and turn brown in concentric zones (Figure 7.41). Dried leaf laminas fall, leaving behind dried petioles having burnt-appearing tips, called **matchsticks**, attached to the vines, even after normal leaf fall. The wood of infected canes matures irregularly, forming patches of immature green bark surrounded by mature brown bark and vice versa. Grape clusters on infected vines stop growing, shrivel,

FIGURE 7.41 Pierce's disease of grapevine. (Courtesy of California Department of Food and Agriculture, PD Control Program, www.pdgwss.net.)

and dry up. Leaf symptoms may vary with different grape cultivars. In Chenin Blanc and Thompson Seedless, discoloration and scorching of leaves may appear in sectors instead of along the margins (Anonymous 2006).

Infection gradually spreads from the point of infection to the upper side and more slowly toward the base of the vines. The die-back of shoots and roots occurs. Climatic differences of different regions can influence the severity and timing of the symptoms, but not the type. In hot climates, the disease progresses rapidly even under adequate soil moisture conditions due to higher moisture stress. The shoots of infected vines grow slowly and vines deteriorate rapidly.

Infected plants show a delayed and stunted growth in the next spring season. Some canes and spurs may fail to bud out. Newly developed leaves show chlorotic areas between the veins. Late in the season, leaves become scorched, and the scorching occurs first on the oldest leaves. Yellow to brown streaks are present in longitudinal and cross sections of the current season wood of the infected canes. The gum formation in vessels and other cells, and formation of tyloses also occur in this wood (Agrios 2004).

Young vines are more susceptible to disease due to fast movement of the bacteria than the old vines. The length of time for which an infected vine can survive depends on climate, age, and variety of vines. One-year-old wine of a susceptible variety like Pinot Noir can die the same year after infection, while chronically infected 10-year-old vines of Chenin Blanc or Ruby Cabernet can live for more than 5 years. Krell et al. (2006) reported that diagnosis of the disease based on foliar symptoms was the only consistent indicator of infection in *Xyl. fastidiosa*-positive grapevines and samples from the basal portion of a cane increased the probability of its detection. Schaad et al. (2002) has highlighted the use of a real-time polymerase chain reaction, requiring only 1 h, for the on-site diagnosis of PD from asymptomatic vines.

A rapid isothermal assay for specific detection of *Xyl. fastidiosa* using the advanced recombinase-polymerase amplification technology has been developed by Agdia. The assay performs both as a real-time and an endpoint test from a single reaction tube at 39°C for 20 min. Reaction template is simply prepared by soaking 50 mg of petiole cross-sections in 0.5 mL AMP1 extraction buffer for 10 min. The assay reacts to over 27 isolates from a dozen hosts including grapevine and citrus, while consistently detecting 22 copies of spiked *Xyl. fastidiosa* genome in petiole extract (1:10, W/V). No reaction background was observed in host tissue such as grapevine, citrus, olive, almond, coffee, blueberry, and blackberry. No cross-reaction was observed to *Escherichia coli*, *Xanthomonas*, *Erwinia*, and *Pseudomonas*. This assay provides users a fast and reliable tool to monitor the regional and international movements of *Xyl. fastidiosa* (Li et al. 2017).

7.58.3 Causal Organism

Xylella fastidiosa subsp. *fastidiosa* Wells et al. 1987
Syn. *Xyl. fastidiosa*

In 1973, the pathogen reported to be associated with this disease, was referred to as "**rickettsia-like**" because of its similarities in morphology and ultrastructure with members of the *Rickettsiaceae*. Wells et al. (1987) proposed a new genus *Xylella* and included all the strains of fastidious, Gram-stain-negative, xylem-limited bacteria under a single species called *Xyl. fastidiosa*. The species epithet *fastidiosa* refers to nutritional fastidiousness, particularly for primary isolation of the bacterium. It does not grow on most common bacteriological media. However, it can be cultured on specialized media like buffered cysteine-yeast extract (BCYE) agar containing charcoal. The Periwinkle wilt (PW) medium has gained the widest acceptance, and supports the growth of all known strains of *Xyl. fastidiosa*. Almeida et al. (2004) has proposed a simpler solid medium for supporting the growth of PD strains of *Xyl. fastidiosa*. The following description of the bacterium is taken mainly from Wells et al. (1987).

Cells are Gram-stain-negative, non-flagellate, motile via type IV pili-mediated twitching, straight rods measuring 0.25–0.35 × 0.9–3.5 μm. Cells are predominantly single, but sometimes form filamentous strands under certain conditions. It is an obligate aerobe, inhibited by 2.5% carbon dioxide. It is non-pigmented, non-halophilic, Kovac's oxidase negative and catalase positive. The bacterium hydrolyzes gelatin, utilizes hippurate, and most strains produce beta-lactamase. It does not ferment D-glucose, does not produce indole and H_2S, and is also negative for beta-galactosidase, lipase, amylase, coagulase, and phosphatase tests. Optimum temperature for growth is 26°C–28°C and optimum pH is 6.5–6.9. The G + C content of the DNA is 51.0–52.4 mol%.

Although Wells et al. (1987) recognized only one species to include all the strains, there are definitely different pathotypes, most of which have not been characterized and compared with each other (Hopkins and Purcell 2002). Schaad et al. (2004) proposed three new sub species, namely, *Xyl. fastidiosa* subsp. *piercei*, *Xyl. fastidiosa* subsp. *multiplex*, and *Xyl. fastidiosa* subsp. *pauca*. *Xyl. fastidiosa* subsp. *pauca* has not been validated due to the lack of deposition of a culture of the type strain. As the type strain designated for subsp. *piercei* is the type strain of the species, the name *Xyl. fastidiosa* subsp. *fastidiosa* takes priority over *Xyl. fastidiosa* subsp. *piercei*. The study of isolates on the basis of which Schaad et al. (2004) proposed three new sub species of *Xyl. fastidiosa* included: strains from grape, alfalfa, almond, and maple classified as *Xyl. fastidiosa* subsp. *piercei*, those from peach, plum, elm, pigeon grape, sycamore, and almond as *Xyl. fastidiosa* subsp. *multiplex,* and those only from citrus as *Xyl. fastidiosa* subsp. *pauca*.

Van Sluys et al. (2003) reported the genome sequence of *Xyl. fastidiosa* (Temecula strain), isolated from a naturally infected grapevine with PD in a wine-grape-growing region of California. Comparative analyses with a previously sequenced *Xyl. fastidiosa* strain 9a5c causing CVC (Simpson et al. 2000) revealed that 98% of the PD *Xyl. fastidiosa* Temecula genes are shared with those of CVC *Xyl. fastidiosa*. Genomic differences are limited to phage-associated chromosomal rearrangements and deletions that also account for the strain-specific genes present in each genome. Genomic islands, one in each genome, were identified, and their presence in other *Xyl. fastidiosa* strains was analyzed. They concluded that these two organisms have identical metabolic functions and are likely to use a common set of genes in plant colonization and pathogenesis, permitting convergence of functional genomic strategies. Comparative analyses of the complete sequences and annotations of PD *Xyl. fastidiosa* and a *Xyl. fastidiosa* representative of the CVC group isolated from Brazil revealed that these strains exhibit remarkably limited genomic variability and share 95.7% amino acid identity in equivalent regions. There are only three genomic rearrangements, two identified genomic islands, 41 PD *Xyl. fastidiosa*, and 152 CVC *Xyl. fastidiosa* strain-specific genes, and some genes harboring frame shifts.

Kandel (2015) described natural competence (NC), a mode of horizontal gene transfer in which bacterial cells take up free DNA from the environment and recombine it, in *Xyl. fastidiosa*. To more closely approximate in planta conditions, the author used a microfluidic chamber setting (MC) that resembles the plant xylem vessels that *Xyl. fastidiosa* naturally inhabits and culture medium supplemented with different concentrations of grapevine sap to create a natural environment mimic system (NEMS). Using combinations of live and dead antibiotic marker-tagged *Xyl. fastidiosa* mutant cells, it was demonstrated that NC occurs in the NEMS and the recombination frequency is comparable to that obtained from experiments conducted on agar plates and in MC containing only artificial media. These results suggest that NC can occur in nature and might be responsible for the recently described host and geographical shifts in *Xyl. fastidiosa*.

7.58.4 Disease Cycle

After penetration into grapevines, the bacteria multiply for several months in water-conducting system of plants and spread through the xylem system. The multiplication of the bacteria causes blocking of xylem vessels, and this blockade is responsible for most of the external symptoms of the disease. Roper et al. (2007), on the basis of presence of genes in the genome of *Xyl. fastidiosa*, hypothesized that several cell wall degrading enzymes like beta-1,4 endoglucanases, xylanases, xylosidases, and polygalacturonase (PG) produced by *Xyl. fastidiosa* breakdown pit membranes. They further demonstrated that among them, PG encoded by *pglA* gene is a critical virulence factor and is required by *Xyl. fastidiosa* for successful infection of grapevines because a mutant in *pglA* lost the pathogenicity.

Wallis and Chen (2012) examined induction of phenolic compounds in grapevines of cv. Thompson Seedless infected with *Xyl. fastidiosa* over a 6-month period. Two months post-inoculation with the bacterium, catechin, digalloylquinic acid, and astringin were found at greater levels in xylem sap; multiple catechins, procyanidins, and stilbenoids were found at greater levels in xylem tissues; and precursors to lignin and condensed tannins were found at greater levels in xylem cell walls. However, such large-scale inductions of phenolic compounds were not observed 4 months after inoculation. After 6 months of inoculation, infected plants showed significantly reduced phenolic levels in xylem sap and tissues, including lowered levels of lignin and condensed tannins compared to control plants. At 6 months, infected plants showed severe disease symptoms and most photosynthetic tissue was abscised. It is evident, even though grapevine plants may initially respond to *Xyl. fastidiosa* infection with increased production of phenolic compounds, ultimately, the bacterium renders the grapevine plants to enter a state of decline, whereby diseased plants no longer have the ability to support secondary metabolite production, including defense-associated phenolic compounds.

Burbank and Stenger (2017) characterized the role of DinJ/RelE toxin-antitoxin system in controlling bacterial proliferation and population size during plant colonization. The ratio of antitoxin/toxin expressed is dependent on bacterial growth conditions, with lower amounts of antitoxin present under conditions designed to mimic grapevine xylem sap. A knockout mutant of DinJ/RelE exhibits a hypervirulent phenotype, with higher bacterial populations and increased symptom development and plant decline. Probably DinJ/RelE acts to prevent excessive population growth, thereby, contributing to the ability of the pathogen to spread systemically without completely blocking the xylem vessels and increasing probability of acquisition by the insect vector.

Shi et al. (2013) reported that *Xyl. fastidiosa* when grown in xylem fluid from the susceptible species, the bacterial cells formed a heavier biofilm compared to those in xylem fluid from the resistant species. Compared with xylem fluid of *V. smalliana* (resistant species), xylem fluid of *V. vinifera* (susceptible species) stimulated the expression of *Xyl. fastidiosa* genes *rpfC*, *gacA*, *xrvA*, *gcvR*, and *cysB*, involved in virulence regulation, and genes *pilI*, *pilU*, *pilE*, and *pilG* involved in biogenesis of pili and twitching motility. Increased expression of virulence genes likely contributes to the expression of PD symptoms in the susceptible grapevines, whereas reduced expression of these genes may lead to limitation of symptoms in resistant grapevines. Shi and Lin (2018) elucidate the roles of *pilG* in pathogenicity of *Xyl. fastidiosa* by generating the *pilG*-deletion mutant

XfΔpilG and complemented strain *XfΔpilG*-C. While all the strains showed similar growth curves *in vitro*, *XfΔpliG* showed significant reduction in cell-matrix adherence and biofilm production compared to wild-type *Xyl. fastidiosa* and *XfΔpilG*-C. The genes *pilE*, *pilU*, *pilT*, and *pilS* were down-regulated in *XfΔpliG* compared to its complemented strain and wild-type *Xyl. fastidiosa*. No symptoms of Pierce's disease appeared in grapevines inoculated with *XfΔpilG,* while the grapevines inoculated with the wild-type *Xyl. fastidiosa* and complemented strain *XfΔpilG*-C produced typical disease symptoms. The results indicate that *pilG* has a role in *Xyl. fastidiosa* virulence in grapevines.

The multiplication and movement of the bacterium varies in different plant species. In *V. vinifera* and wild grapes, the bacterium multiplies and reaches a concentration of 10 million to 1 billion bacterial cells per gram of host tissue, and it spreads systemically inside the host. In blackberry, the infection is systemic, but the multiplication (1/100–1/1000 times less) and movement of the bacterium is slower than in grapevines. In mugwort, the bacterium multiplies without causing any systemic infection.

Baccari and Lindow (2011) examined the movement and multiplication of a green fluorescent protein-marked strain of *Xyl. fastidiosa* in the stems and petioles of Cabernet Sauvignon, Chenin Blanc, Roucaneuf, and Tampa grape cultivars that differ in their susceptibility to the pathogen. The bacterium attained very low population sizes and colonized fewer xylem vessels in the stem of resistant cultivars compared with more susceptible cultivars. In contrast, *Xyl. fastidiosa* achieved similarly high population sizes and colonized a similar proportion of the vessels in petioles of susceptible and resistant cultivars, suggesting that, compared with the stem, *Xyl. fastidiosa* is relatively unrestricted in its movement and growth within the petiole. The high population sizes of *Xyl. fastidiosa* in stems of susceptible genotypes were associated with both a high number of infected vessels and a much higher extent of colonization of those vessels that become infected than in more resistant cultivars. The formation of large cellular aggregates in vessels is not required for *Xyl. fastidiosa* to move laterally in the stem to adjacent vessels, because most vessels harbored only small assemblages, particularly in resistant cultivars such as Roucaneuf, in which some intervessel movement was detected. Resistance to Pierce's disease is apparently not due to inhibitory compounds that circulate in the xylem because they might be expected to operate similarly in all tissues.

The bacterium is not seed-borne. The pathogen is transmitted by grafting and insects. The sharpshooters and spittlebugs belonging to sub family Cicadellinae and family Cercopidae, respectively, are the insect vectors of the pathogen. These insects acquire the bacteria from the xylem vessels of the infected vines and transmit them to the healthy plants. Four species of sharpshooters, namely, blue-green (*Graphocephala atropunctata*), green (*Draeculacephala minerva*), red-headed (*Carneocephala fulgida*), and glassy-winged sharpshooter (*Homalodisca vitripennis*, earlier name *H. coagulata*) are the main vectors of *Xyl. fastidiosa* subsp. *fastidiosa* in North America. In southeast Europe, *Cicadella viridis* and *Philaenus spumarius* (meadow spittlebug) are the main vectors. As glassy-winged sharpshooter moves faster and flies longer distances in grapevine yards than the other species of sharpshooters, it is a serious threat to vineyards. It was found in high numbers in 1998–1999 in Kern County vineyards of California, and in 2001, hundreds of vines were found infected with Pierce's disease.

Some vectors are highly efficient in transmitting the Pierce's disease bacterium as they require less than 100 bacteria per insect for transmission and transmit it immediately after acquiring it from the host. An infectious blue-green sharpshooter has more than 90% chances of transmission. An infectious adult is capable of transmitting the bacterium throughout its life. However, the infectious nymphs transmit the bacterium until they molt, and they have to reacquire the bacteria after every molt. *Xyl. fastidiosa* is transmitted from the foregut and foregut lining is shed during molting. Xylem feeding insects require succulent plant tissue for feeding. As grapevines are pruned every year and produce a succulent new growth, they serve as good feeding host (Varela, L.G.). Grape leafhoppers are not vectors as they feed on mesophyll cells and phloem tissue.

Xyl. fastidiosa is the only known arthropod-transmitted bacterial plant pathogen that does not circulate in the vector's hemolymph. To determine how a sharpshooter vector inoculates bacteria from foregut acquisition sites, Backus and Morgan (2011) used confocal laser-scanning microscopy to identify locations in non-dissected, anterior foreguts of the *H. vitripennis* colonized by green fluorescent protein-expressing *Xyl. fastidiosa* and daily examined spatial and temporal distributions of colonizing *Xyl. fastidiosa* over acquisition access periods of 1–6 days for both contaminated field collected and clean laboratory reared *H. vitripennis*. It was found that established populations of *Xyl. fastidiosa* can disappear from vector foreguts over time. Taking into account the existing knowledge on behavior, physiology, and functional anatomy of sharpshooter feeding, the present results support the idea that the disappearance is caused by outward fluid flow (egestion) and not inward flow (ingestion, i.e., swallowing). Therefore, the egestion is a critical part of the *Xyl. fastidiosa* inoculation mechanism. Moreover, the results suggest a cyclical, spatiotemporal pattern of microbial colonization, disappearance, and recolonization in the precibarium. Colonization patterns also support two types of egestion, termed rinsing and discharging egestion herein. The comparison of acquisition results for field collected versus laboratory reared sharpshooters suggest that there may be competitive binding for optimum acquisition sites in the foregut. Hence, the successful inoculation of *Xyl. fastidiosa* may depend, in large part, on vector load in the precibarium.

Backus et al. (2015) traced green fluorescent protein-expressing *Xyl. fastidiosa* or fluorescent nanoparticles acquired from artificial diets by glassy-winged sharpshooters, *H. vitripennis*, as they were egested into simultaneously secreted saliva. *Xyl. fastidiosa* or nanoparticles were shown to mix with gelling saliva to form fluorescent deposits and salivary sheaths on artificial diets, providing the first direct and conclusive evidence of egestion by any hemipteran insect. The present findings strongly support an egestion-salivation mechanism of *Xyl. fastidiosa* inoculation and also that a column of fluid is transiently held in the foregut without being swallowed. The findings also support, though do not definitively prove, that bacteria were suspended in the column of fluid during the vector's transit from diet to diet, and were egested with the held fluid. Therefore, the authors hypothesized that sharpshooters could be true "flying syringes," especially when inoculation occurs very soon after uptake of bacteria, suggesting the new paradigm of a non-persistent *Xyl. fastidiosa* transmission mechanism.

Cornara et al. (2016) studied the transmission of *Xyl. fastidiosa* to grapevines by spittlebugs, *Philaenus spumarius* (Hemiptera, Cercopoidea). It was found that transmission efficiency of *Xyl. fastidiosa* by *P. spumarius* increased with plant access time, similarly to insect vectors in another family (Hemiptera, Cicadellidae). Moreover, a positive correlation between pathogen populations in *P. spumarius* and transmission to plants was observed. However, bacterial populations in *P. spumarius* were one to two orders of magnitude lower than those observed in leafhopper vectors, and population size peaked within 3 days of plant access period. The results indicate that *P. spumarius* has either a limited number of sites in the foregut that may be colonized, or that fluid dynamics in the mouthparts of these insects is different from that in leafhoppers. It seems that *Xyl. fastidiosa* transmission by spittlebugs is similar to that by leafhoppers.

Some vines may recover from the disease in the first winter following infection, but the recovery depends on the variety and time of infection. The vines in which the bacteria have survived the first winter, i.e., remained infected for over a year, their chances of recovery are remote. Infections that occur until June have the highest chances to continue in the next year. However, late infections, i.e., after June by blue-green, green, and red-headed sharpshooters are less likely to persist in the following season. However, the late infections caused by glassy-winged sharpshooters survive better because the insect feeds on leaves near the base of cane and also on 2-year-old dormant wood, and these infected parts are not removed by pruning.

Francis et al. (2008) reported *Nicotiana tabacum* cv. SR1 (Petite Havana) as an excellent bioassay host to detect the bacterium from infected plants and to investigate the host pathogen interactions. It was a better bioassay host than *N. tabacum* cvs. Havana, RP1, and TNN reported previously.

Rootstock species and hybrids vary greatly in susceptibility to the pathogen. Many rootstock species are resistant to the pathogen, but a resistant rootstock does not impart resistance to a susceptible grape variety grafted on it.

Lieth et al. (2011) developed a mathematical model for cold curing of grapevines inoculated with *Xyl. fastidiosa* and calibrated the model with cold-curing data collected in a field study. Parameter estimation resulted in lowest sum of squared differences across all ten trials to be low temperature below which the organism is killed (T_0) = 6°C, number of hours to achieve 100% cure (N_{100}) = 195 h, number of hours to achieve 10% cure (N_{10}) = 20 h, and killing index (K_x) = 0.45 for cv. Pinot Noir and T_0 = 6°C, N_{100} = 302 h, N_{10} = 170 h, and K_x = 0.41 for cv. Cabernet Sauvignon. After the optimization of parameter estimates by model calibration, the simulation model was effective at predicting cold curing in four locations during the experiment, although there were some differences between Hopland location for Pinot Noir and Davis location for Cabernet Sauvignon. Using historical temperature data, the model accurately predicted the known severity of Pierce's disease in other grape-growing regions of California, suggesting that it may have utility in assessing the relative risk of developing Pierce's disease in proposed new vineyard sites.

7.58.5 Management

The management of the disease is difficult due to the systemic nature of the pathogen and wide habitats of the vectoring insects. However, the following measures will help in reducing the disease incidence.

1. Insecticidal control of insect vectors in habitats adjacent to the vineyards helps in reducing the disease incidence, but it is not viable in highly susceptible cultivars or in vineyards having less than 3-year-old vines
2. Plant less susceptible or tolerant cultivars in vineyards that are near a known or suspected hot spot of the disease
3. Use systemic insecticides to control insect vectors on vines and also on a 50–100 feet wide strip of natural vegetation along the vineyard edges. Tubajika et al. (2007) obtained 88%, 78%, and 78% disease control with the application of kaolin, harpin, and imidacloprid insecticides, respectively, in Kern county of California
4. Insect vectors breeding on irrigated grasses should be controlled by weed control rather than insecticide control
5. Remove and destroy chronically infected vines, i.e., vines showing disease symptoms for more than 1 year
6. Also remove vines showing extensive foliar symptoms and severe die-back even if it is their first year of infection
7. Early disease detection and vine removal is recommended in vineyards that experience influxes of glassy-winged sharpshooters
8. Hot water treatment of cuttings and propagation buds at 45°C for 3 h
9. *Xyl. fastidiosa* subsp. *fastidiosa* is a quarantined pest (EPPO A1) in many countries. Enforcement of quarantine regulations will prevent the entry of the pathogen in disease-free areas
10. Tetracycline is effective in reducing the disease incidence, but its efficacy and viability needs to be tested at the field level

11. Meyer and Kirkpatrick (2011) reported that foliar and drench applications of the plant growth regulator, abscisic acid to 1-year-old potted *V. vinifera* "Pinot Noir" and "Cabernet Sauvignon" vines infected with *Xyl. fastidiosa* showed some effectiveness in cure of affected grapevines. Pinot Noir grapevines treated with drench applications of abscisic acid showed significantly greater disease cure than the unsprayed control plants. As plant phenolics have antimicrobial properties, a positive correlation between effective abscisic acid treatments and the total phenolic compound content of xylem sap extracted from Pinot Noir vines was established
12. Hopkins (2005) reported that biological control by inoculation of susceptible grapevines with benign strains of *Xyl. fastidiosa*, especially strain EB92-1, appears to have the potential to control PD in commercial vineyards in Florida and other areas where the disease occurs
13. Fletcher and Wayadande (2002) have suggested a novel approach of paratransgenesis to control plant diseases caused by *Xyl. fastidiosa*. In this process, non-*Xylella* enteric bacteria, i.e., normally gut bacteria of leafhoppers will be transformed to secrete toxin(s) lethal to *Xyl. fastidiosa* subsp. *fastidiosa*. These transformed bacteria will pass from insect to insect via normal copraphagus activity (ingestion of fecal matter), and these paratransgenic insects will harm the *Xyl. fastidiosa* subsp. *fastidiosa* cells acquired by them

REFERENCES

Agrios, G. N. 2004. *Plant Pathology*. 5th ed. Elsevier Academic Press, San Diego, CA, 922 pp.

Almeida, R. P. P., Mann, R., and Purcell, A. H. 2004. *Xylella fastidiosa* cultivation on a minimal solid defined medium. *Curr. Microbiol.* 48: 368–372.

Anonymous. 2000. Pierce's disease. Hortguard threat data sheet. Department of Agriculture, Government of Western Australia, p. 10.

Anonymous. 2006. Grape Pierce's disease. IPM Programme, University of California [http://www.ipm.ucdavis.edu/PMG/r302101211.html]

Baccari, C., and Lindow, S. E. 2011. Assessment of the process of movement of *Xylella fastidiosa* within susceptible and resistant grape cultivars. *Phytopathology* 101: 77–84.

Backus, E. A., and Morgan, D. J. W. 2011. Spatiotemporal colonization of *Xylella fastidiosa* in its vector supports the role of egestion in the inoculation mechanism of foregut-borne plant pathogens. *Phytopathology* 101: 912–922.

Backus, E. A., Shugart, H. J., Rogers, E. E., Morgan, J. K., and Shatters, R. 2015. Direct evidence of egestion and salivation of *Xylella fastidiosa* suggests sharpshooters can be "flying syringes." *Phytopathology* 105: 608–620.

Burbank, L. P., and Stenger, D. C. 2017. The DinJ/RelE toxin-antitoxin system suppresses bacterial proliferation and virulence of *Xylella fastidiosa* in grapevine. *Phytopathology* 107: 388–94.

Chatterjee, S., Almeida, R. P. P., and Lindow, S. 2008. Living in two worlds: The plant and insect lifestyles of *Xylella fastidiosa*. *Annu. Rev. Phytopathol.* 46: 243–271.

Cornara, D., Sicard, A., Zeilinger, A. R., Porcelli, F., Purcell, A. H., et al. 2016. Transmission of *Xylella fastidiosa* to grapevine by the meadow spittlebug. *Phytopathology* 106: 1285–1290.

Fletcher, J., and Wayadande, A. 2002. Fastidious vascular-colonizing bacteria. APS Education Center [http://www.apsnet.org/education/introplantpath/pathogengroups/fastidious]

Francis, M., Civerolo, E. L., and Bruening, G. 2008. Improved bioassay of *Xylella fastidiosa* using *Nicotiana tabacum* cultivar SR1. *Plant Dis.* 92: 14–20.

Hill, B. L., and Purcell, A. H. 1997. Populations of *Xylella fastidiosa* in plants required for transmission by an efficient vector. *Phytopathology* 87: 1197–1201.

Hopkins, D. L. 1989. *Xylella fastidiosa*: Xylem-limited bacterial pathogen of plants. *Annu. Rev. Phytopathol.* 27: 271–290.

Hopkins, D. L. 2005. Biological control of Pierce's disease in the vineyard with strains of *Xylella fastidiosa* benign to grapevine. *Plant Dis.* 89: 1348–1352.

Hopkins, D. L., and Purcell, A. H. 2002. *Xylella fastidiosa*: Cause of Pierce's disease of grapevine and other emergent diseases. *Plant Dis.* 86: 1056–1066.

Janse, J. D., and Obradovic, A. 2010. *Xylella fastidiosa*: Its biology, diagnosis, control and risks. *J. Plant Pathol.* 92: S1.35–48.

Kandel, P. P. 2015. Genetic recombination of *Xylella fastidiosa* cultured in microfluidic chambers with grapevine sap. *Phytopathology* 105: S2.6.

Krell, R. K., Perring, T. M., Farrar, C. A., Park, Y.-L., and Gispert, C. 2006. Intraplant sampling of grapevines for pierce's disease diagnosis. *Plant Dis.* 90: 351–357.

Li, R., Russell, P., Zhang, S., Davenport, B., and Eads, A. 2017. Rapid isothermal detection of *Xylella fastidiosa* through recombinase polymerase amplification (Abstr.). *Phytopathology* 107: S4.12.

Lieth, J. H., Meyer, M. M., Yeo, K.-H., and Kirkpatrick, B. C. 2011. Modeling cold curing of Pierce's disease in *Vitis vinifera* "Pinot Noir" and "Cabernet Sauvignon" grapevines in California. *Phytopathology* 101: 1492–1500.

Meyer, M. M., and Kirkpatrick, B. C. 2011. Exogenous applications of abscisic acid increase curing of Pierce's disease-affected grapevines growing in pots. *Plant Dis.* 95: 173–177.

Purcell, A. H. 2006. *Xylella fastidiosa*. A scientific and community internet resource on plant diseases caused by the bacterium *Xylella fastidiosa* [http://nature.berkeley.edu/xylella/central.hosts.htm]

Roper, M. C., Greve, L. C., Warren, J. G., Labavitch, J. M., and Kirkpatrick, B. C. 2007. *Xylella fastidiosa* requires polygalacturonase for colonization and pathogenicity in *Vitis vinifera* grapevines. *Mol. Plant-Microbe Interact.* 20: 411–419.

Schaad, N. W., Opgenorth, D., and Gaush, P. 2002. Real-time polymerase chain reaction for one-hr on-site diagnosis of Pierce's disease of grape in early season asymptomatic vines. *Phytopathology* 92: 721–728.

Schaad, N. W., Postnikova, E., Lacy, G., Fatmi, M., and Chang, C.-J. 2004. *Xylella fastidiosa* subspecies: *X. fastidiosa* subsp. *fastidiosa*, subsp. nov., *X. fastidiosa* subsp. *multiplex* subsp. nov. and *X. fastidiosa* subsp. *pauca* subsp. nov. *Syst. Appl. Microbiol.* 27: 290–300.

Shi, X., Bi, J., Morse, J. G., Toscano, N. C., and Cooksey, D. A. 2013. Effect of xylem fluid from susceptible and resistant grapevines on developmental biology of *Xylella fastidiosa*. *Eur. J. Plant Pathol.* 135: 127–135.

Shi, X., and Lin, H. 2018. The chemotaxis regulator *pilG* of *Xylella fastidiosa* is required for virulence in *Vitis vinifera* grapevines. *Eur. J. Plant Pathol.* 150: 351–362.

Simpson, A. J., Reinach, F. C., Arruda, P., Abreu, F. A., Acencio, M., et al. 2000. The genome sequence of the plant pathogen *Xylella fastidiosa*. The *Xylella fastidiosa* Consortium of the Organization for Nucleotide Sequencing and Analysis. *Nature* 406: 151–157.
Su, C.-C., Chang, C. J., Chang, C.-M., Shih, H.-T., Tzeng, K.-C., et al. 2013. Pierce's disease of grapevines in Taiwan: Isolation, cultivation and pathogenicity of *Xylella fastidiosa*. *J. Phytopathol.* 161: 389–396.
Tubajika, K. M., Civerolo, E. L., Puterka, G. J., Hashim, J. M., Luvisi, D. A., et al. 2007. The effects of kaolin, harpin and imidacloprid on development of Pierce's disease in grape. *Crop Prot.* 26: 92–99.
Van Sluys, M. A., de Oliveira, M. C., Monteiro-Vitorello, C. B., Miyaki, C. Y., and Furlan, L. R. 2003. Comparative analyses of the complete genome sequences of pierce's disease and citrus variegated chlorosis strains of *Xylella fastidiosa*. *J. Bacteriol.* 185: 1018–1026.
Varela, L. G. Pierce's disease in the North coast. University of California. Cooperative Extension and Statewide IPM Project [http://www.cnr.berkeley.edu/Xylella/pd97.html]
Wallis, C. M., and Chen, J. 2012. Grapevine phenolic compounds in xylem sap and tissues are significantly altered during infection by *Xylella fastidiosa*. *Phytopathology* 102: 816–826.
Wells, J. M., Raju, B. C., Hung, H. Y., Weisburg, W. G., Mandelco-Paul, L., et al. 1987. *Xylella fastidiosa* gen. nov., sp. nov: Gram-negative, xylem-limited, fastidious plant bacteria related to *Xanthomonas* spp. *Int. J. Syst. Bacteriol.* 37: 136–143.
Wistrom, C., and Purcell, A. H. 2005. The fate of *Xylella fastidiosa* in vineyard weeds and other alternate hosts in California. *Plant Dis.* 89: 994–999.

7.59 BACTERIAL BLIGHT OF GRAPEVINE

Bacterial blight or bacterial necrosis of grapevine (*Vitis vinifera* L.) is a chronic and destructive vascular disease of grapevine, affecting commercially important cultivars. Some cultivars are more susceptible than others. Losses arise from reduced productivity and shortened life span of diseased vines.

The disease was first described from Crete, Greece (Panagopoulos 1969). It also occurs in Asia, France, Italy, Moldova, Portugal, Slovenia, Spain, Turkey (now eradicated), South Africa, South America, and USSR. It is widespread in the Mediterranean region and South Africa. Severe damage was reported from South Africa (Matthee et al. 1970) and France in the past, but nowadays the disease appears to be of minor importance. The "maladie d'Oléron" of grapevine, described in France in 1895 (Ravaz 1895) was first thought to be caused by *Erwinia vitivora*, but Prunier et al. (1970) showed it to be caused by *Xylophilus ampelinus*.

7.59.1 Host Range

The grapevine is the only host of *Xyp. ampelinus*.

7.59.2 Symptoms

The disease is characterized by typical symptoms such as cankers on stems and petioles, necrotic foliar spots, and bud death. In early spring, buds on infected spurs fail to open or show stunted growth and eventually die. Affected spurs often appear slightly swollen due to hyperplasia of the cambial tissue. Cracks appear along such spurs, which become deeper and longer, forming cankers. Immature flowers turn black and die-back. Infected grape bunch stalks show symptoms similar to the infected shoots. Cankers on bunch stalks cause partial or total death of a fruit bunch.

The pale yellowish-green spots usually develop on the lowest internodes of young shoots. These spots extend upwards forming reddish-brown to black streaks, which crack and develop into cankers. Later cracks and cankers reach up to the pith of the shoot. Shoots subsequently wilt, droop, and dry up. Discoloration is less common on very young shoots, but the whole shoot dies-back. In severe infection, a large number of adventitious buds develop, but these quickly die-back. Infected shoots are shorter, giving the vine a stunted appearance. Tissue browning is revealed in stem cross-sections. Later in spring, cracks and later on cankers also form on more woody branches (Figure 7.42a and b).

In summer, cankers are often seen on the sides of petioles, causing a characteristic one-sided necrosis of the leaf. They may also appear on main and secondary flower and fruit stalks. The leaf infection through stomata results in the development of angular, reddish-brown lesions, while the infection through hydathodes results in reddish-brown discoloration of the leaf

(a)

(b)

FIGURE 7.42 Bacterial blight of grapevine: (a) Leaf symptoms. (b) Symptoms on stem. (Courtesy of C.G. Panagopoulos, Agricultural University, Bugwood.org.)

margins. When leaves are penetrated via the petiole, they become brown, necrotic, and dry. Pale yellow bacterial ooze may be seen on infected leaves when humidity is high. Roots may also be attacked and show brown internal discoloration, resulting in retardation of shoot growth. The ability of the bacterium to survive for several years inside plants without inducing symptoms may result in a latency period. Panagopoulos (1987) found that 50% of apparently healthy canes from diseased vineyards in Crete, Greece were latently infected.

The bacterium can be detected by immunofluorescent microscopy in washings from diseased leaves or in homogenates of woody tissues, as well as in ooze or on pruning shears. Molecular methods have also been developed to detect the bacterium from the grapevine tissue. Botha et al. (2001) developed a nested PCR, using pathogen specific primers, to detect the bacterium from grapevine cuttings. Dreo et al. (2007) developed a real-time PCR, which was ten times more sensitive than the abovementioned nested PCR. Manceau et al. (2005) developed a sensitive, specific, and reliable technique for the detection of *Xyp. ampelinus* employing a PCR and Microwell plate system.

7.59.3 Causal Organism

Xylophilus ampelinus (Panagopoulos 1969) Willems et al. 1987
Syn. *Xanthomonas ampelina*
Xyp. ampelinus is an EPPO A2 quarantine pest.

It is difficult to isolate the bacterium from infected tissue due to its slow growth on nutrient agar. Serfontein et al. (1997) reported that incubation of naturally infected cuttings at 15°C for 3 days under moist conditions significantly increased the number of isolated colonies. Using this enrichment method, the organism has also been detected in shoots from vine plantings without symptoms, previously thought to be free of the disease.

Xyp. ampelinus is a strictly aerobic, non-spore forming, Gram-stain-positive rod measuring 0.5–0.8 × 1.6–3.2 μm and motile by one polar flagellum. Occasionally isolates have some filamentous cells 8–10 times longer than usual. Cells occur singly or in pairs. It produces a yellow insoluble pigment and metabolizes sugars oxidatively.

Growth on artificial media is very slow. On nutrient agar, the colonies are circular, entire, glistening, translucent, and pale yellow, and measure 0.2–0.3 mm in diameter after 6 days and 0.6–0.8 mm in diameter after 15 days of incubation at 26°C. On nutrient agar with 5% sucrose or on yeast glucose chalk agar (containing 0.5% yeast extract, 1% glucose, 3% calcium carbonate, and 2% agar), growth is faster and deep yellow. Best growth is obtained on 2% galactose, 1% yeast extract, 2% calcium carbonate, and 2% agar; which is deep yellow with a diffusible brown pigment. No fluorescent pigment is produced on King's medium B. Growth fails or is very poor in presence of 0.02% triphenyltetrazolium chloride. Maximum tolerance of NaCl is 1%. Minimum temperature for growth is 6°C, while maximum is 30°C.

No strain hydrolyzes aesculin, arbutin, casein, gelatin, sodium hippurate, or starch, or shows pectolytic activity on potato tissue, reduces nitrate to nitrite, produces ammonia, indole, arginine dihydrolase, ornithine and lysine decarboxylases, or phenylalanine deaminase. H_2S is produced from cysteine and thiosulphate by all strains, but from peptone only by some. The malonate and gluconate tests are negative, while tyrosinase activity varies. Lipolysis is positive on Tween 80, but negative or weak on tributyrin. Urease is strongly positive, catalase positive, but oxidase is negative. Metabolism is strictly aerobic. The G + C content of the DNA is 68%–69% by thermal denaturation (Bradbury 1991).

Komatsu et al. (2016) reported that genomic fingerprints generated from 43 strains of *Xyp. ampelinus*, collected from Europe and Hokkaido, Japan, revealed four DNA types (A–D) based on the combined results of Rep-, ERIC-, and Box-PCR. Genetic variation was found among the strains examined; strains collected from Europe belonged to DNA types A or B, and those collected from Hokkaido belonged to DNA types C or D. However, the strains belonging to different DNA types did not differ in pathogenicity and virulence on grapevine cultivar Kerner.

7.59.4 Disease Cycle

The bacterium can occur as an epiphyte and survives in vascular tissue of planting material. Using green fluorescence protein-labeled, GFP-marked strains of *Xyp. ampelinus* in histological studies, Grall and Manceau (2003) reported that after stem inoculation by wounding, the bacteria progressed down to the crown through the xylem vessels and formed biofilms. When the bacteria were forced into woody cuttings, they rarely colonized the emerging plantlets. Xylem vessels could play a key role in the multiplication and conservation of the bacteria, rather than being a route for plant colonization. When bacterial suspension was sprayed on the plants, the bacteria progressed in two directions, i.e., to the emerging organs and down to the crown, thus displaying the importance of epiphytic colonization in disease development.

Infected propagating material is a major source of primary infection. The bacterium emerges in spring and is carried to healthy shoots mainly by rain splash. Shoots are susceptible to infection during autumn and winter and non-susceptible during spring and summer. The disease is favored by warm moist weather. Secondary spread occurs through grafting, wind-splashed rain, overhead/sprinkler irrigation, pruning tools, and bleeding sap. Contaminated bleeding sap is an important source of inoculum for external contamination due to the high susceptibility of young merging shoots to the pathogen (Grall et al. 2005). They further reported that the old wood was contaminated throughout the year and constituted a stock inoculum for endophytic contamination of crude sap during the winter and the spring.

Wounds may facilitate entry, but infection may also occur without wounding (Bradbury 1991). From initial disease foci, local spread in vineyards tends to occur along rows. Illegally imported plants pose the greatest risk and if such material is infected, the disease is likely to become established. The varieties Sultana and Barlinka are very susceptible. Hermitage is moderately susceptible and Green Grape and Riesling are moderately resistant.

Komatsu and Kondo (2015) investigated the distribution of *Xyp. ampelinus* in grapevines and its winter habitat in Hokkaido, Japan and found that the bacterium is possibly not maintained inside tissues or sap and that it might not survive on plant surfaces; rather, bacteria can be found on the underside surface of the bract and bud wool during winter in Hokkaido.

Sevillano et al. (2014), using transposon insertion mutagenesis, found that MntH protein plays a significant role in the virulence of *Xyp. ampelinus*. This is the first report showing that transposon mutagenesis is an effective strategy for the isolation of *Xyp. ampelinus* mutants. It is also the first report characterizing a gene encoding a protein involved in virulence in this grapevine pathogen.

7.59.5 Management

The following good viticultural practices can prevent or control the disease.

1. Use of healthy planting material and indexing of mother plants
2. Pruning in dry weather and disinfection of pruning tools
3. Avoidance of overhead/sprinkler irrigation

REFERENCES

Botha, W. J., Serfontein, S., Greyling, M. M., and Berger, D. K. 2001. Detection of *Xylophilus ampelinus* in grapevine cuttings using a nested polymerase chain reaction. *Plant Pathol.* 50: 515–526.

Bradbury, J. F. 1991. *Xylophilus ampelinus. IMI Descriptions of Fungi and Bacteria.* No. 1050. *Mycopathologia* 115: 63–64.

Dreo, T., Gruden, K., Manceau, C., Janse, J. D., and Ravnikar, M. 2007. Development of a real-time PCR-based method for detection of *Xylophilus ampelinus. Plant Pathol.* 56: 9–16.

Grall, S., and Manceau, C. 2003. Colonization of *Vitis vinifera* by a green fluorescence protein-labeled, gfp-marked strain of *Xylophilus ampelinus*, the causal agent of bacterial necrosis of grapevine. *Appl. Environ. Microbiol.* 69: 1904–1912.

Grall, S., Roulland, C., Guillaumes, J., and Manceau, C. 2005. Bleeding sap and old wood are two main sources of contamination of merging organs of vine plants by *Xylophilus ampelinus*, the causal agent of bacterial necrosis. *Appl. Environ. Microbiol.* 71: 8292–8300.

Komatsu, T., and Kondo, N. 2015. Winter habitat of *Xylophilus ampelinus*, the cause of bacterial blight of grapevine, in Japan. *J. Gen. Plant Pathol.* 81: 237–242.

Komatsu, T., Shinmura, A., and Kondo, N. 2016. DNA type analysis to differentiate strains of *Xylophilus ampelinus* from Europe and Hokkaido, Japan. *J. Gen. Plant Pathol.* 82: 159–164.

Manceau, C., Grall, S., Brin, C., and Guillaumes, J. 2005. Bacterial extraction from grapevine and detection of *Xylophilus ampelinus* by a PCR and Microwell plate detection system. *Bull. OEPP/EPPO Bull.* 35: 55–60.

Matthee, F. N., Heyns, A. J., and Erasmus, H. D. 1970. Present position of bacterial blight (Vlamsiekte) in South Africa. *Deciduous Fruit Grower* 20: 81–84.

Panagopoulos, C. G. 1969. The disease "Tsilik marasi" of grapevine: Its description and identification of the causal agent (*Xanthomonas ampelina* sp. nov.). *Ann. Inst. Phytopathol. Benaki* 9: 59–81.

Panagopoulos, C. G. 1987. Recent research progress on *Xanthomonas ampelina. EPPO Bull.* 17: 225–230.

Prunier, J. P., Ridé, M., Lafon, R., and Bulit, J. 1970. La nécrose bactérienne de la vigne. *C. R. Acad. Agric. France* 56: 975–82.

Ravaz, L. 1895. La maladie d'Oléron. *Annales de l'École Nationale d'Agriculture, Montpellier* 9: 299–317.

Serfontein, S., Serfontein, T. J., Botha, W. J., and Staphorst, J. L. 1997. The isolation and characterisation of *Xylophilus ampelinus. Vitis* 36: 209–210.

Sevillano, S., Cobos, R., García-Angulo, P., Alonso-Monroy, A., Álvarez-Rodríguez, M. L., et al. 2014. Manganese transporter protein MntH is required for virulence of *Xylophilus ampelinus*, the causal agent of bacterial necrosis in grapevine. *Aust. J. Grape and Wine Res.* 20: 442–450.

Willems, A., Gillis, M., Kersters, K., Van den Broecke, L., and De Ley, J. 1987. Transfer of *Xanthomonas ampelina* Panagopoulos 1969 to a new genus, *Xylophilus* gen. nov., as *Xylophilus ampelinus* (Panagopoulos 1969) comb. nov. *Int. J. Syst. Bacteriol.* 37: 422–430.

7.60 BACTERIAL CANKER OF GRAPEVINE

Grapevine (*Vitis vinifera* L.) suffers from several bacterial diseases, namely, bacterial blight (*Xylophilus ampelinus*), bacterial necrosis (*Pseudomonas syringae* pv. *syringae*), Pierce's disease (*Xylella fastidiosa* subsp. *fastidiosa*), crown gall (*Agrobacterium vitis*), and bacterial canker (*Xanthomonas campestris* pv. *viticola*). All these diseases except, bacterial necrosis are described in this book. Bacterial canker is also called by another name, i.e., **bacterial leaf spot of grapevine**.

The disease was first observed by Nayudu (1972) in Andhra Pradesh, India. Subsequently, the disease has been reported from other states of India, namely, Punjab, Karnataka, and Maharashtra, and it also occurs in other grape growing areas of the country. In Punjab, the disease is severe on commercial grown cultivars like Perlette, Anabe-e-shahi, and Thompson Seedless. In certain areas of Punjab, the disease incidence as high as 93% and disease intensity up to 29% has been reported. Chand and Kishun (1990) reported an outbreak of the disease from India and concluded that primarily the more extensive cultivation of the susceptible seedless vareities was the cause of the outbreak.

The disease also occurs in Brazil, where it was first observed in 1998 in vineyards in the Sumedio of the Sao Francisco River basin area of Pernambuco state. The disease incidence was as high as 100%, and in some cases the yield losses were nearly total (Lima et al. 1999). The disease also occurs in other states of Brazil, namely, Piaui, Bahia, Ceara, and Roraima. In Brazil, it is a quarantined disease, officially restricted to these states including Pernambuco (Halfeld-Vieira and Nechet 2006).

7.60.1 Host Range

Chand and Kishun (1990) tested 16 plant species and found that only mango was infected. Although mango does not show infection under natural conditions, its infection under artificial conditions indicates that it may serve as an additional source of inoculum of the bacterium. The neem (*Azadirachta indica* A. Juss.) reported as a colateral host of the bacterium by Nayudu (1972) also did not show any infection.

7.60.2 Symptoms

The symptoms are found on all the aboveground plant parts. The symptoms first appear on young leaves as small, water-soaked lesions on the lower surface surrounded by yellow haloes. Lesions enlarge; become angular, cankerous, and dark brown. Lesions then also appear on the upper leaf surface. Sometimes, the lesions coalesce and cause death of extensive leaf areas. Spots are rough to touch due to drying of gummy bacterial ooze. Cankers on the veins are also common. Vein infection may range from 1 mm long to entire length of the vein. On petioles, the lesions are brown, elongated, and cankerous, and measure 1–5 mm in length (Figure 7.43).

On stems, brown to black, elongated, raised cankers appear. In advanced stage, cankers split longitudinally in the center exposing the vascular discoloration. Severe infection on young twigs and branches results in drying and killing of twigs. Brown to black, irregular to circular, raised, skin-deep cankers appear on the berries. The cankers are mainly formed on the inner berries of the bunch. The symptoms on berries appear near maturity.

In Punjab, India, the disease appears in 3rd or 4th week of March, and the progress of the disease continues on the new plant growth until middle of May. From mid-May to mid-July, the fresh infections do not occur on young plant parts. With the onset of rains, the fresh infections take place from mid-July, and the disease continues to progress up to September.

FIGURE 7.43 Bacterial canker of grapevine: Leaf symptoms.

7.60.3 Causal Organism

Xanthomonas campestris pv. *viticola* (Nayudu 1972) Dye 1978
Syn. *Pseudomonas viticola*

The colonies are white, convex, and mucoid, and no fluorescent pigment is produced in culture media. Cells are Gram-stain-negative rods, non-sporing, and grow aerobically. The bacterium hydrolyzes esculin and starch and produces acid from trehalose, glucose, cellobiose, melibiose, and arabinose; weakly positive for glycerol. Acid is not produced from dulcitol, and the tests for oxidase and urease are also negative. Asparagine is not utilized as a sole source of carbon and nitrogen. The bacterium produces hypersensitive reaction on tomato and tobacco in 48 h.

On the basis of combined analysis of PCR patterns obtained with primers REP, ERIC, and BOX, Trindade et al. (2005) found a high degree of similarity between Brazilian strains and Indian type strain NCPPB 2475. Trindade et al. (2007) designed primer Xcv1F/Xcv3R and RST2/Xcv3R based on the partial sequence of *hrpB* gene for the detection and identification of *Xa. campestris* pv. *viticola* (Xcv). The detection limit of primer RST2/Xcv3R was 10^4 cfu mL^{-1}, and the sensitivity of Xcv1F/Xcv3R and RST2/Xcv3R was 10 pg and 1 pg of purified bacterial DNA, respectively. The presence of the bacterium in tissues of grapevine petioles, previously inoculated with it, could not be detected by PCR using macerated extract added directly in the reaction. However, amplification was positive with the introduction of an agar-plating step prior to PCR.

Keeping in view the doubtful taxonomic position of *Xa. campestris* pv. *viticola* and pigmented isolates pathogenic to cashew plant, da Gama et al. (2018) used the multi-locus sequence analysis technique, average nucleotide identity values, and tetranucleotide frequency correlation coefficients to analyze their phylogenetic relationship in relation to other *Xanthomonas* species. On the basis of results obtained, they suggested the taxonomic repositioning of *Xa. campestris* pv. *viticola* (Nayudu 1972) Dye 1978 as *Xa. citri* pv. *viticola* (Nayudu 1972) Dye 1978 comb. nov. and emendation of the description of *Xa. citri* pv. *anacardii* to include pigmented isolates pathogenic to cashew plant.

Kamble et al. (2017) evaluated *in vitro* 11 chemicals along with three biocontrol agents against *Xa. campestris* pv. *viticola*. Among the chemicals, Streptocycline, mancozeb, and bronopol showed significant inhibition of the pathogen, while kasugamycin, copper oxychloride, and copper hydroxide exhibited low efficacy. Copper sulphate, validamycin, difenoconazole, carbendazim, and potassium phosphate showed no inhibitory effect at 50, 100, 500, 1,000, 2,000, and 3,000 μg mL^{-1}. Among the biocontrol agents, *Bacillus subtilis* and *Trichoderma asperolloides* showed the inhibitory effect against the bacterium.

7.60.4 Disease Cycle

The bacterium survives mainly in the stem cankers. The survival also occurs in the infected crop residue. The local spread of the bacterium occurs through wind-splashed rain and harvesting and pruning tools. The long distance dissemination of the pathogen occurs through infected propagating material.

Naue et al. (2014) evaluated the efficacy of seven sanitizing agents to disinfect grapevine cutting tools and water contaminated with *Xa. campestris* pv. *viticola* 2Rif. Sodium hypochlorite (20,000 μg mL^{-1}) and dodecyl dimethyl ammonium chloride (1,140 μg mL^{-1}) produced the largest inhibition zones against four isolates of the bacterium and were 100% effective in disinfecting contaminated harvest shears. These two sanitizing agents maintained their efficacy for 8 h. The bacterium survived for 24 h on harvest shears and was spread by contaminated harvest shears until the 24th cut, on an average. Water disinfection was successfully achieved with dodecyl dimethyl ammonium chloride (570 μg mL^{-1}), sodium hypochlorite (5,000 μg mL^{-1}), and benzalkonium chloride (122.5 μg mL^{-1}).

7.60.5 Management

1. Use of disease-free propagating material. Hot water treatment of cuttings at 52°C for 30 min reduces the infection
2. Rouging and drastic pruning of the infected vines. The time of pruning also affects the disease management. Chand et al. (1991) found that disease intensity was 76.8% at three locations in Karnataka, India when vines were pruned between September 1 and 15, while it was 5.6% in vineyards pruned after October 10. Pruning the vines from the second week of October and onwards gave the best results in the grape producing areas of Maharashtra and North Karnataka
3. Disinfection of harvesting and pruning tools
4. Use of windbreaks to reduce pathogen dissemination
5. Give protective sprays of copper and thiocarbamate fungicides (Nascimento and Mariano 2004). Under Punjab (India) conditions, four sprays at 15 days intervals starting from July 15 gave an effective control of the disease. Bordeaux mixture (0.8%) proved most effective followed by Streptocycline (200 μg mL^{-1})

Shivananda et al. (2011) reported that application of Streptocycline (500 μg mL^{-1}) + copper oxychloride (2,000 μg mL^{-1}) thrice at 20 days interval was most effective, as it gave minimum percent disease index (PDI) (29.86%), maximum yield (26.95 t ha^{-1}), more number of bunches produced (20.03/plant), lowest infected bunches (3.40/plant), and maximum single bunch weight (925 g), followed by Streptocycline (500 μg mL^{-1}) alone, which gave PDI (35.35%), yield (23.50 t ha^{-1}), bunches produced (17.13/plant), number of infected bunches (5.05/plant), and single bunch weight (862 g). Among the biocontrol agents, *B. subtilis* at 5,000 μg mL^{-1} proved most effective as it gave PDI (45.48%), yield (14.23 t ha^{-1}), bunches produced (9.35/plant), number of infected bunches (7.10/plant), and single bunch weight (413 g), followed by *Pseudomonas fluorescens* at 5,000 μg mL^{-1}, which gave PDI (47.09%), yield (11.35 t ha^{-1}), bunches produced (7.43/plant), infected bunches (7.08/plant), and single bunch weight (318 g).

REFERENCES

Chand, R., and Kishun, R. 1990. Outbreak of grapevine bacterial canker disease in India. *Vitis* 29: 183–188.

Chand, R., Patil, B. P., and Kishun, R. 1991. Management of bacterial canker disease (*Xanthomonas campestris* pv. *viticola*) of grapevine (*Vitis vinifera*) by pruning. *Indian J. Agri. Sci.* 61: 220–222.

da Gama, M. A. S., Mariano, R., de Lima R., da Silva Jr., W. J., de Farias, A. R. G., Barbosa, M. A. G., et al. 2018. Taxonomic repositioning of *Xanthomonas campestris* pv. *viticola* (Nayudu 1972) Dye 1978 as *Xanthomonas citri* pv. *viticola* (Nayudu 1972) Dye 1978 comb. nov. and emendation of the description of *Xanthomonas citri* pv. *anacardii* to include pigmented isolates pathogenic to cashew plant. *Phytopathology* 108: 1143–1153.

Dye, D. W. 1978. Genus IX. *Xanthomonas* Dowson 1939. In Young, J. M., Dye, D. W., Bradbury, J. F., Panagopoulos, C. G., and Robbs, C. F. 1978. A proposed nomenclature and classification for plant pathogenic bacteria. *N. Z. J. Agric. Res.* 21: 162–166.

Halfeld-Vieira, B. A., and Nechet, K. de Lima. 2006. Bacterial canker of grapevine in Roraima, Brazil. *Fitopatologia Brasileira* 31: 604.

Kamble, A. K., Sawant, S. D., Saha, S., and Sawant, I. S. 2017. *In vitro* efficacy of different chemicals and biological agents against *Xanthomonas campestris* pv. *viticola* causing bacterial leaf spot of grapes. *Int. J. Agric. Sci.* 9: 4427–4430.

Lima, M. F., Ferreira, M. A. S. V., Moreira, W. A., and Dianese, J. C. 1999. Bacterial canker of grapevine in Brazil. *Fitopatologia Brasileira* 24: 440–443.

Nascimento, A. R. P., and Mariano, R. de L. R. 2004. Bacterial canker of grapevine: Etiology, epidemiology and control strategies. *Ciencia Rural* 34: 301–307.

Naue, C. R., Costa, V. S. O., Barbosa, M. A. G., Batista, D. C., Souza, E. B., et al. 2014. *Xanthomonas campestris* pv. *viticola* on grapevine cutting tools and water: Survival and disinfection. *J. Plant Pathol.* 96: 451–458.

Nayudu, M. V. 1972. *Pseudomonas viticola* sp. nov., incitant of a new bacterial disease of grapevine. *Phytopathol. Z.* 73: 183–186.

Shivananda, J., Ravikumar, M. R., and Hiremani, N. 2011. Evaluation of different chemicals and bioagents against bacterial leaf spot of grapevine and their effect on yield and yield parameters. *Int. J. Plant Prot.* 4: 377–380.

Trindade, L. C., Lima, M. F., and Ferreira, M. A. S. V. 2005. Molecular characterization of Brazilian strains of *Xanthomonas campestris* pv. *viticola* by rep-PCR fingerprinting. *Fitopatologia Brasileira* 30: 46–54.

Trindade, L. C. da, Marques, E., Lopes, D. B., and Ferreira, M. A. da S. V. 2007. Development of a molecular method for detection and identification of *Xanthomonas campestris* pv. *viticola*. *Summa Phytopathologica* 33: 16–23.

7.61 BACTERIAL CANKER OF KIWI

Bacterial canker of kiwi (staminate kiwi, *Actinidia deliciosa*) was first reported from California, USA in 1980. Later on in 1983, the causal bacterium was identified as *Pseudomonas syringae*. Serizawa et al. (1989) reported the disease on Chinese gooseberry (*A. chinensis*) from Japan in early 1980s, where its severity is one of the limiting factors in kiwifruit production. According to Goto (1992), the disease found in Japan is more destructive and such form of disease has not been found in other regions of the world. In the EPPO region, the disease was first noticed in northern Italy in 1992 where it remained sporadic and with a low incidence for 15 years. However, in 2007/2008 economic losses started occurring, particularly in the Lazio region, and the possible spread of the disease to other kiwifruit producing regions in Italy began to raise concerns. Due to the emerging problem of *Psm. syringae* pv. *actinidiae* (hereafter referred to as *Psa*) in the Mediterranean region, the EPPO Secretariat decided to add it to the EPPO Alert List. Ferrante and Scortichini (2010) reported a severe epiphytotic of the disease during 2008–2009 from central Italy on *A. chinensis* (yellow kiwifruit). In Italy, it is estimated that the economic losses, including impact on trade, due to *Psa* have reached 2 million euros (Balestra et al. 2009). The disease has also been reported from China, New Zealand, Australia, Spain, Chile, and Korea. Kim et al. (2017) reported the appearance of biovar *Psa*3 in Korea in 2011 (the biovar *Psa*2 was already present there since 1988) that caused tremendous economic losses by destroying many vines or orchards of yellow-fleshed kiwifruit cultivars in one or several growing seasons. Bacterial canker epiphytotics caused by both *Psa*2 and *Psa*3 biovars were prevalent in Korea in recent years. In this review, the authors also summarized the symptomatology, etiology, disease cycle, diagnosis, and epidemiology of kiwifruit bacterial canker in Korea.

Koh et al. (2012) reported *Pectobacterium carotovorum* subsp. *actinidiae* as a new bacterial pathogen causing kiwifruit canker-like symptoms in Korea. Wu et al. (2017) reported this bacterium causing kiwifruit bacterial canker-like symptoms in China, this being the first report of this disease from China and also outside of Korea.

7.61.1 Host Range

Psa naturally infects only kiwi plants, both cultivated and wild. It can infect peach and Mume on artificial inoculation. *Psm. syringae* pv. *syringae* is pathogenic to cowpea and carrots in addition to kiwi.

7.61.2 Symptoms

The symptoms appear on almost all the aboveground plant parts. The symptoms appear on trunks and limbs from late winter to early spring, after the break in dormancy of vines. The most conspicuous symptom is the red-rusty exudation that appears on buds, joints of branches and leaf, and pruning scars. The bark of affected parts deeply shrivels and dries, often leading to the formation of cracks (Figure 7.44b).

On leaves, the infection is generally limited to young expanding leaves. Initially, small, water-soaked, light green, scattered spots appear on leaf lamina. The spots increase in size (2–3 mm in diameter), turn brown to dark brown, and become angular in shape due to the restrictions imposed by veins. The spots are surrounded by bright yellow haloes. Bacterial ooze is generally formed on the lower surface of leaves. In severe infection, the entire leaf may be blighted and shriveled (Figure 7.44a).

Water-soaked, dark green, elongated lesions appear on the infected canes. As the infection advances, longitudinal cracks often appear leading to wilting and blightening of the whole shoot and exudation of bacterial ooze. Removal of the bark usually reveals a brown discoloration of the external vascular tissues and reddening of the tissues beneath lenticels. Infected flowers turn brown and often wither before opening. Infected flowers that open generally have under-developed petals. Necrotic lesions are also formed on the sepals (Goto 1992).

Another disease of kiwi, called "**bacterial blossom blight**" or "**blossom rot**" has some similarities with bacterial canker. Blossom blight occurs only on young flowers and is

(a)

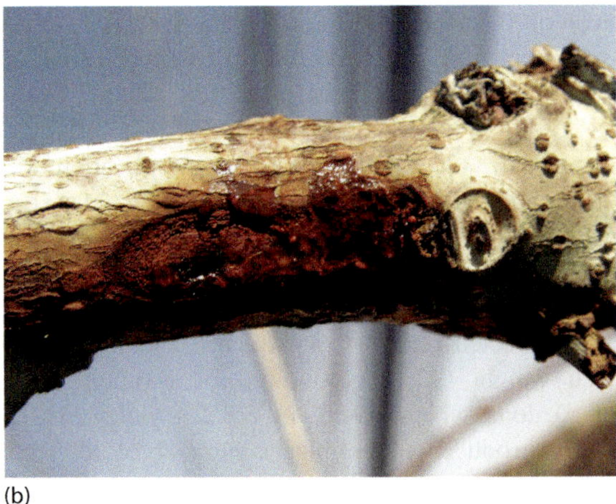

(b)

FIGURE 7.44 Bacterial canker of kiwi: (a) Symptoms on leaf. (b) Stem canker. (Courtesy of Riccardo Bugiani, Plant Protection Service of Emilia-Romagna region, Italy.)

characterized by browning of petals and shedding of premature flowers. The severity of the disease increases with rainfall during the flowering period. *Psm. syringae*, *Psm. viridiflava*, and *Psa* form the causal complex, which causes blossom blight.

7.61.3 Causal Organisms

Pseudomonas syringae pv. *actinidiae* Takikawa et al. 1989
Syn. *Psm. syringae* pv. *avellanae*
Psm. syringae pv. *syringae* van Hall 1902
Syn. *Psm. syringae* pv. *japonica*
Psm. syringae pv. *actinidifoliorum* pv. nov. Cunty et al. 2015

7.61.3.1 *Psm. syringae* pv. *actinidiae*

It was first identified by Takikawa et al. (1989) as a new pathovar of *Psm. syringae*. Ushiyama et al. (1992) are of the view that *Psa* might have originated from *Actinidia* wild species, namely, *A. arguta* and *A. kolomikta*, which are distributed in northern areas of Japan and show natural infection.

Bacterial cells are Gram-stain-negative, non-sporing rods, motile by polar flagella. It does not produce fluorescent pigment on King's medium B. Neither β-galactosidase and urease are produced, nor is gelatin liquefied. The bacterium does not utilize DL-lactate, caprate, erythritol, xylose, L-histidine, and trigonelline (Goto 1992). Some strains of the bacterium produce phaseolotoxin, as does *Psm. syringae* pv. *phaseolicola*, the causal agent of bean (*Phaseolus vulgaris*) halo blight. However, the Korean strains of the bacterium were found to produce coronatine, but not phaseolotoxin (Han et al. 2003). Ferrante and Scortichini (2010) reported that none of the strains isolated from central Italy during 2008–2009 epiphytotic possessed genes coding for phaseolotoxin or coronatine, but all had an effector protein, namely, *hopA1*, differentiating them from the strains causing past outbreaks in Japan and Italy. They further found that strains, which caused outbreak in central Italy, appear to be distinct from those found previously in Japan, South Korea, and Italy.

Four populations of *Psa*, namely, *Psa*1, *Psa*2, *Psa*3 and *Psa*4 have been described and *Psa*4 is less virulent than the other three populations. Also, multi-locus sequence typing of housekeeping genes or broad *Psa* strain genome comparisons revealed that *Psa* 4 strains cluster separately from other three *Psa* populations. Furthermore, repetitive-sequence PCR fingerprinting, type III secretion system effector protein genes detection, and colony morphology clearly indicate the distinctiveness of *Psa* 4 strains from the other three *Psa* populations. Rep-PCR molecular typing revealed a high similarity of the *Psa* 4 strains with members of *Psm. avellanae* species. It was concluded that only *Psa*1, *Psa*2, and *Psa*3 populations, pathogenic to *Actinidia* spp. should be included in *Psa*, and the *Psa*4 strains isolated in New Zealand do not belong to the pathovar *actinidiae*, and most probably, belong to a new still unnamed pathovar (Ferrante and Scortichini 2015). Cunty et al. (2015) performed a multi-locus sequence analysis based on four housekeeping genes (*gapA*, *gltA*, *gyrB*, and *rpoD*) on 72 strains representative of the French outbreak that occurred since the beginning of 2010, and found that all the strains fell into two phylogenetic groups; one clonal corresponding to biovar 3 and the other corresponding to biovar 4. Strains of biovar 4 were found substantially different from those of the other biovars as they were less aggressive and caused only leaf spots, whereas *Psa* biovars 1, 2, and 3 also caused canker and shoot die-back. Based on these pathogenic differences, supported by phenotypic, genetic, and phylogenetic differences, the authors proposed that *Psa* biovar 4 be renamed as *Psm. syringae* pv. *actinidifoliorum* pv. nov.

Psa has an extensive global distribution, yet the isolates causing widespread losses to the kiwifruit industry can all be traced to a single MLSA group, *Psa*3 (Chapman et al. 2012).

Patel et al. (2017) performed a genetic screen in order to identify transposon mutants altered in the lipolytic activity and aimed to identify the set of secretion and global regulatory loci that control lipolytic activity and also play important roles in *in planta* fitness. The screen for altered lipolytic activity phenotype identified a total of 58 Tn*5* transposon mutants. Mapping all these Tn*5* mutants revealed that the transposons were inserted in genes that play roles in cell division, chemotaxis, metabolism, movement, recombination, regulation, signal transduction, and transport as well as a few unknown functions. Several of these identified *Psa* Tn*5* mutants, notably the functions affected in phosphomannomutase AlgC, lipid A biosynthesis acyltransferase, glutamate-cysteine ligase, and the type IV pilus protein PilI, were also found affected in *in planta* survival and/or growth in kiwifruit plants.

7.61.3.2 *Psm. syringae* pv. *syringae*

Psm. syringae pv. *syringae*, the bacterium first isolated from kiwi canker in 1983, differs from *Psa* in liquefying gelatin and producing fluorescent pigment.

Strains of both the abovementioned pathogens resistant to streptomycin occur in Korea and Japan (Han et al. 2004). Strains of *Psa* resistant to copper sulphate occur in Japan and the resistance to copper sulphate operates through copper-trapping system (Masami et al. 2004). Strains of *Psm. syringae* resistant to copper have also been reported from New Zealand (Vanneste and Voyl 2003).

7.61.3.3 *Psm. syringae* pv. *actinidifoliorum* pv. nov

The strains of this bacterium, i.e., biovar 4 are substantially different from those of *Psa* biovars 1, 2, and 3. Strains of biovar 4 are less aggressive and cause only leaf spots, whereas strains of *Psa* biovars 1, 2, and 3 also cause canker and shoot die-back.

7.61.4 Disease Cycle

The bacteria survive in cankers formed on the infected vines. In Japan, infected *Actinidia* wild spp. (*A. arguta*, *A. kolomikta*) also help in the perpetuation of the bacterium and act as a source of infection for cultivated vines. The bacterial cells ooze out from cankers formed on trunks, limbs, and canes, and are disseminated by wind-splashed rain and overhead-irrigation. Long distance spread occurs through infected planting material. The bacteria enter the host through stomata, lenticels, and wounds caused by pruning and cultural operations. Mature leaves and canes are not infected in summer and even the penetration through wounds does not result in infection. The disease is severe when the mean temperature in early spring remains below 15°C and is coupled with frequent rains. Winter cankers cause more yield loss as they reduce the size of productive vines by damaging the main vine structure. Spring cankers have less direct effect on the yield, but are important for the formation of winter cankers in the next year. Four species of *Actinidia*, namely, *A. deliciosa*, *A. chinensis*, *A. arguta*, and *A. kolomikta* are susceptible to the pathogen. There is no data on the susceptibility of other *Actinidia* species. Observations made in Italy suggest that the damage is more severe on yellow-fleshed kiwifruit, namely, *A. chinensis* cvs. Hort 16A and Jin Tao than on the more widely grown green-fleshed cv. Hayward of *A. deliciosa*.

Ferrante and Scortichini (2014) reported that both autumn and winter frosts promote *Psa* multiplication in the inoculated twigs of both *A. chinensis* and *A. deliciosa*. The damage from frost, freeze thawing, and the accumulation of *Psa* in *Actinidia* twigs promotes the migration of the pathogen within and between the orchards. The results obtained in this study confirmed that *A. deliciosa* is more frost tolerant than *A. chinensis*, autumn frosts are more dangerous to these hosts than winter frosts, and in the absence of *Psa*, young kiwifruit plants remain sensitive to frost.

Li et al. (2005) reported that the resistance of cultivars was related to the contents of lignin and soluble sugars in annual twigs and foliage. The contents of both the compounds were higher in resistant cultivars than those of susceptible cultivars. After infection, soluble sugar decreased in both the cultivars, but the decrease was much more in susceptible cultivar compared to resistant cultivar. In contrast, the lignin content increased in both the cultivars and the increase in resistant cultivar was much higher in comparison to susceptible cultivar. Balestra et al. (2001) reported that staminate kiwi cv. GTH, highly resistant to *Psm. syringae* pv. *syringae*, acquired susceptibility when transformed with *rolABC* of *Agrobacterium rhizogenes*. The susceptibility was inherited by the T1 offspring containing *rolABC* genes, derived by crossing pollen of transgenic *rolABC* GTH with the pistilate cv. Hayward (highly susceptible). The susceptibility was probably correlated to high nitrogen content in the leaves of *rolABC* plants. The high resistance of the staminate GTH was inherited by all the offsprings lacking *rolABC* genes.

Beatrice et al. (2017) evaluated the chitosan's elicitation effect to induce SAR in kiwi plants for the control of *Psa* infection. Chitosan elicited a systemic response in kiwi plants with intensity comparable to other well-known signalling compounds such as salicylic acid, methyl jasmonate, or ethylene. The data obtained by chitosan treatments in *in vitro* cultures were confirmed in plants grown in greenhouse, in which, the combination of chitosan treatment and the bacterial inoculum had the greatest effect on pathogenesis related proteins. The study also proved that chitosan, leading to an increased expression of both pathogenesis related proteins, has a role in kiwifruit defense reactions.

Rees-George et al. (2010) found that several PCR primers designed earlier to identify *Psa* were not specific. Therefore, they designed two new sets of PCR primers, PsaF1/R2 and PsaF3/R4 to be complementary to a portion of the 16S–23S rDNA intertranscribed spacer regions. These primers amplified a DNA fragment from strains of *Psa* and a strain of the tea pathogen, *Psm. syringae* pv. *theae*, but not from 56 strains of bacteria belonging to six genera and 17 species. It was not possible to distinguish *Psa* from the phylogenetically similar *Psm. syringae* pv. *theae* using the ITS, *gyrB*, *acnB*, *rpoD*, *pgi*, or *cts* gene regions to design PCR primers. As *Psm. syringae* pv. *theae* is unlikely to be present in kiwifruit, primers PsaF1/R2 and PsaF3/R4 can be used to detect *Psa* from kiwifruit tissue.

Balestra et al. (2013) developed and validated a new multiplex PCR assay using 32 *Psa* isolates and 15 non-*Psa* strains and correctly assigned *Psa* strains to three different haplotypes: a Japanese/Korean group, a European group, and a Chinese group. The described PCR assays has a limit of detection of approximately 5–50 pg of purified DNA or of 5×10^2 bacteria/PCR and were shown to work with both artificially and naturally infected plant tissues. Hence, the described method offers a suitable tool for detection of *Psa* and haplotype attribution, in particular, when testing a high number of samples during surveillance and prevention activities.

Ruinelli et al. (2017) developed a rapid, sensitive, and reliable field-based assay for the detection of *Psa*, using a comparative genomic approach on the publicly available *Psa* genomic data to select unique target regions for the development of two LAMP assays able to detect *Psa* and to discriminate strains belonging to the highly virulent and globally spreading *Psa* biovar 3. Both the LAMP assays showed specificity in accordance with their target and were able to detect reliably 125 cfu per reaction in less than 30 min. The developed assays were also able to detect the presence of *Psa* in naturally infected kiwifruit material with and without symptoms, thereby increasing the potential of the LAMP assays for phytosanitary use.

Chen et al. (2018) raised a polyclonal antiserum to the bacterial effector-hopz5 (PAb:hopz5) and tested its specificity. PAb:hopz5 was able to detect *Psa* from culture and infected plant samples, thus representing a suitable tool for the immunodetection of the agent of kiwifruit bacterial canker in field samples. The detection sensitivity of a colloidal gold immunochromatographic strip was high, up to 2.2×10^3 cfu mL^{-1}. Therefore, PAb:hopz5 constitutes a suitable and accurate tool for detection of *Psa* and haplotype distribution.

Kim et al. (2018) standardized an infection risk model used in New Zealand by modifying the temperature and rainfall functions of the existing model for use in Korea. Analyses using statistical correlation and prediction-realization tables revealed that the modified model is valid with high agreement (a correlation coefficient of 0.85 and an accuracy of 85.7%, respectively) between the observed disease incidence and simulated disease risk from the model. The model was also found to be highly sensitive to the presence or absence of rainfall than any other weather variable inputs. The modified *Psa* risk model can be used to provide practical and applicable information for timely disease control to the kiwifruit growers in Korea.

7.61.5 Management

The following measures are helpful in minimizing the disease incidence.

1. Use of disease-free nursery plants
2. Use of windbreaks to prevent injuries to the vines from strong winds
3. During pruning, infected vines show bacterial ooze at the cut surface. In such cases, these should be cut back to the healthy portion
4. During pruning, the scissors should be disinfected regularly with disinfectants like sodium hypochlorite solution
5. Avoidance or minimum use of overhead irrigation
6. Streptomycin and copper sulphate are generally used to control the disease in orchards. However, their continuous use has resulted in the development of pathogen strains resistant to both the chemicals and this has created a serious problem in the management of the disease. The strains of *Psa* resistant to streptomycin and copper sulphate occur in Japan (Goto et al. 1994; Lee et al. 2005) and those of *Psm. syringae* in New Zealand (Vanneste and Voyl 2003). Han et al. (2004) also showed that Japanese and Korean isolates of *Psa* and *Psm syringae* pv. *syringae* harbored *strA-strB* streptomycin resistance genes for a streptomycin-resistance determinant

In case of the appearance of copper/antibiotic-resistant strains of the pathogen, Scortichini (2014) compared the field efficacy of a chitosan-based compound with a copper compound in a 3-year trial. Chitosan did not cause any phytotoxic effect on plants and fruits and showed an overall higher level of performance than the copper compound in reducing the disease severity throughout the trial. Chitosan also significantly reduced the presence of exudates on trunk and leaders recorded at the end of winter.

Scortichini (2016) tested the field efficacy of a zinc (4.7%)-copper (2.6%)-hydracid of citric acid (21.4%) biocomplex compound (applied three times, at a dose of 1.0%) to reduce the occurrence of bacterial exudates oozing out from the main trunk and leaders in early spring, during winters of 2013–2014 and 2014–2015 at four sites in Italy. The biocomplex compound was more effective than 1% Bordeaux mixture used under the same conditions. This biocomplex formulation could be effectively used to reduce the amount of exudates produced by the bacterium in areas where winter frost occurs regularly.

REFERENCES

Balestra, G. M., Mazaglia, A., Quattrucci, A., Renzi, M., and Rossetti, A. 2009. Occurrence of *Pseudomonas syringae* pv. *actinidiae* in Jin Tao kiwi plants in Italy. *Phytopathol. Mediterr.* 48: 299–301.

Balestra, G. M., Rugni, E., and Varvaro, L. 2001. Increased susceptibility to *Pseudomonas syringae* pv. *syringae* and *Pseudomonas viridiflava* of kiwi plants having transgenic *rolABC* genes and its inheritance in the T1 offspring. *Phytopathol. Z.* 149: 189–194.

Balestra, G. M., Taratufolo, M. C., Vinatzer, B. A., and Mazzaglia, A. 2013. A multiplex PCR assay for detection of *Pseudomonas syringae* pv. *actinidiae* and differentiation of populations with different geographic origin. *Plant Dis.* 97: 472–478.

Beatrice, C., Linthorst, J. M. H., Cinzia, F., and Luca, R. 2017. Enhancement of *PR1* and *PR5* gene expressions by chitosan treatment in kiwifruit plants inoculated with *Pseudomonas syringae* pv. *actinidiae*. *Eur. J. Plant Pathol.* 148: 163–179.

Chapman, J. R., Taylor, R. K., Weir, B. S., Romberg, M. K., Vanneste, J. L., et al. 2012. Phylogenetic relationships among global populations of *Pseudomonas syringae* pv. *actinidiae*. *Phytopathology* 102: 1034–1044.

Chen, H., Hu, Y., Qin, K., Yang, X., Jia, Z., et al. 2018. A serological approach for the identification of the effector *hopz5* of *Pseudomonas syringae* pv. *actinidiae*: A tool for the rapid immunodetection of kiwifruit bacterial canker. *J. Plant Pathol.* 100: 171–177.

Cunty, A., Poliakoff, F., Rivoal, C., Cesbron, S., Fischer-Le Saux, M., et al. 2015. Characterization of *Pseudomonas syringae* pv. *actinidiae* (Psa) isolated from France and assignment of Psa biovar 4 to a *de novo* pathovar: *Pseudomonas syringae* pv. *actinidifoliorum* pv. nov. *Plant Pathol.* 64: 582–596.

Ferrante, P., and Scortichini, M. 2010. Molecular and phenotypic features of *Pseudomonas syringae* pv. *actinidiae* isolated during recent epidemics of bacterial canker on yellow kiwifruit (*Actinidia chinensis*) in central Italy. *Plant Pathol.* 59: 954–962.

Ferrante, P., and Scortichini, M. 2014. Frost promotes the pathogenicity of *Pseudomonas syringae* pv. *actinidiae* in *Actinidia chinensis* and *A. deliciosa* plants. *Plant Pathol.* 63: 12–19.

Ferrante, P., and Scortichini, M. 2015. Redefining the global populations of *Pseudomonas syringae* pv. *actinidiae* based on pathogenic, molecular and phenotypic characteristics. *Plant Pathol.* 64: 51–62.

Goto, M. 1992. *Fundamentals of Bacterial Plant Pathology*. Academic Press, San Diego, CA, 342 pp.

Goto, M., Hikota, T., Nakajima, M., Takikawa, Y., and Tsuyumu, S. 1994. Occurrence and properties of copper-resistance in plant pathogenic bacteria. *Ann. Phytopathol. Soc. Japan* 60: 147–153.

Han, H. S., Koh, Y. J., Hur, J.-S., and Jung, J. S. 2003. Identification and characterization of coronatine-producing *Pseudomonas syringae* pv. *actinidiae*. *J. Microbiol. Biotechnol.* 13: 110–118.

Han, H. S., Koh, Y. J., Hur, J.-S., and Jung, S. 2004. Occurrence of the *strA-strB* streptomycin resistance genes in *Pseudomonas* species isolated from kiwifruit plants. *J. Microbiol.* 42: 365–368.

Kim, G. H., Jung, J. S., and Koh, Y. J. 2017. Occurrence and epidemics of bacterial canker of kiwifruit in Korea. *The Plant Pathol. J.* 33: 351–361.

Kim, K.-H., Son, K. I., and Koh, Y. J. 2018. Adaptation of the New Zealand Psa risk model for forecasting kiwifruit bacterial canker in Korea. *Plant Pathol.* 67: 1208–1219.

Koh, Y. J., Kim, G. H., Lee, Y. S., Sohn, S. H., Koh, H. S., et al. 2012. *Pectobacterium carotovorum* subsp. *actinidiae* subsp. nov., a new bacterial pathogen causing canker-like symptoms in yellow kiwifruit, *Actinidia chinensis*. *N. Z. J. Crop Hortic. Sci.* 40: 269–279.

Lee, J. H., Kim, J. H., Kim, G. H., Jung, J. S., Hur, J.-S., et al. 2005. Comparative analysis of Korean and Japanese strains of *Pseudomonas syringae* pv. *actinidiae* causing bacterial canker of kiwifruit. *Plant Pathol. J.* 21: 119–126.

Li, M., Tan, G., Li, Y., Cheng, H., and Zhou, Z. 2005. Relationship between contents of lignin and soluble sugar in plants of kiwifruit cultivars and their resistance to kiwifruit bacterial canker infected by *Pseudomonas syringae* pv. *actinidiae*. *Acta Phytophylacica Sin.* 32: 138–142.

Masami, N., Masao, G., Katsumi, A., and Tadaaki, H. 2004. Nucleotide sequence and organization of copper resistance genes from *Pseudomonas syringae* pv. *actinidiae*. *Eur. J. Plant Pathol.* 110: 223–226.

Patel, H. K., Ferrante, P., Xianfa, M., Javvadi, S. G., Subramoni, S., et al. 2017. Identification of loci of *Pseudomonas syringae* pv. *actinidiae* involved in lipolytic activity and their role in colonization of kiwifruit leaves. *Phytopathology* 107: 645–653.

Rees-George, J., Vanneste, J. L., Cornish, D. A., Pushparajah, I. P. S., Yu. J., et al. 2010. Detection of *Pseudomonas syringae* pv. *actinidiae* using polymerase chain reaction (PCR) primers based on the 16S-23S rDNA intertranscribed spacer region and comparison with PCR primers based on other gene regions. *Plant Pathol.* 59: 453–464.

Ruinelli, M., Schneeberger, P. H. H., Ferrante, P., Bühlmann, A., Scortichini, M., et al. 2017. Comparative genomics-informed design of two LAMP assays for detection of the kiwifruit pathogen *Pseudomonas syringae* pv. *actinidiae* and discrimination of isolates belonging to the pandemic biovar 3. *Plant Pathol.* 66: 140–149.

Scortichini, M. 2014. Field efficacy of chitosan to control *Pseudomonas syringae* pv. *actinidiae*, the causal agent of kiwifruit bacterial canker. *Eur. J. Plant Pathol.* 140: 887–892.

Scortichini, M. 2016. Field efficacy of a zinc-copper-hydracid of citric acid biocomplex compound to reduce oozing from winter cankers caused by *Pseudomonas syringae* pv. *actinidiae* to *Actinidia* spp. *J. Plant Pathol.* 98: 651–655.

Serizawa, S., Ichikawa, T., Takikawa, Y., Tsuyumu, S., and Goto, M. 1989. Occurrence of bacterial canker of kiwifruit in Japan: Description of symptoms, isolation of the pathogen and screening of bactericides. *Ann. Phytopathol. Soc. Japan* 55: 427–436.

Takikawa, Y., Serizawa, S., Ichikawa, T., Tsuyumu, S., and Goto, M. 1989. *Pseudomonas syringae* pv. *actinidiae* pv. nov.: The causal bacterium of canker of kiwifruit in Japan. *Ann. Phytopathol. Soc. Japan* 55: 437–444.

Ushiyama, K., Kita, N., Aono, N., Ogawa, J., and Fujii, H. 1992. Bacterial canker disease of wild actinidia plants as the infection source outbreak of bacterial canker of kiwifruit caused by *Pseudomonas syringae* pv. *actinidae*. *Ann. Phytopathol. Soc. Japan* 58: 426–430.

van Hall, C. J. J. 1902. *Bijdragen tot de kennis der Bakterieele Plantenzeikten*. Coöperatieve Drukkerij-vereeniging "Plantijn," Inaugural dissertation, Amsterdam, the Netherlands.

Vanneste, J. L., and Voyl, M. D. 2003. Genetic basis of copper resistance in New Zealand strains of *Pseudomonas syringae*. *N. Z. Plant Prot.* 56: 109–112.

Wu, W. X., Liu, Y., Huang, X. Q., and Zhang, L. 2017. First report of summer canker caused by *Pectobacterium carotovorum* subsp. *actinidiae* on kiwifruit in China's Sichuan Province. *Plant Dis.* 101: 1540.

7.62 ANGULAR LEAF SPOT OF STRAWBERRY

Angular leaf spot (also called **bacterial angular leaf spot**) of strawberry (*Fragaria* × *ananassa* Duch.) was first reported from Minnesota, USA in 1960 and subsequently it was found in other states, namely, California, Florida, Kentucky, and Wisconsin of the USA. The disease probably spread from there, with infected planting material, to other continents, but this is only a presumption. The disease also occurs in France (Rat 1974), Greece (Panagopoulos et al. 1978), Israel, Italy (including Sicily; Mazzucchi et al. 1973), Portugal (Fernandes and Pinto-Ganhao 1981), Romania (Severin et al. 1985), Spain (Lopez et al. 1985), Switzerland (Grimm et al. 1993), Israel, Taiwan, Ethiopia, Argentina (Alippi et al. 1989), Brazil, Belgium, Ecuador, Paraguay, Uruguay, Venezuela, and New Zealand. Fernández-Pavía et al. (2014) reported its occurrence from Mexico.

Fragaria ananassa, the predominant cultivated strawberry, whose progenitors derive from hybridization between *F. chiloensis* and *F. virginiana*, is the main host, but its many cultivars vary greatly in susceptibility. *F. virginiana*, *F. vesca*, *Potentilla fruticosa*, and *P. glandulosa* have been infected following experimental inoculation. Among *Fragaria* spp., only *F. moschata* is immune (Kennedy and King 1962a; Kennedy 1965; Maas 1984). Cultivated strawberries are the host of concern throughout the European and Mediterranean Plant Protection Organization (EPPO) region. Generally, the disease is not destructive, but heavy losses may occur with frequent overhead/sprinkler irrigation.

7.62.1 Symptoms

The symptoms first appear as minute, water-soaked spots on the undersides of leaves. As the lesions enlarge, they become delineated by small veins and become angular in shape. Symptoms appear translucent when viewed with transmitted light and dark green when viewed with reflected light. This is an important distinguishing characteristic of the disease. The spots enlarge, coalesce, and after about 2 weeks are also visible on the upper surface as water-soaked, angular, and reddish-brown necrotic spots (Figure 7.45). A chlorotic halo may surround the lesion, and at this stage, it is difficult

FIGURE 7.45 Angular leaf spot of strawberry. (Courtesy of Don Ferrin, Louisiana State University Agricultural Center, Bugwood.org.)

to distinguish the symptoms from those of common leaf spot and leaf scorch caused by *Mycosphaerella fragariae* and *Diplocarpon earliana*, respectively. It is also possible that the disease can be confused with the symptoms caused by a new pathovar of *Xanthomonas*, *Xa. arboricola* pv. *fragariae* (Janse et al. 2001). However, this new disease, called bacterial leaf blight, has been recorded only in Italy and the pathogen is considered to be only weakly pathogenic. This bacterium exists more commonly as an epiphyte, and it has not reappeared as a pathogen since it was first observed in 1993–1995 (M. Scortichini, personal communication). Under humid conditions, the lesions on the lower surface of leaves may exude viscous bacterial ooze that dries to a cream to brown colored crusty film. Spots coalesce more frequently along the primary and secondary veins. Entire leaves may die if major veins become infected. The dead tissues tear and break off, and the diseased leaf may assume a ragged appearance.

While leaf lesions are the most common symptoms, the bacterium can also infect other plant parts. Infected calyx tissue first appears dark green and water-soaked, but later turns brown or black and dry up. Fruit tissue nearest the infected calyx becomes water-soaked, resulting in unmarketability of the fruits. The bacterium can move into the roots, crowns, and stolons without causing any apparent symptoms. In the most severe cases, infected pockets may be seen inside the crown tissue after dissection. They appear as localized, water-soaked zones, frequently confined to one side of the crown.

Primary infection by *Xa. fragariae* occurs through stomata on the underside of the leaves. Healthy, young rapidly growing leaves are the most susceptible. Emerging leaves and older leaves are not usually attacked. Although primarily a disease of the leaves, it can also affect petioles, runners, and flowers. Severe infections result when bacteria invade the crown tissue and cause collapse of the vascular system. All types of cells in the vascular system may be affected when the infection is systemic.

Apart from symptoms, the presence of the disease can be ascertained by immunofluorescence antibody staining technique (Mazzucchi et al. 1973) and enzyme linked immunosorbent assay (Lopez et al. 1987).

Gétaz et al. (2017) designed a reliable and sensitive LAMP assay using a unique marker, providing a highly specific and rapid detection technique, convenient for on-site detection. The specificity of the designed assay was tested on 37 strains from a culture collection of *Xa. fragariae*, 82 strains of other *Xanthomonas* species and pathovars, and 11 strains of other bacterial genera isolated from strawberry leaves. A detection limit of 10^2 fg was achieved, approximating to 20 genome copies per reaction. A consistent lower detection efficiency of 10^2 cfu mL^{-1} was achieved, when the analysis was performed with crude plant material. The LAMP assay designed in this study was adapted to work on crude plant material without any prior extensive extraction steps or incubation period. Moreover, it does not require advanced analytical knowledge or a fully equipped laboratory. Results were achieved within 7–20 min, depending on the pathogen concentration, thus providing a high-throughput and user-friendly method for detection and screening of plant material for quarantine regulations.

7.62.2 Causal Organism

Xanthomonas fragariae Kennedy and King 1962b

Xa. fragariae is slow growing and fastidious in culturing, hence, its isolation is difficult. For isolation, choose leaves with young lesions and these should not be surface sterilized as this may kill or significantly reduce the number of pathogen's cells. Surface debris can be washed off with tap water followed by sterile distilled water. It is preferable not to macerate the tissue for isolation as this can increase the dispersion of saprophytic bacteria. For each sample, use a dissecting microscope and excise very small pieces of tissue, not more than 1 mm wide, from the margin of the lesions. Place a few pieces of freshly cut tissue into a few drops of sterile distilled water or phosphate saline buffer. Allow approximately 10–20 min for the bacteria to egress and streak the suspension on to Wilbrinks-N medium plates. Yeast peptone glucose agar or sucrose peptone agar can also be used, but these are reported to be less successful. Use of purified microbiological grade agar is recommended in all media because impurities from other commercial agars can inhibit the growth of *Xa. fragariae*.

Incubate the plates at 25°C–28°C and examine after 3 days. Use a dissecting microscope to examine the streaked plates for the presence of the slow growing *Xa. fragariae* colonies, which may arise along the streaks between the larger colonies of saprophytes. If possible use a reference strain for comparison. If colonies resembling *Xanthomonas* appear macroscopically after 3 days of incubation, these are not of *Xa. fragariae*. Mark the colonies and continue to incubate. Individual colonies of *Xa. fragariae* will be visible approximately after 7 days of incubation.

The colonies are circular, 0.5–1.0 mm in diameter, yellow-pigmented, and dome-shaped with entire edges. On beef extract peptone agar, or similar media without added carbohydrate, colonies are circular, entire, convex, glistening, and translucent to pale-yellow (Bradbury 1977). The pure culture is distinguishable from other phytopathogenic xanthomonads by at least seven characteristics, i.e., no growth at 33°C, no hydrolysis of aesculin, no acid from arabinose, galactose, trehalose, and cellobiose, and tolerance to 0.5%–1.0% NaCl (Kennedy and King 1962b; Bradbury 1984).

Xa. fragariae is an aerobic, Gram-stain-negative, nonsporing, non-capsulated, rod-shaped bacterium, measuring 0.4 × 1.3 μm, with a single polar flagellum. It prefers growth media containing high levels of available carbohydrate and colonies are shiny, pale yellow, circular, entire, convex, and mucoid. Sub cultured colonies are just visible after 3 days of incubation at 25°C.

Xa. fragariae has the common characteristics of all xanthomonads: Gram-stain-negative, aerobic rods, with a single polar flagellum, nitrates not reduced, catalase positive, asparagine not used as a sole source of carbon and nitrogen, produces xanthomonadin, and weak production of acids from carbohydrates. Colonies are mucoid, convex, and shiny on yeast peptone glucose agar and Wilbrinks-N media (Dye 1962; van den Mooter and Swings 1990; Swings et al. 1993). Growth of *Xa. fragariae* is poor on nutrient agar, but much better when supplemented with 5% glucose.

Vandroemme et al. (2013), on the basis of polyphasic analysis, reported: (i) no clear criteria exist for the identification of strains as *Xa. arboricola* pv. *fragariae*, (ii) the name *Xa. arboricola* pv. *fragariae* is currently used for a genetically diverse assortment of strains, and (iii) the species *Xa. arboricola* holds many undetermined plant-associated bacteria besides the described pathovars.

PCR is used in the United States for detection of *Xa. fragariae* from symptomatic plants. A nested PCR has been used to detect *Xa. fragariae* in asymptomatic plants.

EPPO lists *Xa. fragariae* as an A2 quarantine pathogen, while Inter-African Phytosanitary Council also considers it as of quarantine significance.

7.62.3 Disease Cycle

The bacterium does not survive freely in the soil, but instead survives in infected dead leaves on or in the soil. This bacterium is resistant to adverse conditions such as desiccation and can survive for long periods in dry leaf debris or buried leaves in soil. In dried infected leaves, kept in the laboratory, the bacterium may survive for at least 2.5 years. Bacterial cells in the residue infect the young leaves at the beginning of the growing season. Crowns of systemically infected (live) plants are another source for overwintering inoculum. Systemically infected plants have the bacteria present in their vascular system. The bacterium is generally introduced into a planting on symptomless systemically infected nursery plants. From crown infection pockets, the bacterium causes lesions along the veins at the base of the youngest leaves, which develop in the apical crown region. Commercial strawberry runners used for planting may spread the bacterium over short and long distances. They may still bear old, whole, or torn, infected leaves or have crown infection pockets. Moreover, almost invisible fragments of infected leaves may be hidden in the apical crown region or between the roots (Kennedy and King 1962a).

The penetration of the bacterium occurs through the stomata. Infections of the crowns occur through local wounds or downwards from the affected leaves. During the growing season, several cycles of secondary infections may occur. The bacterial cells exuded from primary lesions on the undersides of leaves serve as the source of secondary spread in plantings. They are spread by rain, wind-splashed rain, and overhead/sprinkler irrigation. van der Wolf et al. (2018) reported that there is a considerable risk of infections of strawberry plants exposed to aerosolized inoculum.

During epiphytotics, when environmental conditions favor exudation and spread, the bacterium may cause systemic infections associated with crown pockets. Systemic infections may arise under damp nursery conditions. The favorable conditions for infection are cool to moderate daytime temperature (18°C–21°C), cold night time temperature (near freezing), and high relative humidity (Maas 1984). Long periods of precipitation, sprinkler irrigation, and heavy dews in the spring also favor disease development. Succulent leaves and calyx tissue are most susceptible to infection.

Kastelein et al. (2014), on the basis of greenhouse studies using a *Xa. fragariae* isolate tagged with a green fluorescent protein reported that during the first 3 days after inoculation, bacterial populations washed off leaves rapidly decreased by at least a factor of 1,000, after which populations remained stable until 14 days post-inoculation, when symptoms first started to appear. After that, populations increased to a level of 10^{12} cfu g^{-1} of leaf material or higher. The densities in leaf extracts were also low (100–1,000 cfu g^{-1} of leaf tissue) during the first 3 days after inoculation. Gradually, populations increased to a level of 10^{9-12} cfu g^{-1} at 28 days post-inoculation. Higher densities of epiphytic populations were found on the abaxial side than on the adaxial leaf side during the first 2 weeks after inoculation. After spray-inoculation of leaves, bacterial populations released from infected plants remained low until symptoms appeared, after which plants became highly infectious, particularly under high humidity.

Bestfleisch et al. (2015) investigated the systemic dispersal of the pathogen in strawberry plants after inoculation of leaves with *Xa. fragariae* strain XF3.9.C and the GFP-tagged strain XF3.9.$C_{(pKAN)}$. The observations were taken 3, 7, 14, and 28 days after inoculation using nested PCR and fluorescence microscopy. After 3 days post-inoculation, *Xa. fragariae* could be found in all tissues tested including the inoculated leaf, its petiole, the rhizome, the heart bud up to the youngest fully expanded leaf, and its petiole. The systemic spread was also detectable in partially resistant genotypes.

Hartung et al. (2003) categorized two genotypes, a native *F. virginiana* from Minnesota and a hybrid between *F. virginiana* from Georgia and *F.* × *ananassa* "Earliglow," resistant to *Xa. fragariae* after screening 81 *Fragaria* genotypes, including both diploid and octoploid accessions. These two resistant genotypes designated as US 4808 and US 4809 were made available to the public as germplasm releases. In controlled crosses between the susceptible variety "Sweet Charlie" and these two resistant genotypes, resistance to *Xa. fragariae* was transmitted to 8%–12% of the progeny of the US 4808 cross and to 4%–18% of the progeny of the US 4809 cross. Lewers et al. (2003) reported that segregation ratios of 7S:1R and 15S:1R obtained suggest that three or four unlinked loci could explain the inheritance of resistance in F_1 seedlings of crosses of susceptible and resistant genotypes.

7.62.4 Management

1. The use of disease-free transplants is extremely important in preventing the introduction of the disease. Always use certified planting materials. EPPO recommends, in its specific quarantine requirements

(OEPP/EPPO 1990), that strawberry planting material from infested countries should be derived from mother plants kept free from *Xa. fragariae* as part of a certification scheme, and in addition, the place of production should have been found free from the disease during the last five growing seasons. In addition, visual inspections during the dormant period can be useful. Inspectors should look for typical angular spots on old leaves or on their remains still attached to the runners. Samples from lots kept in cold storage must be inspected immediately after the runners are taken out and thawed because the spots cannot be seen after keeping them at room temperature even for a day
2. Follow crop rotation and avoid overhead-irrigation when possible
3. To minimize spread of angular leaf spot within infected fields, avoid moving equipment or harvesting when plants are wet. This also reduces wounding of plants when the bacteria are more likely to spread
4. Use of tolerant or less susceptible cultivars will be helpful in managing the disease
5. Use of chemicals to control the disease is generally not very effective. Protectant applications of fixed copper fungicides have been used with varying success. Low doses of copper fungicides are recommended since repeated applications can result in phytotoxicity. Phytotoxicity has been reported after 6–7 successive copper applications. Moreover, the applications of copper after the initiation of bloom may result in more damage than the disease, especially under cool, slow drying conditions. Turechek et al. (2013) reported that a strategic combination of control practices that includes heat treatment should reduce significantly the initial amount of bacteria introduced into a field as none of the treatments completely eliminated *Xa. fragariae* from the planting stock.

REFERENCES

Alippi, A. M., Ronco, B. L., and Carranza, M. R. 1989. Angular leaf spot of strawberry, a new disease in Argentina. Comparative control with antibiotics and fungicides. *Adv. Hort. Sci.* 1: 3–6.

Bestfleisch, M., Richter, K., Wensing, A., Wünsche, J. N., Hanke, M.-V., et al. 2015. Resistance and systemic dispersal of *Xanthomonas fragariae* in strawberry germplasm (*Fragaria* L.). *Plant Pathol.* 64: 71–80.

Bradbury, J. F. 1977. *Xanthomonas fragariae. CMI Descriptions of Pathogenic Fungi and Bacteria No. 558.* CAB International, Wallingford, UK.

Bradbury, J. F. 1984. *Xanthomonas.* in: *Bergey's Manual of Systematic Bacteriology*, vol. 1, N. R. Krieg, and J. G. Holt eds. Williams & Wilkins, Baltimore, MD.

Dye, D. W. 1962. The inadequacy of the usual determinative test for the identification of *Xanthomonas* spp. *N. Z. J. Sci.* 5: 393–416.

Fernandes, A. M. M., and Pinto-Ganhao, J. F. 1981. *Xanthomonas fragariae* Kennedy & King—A new bacterial disease in Portugal. *Agros* 64: 5–8.

Fernández-Pavía, S. P., Rodríguez-Alvarado, G., Garay-Serrano, E., and Cárdenas-Navarro, R. 2014. First report of *Xanthomonas fragariae* causing angular leaf spot on strawberry plants in México. *Plant Dis.* 98: 682.

Gétaz, M., Bühlmann, A., Schneeberger, P. H. H., Van Malderghem, C., Duffy, B., et al. 2017. A diagnostic tool for improved detection of *Xanthomonas fragariae* using a rapid and highly specific LAMP assay designed with comparative genomics. *Plant Pathol.* 66: 1094–1102.

Grimm, R., Lips, T., and Vogelsanger, J. 1993. Angular leaf spot of strawberry. *Schweiz. Zeitsch. Obst Weinbau* 128: 130–31.

Hartung, J. S., Gouin, C. C., Lewers, K. S., Maas, J. L., and Hokanson, S. 2003. Identification of sources of resistance to bacterial angular leaf spot disease of strawberry. *Acta Hortic.* (*ISHS*) 626: 155–159.

Janse, J. D., Rossi, M. P., Gorkink, R. F. J., Derks, J. H. J., Swings, J., et al. 2001. Bacterial leaf blight of strawberry (*Fragaria* (x) *ananassa*) caused by a pathovar of *Xanthomonas arboricola*, not similar to *Xanthomonas fragariae* Kennedy & King. Description of the causal organism as *Xanthomonas arboricola* pv. *fragariae* (pv. nov., comb. nov.). *Plant Pathol.* 50: 653–665.

Kastelein, P., Krijger, M., Czajkowski, R., van der Zouwen, P. S., van der Schoor, R., et al. 2014. Development of *Xanthomonas fragariae* populations and disease progression in strawberry plants after spray-inoculation of leaves. *Plant Pathol.* 63: 255–263.

Kennedy, B. W. 1965. Infection of *Potentilla* by *Xanthomonas fragariae. Plant Dis. Reptr* 49: 491–492.

Kennedy, B. W., and King, T. H. 1962a. Studies on epidemiology of bacterial angular leaf spot on strawberry. *Plant Dis. Reptr* 46: 360–363.

Kennedy, B. W., and King, T. H. 1962b. Angular leaf spot of strawberry caused by *Xanthomonas fragariae* sp. nov. *Phytopathology* 52: 873–875.

Lewers, K. S., Maas, J. L., Hokanson, S. C., Gouin, C., and Hartung J. S. 2003. Inheritance of resistance in strawberry to bacterial angular leaf spot disease caused by *Xanthomonas fragariae. J. Amer. Soc. Hort. Sci.* 128: 209–212.

Lopez, M. M., Aramburu, J. M., Cambra, M., and Borras, V. 1985. Detection and identification of *Xanthomonas fragariae* in Spain. *Anales del Instituto Nacional de Investigaciones Agrarias. Serie Agricola* 28 (Suppl.): 245–260.

Lopez, M. M., Cambra, M., Aramburu, J. M., and Bolinches, J. 1987. Problems of detecting phytopathogenic bacteria by ELISA. *Bull.OEPP/EPPO Bull.* 17: 113–117.

Maas, J. L. 1984. *Compendium of Strawberry Diseases*. American Phytopathology Society, St. Paul, MN.

Mazzucchi, U., Alberghina, A., and Dalli, A. 1973. Occurrence of *Xanthomonas fragariae* Kennedy & King in Italy. *Phytopathol. Z.* 76: 367–370.

OEPP/EPPO. 1990. Specific quarantine requirements. EPPO Technical Documents No. 1008.

Panagopoulos, C. G., Psallidas, P. G., and Alivizatos, A. S. 1978. A bacterial leaf spot of strawberry in Greece caused by *Xanthomonas fragariae. Phytopathol. Z.* 91: 33–38.

Rat, B. 1974. Présence en France de la maladie des taches angulaires du fraisier. *Ann. Phytopathol.* 6: 223.

Severin, V., Stancescu, C., and Zambrowicz, E. 1985. Angular leaf spot, a new bacterial disease of strawberry in Romania. Buletinul de Protectia Plantelor Nos. 1–2, 21–23.

Swings, J., Vauterin, L., and Kersters, K. 1993. The bacterium *Xanthomonas*. pp. 138–44 in: *Xanthomonas*. J. Swings, and E. L. Civerolo eds. Chapman & Hall, London, UK.

Turechek, W. W., Wang, S., Tiwari, G., and Peres, N. A. 2013. Investigating alternative strategies for managing bacterial angular leaf spot in strawberry nursery production. *Int. J. Fruit Sci.* 13: 234–245.

van den Mooter, M., and Swings, J. 1990. Numerical analysis of 295 phenotypic features of 266 *Xanthomonas* strains and related strains and an improved taxonomy of the genus. *Int. J. Syst. Bacteriol.* 40: 348–369.

van der Wolf, J. M., Evenhuis, A., Kastelein, P., Krijger, M. C., Funke, V. Z., et al. 2018. Risks for infection of strawberry plants with an aerosolized inoculum of *Xanthomonas fragariae*. *Eur. J. Plant Pathol.* 152: 711–722.

Vandroemme, J., Cottyn, B., Pothier, J. F., Pflüger, V., Duffy, B., et al. 2013. *Xanthomonas arboricola* pv. *fragariae*: What's in a name? *Plant Pathol.* 62: 1123–1131.

7.63 BACTERIAL FRUIT BLOTCH OF WATERMELON

The disease was first observed in 1965 on watermelon at the USDA Plant Introduction Station, Griffin, Georgia, where it caused seedling blight symptoms. The disease was confined to seedlings at the station, suggesting that it was introduced by seeds of plant introductions. However, no outbreaks were reported in commercial watermelon fields during this period. Crall and Schenck (1969) described a bacterial fruit rot of watermelon that they observed at Leesburg, Florida experiment station during the previous 2 years. They described the symptoms of the disease, but did not determine the etiology. Wall (1989) for the first time showed the relationship between the seedling disease and fruit symptoms. Latin and Hopkins (1995) have given an elaborate account of the disease chronology.

By 1988, the disease outbreaks were observed in commercial watermelon fields in the Mariana Islands (in the North Pacific Ocean). It is apparent that the source of inoculum in the commercial watermelons fields was independent of the seedling blight outbreak at the USDA Plant Introduction Station in Georgia. In spring 1989, a bacterial fruit rot of watermelon appeared in commercial watermelon fields in Florida and, as the season progressed, the disease was observed in South Carolina, North Carolina, Maryland, Delaware, and Indiana. In some fields, the losses were more than 90% of the total marketable fruit. Since then, the disease has also been found in Alabama, Arkansas, California, Georgia, Iowa, Mississippi, Missouri, Oklahoma, and Texas. Since 1989, bacterial fruit blotch has occurred in one or more states in the eastern US every year. Fortunately, in most of these years, the disease occurred in relatively few fields, though it was devastating in some of the fields. Fruit blotch was widespread in Georgia in 1992. Most notably, the disease caused significant losses in 1994 when the seed-borne inoculum of the pathogen, along with the lack of adequate seed health assays, resulted in its widespread outbreaks in thousands of acres distributed over at least ten states, including Florida, Georgia, Indiana, South Carolina, and Texas.

These outbreaks created a national scare and highlighted the general threat posed by seed-borne diseases. The repercussions of these outbreaks resulted in direct economic losses to the growers, as well as costly lawsuits against seed and transplant producers. The economic magnitude of these lawsuits forced some seed companies out of business, while others suspended watermelon seed sales in certain "high risk" states. After the implementation of routine seed health testing, many companies resumed the sale of watermelon seed with reduced risk of *Acidovorax citrulli* transmission. In spite of the implementation of routine seed health testing, sporadic outbreaks of the disease continued to occur, and widespread outbreaks occurred in 2000 and 2001 across the US (Walcott 2005). The disease has also been reported from Australia in 1986, but has not reoccurred. Bacterial fruit blotch also occurs in melon (cantaloupe and honeydew), cucumber, pumpkin, squash, gourds, citron melon, and other cucurbits, but watermelon is the preferred host.

The widespread geographical distribution of the disease indicates that seed is still an important primary source of inoculum for the disease. To guard against costly lawsuits, seed producers in the US include a disclaimer with each watermelon seed package, stating that even though seeds have been tested, there is no guarantee that these are free of *Aci. citrulli*. At the time of seed purchase, growers are also required to waive their rights to pursue legal action against seed producers in the event of a disease outbreak (Walcott 2005).

Burdman and Walcott (2012) have summarized the current knowledge on *Aci. citrulli*, with emphasis on its epidemiology and the factors involved in its pathogenicity and virulence.

7.63.1 Symptoms

The symptoms are generally found on seedling transplants, mature leaves, and fruits. The first symptoms in watermelon seedlings appear as dark, water-soaked "oily" areas on the lower surface of cotyledons and leaves. As cotyledons expand, lesions become dark brown and often extend along the length of midrib. In young seedlings, lesions can occur in the hypocotyl, resulting in collapse and death of the plant (Figure 7.46a). Foliar symptoms develop throughout the growing season, but in many cases, they are neither abundant nor very distinctive and may easily go unnoticed. Leaf lesions are small, dark brown, somewhat angular, and generally inconspicuous. When viewed from the underside of the leaf, the margins of the lesions appear water-soaked, especially in humid weather. Leaf lesions in the field do not result in defoliation, but serve as an important source of inoculum for fruit infection. The symptoms generally do not develop on stems, petioles, and roots. Under certain conditions, foliar symptoms are not very conspicuous, and the growers fail to realize the gravity of the problem until fruit infection renders the crop unmarketable. Therefore, close inspection of plants is very essential for early detection of the disease.

On cantaloupe leaves, symptoms include tan to reddish-brown lesions along leaf veins, but also include V-shaped lesions that extend from the margin to the base of the leaf.

(a)

(b)

FIGURE 7.46 Bacterial fruit blotch of watermelon: (a) Leaf lesions. (Courtesy of Jason Brock, University of Georgia, Bugwood.org.) (b) Symptoms on fruit. (Courtesy of Rebecca A. Melanson, Mississippi State University Extension, Bugwood.org.)

Infection of watermelon fruits occurs at flowering and early fruit set. Two- to three-week-old developing fruits that have not formed a wax layer are most susceptible to infection. It is more difficult for the bacterium to invade a fruit that matures and develops a wax layer. The diagnostic symptom on fruit is a dark olive green stain or "**blotch**" on the upper surface of a developing fruit. Initially, the blotch may be ½ inch in diameter, but it enlarges rapidly to become dark green, water-soaked lesion several centimeters in diameter having irregular margins. If environmental conditions are favorable for infection, these lesions may rapidly expand to cover much of the upper fruit surface within 7–10 days (Figure 7.46b). In fruits of watermelon cultivars with light and dark green striped rinds, the water-soaking may be restricted, resulting in smaller lesions on the lighter green stripe. Typical water-soaking may not occur under certain environmental conditions and in infection of very young fruits. As the blotch increases in size, the initial infection site turns necrotic. Fruit lesions rarely extend into the flesh of the melon. In advanced stage of lesion development, the periderm of the fruit may crack and a total fruit rot may develop; secondary organisms are responsible for ultimate decay and collapse of the fruit. After cracking, sticky, white bacterial ooze frequently exudes from the fruits.

7.63.2 Causal Organism

Acidovorax citrulli (Schaad et al. 1978) Schaad et al. 2009 comb. nov.

Syn. *Aci. avenae* subsp. *citrulli*, *Pseudomonas pseudoalcaligenes* subsp. *citrulli*, *Psm. avenae* subsp. *citrulli*

Schaad et al. (2008), on the basis of genetic and phenotypic results, proposed an emendation of the species *Aci. avenae* and elevated *Aci. avenae* subsp. *citrulli* to species rank as *Acidovorax citrulli*.

The cells are Gram-stain-negative rods with average dimensions of 0.5 × 1.7 µm, have one polar flagellum, and obligately aerobic. Colonies on nutrient glucose agar are white. On King's medium B, the colonies are smooth, round, creamy, and non-fluorescent after 48 h of incubation. The bacterium shows positive lipolytic and oxidase activity, but tests for arginine dihydrolase, nitrate reduction, levan production, and starch hydrolysis are negative. Gelatin hydrolysis is very mild, no pitting occurs on CVP, and hypersensitive reaction on tobacco and tomato is positive in 24 h.

Aci. citrulli strains can be divided into two genetically distinct groups, group I strains infecting a range of cucurbit species and group II strains being predominantly associated with watermelon. Silva et al. (2016) reported that Brazilian population of *Aci. citrulli* comprised predominantly of group I strains (98%), regardless of the geographical region and the host. The greatest diversity was found among strains collected from the northeastern region of Brazil, which accounts for more than 90% of the country's melon production. The study also provided the first evidence to suggest that temperature might be a driver in the ecological adaptation of *Aci. citrulli* populations.

Feng et al. (2009) developed a multi-locus sequence-typing scheme for *Aci. avenae* subsp. *citrulli* using seven conserved loci and identified two major clonal complexes, CC1 and CC2. In China, the predominant strains were CC1 and were found nearly equally on melon and watermelon. Yan et al. (2013) reported genetic differentiation among *Aci. citrulli* strains obtained from watermelon and melon, and it was the first attempt to compare pulsed field gel electrophoresis and MLST for analyzing genetic diversity of *Aci. citrulli* strains. The study showed MLST could better distinguish *Aci. citrulli* strains. PCR results showed that the primers, based on the *pil*L gene of a group II strain, were able to detect group II strains of *Aci. citrulli* and distinguish between strains of groups I and II rapidly and accurately (Zhong et al. 2016). Comparative analysis of the 11 effector genes from a collection of 22 *Aci. citrulli* strains led to the identification of a third *Aci. citrulli* group, which was supported by DNA:DNA hybridization, DNA fingerprinting, multi-locus sequence analysis of conserved genes, and virulence assays (Eckshtain-Levi et al. 2014).

7.63.3 Disease Cycle

Infested seed is the main source of perpetuation of the bacterium, and it has been found to survive for >34 years in stored melon and watermelon seeds. Wall et al. (1990) established the seed-borne nature of the bacterium confirming the previous reports. Dutta et al. (2012) reported that invasion of pericarp by *Aci. citrulli* resulted in superficial contamination of the testae and perisperm-endosperm layers, while pistil invasion resulted in the deposition of the bacterium in seed embryos. Dutta et al. (2015) showed that stigma inoculation resulted in faster colonization of watermelon ovules by *Aci. citrulli* than pericarp inoculation, though there was no difference in the levels of infection in mature seeds. It was also observed that pollen germ tubes played an important role in *Aci. citrulli* ingress into watermelon seeds via stigmas. Dutta et al. (2016) reported that *Aci. citrulli* cells are not intrinsically tolerant to desiccation, and the localization of the bacterium to testa tissues did not enhance the survival of the bacterium. In contrast, it was likely that embryo/endosperm localization enhanced its survival in the seeds. Dutta et al. (2014) reported that *Aci. citrulli* survived for 7 years in citron melon seeds, and the survival was significantly greater in seed lots generated via pistil (stigma) inoculation than via pericarp (ovary wall) inoculation.

Watermelon blossoms are a potential site of ingress by the bacterium for fruit and seed infestation. Walcott et al. (2003) found that approximately 98% (84/87) of fruits developed from blossoms, inoculated with 10^7 or 10^9 cfu of *Aci. avenae* subsp. *citrulli* per blossom, were asymptomatic. However, from these asymptomatic fruits the bacterium was recovered from 31% of the seed lots and 27% of these lots produced diseased seedlings. Further, the bacterium was detected and recovered from the pulp of 33% and 19% of these symptomless fruits, respectively. Carvalho et al. (2013) reported that the seeds from symptomatic and asymptomatic fruits of seven tested genotypes showed transmission rates of *Aci. citrulli* up to 35.3% and 8.7%, respectively, suggesting that asymptomatic fruits contain contaminated seeds that are responsible for the transmission of the pathogen. Johnson and Walcott (2013) investigated the role of quorum sensing in watermelon seed colonization and seed-to-seedling transmission of *Aci. citrulli* and found that quorum sensing played a role in regulation of genes involved in seed-to-seedling transmission of the bacterium.

Wu et al. (2017) designed the specific primers and probe based on a specific DNA fragment from the genome of *Aci. citrulli* and developed a specific, rapid, and sensitive method based on TaqMan probe-based insulated isothermal PCR (TiiPCR) to detect successfully bacterial cells of *Aci. citrulli* from watermelon seeds. Results revealed that ten copies of plasmid DNA were detectable within the modified reagents by TiiPCR and ten bacterial cells in each reaction tube were detectable at a 100% detection rate in this condition with fluorescent signal intensification over 1.8.

The other sources of perpetuation of the bacterium are infected volunteer watermelon seedlings (from infected fruits left in field last year) and infected wild cucurbits. Wild citron, a widespread weed in the southern US, is very susceptible to both foliage and fruit infection. Seed–plant transmission of the bacterium has been demonstrated in citron. Citron and other wild cucurbits are potential over seasoning hosts of the fruit blotch bacterium. The pathogen may also survive from one season to another on infested watermelon rind.

The disease development appears to be favored by high humidity and high temperature. Symptom development and spread of bacterium on foliage and fruit is most rapid during periods when the weather is hot, humid, and sunny with afternoon thundershowers. In Florida, this usually occurs in May–June with the spring crop, and in the more northerly states such as Georgia, South Carolina, and Indiana, it is in June–July. Optimum conditions for disease development may also occur early with the fall crop in Florida (August–September). Symptoms do not develop as severely during cool, rainy weather. In favorable weather, the bacterium spreads rapidly, and the spread occurs by wind-splashed rain and mechanical means. A few primary infection sites in a field can result in infection of all plants by the time of harvest.

Infected transplants represent the most important means of disease spread because the bacterium can disseminate throughout the transplant operation, can be asymptomatic on transplants, and can lead to higher percentage of infected transplants (due to common use of sprinkler irrigation in greenhouses) entering field plantings. The asymptomatic transplants constitute a dangerous source of hidden spread of the bacterium. Chalupowicz et al. (2015) reported that overhead-irrigation dispersed the bacterium from infected seedlings to 95% of the neighboring healthy seedlings, with 80% of them showing disease severity. In contrast, when sub irrigation by floating was employed, the neighboring plants of the infected ones did not display disease symptoms and were not colonized by the bacterium. The assessment of *Aci. citrulli* populations in different parts of the seedlings revealed that cotyledons were the most colonized part of the plant. Images of fluorescent binocular and confocal laser-scanning microscopy of seedlings infected with a GFP-labeled strain of the bacterium showed that the pathogen forms abundant aggregates on the surface of cotyledons, extensively colonizes the intercellular spaces of the parenchymatic tissues, and moves through the vascular system of the hypocotyls, leading to infection of emerging leaves.

Bacteria produced in leaf lesions can spread and infect developing fruit. The fruit blotch bacterium enters fruit through stomata on the fruit surface and a small, water-soaked lesion develops after 3–7 days. Mature fruit are covered with a wax layer that plugs the stomata and prevents entry of the bacteria into the fruit. After the formation of wax layer, mature fruits can be invaded by the fruit blotch bacterium only after wounding.

Zimerman-Lax et al. (2016) reported that nitrogen fertilization, based on nitrate only, led to reduced disease severity and bacterial numbers in melon leaves, as compared with two combinations of nitrate and ammonium. The reduction of

Aci. citrulli establishment in the plant foliage is of particular importance because the establishment of the pathogen during the growing season is assumed to increase the incidence of fruit infection, leading to serious yield losses.

Yan et al. (2017) described an assay using immature, detached melon fruit of "Joaquin Gold" melon (Syngenta, Rogers brand) that clearly indicates differences in host specificity between group I and II *Aci. citrulli* strains. In this assay, four strains of group I induced typical water-soaked lesions in melon fruit rind tissue 7–10 days after pinprick inoculation, while four strains of group II did not induce water-soaked lesions on detached melon fruit rinds during the same period. It is evident that group I *Aci. citrulli* strains have a specific capacity to infect immature Joaquin Gold melon fruit, whereas group II strains do not. Interestingly, this differential pathogenicity phenotype was not observed on foliar seedling tissues of the same melon cultivar, suggesting that host preference of *Aci. citrulli* strains is specific to immature fruit tissues. In another immature melon fruit inoculation assay, TIIIS mutant of the group I *Aci. citrulli* strain, M6 (M6Δ*hrcV*), failed to induce water-soaking. It suggests that TIIIS effectors are involved in *Aci. citrulli* cucurbit host preference, and that this assay is suitable for future studies of unique TIIIS effectors that distinguish group I and II strains.

At present, none of the commercial cultivars of watermelon is completely resistant to the bacterium. However, some cultivars are more susceptible to the pathogen than others. Cultivars with dark green fruit rind are less susceptible than light green rind cultivars. Triploid seedless watermelons are less susceptible than seeded diploid melons. Seedless melons can sustain high levels of leaf infection with low levels of fruit infection.

Carvalho et al. (2013) evaluated 110 watermelon genotypes at different plant developmental stages, namely, seeds (74), seedlings and plants before flowering (29), as well as plants during flowering and fruiting (7) and found that none of the genotypes was immune to the pathogen, but genotypes BGCIA 979, BGCIA 34, and Sugar Baby showed high level of resistance at most stages of plant development.

7.63.4 Management

The ideal management strategy is the exclusion of the pathogen by using pathogen-free seed and seedlings. Additionally, effective disease management must be practiced in seed and transplant production.

7.63.4.1 Management in Seed Production Fields

Use certified pathogen-free seed. To reduce the risk of seed infestation, the seed should be produced in dry and cool climatic regions of countries with no history of disease and a crop rotation with non-cucurbit hosts should be practiced for 3–5 years. Seed fields should be visually inspected, and only seed from disease-free fields should be used. Wall (1989) reported effective control of the disease in watermelon seedlings by hot water treatment of infested seed at 50°C for 20 min. Kubota et al. (2012) achieved complete disinfection of seeds of melon, cucumber, and small-seeded squash with dry heat treatment at 85°C for 3–5 days without any loss of germination.

7.63.4.2 Management in Transplant Production Areas

1. Use certified pathogen-free transplants. Transplant growers should carefully inspect their seedlings; destroy any flats containing seedlings with suspicious symptoms, and immediately separate flats adjacent to those with symptoms, from healthy plants
2. Cultural practices in the transplant house should include minimal manipulation of plants; closing sides of the transplant house during storms or windy periods; and destroying discarded plant material
3. Spread of the bacterium in the greenhouse can be minimized by low humidity, low temperatures, and by following practices that minimize or eliminate long periods of leaf wetness. If possible, ebb and flow irrigation should be used instead of overhead-irrigation to reduce splash dispersal of *Aci. citrulli*. However, this is cost prohibitive and rarely implemented. With overhead irrigation, watering should be done at mid day to facilitate rapid drying of plant surfaces, and the water delivery pressure should be low to limit aerosol generation. It would also be advisable to segregate different seed lots in the transplant house to reduce the chance of cross contamination
4. Disinfect a greenhouse that had infected seedlings and wait at least for 2 to 3 weeks to plant cucurbits in it again. Use 10% bleach solution or Physan 20% or 0.525% sodium hypochlorite solution for disinfection. Also dip all tools, plant containers, and recycled plastic trays for at least 10 min for disinfection.

7.63.4.3 Management in Fruit Production Fields

1. Deep plough and burry the infected crop residue including watermelon culls to reduce the overseasoning inoculum
2. Follow a strict 3- to 5-year crop rotation with non-cucurbit crops
3. Destroy immediately volunteer watermelon seedlings from previous crops by planting subsequent crops that utilize herbicides that will kill volunteer watermelon seedlings. This practice is followed in northern growing areas in the United States
4. Eliminate wild cucurbits and volunteer cucurbits near transplant houses and production fields
5. Do not work in an infested field if the foliage is wet. Decontaminate irrigation or mechanical equipment before moving it from an infested field to a healthy field

6. Grow less susceptible cultivars like seedless triploids and cultivars having darker green rind fruits
7. Applications of copper-containing fungicides, initiated prior to fruit set, reduce the incidence of the disease. A minimum of two to three copper applications and thorough coverage of the foliage are essential for an effective disease management. Applications should begin at first flowering, or earlier, and continue until all fruit are mature. Application of copper often causes some marginal yellowing of watermelon foliage. Under certain conditions, slight stunting of the vines may also occur. However, field studies have shown that the stunting and marginal yellowing have no detrimental effect on yield. In South Carolina in 1994, weekly applications of copper-containing fungicides at half the recommended rate gave slightly better control of fruit blotch than applications after every 14 days at the recommended rate. Marginal chlorosis was noted after the first application of copper, but later applications appeared to have no phytotoxic effect. Stunting was not observed with any of the materials or rates
8. The application of Kocide, *Psm. fluorescens* A506, and *Aci. avenae* subsp. *avenae* AAA 99-2 to blossoms, 5 h prior to inoculation, resulted in seed lot infestation of 21.1%, 24.1%, and 13.8%, respectively, and the effect was statistically significant. The seed treatment with *Aci. avenae* subsp. *avenae* AAA 99-2 suppressed the disease and reduced the pathogen transmission by 96.5% (Fessehaie and Walcott 2005)
9. Chitosan A application at 0.40 mg mL^{-1} significantly reduced disease index of watermelon seedlings compared to pathogen control (Li et al. 2013)
10. Preventing secondary spread of the pathogen in melon nurseries by sub irrigation combined with a bactericidal (Kocide) spray at the cotyledon stage reduced the disease incidence to 40% (Chalupowicz et al. 2015)

REFERENCES

Burdman, S., and Walcott, R. 2012. *Acidovorax citrulli*: Generating basic and applied knowledge to tackle a global threat to the cucurbit industry. *Mol. Plant Pathol.* 13: 805–815.

Carvalho, F. C. Q., Santos, L. A., Dias, R. C. S., Mariano, R. L. R., and Souza, E. B. 2013. Selection of watermelon genotypes for resistance to bacterial fruit blotch. *Euphytica* 190: 169–180.

Chalupowicz, L., Dror, O., Reuven, M., Burdman, S., and Manulis-Sasson, S. 2015. Cotyledons are the main source of secondary spread of *Acidovorax citrulli* in melon nurseries. *Plant Pathol.* 64: 528–536.

Crall, J. M., and Schenck, N. C. 1969. Bacterial fruit rot of watermelon in Florida. *Plant Dis. Reptr* 53: 74–75.

Dutta, B., Avci, U., Hahn, M. G., and Walcott, R. R. 2012. Location of *Acidovorax citrulli* in infested watermelon seeds is influenced by the pathway of bacterial invasion. *Phytopathology* 102: 461–468.

Dutta, B., Ha, Y., Lessl, J. T., Avci, U., Sparks, A. C., et al. 2015. Pathways of bacterial invasion and watermelon seed infection by *Acidovorax citrulli*. *Plant Pathol.* 64: 537–544.

Dutta, B., Sanders, H., Langston, D. B., Booth, C., Smith, S., et al. 2014. Long-term survival of *Acidovorax citrulli* in citron melon (*Citrullus lanatus* var. *citroides*) seeds. *Plant Pathol.* 63: 1130–1137.

Dutta, B., Schneider, R. W., Robertson, C. L., and Walcott, R. R. 2016. Embryo localization enhances the survival of *Acidovorax citrulli* in watermelon seeds. *Phytopathology* 106: 330–338.

Eckshtain-Levi, N., Munitz, T., Živanović, M., Traore, S. M., Spröer, C., et al. 2014. Comparative analysis of type III secreted effector genes reflects divergence of *Acidovorax citrulli* strains into three distinct lineages. *Phytopathology* 104: 1152–1162.

Feng, J., Schuenzel, E. L., Li, J., and Schaad, N. W. 2009. Multilocus sequence typing reveals two evolutionary lineages of *Acidovorax avenae* subsp. *citrulli*. *Phytopathology* 99: 913–920.

Fessehaie, A., and Walcott, R. R. 2005. Biological control to protect watermelon blossoms and seed from infection by *Acidovorax avenae* subsp. *citrulli*. *Phytopathology* 95: 413–419.

Johnson, K. L., and Walcott, R. R. 2013. Quorum sensing contributes to seed-to-seedling transmission of *Acidovorax citrulli* on watermelon. *J. Phytopathol.* 161: 562–573.

Kubota, M., Hagiwara, N., and Shirakawa, T. 2012. Disinfection of seeds of cucurbit crops infested with *Acidovorax citrulli* with dry heat treatment. *J. Phytopathol.* 160: 364–68.

Latin, R. X., and Hopkins, D. L. 1995. Bacterial fruit blotch of watermelon. The hypothetical exam question becomes reality. *Plant Dis.* 79: 761–765.

Li, B., Shi, Y., Shan, C., Zhou, Q., Ibrahim, M., et al. 2013. Effect of chitosan solution on the inhibition of *Acidovorax citrulli* causing bacterial fruit blotch of watermelon. *J. Sci. Food Agric.* 93: 1010–1015.

Schaad, N. W., Postnikova, E., Sechler, A., Claflin, L. E., Vidaver, A. K., et al. 2009. List of new names and new combinations previously effectively but not validly, published. *Int. J. Syst. Evol. Microbiol.* 59: 923–925. Effective publication: Schaad, N. W., Postnikova, E., Sechler, A., Claflin, L. E., Vidaver, A. K., et al. 2008. Reclassification of subspecies of *Acidovorax avenae* as *A. Avenae* (Manns 1905) emend., *A. cattleyae* (Parvarino, 1911) comb. nov., *A. citrulli* (Schaad et al., 1978) comb. nov., and proposal of *A. oryzae* sp. nov. *Syst. Appl. Microbiol.* 31: 434–446.

Schaad, N. W., Sowell, G., Goth R. W., Colwell, R. R., and Webb, R. E. 1978. *Pseudomonas pseudoalcaligenes* subsp. *citrulli* subsp. nov. *Int. J. Syst. Bacteriol.* 28: 117–125.

Silva, G. M., Souza, R. M., Yan, L., Júnior, R. S., Medeiros, F. H. V., et al. 2016. Strains of the group I lineage of *Acidovorax citrulli*, the causal agent of bacterial fruit blotch of cucurbitaceous crops, are predominant in Brazil. *Phytopathology* 106: 1486–1494.

Walcott, R. R. 2005. Bacterial fruit blotch of cucurbits. The Plant Health Instructor. DOI: 10.1094/PHI-I-2005-1025-02.

Walcott, R. R., Gitaitis, R. D., and Castro, A. C. 2003. Role of blossoms in watermelon seed infestation by *Acidovorax avenae* subsp. *citrulli*. *Phytopathology* 93: 528–534.

Wall, G. C. 1989. Control of watermelon fruit blotch by seed heat-treatment (Abstr.). *Phytopathology* 79: 1191.

Wall, G. C., Santos, V. M., Cruz, F. J., Nelson, D. A., and Cabrera, I. 1990. Outbreak of watermelon fruit blotch in the Mariana Islands. *Plant Dis.* 74: 80.

Wu, P.-Y., Ho, L.-C., Chang, J.-J., Tzeng, K.-C., Deng, W.-L., et al. 2017. Development of a TaqMan probe-based insulated isothermal PCR (TiiPCR) for the detection of *Acidovorax citrulli*, the bacterial pathogen of watermelon fruit blotch. *Eur. J. Plant Pathol.* 147: 869–875.
Yan, S., Yang, Y., Wang, T., Zhao, T., and Schaad, N. W. 2013. Genetic diversity analysis of *Acidovorax citrulli* in China. *Eur. J. Plant Pathol.* 136. 171–181.
Yan, L., Hu, B., Chen, G., Zhao, M., and Walcott, R. R. 2017. Further evidence of cucurbit host specificity among *Acidovorax citrulli* groups based on a detached melon fruit pathogenicity assay. *Phytopathology* 107: 1305–1311.
Zhong, J., Lin, Z-Y., Ma, Y-M., Gao, B-D., Liu, H-Q., et al. 2016. Rapid discrimination between groups I and II of *Acidovorax citrulli* using a primer pair specific to a *pil*L gene. *J. Phytopathol.* 164: 558–562.
Zimerman-Lax, N., Shenker, M., Tamir-Ariel, D., Perl-Treves, R., and Burdman, S. 2016. Effects of nitrogen nutrition on disease development caused by *Acidovorax citrulli* on melon foliage. *Eur. J. Plant Pathol.* 145: 125–137.

7.64 BACTERIAL BLIGHT OF WALNUT

Walnut blight caused by *Xanthomonas arboricola* pv. *juglandis* is the most important bacterial disease of Persian (English) walnuts (*Juglans regia* L.) and other *Juglans* species, which affects a high percentage of pistillate flowers and fruits. The disease reduces the yield and frequently lowers quality of harvested nuts. Although none of the English walnut cultivars appear to be resistant to the disease, it is most severe on early varieties whose bloom periods coincide with spring rains. In years with extended spring rains, blight can reduce nut yield from 50% to 80%.

Pierce (1901) described the disease for the first time from California. Heavy losses up to 30%, exceptionally up to 100% have been reported from the USA (Rudolph 1933). Lang and Evans (2010) found disease incidence ranging from 19% to 100% on fruits of cultivars Franquette and Vina near harvest, in three growing years in Tasmania. In other parts of the world, the disease is apparently less damaging, probably due to variation in germplasm resistance. Californian cultivars and a Hungarian selection A 117 show high susceptibility, French cultivars medium susceptibility, and the Italian cultivar Sorrento is tolerant (Tamponi and Donati 1990). Lamichhane (2014), in a review article, has provided an overview of *Xa. arboricola* diseases of stone fruit, almond, and walnut trees, and has discussed the current and future management strategies. *Xa. arboricola* pv. *juglandis* has also been isolated from tissues affected by bacterial apical necrosis (Belisario et al. 2002).

Walnut blight has been recorded from Europe (Austria, Bulgaria, Denmark, France, Germany, Greece, Italy, Moldova, Netherlands, Poland, Portugal, Romania, Russia, southern Russia, Slovenia, Switzerland, Spain, UK, Ukraine, and former Yugoslavia), Asia [Azerbaijan, China (Hebei, Henan, Jiangsu, Shaanxi, Shandong), Georgia, India (Himachal Pradesh and Uttar Pradesh states), Iran (Qazvin, Hamadan, Tehran, Markazi, Aradabil, Guilan, Mazandaran, and Golestan provinces, Golmohammadi et al. 2002), Iraq, Israel, Lebanon and Uzbekistan], Africa (South Africa and Zimbabwe), North America [Canada, Mexico, and USA (Alabama, Arkansas, California, Connecticut, Delaware, Georgia, Indiana, Kansas, Louisiana, Maryland, Michigan, Minnesota, Missouri, Maine, New Jersey, New York, Ohio, Oregon, Pennsylvania, Texas, Virginia, and Washington states)], South America (Argentina, Uruguay, and Chile), and Oceania [Australia (New South Wales, Victoria, Queensland, South Australia, Tasmania, western Australia) and New Zealand] (CABI-EPPO, 2001).

Xa. arboricola pv. *juglandis*, like most other xanthomonads, is host specific, infecting only *Juglans* species: *J. regia* (Persian walnut), *J. nigra* (eastern black walnut), *J. ailantifolia* (Japanese walnut), *J. ailantifolia* var. *cordiformis*, *J. californica* (southern California black walnut), *J. cinerea* (butternut), and *J. hindsii* (northern California black walnut.

7.64.1 Symptoms

The disease develops usually in spring under wet weather conditions from bacteria that survive in buds and catkins and to a lesser extent in holdover cankers on twigs. Water-soaked spots appear on leaves and petioles, that later coalesce into large brown-black necrotic lesions. Blighted leaves, irrespective of their time of infection, remain in the canopy through the rest of the growing season and are the major source of inoculum for secondary spread of the pathogen. Tips of twigs become black and subsequently the whole twig will die. The bacterium invades primarily the parenchymatous tissues and wood older than 1 year is not attacked. Water-soaked black spots can also develop on the male and female catkins and cause severe damage. The pollens of such catkins are contaminated with the bacteria and these pollens can spread the pathogen. Catkins, whether healthy or diseased, drop from the leaf canopy following elongation. Water-soaked spots are formed on fruits and these spots turn black and become slightly sunken. The infected black tissue enlarges, can extend from the epicarp into the nut and destroys it. Nutlets infected in the early stages of development generally drop a few weeks after infection. Late season infections on maturing nuts can discolor the shell and nutmeats, lowering their quality or making them unmarketable. Trees are not killed.

Symptoms on leaves and fruits may be confused with those caused by the fungus, *Marssonina juglandis*, but in case of fungal attack acervulus fruiting bodies are generally found.

Two *Brenneria* species, namely, *Br. rubrifaciens* and *Br. nigrifluens*, cause **deep bark canker** and **shallow bark canker**, respectively. Deep bark canker is characterized by cracks in scaffold branches and trunks, where infection is first established. The symptoms of shallow bark canker consist of brownish to black round spots, which appear on trunk or on lower scaffold limbs.

7.64.2 Causal Organism

Xanthomonas arboricola pv. *juglandis* (Pierce 1901) Vauterin *et al.* 1995

Syn. *Xa. juglandis, X. campestris* pv. *juglandis, X. juglandis* pv. *juglandis*

Mulrean and Schroth (1981) developed a selective medium with a dye (brilliant cresyl blue) and potato starch as its major selective agents; which allow identification of the walnut blight bacterium on the basis of a distinctive colony character. Preferential utilization of starch by the bacterium creates a distinctive hydrolytic zone around the pathogen colonies, making them easily distinguishable from some pseudomonads that also grow in the medium. The selective medium provides a simple and an effective method of isolating *Xa. arboricola* pv. *juglandis* from walnut tissue while eliminating 97 percent of the contaminating bacteria that grow on ordinary media. The selective medium was also useful for isolating six other *Xa. campestris* pathovars, namely, pv. *begoniae*, pv. *campestris*, pv. *incanae*, pv. *malvacearum*, pv. *phaseoli*, and pv. *vesicatoria*.

Giovanardi et al. (2016) reported that 77 isolates of *Xa. arboricola* pv. *juglandis* obtained from Romagna, Italy were able to grow on mannitol glutamate yeast agar containing 50 μg mL^{-1} of copper sulphate. However, several isolates were found to be highly resistant to copper, even up to 500 μg mL^{-1}. They also found high phenotypic and genotypic variability among these isolates and attributed this variability to the different origin of the propagation material. Information provided in the study on the Italian collection allows a better understanding of the walnut bacterial blight epidemiology.

Loreti et al. (2001) and Scortichini et al. (2001) reported the high genetic diversity among strains of *Xa. arboricola* pv. *juglandis* isolated from walnut, and there was relationship among strains with the geographic origin. This could be because Persian walnut is propagated by locally grown seeds adapted to different environments, allowing the selection of different *Xa. arboricola* pv. *juglandis* populations (Scortichini et al. 2001). Hajri et al. (2010) reported a distinct genetic lineage of *Xa. arboricola* pv. *juglandis* strains pathogenic to walnut, highlighting an even wider diversity among these strains than what was known earlier.

Multilocus sequence analysis (MLSA) of partial sequences of the *fyuA*, *gyrB*, and *rpoD* genes of 18 isolates of *Xa. arboricola* pv. *juglandis* collected from six locations in Poland, and respective sequences from GenBank of pathotype strains of other pathovars of *Xa. arboricola* showed that the *Xa. arboricola* pv. *juglandis* isolates consisted of different phylogenetic lineages. An incongruence among MLSA gene phylogenies and traces of intergenic recombination events were proved. These data suggest that the sequence analysis of several housekeeping genes is necessary for the proper identification of *Xa. arboricola* pathovars (Kałużna et al. 2014).

Ivanović et al. (2015) examined the molecular diversity of 59 isolates of *Xa. arboricola* pv. *juglandis* collected from different geographic locations in Serbia. Genomic variability was assessed by using repetitive PCR, *Spe*I macrorestriction analysis of genomic DNAs by pulsed-field gel electrophoresis, and partial sequencing of the *gyrB* gene. Molecular analyses showed substantial genetic diversity among strains and existence of diverse populations of *Xa. arboricola* pv. *juglandis* in Serbia.

7.64.3 Disease Cycle

The bacterium overwinters on dormant buds, primarily under the outer bud scales or cataphylls. Bud population evaluations have shown that the inner buds where the immature walnut flowers are located are relatively free of bacteria. The bacterium also overwinters in the twig cankers. The disease severity increases with earliness of flowering (Aletà et al. 2001; Piccirillo and Petriccione 2006). The bacterium enters the host mainly through stomata or wounds. Rain is the most important dispersal agent (Piccirillo 2003) but insects and mites [black walnut blister mite or walnut leaf gall mite *Eriophyes erineus* (=*Aceria tristriatus*) (Rudolph 1943)] may also carry the bacteria. The presence of aphids (*Chromaphis juglandicola*) did not increase the percentage of affected fruits, but it resulted in a higher disease incidence on the leaves (Arquero et al. 2006). Pollens from infected catkins can carry the bacteria. The pollen transmission is considered not to be important in Oregon (USA), although it does occur in other areas. Ark (1944) showed pollens as a source of walnut bacterial blight infection in California.

Cultivar blooming habits, rainfall patterns, and orchard design influence the development and severity of blight and the subsequent infestation of buds and catkins. Buds infestation is generally higher than catkins. The incidence of bud infestation is generally more in the early-blooming varieties. Middle- and late-blooming varieties usually bloom after spring rains have subsided and have low nut and leaf blight incidence during the season. Inoculum from infected leaves, spreads through sprinkler irrigation, dew, and rain, and can infect other leaves or maturing nuts and also contributes to the infestation of next season's buds and catkins. An adequate amount of inoculum on the infested dormant buds and catkins survives the winter to potentially initiate another blight epidemic in the coming spring.

Lindow et al. (2014) reported that *Xa. arboricola* pv. *juglandis* populations on developing nuts were strongly related to that on the shoots on which they were borne. The inoculum of the bacterium present on dormant buds is the primary determinant of nut infections and the risk of disease can be predicted from both the numbers of the bacteria in buds and the incidence of early spring rain.

Soltani and Aliabadi (2010) reported that genotype 94 showed the highest resistance to both disease incidence and its progress after 4–6 weeks of infection. Genotype 65 showed high susceptibility to disease while genotype 69 showed high susceptibility both to disease incidence and its progress after 4–6 weeks of infection.

Fernandes et al. (2017) proposed nine novel specific DNA markers (XAJ1 to XAJ9), selected by dedicated in silico approaches, to identify *Xa. arboricola* pv. *juglandis* isolates and detect these bacteria in infected plant material. Dot blot hybridization was carried out across a large set of xanthomonads, which confirmed the pathovar specificity of these markers, allowed to identify four broad-range markers (XAJ1, XAJ4, XAJ6, and XAJ8) and five narrow-range markers (XAJ2, XAJ3, XAJ5, XAJ7, and XAJ9), originating 12 hybridization patterns (HP1 to HP12). Four isolates that clustered together according to the *gyrB* phylogenetic analysis (CPBF 1507, 1508, 1514, and 1522) presented the same hybridization pattern (HP11), suggesting that these nine markers might be informative to rapidly discriminate and identify different *Xa. arboricola* pv. *juglandis* lineages. A multiplex PCR was optimized using markers XAJ1, XAJ6, and XAJ8 and this triplex PCR, besides confirming the dot blot data for each of the 52 *Xa. arboricola* pv. *juglandis*, was able to detect *Xa. arboricola* pv. *juglandis* in field infected walnut leaves and fruits. On the whole, these nine *Xa. arboricola* pv. *juglandis*-specific markers allow conciliating the specificity of DNA-detection assays with typing resolution, contributing to rapid detection and identification of potential emergent and acutely virulent *Xa. arboricola* pv. *juglandis* genotypes, infer their distribution, disclose the presence of this phytopathogen on potential alternative host species and improve phytosanitary control.

7.64.4 Management

1. Prune the infected plant parts a good way back from visible damage and burn.
2. Maintain the soil pH above 6.0.
3. Avoid wetting of foliage, especially during bloom, by preventing sprinkler irrigation.
4. Give the balanced nitrogen fertilization.
5. Create an open structure for aeration by pruning.
6. Multiple protective copper sprays are generally given to reduce the amount of overwintering inoculum. Kocide (0.2%) treatment at bud break substantially reduces the disease incidence either alone or together with Silwet 77 (0.1%). The addition of Silwet 77 to copper helps the penetration of the active ingredient into buds. A good-quality copper product mixed at the label rate with Manex also provides good protection. A good-quality adjuvant will wet cataphylls and encourage the copper/Manex mix to penetrate between bud scales and kill overwintering bacteria. Under severe walnut blight pressure, the second application should be given 7–10 days after the first spray. Additional spray applications are based upon the weather and the risk of infection. Copper resistance has been observed in California, France and Italy (Teviotdale and Schroth 1998; Scortichini et al. 2001). Lang and Evans (2010) reported that in a dry year, two sprays of copper-based biocide (300 g kg^{-1} of copper hydroxide formulated with 150 g kg^{-1} mancozeb) applied one week apart from 5% terminal bud-burst provided commercially acceptable control whereas up to nine sprays in a wetter year failed to control the disease. The inability of copper materials to penetrate walnut buds and catkins results in unsatisfactory control of the disease. Until bactericides with eradicative or systemic activities are developed, walnut blight will continue to be a difficult disease to control.

   Ninot et al. (2002), on the basis of a three-year trial, worked out a reduced spray schedule of three sprays to replace the standard schedule of seven copper sprays from bud break until harvest followed in western Europe. The disease control on nuts was similar or even better with the reduced spraying schedule than with the standard schedule, along with significantly lower amount of copper accumulation (about half) in the soil.
7. Three strains of *Pseudomonas fluorescens*, isolated from phylloplane of healthy walnut trees, reduced the disease incidence ranging from 42% to 82% for two consecutive years. Antagonistic strains were sprayed in a suspension of 10^9 cfu mL^{-1} on the leaves, which were subsequently inoculated with suspension of pathogen strain W7/1 (10^8 cfu mL^{-1}) 24 h after the antagonist's application (Ozaktan et al. 2012).

A walnut blight forecast system that may reduce chemical sprays has been developed in the USA (http://www.fieldwise.com).

REFERENCES

Aletà, N., Ninot, A., Moragrega, C., Llorente, I., and Montesinos, E. 2001. Blight sensitivity of Spanish selections of *Juglans regia*. *Acta Hortic*. (*ISHS*) 544: 353–362.

Ark, P. A. 1944. Pollen as a source of Walnut bacterial blight infection. *Phytopathology* 34: 329–334.

Arquero, O., Lovera, M., Rodriguez, R., and Trapero, A. 2006. Walnut blight (*Xanthomonas arboricola* pv. *juglandis*): Factors influencing disease incidence. *Acta Hortic*. (*ISHS*) 705: 443–446.

Belisario, A., Maccaroni, M., Corazza, L., Balmas, V., and Valier, A. 2002. Occurrence and etiology of brown apical necrosis on Persian (English) walnut fruit. *Plant Dis*. 86: 599–602.

CABI-EPPO. 2001. *Xanthomonas arboricola* pv. *juglandis*. Distribution Maps of Plant Diseases, No. 133.

Fernandes, C., Albuquerque, P., Sousa, R., Cruz, L., and Tavares, F. 2017. Multiple DNA markers for identification of *Xanthomonas arboricola* pv. *juglandis* isolates and its direct detection in plant samples. *Plant Dis*. 101: 858–865.

Giovanardi, D., Bonneau, S., Gironde, S., Saux, M. F-Le, Manceau, C., et al. 2016. Morphological and genotypic features of *Xanthomonas arboricola* pv. *juglandis* populations from walnut groves in Romagna region, Italy. *Eur. J. Plant Pathol*. 145: 1–16.

Golmohammadi, M., Alizadeh, A., and Rahimian, H. 2002. Homogeneity of the strains inciting bacterial blight of walnut in the central and northern provinces of Iran. *Iranian J. Plant Pathol.* 38: 11–20.

Hajri, A., Meyer, D., Delort, F., Guillaumes, J., Brin, C., et al. 2010. Identification of a genetic lineage within *Xanthomonas arboricola* pv. *juglandis* as the causal agent of vertical oozing canker of Persian (English) walnut in France. *Plant Pathol.* 59: 1014–1022.

Ivanović, Ž., Popović, T., Janse, J., Kojić, M., Stanković, S., et al. 2015. Molecular assessment of genetic diversity of *Xanthomonas arboricola* pv. *juglandis* strains from Serbia by various DNA fingerprinting techniques. *Eur. J. Plant Pathol.* 141: 133–145.

Kałużna, M., Pulawska, J., Waleron, M., and Sobiczewski, P. 2014. The genetic characterization of *Xanthomonas arboricola* pv. *juglandis*, the causal agent of walnut blight in Poland. *Plant Pathol.* 63: 1404–1416.

Lamichhane, J. R. 2014. *Xanthomonas arboricola* diseases of stone fruit, almond, and walnut trees: Progress toward understanding and management. *Plant Dis.* 98: 1600–1610.

Lang, M. D., and Evans, K. J. 2010. Epidemiology and status of walnut blight in Australia. *J. Plant Pathol.* 92: S1.49–55.

Lindow, S., Olson, W., and Buchner, R. 2014. Colonization of dormant walnut buds by *Xanthomonas arboricola* pv. *juglandis* is predictive of subsequent disease. *Phytopathology* 104: 1163–1174.

Loreti, S., Gallelli, A., Belisario, A., Wajnberg, E., and Corazza, L. 2001. Investigation of genomic variability of *Xanthomonas arboricola* pv. *juglandis* by AFLP analysis. *Eur. J. Plant Pathol.* 107: 583–591.

Murlean, E. N., and Schroth, M. N. 1981. Bacterial blight on Persian walnuts. California Agriculture, September–October, pp. 11–13.

Ninot, A., Aletà, N., Moragrega, C., and Montesinos, E. 2002. Evaluation of a reduced copper spraying program to control bacterial blight of walnut. *Plant Dis.* 86: 583–587.

Ozaktan, H., Erdal, M., Akkopru, A., and Aslan, E. 2012. Biological control of bacterial blight of walnut by antagonistic bacteria. *J. Plant Pathol.* 94: S1.53–56.

Piccirillo, P. 2003. The phytopathological situation of walnuts (*Juglans regia* L.) according to observations made by the Fruit Research Institute of Caserta. *Rivista di Frutticoltura e Ortofloricoltura* 65: 39–43.

Piccirillo, P., and Petriccione, M. 2006. Walnut collection of the Italian Research Council of Agriculture in Caserta. *Acta Hortic.* (*ISHS*) 705: 177–181.

Pierce, N. B. 1901. Walnut bacteriosis. *Bot. Gaz.* 31: 272–273.

Rudolph, B. A. 1933. Bacteriosis (blight) of the English walnut and its control. California Agricultural Experiment Station Bulletin 564, pp. 1–88.

Rudolph, B. A. 1943. The walnut erinose mite, a carrier of walnut blight. *Diamond Walnut News* 11, pp. 1–2.

Scortichini, M., Marchesi, U., and Di Prospero, P. 2001. Genetic diversity of *Xanthomonas arboricola* pv. *juglandis* (synonyms: *X. campestris* pv. *juglandis*; *X. juglandis* pv. *juglandis*) strains from different geographical areas shown by repetitive polymerase chain reaction genomic fingerprinting. *J. Phytopathol.* 149: 325–332.

Soltani, J., and Aliabadi, A. A. 2010. Genetic resistance to bacterial blight disease in Persian walnut. *Eur. J. Plant Pathol.* 128: 65–70.

Tamponi, G., and Donati, G. P. 1990. Walnut cultivars susceptibility to *Xanthomonas juglandis*. *Acta Hortic.* (*ISHS*) 284, 301–302.

Teviotdale, B. L., and Schroth, M. N. 1998. Bark, fruit, and foliage diseases. pp. 242–46 in: *Walnut Production Manual.* D. E. Ramos, ed. Publ. 3373. Division of Agriculture and Natural Sciences, University of California, Oakland, CA.

Vauterin, L., Hoste, B., Kersters, K., and Swings, J. 1995. Reclassification of *Xanthomonas*. *Int. J. Syst. Bacteriol.* 45: 472–489.

7.65 BACTERIAL CANKER AND DECLINE OF HAZELNUT

Bacterial canker and decline of European hazelnut or filbert (*Corylus avellana* L.) caused by *Pseudomonas avellanae* was first observed in northern Greece in 1976 by Psallidas and Panagopoulos (1979). Within a few years, the disease caused substantial losses in northern Greece by destroying young plantations of Turkish cultivar Palaz (Psallidas 1987). Subsequently, the same disease, referred to locally as "moria," was observed in plantations throughout a 20,000 ha area in the Latium region of central Italy of Viterbo province (Scortichini and Tropiano 1994). Since its first appearance in the Latium area of central Italy, it has caused death of more than 40,000 trees spread over an area of approximately 1,000 ha. On an annual basis, losses are estimated to be approximately $1.5 million and the disease is considered a serious problem (De Castro 1999). Bacterial canker has been found also in wild European hazelnut trees growing in forests adjacent to commercial orchards in Italy. Mahdavian and Hasanzadeh (2012) reported decline of hazelnut from Mazandaran province of Iran due to *Psm. syringe*.

A less destructive bacterial disease of European hazelnut, causing limited twig and branch wilting, but without canker formation and death of trees, has been found in northern Italy and is caused by strains closely related to *Psm. syringae* pv. *syringae* (Scortichini et al. 2002), but Scortichini et al. (2005) have identified it as *Psm. syringae* pv. *coryli*.

7.65.1 Host Range

The hosts range of the bacterium includes hazelnut (*C. avellana*), wild and cultivated trees.

7.65.2 Symptoms

The first symptom may occur as wilting of the male catkins during winter. Dead catkins as well as dead leaves usually remain attached to the twig. Also female catkins may show necrosis in February and March. In spring, early symptoms may be delayed bud break and leaf emergence. Emerging leaves wilt and die rapidly, but sometimes trees exhibit only some pale green foliage in spring. In severe infections, trees are completely killed in one season. The most obvious symptoms are observed in summer, i.e., rapid wilting of leaves on a part of the tree. Apart from necrotic leaves; the dead immature fruits also remain attached to the twigs for many weeks. During summer, longitudinal cankers on branches and the trunk may be formed. Diseased bark turns reddish brown, and a brown discoloration of the sapwood is visible when the bark

FIGURE 7.47 Bacterial canker and decline of hazelnut. (Courtesy of J. Lamichhane, characteristic decline symptoms on hazelnut plants of cv. Tonda Gentile Romana [a], decline and dieback affected branch [b], and excised branch showing discoloration of cambium [c])

is removed from the branch (Figure 7.47a, b and c). Root necrosis may also occur. Infected trees that survive the winter often die the following summer (Scortichini 2002).

Very similar symptoms, but with a less aggressive disease development are caused by strains that are not *Psm. avellanae,* but closely related to *Psm. syringae* pv. *syringae* (Scortichini et al. 2002). However, Scortichini et al. (2005) identified these strains as *Psm. syringae* pv. *coryli.*

7.65.3 Causal Organism

Pseudomonas avellanae Janse et al. 1997
Syn. *Psm. syringae* pv. *avellanae*

The bacterium causing bacterial canker of hazelnut was originally described as *Pseudomonas syringae* pv. *avellanae* (Psallidas 1984, 1993). Janse et al. (1996) reclassified it as *Pseudomonas avellanae* on the basis of fatty acids methyl ester analysis, whole-cell protein profiles, and sequence comparisons of 16S rRNA gene. Based on DNA relatedness studies of *Psm. syringae* and related species, *Psm. avellanae* was included in the genomospecies 8, together with *Psm. syringae* pv. *theae* (Gardan et al. 1999).

Scortichini et al. (2016) reported that the studies based on multiple comparative taxonomic analyses, including MLSA and average nucleotide identity approaches; reveal the clear existence of strains associated with bacterial canker and decline of European hazelnut trees belonging to two different genomospecies: *Psm. avellanae* and *Psm. syringae*. Strains of the latter were tentatively named as *Psm. syringae* pv. *avellanae*. The distinctiveness of these two phytopathogens associated with the same symptoms on *C. avellana* requires the formal emended description of *Psm. syringae* pv. *avellanae*. Hence, the authors provided the emended description of *Psm. syringae* pv. *avellanae*.

The bacterium can be isolated on King's medium B and nutrient sucrose agar containing 5% sucrose. Useful discriminative tests have been described. SDS-PAGE comparison of protein extracts and fatty acid methyl esters analysis have proven effective. Repetitive-PCR using ERIC primers is also reliable. Using these techniques, along with some preliminary key biochemical tests, i.e., absence of oxidase and levan production, *Psm. avellanae* isolates can be identified in 4–6 days. Completion of Koch's postulates involves inoculation of leaf scars of young hazelnut trees and may take up to 7 months.

Psm. avellanae is differentiated from a distinct taxon closely related to *Psm. syringae* pv. *syringae* from European hazelnut by a repetitive PCR analysis with enterobacterial repetitive intergenic consensus, box element, and repetitive extragenic palindromic primer sets, sodium dodecyl sulfate-polyacrylamide gel electrophoresis of whole-cell protein extracts, a carbon compound utilization analysis, and an analysis to determine the presence of the *syrB* gene (Scortichini et al. 2002). All *Psm. avellanae* strains lack the *syrB* gene, whereas most of the *Psm. syringae* pv. *syringae* strains from hazelnut have this gene encoding for the production of phytotoxins (Sorensen et al. 1998).

Psm. avellanae strains possess *hrp* genes encoding for the harpin proteins that are involved in the elicitation of the hypersensitivity reaction in leaf tissues (Loreti et al. 2001). Repetitive polymerase chain reaction using ERIC, BOX, and REP primers sets clearly differentiated *Psm. avellanae* strains from Greece and central Italy (Scortichini et al. 1998), and Scortichini (2012) further divided these strains into two lineages. Similarly monoclonal antibodies raised toward cell wall polysaccharides of strains isolated in Greece and central Italy clearly differentiated two serotypes (Ovod et al. 1999).

A sensitive and rapid detection procedure uses 16s rRNA targeted primers PAV 1 and PAV 2, amplifying a 762-bp product present in both Greek and Italian isolates of *Psm. avellanae,* but absent in other pseudomonads. This technique allows for direct amplification from infected hazelnut tissue (extracts from twigs, branches, or roots) and also allows for detection of *Psm. avellanae* in asymptomatic trees (Scortichini and Marchesi 2001). The addition of 4% BLOTTO (10% skim milk powder and 0.2% $NaNo_3$) to the PCR mixture, as suggested by De Boer et al. (1995), is essential to avoid interference of *C. avellana* tissue compounds in the extracts during amplification. This method is very useful when sanitizing diseased plantings or screening propagative material. Other specific primers amplify the *hrpW* gene of the bacterium and can be used for the rapid detection (Loreti and Gallelli 2002).

Gervasi and Scortichini (2009) developed a real-time PCR, based on primers and probe designed within the 16S rRNA gene of the pathogen, to detect *Psm. avellanae* from hazelnut. The addition of bovine serum albumin was essential to prevent inhibition by compounds released from the hazelnut tissue. The assay detected the pathogen when inoculated in low numbers (80 cells) in a leaf scar up to 9 days after inoculation. The other bacterial pathogens of hazelnut, namely, *Psm. syringae* pv. *coryli*, *Psm. syringae* pv. *syringae*, and *Xanthomonas arboricola* pv. *corylina*, did not react in the assay. Loreti et al. (2009) reported that a PCR method based on the amplification of a 350 bp fragment of the *hrpW* gene of *Psm. avellanae*, encoding a harpin protein, proved to be more reliable than isolation for the detection of *Psm. avellanae*.

7.65.4 Disease Cycle

Latently infected suckers used for propagation may be the main source for widespread dispersal of the pathogen. Regenerating suckers on stumps of destroyed diseased trees show infection within 1–3 years. Scortichini and Tropiano (1994) demonstrated introduction of the disease with latently infected propagating material from northern to southern orchards in the Latium region of Italy. The bacterium infects mainly through not yet suberized leaf scars in early autumn. After leaf scar infection, the bacterium may overwinter under the bark in the parenchymatous and xylem tissue, and from there in spring, the pathogen can move systemically within the plant, with the potential to reach the roots and cause cankers on the trunk and branches. Secondary dissemination in an orchard is mainly by wind-driven rain (Martins and Scortichini 1998). The pathogen's spread to the apparently highly susceptible wild European hazelnut population is a matter of concern (Scortichini et al. 2000a). Acidic soils and spring frost probably are important for predisposing trees to infection. Frosty weather, causing the bark of branches and trunk to crack, facilitates colonization by *Psm. avellanae*. Epiphytic populations of the bacterium could not be demonstrated, and it has never been found in fruits or seeds. The bacterium may survive in branches and roots (Scortichini 1998).

Insects may play a role in disseminating *Psm. avellanae* from an infected tree to a nearby healthy tree. Several scolytid beetles, including *Xyleborus* (*Anisandrus*) *dispar* L. and *X. saxesenii* Ratz. are attracted by terpenes released from diseased trees. Adult insects may come into contact with the bacterium during oviposition, and larvae may be contaminated during tunneling. Although the bacterium has been isolated from larvae and adult scolytid beetles, conclusive evidence for insect transmission is still lacking.

7.65.5 Management

1. Production of disease-free planting material and avoidance of introduction of latently infected plants are the key factors in the prevention of the disease
2. Monitor for disease symptoms during early spring and summer and remove the infected plant parts far below the infection site and in case of severely infected trees, remove roots and suckers also. All this material should be burnt. Avoid pruning and/or sucker removal during humid periods. After pruning, it is advisable to seal the wounds with Bordeaux paint
3. If the orchard soil is very acidic, apply lime to increase the soil pH
4. Sprays of copper-based compounds are not very effective due to the systemic nature of the pathogen. However, when applied immediately after pruning, spring frost, hail, and windy-storms in early autumn and at the beginning and middle of leaf drop, this treatment may reduce the possibility of wound colonization by the bacterium
5. Control the scolytid beetles by using chromotropic traps
6. Acibenzolar-*S*-methyl (registered as Bion and Actigard in Europe and the United States, respectively) has been found to control plant diseases by inducing systemic acquired resistance in plants. Scortichini et al. (2000b) obtained 25%–30% disease reduction with five applications of Bion at 25 g a.i. ha^{-1} when applied once a month from late April to July including the final application in September after harvest
7. An integrated approach including destruction of wilted twigs, branches, and trees, copper treatments at critical times, insect control, and application of acibenzolar-*S*-methyl may result in satisfactory management of the disease

REFERENCES

De Boer, S. H., Ward, L. J., Li, X., and Chittaranjan, S. 1995. Attenuation of PCR inhibition in the presence of plant compounds by addition of BLOTTO. *Nucleic Acids Res.* 23: 2567–2568.

De Castro, P. 1999. Disposizioni a favour delle colture corilicole colpite dalla moria del nocciolo. Legge della Repubblica Italiana n° 307, del 17 Agosto 1999. Gazzetta Ufficiale n° 210.

Gardan, L., Shafik, H., Belouin, S., Brosch, R., Grimont, F., et al. 1999. DNA relatedness among the pathovars of *Pseudomonas syringae* and description of *Pseudomonas tremae* sp. nov. and *Pseudomonas cannabina* sp. nov. (ex Sutic and Dowson 1959). *Int. J. Syst. Bacteriol.* 49: 469–478.

Gervasi, F., and Scortichini, M. 2009. Detection of *Pseudomonas avellanae* from hazelnut twigs by taqman real-time PCR. *J. Plant Pathol.* 91: 573–578.

Janse, J. D., Rossi, M. P., Angelucci, L., Scortichini, M., Derks, J. H. J., et al. 1997. Validation of the publication of new names and new combinations previously effectively published outside the IJSB. *Int. J. Syst. Bacteriol.* 47: 601–602. Effective publication: Janse, J. D., Rossi, M. P., Angelucci, L., Scortichini, M., Derks, J. H. J., et al. 1996. Reclassification of *Pseudomonas syringae* pv. *avellanae* as *Pseudomonas avellanae* (spec. nov.), the bacterium causing canker of hazelnut (*Corylus avellana* L.). *Syst. Appl. Microbiol.* 19: 589–595.

Loreti, S., and Gallelli, A. 2002. Rapid and specific detection of virulent *Pseudomonas avellanae* strains by PCR amplification. *Eur. J. Plant Pathol.* 108: 237–244.

Loreti, S., Gallelli, A., De Simone, D., and Bosco, A. 2009. Detection of *Pseudomonas avellanae* and the bacterial microflora of hazelnut affected by "moria" in central Italy. *J. Plant Pathol.* 91: 365–373.

Loreti, S., Sarrocco, S., and Gallelli, A. 2001. Identification of *hrp* genes, encoding harpin protein, in *Pseudomonas avellanae* (Psallidas) Janse et al. *J. Phytopathol.* 149: 219–226.

Mahdavian, S. E., and Hasanzadeh, N. 2012. Isolation of *Pseudomonas syringae* from hazelnut trees in west part of Mazandaran province. *Int. J. Innov. Res. Sci. Eng. Technol.* 1: 226–229.

Martins, J. M. S., and Scortichini, M. 1998. Spatio-temporal spread of a bacterial disease in a hazelnut orchard. pp. 82–88 in: *Actas 2a Reuniao bienal de Sociedade Portuguesa de Fitopatologia*, Oeiras, Portugal.

Ovod, V. V., Knirel, Y. A., Samson, R., and Krohn, K. J. 1999. Immunochemical characterization and taxonomic evaluation of the O polysaccharides of the lipopolysaccharides of *Pseudomonas syringae* serogroup O1 strains. *J. Bacteriol.* 181: 6937–6947.

Psallidas, P. G. 1984. Bacterial canker of *Corylus avellana*: The taxonomic position of the causal agent. *Proceedings 2nd Working Group on Pseudomonas syringae Pathovars*, Sounion, Greece, pp. 53–55.

Psallidas, P. G. 1987. The problem caused by *Pseudomonas syringae* pv. *avellanae* in Greece. *EPPO Bull.* 94:103–111.

Psallidas, P. G. 1993. *Pseudomonas syringae* pv. *avellanae* pathovar nov., the bacterium causing canker disease on *Corylus avellana*. *Plant Pathol.* 42: 358–363.

Psallidas, P. G., and Panagopoulos, C. G. 1979. A bacterial canker of hazelnut in Greece. *Phytopathol. Z.* 94: 103–111.

Scortichini, M. 1998. Response of *Corylus avellana* germplasm to artificial inoculation with *Pseudomonas avellanae* (Psallidas) Janse et al. *Agricultura Mediterranea* 128: 153–156.

Scortichini, M. 2002. Bacterial canker and decline of European hazelnut. *Plant Dis.* 86: 704–709.

Scortichini, M. 2012. Detection and identification methods and new tests as developed and used in the framework of cost 873 for bacteria pathogenic to stone fruits and nuts *Pseudomonas avellanae*. *J. Plant Pathol.* 94: S1.115–16.

Scortichini, M., Dettori, M. T., Rossi, M. P., Marchesi, U., and Palombi, M. A. 1998. Differentiation of *Pseudomonas avellanae* strains from Greece and Italy by rep-PCR genomic fingerprinting. *J. Phytopathol.* 146: 417–420.

Scortichini, M., Ferrante, P., Cozzolino, L., and Zoina, A. 2016. Emended description of *Pseudomonas syringae* pv. *avellanae*, causal agent of European hazelnut (*Corylus avellana* L.) bacterial canker and decline. *Eur. J. Plant Pathol.* 144: 213–215.

Scortichini, M., Liguori, R., Nobile, M., Moretti, D., Valentini, B., et al. 2000a. Moria del nocciolo: un triennio di prove di campo con acibenzolar-Smetil, induttore di resistenza sistemica acquisita. *Atti Giornate Fitopatologico* 2: 379–384.

Scortichini, M., and Marchesi, U. 2001. Sensitive and specific detection of *Pseudomonas avellanae* strains using primers based on 16S rRNA gene sequences. *J. Phytopathol.* 149: 527–532.

Scortichini, M., Marchesi, U., Angelucci, L., Rossi, M. P., and Dettori, M. T. 2000b. Occurrence of *Pseudomonas avellanae* (Psallidas) Janse et al. and related pseudomonads on wild *Corylus avellana* trees and genetic relationships with strains isolated from cultivated hazelnut. *J. Phytopathol.* 148: 523–532.

Scortichini, M., Marchesi, U., Rossi, M. P., and Di Prospero, P. 2002. Bacteria associated with hazelnut (*Corylus avellana* L.) decline are of two groups: *Pseudomonas avellanae* and strains resembling *P. syringae* pv. *syringae*. *Appl. Environ. Microbiol.* 68: 476–484.

Scortichini, M., Rossi, M. P., Loreti, S., Bosco, A., Fiori, M., et al. 2005. *Pseudomonas syringae* pv. *coryli*, the causal agent of bacterial twig dieback of *Corylus avellana*. *Phytopathology* 95: 1316–1324.

Scortichini, M., and Tropiano, F. G. 1994. Severe outbreak of *Pseudomonas syringae* pv. *avellanae* on hazelnut in Italy. *J. Phytopathol.* 140: 65–70.

Sorensen, K. W., Kim, K. H., and Takemoto, J. Y. 1998. PCR detection of cyclic lipodepsinonapeptide-producing *Pseudomonas syringae* pv. *syringae* and similarity of strains. *Appl. Environ. Microbiol.* 64: 226–230.

7.66 WILDFIRE OF TOBACCO

Wildfire of tobacco (*Nicotiana tabacum* L.) is also known by another name, i.e., **"angular leaf spot."** It can affect tobacco in both the seedbeds/float trays and the field, although wildfire tends to be more of a problem in the seedbed and angular leaf spot in the field. It is not a major problem in many tobacco-producing areas, such as the USA, Brazil, and Europe. In Africa, these diseases are of major economic importance and can cause devastating losses, especially in wet seasons. Disease incidence up to 76% was observed on tobacco in Zimbabwe (Fisher and Mpofu 1992). However, the availability of resistant cultivars has reduced concern for wildfire and angular leaf spot (also called **"black fire"**) worldwide, and there are few reports of severe epiphytotics in recent years.

The bacteria that cause wildfire and angular leaf spot are identical in all respects except that the wildfire bacteria produce a toxin and the angular leaf spot bacteria do not. Wildfire is, therefore, caused by the "**tox**$^{+}$" strain and angular leaf spot by the "**tox**$^{-}$" strain of *Pseudomonas syringae* pv. *tabaci*.

Destéfano et al. (2010) have reported the occurrence of bacterial leaf spot of coffee caused by *Psm. syringae* pv. *tabaci* in Brazil. It was the first report of this species causing disease on coffee.

7.66.1 Host Range

Deadly nightshade (*Atropa belladonna*), oats (*Avena sativa*), bell pepper (*Capsicum annuum*), fat hen (*Chenopodium album*), arabica coffee (*Coffea arabica*), cucumber (*Cucumis sativus*), jimson weed/stinkblaar (*Datura stramonium*), soybean (*Glycine max*), apple of Peru (*Nicandra physaloides*), common bean (*Phaseolus vulgaris*), lima bean (*P. lunatus*), pea (*Pisum sativum*), tomato (*Solanum lycopersicum*), aubergine (*S. melongena*), black nightshade (*S. nigrum*), potato (*S. tuberosum*), dandelion (*Taraxacum officinale* complex), and cowpea (*Vigna unguiculata*) also act as natural hosts.

7.66.2 Symptoms

The symptoms of wildfire (tox^{+} or toxin producing) and angular leaf spot (tox^{-} or non-toxin producing) forms or phases of this disease are quite different.

7.66.2.1 Wildfire

The name wildfire denotes the burnt appearance of heavily infected tobacco plants. Initially, wildfire symptoms are small, circular, pale-green areas (ca 0.5 cm in diameter), which turn brown in the center owing to necrosis of parenchymatic tissues. The necrotic process advances rapidly and the chlorotic halo extends; within a few days spots enlarge up to 3 cm in diameter, with a water-soaked zone surrounding the necrotic center. Spots may coalesce and dead tissue expands irregularly until the diseased tissue eventually falls out leaving ragged holes. Heavily infected leaves grow poorly and become distorted. The formation of the chlorotic halo at the border of the necrotic area results from the diffusion of toxins released by the pathogen. Wildfire can be systemic in seedlings, causing distortion of the apical bud and leaves.

In seedbeds, the disease develops at irregularly distributed sites. Under high moisture or crowded conditions, leaves can be completely infected and rot rapidly. In the field, wet weather allows the formation of large necrotic spots with a narrow halo. Although leaves are the preferred sites for the disease, lesions can occasionally be observed on stalks, flowers, and capsules.

7.66.2.2 Angular Leaf Spot

Necrotic areas are brown, dark brown, or almost black and much larger than the wildfire lesions. These are angular in shape due to their confinement by the lateral veins and the chlorotic halos are absent. Necrotic tissues in the center of the spots may break away. Deformation of the leaves occurs due to the uneven growth of healthy and infected areas (Figure 7.48). During epiphytotics, diffused dropping of necrotic tissues occurs and only veins may remain in place of the leaves. On a number of cultivars, leaf spots may be circular and show concentric circles. In Africa, both phases tend to be more severe at the top of the plant.

Deall and Cole (1986) recorded significant differences between strains causing wildfire or angular leaf spot in localization of spots with reference to plant age and stalk position; lower leaves of young plants being more susceptible to wildfire, while top leaves of older plants were more susceptible to angular leaf spot. However, at the end of the season, it is often difficult to distinguish between the two described syndromes, which can occur in the same field at the same time.

7.66.3 Causal Organism

Pseudomonas syringae pv. *tabaci* tox$^+$ and tox$^-$ (Wolf and Foster 1917) Young et al. 1978
Syn. *Psm. tabaci*, *Psm. angulata*, *Psm. syringae* pv. *tabaci*, *Psm. syringae* pv. *angulata*

Wang et al. (2015) determined the draft genome sequence of *Psm. syringae* pv. *tabaci* strain yuexi-1 isolated from tobacco sample of yuexi-1, Sichuan province, China. The genome contains a single circular chromosome of 6,232,497 bp with a GC content of 58.2 mol%. Total coding genes were 5,354,916 bp, and 5,701 protein-coding genes were identified in the genome, with an average length of 939 bp. A total of 3,261 (57.2%) genes were classified into clusters of orthologous-group families comprising 21 functional categories.

FIGURE 7.48 Wildfire of tobacco. (Courtesy of R.J. Reynolds Tobacco Company Slide Set, R.J. Reynolds Tobacco Company, Bugwood.org.)

7.66.4 Disease Cycle

Both the bacteria are seed-borne and can be transmitted through infected seed. Diseased plant debris and tobacco regrowth are other important sources of perpetuation of the bacteria and these also provide inoculum to infect overwintering weed hosts. Many solanaceous weeds including apple of Peru and jimson weed/stinkblaar act as collateral hosts of the bacteria. In the semi-tropical areas, where these diseases are a problem, winters are seldom cold enough to kill overwintering weeds and tobacco regrowth. The short distance dissemination of the bacteria occurs primarily through wind-driven rains and overhead irrigation. Wind-driven rains and sand blasting winds exacerbate the problem considerably. The long distance dissemination can occur through infected seed and seedlings and contaminated irrigation water. Wildfire and angular leaf spot are favored by cloudy wet weather.

Host resistance: Many species in the genus *Nicotiana*, namely, *N. acaulis*, *N. alata*, *N. arentsii*, *N. attenuata*, *N. bigelovii*, *N. bonariensis*, *N. cavicola*, *N. debneyi*, *N. fragrans*, *N. glauca*, *N. langsdorffii*, *N. longiflora*, *N. nudicaulis*, *N. plumbaginifolia*, *N. repanda*, *N. rustica*, *N. suaveolens*, *N. trigonophylla*, *N. undulata*, and *N. wigandioides* possess some level of resistance to the wildfire pathogen (Stavely 1979). Considerable efforts have been made to develop tobacco varieties resistant to wildfire. *N. longiflora* was the first species whose monogenic resistance was transferred to *N. tabacum*; an interspecific cross allowed the release of TL 106, a widely employed breeding line. However, mutant strains of the pathogen able to overcome this resistance were reported (Valleau et al. 1962; Fulton 1980). Race 1 is capable of overcoming resistance of all *Nicotiana* species except *N. nudicaulis*, *N. repanda*, *N. rustica*, and *N. undulata* (Stavely 1979). Monogenic resistance to race 1 from *N. rustica* var. *brasilea* has been introduced into *N. tabacum*, but there are problems in its transmission to successive generations (Smeeton and Ternouth 1992; Woodend and Mudzengerere 1992). Attempts to incorporate resistance from *N. repanda* have also been made (Gwynn et al. 1986).

Burley 21, the first tobacco variety resistant to wildfire was released for planting in 1955. Its high resistance to wildfire was derived from *N. longiflora* and resistance to tobacco mosaic of the local-lesion type from *N. glutinosa* (Heggestad et al. 1960). Resistance to wildfire pathogen race 0, derived from *N. longiflora* via Burley 21, is monogenic, complete, and fully dominant. This resistance has proved very durable in most parts of the world, but in Africa, it broke down in the 1970s with the emergence of race 1. *N. rustica* derived resistance to races 0 and 1 of wildfire and angular leaf spot pathogens is also monogenic, complete, and fully dominant. Resistant varieties to race 1 were released in Zimbabwe in the early 1990s, but these varieties succumbed to the pathogen attack in the late 1990s due to the emergence of race 2 of both the pathogens. No source of resistance to race 2 has been identified. Polygenic resistance is generally low, but some of the newer multi-resistant Zimbabwean varieties have some level of polygenic resistance to race 2.

Lovrekovich and Farkas (1965) reported that pretreatment of tobacco leaves by heat-killed bacteria induced protection against infection by *Psm. tabaci*. Thilmony et al. (1995) found that expression of the tomato *Pto* gene in tobacco enhances resistance to *Psm. syringae* pv. *tabaci* expressing *avrPto* and the results indicate that essential components of a *Pto*-mediated signal transduction pathway are conserved in tobacco and should prompt examination of resistance gene function across even broader taxonomic distances. Huang et al. (1997) demonstrated that the pathogen-induced promoter and the secretory sequence were competent elements for transforming a cecropin gene into an effective disease-control gene for plants. Bacterial multiplication in leaves of MB39-transgenic plants was suppressed more than 10-fold compared to control plants, and absence of disease symptom development was associated with this growth suppression. Anzai et al. (1990) cloned the tabtoxin resistance gene (*ttr*) coding for an acetyltransferase from *Psm. syringae* pv. *tabaci*. Transgenic tobacco plants expressing *ttr* gene showed no chlorotic symptoms after treatment with tabtoxin and infection by *Psm. syringae* pv. *tabaci*. Lee et al. (2013) reported that *Psm. syringae* pv. *tabaci* extracellular metabolite extracts suppressed plant defense responses such as stomatal closure and non-host hypersensitive response cell death induced by the non-host bacterial pathogen, *Psm. syringae* pv. *tomato* T1 in *N. benthamiana*.

7.66.5 Management

An integrated approach to the management and control of wildfire and angular leaf spot includes:

1. Use certified pathogen-free tobacco seed
2. Seed treatment with silver nitrate, ideally as part of the seed certification requirement, is essential
3. Rotation of seedbed sites and proper fumigation of the seedbed areas
4. Apply preventative sprays of a combination of a copper-based compound (e.g., Kocide 101, copper oxychloride) and ASM in seedbed/float bed. ASM should be used on cv. Burley 21 seedlings with caution and only when necessary
5. Regularly scout seedbeds, remove and destroy all seedlings within 1 m area of an infected plant. Drench the surrounding area with disinfectant (e.g., bleach). It should be done more regularly particularly when weather conditions are more conducive to disease development, i.e., wet and cloudy weather and after rains
6. Eradicate natural weed hosts from fields and particularly, those near the seedbeds/float beds. This is particularly important in areas, which do not have killing winter frosts and weed hosts survive
7. Sterilize all seedbed tools, particularly those used for clipping/mowing, with bleach or a copper-based compound
8. Select site to avoid later planted crops downwind of earlier planted ones
9. Apply correct doses of fertilizers and maintain correct pH; avoid excessive doses of nitrogen and low dose of potash
10. Avoid or minimize leaf injury by avoiding leaf breakage and not mowing too close to the plants
11. Give preventative sprays of acibenzolar-*S*-methyl (trade name, Actiguard and Bion) in the field or immediately after the appearance of the disease (Cole 1997)
12. Minimize pathogen spread by reaping clean fields before infected ones
13. Deep plough and burry the diseased plant debris at the end of the season and also destroy all re-growth. To make it more effective, all the growers in an area should follow this practice
14. Use of resistant varieties is probably the most effective control measure of the disease and this has been widely exploited in case of tobacco

REFERENCES

Anzai, H., Yoneyama, K., and Yamaguchi, I. 1990. The nucleotide sequence of tabtoxin resistance gene (*ttr*) of *Pseudomonas syringae* pv. *tabaci*. *Nucleic Acids Res*. 18: 1890.

Cole, D. L. 1997. The efficacy of a plant activator CGA 245704 against field diseases of tobacco. *Presented: CORESTA Agronomy & Phytopathology Group Meeting*, Montreux, Switzerland.

Deall, M. W., and Cole, J. S. 1986. A comparative study of the pathogenicity and epidemiology of strains of *Pseudomonas syringae* pv. *tabaci* that cause wild fire and angular leaf spot diseases of tobacco in Zimbabwe. *Plant Pathol*. 35: 74–81.

Destéfano, S. A. L., Rodrigues, L. M. R., Beriam, L. O. S., Patrício, F. R. A., Thomaziello, R. A., et al. 2010. Bacterial leaf spot of coffee caused by *Pseudomonas syringae* pv. *tabaci* in Brazil. *Plant Pathol*. 59: 1162–1163.

Fisher, C. R., and Mpofu, S. I. 1992. *Plant Pathology. Annual Report and Accounts—Tobacco Research Board (Harare) for the Year Ended 30 June 1992*. Tobacco Research Board, Harare, Zimbabwe, pp. 17–18.

Fulton, R. W. 1980. Tobacco blackfire disease in Wisconsin. *Plant Dis*. 64: 100.

Gwynn, G. R., Barker, K. R., Reilly, J. J., Komm, D. A., Burk, L. G., et al. 1986. Genetic resistance to tobacco mosaic virus, cyst nematodes, root-knot nematodes, and wildfire from *Nicotiana repanda* incorporated into *N. tabacum*. *Plant Dis*. 70: 958–962.

Heggestad, H. E., Clayton, E. E., Neas, M. O., and Skoog, H. A. 1960. Development of Burley 21, the first wildfire-resistant tobacco variety, including results of variety trials. University of Tennessee Agricultural Experiment Station, *Bulletins* [http://trace.tennessee.edu/utk_agbulletin/265]

Huang, Y., Nordeen, R. O., Di, M., Owens, L. D., and McBeath, J. H. 1997. Expression of an engineered cecropin gene cassette in transgenic tobacco plants confers disease resistance to *Pseudomonas syringae* pv. *tabaci*. *Phytopathology* 87: 494–499.

Lee, S., Yang, D. S., Uppalapati, S. R., Sumner, L. W., and Kirankumar, S. M. 2013. Suppression of plant defense responses by extracellular metabolites from *Pseudomonas syringae* pv. *tabaci* in *Nicotiana benthamiana*. *BMC Plant Biol*. 13: 65.

Lovrekovich, L., and Farkas, G. L. 1965. Induced protection against wildfire disease in tobacco leaves treated with heat-killed bacteria. *Nature* 205: 823–824.

Smeeton, B. W., and Ternouth, R. A. F. 1992. Sources of resistance to powdery mildew, wildfire, angular leaf spot, and *Alternaria*. *Bulletin d'Information—CORESTA* 3–4: 127–135.

Stavely, J. R. 1979. Disease resistance. pp. 87–110 in: *Nicotiana: Procedures for Experimental Use*. R. D. Durbin ed. US Department of Agriculture Technical Bulletin 1586.

Thilmony, R. L., Chen, Z., Bressan, R. A., and Martina, G. B. 1995. Expression of the tomato *Pto* gene in tobacco enhances resistance to *Pseudomonas syringae* pv. *tabaci* expressing *avrPto*. *Plant Cell* 7: 1529–1536.

Valleau, W. D., Litton, C. C., and Johnson, E. M. 1962. Susceptibility of wildfire resistant tobacco varieties to certain strains of *Pseudomonas tabaci* and *Pseudomonas angulata*. *Plant Dis. Reptr* 46: 36–37.

Wang, T., Yang, Y., and Zhao, T. 2015. Genome sequence of a *Pseudomonas syringae* pv. *tabaci* strain, yuexi-1, causing wildfire disease in tobacco. *Genome Announc.* 3: e00180–e00215.

Wolf, F. A., and Foster, A. C. 1917. Bacterial leaf spot of tobacco. *Science, Washington* 46: 361–362.

Woodend, J. J., and Mudzengerere, E. 1992. Inheritance of resistance to wildfire and angular leaf spot derived from *Nicotiana rustica* var. *brasilea*. *Euphytica* 64: 149–156.

Young, J. M., Dye, D. W., and Wilkie, J. P. 1978. Genus *Pseudomonas* Migula 1894. In: Young, J. M., Dye, D. W., Bradbury, J. F., Panagopoulos, C. G., and Robbs, C. F. 1978. A proposed nomenclature and classification for plant pathogenic bacteria. *N. Z. J. Agric. Res.* 21: 153–177.

7.67 BACTERIAL BLIGHT OF CASSAVA

Bacterial blight of cassava (*Manihot esculenta* Crantz) caused by a vascular pathogen, *Xanthomonas axonopodis* pv. *manihotis* is the most common biotic constraint to its production worldwide. The disease was first described in 1912 from Brazil (Bondar 1912). During the 1970s, it was found in most of the cassava growing areas of central and South America, the Caribbean, Africa, and Asia (Bradbury 1986). Despite its widespread occurrence, there are still areas which are disease-free, including the southern Pacific. Quarantine measures in this area are particularly relevant. Yield losses in different areas of Africa and South America may be anywhere between 12% and 90% (Persley 1977; Lozano & Laberry 1982). During the epiphytotic years of 1970–1975 when losses in central Africa were as high as 80%, the disease contributed immensely to starvation in Zaire (Lozano 1986). Crop losses as high as 90%–100% were observed in some parts of Uganda, 2 years after the disease was first recorded (Otim-Nape 1980).

7.67.1 Host Range

Besides cassava (*M. esculenta*), the bacterium attacks the wild species, namely, *M. aipi*, *M. glaziovii*, and *M. palmata*. The symptoms similar to those observed on cassava were induced on *Euphorbia pulcherrima* (poinsettia) and *Pedilanthus tithymaloides* following artificial inoculation with *Xa. axonopodis* pv. *manihotis* (Dedal et al. 1980). However, Van den Mooter et al. (1987) did not observe typical symptoms, but found only very restricted lesions on *E. pulcherrima* inoculated with a wide range of *Xa. axonopodis* pv. *manihotis* isolates.

7.67.2 Symptoms

Considerable variations in the predominance and severity of the various symptoms described here are observed depending on the location, season, aggressiveness of the occurring bacterial strains, and the cassava cultivars (Maraite 1993). The initial symptoms are wilting of young germinating sprouts followed by their die-back, resulting from the sowing of infected planting material. Then the other symptoms like, angular leaf spots, leaf blightening, defoliation, wilting of the immature shoot, and finally die-back appear.

Infected leaves show dark-green to blue, water-soaked, angular spots (1–4 mm diameter), limited by veinlets, and irregularly distributed on the lamina. These spots frequently extend and coalesce along the veins or the edges of the leaf; the central portion turns brown and the water-soaked part often becomes surrounded by a chlorotic halo. The lesions appear as translucent spots when viewed against the light. Under a magnifying glass, small droplets of exudate oozing from the central portion of the lesion are visible on the lower surface of the leaves. The droplets, which first glisten creamy-white and later yellow, are easily dissolved by rain or dew, and on drying they form a thin scale. Under favorable conditions, water-soaked pinpoint spots develop, scattered around young angular spots. The surrounding part of the lamina turns light brown and within 2–3 days extensive areas of the leaflets become withered not only toward the tip or border of the leaflet, but also toward the base. The affected parts show light-brown and green zonations as if they had been superficially burnt. These necrotic areas are not translucent, no bacterial exudate is observed, and bacteria are absent or present in very limited numbers at the borders of the extending blight lesions. A severe attack leads to premature drying and shedding of the leaves. Under conditions of high humidity, infection may spread through the vascular bundles from the leaflets to the petioles and twigs or stems, with the formation of dark-brown to black streaks as well as exudation drops along the pathway of progression. These parts may also become infected directly through wounds, which may be due to removal of leaves for consumption or insect punctures.

On the unlignified twigs or stems, a dark-green to black water-soaked area develops around the infection point. Large gummy exudation drops appear, some distance away from the point of infection, in the axis of vascular bundles, and one or a small number of leaves, located on the same side, show a sudden loss of turgidity, followed by rapid wilting and shriveling. Afterwards, the base of the petiole collapses, but the dried leaves generally remain attached for some time. All leaves located above those showing the first symptoms wilt progressively. Finally, the unlignified tip dies, appearing as a wick on the withered stem end, giving the "**candle**" symptom, while new shoots grow out lower down the stem. As the infection progresses toward the base of the stem, these shoots often also become wilted leading to plant die-back. In the infected shoots, xylem vessels are brownish.

In the microscopic examination, the infected vessels appear obstructed by bacteria, tyloses, and mucilaginous substances. Lytic pockets develop around the protoxylem. The spread of these pockets causes rupture of the xylem ring, development of lytic pockets in the phloem, and later rupture of the fiber ring in the cortical collenchyma. Externally, the latter pockets become visible as dark-green, water-soaked spots and small black streaks, corresponding to altered laticifers. These spots swell, rupture, and extrude a sticky white-yellow gum. In fully lignified stems or branches, only internal vascular browning is visible. Infection may spread more than 50 cm below any external visual symptom.

Water-soaked expanding spots appear on fruit or seed balls. Severely infected seeds from such fruit may be deformed, with necrotic areas on cotyledons and endosperm and corrugation of the testa. Very few of such seeds germinate. Generally roots of infected plants do not show symptoms. However, in some susceptible cultivars, swollen roots may show dry, rotted spots around the dead vascular strands. However, the rotting is usually restricted to the vascular tissues and the root tissues remain apparently healthy.

The infection of young plants may destroy their aerial parts. The infected plants usually produce new shoots from the stem, either above or below the ground. These young shoots are extremely susceptible and rapidly get infected during the rainy season, thereby providing inoculum for the secondary infection.

Xa. cassavae causes angular leaf spots very similar to those induced by *Xa. axonopodis* pv. *manihotis*. However, these spots do not evolve into blight areas, and produce very tiny bright yellow exudates (Maraite and Perreaux 1979).

7.67.3 Causal Organism

Xanthomonas axonopodis pv. *manihotis* (Bondar 1915) Vauterin et al. 1995
Syn. *Xa. manihotis*, *Xa. campestris* pv. *manihotis*

The bacterium was first named as *Bacillus manihotis* Arthaud-Berthet and then *Phytomonas manihotis* (Arthaud-Berthet and Bondar) Viegas. In a study on the nutrition of phytopathogenic bacteria in 1946, Starr listed this bacterium as *Xanthomonas manihotis*. After the adoption of pathovar system, Dye (1978) renamed it as *Xa. campestris* pv. *manihotis* (Arthaud-Berthet and Bondar) Dye. On the basis of their phenetic and genomic characteristics, Vauterin et al. (1995) proposed the name *Xa. axonopodis* pv. *manihotis*, which is now used as reference name in the list of plant pathogenic bacteria (Young et al. 1996).

Cells are Gram-stain-negative rods, measure 0.4 × 1.3 µm (0.28–0.6 × 1.0–1.75 µm), occur singly or in pairs and motile with one polar flagellum. Individual colonies on nutrient agar become visible after 24 h of incubation at 28°C and measure about 1 mm in diameter after 48 h. On nutrient agar, the colonies are whitish-grey to creamy, raised, convex, smooth, shiny, with entire edges; at first hyaline, then opaque and turbid, and of viscous consistency.

Growth on potato dextrose agar and tryptone soya agar is faster than on nutrient agar. Colonies on tetrazolium medium (Kelman 1954) are 8 mm in diameter after 6 days of incubation, and are round, smooth with bright red centers, and a narrow edge. They resemble the colony type of weakly pathogenic mutants of *Pseudomonas solanacearum*, recent name *Ralstonia solanacearum* (Maraite and Meyer,1975). On glucose yeast saline agar, colonies are distinctly convex and shiny. For some strains, rough variants are easily detected on this medium (Maraite et al. 1981). The Kovacs' oxidase test is most useful for distinguishing *Xa. axonopodis* pv. *manihotis* (negative or slow positive) from *Ral. solanacearum* (rapidly positive). Except for lack of pigmentation, most physiological and biochemical characteristics of *Xa. axonopodis* pv. *manihotis* are those of xanthomonads.

Variation was observed among *Xa. axonopodis* pv. *manihotis* strains in biochemical and physiological (Fessehaie 1997), serological (Wydra et al. 1998) and genomic characteristics (Assigbétsé et al. 1998; Restrepo et al. 1999), and strain × genotype interactions have been reported. Differences in virulence among *Xa. axonopodis* pv. *manihotis* isolates were first described by Robbs et al. (1972) in South American countries. In Colombia, 189 isolates were grouped into seven clusters following stem-inoculation (Restrepo and Verdier 1997), and a set of 27, four, and one pathotype was identified among bacterial strains from different Colombian ecozones (Restrepo et al. 2000, 2004). Ogunjobi et al. (2007) determined the pathogenic variation in the population structure of 72 strains of *Xa. axonopodis* pv. *manihotis* isolated from Nigeria. Out of 72 strains examined, 12.9% were classified as highly virulent, 71.9% as virulent, 12.5% as less virulent, and 2.8% were non-pathogenic.

Chege et al. (2017) determined the diversity of 33 isolates of *Xa. axonopodis* pv. *manihotis* obtained from different cassava growing regions of Kenya using phenotypic characteristics and repetitive DNA polymerase chain reaction-based fingerprinting. The dendrogram generated from analysis of phenotypic characteristics of the isolates produced two major clusters at 75% similarity level. Analysis of 19 isolates with repetitive extragenic palindromic primers yielded characteristic fingerprint pattern with bands ranging between 400 and 2,000 bp in size and their numbers ranged from 1 to 6 bands per isolate. The cluster analysis using unweighted pair group method with arithmetic averages did not reveal any significant differences in clustering and relationship to the geographical origin, with exception of a single isolate that had unique fingerprints. These findings suggest that *Xa. axonopodis* pv. *manihotis* population in Kenya is a uniform population that evolved from the same origin, and this information may prove useful in future breeding programs.

Arrieta-Ortiz et al. (2013) generated the first manually annotated high-quality draft genome sequence of *Xa. axonopodis* pv. *manihotis* strain CIO151 and found 126 genes potentially unique to the bacterium, as well as potential horizontal transfer events in the history of the genome. They further reported that the relation of these regions with virulence and pathogenicity could explain several aspects of the biology of this pathogen, including its ability to colonize both vascular and non-vascular tissues of cassava plants.

7.67.4 Disease Cycle

The bacterium survives for many months in infected stems and bacterial exudate, reviving multiplication in wet periods. The survival rate of the bacterium is higher in dry than in moist conditions (Persley 1979b) and better in cassava debris than in rhizosphere soil of cassava weeds (Thaveechai et al. 1993). Daniel and Boher (1985) found large number of pathogen cells on symptomless cassava leaves in farm crops and cultivar collections in the Congo during the rainy season, when spread of the disease occurs. In the dry season, the cell numbers fell to undetectable levels, but the presence of *Xa. axonopodis* pv. *manihotis* a few weeks before new symptoms appeared at the beginning of the next season suggested that it could survive as an epiphyte.

Amaranthus species, *Panicum fasciculatum*, Sida species, *Sorghum halepense*, and several species belonging to the *Euphorbiaceae* have been identified as possible alternative hosts of the bacterium in Venezuela (Marcano and Trujillo 1982). Marcano and Trujillo (1984) demonstrated epiphytic survival of the bacterium on several plant species under high relative humidity, heavy rainfall, and thick cloud cover, but not under dry conditions. However, Ikotun (1981) concluded from studies in Nigeria and Colombia that *Xa. axonopodis* pv. *manihotis* does not survive naturally on alternative hosts in and around cassava plants, but observed a hypersensitive reaction on *E. repanda* after inoculation.

The bacterium can also perpetuate through the infected seed. The possibility of seed-plant transmission of *Xa. axonopodis* pv. *manihotis* has been demonstrated by Persley (1979a) by a leaf-infiltration technique, using the supernatant from seeds soaked in sterile water at 30°C for 2–4 h as inoculum. Elango and Lozano (1980) confirmed seed-plant transmission, with embryo infection ranging from 0% to 40% depending on the sampling period by immunofluorescence (IF) and ELISA on enriched embryonic extracts. Clear positive reactions were obtained with 100,000 cells mL^{-1} in IF and with 10,000 cells mL^{-1} in ELISA. Daniel and Boher (1981) determined pathogen population on and within the seeds (testa, caruncle, endosperm, cotyledon, and embryo) up to 1 million bacteria per seed in histological studies combined with IF. Infected seeds are recognized as an important means of transmission of the pathogen (Lozano 1986).

The bacterium spreads to new areas in infected symptomless cuttings and seed. Because cassava is vegetatively propagated, infected propagative planting material forms a significant vehicle for dissemination of the pathogen. The movement of infected vegetative planting material from diseased plantations has been the main mode of disseminating *Xa. axonopodis* pv. *manihotis* over long distances into Africa and Asia. It is also the source of primary infections in newly established plantations. This is mainly due to the lack of visible symptoms on lignified stems and the ability of the pathogen to survive in invaded tissue for a very long time (Lozano and Nolt 1989). Within the field, spread is mostly by rain splashes (Persley 1979b) and epidemics build up during the rainy season. Contaminated tools are also an important means of spread, because planting material is prepared simultaneously with the harvest, and the pathogen may be spread to uninfected cuttings when planting stakes are prepared from apparently healthy mature stems harboring the pathogen. Movement of man and animals through plantations, especially during or after rain, may also contribute to pathogen spread (Lozano 1986).

The entry of the pathogen into the host occurs through stomata or epidermal wounds. In central and western Africa, stem infections have often been found to originate from punctures by the bug *Pseudotheraptus devastans*, an occasional vector of the bacterium (Maraite and Meyer 1975). Bouriquet (1939) found a similar role for *Anoplecnemis madagascariensis* and the grasshopper, *Zonocerus variegatus*.

A relative humidity of 90%–100% for 12 h along with temperature range of 22°C–26°C is required for infection (Lozano 1986). The optimum temperature for disease development in growth chambers is around 30°C (Maraite and Perreaux 1979). In South America, temperature is considered to be a major factor affecting the disease severity. In regions such as the Amazon, with average minimum and maximum temperatures above 20°C and 30°C, respectively, cassava bacterial blight is not an important disease, despite the high rainfall (Takatsu et al. 1979). Fluctuating day/night temperatures exceeding 10°C during the rainy season were found to enhance disease severity (Lozano 1986; Joseph and Elango 1991), which may explain the low incidence in forest zones with buffered temperature variations. In Africa, cassava bacterial blight is also more prevalent in the savannah and the forest-savannah transition zones than in the forest zones, or even the forest galleries within the savannahs (Maraite and Meyer 1975; Persley 1980). Apart from temperature fluctuation effect, this may also be due to increased susceptibility of cassava on the leached savannah soil: potassium fertilization to increase the potassium content of the leaves tends to reduce disease severity and enhances yield (Odurukwe and Arene 1980).

Bacterial multiplication in the intercellular spaces of parenchymatous leaf or stem tissues occurs together with cell-wall degradation, possibly due to production of pectinases and cellulases (Maraite and Weyns 1979; Dianese 1985; Boher et al. 1995), and leads to the formation of lytic pockets filled with mucilaginous substances, including bacterial extracellular polysaccharide (Ikotun 1984). Either directly or after disruption, the lytic pockets provide the bacteria with access to the xylem which results in systemic spread throughout the plant. The appearance and extent of the blight symptoms are associated with an increased leaf concentration of the blight-inducing toxin, 3-methylthiopropionic acid (Perreaux et al. 1986) and other phytotoxic carboxylic acids produced by the bacteria (Ewbank 1992). These toxins result from the induction of transamination and decarboxylation catabolism of methionine (Ewbank and Maraite 1990) or of other amino acids in the bacteria (Ewbank 1992).

Resistance in *M. esculenta*, which was introgressed from the wild relative *M. glaziovii*, is polygenic and additively inherited. Genetic diversity and resistance to bacterial blight revealed a high level of polymorphism among cassava varieties (Sãnchez et al. 1999). Jorge et al. (2000) identified six regions of the cassava genome controlling resistance to *Xa. axonopodis* pv. *manihotis* strains, confirming the polygenic character of the resistance. A specific interaction between the host and the pathogen was suggested, and resistance markers specific for African bacterial strains were identified (Wydra et al. 2003). Resistance to cassava bacterial blight appears to be due to several genes, mainly with additive effects, but also to some extent with non-additive effects; resistance appears to be recessive to susceptibility (Hahn 1979; Uremura and Kawano 1983). Breeding work at the International Institute of Tropical Agriculture, Nigeria and the Centro Internacional de Agricultura Tropical, Colombia led to the development of cassava lines resistant to the pathogen and also showed that resistance is multigenic and maintained over time and locations. Banito et al. (2010) screened 24 improved and local genotypes of cassava from Togo against four highly virulent *Xa. axonopodis* pv. *manihotis* strains from different African origins and found six genotypes resistant to four strains, the local genotype Gbazékouté and the improved CVTM4 from the International Institute of Tropical Agriculture, Nigeria being the most resistant. Six groups of genotypes with differential reactions to the strains were identified, which allowed the strains to be defined as pathotypes.

Using an extensive database of effector proteins from animal and plant pathogens, Bart et al. (2012) identified the effector repertoire for each of the 65 sequenced strains of *Xa. axonopodis* pv. *manihotis* and used a comparative sequence analysis to deduce the least polymorphic of the conserved effectors. To understand the roles of pathogenesis-related (PR) protein genes for hevamine, chitinase, and sulphite reductase, induced in a resistant cassava cultivar during the response to *Xa. axonopodis* pv. *manihotis* in defense responses, Román et al. (2014) generated a protein-protein interaction map based on yeast two-hybrid assays of these candidate PR proteins. They showed that the cassava PRs interacted with each other and, although hevamine and chitinase belong to the same class of chitinases, they shared only four interactors. Co-regulated expression of PRs and their interactors was observed and similar ontology terms were identified for interactors. Moreover, an overlap between immune and other metabolic pathways was noticed based on other bioinformatics studies.

7.67.5 Management

An integrated approach including regulatory control, sanitary measures, cultural practices, varietal resistance, chemical, and biological control have greatly reduced the losses caused by the disease (Lozano 1986).

1. Regulatory control: In areas where cassava bacterial blight does not yet occur, great care must be taken in the introduction of germplasm. Vegetative propagated material must be introduced as meristem culture multiplied *in vitro* and certified disease-free. Botanical seed should originate from areas unfavorable for disease development, be heat-treated and planted in quarantine. The FAO/IBPGR (International Board for Plant Genetic Resources) technical guidelines for the safe movement of cassava germplasm recommend, visual inspection of the seed, density selection, followed by treatment of the seed by immersing them in water and heating in a microwave oven at full power until the water temperature reaches 73°C, and then immediately pouring the water off. If a microwave oven is not available, a dry heat treatment for 2 weeks at 60°C is recommended (Frison and Feliu 1991). A subsequent thiram dust treatment reduces seed re-infestation. Lozano and Nolt (1989) mentioned 77°C instead of 73°C for the microwave treatment
2. Cultural control: In areas where cassava bacterial blight is already widespread, disease incidence can be reduced by the use of clean planting material. Cuttings should be taken only from plantations that have been found to be free from the disease by inspections at the end of the rainy season. In cases of sporadic occurrence of the disease, great care must be exercised in collecting cuttings only from healthy plants and from the most lignified portion of the stem, i.e., up to 1 m from the base, combined with visual inspection for the absence of vascular browning
3. Sanitary measures: The following sanitary measures should be practiced
   a. Tools should be regularly disinfected using a bactericide
   b. Infected clones can be cleaned by rooting bacteria-free stem tips in conditions unfavorable for infection or by meristem cultures *in vitro*
   c. Crop rotation and fallowing proves very successful when the new crop is planted with uninfected cuttings. Rotation or fallowing should be practiced at least for 1 year
   d. All infected plant debris and weeds harboring epiphytic population of the pathogen should be removed and burnt or incorporated into the soil
   e. In some areas, planting toward the end of the rainy season instead of at the beginning will delay epiphytotic development during the growing period and so reduce yield loss
   f. Intercropping cassava with maize or melon has been reported to reduce the disease significantly (Ene 1977)
   g. In potassium-deficient soils, increasing the potassium content of the leaves by fertilization tends to reduce disease severity (Odurukwe and Arene 1980).

4. Host-plant resistance: Clear differences in host-plant resistance occur, especially with regard to stem infection and wilt; use of resistant genotypes is a major control strategy
5. Chemical control: Soaking of infested botanical seed in hot water at 60°C for 20 min, followed by drying in shallow layers at 30°C overnight or at 50°C for 4 h, reduced the number of bacteria to less than the minimum detectable level without appreciably reducing germination (Persley 1979a)
6. Biological control: Foliar application of *Psm. fluorescens* and *Psm. putida* has been shown to significantly reduce the leaf infection by *Xa. axonopodis* pv. *manihotis* (Lozano 1986). However, biological control has not yet gained practical acceptance

REFERENCES

Arrieta-Ortiz, M. L., Rodríguez-R, L. M., Pérez-Quintero, Á. L., Poulin, L., Díaz, A. C., et al. 2013. Genomic survey of pathogenicity determinants and VNTR markers in the cassava bacterial pathogen *Xanthomonas axonopodis* pv. *manihotis* Strain CIO151. *PLoS ONE* 8(11): e79704.

Assigbétsé, K., Verdier, V., Wydra, K., Rudolph, K., and Geiger, J. P. 1998. Genetic variation of the cassava bacterial blight pathogen, *Xanthomonas campestris* pv. *manihotis*, originating from different ecoregions in Africa. pp. 223–229 in: *Proceedings of 9th International Conference on Plant Pathogenic Bacteria*. A. Mahadevan ed. University of Madras.

Banito, A., Kpémoua K. E., and Wydra, K. 2010. Screening of cassava genotypes for resistance to bacterial blight using strain × genotype interactions. *J. Plant Pathol.* 92: 181–186.

Bart, R., Cohn, M., Kassen, A., McCallum, E. J., Shybut, M., et al. 2012. High-throughput genomic sequencing of cassava bacterial blight strains identifies conserved effectors to target for durable resistance. *Proc. Nat. Acad. Sci.* E1972–979.

Boher, B., Kpemoua, K., Nicole, M., Luisetti, J., and Geiger, J. P. 1995. Ultrastructure of interactions between cassava and *Xanthomonas campestris* pv. *manihotis*: Cytochemistry of cellulose and pectin degradation in a susceptible cultivar. *Phytopathology* 85: 777–788.

Bondar, G. 1912. Una nova molestia bacteriana das hastas da mandioca. *Chacaras e Quintaes, São Paulo* 5: 15–18.

Bondar, G. 1915. Molestia bacteriana da mandioca. *Boletim de Agricultura, São Paulo* 16a: 513–524.

Bouriquet, G. 1939. Mosaïque et maladie bactérienne du manioc. *Recherches systématiques, biologiques et cytologiques sur les maladies des plantes cultivées à Madagascar*. Chevalier, Paris, France, pp. 83–113.

Bradbury, J. F. 1986. *Guide to Plant Pathogenic Bacteria*. CAB International, Wallingford, UK.

Chege, M. N., Wamunyokoli, F., Kamau, J., and Nyaboga, E. N. 2017. Phenotypic and genotypic diversity of *Xanthomonas axonopodis* pv. *manihotis* causing bacterial blight disease of cassava in Kenya. *J. Appl. Biol. & Biotechn.* 5: 038–044.

Daniel, J. F., and Boher, B. 1981. Contamination of cassava flowers, fruits and seeds by *Xanthomonas campestris* pv. *manihotis*. pp. 358–68 in: *Proceedings of 5th International Conference on Plant Pathogenic Bacteria*. J. C. Lozano, ed. Centro Internacional de Agricultura Tropical, Cali, Colombia.

Daniel, J.-F., and Boher, B. 1985. Epiphytic phase of *Xanthomonas campestris* pathovar *manihotis* on aerial parts of cassava. *Agronomie* 5: 111–115.

Dedal, O. I., Palomar, M. K., and Napiere, C. M. 1980. Host range of *Xanthomonas manihotis* Starr. *Ann. Trop. Res.* 2: 149–55.

Dianese, J. C. 1985. Atividade pectolitica das patovares de *Xanthomonas campestris* que afetam a mandioca. *Rev. Microbiol.* 16: 195–202.

Dye, D. W. 1978. Genus IX. *Xanthomonas* Dowson 1939. In: Young, J. M., Dye, D. W., Bradbury, J. F., Pangopoulos, C. G., Robbs, C. F. eds. A proposed nomenclature and classification for plant pathogenic bacteria. *N. Z. J. Agric. Res.* 21:162–166.

Elango, F., and Lozano, J. C. 1980. Transmission of *Xanthomonas manihotis* in seed of cassava (*Manihot esculanta*). *Plant Dis.* 64: 784–786.

Ene, L. S. O. 1977. Control of cassava bacterial blight (CBB). *Tropical Root and Tuber Crops Newsletter* 10: 30–31.

Ewbank, E. 1992. Etude du catabolisme d'acides aminés conduisant à la formation de phytotoxines par *Xanthomonas campestris* pv. *manihotis* (agent pathogFne de la bactériose du manioc). PhD thesis, Université Catholique de Louvain, Louvain-la-Neuve, Belgium.

Ewbank, E., and Maraite, H. 1990. Conversion of methionine to phytotoxic 3-methylthio- propionic acid by *Xanthomonas campestris* pv. *manihotis*. *J. Gen. Microbiol.* 136: 1185–1189.

Fessehaie, A. 1997. Biochemical/physiological characterization and detection methods of *Xanthomonas campestris* pv. *manihotis* (Berthet-Bondar) Dye 1978, the causal organism of cassava bacterial blight. PhD thesis, University of Göttingen, Germany.

Frison, E. A., and Feliu, E. 1991. *FAO/IBPGR Technical Guidelines for the safe Movement of Cassava Germplasm*. Food and Agriculture Organization of the United Nations/International Board for Plant Genetic Resources, Rome, Italy.

Hahn, S. K. 1979. Breeding of cassava for resistance to cassava mosaic disease (CMD) and cassava bacterial blight (CBB) in Africa. pp. 211–219 in: *Diseases of Tropical Food Crops*. H. Maraite, and J. A. Meyer eds. *Proceedings of an International Symposium, UCL*, 1978, Université Catholique de Louvain, Louvain-la-Neuve, Belgium.

Ikotun, T. 1981. Studies on the host range of *Xanthomonas manihotis*. *Fitopatol. Bras.* 6: 15–21.

Ikotun, T. 1984. The nature and function of the extracellular polysaccharide produced by *Xanthomonas campestris* pathovar *manihotis*. *Fitopatol. Bras.* 9: 467–473.

Jorge, V., Fregene, M. A., Duque, M. C., Bonierbale, M. W., Tohme, J., et al. 2000. Genetic mapping of resistance to bacterial blight disease in cassava (*Manihot esculenta* Cranz). *Theoret. Appl. Genet.* 101: 865–872.

Joseph, J., and Elango, F. 1991. The status of cassava bacterial blight caused by *Xanthomonas campestris* pv. *manihotis* in Trinidad. *J. Phytopathol.* 133: 320–326.

Kelman, A. 1954. The relationship of pathogenicity of *Pseudomonas solanacearum* to colony appearance in tetrazolium medium. *Phytopathology* 44: 693–695.

Lozano, J. C. 1986. Cassava bacterial blight: A manageable disease. *Plant Dis.* 70: 1089–1093.

Lozano, J. C., and Laberry R. 1982. Screening for resistance to cassava bacterial blight. *Plant Dis.* 66: 316–318.

Lozano, J. C., and Nolt, B. L. 1989. Pests and pathogens of cassava. pp. 169–182 in: *Plant Protection and Quarantine*. Vol. II. Selected Pests and Pathogens of Quarantine Significance. R. P. Kahn ed. CRC Press, Boca Raton, FL.

Maraite, H. 1993. *Xanthomonas campestris* pathovars on cassava: Cause of bacterial blight and bacterial necrosis. pp. 18–24 in: *Xanthomonas*. J. G., Swing, and E. L. Civerolo eds. Chapman & Hall, London, UK.

Maraite, H., and Meyer, J. A. 1975. *Xanthomonas manihotis* (Arthaud-Berthet) Starr, causal agent of bacterial wilt, blight and leaf spots of cassava in Zaire. *PANS* 21: 27–37.

Maraite, H., and Perreaux, D. 1979. Comparative symptom development in cassava after infection by *Xanthomonas manihotis* or *X. cassavae* under controlled conditions. pp. 17–24 in: *Cassava Bacterial Blight in Africa; Past, Present and Future*. E. R. Terry, G. J. Persley, and S. C. A. Cook eds. Report of an interdisciplinary workshop, IITA, Ibadan, Nigeria, 1978, Centre for Overseas Pest Research, London, UK.

Maraite, H., and Weyns, J. 1979. Distinctive physiological, biochemical and pathogenic characteristics of *Xanthomonas manihotis* and *X. cassavae*. pp. 103–17 in: *Diseases of Tropical Food Crops*. H. Maraite, and J. A. Meyer eds. *Proceedings of an International Symposium, UCL*, 1978. Université Catholique de Louvain, Louvain-la-Neuve, Belgium.

Maraite, H., Weyns, J., Yimkwan, O., Lipembra, P., and Perreaux, D. 1981. Physiological and pathogenic variations in *Xanthomonas campestris* pv. *manihotis*. pp. 358–68 in: *Proceedings of 5th International Conference on Plant Pathogenic Bacteria*. J. C. Lozano, and P. Gwin eds. Centro Internacional de Agricultura Triopical (CIAT), Cali, Colombia.

Marcano, M., and Trujillo, G. 1982. Bacteriosis de la yuca; identificados algunos aspectos epidemiologicos del pathogeno que pueden utilizarse para disminuir los danos ocasionados par la enfernedad. *Fonaiap Divulga (Venezuela)* 1: 12–13.

Marcano, M., and Trujillo, G. 1984. Role of weeds in relation to survival of bacterial blight of cassava. Revista de la Facultad de Agronomia. *Universidad Central de VeneZuela* 13(1/4): 167–81.

Odurukwe, S. O., and Arene, O. B. 1980. Effect of N, P, K fertilizers on cassava bacterial blight and root yield of cassava. *Trop. Pest Manag*. 26: 391–395.

Ogunjobi, A. A., Fagade, O. E., Dixon, A. G. O., and Amusa, N. 2007. Pathological variation in cassava bacterial blight (cbb) isolates in Nigeria. *World Appl. Sci. J*. 2: 587–593.

Otim-Nape, G. W. 1980. Cassava bacterial blight in Uganda. *Trop. Pest Manag*. 26: 274–277.

Perreaux, D., Maraite, H., and Meyer, J. A. 1986. Detection of 3(methylthio) propionic acid in cassava leaves infected by *Xanthomonas campestris* pv. *manihotis*. *Physiol. Mol. Plant Pathol*. 28: 323–328.

Persley, G. J. 1977. Distribution and importance of cassava bacterial blight in Africa. pp. 9–14 in: *Cassava Bacterial Blight*. G. J. Persley, E. R. Terry, and R. F. MacIntyre eds. (IITA), Idaban, Nigeria.

Persley, G. J. 1979a. Studies on the survival and transmission of *Xanthomonas manihotis* on cassava seed. *Ann. Appl. Biol*. 93: 159–166.

Persley, G. J. 1979b. Studies on the epidemiology and ecology of cassava bacterial blight. pp. 5–7 in: *Cassava Bacterial Blight in Africa; Past, Present and Future*. E. R. Terry, J. G. Persley, and S. C. A. Cook eds. *Report of an Inter-disciplinary Workshop*, IITA, Ibadan, Nigeria, 1978. Centre for Overseas Pest Research, London, UK.

Persley, G. J. 1980. Studies on bacterial blight of cassava in Africa. PhD thesis, University of Queensland, Australia.

Restrepo, S., Valle, T., Duque, M. C., and Verdier, V. 1999. Assessing genetic variability among Brazilian isolates of *Xanthomonas axonopodis* pv. *manihotis* through RFLP and AFLP analyses. *Canad. J. Microbiol*. 45: 754–763.

Restrepo, S., and Verdier, V. 1997. Geographical differentiation of the population of *Xanthomonas axonopodis* pv. *manihotis* in Colombia. *Appl. Environ. Microbiol*. 63: 4427–4434.

Restrepo, S., Duque, M. C., and Verdier, V. 2000. Characterization of pathotypes among isolates of *Xanthomonas axonopodis* pv. *manihotis* in Colombia. *Plant Pathol*. 49: 680–687.

Restrepo, S., Velez, C. M., Duque, M. C., and Verdier, V. 2004. Genetic structure and population dynamics of *Xanthomonas axonopodis* pv. *manihotis* in Colombia from 1995 to 1999. *Appl. Environ. Microbiol*. 70: 255–261.

Robbs, C. F., Ribeiro, R., Kimura, O., and Akiba, F. 1972. Variasks en *Xanthomonas manihotis* (Arthaud-Berthet) Starr. *Rev. Soc. Bras. Fitopatol*. 5: 67–75.

Román, V., Bossa-Castro, A. M., Vásquez, A., Bernal, V., Schuster, M., et al. 2014. Construction of a cassava PR protein-interacting network during *Xanthomonas axonopodis* pv. *manihotis* infection. *Plant Pathol*. 63: 792–802.

Sãnchez, G., Restrepo, S., Duque, M. C., Bonierbale, M., and Verdier, V. 1999. AFLP assessment of genetic variability in cassava accessions (*Manihot esculenta*) resistant and susceptible to the cassava bacterial blight (CBB). *Genome* 42: 163–172.

Takatsu, A., Fukuda, S., and Perrin, S. 1979. Epidemiological aspects of bacterial blight of cassava in Brazil. pp. 141–150 in: *Proceedings of the International Symposium on Diseases of Tropical Food Crops*, 1978. H. Maraite, and J. A. Meyer, eds. Université Catholique de Louvain, Belgium, Louvain-la-Neuve, Belgium.

Thaveechai, N., Leksomboon, C., Kositratana, W., Paradornuwat, A., and Rojanaridpiched, C. 1993. Survival of *Xanthomonas campestris* pv. *manihotis* under natural field conditions. *Kasetsart J. Nat. Sci*. 27: 25–32.

Uremura, Y., and Kawano, K. 1983. Field assessment and inheritance of resistance to cassava bacterial blight. *Crop Sci*. 23: 1127–1132.

Van den Mooter, M., Maraite, H., Meiresonne, L., Swings, J., Gillis, M., et al. 1987. Comparison between *Xanthomonas campestris* pv. *manihotis* (ISPP list 1980) and *X. campestris* pv. *cassava* (ISPP list 1980) by means of phenotypic, protein electrophoretic, DNA hybridization and phytopathological techniques. *J. Gen. Mirobiol*. 133: 57–71.

Vauterin, L., Hoste, B., Kersters, K., and Swings, J. 1995. Reclassification of *Xanthomonas*. *Int. J. Syst. Bacteriol*. 45: 472–489.

Wydra, K., Agbicodo, E., Ahohuendo, B., Banito, A., Cooper, R. M. C., et al. 2003. Integrated control of cassava bacterial blight by (1) combined cultural control measures and (2) host plant resistance adapted to agro-ecological conditions, and (3) improved pathogen detection. pp. 506–515 in: *Proceedings of the 8th Triennial Symposium of the International Society for Tropical Root Crops—Africa Branch*. Akoroda M. ed. ISTRC-AB, Ibadan, Nigeria.

Wydra, K. A., Fessehaie, A., Fanou, A., Sikirou R., Janse, J., et al. 1998. Variability of isolates of *Xanthomonas campestris* pv. *manihotis* (*Xcm*), incitant of cassava (*Manihot esculenta* Crantz) bacterial blight, from different geographic origins in pathological, biochemical and serological characteristics. pp. 317–323 in: *Proceedings of 9th International Conference on Plant Pathogenic Bacteria*. A. Mahadevan ed. University of Madras, India.

Young, J. M., Saddler, G. S., Takikawa, Y., de Boer, S. H., Vauterin, L., et al. 1996. Names of plant pathogenic bacteria 1864–1995. *Rev. Plant Pathol.* 75: 721–763.

7.68 YELLOW DISEASE OF HYACINTHS

Yellow disease of hyacinth (*Hyacinthus orientalis* L.) was the first disease caused by a xanthomonad to be described and one of the first plant diseases recognized to be caused by a bacterium. The disease was originally described from the Netherlands by Wakker (1883). After its first appearance, the disease soon assumed severe and even epiphytotic proportions in the Netherlands, but was subsequently brought under control. It is now a regular, but normal constraint on intensive hyacinth bulb production in that country. It also occurs in Australia, Finland, France, Hungary, Ireland, Italy, Japan, Poland, Romania, Russian federation, Sweden, UK, USA, and former Yugoslavia. Heavy losses may occur in susceptible cultivars under intensive cultivation. In all other cases, the disease appears to be of minor importance.

7.68.1 Host Range

Hyacinth is the main host. Natural infection of *Xanthomonas hyacinthi* has also been reported on *Scilla tubergeniana*, *Eucomis autumnalis*, *Puschkinia scilloides*, and *P. libanotica* (Janse and Miller 1983; Van der Tuin et al. 1995; van Doorn et al. 2009). Artificial infection occurs on *S. mischtschenkoana*, *S. siberica*, and *Muscari azureum* (van Doorn et al. 2009).

7.68.2 Symptoms

The mildly infected bulbs do not reveal external symptoms. However, when such bulbs are cut across, small yellow spots become visible in the scale tissue. In longitudinal sections, these lesions appear as vertical yellow stripes leading toward the base of the bulb. Such lesions may enlarge and coalesce with time. When such bulbs are planted, leaf tips and edges may wilt or the leaves may develop normally, but subsequently show dark green, water-soaked stripes coming from the nose of the bulb. The sprouts from severely infected bulbs may not develop normally, remaining small, and showing extensive water-soaked areas. Very severely infected bulbs usually decay before planting or do not sprout.

The primary infection of the leaves in the field generally results in the formation of water-soaked spots at the leaf tip and dark streaks along the veins at the margin (resulting from hydathodal infection) or other parts of the leaf. These spots elongate longitudinally as the bacteria progress downwards and, especially under wet weather conditions new infections may develop on lower parts of the leaves. Leaf margins may be infected through wounds inflicted by rubbing of adjacent leaves. The lesions desiccate and turn yellowish-brown, the unaffected portions of the leaf remaining green. Similar symptoms may appear on pedicels.

Often *Erwinia carotovora* pv. *atroseptica* is found in hyacinth bulbs with *Xa. hyacinthi* infection. This may hamper diagnosis of the disease by obscuring the characteristic yellow disease symptoms with its glassy, slimy, and resinous growth (van Doorn and Roebroeck 2012).

van Doorn et al. (2001) developed a sensitive and specific PCR-mediated detection method based on amplification of a subsequence of the type IV fimbrial-sub unit gene *fimA*. The *fimA* gene was amplified by PCR with degenerate DNA primers designed by using the *N*-terminal and *C*-terminal amino acid sequences of trypsin fragments of FimA. The method proved to be pathovar specific, as assessed by testing 71 *Xanthomonas* pathovars and bacterial isolates belonging to other genera, such as *Erwinia* and *Pseudomonas*. In inoculated leaves of hyacinths, the threshold was 5,000 cfu mL^{-1}. The results indicated that infected hyacinths with early symptoms could be successfully screened for *Xa. hyacinthi* with this method.

7.68.3 Causal Organism

Xanthomonas hyacinthi (*ex* Wakker 1883) Vauterin et al. 1995
Syn. *Xanthomonas campestris* pv. *hyacinthi*

The bacterium was first named as *Bacterium hyacinthi* by Wakker in 1883. After the recognition of *Xanthomonas* genus in 1939, it was changed to *Xanthomonas hyacinthi*. After the introduction of pathovar concept, it was changed to *Xa. campestris* pv. *hyacinthi*. Vauterin et al. (1995) changed it to *Xa. hyacinthi*.

The cells are 1.5 μm long rods, Gram-stain-negative having one polar flagellum. The optimum growth occurs at 28°C–30°C, colonies are yellow and shiny, and metabolism is strictly aerobic. Gelatin and starch are hydrolyzed; hydrogen sulphide, oxidase, catalase, casein, and esculin are produced, and lipolytic activity is positive. Acid is produced from L(+) arabinose, cellobiose, D(+) galactose, glycerol, maltose, D(+) mannose, and sucrose, but not from dulcitol, inositol, lactose, D(−) mannitol, L(+) rhamnose, D(−) ribose, salicin, and D(−) sorbitol. Nitrate reduction, urease production, pectolytic activity, utilization of L(+) tartrate and lactate, and hypersensitive reaction in tobacco are negative.

7.68.4 Disease Cycle

The bacterium survives mainly in the infected bulbs. In February and March, shoots from infected bulbs carrying the bacterium act as foci of infection. The penetration of the pathogen occurs through wounds, hydathodes, and stomata. The dissemination of the pathogen occurs through wind-splashed rains and overhead-irrigation. In addition, dispersal occurs through movement of animals through wet fields and contact of infected leaves with adjacent plants. Machines used for lifting and storing of bulbs and knives used for scooping and scoring can spread the bacteria from diseased bulbs to healthy ones. The bacterium is not known to survive as an epiphyte on non-host plants and apparently it cannot survive in soil. No vectors have been described, but mechanical transmission by man is easily possible (Kamerman 1975).

An analysis of an incomplete diallel design with 14 hyacinth cultivars for resistance to the disease showed a significant general combining ability (GCA) and a heritability of 0.75. A correlation appeared to exist between flowering date and degree of resistance to the disease; early flowering seedlings being generally less resistant (Van Tuyl and Toxopeus 1980). Tuyl (1982) also reported GCA for resistance to the disease. The late flowering hyacinth cultivars King of the Blues and Marconi and the early flowering *H. orientalis* 70129 were found to be the best combiners for disease resistance.

van Tuyl et al. (1986) found that on the basis of degree of leaf infection, the cultivars King of the Blues and Pink Pearl showed a high degree of resistance, while Blue Giant and Carnegie were found to be very susceptible. This leaf infection corresponds with differences in resistance known in practice. They also reported significant GCA component for resistance to the disease. The late flowering cultivars King of the Blues and Marconi and the early flowering *H. orientalis* were found to be the best combiners for the disease. Within a cultivar, the degree of earliness itself was not associated with degree of resistance. Correlation coefficients of leaf characteristics and degree of resistance showed that hyacinths with short and/or narrow leaves are generally more resistant than those with long and/or broad leaves. Tetraploid cultivars with few, large stomata tend to be susceptible, diploid ones with many small stomata tend to be the more resistant. In 1983, the first yellow disease resistant cultivar out of this program, Nestor, originating from a cross with *H. orientalis*, was released to five hyacinth growers in the Netherlands.

7.68.5 Management

1. Inspect bulbs carefully before planting
2. Remove volunteer plants in early spring
3. Do not allow the men and animals to enter wet fields
4. Carry out field inspections of mother bulb crops to trace the diseased plants. Diseased plants and those in the immediate vicinity (within a radius of about 1.5 m) should be rouged or killed *in situ* to prevent bacterial spread from these foci. Treat the surrounding soil with diquat or 5% formalin and the surrounding plants with kasugamycin (if permitted)
5. Apply alkyl dimethyl benzyl ammonium chloride (if permitted) when leaf symptoms appear, but it is not completely effective
6. Give foliar sprays of Groen-ex in the field and it should be applied early when leaf spotting is minimal (Koster et al. 1988). Captan was effective under glasshouse conditions, but not in the field (Kamerman 1983)
7. Harvest bulbs from diseased spots lastly and separately and destroy them or give the heat treatment. The following heat treatment has been found to be quite effective. Store bulbs first for a period of 4 weeks at 30°C, thereafter for 2 weeks at 38°C, with a final treatment for 3 days at 44°C. Subsequently bulbs are stored at 28°C–30°C (Kruyer and Vreeburg 1981). Ensure sufficient ventilation to avoid harmful effects of the heat treatment and dry the bulbs rapidly. This treatment should be carried out within the 10 weeks after harvesting. It may have undesirable side-effects on the stored bulbs (Vreeburg et al. 1986; van der Hulst and Munk,1992) and should only be applied when necessary
8. Disinfect machines and implements after harvest
9. Follow the classification system to produce yellow disease-free hyacinths developed by OEPP/EPPO (1998)

REFERENCES

Janse, J. D., and Miller, H. 1983. Yellow disease in *Scilla tubergeniana* and related bulbs caused by *Xanthomonas campestris* pv. *hyacinthi*. *Netherlands J. Plant Pathol.* 89: 203–206.

Kamerman, W. 1975. Biology and control of *Xanthomonas hyacinthi* in hyacinths. II International Symposium on Flower Bulbs. *Acta Hortic.* (*ISHS*) 47: 99–106.

Kamerman, W. 1983. Captan is unsuitable for controlling hyacinth yellows in the field. *Bloembollen Cultuur* 93(33): 845.

Koster, A. T. J., Vreeburg, P. J. M., Conijn, C. G. M., and Kamerman, W. 1988. Pay attention to the control of yellow disease in the field. *Bloembollen Cultuur* 99(7): 12–13.

Kruyer, C. J., and Vreeburg, P. J. M. 1981. How can hyacinths best be stored after heat treatment? *Bloembollen Cultuur* 92: 224–225.

OEPP/EPPO. 1998. EPPO Standards PM 4/23 (1) Classification scheme for hyacinth. *OEPP/EPPO Bull.* 28: 235–242.

Tuyl, J. M. 1982. Breeding for resistance to yellow disease of hyacinths. II. Influence of flowering time, leaf characters, stomata and chromosome number on the degree of resistance. *Euphytica* 31: 621–628.

van der Hulst, C. T. C., and de Munk, W. J. 1992. Non-parasitic disorders in hyacinths caused by a high-temperature treatment to control yellow disease (*Xanthomonas campestris* pv. *hyacinthi*), and the analysis of gaseous emanations from the bulbs. *Acta Hortic.* (*ISHS*) 325: 149–155.

Van der Tuin, W. R., Spit, B. E., Nahumury, T., and Janse, J. D. 1995. *Xanthomonas campestris* pv. *hyacinthi* in *Eucomis autumnalis* and *Puschkinia* species. Verslagen en Mededelingen nr. 177, Annual Report 1994, Plant Protection Service, Wageningen, The Netherlands, pp. 22–23.

van Doorn, J., Hollinger, T. C., and Oudega, B. 2001. Analysis of the type IV fimbrial-subunit gene*fimA* of *Xanthomonas hyacinthi*: Application in PCR-mediated detection of yellow disease in hyacinths. *Appl. Environ. Microbiol.* 67: 598–607.

van Doorn, J., and Roebroeck, E. J. A. 2012. *Xanthomonas campestris* pv. *hyacinthi*: Cause of yellow disease in hyacinths. pp. 83–91 in: *The Hosts of Xanthomonas*. J. Swings, and L. Civetta eds. Springer Science and Business Media, 399 pp.

van Doorn, J., van Leeuwen, P., and Miglino, R. 2009. Yellow disease in hyacinth and in rare bulbs: Detection and spread of *Xanthomonas hyacinthi*. *Gewasbescherming* 40: 196–199.

Van Tuyl, J. M., and Toxopeus, S. J. 1980. Breeding for resistance to yellow disease of hyacinths. I. Investigations on F_1's from diallel crosses. *Euphytica* 29: 555–60.

van Tuyl, J. M., van Eijk, J. P., and Kwakkenbos, T. A. M. 1986. Breeding for resistance to yellow disease of hyacinths. IV International Symposium on Flower Bulbs. *Acta Hortic. (ISHS)* 177: 585–589.

Vauterin, L., Hoste, B., Kersters, K., and Swings, J. 1995. Reclassification of *Xanthomonas*. *Int. J. Syst. Bacteriol.* 45: 472–489.

Vreeburg, P. J. M., Korsuize, C. A., and Hof, N. A. A. 1986. Yellows disease in bulbs for propagation remains a problem. Only give heat treatment if it is necessary. *Bloembollencultuur* 97(31): 12–13.

Wakker, J. H. 1883. Vorläufige Mittheilungen über Hyacinthenkrankheiten. *Botanisches Centralblatt Bd.* XIV(23): 315–316.

7.69 BACTERIAL CANKER OF POPLAR

Bacterial or oozing canker caused by *Xanthomonas populi* pv. *populi* is probably the most damaging bacterial disease of poplar in Europe where it has been known for over 100 years. It has become an endemic disease because of the existence of infected plantations of old susceptible cultivars such as *Populus euramericana* cvs. Regenerata, Grandis, or Brabantica, and the inconspicuous presence of the disease in spontaneous populations of aspen (*P. tremula*). Moreover, the recent distribution of polyclonal mixtures not tested against the bacterium and of some clones (*P. euramericana* cv. Dorskamp, *P. trichocarpa* × *P. deltoides* Barn), whose susceptibility to bacterial canker was not sufficiently known, have created good conditions for new infection sources.

Bacterial canker occurs in Austria, Belgium, France, Czechoslovakia, USSR, the Netherland, Germany, Ireland, Poland, the United Kingdom, Denmark, Hungary, Romania, and Yugoslavia. The disease is not present in North America and Canada. **Bark canker** of poplar caused by *Lonsdalea quercina* subsp. *populi* has also been reported from Spain by Berruete et al. (2016) and from Henan and Shandong provinces of China by Li et al. (2014).

Due to sanitary measures now being adopted in the main European poplar growing countries, the disease is not present in nurseries. However, it affects young industrial poplar plantations when susceptible cultivars are introduced near wild cankered aspen or old cankered plantations. One such situation occurred when *P. euramericana* cv. Blanc de Poitou was introduced in 1958–1960 into a high risk area of northern France. By 1964, 30% of the trees were affected with the disease and 10%–28% of the trees had trunk cankers. The losses in commercial value of the timber in cankered trees ranged from 5% to 60%.

Besides *Xa. populi* pv. *populi*, the other bacterial pathogens have also been shown to cause canker on poplars. In Britain, *Pseudomonas syringae* plays a minor role in canker lesions. It may cause shoot blight in spring, a distinct pathological condition frequently associated with bacterial canker (Whitbread 1967). Bacterial canker of poplar caused by *Erwinia* sp. is a serious disease spreading in most parts of northeast China, where poplars are cultivated. The disease seriously infects those poplars belonging to section *Aigeiros* or the intersection hybrids of section *Tacamahaca* and section *Aigeiros* and their clones (Wu et al. 1991). The authors also identified the general combining ability (GCA) of five fine poplar cultivars, resistant to bacterial canker, by screening tests. Xiang et al. (1991) reported **bacterial swollen stem canker** of poplar prevalent in planted poplars of northeast China. Serious damages occur in the form of necrosis, cankers, swollen stems, and discolored wood of trunks and branches of young and mature trees. On the basis of pathogenicity test, Gram-stain reaction, and cultural characteristics, the bacterium was identified as *Erwinia* sp.

Since 2009, a specific symptom has been detected in poplar (*Populus* × *euramericana*) stands in the central part of Hungary. The bark of symptomatic trees is vertically cracked, and a sticky, brown-colored fluid oozes from the canker. During 2011, many bacterial strains isolated from oozing bark cracks and creamy slime under the bark of poplar trees were tentatively identified as *Lo. quercina* (formerly *Brenneria quercina*) based on partial 16S rRNA sequencing, but it differed from *Lo. quercina* in several phenotypic traits. Tóth et al. (2013) investigated the taxonomic position of seven Gram-stain-negative bacterial strains isolated from oozing bark canker of poplar (*Populus* × *euramericana*) trees in Hungary using multi-locus sequence analysis of three housekeeping genes (*gyrB*, *atpD*, and *infB*), DNA-DNA hybridization analysis, and phenotypic and physiological characteristics and classified them into a new sub species, *Lo. quercina* subsp. *populi*.

7.69.1 Host Range

The main hosts infected by *Xa. populi* pv. *populi* are cottonwoods, poplars, and aspens. The bacterium attacks *Populus* spp., which are widely grown throughout the European and Mediterranean Plant Protection Organisation (EPPO) region. It occurs naturally in wild *P. tremula*, but mainly damages plantings of hybrid cultivars of *P. canadensis*. Provenances of *P. trichocarpa*, *P. deltoids*, and *P. tremuloides* from North America, used in European breeding programs, are highly susceptible (Ridé 1988). Willow is also susceptible to this bacterial pathogen.

7.69.2 Symptoms

Symptoms of the disease vary with the affected clone, ecological conditions, and the physiological state of the diseased tree. The first symptoms are observed in spring on the previous year's shoots, shortly before bud-burst. Many buds fail to open on young branches, and symptoms are visible at bud bases as slightly swollen areas with superficial cracks that may exude dense, greyish, mucoid ooze. The bacterial exudate may also appear as droplets through the lenticels or cracks on the internodes of 1-year-old shoots as well as from the edges of callus of perennial cankers. In the absence of this exudate, the bacterium may be located by cutting into the necrotic region, which will reveal translucent, vitreous, sticky tissues beneath. Highly susceptible clones are often girdled before the end of the second summer following infection, or they may develop highly erumpent cankers. The buds infected during the previous year are generally killed, but sometimes produce leaves that may exhibit blackened or discolored areas due to petiole infection by the bacteria. Many tender green shoots developing from such buds are killed.

On susceptible cultivars, lignified young twigs either dieback at the end of the season or exhibit cracks that develop into perennial cankers. Active cankers develop year by year on twigs, main branches, or trunk and may girdle and kill them, but can also result in stem breakage and subsequent sprouting. Active cankers also cause a massive degradation of wood structure, even when they are healed (Ridé 2012). In less susceptible clones, small lesions or only a roughened healing callus over the wound is found. In highly resistant clones, there is no canker development, and complete healing of the wound occurs.

The symptoms appear on 1- to 2-month-old leaves as youngest and oldest leaves are resistant to infection. The new lesions appear frequently along the main veins on the lower surface of the leaves. The lesions increase in size and subsequently the spots coalesce and become visible on the upper surface of leaves. The death of the tissue invaded by bacteria results in reddish-brown necrotic irregular spots. Finally, the leaflets shrivel completely. No symptoms appear on fruits.

7.69.3 Causal Organism

Xanthomonas populi pv. *populi*
Syn. *Aplanobacterium populi*, *Aplanobacter populi*

The bacterial cells are short, ovoid, Gram-stain-negative, nonsporing, and non-motile rods, which contain inclusion bodies. Colonies on yeast extract peptone glucose agar are creamy, mucoid, and slow-growing; no growth occurs on nutrient agar (Bradbury 1970).

EPPO has listed *Xa. populi* pv. *populi* as an A2 quarantine pest (OEPP/EPPO 1978) and it is also of quarantine significance for Interafrican Phytosanitary Council (IAPSC) and North American Plant Protection Organisation. In fact, the bacterium is fairly widely distributed throughout the EPPO region, but is absent only from Mediterranean poplar-growing countries, particularly Italy and Spain (Cadahia 1986).

Nesme et al. (1994) identified five physiological races of *Xa. populi* pv. *populi* after inoculating 19 strains of the bacterium on five poplar clones and one willow clone. Analysis of variance indicated a significant poplar clone-strain interaction and no significant clone-strain-inoculation date interaction for most strains, using the two variables. Race 3 was totally avirulent on the clone Italica, although the other races gave relatively small differences in canker severity. Strains belonging to races 1, 2, 3, and 5 were isolated in continental Europe (Belgium, France, and the Netherlands), whereas race 4, more virulent to *P. trichocarpa* clones, was isolated in Britain or close to the Belgian coast.

7.69.4 Disease Cycle

The bacterium survives in infected nursery plants from where it spreads to near and far of places. The progressive spread of the disease to the pathogen-free countries probably reflects the ease with which the pathogen is carried on planting material. In spring and summer, the massive bacterial slime produced by newly developing cankers is spread extensively by the wind-splashed rain. The bacterium can also spread through contaminated tools, insects, and animals. Larvae of the cambial miner, *Phytobia carbonaria* (Diptera: Agromyzidae) make tunnels under the bark of young stems and the bacterium may spread through the tunnels to form elongated cankers. It is possible, though not proven, that this insect could also transmit the disease. Feeding sites and larval galleries of the sessiid moth *Paranthrene tabaniformis* are also reported to permit entry of the bacterium (Bertucci 1986).

Leaf scars provide the main avenues for infection, but their infectibility declines rapidly during October (Whitbread 1967). The penetration can also occur through fresh wounds on branches and stems. Wounds and leaf scars are present throughout the growing season. In young stems, infection spreads through the cortical tissue, which subsequently splits open. The extent of bark and stem infection depends on the tree's susceptibility to the pathogen. The cankers resulting from early-season infection produce slime during autumn for the infection of scars resulting from normal leaf fall.

After inoculation of poplar trees with ice-nucleation active bacteria, the moisture content of bark decreased, but relative turgidity increased; electrolyte effusion rate increased and was at peak at temperature ranging from −4°C to 5°C. The lignin content also increased and positively correlated with poplars' disease resistance and the increase in plenylalanine ammonialyase activity also showed a significant positive correlation with poplars' disease resistance (Dong et al. 2012).

7.69.5 Management

The following measures will help in managing or reducing the incidence of the disease.

1. Use of pathogen-free planting material is the most effective method of disease management
2. Planting highly resistant clones is the best way to control the disease. Selection and breeding programs have produced many resistant clones. The disease has apparently become less damaging following the planting of more resistant poplar clones
3. The EPPO specific quarantine program (OEPP/EPPO, 1990) recommends a growing season inspection and also suggests that all poplar clones moving in international trade in the EPPO region should be resistant to *Xa. populi* pv. *populi*
4. Wu et al. (1991) reported the effectiveness of 50% DT germicide, 40% XF-136, 10% C.C.M.A. or 12.5% ramphencol, and validamycin 50,000 μg mL^{-1}, when applied in water on stems or the roots for the control of the disease

REFERENCES

Berruete, I. M., Cambra, M. A., Collados, R., Monterde, A., López, M. M., et al. 2016. First report of bark canker disease of poplar caused by *Lonsdalea quercina* subsp. *populi* in Spain. *Plant Dis.* 100: 2159.

Bertucci, B. M. 1986. The poplar wasp moth. *Inf. Fitopatol.* 36: 29–34.

Bradbury, J. F. 1970. *Aplanobacter populi. CMI Descriptions of Pathogenic Fungi and Bacteria* No. 231. CAB International, Wallingford, UK.

Cadahia, D. 1986. Problèmes phytosanitaires forestiers dus aux organismes de quarantaine OEPP. *Bull. OEPP/EPPO Bull.* 16: 537–541.

Dong, A.-r., Zhang, X.-y., Wang, Y.-t., Zheng, Q.-z., and Li, J. 2012. Pathogenetic and physiological mechanisms of poplar ice nucleation active bacterial canker. *J. Forestry Res.* 12: 253–256.

Li, Y., He, W., Ren, F., Guo, L., Chang, J., et al. 2014. A canker disease of *Populus × euramericana* in China caused by *Lonsdalea quercina* subsp. *populi*. *Plant Dis.* 98: 368–378.

OEPP/EPPO. 1978. Data sheets on quarantine organisms No. 47, *Aplanobacter populi*. *Bull. OEPP/EPPO Bull.* 8(2).

OEPP/EPPO. 1990. Specific quarantine requirements. EPPO Technical Documents No. 1008.

Nesme, X., Steenackers, M., Steenackers, V., Picard, Ch., Menard, M., et al. 1994. Differential host-pathogen interactions among clones of poplar and strains of *Xanthomonas populi* pv. *populi*. *Phytopathology* 84: 101–107.

Ridé, J. 1988. *Xanthomonas populi*. pp. 169–71 in: *European Handbook of Plant Diseases*. I. M. Smith, J. Dunez, R. A. Lelliot, D. H. Phillips, S. A. Archer eds. Blackwell Scientific Publications, Oxford, UK.

Ridé, M. 2012. *Xanthomonas populi*: Cause of bacterial canker of poplar. pp. 64–70 in: *Xanthomonas*. J. Swings, and L. Civetta eds. Springer, 399 pp.

Tóth, T., Lakatos, T., and Koltay, A. 2013. *Lonsdalea quercina* subsp. *populi* subsp. nov., isolated from bark canker of poplar trees. *Int. J. Syst. Evol. Microbiol.* 63: 2309–2313.

Whitbread, R. 1967. Bacterial canker of poplars in Britain: The cause of the disease and the role of leaf-scars in infection. *Ann. Appl. Biol.* 59: 123–131.

Wu, X., Zhang, J., Xiang, C., Li, G., Shao, Z., et al. 1991. Control technique of bacterial canker of poplar. *J. Northeast Forestry Univ.* 2: 7–16.

Xiang, C., Sun, H., Li, G., Wu, X., Li, Z., et al. 1991. Bacterial swollen stem canker on poplar: The disease and its causes. *J. Northeast Forestry Univ.* 2: 48.

7.70 BACTERIAL BLIGHT OF MULBERRY

Bacterial blight of mulberry (*Morus alba* L.) was first reported from the USA in 1983. Prior to it, it was incompletely described by Boyer and Lambert from France in 1893. It is the earliest record of a disease caused by a *Pseudomonas* species. The occurrence of the disease in Japan was confirmed in 1901. The disease has also been reported from eastern Anatolia region of Turkey where its incidence was nearly 100% in spring 1999 (Sahin et al. 1999). The disease also occurs in Pakistan, Korea DPR, Republic of Korea, China, Brazil, Bulgaria, Canada, Australia, Germany, Hungary, New Zealand, Romania, South Africa, UK, Uganda, Iran, Italy, and India. In India, the disease is widespread in Karnataka state where mulberry is widely grown due to the presence of sericulture industry. The disease incidence was 45.9% and the percentage of infested gardens ranged from 80 to 85 in Kollegal area of Karnataka state in rainy season of 1998 (Sharma et al. 2000).

Initially, the symptoms of the disease appear in cool, wet weather of spring season, but the severity of the disease increases in rainy season during June to July. The severity of the disease decreases after the rainy season, but it appears again in autumn in a mild form.

Bacterial leaf spot incited by *Xanthomonas campestris* pv. *mori* is another devastating foliar disease of mulberry reported globally (Banerjee et al. 2012). Allahverdi et al. (2016) reported bacterial canker of white mulberry caused by *Citrobacter freundii* from northwestern Iran. The symptoms of disease were dark brown to black, elongated lesions on the trunk. Dead barks were easily removed from the underlying tissues and a dark brown gum exuded from the shoot and trunk cankers.

7.70.1 Host Range

Mulberry is the only naturally infected host. In addition to *M. alba*, *M. alba* var. *taratarica*, *M. bombycis* (Japanese mulberry), *M. kagayamae*, *M. latifolia*, *M. nigra* (black mulberry), and *M. ruba* (red mulberry) have been reported natural hosts of *Psm. syringae* pv. *mori* (Janse 2005). On artificial inoculation, the bacterium infects kidney beans (Goto 1992).

7.70.2 Symptoms

The symptoms appear on leaves, twigs, and buds. On leaves, small, water-soaked, irregular, dark green, sunken spots appear. The spots turn brown to black with yellow borders and translucent edges as they age. The spots enlarge, become angular in shape, and often fall out. Elongated, dark brown and slightly sunken spots appear on large veins, midribs, and petioles of the leaves (Figure 7.49). The occurrence of these spots causes curling and distortion of the leaves due to the unbalanced growth of lamina. Diseased leaves at the tips of twigs wilt and later on die. Infected buds may become disfigured as they swell.

FIGURE 7.49 Bacterial blight of mulberry. (Courtesy of William M. Brown Jr., Bugwood.org.)

Dark brown to black, elongated, more or less sunken streaks appear on twigs. In humid weather, creamy-yellowish bacterial ooze exudes from lenticels of the affected parts. The streaks may coalesce to girdle the shoots resulting in the die-back symptoms. Severely infected young trees often appear stunted.

7.70.3 Causal Organism

Pseudomonas syringae pv. *mori* (Boyer and Lambert 1893) Young et al., 1978
Syn. *Psm. mori*

Cells are Gram-stain-negative, aerobic rods measuring 0.9–1.3 × 1.8–4.5 μm and motile with a polar flagellum. Green fluorescent pigment is produced on King's medium B. Colonies on agar media are white, slow growing, smooth, flat with entire margins, and becoming undulate. The tests for gelatin liquefaction, nitrate reduction, hydrogen sulphide, and oxidase production are negative, while those for catalase and levan production are positive. Milk becomes clear with alkaline reaction. Arbutin is hydrolyzed and D-gluconate, 2-ketoglutarate, inositol, mannitol, sorbitol, and trigonelline are used as a sole source of carbon. DL-lactate, quinate, L-tartrate, erythritol, and DL-homoserine are not used as a sole source of carbon. On the basis of fatty acid methyl ester analysis by gas chromatography, the bacterium confirms to *Psm. syringae* identification, with a similarity index ranging from 0.82 to 0.94 (Sahin et al. 1999).

Sharma et al. (2000) reported that both *Psm. syringae* pv. *mori* and *Xa. campestris* pv. *mori* were responsible for causing bacterial blight of mulberry, and they also reported the synergistic effect of both the bacteria on disease development. There is a need to study the diseases caused by these bacteria individually and to distinguish these diseases on the basis of symptoms. However, *Xa. campestris* pv. *mori* has been reported to cause bacterial leaf spot of mulberry, as mentioned above. Another report by Nagaraj et al. (2001), mentioning *Xa. campestris* pv. *moricola* as the cause of bacterial blight of mulberry from Karnataka, India has further added to the confusion regarding the causal agent(s) of the disease. *Xa. campestris* pvs. *mori* and *moricola* are not included in the validly published names of plant pathogenic bacteria enlisted by Young et al. (2004). However, Bull et al. (2010) have included *Xa. campestris* pv. *mori* in the comprehensive list of names of plant pathogenic bacteria published during 1980 to 2007.

7.70.4 Disease Cycle

The bacterium survives in the cankers formed on the twigs. Pruned shoots and fallen leaves on ground, if not removed and destroyed, also serve as a source of inoculum. The bacterium is not transmitted through seed. Bacterial cells oozing from twig lesions are disseminated by wind-splashed rain. Secondary spread of the bacterium occurs through rain from freshly formed lesions.

The increased severity of the disease during rainy season is due to the high humidity and dissemination of the bacterium caused by wind-splashed rain. Gangwar and Thangavelu (1995) found a highly significant positive correlation between disease severity and rainfall and relative humidity, but a non-significant correlation with minimum and maximum temperatures. High doses of nitrogenous fertilizers, close spacing of trees, poor field sanitation, and wounding of trees also increase the disease severity.

Among the indigenous mulberry varieties, grown in India, S 3, S 13, S 14, and S 1531 are highly resistant and K 2, MR 2, S 54, S 25, S 34, S 1, CSRS 1, Sujanpur, and SLM 1 are resistant to the pathogen. Some of the exotic germplasms, namely, English Black and *M. australis* are also resistant (Gangwar and Thangavelu 1995).

7.70.5 Management

1. Use of resistant varieties and healthy planting material are the most effective management practices
2. Minimize wounds to the limbs and new shoots
3. Prune the affected shoots and branches in the late dormant season and destroy them. Pruning should be done back to healthy wood and the pruning tools should be disinfected with 10% bleach or 70% alcohol between cuts
4. Do not apply excessive doses of nitrogenous fertilizers
5. Plant the trees at proper spacing to provide good air circulation
6. Spray Kocide 2000 T/N/O at 0.75–3 lb per acre as a dormant spray, and at 0.75–2 lb per acre when new growth starts. The application of Kocide is not recommended on trees grown for edible fruit

REFERENCES

Allahverdi, T., Rahimian, H., and Ravanlou, A. 2016. First report of bacterial canker in mulberry caused by *Citrobacter freundii* in Iran. *Plant Dis*. 100: 1774.

Banerjee, R., Das, N. K., Doss, S. G., Saha, A. K., Bajpai, A. K., et al. 2012. Narrow sense heritability estimates of bacterial leaf spot resistance in pseudo $F_2(F_1)$ population of mulberry (*Morus* spp.). *Eur. J. Plant Pathol*. 133: 537–544.

Boyer, G., and Lambert, F. 1893. Sur deux nouvelles maladies du mûrier. *C. R. Hebd. Séanc. Aced. Sci*. 117: 342–343.

Bull, C. T., De Boer, S. H., Denny, T. P., Firrao, G., Fischer-Le Saux, M., et al. 2010. Comprehensive list of names of plant pathogenic bacteria, 1980–2007. *J. Plant Pathol*. 92: 551–92.

Gangwar, S. K., and Thangavelu, K. 1995. Varietal and seasonal incidence of bacterial leaf blight disease of mulberry caused by *Pseudomonas mori* (Boyer *et* Lambert) Stevens. *Indian J. Sericult*. 34: 110–113.

Goto, M. 1992. *Fundamentals of Bacterial Plant Pathology*. Academic Press, San Diego, CA, 342 pp.

Janse, J. D. 2005. *Phytobacteriology: Principles and Practice*. CABI Publishing, Wallingford, UK, 360 pp.

Nagaraj, M. S., Srinivasachary, and Khan, A. N. A. 2001. Studies on the management of *Xanthomonas campestris* pv. *moricola* of mulberry. *J. Mycopathol. Res.* 39: 87–90.

Sahin, F., Kotan, R., and Dönmez, M. F. 1999. First report of bacterial blight of mulberries caused by *Pseudomonas syringae* pv. *mori* in the eastern Anatolia region of Turkey. *Plant Dis.* 83: 1176.

Sharma, D. D., Baqual, M. F., Gupta, V. P., and Chandrashekar, D. S. 2000. A survey on the occurrence of bacterial blight disease complex in mulberry. *Indian J. Sericult.* 39: 113–116.

Young, J. M., Bull, C. T., De Boer, S. H., Firrao, G., Saddler, G. E., et al. 2004. Names of plant pathogenic bacteria, 1864–2004 [http://www.isppweb.org/names_bacterial _revised.asp]. Accessed December 20, 2018.

Young, J. M., Dye, D. W., and Wilkie, J. P. 1978. Genus *Pseudomonas* Migula 1894. In: Young, J. M., Dye, D. W., Bradbury, J. F., Panagopoulos, C. G., and Robbs, C. F. 1978. A proposed nomenclature and classification for plant pathogenic bacteria. *N. Z. J. Agric. Res.* 21: 153–177.

Index

A

B

C

D

E

F

G

H

Q

R

S

Y